More information about this series at http://www.springer.com/series/1244

Lecture Notes in Artificial Intelligence 9078

Subseries of Lecture Notes in Computer Science

LNAI Series Editors

Randy Goebel
  University of Alberta, Edmonton, Canada
Yuzuru Tanaka
  Hokkaido University, Sapporo, Japan
Wolfgang Wahlster
  DFKI and Saarland University, Saarbrücken, Germany

LNAI Founding Series Editor

Joerg Siekmann
  DFKI and Saarland University, Saarbrücken, Germany

Tru Cao · Ee-Peng Lim
Zhi-Hua Zhou · Tu-Bao Ho
David Cheung · Hiroshi Motoda (Eds.)

Advances in Knowledge Discovery and Data Mining

19th Pacific-Asia Conference, PAKDD 2015
Ho Chi Minh City, Vietnam, May 19–22, 2015
Proceedings, Part II

 Springer

Editors
Tru Cao
Ho Chi Minh City University of Technology
Ho Chi Minh City
Vietnam

Ee-Peng Lim
Singapore Management University
Singapore
Singapore

Zhi-Hua Zhou
Nanjing University
Nanjing
China

Tu-Bao Ho
Japan Advanced Institute of Science and
    Technology
Nomi City
Japan

David Cheung
The University of Hong Kong
Hong Kong
Hong Kong SAR

Hiroshi Motoda
Osaka University
Osaka
Japan

ISSN 0302-9743
ISSN 1611-3349   (electronic)
Lecture Notes in Artificial Intelligence
ISBN 978-3-319-18031-1
ISBN 978-3-319-18032-8   (eBook)
DOI 10.1007/978-3-319-18032-8

Library of Congress Control Number: 2015936624

LNCS Sublibrary: SL7 – Artificial Intelligence

Printed on acid-free paper

Springer International Publishing AG Switzerland is part of Springer Science+Business Media
(www.springer.com)

Preface

After ten years since PAKDD 2005 in Ha Noi, PAKDD was held again in Vietnam, during May 19–22, 2015, in Ho Chi Minh City. PAKDD 2015 is the 19th edition of the Pacific-Asia Conference series on Knowledge Discovery and Data Mining, a leading international conference in the field. The conference provides a forum for researchers and practitioners to present and discuss new research results and practical applications.

There were 405 papers submitted to PAKDD 2015 and they underwent a rigorous double-blind review process. Each paper was reviewed by three Program Committee (PC) members in the first round and meta-reviewed by one Senior Program Committee (SPC) member who also conducted discussions with the reviewers. The Program Chairs then considered the recommendations from SPC members, looked into each paper and its reviews, to make final paper selections. At the end, 117 papers were selected for the conference program and proceedings, resulting in the acceptance rate of 28.9%, among which 26 papers were given long presentation and 91 papers given regular presentation.

The conference started with a day of six high-quality workshops. During the next three days, the Technical Program included 20 paper presentation sessions covering various subjects of knowledge discovery and data mining, three tutorials, a data mining contest, a panel discussion, and especially three keynote talks by world-renowned experts.

PAKDD 2015 would not have been so successful without the efforts, contributions, and supports by many individuals and organizations. We sincerely thank the Honorary Chairs, Phan Thanh Binh and Masaru Kitsuregawa, for their kind advice and support during preparation of the conference. We would also like to thank Masashi Sugiyama, Xuan-Long Nguyen, and Thorsten Joachims for giving interesting and inspiring keynote talks.

We would like to thank all the Program Committee members and external reviewers for their hard work to provide timely and comprehensive reviews and recommendations, which were crucial to the final paper selection and production of the high-quality Technical Program. We would also like to express our sincere thanks to the following Organizing Committee members: Xiaoli Li and Myra Spiliopoulou together with the individual Workshop Chairs for organizing the workshops; Dinh Phung and U Kang with the tutorial speakers for arranging the tutorials; Hung Son Nguyen, Nitesh Chawla, and Nguyen Duc Dung for running the contest; Takashi Washio and Jaideep Srivastava for publicizing to attract submissions and participants to the conference; Tran Minh-Triet and Vo Thi Ngoc Chau for handling the whole registration process; Tuyen N. Huynh for compiling all the accepted papers and for working with the Springer team to produce these proceedings; and Bich-Thuy T. Dong, Bac Le, Thanh-Tho Quan, and Do Phuc for the local arrangements to make the conference go smoothly.

We are grateful to all the sponsors of the conference, in particular AFOSR/AOARD (Air Force Office of Scientific Research/Asian Office of Aerospace Research and Development), for their generous sponsorship and support, and the PAKDD Steering

Committee for its guidance and Student Travel Award and Early Career Research Award sponsorship. We would also like to express our gratitude to John von Neumann Institute, University of Technology, University of Science, and University of Information Technology of Vietnam National University at Ho Chi Minh City and Japan Advanced Institute of Science and Technology for jointly hosting and organizing this conference. Last but not least, our sincere thanks go to all the local team members and volunteering helpers for their hard work to make the event possible.

We hope you have enjoyed PAKDD 2015 and your time in Ho Chi Minh City, Vietnam.

May 2015

Tru Cao
Ee-Peng Lim
Zhi-Hua Zhou
Tu-Bao Ho
David Cheung
Hiroshi Motoda

Organization

Honorary Co-chairs

Phan Thanh Binh — Vietnam National University, Ho Chi Minh City, Vietnam

Masaru Kitsuregawa — National Institute of Informatics, Japan

General Co-chairs

Tu-Bao Ho — Japan Advanced Institute of Science and Technology, Japan

David Cheung — University of Hong Kong, China

Hiroshi Motoda — Institute of Scientific and Industrial Research, Osaka University, Japan

Program Committee Co-chairs

Tru Hoang Cao — Ho Chi Minh City University of Technology, Vietnam

Ee-Peng Lim — Singapore Management University, Singapore

Zhi-Hua Zhou — Nanjing University, China

Tutorial Co-chairs

Dinh Phung — Deakin University, Australia

U. Kang — Korea Advanced Institute of Science and Technology, Korea

Workshop Co-chairs

Xiaoli Li — Institute for Infocomm Research, A*STAR, Singapore

Myra Spiliopoulou — Otto-von-Guericke University Magdeburg, Germany

Publicity Co-chairs

Takashi Washio — Institute of Scientific and Industrial Research, Osaka University, Japan

Jaideep Srivastava — University of Minnesota, USA

Proceedings Chair

Tuyen N. Huynh   John von Neumann Institute, Vietnam

Contest Co-chairs

Hung Son Nguyen   University of Warsaw, Poland
Nitesh Chawla   University of Notre Dame, USA
Nguyen Duc Dung   Vietnam Academy of Science and Technology,
                                 Vietnam

Local Arrangement Co-chairs

Bich-Thuy T. Dong   John von Neumann Institute, Vietnam
Bac Le   Ho Chi Minh City University of Science, Vietnam
Thanh-Tho Quan   Ho Chi Minh City University of Technology,
                                 Vietnam
Do Phuc   University of Information Technology, Vietnam
                                 National University at Ho Chi Minh City,
                                 Vietnam

Registration Co-chairs

Tran Minh-Triet   Ho Chi Minh City University of Science,
                                 Vietnam
Vo Thi Ngoc Chau   Ho Chi Minh City University of Technology,
                                 Vietnam

Steering Committee

Chairs

Tu-Bao Ho (Chair)   Japan Advanced Institute of Science and
                                 Technology, Japan
Ee-Peng Lim (Co-chair)   Singapore Management University, Singapore

Treasurer

Graham Williams   Togaware, Australia

Members

Tu-Bao Ho	Japan Advanced Institute of Science and Technology, Japan (Member since 2005, Co-chair 2012–2014, Chair 2015–2017, Life Member since 2013)
Ee-Peng Lim (Co-chair)	Singapore Management University, Singapore (Member since 2006, Co-chair 2015–2017)
Jaideep Srivastava	University of Minnesota, USA (Member since 2006)
Zhi-Hua Zhou	Nanjing University, China (Member since 2007)
Takashi Washio	Institute of Scientific and Industrial Research, Osaka University, Japan (Member since 2008)
Thanaruk Theeramunkong	Thammasat University, Thailand (Member since 2009)
P. Krishna Reddy	International Institute of Information Technology, Hyderabad (IIIT-H), India (Member since 2010)
Joshua Z. Huang	Shenzhen Institutes of Advanced Technology, Chinese Academy of Sciences, China (Member since 2011)
Longbing Cao	Advanced Analytics Institute, University of Technology, Sydney, Australia (Member since 2013)
Jian Pei	School of Computing Science, Simon Fraser University, Canada (Member since 2013)
Myra Spiliopoulou	Otto-von-Guericke-University Magdeburg, Germany (Member since 2013)
Vincent S. Tseng	National Cheng Kung University, Taiwan (Member since 2014)

Life Members

Hiroshi Motoda	AFOSR/AOARD and Institute of Scientific and Industrial Research, Osaka University, Japan (Member since 1997, Co-chair 2001–2003, Chair 2004–2006, Life Member since 2006)
Rao Kotagiri	University of Melbourne, Australia (Member since 1997, Co-chair 2006–2008, Chair 2009–2011, Life Member since 2007)
Huan Liu	Arizona State University, USA (Member since 1998, Treasurer 1998–2000, Life Member since 2012)

Ning Zhong	Maebashi Institute of Technology, Japan (Member since 1999, Life member since 2008)
Masaru Kitsuregawa	Tokyo University, Japan (Member since 2000, Life Member since 2008)
David Cheung	University of Hong Kong, China (Member since 2001, Treasurer 2005–2006, chair 2006–2008, Life Member since 2009)
Graham Williams	Australian National University, Australia (Member since 2001, Treasurer since 2006, Co-chair 2009–2011, Chair 2012–2014, Life Member since 2009)
Ming-Syan Chen	National Taiwan University, Taiwan, ROC (Member since 2002, Life Member since 2010)
Kyu-Young Whang	Korea Advanced Institute of Science and Technology, Korea (Member since 2003, Life Member since 2011)
Chengqi Zhang	University of Technology, Sydney, Australia (Member since 2004, Life Member since 2012)

Senior Program Committee Members

Arbee Chen	National Chengchi University, Taiwan
Bart Goethals	University of Antwerp, Belgium
Charles Ling	University of Western Ontario, Canada
Chih-Jen Lin	National Taiwan University, Taiwan
Dacheng Tao	University of Technology, Sydney, Australia
Dou Shen	Baidu, China
George Karypis	University of Minnesota, USA
Haixun Wang	Google, USA
Hanghang Tong	City University of New York, USA
Hui Xiong	Rutgers Univesity, USA
Ian Davidson	University of California Davis, USA
James Bailey	University of Melbourne, Australia
Jeffrey Yu	The Chinese University of Hong Kong, Hong Kong
Jian Pei	Simon Fraser University, Canada
Jianyong Wang	Tsinghua University, China
Jieping Ye	Arizona State University, USA
Jiuyong Li	University of South Australia, Australia
Joshua Huang	Shenzhen Institutes of Advanced Technology, Chinese Academy of Sciences, China
Kyuseok Shim	Seoul National University, Korea
Longbing Cao	University of Technology, Sydney, Australia
Masashi Sugiyama	University of Tokyo, Japan
Michael Berthold	University of Konstanz, Germany

Ming Li   Nanjing University, China
Ming-Syan Chen   National Taiwan University, Taiwan
Min-Ling Zhang   Southeast University, China
Myra Spiliopoulou   Otto-von-Guericke-University Magdeburg,
                              Germany
Nikos Mamoulis   University of Hong Kong, Hong Kong
Ning Zhong   Maebashi Institute of Technology, Japan
Osmar Zaiane   University of Alberta, Canada
P. Krishna Reddy   International Institute of Information Technology,
                              Hyderabad, India
Peter Christen   Australian National University, Australia
Sanjay Chawla   University of Sydney, Australia
Takashi Washio   Institute of Scientific and Industrial Research,
                              Osaka University, Japan
Vincent S. Tseng   National Cheng Kung University, Taiwan
Wee Keong Ng   Nanyang Technological University, Singapore
Wei Wang   University of California at Los Angeles, USA
Wen-Chih Peng   National Chiao Tung University, Taiwan
Xiaofang Zhou   University of Queensland, Australia
Xiaohua Hu   Drexel University, USA
Xifeng Yan   University of California, Santa Barbara, USA
Xindong Wu   University of Vermont, USA
Xing Xie   Microsoft Research Asia, China
Yanchun Zhang   Victoria University, Australia
Yu Zheng   Microsoft Research Asia, China

Program Committee Members

Aijun An   York University, Canada
Aixin Sun   Nanyang Technological University, Singapore
Akihiro Inokuchi   Kwansei Gakuin University, Japan
Alfredo Cuzzocrea   ICAR-CNR and University of Calabria, Italy
Andrzej Skowron   University of Warsaw, Poland
Anne Denton   North Dakota State University, USA
Bettina Berendt   Katholieke Universiteit Leuven, Belgium
Bin Zhou   University of Maryland, Baltimore County, USA
Bing Tian Dai   Singapore Management University, Singapore
Bo Zhang   Tsinghua University, China
Bolin Ding   Microsoft Research, USA
Bruno Cremilleux   Université de Caen Basse-Normandie, France
Carson K. Leung   University of Manitoba, Canada
Chandan Reddy   Wayne State University, USA
Chedy Raissi   Inria, France
Chengkai Li   The University of Texas at Arlington, USA

Chia-Hui Chang   National Central University, Taiwan
Chiranjib Bhattacharyya   Indian Institute of Science, India
Choochart Haruechaiy   National Electronics and Computer Technology
                              Center, Thailand
Chun-Hao Chen   Tamkang University, Taiwan
Chun-hung Li   Hong Kong Baptist University, Hong Kong
Clifton Phua   NCS, Singapore
Daoqiang Zhang   Nanjing University of Aeronautics and
                              Astronautics, China
Dao-Qing Dai   Sun Yat-Sen University, China
David Taniar   Monash University, Australia
David Lo   Singapore Management University, Singapore
De-Chuan Zhan   Nanjing University, China
Dejing Dou   University of Oregon, USA
De-Nian Yang   Academia Sinica, Taiwan
Dhaval Patel   Indian Institute of Technology, Roorkee, India
Dinh Phung   Deakin University, Australia
Dragan Gamberger   Ruđer Bošković Institute, Croatia
Du Zhang   California State University, Sacramento, USA
Duc Dung Nguyen   Institute of Information Technology, Vietnam
Enhong Chen   University of Science and Technology of China,
                              China
Fei Liu   Carnegie Mellon University, USA
Feida Zhu   Singapore Management University, Singapore
Florent Masseglia   Inria, France
Geng Li   Oracle Corporation, USA
Giuseppe Manco   Università della Calabria, Italy
Guandong Xu   University of Technology, Sydney, Australia
Guo-Cheng Lan   Industrial Technology Research Institute, Taiwan
Gustavo Batista   University of São Paulo, Brazil
Hady Lauw   Singapore Management University, Singapore
Harry Zhang   University of New Brunswick, Canada
Hiroshi Mamitsuka   Kyoto University, Japan
Hong Shen   University of Adelaide, Australia
Hsuan-Tien Lin   National Taiwan University, Taiwan
Hua Lu   Aalborg University, Denmark
Hui Wang   University of Ulster, UK
Hung Son Nguyen   University of Warsaw, Poland
Hung-Yu Kao   National Cheng Kung University, Taiwan
Irena Koprinska   University of Sydney, Australia
J. Saketha Nath   Indian Insitiute of Technology, India
Jaakko Hollmén   Aalto University, Finland
Jake Chen   Indiana University–Purdue University Indianapolis,
                              USA

James Kwok   Hong Kong University of Science and Technology,
                              China
Jason Wang   New Jersey Science and Technology University,
                              USA
Jean-Marc Petit   Université de Lyon, France
Jeffrey Ullman   Stanford University, USA
Jen-Wei Huang   National Cheng Kung University, Taiwan
Jerry Chun-Wei Lin   Harbin Institute of Technology Shenzhen,
                              China
Jia Wu   University of Technology, Sydney, Australia
Jialie Shen   Singapore Management University, Singapore
Jiayu Zhou   Samsung Research America, USA
Jia-Yu Pan   Google, USA
Jin Soung Yoo   Indiana University–Purdue University
                              Indianapolis, USA
Jingrui He   IBM Research, USA
Jinyan Li   University of Technology, Sydney, Australia
John Keane   University of Manchester, UK
Jun Huan   University of Kansas, USA
Jun Gao   Peking University, China
Jun Luo   Huawei Noah's Ark Lab, Hong Kong
Jun Zhu   Tsinghua University, China
Junbin Gao   Charles Sturt University, Australia
Junjie Wu   Beihang University, China
Junping Zhang   Fudan University, China
K. Selcuk Candan   Arizona State University, USA
Keith Chan   Hong Kong Polytechnic University, Hong Kong
Khoat Than   Hanoi University of Science and Technology,
                              Vietnam
Kitsana Waiyamai   Kasetsart University, Thailand
Krisztian Buza   Semmelweis University, Budapest, Hungary
Kun-Ta Chuang   National Cheng Kung University, Taiwan
Kuo-Wei Hsu   National Chengchi University, Taiwan
Latifur Khan   University of Texas at Dallas, USA
Ling Chen   University of Technology, Sydney, Australia
Lipo Wang   Nanyang Technological University, Singapore
Manabu Okumura   Japan Advanced Institute of Science and
                              Technology, Japan
Marco Maggini   Università degli Studi di Siena, Italy
Marian Vajtersic   University of Salzburg, Austria
Marut Buranarach   National Electronics and Computer Technology
                              Center, Thailand
Mary Elaine Califf   Illinois State University, USA
Marzena Kryszkiewicz   Warsaw University of Technology, Poland

Tuyen N. Huynh	John von Neumann Institute, Vietnam
Tzung-Pei Hong	National University of Kaohsiung, Taiwan
Van-Nam Huynh	Japan Advanced Institute of Science and Technology, Japan
Vincenzo Piuri	Università degli Studi di Milano, Italy
Wai Lam The	Chinese University of Hong Kong, Hong Kong
Walter Kosters	Universiteit Leiden, The Netherlands
Wang-Chien Lee	Pennsylvania State University, USA
Wei Ding	University of Massachusetts Boston, USA
Wenjie Zhang	University of New South Wales, Australia
Wenjun Zhou	University of Tennessee, Knoxville, USA
Wilfred Ng	Hong Kong University of Science and Technology, Hong Kong
Wu-Jun Li	Nanjing University, China
Wynne Hsu	National University of Singapore, Singapore
Xiaofeng Meng	Renmin University of China, China
Xiaohui (Daniel) Tao	University of Southern Queensland, Australia
Xiaoli Li	Institute for Infocomm Research, A*STAR, Singapore
Xiaowei Ying	Bank of America, USA
Xin Wang	University of Calgary, Canada
Xingquan Zhu	Florida Atlantic University, USA
Xintao Wu	University of Arkansas, Arkansas
Xuan Vinh Nguyen	University of Melbourne, Australia
Xuan-Hieu Phan	University of Engineering and Technology–Vietnam National University, Hanoi, Vietnam
Xuelong Li	University of London, UK
Xu-Ying Liu	Southeast University, China
Yang Yu	Nanjing University, China
Yang-Sae Moon	Kangwon National University, Korea
Yasuhiko Morimoto	Hiroshima University, Japan
Yidong Li	Beijing Jiaotong University, China
Yi-Dong Shen	Chinese Academy of Sciences, China
Ying Zhang	University of New South Wales, Australia
Yi-Ping Phoebe Chen	La Trobe University, Australia
Yiu-ming Cheung	Hong Kong Baptist University, Hong Kong
Yong Guan	Iowa State University, USA
Yonghong Peng	University of Bradford, UK
Yue-Shi Lee	Ming Chuan University, Taiwan
Zheng Chen	Microsoft Research Asia, China
Zhenhui Li	Pennsylvania State University, USA
Zhiyuan Chen	University of Maryland, Baltimore County, USA
Zhongfei Zhang	Binghamton University, USA
Zili Zhang	Deakin University, Australia

External Reviewers

Ahsanul Haque	University of Texas at Dallas, USA
Ameeta Agrawal	York University, Canada
Anh Kim Nguyen	Hanoi University of Science and Technology, Vietnam
Arnaud Soulet	Université François Rabelais, Tours, France
Bhanukiran Vinzamuri	Wayne State University, USA
Bin Fu	University of Technology, Sydney, Australia
Bing Tian Dai	Singapore Management University, Singapore
Budhaditya Saha	Deakin University, Australia
Cam-Tu Nguyen	Nanjing University, China
Cheng Long	Hong Kong University of Science and Technology, Hong Kong
Chung-Hsien Yu	University of Massachusetts Boston, USA
Chunming Liu	University of Technology, Sydney, Australia
Dawei Wang	University of Massachusetts Boston, USA
Dieu-Thu Le	University of Trento, Italy
Dinusha Vatsalan	Australian National University, Australia
Doan V. Nguyen	Japan Advanced Institute of Science and Technology, Japan
Emmanuel Coquery	Université Lyon1, CNRS, France
Ettore Ritacco	ICAR-CNR, Italy
Fan Jiang	University of Manitoba, Canada
Fang Yuan	Institute for Infocomm Research A*STAR, Singapore
Fangfang Li	University of Technology, Sydney, Australia
Fernando Gutierrez	University of Oregon, USA
Fuzheng Zhang	University of Science and Technology of China, China
Gensheng Zhang	University of Texas at Arlington, USA
Gianni Costa	ICAR-CNR, Italy
Guan-Bin Chen	National Cheng Kung University, Taiwan
Hao Wang	University of Oregon, USA
Heidar Davoudi	York University, Canada
Henry Lo	University of Massachusetts Boston, USA
Ikumi Suzuki	National Institute of Genetics, Japan
Jan Bazan	University of Rzeszów, Poland
Jan Vosecky	Hong Kong University of Science and Technology, Hong Kong
Javid Ebrahimi	University of Oregon, USA
Jianhua Yin	Tsinghua University, China
Jianmin Li	Tsinghua University, China
Jianpeng Xu	Michigan State University, USA
Jing Ren	Singapore Management University, Singapore
Jinpeng Chen	Beihang University, China

Jipeng Qiang	University of Massachusetts Boston, USA
Joseph Paul Cohen	University of Massachusetts Boston, USA
Junfu Yin	University of Technology, Sydney, Australia
Justin Sahs	University of Texas at Dallas, USA
Kai-Ho Chan	Hong Kong University of Science and Technology, Hong Kong
Kazuo Hara	National Institute of Genetics, Japan
Ke Deng	RMIT University, Australia
Kiki Maulana Adhinugraha	Monash University, Australia
Kin-Long Ho	Hong Kong University of Science and Technology, Hong Kong
Lan Thi Le	Hanoi University of Science and Technology, Vietnam
Lei Zhu	Huazhong University of Science and Technology, China
Lin Li	Wuhan University of Technology, China
Linh Van Ngo	Hanoi University of Science and Technology, Vietnam
Loc Do	Singapore Management University, Singapore
Maksim Tkachenko	Singapore Management University, Singapore
Marc Plantevit	Université de Lyon, France
Marian Scuturici	INSA de Lyon, CNRS, France
Marthinus Christoffel du Plessis	University of Tokyo, Japan
Md. Anisuzzaman Siddique	Hiroshima University, Japan
Min Xie	Hong Kong University of Science and Technology, Hong Kong
Ming Yang	Binghamton University, USA
Minh Nhut Nguyen	Institute for Infocomm Research A*STAR, Singapore
Mohit Sharma	University of Minnesota, USA
Morteza Zihayat	York University, Canada
Mu Li	University of Technology, Sydney, Australia
Naeemul Hassan	University of Texas at Arlington, USA
NhatHai Phan	University of Oregon, USA
Nicola Barbieri	Yahoo Labs, Spain
Nicolas Béchet	Université de Bretagne Sud, France
Nima Shahbazi	York University, Canada
Pakawadee Pengcharoen	Hong Kong University of Science and Technology, Hong Kong
Pawel Gora	University of Warsaw, Poland
Peiyuan Zhou	Hong Kong Polytechnic University, Hong Kong
Peng Peng	Hong Kong University of Science and Technology, Hong Kong
Pinghua Gong	University of Michigan, USA

Qiong Fang	Hong Kong University of Science and Technology, Hong Kong
Quan Xiaojun	Institute for Infocomm Research A*STAR, Singapore
Riccardo Ortale	ICAR-CNR, Italy
Sabin Kafle	University of Oregon, USA
San Phyo Phyo	Institute for Infocomm Research A*STAR, Singapore
Sang The Dinh	Hanoi University of Science and Technology, Vietnam
Shangpu Jiang	University of Oregon, USA
Shenlu Wang	University of New South Wales, Australia
Shiyu Yang	University of New South Wales, Australia
Show-Jane Yen	Ming Chuan University, Taiwan
Shuangfei Zhai	Binghamton University, USA
Simone Romano	University of Melbourne, Australia
Sujatha Das Gollapalli	Institute for Infocomm Research A*STAR, Singapore
Swarup Chandra	University of Texas at Dallas, USA
Syed K. Tanbeer	University of Manitoba, Canada
Tenindra Abeywickrama	Monash University, Australia
Thanh-Son Nguyen	Singapore Management University, Singapore
Thin Nguyen	Deakin University, Australia
Tiantian He	Hong Kong Polytechnic University, Hong Kong
Tianyu Kang	University of Massachusetts Boston, USA
Trung Le	Deakin University, Australia
Tuan M. V. Le	Singapore Management University, Singapore
Xiaochen Chen	Google, USA
Xiaolin Hu	Tsinghua University, China
Xin Li	University of Science and Technology, China
Xuhui Fan	University of Technology, Sydney, Australia
Yahui Di	University of Massachusetts Boston, USA
Yan Li	Wayne State University, USA
Yang Jianbo	Institute for Infocomm Research A*STAR, Singapore
Yang Mu	University of Massachusetts Boston, USA
Yanhua Li	University of Minnesota, USA
Yanhui Gu	Nanjing Normal University, China
Yathindu Rangana Hettiarachchige	Monash University, Australia
Yi-Yu Hsu	National Cheng Kung University, Taiwan
Yingming Li	Binghamton University, USA
Yu Zong	West Anhui University, China
Zhiyong Chen	Singapore Management University, Singapore
Zhou Zhao	Hong Kong University of Science and Technology, Hong Kong
Zongda Wu	Wenzhou University, China

Contents – Part II

Outlier and Anomaly Detection

Mining Uncertain and Imprecise Data

Mining Temporal and Spatial Data

Feature Extraction and Selection

Mining Heterogeneous, High Dimensional, and Sequential Data

Entity Resolution and Topic Modelling

Itemset and High Performance Data Mining

Recommendation

Contents – Part I

Classification

Machine Learning

Applications

Novel Methods and Algorithms

Opinion Mining and Sentiment Analysis

Emotion Cause Detection for Chinese Micro-Blogs Based on ECOCC Model

Kai Gao[1,2], Hua Xu[1](✉), and JiushuoWang[1,2]

[1] State Key Laboratory of Intelligent Technology and Systems, Tsinghua National Laboratory for Information Science and Technology, Department of Computer Science and Technology, Tsinghua University, Beijing, China
xuhua@tsinghua.edu.cn
[2] School of Information Science and Engineering, Hebei University of Science and Technology, Shijiazhuang, Hebei, China
gaokai68@139.com, wangjiushuo@126.com

Abstract. Micro-blog emotion mining and emotion cause extraction are essential in social network data mining. This paper presents a novel approach on Chinese micro-blog emotion cause detection based on the ECOCC model, focusing on mining factors for eliciting some kinds of emotions. In order to do so, the corresponding emotion causes are extracted. Moreover, the proportions of different cause components under different emotions are also calculated by means of combining the emotional lexicon with multiple characteristics (e.g., emoticon, punctuation, etc.). Experimental results show the feasibility of the approach. The proposed approaches have important scientific values on social network knowledge discovery and data mining.

Keywords: Micro-blog · Text mining · Emotion cause detection · Emotion analysis

1 Introduction

Nowadays, millions of people present and share their opinions or sentiments in micro-blogs, and as a result, to produce a large number of social network data, containing almost all kinds of opinions and sentiment. These micro-blog data usually includes the description of emergencies, incidents, disasters and some other hot events, and some of them have some kinds of emotions and sentiments. If we can mine or discover the hidden emotions behind the big data, it is essential for public opinion surveillance. Meanwhile, individual emotion generation, expression and perception are influenced by many factors, so emotion cause feature extraction and analysis are necessary.

According to the related works, the cause events are usually composed of verbs, nominalizations and nouns. They evoke the presence of the corresponding emotions by some linguistic cues, which are based on Chinese emotion cause annotated corpus [1]. For example, as for the Chinese micro-blog post *"Ta1*

© Springer International Publishing Switzerland 2015
T. Cao et al. (Eds.): PAKDD 2015, Part II, LNAI 9078, pp. 3–14, 2015.
DOI: 10.1007/978-3-319-18032-8_1

Men Ying2 De2 Le Bi3 Sai4 Shi3 Wo3 Hen3 Kai1 Xin1." ("They won the game makes me very happy."), the causative verb is *"Shi3"* ("makes"), and the emotion keyword is *"Kai1 Xin1"* ("happy"). With the help of the corresponding linguistic rules, we can infer that the emotion cause event is *"Ta1 Men Ying2 De2 Le Bi3 Sai4"* ("they won the game"). Meanwhile, the emotion causes also were detected by extracting cause expressions and constructions in [2]. And Li et al. [3] proposed and implemented a novel method for identifying emotions of micro-blog posts, and tried to infer and extract the emotion causes by using knowledge and theories from other fields such as sociology. On the one hand, Rao et al. [4] proposed two sentiment topic models to extract the latent topics that evoke emotions of readers, then the topics were seen as the causes of emotions as well.

In this paper, an emotion model with cause events is proposed, it describes the causes that trigger bloggers' emotions in the progress of cognitive evaluation. Meanwhile, all of the sub-events are extracted in micro-blogs. This paper detects the corresponding cause events on the basis of the proposed rule-based algorithm. Finally, the proportions of different cause components under different emotions are calculated by constructing the emotional lexicon from the corpus and combining multiple features of Chinese micro-blogs.

2   ECOCC Model Construction

According to the research on cognitive theory, this paper improves the structure about the eliciting conditions of emotions on the basis of the OCC model referred in [5], and presents an emotion model named as ECOCC (Emotion-Cause-OCC) model for micro-blog posts. It describes the combination of psychology and computer science to analyze the corresponding emotion cause events. The improved model describes a hierarchy that classifies 22 fine-grained emotions, i.e., "hope", "fear", "joy", "distress", "pride", "shame", "admiration", "reproach", "liking", "disliking", "gratification", "remorse", "gratitude", "anger", "satisfaction", "fears-confirmed", "relief", "disappointment", "happy-for", "resentment", "gloating" and "pity". This hierarchy contains three main branches (i.e., results of events, actions of agents and aspects of objects), and some branches are combined to form a group of the compound and extended emotions.

According to the three main branches, the components of the model matching the emotional rules in the model are divided into six sub-classes (i.e., event_state, event_norm, action_agent, action_norm, object_entity and object_norm). Within the components, the evaluation-schemes can be defined from event_state (i.e., the state that something will happen, the state that something has happened, the state that something didn't happen), action_agent (i.e., the emotional agent or others) and object_entity (i.e., the elements of objects). They are also divided into six classes (i.e., prospective, confirmation, disconfirmation, main-agent, other-agent and entity). Meanwhile, the corresponding evaluation-standards are defined from the aspects of event_norm (i.e., being satisfactory or unsatisfactory for the event), action_norm (i.e.,being approval or disapproval for the agent) and object_norm (i.e., being attractive or unattractive to the entity). And they are separated into six types (i.e., desirable, undesirable, praiseworthy, blameworthy, positive and negative).

As for the production process of the 22 types of emotions, the following rules (see Table 1) are used to describe them according to the components in the proposed ECOCC model. Here, "s" means the micro-blog post, "C" represents the cause components that trigger emotions, "→" represents the state that changes from one to the other, "∩" has the same meaning as "and" and "∪" has the same meaning as "or".

Table 1. The emotional rules

Classes	The emotional rules
The emotions in results of events	$Hope(C) \stackrel{def}{=} Prospective(s) \cap Desirable(s)$
	$Fear(C) \stackrel{def}{=} Prospective(s) \cap Undesirable(s)$
	$Joy(C) \stackrel{def}{=} [Confirmation(s) \cup Disconfirmation(s)] \cap Desirable(s)$
	$Distress(C) \stackrel{def}{=} [Confirmation(s) \cup Disconfirmation(s)] \cap Undesirable(s)$
The emotions in actions of agents	$Pride(C) \stackrel{def}{=} main\text{-}agent \cap Praiseworthy(s)$
	$Shame(C) \stackrel{def}{=} main\text{-}agent \cap Blameworthy(s)$
	$Admiration(C) \stackrel{def}{=} other\text{-}agent \cap Praiseworthy(s)$
	$Reproach(C) \stackrel{def}{=} other\text{-}agent \cap Blameworthy(s)$
The emotions in aspects of objects	$Liking(C) \stackrel{def}{=} entity \cap Positive(s)$
	$Disliking(C) \stackrel{def}{=} entity \cap Negative(s)$
Compound emotions	$Gratification(C) \stackrel{def}{=} Pride(C) \cap Joy(C)$
	$Remorse(C) \stackrel{def}{=} Shame(C) \cap Distress(C)$
	$Gratitude(C) \stackrel{def}{=} Admiration(C) \cap Joy(C)$
	$Anger(C) \stackrel{def}{=} Reproach(C) \cap Distress(C)$
Extended emotions	$Satisfaction(C) \stackrel{def}{=} Joy(C) \cap [Prospective(s) \rightarrow Confirmation(s)] \cap Hope(C)$
	$Fears\text{-}confirmed(C) \stackrel{def}{=} Distress(C) \cap [Prospective(s) \rightarrow Confirmation(s)] \cap Fear(C)$
	$Relief(C) \stackrel{def}{=} Joy(C) \cap [Prospective(s) \rightarrow Disconfirmation(s)] \cap Fear(C)$
	$Disappointment(C) \stackrel{def}{=} Distress(C) \cap [Prospective(s) \rightarrow Disconfirmation(s)] \cap Hope(C)$
	$Happy\text{-}for(C) \stackrel{def}{=} Joy(C) \cap [other\text{-}agent \cap Praiseworthy(s) \cap Desirable(s)]$
	$Resentment(C) \stackrel{def}{=} Distress(C) \cap [other\text{-}agent \cap Blameworthy(s) \cap Desirable(s)]$
	$Gloating(C) \stackrel{def}{=} Joy(C) \cap [other\text{-}agent \cap Blameworthy(s) \cap Undesirable(s)]$
	$Pity(C) \stackrel{def}{=} Distress(C) \cap [other\text{-}agent \cap Praiseworthy(s) \cap Undesirable(s)]$

3   Emotion Causes Extraction

In this section, the internal events can be taken into account in the process of detecting emotion cause components, which are usually the direct reasons for triggering the change of individual emotion. They can be extracted from the domain of results of events, actions of agents and aspects of objects based on the ECOCC model.

As for the domain of results of events, the LTP (Language Technology Platform) is used to set up the model of extracting sub-events based on the named entity recognition, dependency parsing, semantic role labeling, and so on [6]. Firstly, we label the parts of speech of Chinese (e.g., nouns, verbs, adjectives) within the micro-blog posts by using ICTCLAS, and identify the person names, place names and institutions by using the named entity recognition. Then the core relation of subject-verbs and verb-objects can be identified by using dependency parsing. Finally, we identify the phrases labeled with the semantic role (i.e., A0) for the actions of the agent, the semantic role (i.e., A1) for actions of the receiver, or other four different core semantic roles (i.e., A2-A5) for different predicates by using the semantic role labeling, respectively.

In detail, this paper first selects the phrases which are labeled as A0, A1, A2-A5 (if it exists) and then combines the above components as the basis of event recognition. Otherwise, the triple $U = (nouns, verbs, nouns)$ will be used as another basis of event recognition. By describing the results of events, it is easy to decide the corresponding emotion and its causes according to the evaluation-schemes and the evaluation-standards.

As for the domain of actions of agents, the feature words can be used to describe agents' actions. Firstly, this paper extracts the class of ACT in HowNet[1] which contains a large number of words describing different kinds of people's actions. If the predicate verb belongs to the class of ACT and there exists an initiative relationship between itself and the agent, then this structure can be as a kind of agent's action. On the other hand, as for main-agent, it contains the explicit-agent and the implicit-agent. The former is highlighted in the text and has the subject-predicate relationship with the predicate verb. The latter does not appear in the text, but it is usually expressed by the context and the act of the verb. Finally, on the basis of the description of the actions of agents, it is easy to decide the corresponding emotion and its causes according to the evaluation-schemes and the evaluation-standards.

As for the domain of aspects of objects, the corresponding emotions "Liking" and "Disliking" describe the reactions of the agent to the corresponding object. For recognizing the characteristic information of aspects of objects, it needs to extract the entities with the help of the HowNet corpus and find the subject-predicate relationship by using dependency parsing. The features of objects are extracted by using the semantic role labeling to confirm the evaluation-standards of object_norm, and then we get the final emotion and its cause components.

---

[1] http://www.keenage.com/

4 Emotion Cause Components Analysis

4.1 Emotional Lexicon Construction

Generally, the emotional lexicon can be constructed manually and automatically from the corpus. Firstly, the standard lexicon can be constructed manually. In this process, the 22 fine-grained emotions based on the ECOCC model are chosen as the basal emotions. Then the intensity scores of the emotional words can be divided into five level ranges (i.e., 0-1.0, 1.0-2.0, 2.0-3.0, 3.0-4.0 and 4.0-5.0). Among them, 0-1.0 represents the word with the weakest emotion intensity; while 4.0-5.0 represents the corresponding word with the strongest emotion intensity. Meanwhile, the standard emotional words which belong to 22 different types of emotions are selected by three different annotators, and those words are from HowNet Dictionary, National Taiwan University Sentiment Dictionary[2], and the Affective Lexicon Ontology [7]. And then we give them the corresponding emotion intensities by the setting of the emotion intensity.

The lexicon will be expanded by acquiring automatically from the corpus for getting the larger capacity. It can be completed by the word2vec[3]. The word2vec provides an efficient implementation for computing vector representation of words by using the continuous bag-of-words and skip-gram architectures [8]. Firstly, a large number of posts are randomly crawled from Sina Micro-blog website (weibo.com) to constitute a 1.5G micro-blog dataset and can be transformed to vector representation of words, and then the synonyms of the standard emotional words are chosen as the candidate words. Secondly, it needs to compute the similarity and choose the maximum between the candidate words and the standard words. Meanwhile, the emotion intensity E_i of the ith selected word can be defined as the formula (1) below, where WD_i represents the ith word in the candidate word list, ST_j represents the jth word in the standard word list, $SIM(WD_i, ST_j)$ represents the maximum similarity between the candidate word and the standard word, and $I(ST_j)$ represents the emotion intensity of the standard word.

$$E_i = SIM(WD_i, ST_j) * I(ST_j) \tag{1}$$

4.2 Multiple Features Recognition of Chinese Micro-Blogs

Emoticons. In this paper, we will combine the emoticon with the corresponding intensity to assist in calculating the proportion of the emotion cause. Firstly, the micro-blogs can be formulated as a triple $U = (C, R, T)$, where C means the emoticon list, R is the emotional keyword list, and T represents a list of micro-blog posts. As for one post, it can be regarded as a triple $u_x = (c_i, r_{kj}, t_n)$, where c_i means the ith emoticon in C, r_{kj} is the jth emotion keyword in R of the kth ($1 \leq k \leq 22$) emotion, t_n represents the nth post in T. If c_i and r_{kj} appear

---

[2] http://nlg18.csie.ntu.edu.tw:8080/opinion/index.html
[3] http://word2vec.googlecode.com/svn/trunk/

within t_n at the same time, the corresponding co-occurrence frequency is called $|CO(c_i, r_{kj})|$.

As for the co-occurrence intensity between the corresponding emoticon and the emotion keyword, it can be represented as $\delta_{ij}(c_i, r_{kj})$, see the formula(2), where $|c_i|$ is the number of c_i appearing in t_n, and $|r_{kj}|$ is the number of r_{kj} appearing in t_n.

$$\delta_{ij}(c_i, r_{kj}) = \frac{|CO(c_i, r_{kj})|}{(|c_i| + |r_{kj}|) - |CO(c_i, r_{kj})|} \tag{2}$$

According to the above definitions, it is easy to construct the co-occurrence graph (see Figure 1). Within the figure, E is the set of center nodes containing emoticons (e.g., c_1, c_2, c_3, etc.), and D is the set of leaf nodes containing emotion keywords (e.g., r_{11}, r_{12}, r_{13}, etc.). P is the side set between E and D, which represents the degree of closeness between the emotion intensities of c_i and r_{kj}. The longer the side is, the closer the emotion intensities of both are [9].

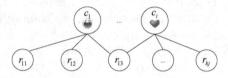

Fig. 1. The co-occurrence graph

As for the weight of the side in the co-occurrence graph (it is defined as $W_{ij}(c_i, r_{kj})$), it can be set to a value equal to the co-occurrence intensity, which is shown in the formula (3):

$$W_{ij}(c_i, r_{kj}) = \delta_{ij}(c_i, r_{kj}) \tag{3}$$

According to the above equations and definitions, the emotion intensity of the ith emoticon (which is called as I_{ICON_i}) can be inferred as follows, see the formula (4), and E_j is the emotion intensity of the jth emotion keyword.

$$I_{ICON_i} = E_j * \max_{1 \le k \le 22} W_{ij}(c_i, r_{kj}) \tag{4}$$

Degree Adverbs. In addition, the modification of degree adverbs is helpful in computing the emotion intensities of cause events. This paper uses the intensifier lexicon including 219 degree adverbs, which are divided into five levels: "Ji2 Qi2 | extreme"; "Hen3 | very"; "Jiao4 | more"; "Shao1 | -ish" and "Qian4 | insufficiently" [10]. Then the influence coefficient is set to x, and the value of x is +0.5, +0.3, -0.1, -0.3 and -0.5 in order. The "-" has the function of weakening the emotion intensities of the corresponding words; and the "+" has the function of strengthening the emotion intensities of the corresponding words. The exponential function e^x is applied to adjust the emotion intensity. Finally, the emotion intensity of the ith emotion keyword (called I_{DA_i}) can be calculated by the following formula (5). Here, γ is the adjustable parameter.

$$I_{DA_i} = \gamma e^x E_i (\gamma \geq 1) \tag{5}$$

Negation Words. As for negation words, they can also impact the negative transformation of emotions or impact the emotion intensity scores. With the location of the negation word changing, the emotion and the corresponding emotion intensity will also change. The following Table 2 describes the influence of the locations of negation words in four situations. "NE" means the negation word, "DA" means the degree adverb, "EWD" means the emotion keyword. The parameter α, η and β stand for the adjustable parameters respectively, and "$-$" is the sign of the transformation of emotion. Here, I_{NEGA_i} is used to express the modified result of the emotion intensity of the ith emotion keyword.

Table 2. The detail of the locations of negation words

The four situations	The formula
NE+NE+EWD	$I_{NEGA_i} = \eta E_i (\eta > 1)$
NE+DA+EWD	$I_{(NEGA_i)} = \beta E_i (0 < \beta < 1)$
NE+EWD	$I_{(NEGA_i)} = -E_i$ or $I_{(NEGA_i)} = 0$
DA+NE+EWD	$I_{(NEGA_i)} = -\alpha e^x E_i (\alpha \geq 1)$

Punctuations. In micro-blogs, the emotional punctuation marks, such as the exclamation mark, the interrogation mark etc., usually strengthen the emotion intensities of sentences in some manners. For example, "You can't be so crazy!!!!!", which expresses the emotion of "anger", and the serial exclamation marks can strengthen the emotional intensities. On the other hand, as the repetitive punctuations strengthen the emotion intensities more obviously than the single punctuation, the former is emphasized. The formula (6) is used to compute the emotion intensity which is influenced by punctuations (called I_{PUNC_i}). ε is the adjustable parameter that can be confirmed by the degree of repeatability which has a direct relationship with the number of punctuations.

$$I_{PUNC_i} = \varepsilon * E_i (\varepsilon \geq 1) \tag{6}$$

Furthermore, with regard to the interrogative sentences, they have different means in different circumstances. Sometimes, the interrogative sentence describes a positive attitude literally, but its actual mean is negative. In another case, it can also strengthen this emotion.

Conjunctions. As we all know, there are different kinds of conjunctions such as coordinating conjunctions, adversative conjunctions, causal conjunctions and so on. Under certain circumstance, conjunctions can play an important role in the emotional expression. For example, the conjunction *"Dan4 Shi4"* ("but") mainly emphasizes the event with a strong emotion which is behind it. And the influencing parameters are divided into two classes. One is that the conjunction impacts on the emotion intensity of its front event, and which is set to F_{before}; the other is that the conjunction impacts on the emotion intensity of its back event, and which is set to F_{after}. If $F_{before} = F_{after}$, it means that the conjunction has the same effect on the front event and the back event, so the values of the two parameters are set to 1; if $F_{before} < F_{after}$, it means that the conjunction has no effect on the front event but strengthens the emotion intensity of the back event, so we set that $F_{before}=1$ and $F_{after} > 1$; and if $F_{before} > F_{after}$, it presents the opposite case with $F_{before} < F_{after}$, so we set that $F_{before} > 1$ and $F_{after} = 1$. Therefore, this paper proposes the formula (7) to express the modified result.

$$I_{CONJ_i} = \begin{cases} F_{before} * E_i & F_{before} > F_{after} \\ E_i & F_{before} = F_{after} \\ F_{after} * E_i & F_{before} < F_{after} \end{cases} \quad (7)$$

4.3   Emotion Cause Components Proportions Calculation

In this section, we combine the characteristics of Bayesian probability model to describe the proportions of emotion causes from the perspective of the prior probability and the conditional probability.

Firstly, this paper constructs an emotion cause component matrix $\rho(s)$ for micro-blog posts, $E(C_m)$ represents the emotion vector with cause components, m is the serial number of 22 types of emotions, E_{nm} represents the nth emotion cause intensity score of the mth emotion, see the formula (8).

$$\rho(s) = (E(C_1), E(C_2), \cdots, E(C_m))^T = \begin{pmatrix} E_{11} & \cdots & E_{1m} \\ \vdots & \ddots & \vdots \\ E_{n1} & \cdots & E_{nm} \end{pmatrix} \quad (8)$$

The proportion of the nth cause component under the mth emotion (which is defined as $P(Emo_m|Cau_n)$) can be computed based on the Bayesian probability, see the formula (9).

$$P(Emo_m|Cau_n) = \frac{P(Cau_n|Emo_m)P(Emo_m)}{\sum_{m=1}^{22} P(Emo_m)P(Cau_n|Emo_m)} \quad (9)$$

In the above formula (9), the parameter Emo_m is the mth emotion and Cau_n is the nth cause component under Emo_m. The prior probability $P(Emo_m)$ is the probability distribution of Emo_m. It can be calculated by the formula (10) and (11), Where $SCORE(Emo_m)$ is the mth emotion intensity score which can be modified by the multiple features in micro-blogs. And I_{ICON_i} is the result modified by emoticons in micro-blogs; I_{DA_i} is the result modified by degree adverbs; I_{NEGA_i} is the result modified by negation words; I_{PUNC_i} is the result modified by punctuations; and I_{CONJ_i} is the result modified by conjunctions which are described in the above sections. If there is no any linguistic feature in micro-blogs, the modified result will be ignored and set to 0.

$$P(Emo_m) = \frac{SCORE(Emo_m)}{\sum_{m=1}^{22} SCORE(Emo_m)} \tag{10}$$

$$SCORE(Emo_m) = \sum_{i=1}(E_i + I_{DA_i} + I_{NEGA_i} + I_{ICON_i} + I_{PUNC_i} + I_{CONJ_i}) \tag{11}$$

Within the above formula (9), $P(Cau_n|Emo_m)$ is the probability density function of the nth cause component in a known condition of emotion. It can be calculated by the formula (12) and (13), where $SCORE(Cau_n)$ is the emotion intensity score of the nth cause component under the mth emotion. It is also influenced by the multiple features in micro-blogs.

$$P(Cau_n|Emo_m) = \frac{SCORE(Cau_n)}{\sum_{n=1} SCORE(Cau_n)} \tag{12}$$

$$SCORE(Cau_n) = \sum_{i=1}(E_{im} + I_{DA_{im}} + I_{NEGA_{im}} + I_{ICON_{im}} + I_{PUNC_{im}} + I_{CONJ_{im}}) \tag{13}$$

5   Experiments and Analysis

5.1   The Experimental Dataset

In this section, some strategies based on simulating browsers' behaviors are used to obtain the micro-blog dataset from Chinese micro-blog website (weibo.com) [11]. In the dataset, the micro-blog posts are short, and most of them are less than 140 characters in each post. After the pre-processing (i.e., removing duplicates, filtering irrelevant results and doing some conversion), 16371 posts are remained in our dataset. Meanwhile, every data is used to label the emotion type and the corresponding cause by some annotators manually. In the process of annotating, the micro-blog posts with obvious emotions are chosen to annotate. If the micro-blog does not belong to any category, it will not be labeled; and if the micro-blog contains both the emotion type and the corresponding cause component, it will be labeled both. If the micro-blog does not contain any cause component, it will be only labeled the emotion type.

5.2   Evaluation Metrics

This paper uses the following three metrics to evaluate the performance: precision (U_P), recall (U_R) and F-score (U_F), see the formula (14), (15) and (16), respectively. S is the set of all posts in the collection; NEC is the number of posts with cause components which are identified through our algorithm; NEM is the total number of posts with cause components.

$$U_P = \frac{s_i \in S | Proportion(s_i) \quad is \quad correct}{NEC} \tag{14}$$

$$U_R = \frac{s_i \in S | Proportion(s_i) \quad is \quad correct}{NEM} \tag{15}$$

$$U_F = \frac{2 * U_P * U_R}{U_P + U_R} \tag{16}$$

As for the correctness of the cause components proportions under different emotions, the following method is used to analyze. Firstly, the proportional range of each cause component is divided into ten levels, and each level is expressed as $\tau_i (1 \leq i \leq 10)$: $\tau_1 \in (0\text{-}10\%)$, $\tau_2 \in (10\%\text{-}20\%)$, $\tau_3 \in (20\%\text{-}30\%)$, $\tau_4 \in (30\%\text{-}40\%)$, $\tau_5 \in (40\%\text{-}50\%)$, $\tau_6 \in (50\%\text{-}60\%)$, $\tau_7 \in (60\%\text{-}70\%)$, $\tau_8 \in (70\%\text{-}80\%)$, $\tau_9 \in (80\%\text{-}90\%)$ and $\tau_{10} \in (90\%\text{-}100\%)$, respectively. The sampled posts with cause components are labeled manually according to the above proportional levels. Secondly, the proportion of the cause component is obtained according to our algorithm and set to x. If the absolute error between x and τ_i is less than φ (φ is a calibration parameter), then the result is correct; otherwise, it is incorrect.

5.3   Experimental Analysis

For examining the effects of emotion cause extraction, this paper conducts the experiments from two aspects. One is to verify the feasibility of our algorithm in the aspect of emotion cause detection; the other is to verify whether using the multiple features can improve the accuracy of calculation of cause component proportion effectively.

To begin with, this paper compares our method with two other methods, Method I is designed on the rule-based system proposed by Lee et al. [1], and Method II is proposed in [3]. The method of ours in extracting emotion causes is based on an emotion model, while the other two methods use some rules and linguistic cues to extract emotion causes. Meanwhile, we use the same metric which is referred in the literature [3], then the results in terms of F-score are shown in Table 3. Obviously, the performance of ours is superior to others, and the F-score improves by 12.95% and 4.21% than Method I and Method II respectively. The experimental result demonstrates the feasibility of this approach and lays a foundation for the calculation of cause component proportion.

Besides, seven experiments are organized for gaining another goal. One is the baseline experiment, we can calculate the proportions only from the emotion intensities of keywords (which is called "EIK"). "EIK_DA" is the experiment which

Table 3. Comparison among the methods

Methods	F-score (%)
Our method	65.51
Method I	52.56
Method II	61.30

calculates the proportions combining "EIK" with degree adverbs. "EIK_ICON" is the experiment which calculates the proportions combining "EIK" with emoticons in micro-blogs. "EIK_NEGA" is the experiment which calculates the proportions combining "EIK" with negation words. "EIK_PUNC" is the experiment which calculates the proportions combining "EIK" with punctuations. "EIK_CONJ" is the experiment which calculates the proportions combining "EIK" with conjunctions. "EIK_ALL" is the experiment which calculates the proportions combining "EIK" with the five features. And the results are shown in Table 4.

Table 4 shows the results of the corresponding baseline experiment and the experiments of the proposed algorithm with multiple features. The precision of "EIK_ALL" is 82.50% which is higher than the other experiments. And the baseline without any language feature has the lowest F-score which is 69.99%. If the experiment is conducted by combining with emoticons, then the F-score increases to 73.09%. By the same token, if we add the other features, the F-score also increases. Obviously, recognizing the linguistic features of emoticons, negation words, punctuations, conjunctions and degree adverbs can be helpful in extracting emotion causes and calculating the proportions of the cause components. We also find that emoticons and negation words have a greater impact on calculating the proportion. And the precisions of the two experiments are 79.91% and 79.55% respectively. That is, people tend to use the two features to express strong emotions in micro-blogs.

Table 4. The results of the experiments

Experiments	Precision (%)	Recall (%)	F-score (%)
Baseline	76.52	64.48	69.99
EIK_DA	77.05	64.94	70.48
EIK_ICON	79.91	67.34	73.09
EIK_NEGA	79.55	67.04	72.76
EIK_PUNC	77.50	65.31	70.89
EIK_CONJ	77.32	65.16	70.72
EIK_ALL	82.50	69.53	75.46

Obviously, by using the method of emotion cause detection based on the ECOCC model, we can extract the cause events that trigger different emotions effectively. Meanwhile, it is possible to find the main cause component on the basis of the proportions of causes under public emotions. Moreover it can also help researchers to study the psychological activity of bloggers, and it has a profound influence on data mining.

6   Conclusions

In this paper, according to the emotional database, we present an ECOCC model which describes the eliciting conditions of emotions. The corresponding cause components under fine-grained emotions are also extracted. Latter, the proportions of cause components in the influence of the multiple features are calculated based on Bayesian probability. The experiment results demonstrate the effectiveness of the approach.

Acknowledgments. This work is sponsored by National Natural Science Foundation of China (Grant No.: 61175110, 61272362) and National Basic Research Program of China (973 Program, Grant No.: 2012CB316301). It is also sponsored by National Science Foundation of Hebei Province and its Academic Exchange Foundation (Grant No.: F2013208105, S2014208001), Key Research Project for University of Hebei Province (Grant No.: ZD2014029) and National Banking Information Technology Risk Management Projects of China.

References

1. Lee, S.Y.M., Chen, Y., Huang, C.-R., Li, S.: Detecting emotion causes with a linguistic rule-based approach. Computational Intelligence **39**(3), 390–416 (2013)
2. Chen, Y., Lee, S.Y.M., Li, S., Huang, C.-R.: Emotion cause detection with linguistic constructions. In: 23rd COLING, pp. 179–187 (2010)
3. Li, W., Xu, H.: Text-based emotion classification using emotion cause extraction. Expert Systems with Applications **41**(4), 1742–1749 (2014)
4. Rao, Y., Li, Q., Mao, X., Liu, W.: Sentiment topic models for social emotion mining. Information Sciences **266**, 90–100 (2014)
5. Steunebrink, B.R., Dastani, M., Meyer, J.-J.C.: A formal model of emotion triggers: an approach for bdi agents. Synthese **185**(1), 83–129 (2012)
6. Che, W., Li, Z., Liu, T.: Ltp: A chinese language technology platform. In: International Conference on Computational Linguistics: Demonstrations, pp. 13–16 (2010)
7. Xu, L., Liu, H., Pan, Y., Ren, H., Chen, J.: Constructing the affective lexicon ontology. Journal of the China Society for Scientific and Technical Information **27**(2), 180–185 (2008)
8. Mikolov, T., Chen, K., Corrado, G., Dean, J.: Efficient estimation of word representations in vector space. In: 1st ICLR (2013)
9. Cui, A., Zhang, M., Liu, Y., Ma, S.: Emotion tokens: bridging the gap among multilingual twitter sentiment analysis. In: Salem, M.V.M., Shaalan, K., Oroumchian, F., Shakery, A., Khelalfa, H. (eds.) AIRS 2011. LNCS, vol. 7097, pp. 238–249. Springer, Heidelberg (2011)
10. Zhang, P., He, Z.: A weakly supervised approach to chinese sentiment classification using partitioned self-training. Journal of Information Science **39**(6), 815–831 (2013)
11. Gao, K., Zhou, E.-L., Grover, S.: Applied methods and techniques for modeling and control on micro-blog data crawler. In: Liu, L., Zhu, Q., Cheng, L., Wang, Y., Zhao, D. (eds.) Applied Methods and Techniques for Mechatronic Systems. LNCIS, vol. 452, pp. 171–188. Springer, Heidelberg (2014)

Parallel Recursive Deep Model
for Sentiment Analysis

Changliang Li[1]($\boxtimes$), Bo Xu[1], Gaowei Wu[1], Saike He[1],
Guanhua Tian[1], and Yujun Zhou[2]

[1] Institute of Automation Chinese Academy of Sciences, Beijing,
People's Republic of China
{changliang.li,xubo,gaowei.wu,saike.he,guanhua.tian}@ia.ac.cn
[2] Jiangsu Jinling Science and Technology Group Co., Ltd, Nanjing, China

Abstract. Sentiment analysis has now become a popular research problem to tackle in Artificial Intelligence (AI) and Natural Language Processing (NLP) field. We introduce a novel Parallel Recursive Deep Model (PRDM) for predicting sentiment label distributions. The main trait of our model is to not only use the composition units, i.e., the vector of word, phrase and sentiment label with them, but also exploit the information encoded among the structure of sentiment label, by introducing a sentiment Recursive Neural Network (sentiment-RNN) together with RNTN. The two parallel neural networks together compose of our novel deep model structure, in which Sentiment-RNN and RNTN cooperate with each other. On predicting sentiment label distributions task, our model outperforms previous state of the art approaches on both full sentences level and phrases level by a large margin.

Keywords: Sentiment analysis · PRDM · Sentiment-RNN

1 Introduction

Sentiment analysis, also called opinion mining, is the field of study that analyzes peoples opinions, sentiments, evaluations, appraisals, attitudes, and emotions towards entities such as products, services, organizations, individuals, issues, events, topics, and their attributes [1].

With the development of Web 2.0, sentiment analysis has now become a popular research problem to tackle. In contrast to Web sites where people are limited to the passive viewing of content, Web 2.0 may allow users to more easily express their views and opinions on social networking sites, such as Twitter and Facebook. The opinion information they leave behind is of great value. For example, by collecting movie reviews, film companies can decide on their strategies for making and marketing movies. Customers can make a better decision which movie is worth watching. Hence, in recent years, sentiment analysis has become a popular topic for many research communities, including artificial intelligence and natural language processing.

© Springer International Publishing Switzerland 2015
T. Cao et al. (Eds.): PAKDD 2015, Part II, LNAI 9078, pp. 15–26, 2015.
DOI: 10.1007/978-3-319-18032-8_2

In this paper, we focus on the task of predicting sentiment label distributions, which aims to predict sentiment label distributions on phrase and sentence level. There has recently been considerable progress in predicting sentiment label distributions. Some existing supervised learning approaches [2] employed annotated corpora of manually labeled documents. Several unsupervised learning approaches have also been proposed [3,4] based on given sentiment lexicons.

Various joint sentiment and topic models [5–7] were proposed to analyze sentiment in detail. Recently, some models based on recursive neural networks have considerable representational power, such as RNN [8], MV-RNN [9] and RNTN [10]. All these models, i.e., RNN and its variant, predict sentiment label of phrase or sentences, based mainly on its vector representation, while missing some valuable local information for the global judgment.

Therefore, we consider that the local information has contributed to the global analysis, i.e. the sentiment distribution of the words in a phrase or sentence is important, and can impact the sentiment analysis for the phrase or sentence. So based on this hypothesis, we propose a Parallel Recursive Deep Model (referred as PRDM), by introducing a sentiment Recursive Neural Networks (sentiment-RNN) to cooperate with RNTN. In our model, each node in neural networks corresponds to a sentiment node, which is represented as sentiment label vector. All the sentiment nodes form a sentiment-RNN. Sentiment-RNN and RNTN composed of a parallel yet interacted structure. As a result, the sentiment of a phrase (or sentence) is calculated by utilizing not only the vector for the phrase or sentence but also the sentiment distribution of its children.

Moreover, we find that our PRDM is able to be seamlessly integrated with Stanford Sentiment Treebank[1], which is the first benchmark with fully labelled parsing trees. This Treebank is a valuable resource for a complete analysis of the compositional effects of sentiment in language [10]. And our PRDM is able to integrate not only the structure information of parsing trees, but also provide a potential network structure over the manually labels in the tree.i.e., PRDM provides a fully supporting platform for Stanford Sentiment Tree bank, to facilitate the analysis of the compositional effects of sentiment in language.

In order to illustrate our models effectiveness in predicting sentiment label distributions, we compare to several models such as RNN [8], MV-RNN [9] and RNTN [10], and baselines such as Naive Bayes (NB), bi-gram NB and SVM. Experimental evaluation demonstrates that our model outperforms previous state of the art.

The rest of the paper is organized as follows: In Section 2, we introduce some related work including word representations, recursive deep learning and Stanford Sentiment Treebank. Section 3 introduces our Parallel Recursive Deep Model and describes the details of parameter learning. Section 4 presents the experiments on predicting sentiment label distributions on both sentences and phrases level. The conclusion and future work are presented in Section 5. All the datasets, code and all relevant parameter specifications are publicly available[2].

---

[1] http://nlp.stanford.edu/sentiment/treebank.html
[2] https://www.dropbox.com/s/w6cqfqlv4ue7jpj/icdm2014.rar

2   Related Work

This work is mainly connected to two areas of NLP research: word representations and recursive deep learning; and one new corpus: Stanford Sentiment Treebank.

2.1   Word Representations

A word representation is a mathematical object associated with each word, often a vector. Each dimensions value corresponds to a feature and might even have a semantic or grammatical interpretation, so we call it a word feature [11].

Each dimension of the embeddings represents a latent feature of the word, hopefully capturing useful syntactic and semantic properties [11]. Many approaches have been proposed to learn good performance word embeddings. There are some embeddings datasets publicly available for evaluation, such as SENNAs embeddings [12], Turians embeddings [11], HLBLs embeddings [13], Huangs embeddings [14].

The training complexity is always a difficult problem of training a variety of language models. It was recently proposed using the distributed Skip-gram or continuous Bag-of-Words (CBOW) models. These models learn word representations using a simple neural network architecture that aims to predict the neighbors of a word. Due to its simplicity, the Skip-gram and CBOW models can be trained on a large amount of text data. There is one parallelized implementation can learn a model from billions of words in hours [15].

2.2   Recursive Deep Learning

Recursive neural networks (RNNs) [8] are able to process structured inputs by repeatedly applying the same neural network at each node of a directed acyclic graph (DAG). The recursive use of parameters is the main difference with standard neural networks. The inputs to all these replicated feed forward networks are either given by using the childrens labels to look up the associated representation or by their previously computed representation. RNNs [8] are related to auto-encoder models such as [16].

The matrix-vector RNNs (MV-RNN) [9] represent a word as both a continuous vector and a matrix of parameters. It assigns a vector and a matrix to every node in a parse tree: the vector captures the inherent meaning of the constituent, while the matrix captures how it changes the meaning of neighboring words or phrases. The composition function employed in MV-RNN can be referred to [17].

In the standard RNN, the input vectors only implicitly interact through the nonlinear function. In MV-RNN, the total number of parameters to learn is large. RNTN [10] aims to build greater interactions between the input vectors. It takes as input phrases of any length. And it utilizes the same tensor-based composition function for all nodes. RNTN has been successfully utilized in predicting phrase or sentence sentiment label.

2.3   Stanford Sentiment Treebank

In order to better express the meaning of longer and variable length phrases, richer supervised training and evaluation resources are necessary. The Stanford Sentiment Treebank is the first corpus with fully labeled parse trees that allows for a complete analysis of the compositional effects of sentiment in language [10].

The corpus consists of 11,855 single sentences, which are movie reviews. The Stanford parser [18] is used to parse all sentences. The Stanford Sentiment Treebank includes a total of 215,154 unique phrases from those parse trees. And each phrase is labeled by Amazon Mechanical Turk. This new dataset allows us to better predict sentiment label of any-length phrases or sentences based on supervised and structured machine learning techniques. The dataset will enable community to train and evaluate compositional models. Stanford Sentiment Treebank has been successfully utilized in predicting movie review [10]. More details about Stanford Sentiment Treebank dataset can be referred to http://nlp.stanford.edu/sentiment/treebank.html.

3   Parallel Recursive Deep Model

In this section, we firstly introduce our Parallel Recursive Deep Model (PRDM). And then we describe the details of parameter learning.

3.1   Parallel Recursive Deep Structure

Our Parallel Recursive Deep Model can strengthen compositional vector representation ability for phrase of any length and sentence, as well as the ability of sentiment label prediction via introducing a sentiment recursive deep structure.

Fig.1 shows our model. The key difference between our model and RNTN is that RNTN use the sentiment label of word or phrase only for local sentiment label training and prediction, while we introduce a sentiment-RNN to propagate local sentiment label information to the whole network, i.e., in the process of training and predicting, any local calculation includes remote linked sentiment label information. In this way, we take more full usage of sentiment labels, and also in this way, we can easily take the RNTN as the special instance of our model. For ease of exposition, we used a tri-gram not very good to explain our model.

We first make some definitions. Each word is represented as a d-dimensional vector. All the word vectors are stacked in the word embedding matrix $L \in R^{d \times |M|}$, $|M|$ is the size of the vocabulary. The word embeddings can be seen as a parameter that is trained jointly with the model. There are two kinds of units in PRDM, term node and sentiment node. Term node, corresponding to a word or phrase or sentence, is represented as a representation vector; sentiment one is represented as the sentiment label vector of word (or phrase, sentence).

When an n-gram is given to the model, it is parsed into a binary tree and each leaf node, corresponding to a word. In Fig.1, term node a, b and c are word vector representations for each word respectively.

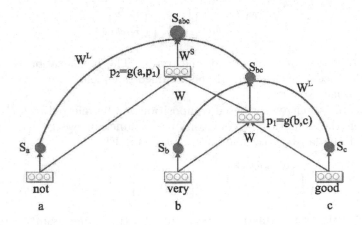

Fig. 1. A tri-gram example "not very good" to show structure of PRDM

Each term node corresponds to a sentiment node, for example, S_a is as sentiment node. The sentiment label vector is C-dimensional. $W^S \in R^{C \times d}$ is the sentiment classification matrix; $W^L \in R^{C \times 2C}$ is the sentiment transformation matrix; $W \in R^{d \times 2d}$ is the transformation matrix. In our model, we omit the bias for simplicity.

Parallel Recursive Deep Model consists of two recursive deep models. One is RNTN and another is sentiment-RNN. RNTN will compute parent vectors in a bottom up fashion using compositionality function g. The parent vectors are again given as features to next layer and sentiment classification matrix.

Meanwhile, sentiment-RNN will take both left and right child sentiment nodes as inputs to compute parent sentiment node. As a result, the final sentiment label of each phrase or sentence is computed through its vector and its children nodes sentiment nodes. From Fig.1, we can see that RNTN and sentiment-RNN are parallel in structure. It is also the reason that we call our model as Parallel Recursive Deep Model.

We use the tensor-based compositionality function g [17]. One advantage of tensor-based compositional function is that it can directly relate input vectors. The output of a tensor product is defined as formula (1).

$$h = \begin{bmatrix} V^{l-child} \\ V^{r-child} \end{bmatrix}^T V^{[1:d]} \begin{bmatrix} V^{l-child} \\ V^{r-child} \end{bmatrix} \tag{1}$$

$V^{[1:d]} \in R^{2d \times 2d \times d}$ represents the tensor that defines multiple bilinear forms. $V^{l-child} \in R^{d \times 1}$ is the left child node, while $V^{r-child} \in R^{d \times 1}$ is the right child node. $\begin{bmatrix} V^{l-child} \\ V^{r-child} \end{bmatrix}$ denotes the concatenation of two childrens column vectors resulting in a $R^{2d \times 1}$ vector. And the output of each slice product $h_i \in R^{C \times 1}$ is computed as formula (2).

$$h_i = \begin{bmatrix} V^{l-child} \\ V^{r-child} \end{bmatrix}^T V^{[i]} \begin{bmatrix} V^{l-child} \\ V^{r-child} \end{bmatrix} \qquad (2)$$

$V^{[i]} \in R^{2d \times 2d}$ is the ith slice of the tensor. The initialization of slices is random. The slices will subsequently be modified to enable a sequence of inputs to compose a new vector.

More details about tensor-based composition can be referred to [17].

Based on the definition of compositionality function, we give the general function as formula (3) to compute a parent vector V^p.

$$V^p = f \left(\begin{bmatrix} V^{l-child} \\ V^{r-child} \end{bmatrix}^T V^{[1:d]} \begin{bmatrix} V^{l-child} \\ V^{r-child} \end{bmatrix} + W \begin{bmatrix} V^{l-child} \\ V^{r-child} \end{bmatrix} \right) \qquad (3)$$

$W \in R^{d \times 2d}$ is the transformation matrix and also the main parameter to learn. When V is set to 0, the compositionality function is the same as used in standard Recursive Neural Networks [9]. $f = \tanh$ is a standard element-wise nonlinearity.

In our example, the vector of p_1 in Fig.1, the parent node of b and c, is computed through formula (4).

$$p_1 = f \left(\begin{bmatrix} b \\ c \end{bmatrix}^T V^{[1:d]} \begin{bmatrix} b \\ c \end{bmatrix} \right) \qquad (4)$$

After computing the first two nodes, the network is shifted by one position and takes as input vectors and again computes a potential parent node. The next parent vector p_2 in Fig.1 will be computed as formula (5).

$$p_2 = f \left(\begin{bmatrix} a \\ p^1 \end{bmatrix}^T V^{[1:d]} \begin{bmatrix} a \\ p^1 \end{bmatrix} + W \begin{bmatrix} a \\ p^1 \end{bmatrix} \right) \qquad (5)$$

Note that the parent vectors must be of the same dimensionality to be recursively compatible and be used as input to the next composition. RNTN model uses the same, tensor based composition function.

When comes to the sentiment-RNN, it repeats the similar process. However, the inputs become the sentiment nodes. Sentiment-RNN model uses standard composition function as formula (6) to compute parent sentiment node S^{bc}.

$$S^{bc} = f \left(W^L \begin{bmatrix} S_b \\ S_c \end{bmatrix} \right) \qquad (6)$$

Where $W^L \in R^{C \times 2C}$ is the sentiment transformation matrix. When W^L is set 0, RNTN [10] is the special case of our Parallel Recursive Deep Model. And similarly we employ formula (7) to compute next parent sentiment node S^{abc}.

$$S^{abc} = f \left(W^L \begin{bmatrix} S_a \\ S_{bc} \end{bmatrix} \right) \qquad (7)$$

We not only use each node as features to a sentiment classification matrix W^s, but also use its children nodes sentiment nodes as inputs to sentiment transformation matrix W^L. Then we use the obtained vectors as inputs to a soft classifier. For classification into C classes, we compute posterior over labels given the node and its corresponding sentiment node. For example, we compute the sentiment label of phrase not very good in Fig.1 as formula (8).

$$y^{S_{abc}} = \text{softmax}(W^s p_2 + S^{abc}) \tag{8}$$

Where $W^s \in R^{C \times d}$ represents sentiment classification matrix. The prediction for other phrases of the given tri-gram, for example very good, is similar to formula (8). We skip description for computing other phrases sentiment labels.

3.2   Backprop Through Parallel Recursive Deep Structure

In this section, we describe how to train our model. Given a sentence, we aim to maximize the probability of correct prediction, or minimize the cross-entropy error between the predicted and target sentiment labels at all nodes. The predicted sentiment label at node i is represented as y^i, and the target sentiment label at node i is represented as t^i, which is labelled by humans.

Let $\theta = (V, W, W^s, W^L, L)$ be our model parameters and β a vector with regularization hyper parameters for all model parameters. The error as a function of the PRDM parameters for a sentence is represented as formula (9).

$$E(\theta) = \sum_i \sum_{j=1}^{Num} t_j^i \log y_j^i + \beta |\theta|^2 \tag{9}$$

j means the jth sentence in corpus. Num is the total number of sentences in corpus.

The derivative for the weights of sentiment transformation W^L is standard. More details about process of compute can be referred to [19]. The derivative for the weights of sentiment transformation for sentiment node S_{abc} in Fig.1 is computed as formula (10).

$$\frac{\partial E^{S_{abc}}}{\partial W^L} = \left((y^{S_{abc}} - t^{S_{abc}}) \circ f'(S_{abc}) \right) \begin{bmatrix} S_a \\ S_{bc} \end{bmatrix}^T \tag{10}$$

$\circ$ is the Hadamard product between the two vectors. f' is the element-wise derivative of function f, which in standard case of using $f = \tanh$, can be computed as $f'(x) = 1 - f^2(x)$.

Similarly, the derivative for the weights of sentiment transformation W^L for sentiment node S_{bc} is computed as formula (11).

$$\frac{\partial E^{S_{bc}}}{\partial W^L} = \left((y^{S_{bc}} - t^{S_{bc}}) \circ f'(S_{bc}) \right) \begin{bmatrix} S_b \\ S_c \end{bmatrix}^T \tag{11}$$

The final result is the sum of all the error. We can compute the final derivative for the weights of sentiment transformation W^L as formula (12).

$$\frac{\partial E(\theta)}{\partial W^L} = \frac{\partial E^{S_{abc}}}{\partial W^L} + \frac{\partial E^{S_{bc}}}{\partial W^L} \tag{12}$$

The derivative for the weights of the sentiment classification matrix W^S at node p_2 is computed as formula (13).

$$\frac{\partial E^{p_2}}{\partial W^S} = \left((y^{S_{abc}} - t^{S_{abc}}) \circ f'(S_{abc})\right)p_2^T \tag{13}$$

The derivative for the weights of the sentiment classification matrix W^S at other nodes is similar to formula (13) and we skip the details here. The final result of the derivative for the weights of the sentiment classification matrix is the sum of all the error from each node. It is computed as formula (14).

$$\frac{\partial E(\theta)}{\partial W^S} = \sum_{i=1}^{T} \frac{\partial E^i}{\partial W^L} \tag{14}$$

T is the total number of nodes in the sentence parse tree.

The full derivative for W and V is the sum of the derivatives at each node. We can use formula (15) to compute the full derivative for V.

$$\frac{\partial E(\theta)}{\partial V^{[k]}} = \delta_k^{p_2,full} \begin{bmatrix} a \\ p_1 \end{bmatrix} \begin{bmatrix} a \\ p_1 \end{bmatrix}^T + \delta_k^{p_1,full} \begin{bmatrix} b \\ c \end{bmatrix} \begin{bmatrix} b \\ c \end{bmatrix}^T \tag{15}$$

$\delta_k^{p_2,full}$ and $\delta_k^{p_1,full}$ are the kth element of $\delta^{p_2,full}$ and $\delta^{p_1,full}$ respectively. $\delta^{p_2,full}$ and $\delta^{p_1,full}$ are full incoming errors for node p_2 and p_1 respectively. We skip the process of computing $\delta^{p_2,full}$ and $\delta^{p_1,full}$ for this tri-gram tree. For the optimization we use AdaGrad [20] to find optimal solution. The process of computing W is similar. More details about learning W and V can be referred to [10].

4   Experiments

For the experiment, we follow the experimental protocols on previous state-of-the-art RNTN model as described in [10]. We employed the same initialized parameters such as learning rate and word embeddings and so on.

The sentences in the Stanford Sentiment Treebank were split into three sets: train set (including 8544 sentences), dev set (including 1101 sentences) and test set (includ-ing 2210 sentences). We use dev set and cross validation over legalization of word vector size, learning rate as well as the weights and minibatch size for AdaGrad.

We compare to the previous state of the art based on RNTN. And we also com-pared to commonly used methods that use bag of words features with Naive Bayes and SVMs, as well as Naive Bayes with bag of bigram features. We abbreviate these with NB, SVM and biNB. We also compare to a model that averages neural word vectors and ignores word order (VecAvg), as well as Recursive Neural Networks (RNN), MV-RNN and RNTN.

4.1   Sentiment Label Prediction

Like other previous work [10], all the sentiment labels are classified into 5 classes. 0 means very negative; 1 means negative; 2 mean neutral; 3 means positive; 4 means very positive.

Fig. 2 shows one example of prediction for semantic label distributions. This ex-ample is A fascinating and fun film.. Note that the full stop . is also taken as a word. The right one is sentiment tree labeled by humans and the left one is predicted result by our model. In this example, we can see that there is only one prediction, (sentiment label for phrase fascinating and), different to sentiment label annotated by humans.

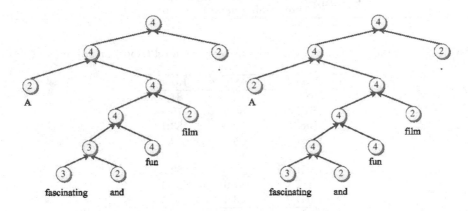

Fig. 2. Example of prediction for semantic label distributions. The left one is our predicted result. The right one is corresponding sentiment tree labeled by humans.

If one nodes sentiment label value $lv > 2$, then it will be classified into positive; If $lv < 2$, then it will be classified into negative; If $lv = 2$, it will be classified into neutral. All neutral sentences are removed from the Treebank, and finally we get 6920 sentences in train set, 872 sentences in dev set and 1821 sentences in test set.

Based on the dataset, like other previous work [10], we analyse binary performance of positive/negative on phrases or sentences ignoring the neutral classes.

Table 1 lists some examples of phrases with different predicted sentiment label values based on our model. The sentiment labels are classified into binary classification of positive/negative according to the values.

Table 2 shows results of this binary classification of positive/negative for both all phrases (All) and full sentences (Root). The results of other models or approaches in Table 2 are cited from [10].

From Table 2, we can see that our model outperforms other approaches by a large margin, especially on all phrases level. The previous state of the art for all phrases was 87.6% based on RNTN [10]. PRDM pushes the previous state of the art on full sentences up to 93.1%. PRDM achieves state of the art on both

Table 1. Phrases with different predicted labels

Label	Sentiment		Phrases
0	negative	very negative	is unrelentingly claustrophobic and unpleasant; a dumb distracted audience;
1		negative	shortcomings; cheesy effects; self-deprecating; flawed; tossing around obscure expressions;
2	neutral	neutral	a journey; below the gloss; one family test boundaries; at any time; the film's open-ended finale; the very end; with a bang;
3	positive	positive	Very true; talent; real talent; treat a subject; super hero; wit and insight;
4		very positive	best examples; visually polished; richly entertaining; predictably heartwarming tale;

Table 2. Accuracy for binary predictions at sentence level (Root) and all nodes (All)

Model	All	Root
NB	82.6	81.8
SVM	84.6	79.4
BiNB	82.7	83.1
VecAvg	85.1	80.1
RNN	86.1	82.4
MV-RNN	86.8	82.9
RNTN	87.6	85.4
PRDM	**93.1**	**86.5**

full sentence and all phrases level. It is expected since that RNTN is the special case of our PRDM when the sentiment transformation matrix W^L is set to 0.

Based on the experimental result, we can make a safe conclusion that our model is more reliable in predicting sentiment label distributions on both phrase and sentence level.

5   Conclusion

We introduce a novel Parallel Recursive Deep Model (PRDM) for predicting senti-ment label distributions. It can express the sentimental semantics of any length phrases and sentences. The main property of our model is to introduce a novel sen-timent-RNN cooperating with RNTN. PRDM not only utilizes the information of the vector for word and phrases, but also utilizes the information of the corresponding sentiment nodes. This strengthens the ability to predict sentiment label on both sen-tence and phrase level. Our model achieves state of the art performance on both full sentences level and all phrases level. Meanwhile, as our model is seamlessly integrat-ed with Stanford Sentiment Tree bank, our PRDM provides a fully supporting plat-form for this Tree bank, to facilitate the analysis of the compositional effects of sen-timent in language. In the future work, we will expand our model for more complicat-ed tasks in sentiment analysis, such as sentiment summarization.

Acknowledgments. This work is supported by National Program on Key Basic Research Project (973 Program) under Grant: 2013CB329302, National Natural Science Foundation of China (NSFC) under Grants No.61175050, No.61203281 and No.61303172.

References

1. Liu, B.: Sentiment analysis and opinion mining. Synthesis Lectures on Human Language Technologies 5(1), 1–167 (2012)
2. Blitzer, J., Dredze, M., Pereira, F.: Biographies, bollywood, boom-boxes and blenders: Domain adaptation for sentiment classification. In: Association for Computa-tional Linguistics (ACL) (2007)
3. Turney, P.D.: Thumbs up or thumbs down?: semantic orientation applied to unsu-pervised classification of reviews. In: Proceedings of the 40th Annual Meeting on Association for Computational Linguistics (2002)
4. Liu, B.: Sentiment Analysis and Subjectivity. To appear in Handbook of natural Language Processing, Second Edition (2010)
5. Mei, Q., Ling, X., Wondra, M., Su, H., Zhai, C.: Topic sentiment mixture: modeling facets and opinions in weblogs. In: Proceedings of the 16th International Conference on World Wide Web, pp. 171–180 (2007)
6. Titov, I., McDonald, R.: Modeling online reviews with multi-grain topic models. In: WWW 08: Proceeding of the 17th International Conference on World Wide Web, pp. 111–120, New York, NY, USA (2008)
7. Lin, C., He, Y.: Joint sentiment/topic model for sentiment analysis. In: Proceedings of the 18th ACM Conference on Information and Knowledge Management (CIKM) (2009)
8. Socher, R., Manning, C.D., Ng, A.Y.: Learning continuous phrase representations and syntactic parsing with recursive neural networks. In: Proceedings of the NIPS-2010 Deep Learning and Unsupervised Feature Learning Workshop (2010)
9. Socher, R., Huval, B., Manning, C.D., Ng, A.Y.: Semantic compositionality through recursive matrix vector spaces. In: EMNLP (2012)
10. Socher, R., Perelygin, A., Wu, J.Y., Chuang, J., Manning, C.D., Ng, A.Y., Potts, C.: Recursive deep models for semantic compositionality over a sentiment treebank. In: Proceedings of EMNLP (2013)
11. Turian, J., Ratinov, L., Bengio, Y.: Word representations: a simple and general method for semi supervised learning. In: Annual Meeting of the Association for Computational Linguistics (ACL) (2010)
12. Collobert, R.: Deep learning for efficient discriminative parsing. In: International Conference on Artificial Intelligence and Statistics (AISTATS) (2011)
13. Mnih, A., Hinton, G.E.: A scalable hierarchical distributed language model. In: NIPS, pp. 1081–1088 (2009)
14. Huang, E.H., Socher, R., Manning, C.D., Ng, A.Y.: Improving word representa-tions via global context and multiple word prototypes. In: Annual Meeting of the Association for Computational Linguistics (ACL) (2012)

15. Mikolov, T., Chen, K., Corrado, G., Dean, J.: Efficient estimation of word representations in vector space (2013). arXiv preprint arXiv:1301.3781
16. Pollack, J.B.: Recursive distributed representations. Artificial Intelligence **46**, November 1990
17. Mitchell, J., Lapata, M.: Composition in distributional models of semantics. Cognitive Science **34**(8), 1388–1429 (2010)
18. Klein, D., Manning, C.D.: Accurate unlexicalized parsing. In: ACL (2003)
19. LeCun, Y.A., Bottou, L., Orr, G.B., Müller, K.-R.: Efficient backprop. In: Orr, G.B., Müller, K.-R. (eds.) NIPS-WS 1996. LNCS, vol. 1524, pp. 9–50. Springer, Heidelberg (1998)
20. Duchi, J., Hazan, E., Singer, Y.: Adaptive subgradient methods for online learning and stochastic optimization. JMLR **12**, July 2011

Sentiment Analysis in Transcribed Utterances

Nir Ofek[✉], Gilad Katz, Bracha Shapira, and Yedidya Bar-Zev

Ben Gurion University, P.O. box 84105, Beer Sheva, Israel, 972-8-6477527
{nirofek,katzgila,bshapira,ybz1984}@bgu.ac.il

Abstract. A single phone call can make or break a valuable customer-organization relationship. Maintaining good quality of service can lead to customer loyalty, which affects profitability. Traditionally, customer feedback is mainly collected by interviews, questionnaires, and surveys; the major drawback of these data collection methods is in their limited scale. The growing amount of research conducted in the field of sentiment analysis, combined with advances in text processing and Artificial Intelligence, has led us be the first to present an intelligent system for mining sentiment from transcribed utterances—wherein the noisiness property and short length poses extra challenges to sentiment analysis. Our aim is to detect and process affective factors from multiple layers of information, and study the effectiveness and robustness of each factor type independently, by proposing a tailored machine learning paradigm. Three types of factors are related to the textual content while two overlook it. Experiments are carried out on two datasets of transcribed phone conversations, obtained from real-world telecommunication companies.

Keywords: Sentiment analysis · Noisy text mining · Customer satisfaction

1 Introduction

Enterprise call-center conversations are one of the most frequent interactions between customers and organizations, as they allow a direct contact to solve product and service-related issues [23]; therefore, they naturally involve negative emotions such as anger and frustration. The main goals of contact centers are to reduce operational cost and to decrease customer dissatisfaction. However, these two goals often have tradeoffs [2]. In decreasing operational cost, organizations have attempted downsizing, outsourcing and automating assistance. Less emphasis has been placed so far on decreasing customer dissatisfaction, in which emotions of customers play a crucial role.

While satisfaction can lead to customer loyalty, which affects profitability [8] customer feedback is mainly collected by interviews, questionnaires, and surveys on a small group of customers. These data collection methods pose three challenges. First, the scale of the data is limited due to time consumption. Second, the sample used is typically biased, since "active customers" who are happy and satisfied are more likely to respond to surveys. Third, these methods often have coarse temporal granularity as they are not analyzed in real-time.

Our system is designed to detect sentiment at the utterance level– the uninterrupted chain of words of the same speaker. Such granular analysis enables detecting change

© Springer International Publishing Switzerland 2015
T. Cao et al. (Eds.): PAKDD 2015, Part II, LNAI 9078, pp. 27–38, 2015.
DOI: 10.1007/978-3-319-18032-8_3

in sentiment [21], or multiple emotional states during a call. However, making meaning out of unstructured information is extremely difficult since text, despite being perfectly suitable for human consumption, is hardly accessible to machines. This problem is aggravated in transcribed text, which is typically generated by using a probabilistic model, which is error-prone due to the absence of punctuation and capitalization, the low quality of voice signals, the usage of irregular language, etc. For these reasons, standard natural language processing (NLP) techniques, such as parsing, are inefficient. Consider the following utterance from Fig. 1: *"no high just   mean in general i mean i'm not,"* which comprises of repetitions and has no sentence boundaries. This fact ultimately renders the sentence grammatically vague – and does not adhere to a coherent syntactic structure.

```
customer: so any of your fair we're  that you can return it
agent: i'm not allowed to recommend anything
customer: no high just  mean in general i mean i'm not
agent: which is all i know you're not authorized
customer: and you know i mean it's not like you know you're porting that
agent: i'm just i'm still not allowed to say anything like that
customer: okay not a problem okay so at least just click on save changes okay
```

Fig. 1. Automatic transcript of a partial conversation

We suggest five types of factors by processing information from multiple layers of conversational data. Since we postulate that sentiment is also expressed in non-textual forms—which are related to the structure and prosody of a conversation—some factor types exploit the textual content, while others overlook it.

The contribution of this study is threefold: first, we suggest designated factors for our specific problem, and elaborate on the lesson learned from utilizing each type. Secondly, our experiments are carried out on two datasets of transcribed phone conversations, obtained from real-world telecommunication companies, each containing thousands of utterances. Therefore, our study faces real-world settings including difficult-to-obtain transcriptions using commercial speech-recognition software and learning from an imbalanced dataset, which contributes greatly to the validity of our results and the conclusions we derive. Thirdly, in addressing the real-world settings, we suggest a seamless machine learning paradigm to utilize all factors and maintain robustness. On the contrary, other works on noisy text analyze the complete call [19], which is too coarse for our task, while there are a handful of works that consider short text [3, 24], but have no conversational context. The next section elaborates on the differences from our work.

2   Related Work

Despite the increased attention to sentiment analysis, little emphasis has been placed on analyzing transcribed text, which is becoming increasingly available with the advent of transcription systems. Transcribed text has a high degree of noise due to the usage of probabilistic models to interpret voice. [23] report an approximate 40% error rate in transcribed text, while [1] report an approximate 20% errors. In general, user-generated-content is hard to parse [16], while this problem is aggravated in transcribed text. Approaches which rely on the syntactic structure of sentences [4, 11, 13] are error-prone in noisy text, due to the various irregularities [24]. For example, the structure of the utterance *"got we can't was may need to delete files,"* differs completely from the original utterance *"well we may need to delete some files."* Thus NLP techniques such as dependency parsing are error-prone; in the example above, the parsed tree of the transcribed sentence considers the verb-noun dependency *got:files*, while it is clear from the original text that the dependency is *delete:files*.

The work of [19] designed a supervised learning scheme, to detect customer satisfaction in transcribed calls. The ground truth values were obtained from customer surveys. Consequently, all calls from customers were regarded as dissatisfied if this was the response of their survey, even if this emotion was not evident in their calls. Our study differs in three major aspects: (1) the data we use has been directly annotated by two professional judges to achieve more direct labels, (2) our method excludes in-domain information, such as ontologies and in-domain lexicons, and (3) since call level analysis may be too coarse, our analysis is performed on a granular level, thus facing the problem of short text mining. Additional studies that detect sentiment in noisy short text mainly use Twitter as their corpus [3], rely heavily on domain specific features (for example, using emoticons and hashtags), or not conversational [24], and therefore have no structure and prosodic flow—which we aim to investigate in this study. The work of [12] studies the effectiveness of utilizing the prosodic of speakers for sentiment analysis, which extracted from speech signals. In the current study, we do not have access to acoustic data.

3   Proposed Approach

To the best of our knowledge, this is the first attempt to detect sentiment in short transcribed noisy utterances. In this study we do not focus upon acoustic data, and rather exploit explicit information in the text, as well as the structural and prosodic layers latent in the conversation. Based on these elements, five types of factors are generated. We then employ a supervised machine learning approach, with the goal of discriminating between sentiment classes. In order to study the merit of each factor type independently, we suggest a machine learning algorithm which is suitable for every type. The main two considerations for our machine learning approach are learning from imbalanced datasets, and high dimensionality of the vector space.

3.1   Factors

In making our approach as generic as possible, we do not exploit information which is domain dependent. Three factor types consider the actual content (i.e., text) of the utterance: contextual, BOW and lexical; the prosodic and structural factor types overlook the textual content. Following, the five types of factors are described.

Prosodic: It is demonstrated that utilizing the prosody of speakers in their speech signals is beneficial to sentiment analysis [12]. However, since we have no access to acoustic data, we generate factors based on non-acoustic signals. Park and Gates [19] suggested that dissatisfied customers tend to "dominate" conversations. We attempt to take this intuition a step further by developing a set of factors dedicated to this issue, such as incorporating the talking time standard deviation. The following prosodic features were computed:

– $AVGTU_A$, $AVGTU_C$ are the average talking time per utterance of the agent and customer respectively—namely, the average utterance's length (in seconds) of each speaker. Long periods of talking time may indicate the confrontation of issues.
– $STDEVTU_A$, $STDEVTU_C$ are the standard deviation of the utterance talking time, for the agent and customer respectively.
– $SPEED$ is the speed of the utterance, namely, the total words divided by the total time. While the talking time may be a matter of style, it also may indicate the emotional state of the speaker.
– TR is the time ratio, i.e., the total time the customer spoke in the conversation, divided by the agent's total speaking time.
– DUR is the duration of the utterance.

Structural: We postulate that sentiment can be expressed in non-textual forms. These features are computed based on the structure of the conversation as a whole, while overlooking the actual content. The following three features were computed:

– UP is the current utterance positions from the beginning of the call. If sentiment is associated with position, this factor may encode relevant information.
– POS_{1stW} is the first word's position, namely, the words offset from the beginning of the call in which the first word of the utterance is positioned. The motivation is given by the previous factor.
– NU is the number of utterances in the call.

Lexical: Lexical information is widely used in sentiment classification. Corpus-based information has a relatively high potential in a supervised paradigm, while external lexicons are widely used in unsupervised, or semi-supervised paradigms. We utilize available sentiment lexicons as well as lexical information. These factors may indicate complexities in a call, such as repeating information, or occurrences of uncommon and informative terms.

The following lexical features were computed:

– $NWEC$, $NWEU$ are the number of Wikipedia entities found in the call and utterance, respectively. We calculate this feature by creating a tri-gram representation of the text of the call and searching for the Wikipedia page titles with an exact match to terms in this representation. The goal of these features is to identify the number of *entities* referred to in the text that may include product and service names. Since customers do not typically call to praise a product, it can imply related problems.

- NWC is the number of words in the call. An emotional call may be longer, in terms of the text spoken.
- NWC_A, NWC_C are the accumulated number of words in the agent's utterances in the call, and in the customer's, respectively. The motivation is explained in the previous feature, while here the information is per speaker.
- $AVGU_A$, $AVGU_C$ are the average number of the agent's and customer's words per utterance, respectively.
- $STDEVU_A$, $STDEVU_C$ is the standard deviation of the number of agent's and customer's words per utterance, respectively. High deviation values may be indicative of interruptions – which may imply an emotionally-driven conversation.
- $AVGWR$ is the average word ratio, i.e., the number of words spoken by the customer, divided by the number of words spoken by the agent during call. High ratio may indicate the presence of problems, which the agent is requested to address in detail.
- NC_{POS}, NC_{NEG} NU_{POS}, NU_{NEG} are the number of positive and negative terms found in a call, and the same feature is computed for the utterance level, respectively. The lexicon contains sets of positive and negative words that were obtained from the work of [9].
- NR is the number of repetitions, i.e., how many words repeat more than once in the utterance. The motivation to use this feature is derived from the re-occurring instances in which dissatisfied customers repeat words in the interest of emphasis.
- $MAXRL$ is the maximum number of times a word is repeated in an utterance. The motivation is given in the previous feature.

Bag-of-words (BOW): Bag-of-words are models where single words or word stems are used in the representation of documents. It is widely used in text classification and provides a strong baseline for sentiment analysis [18, 27]. In our problem, we use the unigram model which was more effective than incorporating bi-grams or three-grams. This finding is in accordance with [18], and may be explained by the fact that many n-grams may contain noisy in some of their word. If each word has the same probability for being miss-transcribed, a bi-gram has twice the chance of being erroneous than a uni-gram.

Contextual: When using the BOW model, a great deal of the information from the original text is discarded. Rather than using single-word information, utilizing contextual indicators aims to integrate information about key terms and the context in which they appear in the model. First, we employ the method proposed by [10] to detect key terms and their relevant contexts in the data. This method, which is based on the language modeling technique [20], was proven efficient for unsupervised keywords extraction. Once detected, we store a frequency table for each key term; each entry is a frequency score which corresponds to the key term's association with any of its context terms. Context terms are identified according to a sliding window of thirty terms in the key terms vicinity. This method is frequently used for word representation by context terms and for word embedding [14]. The mentioned process is repeated separately for each sentiment class, i.e., only utterances pertaining to the specific sentiment class will participate in generating the key terms and frequency tables. Based on the key terms and their context terms, we compute a set of features. Due to space limitations, and since the scope of this paper involves proposing factor *types* and evaluating the merit of each type for our problem, we provide a subset of the used features. The features are divided into two groups, based on their origin:

- *Key terms-based features* are a set of features which provide various aspects of the detected key terms: their proximity to other key terms in the utterance, the number of times each of them appears in the text etc. Some features also model the *absence* of key terms, analyzing the length and frequency of text sequences without key terms.
- *Context terms-based features* are a set of features aimed at representing the context of key terms detected in the document. The generated features include the average number of context terms per key term, the number of context terms shared by several key terms etc.

All factors aim to exploit different aspects of the data that may indicate dissatisfaction or any other form of sentiment. For example, repetition or extremely long utterances may indicate customers who narrate their issues or problems. We do not consider in-domain information, such as a list of the company's products and services, or competitor products' names (as done by [19]). However, we do use Wikipedia in the attempt to identify similar information as explained before.

3.2 Learning Algorithms

We explored machine learning algorithms while placing emphasis on facing our specific problem characteristics. The first is the imbalanced class problem. Induction algorithms tend to misclassify the instances of the less represented classes for a variety of reasons [5]. The second characteristic is the high vector space dimensionality, which is common in text processing and language-model bases techniques. The learning process in this case might be affected by the curse-of-dimensionality, and can lead to overfitting. In this section we describe the algorithms which we found effective to our problem along with the motivation for their use.

Rotation Forest (RF) [22]. is an ensemble classifier that applies rotations on the input data instances through feature selection algorithms. It is considered as an effective algorithm for many classification tasks, and can outperform both single and ensemble benchmark classifiers [28]. The confidence output for classifying instance x is determined as follows: Let N be the number of classifiers in the ensemble, R_j^a (j=1...N) a rotation matrix correspond to the *jth* classifier (review [22] for more details on generating the matrix), $d_{j,i}(xR_j^a)$ the probability that x pertains to class w_i according to classifier D_j. The aggregated confidence for class w_i across N classifiers is determined by the following averaging equation:

$$\mu_i(x) = \frac{1}{N}\sum\nolimits_{j=1}^{N} d_{j,i}(xR_j^a)$$

(1)

Generally, ensemble techniques were found more resistant than base-algorithms in classifying imbalanced datasets [7]; hence, while the use of several classifiers contributes to the complexity, it justifies itself by maintaining robustness to imbalanced class distribution. This principle can be seen in Equation 1; each classifier was trained by using rotation matrix of manipulated features, and a sample of bootstrap instances, to promote the diversity of the classifiers. Averaging the probabilities of diverse classifiers benefits the ensemble learning in imbalanced learning, since implication of misclassifications is reduced. To better maintain robustness, bootstrapping can be performed on only majority class instances. The study of [6] found RF resistant in classifying imbalanced datasets while it uses principal component analysis (PCA) for linear transformation, which increases the diversity among ensemble classifiers.

Support Vector Machine (SVM). aims at finding the hyperplane which splits instances of two classes with the largest distance in between the instances of the two classes [25]. Classifying text potentially involves vector space with high dimensionality. As such, classification performance might be affected by the curse-of-dimensionality, which can lead to overfitting. It is well known that SVMs work well in the case of high dimensionality [15] since they can map complex nonlinear input/output relationships with relatively good accuracy. SVM models are also found to be resistant to noisy text classification tasks [26].

4   Experimental Evaluation

Our task is to detect sentiment in transcribed utterances, which were obtained from contact-center conversations. Therefore, they naturally involve negative sentiment such as anger and frustration. The organizations that contributed the data wish to detect negative sentiment in utterances, to spot problematic sections in conversations. Consequently, in our experiments, the goal is to discriminate between negative and non-negative sentiment classes. This problem is designed in the same manner as the classical negative/positive problem, in which the positive class is regarded as non-negative. The Rotation Forest classifier is employed, with PCA for feature transformation, and each factor type is used independently. In cases involving high dimensionality due to the use of BOW, the SVM is employed for classification; since our system uses all factors (including BOW), both SVM and RF are compared in this case. The BOW model was used as a baseline, since it is considered as a strong baseline for sentiment analysis [12, 18].

4.1   Datasets

The speech recognition system is able to segment utterance boundaries and to distinguish between the speakers. The two datasets used in this study were recorded from a call center of telecom companies and correspond with users making queries regarding various concerns, such as billing, technical issues, and accessing services. In Table 1 we present the statistics of the automatic transcribed call-center calls datasets from two telecommunication companies' (referred to as Tele1 and Tele2). The dataset of Tele1 was the larger dataset comprised of 377 calls and 20,338 utterances compared to 113 calls and 15,183 utterances for Tele2. However, the main difference between the datasets is related to the imbalance ratio of the two classes. Negative utterances in Tele2 comprise 15.39% of the dataset, while 20.31% of the Tele1 dataset.

Table 1. Summary of the datasets

Data-set	Calls	Utterances	Average Utterances per Call	Negative Utterances	Average Words per Utterance
Tele1	377	20338	53.94	862 (8.7%)	14.91
Tele2	113	15182	134.35	1563 (20.3%)	17.11

4.2   Gold Standard

We developed a designated tagging tool to enable the annotators to view a complete call, and to annotate each utterance. The system used in this study is able to automatically identify the speaker—namely, whether an utterance is spoken by the customer, or by the representative agent. Two professional judges were instructed to tag each utterance as either negative or non-negative by accessing only the text. A negative tag is given when emotions are explicitly and clearly expressed, such as: *'I'm I'm very upset mean every it's been its just five years'*. Another negative indicator is when a disagreement between speakers is observed. Since the agreement rate was high (nearly 90%), to avoid the problem of disagreement, in our evaluation we consider only utterances that both judges agreed upon their sentiment.

4.3   Evaluation Measures

The Receiver Operating Characteristic (ROC) curve is a graph produced by plotting the true positives rate versus the false positives rate over all possible acceptance thresholds. In our evaluation we will use the area under the curve (AUC) of the ROC curve measure, since it is not overwhelmed by majority instances of unbalanced datasets and is more robust than other measures [17]. The same rationale applies for precision/recall curves, as they enable evaluation in several recall points and do not involve setting a single threshold. This measure is widely used in information retrieval, and we present this in-depth evaluation for the negative class, which is of interest to our experiments.

4.4   Results

Table 2 provides results on the two datasets (all experiments setups are five-folds c.v.). It is clear that all factor types achieve higher results than a random classifier (which has an AUC of 0.5). The best performance was achieved when using all factors with RF, while it was significantly better than alternatively using SVM as a classifier. This method and the contextual-based method are significantly better than the BOW baseline, in the Tele1 dataset. Fig. 2 provides in-depth evaluation for the negative class. It is clear the contextual and all-factors (using RF) methods dominate in both datasets. However, despite seems minor in Table 2, the difference between all factors model and the contextual model is statistically significant (level of significance 0.1). All other methods are comparable, except the structural-based method which is hardly indicative. Examples of non-negative classifications, using the "all factors" system include: *"Alright just give me about a bit what but it has full keyboard"*, *"okay well then I see you've been good to me then ill do it with you"*. Utterances classified as negative include: *"yes this is ridiculous and I mean it wont be seventy fifty did out until tomorrow it its just crazy dont it show where do changed it September the"*, *"Im im very uoset mean every its been its just five rears"*.

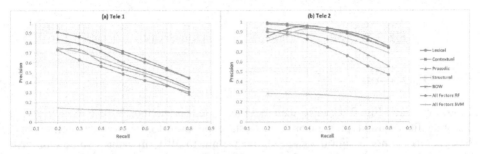

Fig. 2. Precision-recall performance of detecting negative utterances

Table 2. AUC results for the evaluated factors

Dataset	Lexical	Contextual	Prosodic	Structural	BOW	All Factors RF	All Factors SVM
Tele1	0.858	0.927•	0.869	0.557	0.890	**0.932**•○	0.880
Tele2	0.873	0.944	0.900	0.596	0.941	**0.948**○	0.928

○Significantly better than "All Factors SVM" • Significantly better than BOW, level of significance 0.05

4.5   Cross Corpus Experiments

In the previous evaluation we evaluated the usefulness of each factor type for our problem. In these experiments one corpus was used for training, while the other was used as a test set. The cross corpus problem often occurs in our domain since it is expensive to annotate each corpus for training a designated model. Thus this section evaluates to what extent each factor type maintained robustness. Experiments carried out similarly to the five folds cross-validation setup, while here the control set was taken from the other corpus.

Table 3 summarizes the performance of the factor types, while "Tele1 to Tele2" indicates that Tele1 was used for training, while Tele2 for testing. Our system which uses all factors achieves the best results, while the structural type itself is not efficient at all. Fig. 3 presents the in-depth analysis of the negative sentiment class. The contextual factor type and using all factors dominates both graphs.

Table 3. AUC results for the evaluated factors

Experiment	Lexical	Contextual	Prosodic	Structural	BOW	All Factors RF
Tele1 to Tele2	0.840	0.921•	0.911	0.464	0.875	**0.925•**
Tele2 to Tele1	0.858	**0.934•**	0.920	0.469	0.917	**0.934•**

• Significantly better than BOW, level of significance 0.05

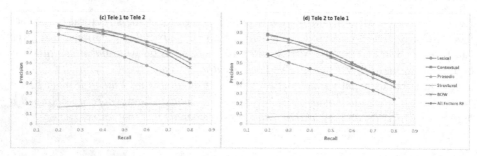

Fig. 3. Precision-recall performance of detecting negative utterances

5   Discussion

From Fig. 2 and Fig. 3 it can be seen that the context is the most effective factor type. The benefits of this type are manifold. First, as opposed to the BOW model, contextual factors do not assign the same initial weight to all terms, since some can have higher values, based on their context words etc. Secondly, the context model considers context in the vicinity of key terms, which is important for sentiment analysis; an intuitive example is negation words which actually can shift the polarity of terms. Composing bi-grams and three-grams encode more context than uni-grams, but are more likely to contain noisy words, as explained on Section 3.1. Structural factors which overlook the textual content are less effective predictors, but still indicative to a minor extent. In Table 2, we can see that structural-based model achieves AUC of 0.55 - 0.6, which is slightly above the random classifier; since this indicates that structural information carries a certain signal to sentiment analysis, and due to the fact that we generated a simple set of factors, an independent study of a structure of a conversation could be a promising direction if further studied. Likewise, the prosodic information overlooks the actual text and did not outperform the BOW model. However, they achieve comparable results and maintain robustness, as we can see in Table 3. This important finding may imply that prosodic characteristics are not corpus specific, and may contribute greatly to the robustness of the system. The lexical factor type carries meaningful signal, as can be seen in Table 2 and Table 3, however, independently it is far less effective than the BOW baseline. The limited contribution can be explained by the fact that spoken language is far less "orderly" than written language, therefore lexicons and knowledge bases have poor coverage. For example, the word "soothingly" appears in the used lexicon, but is not likely to be spoken in a regular conversation. Apart from this, the effectivity of some lexical factors may be limited due to the irregularity trait of spoken text. For example, repetitions may imply an argument, however may also indicate that the customer was not property heard, due to poor voice signals, and repeated herself.

The combined model achieves the best performance, and despite being correlated with the contextual model, it outperforms the contextual model with statistical significance. We found the Rotation Forest classifier to be resistant to the curse of dimensionality and due to its ensemble mechanism, resistant in imbalanced class learning. Table 2 demonstrates that the difference between SVM and RF (by using all factors) is more distinguished in Tele1, in which the imbalance ratio is higher. Fig. 2 further supports this observation. Our finding which SVMs are less resistant to imbalanced learning is further supported by the BOW model. Which it uses SVM, it is less effective in Tele1 (imbalance ratio is 1:10), while in Tele2 (imbalance ratio is 1:4) it achieves relatively better results. By comparing the results of the combined model (in Table 2 and Table 3), we can see that the RF model maintains robustness which indicate that this model is not biased towards a particular dataset and therefore is not over-fitted.

6   Conclusions and Future Work

We studied five types of factors and trained a model to utilize each type independently, and jointly. Our findings highlight the benefit of the contextual factors. For future work, we plan to conduct an intensive discourse analysis, and study the usefulness of utilizing contextual information, while here context refers to adjacent urethanes. Consequently, we will incorporate the content of adjacent utterances to detect sentiment in a specific utterance.

References

1. Abdel-Hamid, O., et al.: Applying convolutional neural networks concepts to hybrid NN-HMM model for speech recognition. In: 2012 IEEE International Conference on Acoustics, Speech and Signal Processing (ICASSP). IEEE (2012)
2. Anderson, E.W., Fornell, C., Rust, R.T.: Customer satisfaction, productivity, and profitability: differences between goods and services. Marketing science 16(2), 129–145 (1997)
3. Barbosa, L., Feng, J.: Robust sentiment detection on twitter from biased and noisy data. In: Proceedings of the 23rd International Conference on Computational Linguistics: Posters. Association for Computational Linguistics (2010)
4. Chan, K.T., King, I.: Let's tango – finding the right couple for feature-opinion association in sentiment analysis. In: Theeramunkong, T., Kijsirikul, B., Cercone, N., Ho, T.-B. (eds.) PAKDD 2009. LNCS, vol. 5476, pp. 741–748. Springer, Heidelberg (2009)
5. Chawla, N.V., Japkowicz, N., Kotcz, A.: Editorial: special issue on learning from imbalanced data sets. ACM Sigkdd Explorations Newsletter 6(1), 1–6 (2004)
6. De Bock, K.W., Poel, D.V.D.: An empirical evaluation of rotation-based ensemble classifiers for customer churn prediction. Expert Systems with Applications 38(10), 12293–12301 (2011)
7. Galar, M., et al.: A review on ensembles for the class imbalance problem: bagging-, boosting-, and hybrid-based approaches. IEEE Transactions on Systems, Man, and Cybernetics, Part C: Applications and Reviews 42(4), 463–484 (2012)
8. Hallowell, R.: The relationships of customer satisfaction, customer loyalty, and profitability: an empirical study. International journal of service industry management 7(4), 27–42 (1996)
9. Hu, M., Liu, B.: Mining and summarizing customer reviews. In: Proceedings of the Tenth ACM SIGKDD (2004)
10. Katz, G., Elovici, Y., Shapira, B.: CoBAn: A context based model for data leakage prevention. Information Sciences 262, 137–158 (2014)

11. Liu, K., et al.: Opinion target extraction using partially-supervised word alignment model. In: Proceedings of the Twenty-Third International Joint Conference on Artificial Intelligence. AAAI Press (2013)

12. Mairesse, F., Polifroni, J., Di Fabbrizio, G.: Can prosody inform sentiment analysis? experiments on short spoken reviews. In: 2012 IEEE International Conference on Acoustics, Speech and Signal Processing (ICASSP). IEEE (2012)

13. Matsumoto, S., Takamura, H., Okumura, M.: Sentiment classification using word subsequences and dependency sub-trees. In: Ho, T.-B., Cheung, D., Liu, H. (eds.) PAKDD 2005. LNCS (LNAI), vol. 3518, pp. 301–311. Springer, Heidelberg (2005)

14. Mikolov, T., et al.: Distributed representations of words and phrases and their compositionality. In: Advances in Neural Information Processing Systems (2013)

15. Morik, K., Brockhausen, P., Joachims, T.: Combining statistical learning with a knowledge-based approach: a case study in intensive care monitoring. Technical Report, SFB 475: Komplexitätsreduktion in Multivariaten Datenstrukturen, Universität Dortmund (1999)

16. Ofek, N., Rokach, L., Mitra, P.: Methodology for connecting nouns to their modifying adjectives. In: Gelbukh, A. (ed.) CICLing 2014, Part I. LNCS, vol. 8403, pp. 271–284. Springer, Heidelberg (2014)

17. Oommen, T., Baise, L.G., Vogel, R.M.: Sampling bias and class imbalance in maximum-likelihood logistic regression. Mathematical Geosciences **43**(1), 99–120 (2011)

18. Pang, B., Lee, L., Vaithyanathan, S.: Thumbs up?: sentiment classification using machine learning techniques. In: Proceedings of the ACL-02 Conference on Empirical Methods in Natural Language Processing, vol. 10. Association for Computational Linguistics (2002)

19. Park, Y., Gates, S.C.: Towards real-time measurement of customer satisfaction using automatically generated call transcripts. In: Proceedings of the 18th ACM Conference on Information and Knowledge Management. ACM (2009)

20. Ponte, J.M., Croft, W.B.: A language modeling approach to information retrieval. In: Proceedings of the 21st Annual International ACM SIGIR Conference on Research and Development in Information Retrieval. ACM (1998)

21. Portier, K., et al.: Understanding topics and sentiment in an online cancer survivor community. JNCI Monographs **2013**(47), 195–198 (2013)

22. Rodriguez, J.J., Kuncheva, L.I., Alonso, C.J.: Rotation forest: A new classifier ensemble method. IEEE Transactions on Pattern Analysis and Machine Intelligence **28**(10), 1619–1630 (2006)

23. Roy, S., Subramaniam, L.V.: Automatic generation of domain models for call centers from noisy transcriptions. In: Proceedings of the 21st International Conference on Computational Linguistics and the 44th Annual Meeting of the Association for Computational Linguistics. Association for Computational Linguistics (2006)

24. Thelwall, M., et al.: Sentiment strength detection in short informal text. Journal of the American Society for Information Science and Technology **61**(12), 2544–2558 (2010)

25. Vapnik, V.: The nature of statistical learning theory. Springer (2000)

26. Vinciarelli, A.: Noisy text categorization. IEEE Transactions on Pattern Analysis and Machine Intelligence **27**(12), 1882–1895 (2005)

27. Xia, R., Zong, C.: Exploring the use of word relation features for sentiment classification. In: Proceedings of the 23rd International Conference on Computational Linguistics: Posters. Association for Computational Linguistics (2010)

28. Zhang, C.-X., Zhang, J.-S.: RotBoost: A technique for combining Rotation Forest and AdaBoost. Pattern Recognition Letters **29**(10), 1524–1536 (2008)

Rating Entities and Aspects Using a Hierarchical Model

Xun Wang$^{(\boxtimes)}$, Katsuhito Sudoh, and Masaaki Nagata

NTT Communication Science Laboratories, Kyoto 619-0237, Japan
{wang.xun,sudoh.katsuhito,nagata.masaaki}@lab.ntt.co.jp

Abstract. Opinion rating has been studied for a long time and recent work started to pay attention to topical aspects opinion rating, for example, the food quality, service, location and price of a restaurant. In this paper, we focus on predicting the overall and aspect rating of entities based on widely available on-line reviews. A novel hierarchical Bayesian generative method is developed for this task. It enables us to mine the overall and aspect ratings of both entity and its reviews at the same time. We conduct experiments on TripAdvisor and results show that we can predict entity-level and review-level overall ratings and aspect ratings well.

1 Introduction

It has becoming a common phenomenon that when we need to make decisions about about traveling, shopping and other consumptions, we turn to user-generated reviews for help. We can easily obtain information about the food, service, location and price of a restaurant without going there by ourselves, as long as we do not mind reading pages and pages of reviews. To save the effort of reading reviews in large quantity, recently researchers started to study text mining on review data. Problems existed for addressing this issue generally include information extraction Ding et al. [2009]; Morinaga et al. [2002]; Popescu and Etzioni [2005], sentiment analysis Kim and Hovy [2004]; Pang and Lee [2005], opinion summarization Hu and Liu [2004]; Jindal and Liu [2006]; Lu and Zhai [2004] and spam detection Li and Cardie [2013a]; Jindal and Liu [2013]; Li et al. [2013b, 2014].

Besides, for a specific entity, usually customers are more concerned about certain aspects than other aspects. Fox example, some customers are price-sensitive and some may be quality-sensitive. The task to predict opinion rating on topical aspects reviews is called Aspect Rating Prediction.

This task aims to help customers to efficiently and accurately grasp the information they need from the huge amount of review data.

Many methods Hu and Liu [2008]; Lu et al. [2011]; Titov and McDonald [2008a]; Wang et al. [2010]; Wang et al [2011] have been proposed to predict opinion ratings on specific aspects directly. While to predict aspect ratings at

© Springer International Publishing Switzerland 2015
T. Cao et al. (Eds.): PAKDD 2015, Part II, LNAI 9078, pp. 39–51, 2015.
DOI: 10.1007/978-3-319-18032-8_4

entity level, as an alternative approach, has not been considered yet. Existing work usually attains aspect rating from reviews directly. However, the entity-level information also has an impact on aspect ratings. For example, a five-star hotel tends to have high quality food and good service. Without considering the entity-level information, the aspect-level rating would suffer greatly. Similarly we also need to consider the review-level ratings for better aspect-level rating.

In this paper, we address the task of Entity Aspect Rating. A novel hierarchical Bayesian generative approach is developed to identify overall and aspect ratings at both review-level and entity-level.

From the review data, we find that (a) reviews on the same entity tend to have similar scores, (b) aspect ratings on the same entity tend to be similar, (c) aspect ratings in one review tend to be similar.

Based on such observations, in the proposed model, we assume (a) the aspect rating of one entity follows a normal distribution with a mean value equal to the overall rating, (b) the review-level rating follows a normal distribution with a mean value of entity-level rating, (c) the aspect rating for each review follows a normal distribution which is a combination of entity-level rating on that aspect and review-level rating. In this way, we successfully model entity overall ratings, entity aspect ratings, review overall ratings and review aspect ratings in a hierarchical model. Text that describes a particular aspect is generated by sampling words from the combination of aspect-word distribution and rating-word distribution within the same aspect.

Experiment results on review data set from Tripadvisor show that our model can well predict both the entity aspect rating score, and aspect ratings in individual reviews. The proposed model can be applied to any collection of review data to find latent topical aspects and predict aspect ratings.

2   Related Work

Aspect review rating has been studied for a long time. Early work Hu and Liu [2004, 2008]; Popescu and Etzioni [2005] usually mined entity features first and then summarized the opinions in each review. Recent work starts to develop systems that are able to detect and aggregate aspects and corresponding opinions Lerman et al. [2009]; Zhuang et al. [2002].

Topic models, due to their ability to uncover latent aspects from text, have been popular in predicting aspect ratings from review data recently.

Several models based on LDA Blei et al. [2003] have been developed to address the aspect rating problem. A typical one is MaxEnt-LDA Zhao et al. [2010] which is a maximum entropy hybrid model and is able to discover both aspects and aspect-specific opinion words from reviews simultaneously. Multi-grain LDA (MG-LDA) Titov and McDonald [2008a] is another LDA based model which models review-specific elements and ratable aspects at the same time. Brody and Elhadad (2008) proposed the Local-LDA which analyzes sentences and discovers ratable aspects in review data.

To predict review rating, supervised LDA (sLDA) model Blei and McAuliffe [2007] is proposed to model the overall rating of each review. Wang et al. [2010]

Wang et al. (2010) proposed a novel method that combined topic models with latent regression technique for aspect ratings in the reviews. Other work include the Good Grief algorithm Synder and Barzilay [2007] which use an online Perceptron Rank (PRank) algorithm to rate aspects for entities and that of Sauper et al. (2010) which integrated the HMM model to improve both multi-aspect rating prediction and aspect-based sentiment summarization.

One problem with these aspect rating approaches mentioned above is that they only consider the review level information and ignore the entity-level information, hence the aspect rating sometimes suffer.

3   Rating Model

3.1   Problem Formulation

For a review date set, let N be the number of entities. Each entity E_i $i \in [1, N]$ is comprised of a collections of reviews, $E_i = \{d_{ij}\}_{j=1}^{j=N_i}$ and N_i denotes the number of reviews for entity i. Each review r is comprised of a collection of sentences $d = \{s_t\}_{t=1}^{t=n_d}$, where n_d denotes the number of sentences in current review. Each sentence is comprised of a collection of words $s = \{w_t\}_{t=1}^{t=n_s}$, where n_s denotes the number of words with sentence s. Assume that there are K aspects for the entity, denoted as $\{A_1, A_2, ..., A_K\}$. V is the vocabulary size. Each entity is associated with an overall rating $R_{i,k}$ and a K-dimensional aspect rating vector $\{R_{ik}\}_{k=1}^{k=K}$. R_{ik} and R_i^{all} can be any positive values. Each review d is associated with an overall rating r_d^{all} and a K-dimensional aspect $\{r_{dk}\}_{k=1}^{k=K}$.

The inputs are $train - data$ and $test - data$, both of which contain a collection of entities. In $train - data$, each review is presented with known values of r_d^{all} and r_{dk}. The output for the proposed model would be 1) the overall rating R_i^{all} and aspect rating $R_{i,k}$ for each entity in $test - data$. 2) the overall rating r_d^{all} and aspect rating $r_{d,k}$ for each review d in $test - data$.

3.2   Hierarchical Model

In the hierarchical model, aspect ratings for the same entity or the same review is not independent from each other. Specifically, for each entity E_i, its overall score is firstly drawn from a Gaussian distribution with mean value μ and deviation Σ^2, as shown in Equ. (1)

$$R_i^{all} \sim N(\mu, \Sigma^2) \tag{1}$$

Then the entity aspect rating R_{ik} and review overall rating r_{dk}, $d \in E_i$ are drawn from a Gaussian distribution with mean value R_i^{all} as follows:

$$R_{ik} \sim N(R_i^{all}, \sigma_1^2) \tag{2}$$

$$r_d^{all} \sim N(R_i^{all}, \sigma_2^2) \tag{3}$$

The aspect rating for review d is drawn from the normal distribution whose the expectation is the combination of overall rating for current review and the aspect rating for the entity as follows:

$$r_{dk} \sim N(\rho r_d^{all} + (1 - \rho)R_{ik}, \sigma_3^2) \tag{4}$$

where ρ is the parameter controls the influence of r_d^{all} and R_{ik}, and is set to 0.5 in our approach.

We assume that words of one sentence are all from the same topic Gruber et al. [2007]; Lu et al. [2011]; Sauper et al. [2010]. The generative story of a reviewer's behavior is as follows: to generate an opinionated review d, the reviewer would first decide the set of aspects he wants to describe. For each aspect, he would choose vocabularies from the word distribution of that aspect. For example, if he wants to comment on food from a restaurant, he can choose vocabularies such as "food", "restaurant", "coffee" etc. Next, he has to decide vocabularies to reflect polarities according to the ratings for such aspect in his minds. If he is satisfied with rooms, he can choose positive words such as "good", "clear" etc. Otherwise, he can choose vocabularies such as "bad", "terrible", "dirty" etc. Since $r_{d,k}$ can be any positive value, we take the rounding value of $r_{d,k}$, denoted as $[r_{d,k}]$ to model rating-word distribution.

In the text modeling, we assume each aspect A_k is characterized by a multinomial word distribution $Multi(\phi_k)$ over vocabulary, where the proportion of aspects θ is drawn from a Dirichlet distribution $Dir(\alpha)$. Each rating score m, $m \in [1, M]$ for aspect k is characterized by a rating-specific word distribution $\psi_{k,m}$. Each review is treated as a mixture over the latent aspects, and the joint probability of observed word contents W, latent aspect assignments $\{z_n\}_{n=1}^{n=n_d}$ and aspect proportion θ is defined as follows:

$$p(\mathbf{W}, z, \theta) = p(\theta_d|\gamma) \prod_{s \in d} p(z_s|\theta_d) \prod_{w \in s} p(w_n|z_s, [r_{d,z_s}]) \tag{5}$$

where $p(w_n|z_s, [r_{d,z_s}])$ is the probability that current word is generated from topic z_s with rating $[r_{z_s}]$. $\phi_{z_s}^{w_n}$ is the probability that word w_n is from aspect z_s and $\psi_{z_s, [r_{d,z_s}]}^{w_n}$ is the probability that word w_n is generated from rating $[r_{d,z_s}]$ in aspect z_s. $p(w_n|z_s, r_{d,z_s})$ can be easily calculated as follows:

$$p(w_n|z_s, r_{d,z_s}) = \phi_{z_n}^{w_n} \cdot \psi_{z_s, [r_{d,z_s}]}^{w_n} \tag{6}$$

The dependency between review text content and the latent aspect ratings is bridged by the aspect assignment z. We try to learn the values of parameters $\Upsilon = \{\mu, \Sigma, \sigma_1^2, \sigma_2^2, \sigma_3^2, \phi, \psi\}$ by maximizing the posterior probability $p(w, r_d^{all}, r_{d,k}|\Upsilon)$ in $train - data$ then use Υ to estimate $\Theta = \{R_i^{all}, \{R_{i,k}\}_{k \in [1,K]}, \{r_d^{all}\}_{d \in E_i}, \{r_{d,k}\}_{d \in E_i}^{k \in [1,K]}\}$ in $test - data$.

For each entity $i \in [1, N]$
  1. draw R_i^{all} and R_{ik} according to Equ.(1)(2).
  2. for each review $d \in E_i$
    2.1 draw r_d^{all} according to Equ.(3).
    2.2 for $k \in [1, K]$:
      draw r_{dk} according to Equ.(4).
    2.3 draw $\theta_d \sim Dir(\alpha)$
    2.4 for each sentence s in d
      2.4.1 draw $z_s \sim Multi(\theta_d)$
      2.4.2 for each word $w \in s$:
        draw $w \sim p(w|z_s, r_{d,z_s}, \phi, \psi)$

Fig. 1. the Generative Story

Combining all the components, the joint probability of training data can thus be defines as follows:

$$p(\mathbf{w}, r_d^{all}, r_{i,k}|\mu, \Sigma, \lambda, \sigma_1, \sigma_2, \sigma_3, \alpha, \phi, \psi)$$

$$= \prod_{i=1}^{N} \int dR_i^{all} \; p(R_i^{all}|\mu, \Sigma)$$

$$\times \sum_{k=1}^{K} \int dR_{ik} \; p(R_{i,k}|R_i^{all}, \sigma_1^2) \prod_{d \in E_i} \int d\theta_d \; p(\theta_d|\alpha) \quad (7)$$

$$\times \; p(r_d^{all}|R_i^{all}, \sigma_2^2) \; p(r_{d,k}|\lambda, R_{i,k}, r_d^{all}, \sigma_2^2)$$

$$\times \prod_{s \in d} \sum_k P(z = k|\theta_d) \sum_{w \in s} P(w|z = k, [r_{d,z}])$$

In order to ensure our discovered aspects are aligned with the pre-defined aspects, in this paper, we use the full set of keywords generated by the bootstrapping method (Wang et al., 2010) as prior to guide the aspect modeling part, which will be discussed in details in Section 4.

3.3 Inference

Since the posterior distribution of latent variables can not be computed efficiently, we use the variational inference method (Jordan et al., 2004) for inference. In variational inference we use variational distribution shown in Equ.(8) to approximate posterior distribution.

$$q(\{R_i^{all}\}, \{R_{i,k}\}, \theta, z|\eta, \Delta, \delta, \beta, \gamma, \varphi)$$

$$= \prod_i q(R_i^{all}|\eta_i, \Delta_i) \prod_k q(R_{i,k}|\beta_{i,k}, \delta_{i,k}^2)$$

$$\times \prod_{d \in E_i} q(\theta_d|\gamma_d) \prod_{s \in d} p(z_s|\varphi_s) \quad (8)$$

where the aspect assignment z for each word is specific by a K-dimensional multinomial distribution $Mul(\varphi_s)$, aspect proportion θ_d is governed by a K-dimensional Dirichlet distribution $Dir(\gamma_d)$, entity overall score R_i^{all} with a normal distribution with mean value η_i and variation Δ_i^2 and entity aspect score $R_{i,k}$ with a normal distribution with mean value $\beta_{i,k}$ and variation $\delta_{i,k}$.

By applying the Jensen's inequality, we obtain a lower bound of the log-likelihood $\log p(\mathbf{w}, r_d^{all}, r_{d,k}|\mu, \Sigma, \lambda, \sigma_1^2, \sigma_2^2, \sigma_3^2, \phi, \psi) \geq L$

$$L = E_q[\log\ p(\mathbf{w}, \{r_d^{all}\}, \{r_{d,k}\}|\mu, \Sigma, \lambda, \sigma^2, \phi, \psi)] + H(q) \tag{9}$$

where $H(q) = -E[\log\ q]$ is the entropy of the variational distribution q.

We use variational EM procedure that maximizes a lower bound with respect to the variational parameters $\{\eta, \Delta, \delta, \beta, \gamma, \varphi\}$, and then, for fixed values of them, maximizes the lower bound with respect to the model parameters z, θ, R_i^{all}, $R_{i,k}, \phi, \psi$.

E-step: Find the optimizing values of the variational parameters $\Phi = \{\eta, \Delta, \beta, \delta, \gamma, \varphi\}$ according to $\frac{\partial L}{\partial \Phi_i} = 0$:

update $\eta, \Delta^2, \beta_k, \delta_k^2$:

$$\eta_i = \frac{\frac{\mu}{\Sigma^2} + \sum_k \frac{\beta_{i,k}}{\sigma_1^2} + \sum_{d \in E_i} \frac{r_d^{all}}{\sigma_2^2}}{\frac{1}{\Sigma^2} + \sum_k \frac{1}{\sigma_1^2} + \sum_d \frac{1}{\sigma_2^2}} \tag{10}$$

$$\Delta_i^2 = \frac{1}{\frac{1}{\Sigma^2} + \sum_k \frac{1}{\sigma_1^2} + \sum_{d \in E_i} \frac{1}{\sigma_2^2}} \tag{11}$$

$$\beta_{i,k} = \frac{\frac{\eta_i}{4\sigma_1^2} + \sum_{d \in E_i} \frac{2r_d^{all} - r_{d,k}}{4\sigma_3^2}}{\frac{1}{\sigma_1^2} + \sum_{d \in E_i} \frac{1}{4\sigma_3^2}} \tag{12}$$

$$\delta_{i,k}^2 = \frac{1}{\frac{1}{\sigma_1^2} + \sum_{d \in E_i} \frac{1}{4\sigma_3^2}} \tag{13}$$

update: γ, φ

$$\varphi_{sk} \propto \exp[\Psi(\gamma_{d,k}) - \Psi(\sum_k \gamma_{d,k})] \cdot \prod_{w \in s} \phi_k^w \cdot \psi_{[r_{d,k}]}^w \tag{14}$$

$$\gamma_{d,k} = \alpha_k + \sum_{s \in d} \varphi_{s,k} \tag{15}$$

where $\Psi(\cdot)$ denotes the first derivative of the $\log \Gamma$ function. Since the derivative with respect to η_i and $\beta_{i,k}$ (shown in Equ.(10)) depends on each other, which is also the case with γ and φ, we employ the strategy shown at Figure 2 to attain their values.

M-step: In M-step, we maximize the resulting lower bound on the log likelihood with respect to the model parameters:

Initialization:
    for all i,d,n,k: $\varphi^0_{i,d,n,k} = 1/K$
    for all i,d,k: $\gamma_{i,d,k} = \alpha_k + n_d/K$
    for all i,k: $\eta = 0$, $\beta_{i,k} = 0$
for all i: calculate Δ^2_i according to Equ.(11)
fro all i,k: calculate $\delta^2_{i,k}$ according to Equ.(13)
Repeat
    for all i,d,n,k: update $\varphi_{i,d,n,k}$ according to (14)
    normalize $\varphi_{i,d,n}$
    for all d,k: update $\gamma_{i,d,k}$ according to (15)
Until Convergence
Repeat
    for all i: update η_i according to Equ. (10)
    for all i,k: update $\beta_{i,k}$ according to Equ.(12)
Until Convergence

Fig. 2. E-step for the training

Update ϕ and ψ:

$$\phi^w_k = \sum_i \sum_d \sum_{s\in d} \varphi_{s,k} \sum_{w\in s} \mathbf{1}(w_n = w) \tag{16}$$

$$\psi^w_{k,m} = \sum_i \sum_d \mathbf{1}([r_{d,k}] = m) \sum_{s\in d} \varphi_{s,k} \sum_{w\in s} \mathbf{1}(w_n = w) \tag{17}$$

Update α

$$\frac{\partial L}{\partial \alpha_k} \propto (\Psi(\sum_k \alpha_k) - \Psi(\alpha_k) + \sum_{d\in E_i}(\Psi(\sum_k \gamma_{d,k}) - \Psi(\gamma_{d,k}))) \tag{18}$$

As shown at Equ.(18) the derivative with respect to α_k depends on α'_k ($k \neq k'$), we therefore must use an iterative method to find the maximum by Newton-Raphson as in Blei et al. (2003) 's work. The Hessian matrix involved can be easily calculated as follows:

$$\frac{\partial^2 L}{\alpha_k \alpha'_k} \propto [\Psi'(\sum_k \gamma_k) - \mathbf{1}(k = k')\Psi'(\gamma_k)] \tag{19}$$

Update $\mu, \Sigma^2, \sigma^2_1, \sigma^2_2, \sigma^2_3$

$$\mu = \frac{1}{N}\sum_i \eta_i \qquad \Sigma^2 = \frac{1}{N}\sum_i [(\eta_i - \mu)^2 + \Delta^2_i] \tag{20}$$

$$\sigma^2_1 = \frac{1}{KN}\sum_i \sum_k (\eta^2_i + \Delta^2_i + \beta^2_{i,k} + \delta^2_{i,k} - 2\beta_{i,k}\eta_i) \tag{21}$$

$$\sigma^2_2 = \frac{1}{\sum_i N_i}\sum_i \sum_{d\in E_i}([r^{all}_d]^2 - 2r^{all}_d \eta_i + \eta^2_i + \Delta^2_i) \tag{22}$$

$$\sigma_3^2 = \frac{1}{\sum_k \sum_i N_i} \sum_i \sum_{d \in E_i} \sum_k (r_{d,k}^2 + \frac{1}{4}(\beta_{i,k}^2 + \delta_k^2)$$

$$+ \frac{1}{4}[r_d^{all}]^2 - r_{d,k}\beta i, k - r_{d,k}r_d^{all} + \frac{1}{2}\beta_{i,k}r_d^{all}) \tag{23}$$

3.4 Prediction

In $test - data$, the prediction for Θ is calculated as follows:

$$\Theta = \arg\max_{\Theta} p(w|\Theta, \Upsilon) \tag{24}$$

The meanings for Θ and Υ have been demonstrated in the previous subsection. $p(w|\Theta, \Upsilon)$ can be calculated by integrating out parameters z_s, α, θ, making it hard to calculate. Here, we employ an alternative method which firstly assign z_s to each of the sentences in $test-data$ which is combined with variational inference in Section 3.3. Then the maximization of $p(w|\Theta, \Upsilon)$ would be transformed to $p(w|\Theta, \Upsilon, z)$ and can be attained as follows:

$$p(w \in E_i|\Theta_i, \Upsilon, z) \propto p(r_i^{all}|\mu, \Sigma^2) \prod_k p(R_{i,k}|R_i^{all}, \sigma_1^2)$$

$$\times \prod_{d \in E_i} p(r_d^{all}|\sigma_2^2, R_i^{all}) \prod_{d \in E_i} \prod_k p(r_{d,k}|\sigma_3^2, R_{i,k}, r_d^{all}) \tag{25}$$

$$\times \prod_{s \in d} \sum_k \mathbf{1}(z_s = k) \sum_{w \in s} \psi_{[r_{d,k}]}^w$$

We set the derivative of $p(w \in E_i|\Theta_i, \Upsilon, z)$ with respect to element in Θ to 0 and get the following results:

$$R_i^{all} = \frac{\frac{\mu}{\Sigma^2} + \sum_k \frac{R_{i,k}}{\sigma_1^2} + \sum_{d \in E_i} \frac{r_d^{all}}{\sigma_2^2}}{\frac{1}{\Sigma^2} + \sum_k \frac{1}{\sigma_1^2} + \sum_{d \in E_i} \frac{1}{\sigma_2^2}} \tag{26}$$

$$R_{i,k} = \frac{\frac{R_i^{all}}{\sigma_1^2} + \sum_{d \in E_i} \frac{2r_{d,k} - r_d^{all}}{4\sigma_3^2}}{\frac{1}{\sigma_1^2} + \sum_{d \in E_i} \frac{1}{4\sigma_3^2}} \tag{27}$$

$$r_d^{all} = \frac{\frac{R_i^{all}}{\Sigma^2} + \sum_{k \in d} \frac{2r_{d,k} - R_{i,k}}{4\sigma_3^2}}{\frac{1}{\Sigma^2} + \sum_{k \in d} \frac{1}{4\sigma_3^2}} \tag{28}$$

$$r_{d,k} = \arg\max_{r_{d,k}} T(r_{d,k}) \tag{29}$$

$$T(r_{d,k}) = exp[-\frac{(r_{d,k}^2 - R_{i,k}r_{d,k} - r_d^{all}r_{d,k})}{2\sigma_3^2}] \times \prod_{s \in d} \mathbf{1}(z_s = k) \prod_{w \in s} \psi_{[r_{d,k}]}^w \tag{30}$$

Since the derivative for one parameter in Θ depends on others, we employ the strategy which first sets all element in Θ to zero. Then parameters are iteratively updated according to Equ.(26)-(30) until convergence.

4   Experiments

4.1   Data Sets and Preprocessing

We collect 878,561 reviews for 3,945 hotels from *TripAdvisor*. As for each hotel, in addition to the overall ratings, reviewers are also asked to provide ratings on 7 predefined aspects such as *service, cleanliness, value, location, room, check in/ frontdesk* and *business service*, ranging from 1 to 5. After necessary filtering, we have 86,134 reviews from 462 hotels.

4.2   Aspect and Rating Evaluation

We compare our hierarchical model HM with existing methods on the quality of identified topical aspects. We compare our model with other three models: **LDA, Local-LDA sLDA**. Our model is similar to LDA except that LDA only use the word co-occurrence. sLDA extends LDA by adding a regression model to capture the overall response and Local-LDA is a sentence-level LDA which is good at discovering aspects. For fair comparison, we do not use any word prior for any models.

For the measurements, we use the metrics of Wang et al. (2011). The full set of keywords (in total 309 words) generated by the bootstrapping method is used as a prior to train a LDA model on this data set. The learned topics are regarded as the ground-truth aspect descriptions. All the three models are then trained without any keyword supervision. To align the topics detected by three models and ground-truth topics, we use the Kuhn-Munkres algorithm and quantitatively measure the quality of the identified aspects using KL divergence.

From the results we can see that our model (HM) and Local-LDA are better than others because the aspects they identified are closer to the ground-truth aspects.

sLDA's performance is the worst. sLDA assumes that the topics directly characterize the overall ratings, so it prefers rating sentiment sensitive words than those general content words. In our model, we have different aspect-word distribution and rating-word distribution, which largely decrease the effect of

Table 1. Top words from 7 aspects

service	cleanliness	value	location	rooms	checkin/frontdesk	business-service
service	clean	price	location	room	check	Internet
hotel	dirty	worth	hotel	rooms	reservation	hotel
food	smell	value	traffic	bed	manager	computer
restaurant	bed	pay	walk	comfortable	help	business
buffet	room	expensive	located	large	staff	wireless
bar	hotel	hotel	close	bedroom	front	stay
nice	good	property	convenient	small	staff	network
staff	rooms	star	subway	spacious	asked	wiki
coffee	bathroom	money	blocks	bathroom	arrived	slow
dinner	area	cheap	minute	floor	checked	experience

Table 2. KL divergence between the align aspects

#Topic	7	10	15	20
HM	5.023	6.464	**8.743**	**11.543**
LDA	5.324	6.687	9.897	13.178
sLDA	13.274	16.562	19.874	22.589
Local-LDA	**4.921**	**6.371**	8.997	11.726

Table 3. Top words from Rating-specific distributions for aspect rooms

5 star	4 star	3 star	2 star	1 star
room (0.0830)	room (0.0797)	room (0.0801)	room (0.0822)	room (0.0843)
clean (0.0564)	nice (0.0647)	rooms (0.0284)	bad (0.0367)	bad (0.0676)
good (0.0359)	rooms (0.0289)	OK(0.0282)	rooms (0.0284)	dirty (0.0427)
rooms (0.0276)	OK (0.0253)	average (0.0214)	dirty (0.0271)	rooms (0.0280)
comfortable (0.0254)	clean (0.0220)	bad (0.0202)	damp (0.0229)	smelly(0.0281)
view (0.00232)	good (0.0204)	good(0.0198)	poor (0.0207)	disgusting (0.0282)
excellent (0.0172)	fine (0.0172)	dirty (0.0182)	stain (0.0192)	bedroom(0.0160)

sentimental word in topic modeling. Local-LDA, as a sentence-level model, is more effective in mining review aspects because usually words in one sentence tends to talk about one aspect.

Table 4. Top words from Rating-specific distributions for "rooms" and "value" after word removal

aspect	5 star	4 star	3 star	2 star	1 star	aspect	5 star	4 star	3 star	2 star	1 star
	clean	nice	OK	bad	bad		value	price	OK	overpriced	bad
rooms	good	OK	average	dirty	dirty	value	worth	pricey	bit	money	overpriced
	comfortable	clean	bad	damp	smelly		excellent	worth	fine	expensive	terrible
	view	good	small	poor	disgusting		quality	fair	cheaper	cheaper	never
	excellent	fine	dirty	stain	dingy		really	pay	overpriced	bad	money
	beach	clean	dislike	below	bed		nice	stay	bad	money	leave
	spacious	bed	bathroom	far	worst		terrific	money	checked	worse	much
	wonderful	large	not	smelly	small		well	deserve	pay	more	worst
	recommend	over	bedroom	finally	broken		pricey	fine	not	asked	poor
	perfect	recommend	stay	carpet	damp		good	recommend	manager	soon	more

Table 1 shows top words in 7 aspects (the topic number is set to 10). We can see that words describing similar aspect tend to show up in the same topic. Table 3 shows the top words in aspect rating word distributions with corresponding probabilities. Aspect words (i.e. hotel, room) mix up with sentimental words (i.e. great, bad) in the rating-word distribution, but aspect words are approximately uniformly distributed (here a word is uniformly distributed if the differences between the probability generated by all rating-specific distributions are less than 10%), while sentimental words usually focus on certain specific aspects. This indicates that aspect words exert little influence for rating prediction in the proposed hierarchical model. For a clearer illustration, we remove the uniformly distributed words and show the remaining top words at Table 4. Due to the space limit, we just show top words with "rooms" aspect and "value" aspect. Top words in different rating-specific topics show a clear pattern of positive and negative attitude except very few exceptions. Such attitudes are much stronger in 5 star and 1 star than 4 and 2 star. It shows the effectiveness of our model.

4.3 Quantitative Prediction

We also quantitatively evaluate the performance of the proposed model in rating prediction. We use the full set of keywords as prior to guide the aspect modeling to align discovered aspects with predefined aspects. For evaluation, we employ measures employed in Wang et al. (2010): (1) Mean Square Error (MSE) of the predicted ratings compared with the ground-truth ratings; (2) Mean Average Precision (MAP_N); (3) $NDCK_N$: a commonly used ranking evaluation algorithm in IR.

For overall rating r_i^{all}, our baselines include **LDA+SVR, sLDA, Bootstrap+SVR** and **LocalLDA+SVR**. Results are presented at Table 5. sLDA and our model (HM) achieves comparative results. Both are better than LDA+SVR and Bootstrap+SVR. Our model and sLDA can leverage the side information (e.g., rating scores) and discover latent topic representations so they outperform the decoupled two-step procedure as adopted in unsupervised topic models (Zhu and Xing, 2001).

As for aspect rating, we include **LDA+SVR,sLDA+SVR, Bootstrap+SVR** and **LocalLDA+SVR** as baselines. Results are presented at Table 6. The proposed model outperforms all other methods in all measures except MAP, for which it attains comparative results with sLDA+SVR. For hotel-level overall rating R_i^{all}, we use the hotel rating scores provided by Northstar[1], as ground-truth ratings. Results are shown at Table 7. The proposed model outperforms other baselines in R_i^{all} prediction. Since ground-truth ratings for hotel aspects

Table 5. Review-level overall rating prediction

	Δ^2	MAP_{5star}	$NDCG_{5star}$
HM	**0.874**	0.439	**0.932**
LDA+SVR	1.548	0.386	0.887
sLDA	1.005	**0.442**	0.845
Bootstrap+SVR	1.342	0.409	0.886
LocalLDA+SVR	1.171	0.408	0.891

Table 6. Review-level aspect rating prediction

	Δ^2	MAP_{5star}	$NDCG_{5star}$
HM	**1.190**	0.409	**0.901**
LDA+SVR	1.784	**0.426**	0.844
sLDA+SVR	2.325	0.262	0.892
Bootstrap+SVR	1.492	0.352	0.861
LocalLDA+SVR	1.488	0.368	0.877

Table 7. Hotel-level overall rating prediction

	Δ^2	MAP_{5star}	$NDCG_{5star}$
HM	**0.678**	**0.511**	**0.947**
LDA+SVR	1.782	0.377	0.861
sLDA	1.320	0.446	0.914
Bootstrap+SVR	1.613	0.389	0.908
LocalLDA+SVR	1.472	0.442	0.817

Table 8. Hotel-level aspect rating prediction

	Δ^2	MAP_{5star}	$NDCG_{5star}$
HM	**1.470**	**0.490**	**0.898**
LDA+SVR	1.982	0.354	0.847
sLDA	2.520	0.349	0.864
Bootstrap+SVR	1.817	0.350	0.858
LocalLAD+SVR	1.784	0.378	0.866

[1] http://www.northstartravelmedia.com/.

are hard to obtain, we use the average aspect ratings for each review in the $test - data$ as ground-truth aspect ratings and compare performances of different models. Each baseline makes review-level aspect predictions and predict hotel-level aspect rating by the average of those predictions. Results are shown at Table 8. We observe that aspect rating prediction (see Table 6 and 8) tends to have a larger value of MSE than overall rating prediction (see Table 5 and 7). This is because aspect information is much noisier and harder to predict. Also the improvement of the proposed model over all the other baselines is greater at hotel-level than at review-level, e.g. Table 7 vs Table 5 and Table 8 vs Table 6. Since reviews within the same hotel tend to have similar ratings, hierarchical structures can help review-level rating predictions. That is why the proposed hierarchical model outperforms others at review level.

5 Conclusion

The hierarchical model proposed in this paper is able to simultaneously predict the review-level and entity-level overall and aspect ratings. Experiments on data crawled from TripAdvisor verify its superiority. In the future, we will develop models that can identify aspects automatically, without the supervision of external keyword input.

References

Blei, D., Andrew, N., Ordan, M.J.: Latent dirichlet allocation. In: JMLR (2003)

Brody, S., Elhadad, N.: An unsupervised aspect-sentiment model for online reviews. In: NAACL 2010 (2010)

Burges, C.: A tutorial on support vector machines for pattern recognition. In: Data Mining and Knowledge Discovery (1998)

Ding, X., Liu, B., Zhang, L.: Entity discovery and assignment for opinion mining applications. In: KDD 2009 (2009)

Gruber, A., Weiss, Y., Rosen-Zvi, M.: Hidden topic Markov models. In: AISTATS 2007 (2007)

Hu, M., Liu, B.: Mining and summarizing customer reviews. In: CIKM 2004 (2004)

Hu, M., Liu, B.: Opinion integration through semi-supervised topic modeling. In: WWW 2008 (2008)

Jindal, N., Liu, B.: Identifying comparative sentences in text documents. In: SIGIR 2006 (2006)

Jordan, M., Ghahramani, Z., Jaakkola, T., Saul, L.: An introduction to variational methods for graphical models. In: Machine Learning (2008)

Lu, B., Ott, M., Cardie, C., Tsou, B.: Multi-aspect sentiment analysis with topic models. In: ICDM 2011 (2011)

Lu, Y., Zhai, C.X., Sundaresan, N.: Rated aspect summarization of short comments. In: WWW 2009 (2009)

Blei, D., McAuliffe, J.: Supervised topic models. In: NIPS 2007 (2007)

Kim, S.-M., Hovy, E.: Determining the sentiment of opinions. In: COLING 2004 (2004)

Lerman, K., Blair-Goldensohn, S., McDonald, R.: Sentiment summarization: evaluating and learning user preferences. In: EACL 2009 (2009)

Lin, C., He, Y.: Joint sentiment/topic model for sentiment analysis. In: CIKM 2009 (2009)

Lu, Y., Zhai, C.: Opinion integration through semi-supervised topic modeling. In: WWW 2004 (2004)

Morinaga, S., Yamanishi, K., Tateishi, K., Fukushima, T.: Mining product reputations on the web. In: KDD 2002 (2002)

Pang, B., Lee, L.: Seeing stars: exploiting class relationships for sentiment categorization with respect to rating scales. In: ACl 2005 (2005)

Popescu, A.-M., Etzioni, O.: Extracting product features and opinions from reviews. In: Natural Language Processing and Text Mining (2007)

Sauper, C., Haghighi, A., Barzilay, R.: Incorporating content structure into text analysis applications. In: EMNLP 2010 (2010)

Titov, I., McDonald, R.: A joint model of text and aspect ratings for sentiment summarization. In: ACL 2008 (2008)

Wang, H., Lu, Y., Zhai, C.X.: Latent aspect rating analysis on review text data: a rating regression approach. In: KDD 2010 (2010)

Wang, H., Lu, Y., Zhai, C.X.: Latent aspect rating analysis without aspect keyword supervision. In: KDD 2011 (2011)

Snyder, B., Barzilay, R.: Multiple aspect ranking using the good grief algorithm. In: HLT-NAACL (2007)

Zhuang, L., Jing, F., Zhu, X.: Movie review mining and summarization. In: CIKM 2006 (2006)

Li, J., Cardie, C.: TopicSpam: a topic-model based approach for spam detection. In: ACL 2013 (2013)

Jindal, N., Liu, B.: Opinion spam and analysis. In: WSDM 2008 (2008)

Li, J., Ott, M., Cardie, C.: Identifying manipulated offerings on review portals. In: EMNLP 2013 (2013)

Li, J., Ott, M., Cardie, C., Hovy, E.: Towards a general rule for identifying deceptive opinion spam. In: ACL 2014 (2014)

Zhao, W., Jiang, J., Yan, H., Li, X.: Jointly modeling aspects and opinions with a MaxEnt-LDA hybrid. In: EMNLP 2010 (2010)

Sentiment Analysis on Microblogging by Integrating Text and Image Features

Yaowen Zhang, Lin Shang$^{(\boxtimes)}$, and Xiuyi Jia

Xianlin Road No.163, Qixia District, Jiangsu Province Nanjing City, China
ken7758521@gmail.com, shanglin@nju.edu.cn, jiaxy@njust.edu.cn

Abstract. Most studies about sentiment analysis on microblogging usually focus on the features mining from the text. This paper presents a new sentiment analysis method by combing features from text with features from image. Bigram model is applied in text feature extraction while color and texture information are extracted from images. Considering the sentiment classification, we propose a new neighborhood classier based on the similarity of two instances described by the fusion of text and features. Experimental results show that our proposed method can improve the performance significantly on Sina Weibo data (we collect and label the data). We find that our method can not only increasingly improve the F values of the classification comparing with only used text or images features, but also outperforms the NaiveBayes and SVM classifiers using all features with text and images.

Keywords: Sentiment analysis · Text features · Image features

1 Introduction

Microblogging has been acknowledged as a source of subjective opinions on a wide range of topics since 2006. Users can post mini messages using various software tools on different electronic devices, such as laptops, mobile phones and tablets without time or space limitation. Because of free format of messages and an easy accessibility of microblogging plateforms, Internet users tend to shift from traditional communication tools (such as traditional blogs or mailing lists) to microblogging services, and share opinions on varieties of topics or discuss current issues.

Mining valuable information from microblogging has drawn many researchers' attention in recent years and sentiment analysis on microblogging has been applied in many real applications [33]. For example, political parties may be interested in knowing whether people support their program or not. The U.S. State Department knew about the importance of this communication tool and asked Twitter to delay a scheduled upgrade which took place 3 days after the election on June 15th, 2009. Social organizations may ask people's opinion on current debates. Jansen et al. [6] showed that 19% of Tweets mention a certain brand, of which 20% contained a sentiment. While reading news about political election [8], it was expected to obtain an overview about the support and

© Springer International Publishing Switzerland 2015
T. Cao et al. (Eds.): PAKDD 2015, Part II, LNAI 9078, pp. 52–63, 2015.
DOI: 10.1007/978-3-319-18032-8_5

opposition about presidential candidates. Analysing broader social and economic trends are also getting popular, such as the relationship between Twitter moods and both stock market fluctuations [10] or the relationship between consumer confidence and political opinion [12].

Various methods about sentiment analysis were explored. The software——SentiStrength is designed for web text using lexicon [25], in [38] a method explores feature definition and selection methods for sentiment polarity analysis. Zagibalov et al. [24] uses a unsupervised knowledge-poor method for domain-independent sentiment analysis.

However, most contributions about sentiment analysis on microblogging focused on the features mining from the texts. Popular social microblogging platforms, such as Twitter and Sina contain massive amounts of visual information in the form of photographs, which are commonly posted with texts. Researchers tend to make advantage of multi resources to mining microblogging. The role of pictures in conveying emotional messages and opinions is recognized as more important than the text. Indeed, a promising research direction is represented by the retrieval of images based on emotional semantics. Some systems based on social tagging such as ALIPR[1] allow users to annotate images using emotional labels. However, emotional mining for only images will ignore the context and background. Taking only one factor (text or image) into account is not enough to classify user's sentiments.

seldom contributions have been proposed for combining text and its corresponding image for sentiment analysis. Our work is inspired by the fusion of text sentiment analysis and image sentiment analysis. We aim to use both text and image information to help improve the result of sentiment analysis. The key issue is how to find a appropriate method to fuse text and image features.

In this paper, we propose a data fusion method which integrates text features and image features satisfactorily. A new neighborhood classifier is proposed to fuse text data and image data properly. The results of the experiments on the Sina microblogging data show that combining text features and image features can improves the average performance of sentiment classification.

2   Related Work

2.1   Sentiment Analysis

With the development of Microblogging and text mining, opinion mining and sentiment analysis has been a field of interest for many researchers. A very broad and excellent overview of the existing work was done by Pang et al. [17].

Diakapolous and Shamma [2] used manual annotations to characterize the sentiment reactions to various issues in political debate. They have found that sentiment was useful as a measure for identifying controversy. Jansen et al. [6] studied the Word-Of-Mouth effect on Twitter using an adjective based sentiment

---

[1] http://alipr.com

classifier, and it was helpful to improve the accuracy of classifying brands on Twitter.

J.Read [23] considered emoticons such as ":-)" and ":-(" to form a training set for the sentiment classification. For this purpose, the author collected texts containing emoticons from Usenet newsgroups. J.Read divided the dataset into "positive" (texts with happy emoticons) and "negative" (texts with sad and angry emotions) samples. Classifiers (SVM and Naive Bayes) trained using emoticons were able to obtain of 70% accuracy on the test set. Go et al. [32] adopted Twitter to collect training data and then performed a sentiment search. The authors have used emoticons to construct corpora and obtained "positive" and "negative" samples, and then various classifiers were tested. The best result was obtained by the Naive Bayes classifier with mutual information measure for feature selection. The author obtained the accuracy of 81% on their test set.

Since emoticons can help analysis sentiment, there is reason to believe that investigating how to combine text features with image features to improve the performance of sentiment analysis is necessary.

2.2   Affective Image

People's understanding and feeling of images is subjective based on the semantic level [36]. As we have no clear definition to give evaluation of fitness other than the one in his mind, most conventional applications are lack of the capability to utilize human intuition and sentiment appropriately in creative applications such as architecture, art, music, and design.

In order to analyse pictures' emotion, we need to extract the effective information from visual signals. In [36], analysis and retrieval of images at the emotional level is studied. Early studies suggested that different semantic classes may need different specific features [34]. Obviously, the key problem is how to choose appropriate features which have close relation with image sentiment or emotion that express the image's semantic.

Colombo et al.[16] designed an image retrieval system which retrieve art paintings through mapping expressive and perceptual features to four emotions. The image was segmented into homogeneous regions, and some features including color, hue, saturation, position, and size from each region was extracted. At last, they used each region's contrasting and harmonious relationships with other regions to capture emotions. Kuroda et al. [21] applied the feature of image regions to extract sky, earth and water which was used in the middle level of the system to derivative impression words and objects in landscape images.

K-DIME [1] built an individual model for each user by using a neural network to judge pictures' sentiment. Wang et al. [5] used specific features of images for affective image classification. Hayashi et al. [26] and Wu et al. [28] applied generic image processing features such as color histograms. In paper of [22], the authors applied Gabor and Wiccest features with SVM to perform emotional valence categorization. Hanjalic [27] and Wang [14] did some work on the affective content analysis in movies.

3   New Microblogging Sentiment Analysis Method with Data Fusion

In this section, we will introduce the new sentiment classification method by considering data fusion. Fig. 1 is the flow chart of our microblogging sentiment analysis method. This method tends to make advantage of text features and image features to help better analyse sentiment.

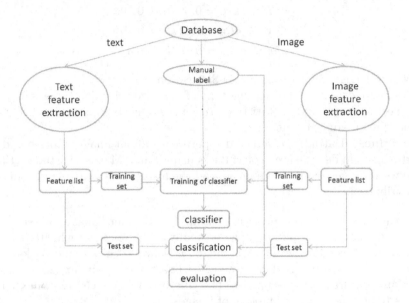

Fig. 1. Our microblogging sentiment analysis procedure

3.1   Text and Image Features

In order to improve the efficiency of sentiment analysis, we considered two groups of features including text features and image features.

1)**Text feature** For text features, bigram as basic unit of bag of words is adopted for its good performance in Chinese document [37]. Some feature measurement method such as MI(Mutual Information), IG(Information Gain) and CHI are choosed to select features.

2)**image feature** Image features are based on relations between induced emotional effects which comes from art theory and color combinations. We will use **color** and **texture** features to conduct our image feature selection.

Artists often use colors to induce emotional effects, but different people from different cultures or backgrounds may have totally different feelings about the same color. Itten [4] and Valdez [30] investigated emotional impact of color and color combinations from different views. For the color features, we adopted several measures which are shown in the follows:

Saturation and Brightness Statistics: According to Osgood dimensional approach [18], saturation and brightness can have direct influence on our pleasure, arousal and dominance. We use the method of Valdez et al. [30] to compute *Pleasure, Arousal* and *Dominance* with saturation and brightness. The result obtained from the analysis gave us a significant relation between brightness, saturation and their emotional impact, which are expressed in the following equation:

$$Pleasure = 0.69Br + 0.22S,$$
$$Arousal = -0.31Br + 0.60S, \qquad (1)$$
$$Dominance = 0.76Br + 0.32S,$$

, where Br is Brightness and S is Saturation.

Hue statistics: Hue expresses the tone of a image. Mardia [7] measured *Hue* by vector-based circular statistics. We compute hue measurements like mean, hue spread, etc as our features.

Colorfulness: Datta [11] conducted experiments about image emotion and colorfulness and he measured colorfulness using Earth Mover's Distance (EMD) between the histogram of an image and the histogram having a uniform color distribution.

Textures are also important to help us analyse the emotional expression of an image. Some artists and photographers usually create pictures which are sharp, or the main object they want to highlight is sharp with a vague background. Blurred images are often used in the category of art photograph images to express fear. Many features are developed to describe texture. We choose some normal features to help us analysis emotion of images.

Tamura textures: Tamura [3] used the Tamura texture features to analysis the emotion of a image. This method is popular among affective image retrieval [28]. We use the first three of the Tamura texture features: *coarseness, contrast* and *directionality*.

Gray Level Cooccurrence Matrix (GLCM): Haralock [9] use GLCM method to analyse the emotion of a image. It is a classic method of measuring textures. We use GLCM to comput *contrast, correlation, energy* and *homegeneity*.

3.2 A Similarity Based Neighborhood Classifier

Consider a common pattern recognition operation based on the K-nearest neighbors algorithm, which is a non-parametric method for classification and regression [29]. It assigns objects "value" or "class" based on K closest training examples in **feature space**. It inspires us to classify sentiment combining texts with images in **text-image space**.

The notion of distance between points in a text-image space can be generalized by introducing the L_p vector norm defined for an n-dimensional vector $(\alpha_1, \alpha_2, ..., \alpha_n)$ as

$$\|\alpha\|_p = \|\alpha_1, \alpha_2, ..., \alpha_n\|_p = (\alpha_1^p + \alpha_2^p + ... + \alpha_n^p)^{\frac{1}{p}} \qquad (2)$$

For text sentiment classification, we use similarity to classify an unlabeled instance into a class. The principle also makes sense in image sentiment field. Thus, we apply cosine similarity as the measure for both text and image features. The cosine similarity can be computed as equation (3) where α, β are the feature vector

$$cos(a, b) = \frac{\alpha_1\beta_1 + \alpha_2\beta_2 + ... + \alpha_n\beta_n}{\|\alpha\|\|\beta\|} \qquad (3)$$

These two kinds of similarities are represented in an coordinate plane, while the X axis represents image similarity and the Y axis represents text similarity in Fig.2.

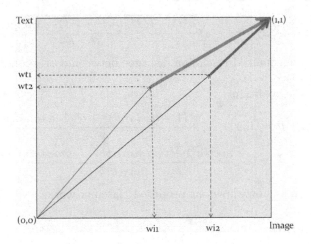

Fig. 2. A similarity based neighborhood classifier

In the text-image plane, each test instance is compared to every training example. We take the distance between the test point and point (1,1) as our matching criteria. The best case is the test instance, which perfectly matches training instance both in text and image axis, that means **(1,1)** point. Usually, we just obtain a point in yellow area in Fig.2, and the closer to the point (1,1), the higher similarity achieved. We take the distance between the test point and point (1,1) as our matching criteria. The closer, the more similar to our labeled instance.

$$sim(doc\_i, doc\_j) = 1 - [\frac{(1 - textsim)^p + (1 - imagesim)^p}{2}]^{\frac{1}{p}} \qquad (4)$$

The equation (4) deals with text feature and image feature. However, the two kinds of features many have different influences on the classification in some cases. Equation (4) is a special case of equation (5) when $a = b = 1$, which means text and image have the same weight.

$$sim(doc\_i, doc\_j) = 1 - [\frac{a^p(1 - textsim)^p + b^p(1 - imagesim)^p}{a^p + b^p}]^{\frac{1}{p}} \quad (5)$$

When p = 1, we can achieve as equation(6) presented:

$$sim(doc\_i, doc\_j) = 1 - [\frac{a(1 - textsim) + b(1 - imagesim)}{a + b}]$$

$$= [textsim, imagesim] \cdot \begin{bmatrix} \frac{a}{a+b} \\ \frac{b}{a+b} \end{bmatrix} \quad (6)$$

That means, when $p = 1$, the similarity is measured by the inner product between document elements(text and image) and the normalized weight represented as $w_1 = \frac{a}{a+b}, w_2 = \frac{b}{a+b}$

When $p = 2$, we obtain the equation(8)

$$sim(doc\_i, doc\_j) = 1 - \sqrt{\frac{a^2(1 - textsim)^2 + b^2(1 - imagesim)^2}{a^2 + b^2}} \quad (7)$$

It becomes a normalized Euclidian distance between corresponding points in the text-image space.

When $p = \infty$, we obtain

$$sim(doc\_i, doc\_j) = \lim_{p \to \infty} 1 - [\frac{a^p(1 - textsim)^p + b^p(1 - imagesim)^p}{a^p + b^p}]^{\frac{1}{p}}$$

$$= 1 - \frac{\max(a(1 - textsim), b(1 - imagesim))}{\max(a, b)} \quad (8)$$

The algorithm 1 shows how our proposed classifier works.

4   Experiments

In this section, we will show how our proposed method works by several experiments and analysis.

4.1   Data Collection

As benchmark data of Sina microblogging with text and image are seldom collected, we collect Sina data by ourselves. 1620 documents with corresponding images are extracted from Sina platform[2]. We assigned a rating score to each microblogging. The microblogging with rating > 3 were labeled positive, and those with rating < 3 were labeled negative. The rest(rating $= 3$) are discarded for their ambiguous polarity and 1000 documents left. There are 575 positive instances and 425 negative instances. Fig.3 shows two examples of the data.

---

[2] weibo.com

Algorithm 1. A similarity based neighborhood classifier

Input:trainset,testset
Output:labeled instance
for $i = 1$ to $testset.length$ **do**
   $distanceset = []$
   for $j = 1$ to $trainset.length$ **do**
      $textsim = cosine(testset\_text[i], trainset\_text[i])$
      $imagesim = cosine(testset\_image[i], trainset\_image[i])$
      distance = minkowskidistance($p, textcosine, imagcosine, a, b$) ▷ # a,b stands
for the weight of text and image, p stands for the value of P normal model
   end for
end for
store k items with lowest distances to distanceset
find the majority class among k items
return the majority class

Fig. 3. Two examples of data

4.2 Experiments Setup

Documents have to be transformed into a suitable representation for classification tasks. We employ vector space model which is widely used in text classification as our feature vector weighting method.

In order to eliminate the unfavorable effect resulting from an unequal dimension, the better solution is to adopt the weighted p-normal form. It turns out that the solution is useful, and there are two ways for text and image weight assignment. The first weight method is based on an intuitive idea. Nearly every Microblogging platform allow users to publish short messages with a limit of 140 characters. Actually, we assign the number of characters to text weight $a = \#characters$ and $b = 140 - a$ for image weight. We call it **Text Proportion** weighting method. The second way is taking the dimensions of each feature as weight according to the feature fusion method [13] and $a = \#dimensions\ of\ text$ $features$, $b = \#dimension\ of\ image\ featuers$. We call this method **Feature dimension** weighting method.

4.3   Comparison and Analysis

In this subsection, we conduct experiments to show the effect of the classifier with the combined text and image through comparing to the common classifers. All experiments are implemented based on WEKA[3] with version 3.6, with all parameters set to their default values.

Tan et al. [19] reported that 5000 to 6000 numbers of Chinese text features would lead to a reasonable performance for almost every common classifier. As SVM and NaiveBayes are commonly used in sentiment analysis, we apply them to our sentiment classifiers. F value is adopted as the measure. Sample values of p = 1,2,5,9 and ∞ were tested experimentally. Table 1 shows the performance of common classifiers considering onlyt text features and text combined images features. Fig.4 is the performances of our proposed classifier considering different p values with different feature selection method. Table 2 are average performance of our experiments based on common classifiers comparing to our proposed method.

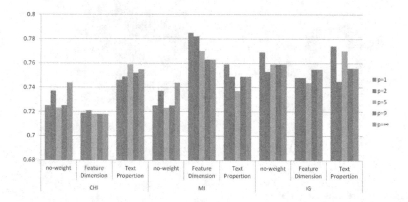

Fig. 4. Our microblogging sentiment analysis procedure

- The results of Table 1 show that simply putting text features and image features together will not improve the performance of sentiment analysis greatyly. Such as taken SVM considered, the F value improves only in CHI mode.
- From Fig.4, we can see that Euclidian distance(p = 2) is not always appropriate for all feature extraction methods. For p-normal model the optimum values appear to occur for p values somewhere between 1 and 9; as the p-values grow larger, the amount of improvement over the common text sentiment analysis cases decreases.

---

[3] http://www.cs.waikato.ac.nz/ml/weka/downloading.html

Table 1. The comparison F-vaules of common classifiers by considering different features

Selection Method	NB text	SVM text	NB text+image	SVM text+image
CHI	0.72	0.707	0.727	0.738
MI	0.727	0.738	0.720	0.707
IG	0.731	0.737	0.726	0.706

Table 2. Average F value of feature weighting model

	NB text	SVM text	NB text + image	SVM text + image	Our Proposed Method(text + image)		
					no-weight	Feature Dimension	Text Proportion
MI	0.739	0.730	0.691	0.705	0.756	**0.789**	0.739
CHI	0.727	0.738	0.720	0.707	0.782	**0.794**	0.761
IG	0.731	0.737	0.726	0.706	0.738	0.775	**0.794**
Aver	0.732	0.735	0.712	0.706	0.759	**0.779**	0.764

- From the experiments results of Table 2 we can see that weighted p-normal performs better than **no-weight** p-normal form. **Feature dimension** is 2.0% higher than **no-weight** model, and **Text Proportion** is 0.5% higher than **no-weight** model. What's more, **Feature dimension** performs best from the experiment tables.
- The average F value of Table 2 prove that our proposed method of data fusion performs better than common sentiment classifiers. For example, the average F values of NaiveBayes for text features and text features with image features are 73.2% and 71.2%, and SVM's F values are 73.5% and 70.6%. None of them are better than anyone of our proposed classifier's performance(The worst average F value is 75.9%).

5   Conclusion and Future Works

In this paper, we proposed a new data fusion method combining text and image to improve the performance of sentiment analysis. We set up a dataset based on microblogging to validate our proposed method. Valued features are selected and sentiment polarity classification (positive or negative) on those messages is addressed. In text feature extraction, we use four typical feature extraction method (CHI, MI, IG) to select the features. In image feature extraction, we extract color features and texture features. Based on the fusion of text and image features, we proposed a new similarity based neighborhood classifier. Several comparison experiments show the efficiency of our proposed method. In the future we will extract more useful text features and image features to analyse microblogging sentiment.

References

1. Bianchi-Berthouze, N.: K-DIME: an affective image filtering system. IEEE on Multimedia **10**(3), 103–106 (2003)
2. Diakopoulos, N.A., Shamma, D.A.: Characterizing debate performance via aggregated twitter sentiment. In: Conference on Human Factors in Computing Systems (CHI 2010) (2010)
3. Tamura, H., Mori, S., Yamawaki, T.: Textural features corresponding to visual perception. IEEE Transactions on Systems, Man and Cybernetics **8**(6), 460–473 (1978)
4. Itten, J.: The art of color: the subjective experience and objective rationale of color. Van Nostrand Reinhold, New York (1973)
5. Wei-ning, W., Ying-lin, Y., Sheng-ming, J.: Image retrieval by emotional semantics: a study of emotional space and feature extraction. In: IEEE International Conference on Systems, Man and Cybernetics, SMC 2006, vol. 4, pp. 3534–3539. IEEE (2006)
6. Jansen, B.J., Zhang, M., Sobel, K., Chowdury, A.: Twitter power: Tweets as electronic word of mouth. Journal of the American Society for Information Science and Technology (2009)
7. Mardia, K.V., Jupp, P.E.: Directional statistics. Wiley (2009)
8. Tumasjan, A., Sprenger, T.O., Sandner, P.G., et al.: Predicting elections with twitter: what 140 characters reveal about political sentiment. In: Proceedings of the fourth international AAAI conference on weblogs and social media, pp. 178–185 (2010)
9. Haralock, R.M., Shapiro, L.G.: Computer and robot vision. Addison-Wesley Longman Publishing Co., Inc (1991)
10. Bollen, J., Mao, H., Zeng, X.: Twitter mood predicts the stock market. Journal of Computational Science **2**(1), 1–8 (2011)
11. Datta, R., Joshi, D., Li, J., Wang, J.Z.: Studying aesthetics in photographic images using a computational approach. In: Leonardis, A., Bischof, H., Pinz, A. (eds.) ECCV 2006. LNCS, vol. 3953, pp. 288–301. Springer, Heidelberg (2006)
12. OConnor, B., Balasubramanyan, R., Routledge, B.R., et al.: From tweets to polls: linking text sentiment to public opinion time series. In: Proceedings of the International AAAI Conference on Weblogs and Social Media, pp. 122–129 (2010)
13. Yang, J., et al.: Feature fusion: parallel strategy vs. serial strategy. Pattern Recognition **36**(6), 1369–1381 (2003)
14. Wang, H.L., Cheong, L.F.: Affective understanding in film. IEEE Transactions on Circuits and Systems for Video Technology **16**(6), 689–704 (2006)
15. Jones, K.S.: A statistical interpretation of term specificity and its application in retrieval. Journal of documentation **28**(1), 11–21 (1972)
16. Colombo, C., Del Bimbo, A., Pala, P.: Semantics in visual information retrieval. IEEE on Multimedia **6**(3), 38–53 (1999)
17. Pang, B., Lee, L.: Opinion mining and sentiment analysis. Found. Trends Inf. Retr. **2**(1–2), 1–135 (2008)
18. Osgood, C.E., Suci, G.J., Tannenbaum, P.H.: The measurement of meaning. University of Illinois Press, Urbana (1957)
19. Tan, S., Zhang, J.: An empirical study of sentiment analysis for chinese documents. Expert Systems with Applications **34**(4), 2622–2629 (2008)
20. Galavotti, L., Sebastiani, F., Simi, M.: Experiments on the use of feature selection and negative evidence in automated text categorization. In: Borbinha, J.L.,

Baker, T. (eds.) ECDL 2000. LNCS, vol. 1923, pp. 59–68. Springer, Heidelberg (2000)

21. Kuroda, K., Hagiwara, M.: An image retrieval system by impression words and specific object names–IRIS. Neurocomputing **43**(1), 259–276 (2002)

22. Yanulevskaya, V., Van Gemert, J.C., Roth, K., et al.: Emotional valence categorization using holistic image features. In: 15th IEEE International Conference on Image Processing, ICIP 2008, pp. 101–104. IEEE (2008)

23. Read, J.: Using emoticons to reduce dependency in machine learning techniques for sentiment classification. In: ACL. The Association for Computer Linguistics (2005)

24. Zagibalov, T.: Unsupervised and knowledge-poor approaches to sentiment analysis. Diss. University of Sussex (2010)

25. Ponomareva, N., Thelwall, M.: Do neighbours help? An exploration of graph-based algorithms for cross-domain sentiment classification. The 2012 Conference on Empirical Methods on Natural Language Processing and Computational Natural Language Learning (EMNLPCoNLL 2012), pp. 655–665 (2012)

26. Hayashi, T., Hagiwara, M.: Image query by impression words-the IQI system. IEEE Transactions on Consumer Electronics **44**(2), 347–352 (1998)

27. Hanjalic, A.: Extracting moods from pictures and sounds: Towards truly personalized TV. IEEE on Signal Processing Magazine **23**(2), 90–100 (2006)

28. Wu, Q., Zhou, C.-L., Wang, C.: Content-based affective image classification and retrieval using support vector machines. In: Tao, J., Tan, T., Picard, R.W. (eds.) ACII 2005. LNCS, vol. 3784, pp. 239–247. Springer, Heidelberg (2005)

29. Altman, N.S.: An introduction to kernel and nearest-neighbor nonparametric regression. The American Statistician **46**(3), 175–185 (1992)

30. Valdez, P., Mehrabian, A.: Effects of color on emotions. Journal of Experimental Psychology: General **123**(4), 394 (1994)

31. Zhang, H.P., Liu, Q., Cheng, X.Q., et al.: Chinese lexical analysis using hierarchical hidden markov model. In: Proceedings of the second SIGHAN workshop on Chinese language processing, vol. 17, pp. 63–70. Association for Computational Linguistics (2003)

32. Go, A., Huang, L., Bhayani, R.: Twitter sentiment analysis. Final Projects from CS224N for Spring 2008/2009 at The Stanford Natural Language Processing Group (2009)

33. Muralidharan, S., Rasmussen, L., Patterson, D., et al.: Hope for Haiti: An analysis of Facebook and Twitter usage during the earthquake relief efforts. Public Relations Review **37**(2), 175–177 (2011)

34. Mojsilovic, A., Gomes, J., Rogowitz, B.: Semantic-friendly indexing and quering of images based on the extraction of the objective semantic cues. International Journal of Computer Vision **56**(1–2), 79–107 (2004)

35. Asur, S., Huberman, B.A.: Predicting the future with social media. In: 2010 IEEE/WIC/ACM International Conference on Web Intelligence and Intelligent Agent Technology (WI-IAT), vol. 1. IEEE (2010)

36. Wang, W., He, Q.: A survey on emotional semantic image retrieval. In: 15th IEEE Int. Conf. on Image Processing, pp. 117–120 (2008)

37. Zhai, Z., et al.: Exploiting effective features for chinese sentiment classification. Expert Systems with Applications **38**(8), 9139–9146 (2011)

38. Mejova, Y., Srinivasan, P.: Exploring feature definition and selection for sentiment classifiers. ICWSM (2011)

TSum4act: A Framework for Retrieving and Summarizing Actionable Tweets During a Disaster for Reaction

Minh-Tien Nguyen[1]([✉]), Asanobu Kitamoto[2], and Tri-Thanh Nguyen[3]

[1] Hung Yen University of Technology and Education (UTEHY),
Hung Yen, Vietnam
tiennm@utehy.edu.vn
[2] National Institute of Informatics, Tokyo, Japan
kitamoto@nii.ac.jp
[3] Vietnam National University, Hanoi (VNU),
University of Engineering and Technology (UET), Hanoi, Vietnam
ntthanh@vnu.edu.vn

Abstract. Social networks (e.g. Twitter) have been proved to be an almost real-time mean of information spread, thus they can be exploited as a valuable channel of information for emergencies (e.g. disasters) during which people need updated information for suitable reactions. In this paper, we present *TSum4act*, a framework designed to tackle the challenges of tweets (e.g. diversity, large volume, and noise) for disaster responses. The objective of the framework is to retrieve actionable tweets (e.g. casualties, cautions, and donations) that were posted during disasters. For this purpose, the framework first identifies informative tweets to remove noise; then assigns informative tweets into topics to preserve the diversity; next summarizes the topics to be compact; and finally ranks the results for user's faster scan. In order to improve the performance, we proposed to incorporate event extraction for enriching the semantics of tweets. *TSum4act* has been successfully tested on Joplin tornado dataset of 230.535 tweets and the completeness of 0.58 outperformed 17%, of the retweet baseline's.

Keywords: Data mining · Event extraction · Tweet summarization · Tweet recommendation

1 Introduction

Twitter provides a new method for spreading information during natural or man-made disasters [13] in the form of tweets (a short text message with the maximum of 140 letters). These tweets mention a wide range of information in all aspects of life, from personal aspects to disaster facts. During a disaster, tweets usually tend to explode in a large volume and high speed. This can challenge the people who seek up-to-date information for making decisions. To seek the information, people can use Twitter's search function. However, this function bases

© Springer International Publishing Switzerland 2015
T. Cao et al. (Eds.): PAKDD 2015, Part II, LNAI 9078, pp. 64–75, 2015.
DOI: 10.1007/978-3-319-18032-8_6

on boolean queries and responses highly redundant tweets in a reverse chronological order. This challenge inspires us to present a framework that incorporates event extraction to generate event graphs in retrieving informative tweets for people in disasters.

Recently, tweet summarization has received a lot of attention from papers [5,8,12]. The authors in these researches have solved the summarization by using term frequency in corporation with inverted document frequency (TF-IDF) or lexical approach. However, these approaches face the noise of tweets and may not exploit entities (e.g. times, locations, numbers, etc) which play an important role in summarization [15]. These limitations inspire us to apply event extraction in our framework. Our contributions are:

- We adapt event extraction for improving the performance of summarization. Our approach has two advantages: (1) exploiting the important role of entities and (2) reducing the impact of lexical representation noise.
- We successfully apply our framework in a real dataset which is collected during Joplin tornado. The completeness measure of 0.58 indicates that information from our framework can be combined with other sources (i.e. TV, online news, or emergency services) to provide important information to people during disasters.

We set our problem as the tweet recommendation by relying on extractive summarization (e.g. based on event extraction). Given a disaster-related query, our model: 1) retrieves tweets containing the query from the source; 2) due to the fact that the tweets are diverse and noisy (e.g. there are a lot of unrelated ones), the model filters out irrelevant ones to get a smaller set of informative tweets; 3) in order to provide finer-grained information to users, the model divides informative tweets into predefined classes (i.e. casualties, cautions, and donations); 4) since the number of tweets in each class is still big, for each class of tweets, the model separates them into topics (in the form of clusters); finally, 5) to make the results compact, the model ranks the tweets and gets the top ones as a summary of each cluster to recommend to users. For improving the semantics and reducing the lexical noise of the tweets, we propose to represent tweets in the form of events (viz. after applying event extraction) for constructing a graph as the input for ranking algorithm.

The rest of this paper is organized as follows: related works are showed in Section 2; we will give our idea to solve informative tweets summarization in Section 3; our contributions are in Section 4.1, 4.2, 4.3, and 4.3; results are showed in Section 5; the last section is the conclusion.

2   Related Work

Ritter et al. [12] proposed a method which automatically extracted open domain events. In this method, events were ranked and classified before being plotted in calender entries. The result increased 14% of F1 over baseline (without NER). However, the ranking mechanism can badly affect in generating result because the author used frequency of entities, thus unimportant entities may generate redundant events.

Chakrabarti and Punera used Hidden Markov Model to identify sub-events of the parent event with 0.5 of precision and 0.52 of recall in football matches [5]. To summarize information, the author combined *tf-idf* and *Cosine similarity*. However, this method might not achieve high precision because the noise of tweets. In our work, we rely on entities rather than single words (terms).

Khan et al. [8] used lexical level underlying topical modeling and graphical model for summarizing tweets in a debating event. The result was about 81.6% of precision and 80% recall with their dataset. However, this method faces the noise from tweets, does not utilize entities, and requires a large of lexicons for generating the summarization.

3   TSum4act Framework

TSum4act solves tweet summarization and recommendation in disaster responses. The problem can be defined as follows:

Input: the keywords related to a natural disaster.

Output: A small set of the most informative tweets which can be used in situational awareness for supporting the making of suitable reaction. The tweets are divided into classes (e.g. casualty, caution, and donation), and each class, in turn, is divided into topics represented by top ranked tweets.

We call this set as *informative tweets* or *actionable tweets* because these tweets provide useful information which help people making suitable decisions in a disaster.

The framework in Fig. 1 consists of four main components: tweet retrieval, informative tweet identification, topic identification, and tweet summarization. The first component receives keywords from users and retrieves relevant tweets from Twitter. The second component identifies informative tweets (i.e. tweets

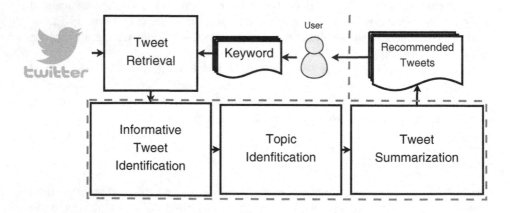

Fig. 1. Overview of TSum4act framework

which help people making decisions in a disaster rather than personal tweets). We solve this problem by using classification because tweets can be divided into *informative* and *not informative*. Another task of this component is to put informative tweets into important classes for users' easy navigation. The third component takes each tweet class to identify the different topics (i.e. tweets of which the meaning is close to each other) as a preliminary step to compress the tweets for reducing the tweet volume size. This component is built based on clustering which has the ability to group items that are close to each other into a cluster. The last component has the responsibility of summarizing the data to produce recommended tweets to users. This task is done by representing tweets in the form of events, and an event graph is constructed for each topic for ranking. After ranking, near-duplicate tweets are removed, and top ranked ones are returned to users as a summary for recommendation.

The first component is rather simple, therefore, in this paper we only focus on the three remaining components.

4   The Process of TSum4act

4.1   Informative Tweet Identification

We follow the approach of [7,14] to apply classification for informative tweet identification. The classification process includes three steps along with three binary classifiers. The first step distinguishes whether tweets are informative or not; the second one identifies whether informative tweets are direct (viz. events that users see or hear directly) or indirect, e.g. a tweet that is forwarded (called *retweet*); and the third step classifies the direct tweets into three classes: casualty/damage, caution/advice, and donation/offer (called valuable classes).

4.2   Topic Identification

The number of tweets is still big after classifying, we propose to assign tweets into topics (in the form of clusters) (to ensure the diversity) for later summarization (to keep the result compact). Since common clustering methods (e.g. K-means) normally base on the frequency of words (e.g. tf-idf), thus they are not suitable for keeping the semantics of tweets whereas hidden topic models are a good selection to solve this issue. Assigning tweets into clusters helps users to easily navigate the information. This component first generates document distributions by LDA and secondly, it uses a clustering algorithm to assign tweets into topic clusters.

LDA: LDA generates a probability distribution of documents (tweets) over topics by using estimated methods. Tweets can be assigned into clusters by a clustering algorithm based on this distribution. We decide to use LDA since it was adapted for short texts (e.g., tweets or messages) [8]. Another interesting aspect is that it not only solves the clustering problem but also generates topical words which can be used for abstractive summarization. For simplicity, we have adopted Latent Dirichlet Allocation (LDA)[1] [1].

Clustering: LDA assigns tweets into hidden topics which are represented by words/phrases. However, the framework expects tweets which belong to clusters. Therefore, a clustering algorithm is used to solve this issue. The idea of assigning tweets into clusters is to find out the closest distance from tweet t to cluster c_i by an iterative algorithm which uses Jensen-Shannon divergence.

4.3   Tweet Summarization

This component first extracts events in tweets in each topic to built a graph. Subsequently, the event graphs are the input for a ranking algorithm to find out important (i.e. highest score) events . Finally, after removing near-duplicate tweets, top ones are returned to users.

Event Extraction: As introduced previously, we use a ranking algorithm to find out informative tweets. In fact, normal ranking algorithms only rely on keywords or even topical statistic can not completely utilize the semantics of tweets, thus, the final result may be not good. We propose to use ranking approach in which a tweet has a correlation with another based on the similarity. The similarity can be denoted by using words or complete tweets. However, using words or complete tweets faces the noise of tweet (i.e. stop words, hashtags, or emoticons); hence, this can badly affect in calculating the similarity. As the contribution, we propose to use event extraction to represent the similarity between two tweets. Using event extraction has two advantages: 1) reducing the noise of tweet and 2) keeping the semantics of tweets. Therefore, the framework can achieve high accuracy in generating informative tweets.

We define an event as a set of entities in a tweet. An event includes *subject*, *event phrase*, *location*, and *number* as follows.

$$event = \{subject, event\ phrase, location, number\} \tag{1}$$

where *subject* answers the question WHAT (e.g. a tornado or a road) which is a cause or result; *event phrase* represents the action/effect of the subject; *location* answers WHERE the event occurs (e.g. Oklahoma); and *number* focuses to answer the question HOW MANY (e.g. the number of victims).

To extract above attributes we use NER tool of [11] which annotates the tweets with predefined tags, then, we parse the result to extract values of the tags corresponding to the attributes. We accept an event which does not have complete attributes. An example of an event from an original tweet is showed as below:

---

[1] http://jgibblda.sourceforge.net/

Table 1. The illustration of two equations

Tweet	Simpson	Cosine
Tornado kills 89 in Missouri		
Tornado kills 89 in Missouri yesterday	0.0	0.912

Original tweet: *"Tornado kills 89 in Missouri yesterday"*
Event: { *Tornado, kills, Missouri, 89* }
In this example, *Tornado* is the subject, *kills* is the event phrase, *Missouri* is the location, and *89* is the number of victims. Events in this step are input for generating event graphs in the next section.

Event Graph Construction: The event graphs require vertices and weights as inputs. In this graph, two vertices (two events) are connected by an edge with a weight. To identify the weight we consider two equations: Simpson and Cosine. However, in this case, Cosine is better. The compared intuition of two equations is showed in Table 1. In this table, the value of *0.0* of Simpson indicates that two tweets are completely similar despite the difference of the word *"yesterday"* between the two sentences. In contrast, the value of Cosine equation is *0.912* giving two tweets are nearly completely similar. Therefore, we follow Cosine to calculate the weight of two events. Given two events A and B, Cosine equation is showed in Eq. (2).

$$sim(A, B) = \frac{\sum_{i=1}^{n} A_i \times B_i}{\sqrt{\sum_{i=1}^{n}(A_i)^2} \times \sqrt{\sum_{i=1}^{n}(B_i)^2}} \qquad (2)$$

The *sim(A, B)* may be equal 0 in some cases indicating the two events are totally different. Therefore, to make the result compact, two events are deemed to have relation if they have the similarity measure above zero, and an event is removed if it has no relation with any other event.

After identifying vertices and weights the framework generates graphs for the raking step. In this paper, a graph is defined as $G = < V, E, W >$ where:

- V: a set of vertices denoted as $V = \{v_1, v_2, ..., v_n\}$ where v_i^{th} corresponds to an *event* i^{th} and n is the number of events in each cluster.
- E: a set of edges denoted as $E = \{e_1, e_2, ..., e_m\}$ where e_j^{th} connects two vertices v_k^{th} and v_h^{th} in V.
- W: a weight matrix holding the weight (in term of similarity) of edges.

In this graph, the weight matrix is calculated by Eq. (2). As this calculation, two vertices are connected when the weight is greater than 0. The graph G is the input for ranking algorithm in the next section.

Ranking: The goal of our method is to retrieve the most useful information. To solve this task a random method can be used after clustering. However, this method retrieves tweets which may not have enough evidence to conclude whether they are informative. An alternative approach is to use a ranking algorithm. It is suitable with our objective because: 1) our goal is to retrieve top k of tweets in each cluster and 2) important tweets converge at central clusters after clustering; hence the ranking algorithm can be easy to find out informative tweets.

The framework uses PageRank [2] for retrieving informative tweets. This is because, firstly, it is a successful algorithm for ranking webpages. Secondly, PageRank ranks vertices to find out important vertices. This is similar with our goal in finding important events. The ranking mechanism is showed in E.q (3).

$$PR(v_j) = (1 - d) + d \sum_{v_i, vj \in E} \frac{W_{i,j} \times PR(v_i)}{\sum_{v_i, v_k \in E} W_{ik}} \tag{3}$$

Equation (3) shows the process to calculate the rank of a node in a recursive mechanism by using power iteration. In this equation, d is the damping factor which is used to reduce the bias of isolated nodes in a graph.

Filtering: The final module retrieves informative tweets corresponding to important events. Though it is possible to apply filtering before ranking, we put it after ranking to avoid the situation where the filtering produces a so sparse event graph that the performance of the ranking is badly affected.

The filtering uses Simpson equation to remove near-duplicate tweets because they provide the same information with others. Two near-duplicate tweets contain a little difference of information (e.g. in Tab 1). The Simpson is showed in Eq. (4).

$$simp(t_1, t_2) = 1 - \frac{|S(t_1) \cap S(t_2)|}{min(|S(t_1)|, |S(t_2)|)} \tag{4}$$

where $S(t)$ denotes the set of words in tweet t.

5   Experiments and Results

5.1   Experimental Setup

Data: Dataset is 230.535 tweets, which were collected during Joplin tornado in the late afternoon of Sunday, May 22, 2011[2] at Missouri. The unique tweets were selected by Twitter Streaming API using the hashtag *#joplin*. The dataset is a part of AIDR project [6].

Parameter Setting: Following [8] we choose $\alpha = \beta = 0.01$ with 1.000 iterations and $k \in [2, 50]$ for LDA. k will be identified in 5.3. In the Section 4.2 we use $\mu = 0.9$. Following [2], we choose $d = 0.85$ for PageRank. Finally, Eq. (4) only keeps tweets which satisfy $simp(t_1, t_2) > 0.25$. The value 0.25 is achieved by running the experiment over many times.

---

[2] http://en.wikipedia.org/wiki/2011_Joplin_tornado

5.2   Retweet Baseline Model

We use retweet model as a baseline [4]. This is because, firstly, it represents the importance of a tweet based on retweet. Intuitively, if a message receives many retweets, it can be considered an informative tweet. Secondly, the model is simple because it has already provided by Twitter.

5.3   Preliminary Results

Classification: To build binary classifiers we use Maximum Entropy (ME)[3] with N-gram features because ME has been a successful method for solving classification problem [10]. Another interesting aspect is that our data is sparse after pre-processing while ME is a suitable approach to deal with sparse data. We do not use hashtags, emoticons, or retweets because these features are usually used in emotional analysis rather than in classification.

The performance of classification is measured by the average of 10-folds cross validation. Results are showed in Table 2.

Table 2. The performance of classification

Class	Precision	Recall	F1
Informative Information	0.75	0.87	0.8
Direct Tweets	0.71	0.83	0.77
Casualty	0.89	0.89	0.89
Caution	0.88	0.91	0.89
Donation	0.87	0.88	0.88

The results in Tab. 2 show that classifiers achieve good performances 0.8, 0.77, and about 0.89 of F1 in the three classification levels, respectively. After classification, the framework only keeps tweets of the valuable classes including casualty/damage, caution/advice, and donation/offer.

Selecting the Number of Clusters: We follow cluster validation to select k [3,9]. We only select $k \in [2 : 50]$ because if $k > 50$ clusters may be overfitting while it is too general if $k < 2$ (only one cluster). The result of cluster validation is illustrated in Fig. 2a. After clustering, informative tweets belong to clusters which will be used for the summarization and recommendation.

Evaluation Method: To evaluate the performance of the framework, we use three annotators to rate retrieved tweets (top 10 tweets). This is because: 1) selecting top k (k=10) is similar to recommendation and 2) our data needs user ratings to calculate performance of this component. The retrieved tweets are given for annotators to rate a value from 1 to 5 (5: very good; 4: good; 3: acceptable; 2: poor; and 1: very poor). Each tweet is rated in three times; the score of

[3] http://www.cs.princeton.edu/maxent

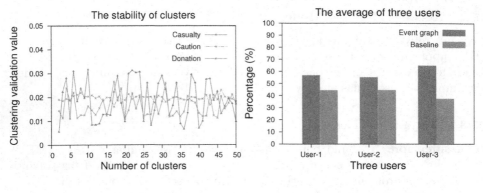

(a) The result of cluster validation   (b) The average of completeness

Fig. 2. Selecting k and the average of completeness of three users

a tweet is the average of rating scores after rounding. The average of inter-rater agreement is 87.6% after cross-checking the results of three annotators.

Tweet Summarization and Recommendation: We define the *completeness* to measure the performance of our method. The completeness measures how well the summary covers the informative content in the recommended tweets by the total rated scores over maximal score in each cluster as in E.q. (5).

$$completeness = \sum (rating\ score)/50 \tag{5}$$

where *rating score* is the users' score, and 50 is the maximum total score (viz. 10 tweets each has a maximum score of 5).

The results of the three annotators are illustrated in Fig. 3, 4, and 5.

indicating that our framework dominates the baseline in Fig. 3a, 3c, 4c, 5a, 5b, and 5c while it is hard to conclude our framework is better in Caution in Fig. 3b and 4b.

The average of completeness of three users is showed in Fig. 6. It is equal in Caution of 1st annotator about 0.42 in Fig. 6a while our framework outperforms baseline on other annotators. In Fig. 6a, the result of our framework is twice higher than baseline in Donation about 0.7 and 0.3. The result of 2nd annotator is higher than other users and the completeness of our method in Fig. 6b is around 0.6.

The average of completeness on three annotators over classes is showed in Fig. 2b. The result shows that the completeness of our method outperforms baseline method about 17% (0.58 and 0.41, respectively). The results in Casualty and Caution are similar about 0.55 with our method and around 0.44 with baseline. In Donation our method outperformed the baseline with the result of 0.64 in comparison with 0.37.

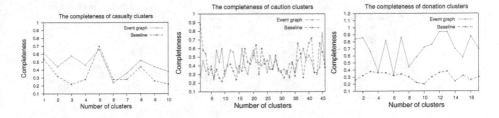

(a) The completeness of casualty

(b) The completeness of caution

(c) The completeness of donation

Fig. 3. The comparison of two methods in tweet recommendation of 1st user

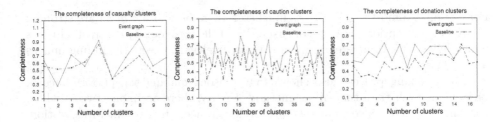

(a) The completeness of casualty

(b) The completeness of caution

(c) The completeness of donation

Fig. 4. The comparison of two methods in tweet recommendation of 2nd user

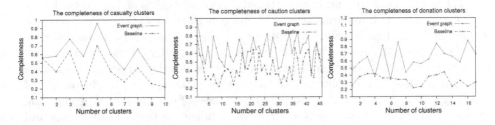

(a) The completeness of casualty

(b) The completeness of caution

(c) The completeness of donation

Fig. 5. The comparison of two methods in tweet recommendation of 3rd user

5.4   Discussion

Fig. 3, 4, and 5 indicate that the results of our method are quite similar with baseline in Fig. 3b, 4a, 4b, and 5b, but in the remaining figures our method prevails.

By checking original tweets we recognize that clusters which have a high completeness contain highly relevant tweets. It appears in both the two methods.

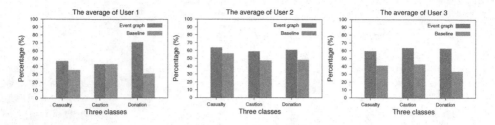

(a) The average of $1^{st} user$ (b) The average of $2^{nd} user$ (c) The average of $3^{rd} user$

Fig. 6. The average of completeness of two methods in tweet recommendation of three users

For example, almost top 10 tweets in cluster 1st in 5b and 5th in 5a are strongly related to the cautions, advices, or casualties of the tornado. By contrast, clusters which have a low completeness include many irrelevant tweets (including images and videos).

The checking process also shows that reasons making the performance of baseline is lower than that of our method. Firstly, there are many tweets which contain images or videos, but they are irrelevant according to our investigation. Secondly, many tweets in donation receive a lot of retweets because it mentions about a famous person (President Obama). Another interesting point is the speech of Eric Cantor, who is a member of GOP of U.S. Eric Cantor opposes disaster relief and saves money. Therefore, his tweets receive many retweets. Finally, a lot of tweets mention to memorial services for victims.

The results indicate that ranking based on event graph is an important factor which making our framework achieves a good performance. If an event receives a high weight, then it has many strong relations with others. It means that this event tends to appear in many tweets; hence, ranking events can retrieve important tweets. This phenomenon also appears in other clusters which have a high completeness. An alternative interesting point appears in cluster 5th of caution in Fig. 3b. It is the lowest of completeness of our method. The checking process shows that many events are irrelevant indicating that the clustering and event extraction are not efficient in this cluster.

6   Conclusion

In this paper, we introduced *TSum4act*, a framework for retrieving informative tweets during a disaster for suitable reaction. Our framework utilizes state-of-the-art machine learning techniques, event extraction, and graphical model to deal with the diversity, large volume and noise of tweets. *TSum4act* was compared with retweet model and the result indicates that our framework can be used in real disasters.

For future works, we plan to first improve the performance of classification by trying multi-class classifiers. Secondly, the evaluation method should be considered because we only select the top 10 tweets. In addition, our method should be compared with other model as lexicons. Finally, we plan to summarize recommended tweets by using abstractive approach.

References

1. Blei, D.M., Ng, A.Y., Jordan, M.I.: Latent dirichlet allocation. The Journal of Machine Learning Research **3**, 993–1022 (2003)
2. Brin, S., Page, L.: The anatomy of a large-scale hypertextual web search engines. Computer Networks and ISDN System **30**(1), 107–117 (1998)
3. Brody, S., Elhadad, N.: An unsupervised aspect-sentiment model foronline reviews. In: Human Language Technologies: The 2010 Annual Conferenceof the North American Chapter of the Association for Computational Linguistics. Association for Computational Linguistics (2010)
4. Busch, M., Gade, K., Larson, B., Lok, P., Luckenbill, S., Lin, J.: Earlybird: real-time search at twitter. In: 2012 IEEE 28th International Conference on Data Engineering (ICDE). IEEE (2012)
5. Chakrabarti, D., Punera, K.: Event summarization using tweets. In: ICWSM (2011)
6. Imran, M., Castillo, C., Lucas, J., Meier, P., Vieweg, S.: Aidr: artificial intelligence for disaster response. In: Proceedings of the Companion Publication of the 23rd International Conference on World Wide Web Companion. International World Wide Web Conferences Steering Committee, pp. 159–162, April 2014
7. Imran, M., Elbassuoni, S., Castillo, C., Diaz, F., Meier, P.: Extracting information nuggets from disaster-related messages in social media. In: ISCRAM, Baden-Baden, Germany (2013)
8. Khan, M.A.H., Bollegala, D., Liu, G., Sezaki, K.: Multi-tweet summarization of real-time events. In: Proc. of ASE/IEEE International Conference on Social Computing. IEEE (2013)
9. Levine, E., Domany, E.: Resampling method for unsupervised estimation of cluster validity. Neural Computation **13**(11), 2573–2593 (2001)
10. Nigam, K., Lafferty, J., McCallum, A.: Using maximum entropy for text classification. In: IJCAI-99 Workshop on Machine Learning for Information Filtering, vol. 1, pp. 61–67 (1999)
11. Ritter, A., Clark, S., Mausam, Etzioni, O.: Named entity recognition in tweets: an experimental study. In: Proceedings of the Conference on Empirical Methods in Natural Language Processing. Association for Computational Linguistics (2011)
12. Ritter, A., Mausam, Etzioni, O., Clark, S.: Open domain event extraction from twitter. In: KDD, pp 1104–1112 (2012)
13. Sakaki, T., Okazaki, M., Matsuo, Y.: Earthquake shakes twitter users: real-time event detection by social sensors. In: Proceedings of the 19th International Conference on World Wide Web. ACM (2010)
14. Vieweg, S.E.: Situational Awareness in Mass Emergency: A Behavioral and Linguistic Analysis of Microblogged Communications. PhD thesis, University of Colorado at Boulder (2012)
15. Xu, W., Grishman, R., Meyers, A., Ritter, A.: A preliminary study of tweet summarization using information extraction. In: NAACL (2013)

Clustering

Evolving Chinese Restaurant Processes for Modeling Evolutionary Traces in Temporal Data

Peng Wang[1,2]([✉]), Chuan Zhou[1], Peng Zhang[3], Weiwei Feng[1],
Li Guo[1], and Binxing Fang[4]

[1] Institute of Information Engineering, Chinese Academy of Sciences, Beijing, China
peng860215@gmail.com, {zhouchuan,guoli}@iie.ac.cn
[2] Institute of Computing Technology, University of Chinese Academy of Sciences,
Beijing, China
[3] QCIS, University of Technology Sydney, Sydney, Australia
peng.zhang@uts.edu.au
[4] Beijing University of Posts and Telecommunications, Beijing, China
fangbx@cae.cn

Abstract. Due to the evolving nature of temporal data, clusters often exhibit complex dynamic patterns like birth and death. In particular, a cluster can branch into multiple clusters simultaneously. Intuitively, clusters can evolve as evolutionary trees over time. However, existing models are incapable of recovering the tree-like evolutionary trace in temporal data. To this end, we propose an Evolving Chinese Restaurant Process (ECRP), which is essentially a temporal non-parametric clustering model. ECRP incorporates dynamics of cluster number, parameters and popularity. ECRP allows each cluster to have multiple branches over time. We design an online learning framework based on Gibbs sampling to infer the evolutionary traces of clusters over time. In experiments, we validate that ECRP can capture tree-like evolutionary traces of clusters from real-world data sets and achieve better clustering results than state-of-the-art methods.

Keywords: Nonparametric bayesian · Chinese restaurant process · Evolutionary clustering

1 Introduction

Cluster analysis groups a set of objects in such a way that objects in the same cluster are more similar with each other than with those in other clusters. In this paper we consider clustering temporal data, and focus on how to model the tree-like evolutionary trace of clusters over time.

Intuitively, temporal dynamics and structural evolution is a universal phenomenon for temporal data, such as dynamic topics in text collections, evolution of species and evolution of communities in social networks [18–20]. Take research publications for example, with the development of scientific research, new topics

© Springer International Publishing Switzerland 2015
T. Cao et al. (Eds.): PAKDD 2015, Part II, LNAI 9078, pp. 79–91, 2015.
DOI: 10.1007/978-3-319-18032-8_7

emerge and obsolete topics die. Moreover, topics can evolve and branch into several topics, e.g. , the topic of *social network* has developed many branches, such as *community detection, social influence* and *social privacy*. Intuitively the evolving trace of clusters over time is like a tree, where a cluster constantly evolves and branches new clusters.

Compared with the traditional clustering tasks, modeling tree-like evolutionary traces meets new challenges. First, evolution of clusters (birth, death and branch) implies that intrinsically the number of clusters over time is not fixed. As a result, parametric clustering models are improper for this task as they require setting the number of clusters manually. Second, the distribution (popularity) and parameters of clusters are dynamic over time. The model should be able to catch the temporal features of clusters and smooth them over time. Third, for the evolutionary tree of clusters, the number of branches is uncertain. But we still need to control the scale of branches; otherwise the evolutionary trace recovered shall be incomprehensible.

A handful of temporal clustering models have been proposed to exploit the temporal dynamics and structural evolution of clusters from temporal data [3,12,16,17]. Non-parametric Bayesian models [2,12] can auto-learn the number of topics over time, so they are widely used for clustering evolutionary temporal data. One of the representative work is Recurrent Chinese Restaurant Process (RCRP) [2,3] proposed by Ahmed et.al. RCRP can capture the birth/death of clusters and similar clusters of adjacent epochs form cluster chains. However, RCRF falls short in modeling nonlinear evolutionary patterns because each topic is allowed to have only one variant. Other nonparametric temporal models [4,12,16] also suffer the same problem. Besides modeling the evolution structure of clusters, there are lots of works on modeling the evolution parameters. Previous schemes for modeling the evolution of cluster parameters are generally designed for chain-like evolutionary trace, e.g., DTM [6] and SNDTM [2]. They are difficult to be extended to more complicated case like trees. In summary, to the best of our knowledge, none of existing methods can handle all the challenges mentioned above.

In this paper, we propose an Evolving Chinese Restaurant Process (ECRP) which can model the tree-like evolutionary trace in temporal data. In the setting of ECRP, temporal collections are divided into epoches. To model cluster branching over time, ECRP lets each cluster in current epoch form a Chinese Restaurant Process (CRP) and uses the combination of cluster-specific CRPs as the prior for clusters in the next epoch. So a cluster can branch into several new clusters. Moreover, ECRP can model dependence of clusters' distribution over time. On the top of ECRP, we adopt conjugate prior cascade scheme to model the clusters' parameter evolution, which can model cluster parameter evolution for tree-like structure. We also design an online learning framework based on Gibbs sampling to infer evolutionary traces of clusters over time.

Clustering with ECRP enjoys the following merits. First, our method can model the birth, branch and death of clusters. Second, our method allows easy inference for posterior of parameters. Third, the evolutionary structure can learning incrementally with the online learning framework. In experiments, we

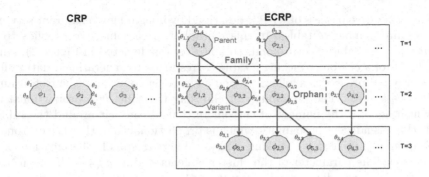

Fig. 1. Illustration of CRP and ECRP. Right: CRP. Note that no temporal concept is evolved. Left: ECRP. Here clusters form three independent evolutionary trees.

demonstrate that ECRP can capture tree-like evolutionary traces of clusters on real-world data sets and outperform the state-of-the-art methods.

The rest of the paper is organized as follows. Section 2 introduces the problem settings and reviews necessary backgrounds. Section 3 introduces the Evolving Chinese Restaurant Process and the learning framework. Section 4 conducts experiments. We survey the related work in Section 5 and conclude the paper in Section 6.

2 Preliminary

In this section, we first introduce the problem settings and then briefly review the Chinese Restaurant Processes (CRP) to make the paper self-contained.

2.1 Problem Settings

In this paper, temporal data are split into epochs. n_t denotes the number of data in the tth epoch, and $x_{t,i}$ denotes the ith datum in epoch t. The temporal data are assumed to be partially exchangeable, i.e., data within an epoch are exchangeable but cannot be exchanged across epochs. we let $\theta_{t,i}$ denote the cluster associated with data $x_{t,i}$. For a cluster k in epoch t, its parameter is denoted as $\phi_{t,k}$. And the set of $\{\phi_{t,k}\}$ define the unique clusters. We assume that the cluster can emerge, branch and die over time. Our goal is to discover the tree-like evolutionary traces of clusters in $\{\mathbf{X_1}, \cdots, \mathbf{X_t}, \cdots\}$.

2.2 Chinese Restaurant Process

CRP is a stochastic process which can model the generation of unbounded number of clusters for static data sets. CRP is specified by a base distribution H and a positive real number α called the concentration parameter. H serves as the prior for clusters' parameters, and α specifies the discretization of clusters.

The generative process of CRP is described by imagining a restaurant serving with infinite number of tables. In this metaphor, a customer corresponds to a datum x; and a table ϕ corresponds to a cluster (as depicted in Figure 1). And the process of a customer picking a table corresponds to generating a datum for a cluster. CRP proceeds as follows. When a customer comes into the restaurant, he chooses a table. The probability of choosing the table k is proportional to n_k, which denotes the number of customers that already sit around the table. While there is nonzero probability, which is proportional to α, that the customer picks an empty table ϕ_{K+}. Picking a new table corresponds the creation of a new cluster. The parameter of the cluster is generated as $\phi_{K+} \sim H$. So in the limit of $\alpha \to 0$, the data are all concentrated on a single cluster, while in the limit of $\alpha \to \infty$, each datum forms a cluster all by itself. The process can be summarized as:

$$\theta_i | \theta_{1:i-1}, H, \alpha \sim \sum_k \frac{n_k^{(i)}}{i - 1 + \alpha} \delta(\phi_k) + \frac{\alpha}{i - 1 + \alpha} H \tag{1}$$

Where θ_i is the cluster chosen by the ith datum. The value of the datum x_i is sampled from ϕ_{θ_i}.

In summary, CRP can describe the generation for static data set with an unbounded number of clusters. In the next section, we present the ECRP process which extends CRP for modeling temporal data and mining tree-like evolutionary trace behind clusters.

3   Model Description

In this section, we first introduce the evolving Chinese restaurant process (ECRP). Then we present how to model the evolution of cluster parameters. Finally, an online learning framework based on Gibbs sampling is presented.

3.1   Evolving Chinese Restaurant Processes

The key to discover tree-like evolutionary traces is to model cluster branching. In ECRP, we let each cluster form a CRP for the next epoch, so a cluster can branch multiple variants in the next epoch. And we adopt the posterior of parent cluster as the base measure for the cluster specific CRP so as to smooth the cluster parameter over time. In the sequel, we present the generation procedure of ECRP.

ECRP is operated in epoches. Let G_t denote the distribution of clusters and their parameters in epoch t. As a time-varying variable, G_t is distributed conditioned on G_{t-1} and H. As a cluster either inherits pre-existing cluster or emerges as a orphan cluster, G_t contains two parts. The first part is $CRP(\gamma, H)$ which generates orphan clusters. The second part is a set of cluster-specific CRPs which can generate branch clusters. For generating a datum $x_{t,i}$, we first pick a cluster $\theta_{t,i}$ based on G_t, then generation its value based on cluster parameter $\phi_{t,\theta_{t,i}}$. In summary, ECRP can be formally defined as follows,

$$G_t | \alpha, \beta, G_{t-1}, H \sim \sum_{k \in G_{t-1}} \omega_{t-1,k} CRP(\beta, H_{\phi_{t-1,k}}) + \omega_{t,O} CRP(\alpha, H),$$

$$\theta_{t,i} | G_t \sim G_t, \qquad x_{t,i} | \theta_{t,i} \sim F(\phi_{t,\theta_{t,i}}) \tag{2}$$

where $\omega_{t-1,k}$ and $\omega_{t,O}$ are the weights of CRPs.

Eq.(2) describe the generation procedure of ECRP. However, two problems remain unsettled. First, how to determine the weight of CRPs, i.e., $\omega_{t-1,k}$ and $\omega_{t,O}$. Second, what's the marginal distribution for a datum's cluster selection $\theta_{t,i}$ based on the G_t.

Weight of CRPs in ECRP. Because the larger a cluster is, the more branches it tends to possess. We assume that the weight of $CRP(\beta, H_{\phi_{t-1,k}})$, i.e. $\omega_{t-1,k}$, is proportional to the number of records which serve dish $\phi_{t-1,k}$ in $t-1$, and the weight of the $DP(\alpha, H)$, i.e. $\omega_{t-1,O}$, is proportional to α. So G_t can be expressed with $\omega_{t-1,k}$ and $\omega_{t,O}$ instantiated, as in Eq.(3).

$$G_t \sim \frac{1}{\alpha + \gamma \sum_k n_{t-1,k}} (\sum_k \gamma n_{t-1,k} CRP(\beta, H_{\phi_{t-1,k}}) + \alpha CRP(\alpha, H)), \tag{3}$$

where $\gamma \in [0,1]$ is a decaying factor which influences the smoothness of cluster popularity over time. Specially, for $t = 0$, G_0 is constructed in the same way as CRP.

Marginal Distribution of $\theta_{t,i}$. The marginal distribution for the data's cluster selection $\theta_{t,i}$ can be calculated with G_t instantiated, which further requires instantiate $CRP(\alpha, H)$ and $CRP(\beta, H_{\phi_{t-1,k}})$. Given the cluster assignment of data in epoch $t-1$ and t, i.e. $\theta_{t-1,\cdot}$ and $\theta_{t,1\dots i-1}$, the $CRP(\alpha, H)$ can be expressed as in Eq.(4)

$$\theta_i | \theta_{1:i-1}, H, \alpha \sim \sum_{o \in O_t} \frac{n_{t,o}}{\sum_{o \in O_t} n_{t,o} + \tau} \delta_{\phi_{t,o}} + \frac{\alpha}{\sum_{o \in O_t} n_{t,o} + \alpha} H, \tag{4}$$

Where O_t is the set of orphan clusters. And $CRP(\beta, H_{\phi_{t-1,k}})$ can be expressed as in Eq.(5):

$$\theta_i | \theta_{1:i-1}, H_{\phi_{t-1,k}}, \beta \sim \sum_{s \in S_{t-1,k}} \frac{n_{t,s}}{\sum_{s \in S_{t-1,k}} n_{t \cdot s} + \beta} \delta(\phi_{t,s}) + \frac{\beta}{\sum_{s \in S_{t-1,k}} n_{t,s} + \beta} H_{\phi_{t-1,k}} \tag{5}$$

where $S_{t-1,k}$ is the set of variants of $\phi_{t-1,k}$.

It is important to notice that $CRP(\beta, H_{\phi_{t-1,k}})$ and $CRP(\alpha, H)$ are different in two aspects. First, the base measure H in $CRP(\gamma, H)$ is a vague prior, while that in $CRP(\tau, H_{\phi_{t-1,k}})$ is determined by $\phi_{t-1,k}$ (we will give describe how to calculate $H_{\phi_{t-1,k}}$ in Section 3.2). Because for a orphan cluster, we often don't have specific knowledge about it, so a vague prior is preferred. While, the parameter of an variant cluster tends to be similar with its parent. Second, the concentration factor α in $CRP(\alpha, H)$ is larger than β in $CRP(\beta, H_{\phi_{t-1,k}})$. Because a larger concentration factor leads to more clusters generated by the CRP. The number of variants for a cluster is generally small, while that of an orphan cluster is often large.

Combining Eq.(3, 4, 5), the marginal distribution of $\theta_{t,i}$ can be written as in Eq.(6),

$$\theta_{t,i} | \{\theta_{t-1}, \cdot\}, \theta_{t,1:i-1}, H, H_{\phi_{t-1,k}}, \alpha, \beta, \gamma \propto$$

$$\sum_{k \in \{\phi_{t-1, \cdot}\}} \frac{\gamma n_{t-1,k}}{R_{t-1}} \left(\sum_{s \in S_{t-1,k}} \frac{n_{t,s}^{(i)}}{\beta + \sum_{s \in S_{t-1,k}} n_{t,s}^{(i)}} \delta_{\phi_{t,s}} + \frac{\beta}{\beta + \sum_{s \in S_{t-1,k}} n_{t,s}^{(i)}} H_{\phi_{t-1,k}} \right)$$

$$+ \frac{\alpha}{R_{t-1}} \left(\sum_{o \in O_t} \frac{n_{t,o}}{\sum_{o \in O_t} n_{t,o} + \alpha} \delta(\phi_{t,o}) + \frac{\alpha}{\sum_{o \in O_t} n_{t,o} + \alpha} H \right)$$

$$(6)$$

where $R_{t-1} = \gamma \sum_{k \in \{\phi_{t-1, \cdot}\}} n_{t-1,k} + \alpha$.

3.2 Modeling Evolving Cluster Parameters

We adopt the conjugate prior cascade to model the evolution of clusters' parameters in ECRP. Because in Bayesian model, the hyper-parameters of a conjugate prior corresponds to having observed a certain number of pseudo observations with properties specified by the parameters [11]. And data in ancestor clusters can serve as pseudo observations for the current cluster. So for a cluster, the posterior distribution of its parent cluster can serve as the conjugate prior. Furthermore, the posterior of the current cluster can be fetched by absorbing observations to the prior. We adopt H as prior for all the orphan clusters. In this manner, given H and observations, the posterior of clusters can be calculated in a cascaded manner. In the conjugate prior cascade, a decaying factor $\eta \in [0, 1]$ is introduced to let the model gradually forget observations in previous clusters.

Evolving Parameter for Temporal Texts. Here we present an evolving topic model for text collections to illustrate the evolution of cluster parameters in ECRP.

For topics (clusters) in text collections, the parameter is the word distribution. We only care about the prior and posterior of the clusters. In the cascade, both of them take the form of Dirichlet distribution. The parameter of Dirichlet corresponds to the count of words. The prior for orphan topics is set to H. And for a variant topic in evolutionary tree, its prior is set as the posterior of its parent with a decaying factor η. So the prior for topic $\phi_{t_h,k}$, i.e., $H_{\phi_{t-1,k}}$, is written as in Eq.(7):

$$p(\phi_{t_h,k} | \mathbf{x}_{t_b}, \cdots, \mathbf{x}_{t_h-1}, H, \eta) = \mathbf{Dir}(\eta^{t_h-t_b} H + \sum_{t=t_b}^{t_h-1} \eta^{t_h-t} \mathbf{x_t}) \qquad (7)$$

Algorithm 1. The online learning framework for ECRP.

Input: Hyperparameters $\{\alpha, \beta, \gamma, \eta, H\}$; temporal data sets $\{X_0, \cdots, X_t, \cdots\}$.
1: $t \longleftarrow 0$, the evolutionary trace is $NULL$.
2: **while** the temporal data does not end **do**
3:    $t \leftarrow t + 1$; Get data $\mathbf{X_t}$;
4:    **while** Not convergence **do**
5:       For each $x_{t,i}$, sampling the cluster assignment θ_{tdi} by Eq.(9).
6:       For each cluster ϕ_{tk}, inferring its parameter by Eq.(8).
7:    **end while**
8:    Update the evolutionary trace according to the sampling result.
9: **end while**

---

where $\{\phi_{t_b,k_b}, \cdots, \phi_{t_h,k_h}\}$ is the path from the root topic to the target topic in the evolutionary tree and $\{\mathbf{x}_{t_b}, \cdots, \mathbf{x}_{t_h}\}$ are the corresponding observations. The posterior of $\phi_{t_h,k}$ can be easily inferred by adding observations to its prior, as in Eq.(8),

$$p(\phi_{t_h,k}|\mathbf{x}_{t_b}, \cdots, \mathbf{x}_{t_h-1}, \mathbf{x}_{t_h}, H, \eta) = \mathbf{Dir}(\eta^{t_h-t_b}H + \sum_{t=t_b}^{t_h} \eta^{t_h-t}\mathbf{x_t}) \qquad (8)$$

3.3   Model Inference

We employ Markov Chain Monte Carlo (MCMC) methodology to estimate the posterior distribution of clusters' states by delivering a Gibbs sampler [14]. The states of the sampler contain both the cluster indicator for every datum $\{\theta_{t,i}\}$ and the posterior of clusters' parameters $\{\phi_{t,k}\}$. In the inference process, we iteratively sample their states until convergence. Note that $\{\phi_{t,k}\}$ can be calculated as described in Section 3.2. So in this part, we focus on sampling $\{\theta_{t,i}\}$.

Sampling Cluster Indicator $\theta_{t,i}$. For a datum $x_{t,i}$, conditioned on parameter of clusters and cluster assignments for other data, its cluster indicator $\theta_{t,i}$ is sampled according to its marginal distribution, which can be written as in Eq.(9).

$$p(\theta_{t,i} = k|\theta_{t-1}, \theta_{t,-i}, x_{t,i}, \{\phi_k\}_{t,t-1}, G_0, \alpha, \gamma) \propto$$

$$\begin{cases} \dfrac{(\gamma n_{t-1,p} n_{t,k}^{(i)}}{\beta + \sum\limits_{s \in S_{t-1,p}} n_{t,k}^{(i)}} F(x_{t,i}|\phi_{k_{f_{t,i}}}) & \text{if } k \in S_{t-1,p} \\[3ex] \dfrac{\gamma n_{t-1,p}\beta}{\beta + \sum\limits_{s \in S_{t-1,p}} n_{t,k}^{(i)}} \int F(x_{t,i}|\phi)d\, p(\phi|\phi_{t-1,p}) & \text{if } k = S_{t-1,p}^+ \\[3ex] n_{k,t}^{(i)} F(x_{t,i}|\phi_{t,o}) & \text{if } k \in O_t \\[2ex] \alpha \int F(x_{t,i}|\phi)dG_0(\phi) & \text{if } k = O_t^+ \end{cases} \qquad (9)$$

where $\theta_{t,-i}$ denotes the set of $\theta_{t,\cdot}$ without $\theta_{t,i}$, and $S_{t-1,p}^+$ denotes new variant of $\phi_{t-1,p}$.

3.4   Online Learning Framework

It is widely admitted that model learning from online data, such as text collections, is computationally demanding. However, ECRP can be incrementally updated and meet the need of online applications.

   In the above Gibbs sampling algorithm, the samplers solely depend on the model states in $t - 1$ and t. To satisfy online learning, we omit the information propagated from epoch $t + 1$. So we can learn an evolutionary trace of $\{\mathbf{X_1}, \cdots, \mathbf{X_t}\}$ without considering $\{\mathbf{X_{t+1}}, \cdots\}$. For $\mathbf{X_t}$, we can update the evolutionary trace of clusters incrementally by branching new clusters from previous clusters and appending new clusters, instead of relearning the model from scratch. The incremental learning framework is presented in Algorithm 1. Although the number of temporal data can be infinitely large, for each epoch t, we only need to handle the newly arrived data. So this framework is efficient for online applications.

4   Experiment

In this section, we test ECRP *w.r.t.* its capability of recovering evolutionary traces from real-world text collections. To this end, we conduct the following tests. First, we compare ECRP with the state-of-the-art clustering models, and demonstrate the superiority of ECRP. Second, we use ECRP to analyze the NIPS data set, and demonstrate that ECRP can recover interesting evolving patterns of data. Note that with minor modification on parameter evolution scheme, ECRP can also analyze various types of temporal data sets.

4.1   Experiment Settings

Measures. We use the metric of *perplexity* for evaluating performance of clustering results. Perplexity is widely used in evaluating clustering models [7,13]. We compute perplexity as in [13]. 10% of each document is randomly sampled to create a test set while the remaining is used for training. The training set is used for learning model parameters while the test set is held out to compute the perplexity. Low perplexity indicates that model can precisely model user behaviors. Formally, the perplexity is defined as:

$$perplexity(D_{test}) = exp\{-\frac{\sum_{i \in D_{test}} \log p(\mathbf{x}_i)}{\sum_{i \in D_{test}} |\mathbf{x}_i|}\} \tag{10}$$

Benchmarks. We compare ECRP with the following clustering models.

  - **CRP [15]:** CRP is a nonparametric clustering model. CRP can only analyze static data set.
  - **Dynamic Topic Model (DTM) [6]:** DTM is an extension of LDA for modeling temporal data;
  - **Recurrent CRP (RCRP) [2]:** RCRP is an extension of CRP for temporal data, where the clusters over time are organized as linear chains;

Table 1. Statistics of Data sets

Data sets	# docs	# terms	# epochs
NIPS[1]	1740	2000	13
DBLP[2]	20480	4000	10
NSF-Awards[2]	30287	4000	14

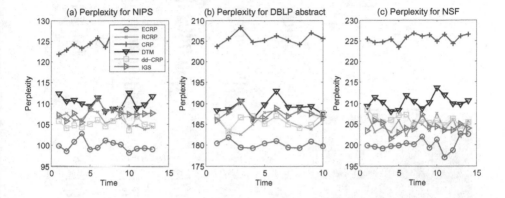

Fig. 2. Comparison for real-world dataset *w.r.t.* perplexity

- **Distance Dependent CRP (dd-CRP)** [4]: dd-CRP is a nonparametric clustering model which can model temporal dependent data;
- **Incremental Gibbs Sampler (IGS)** [10]: IGS can infer connection of clusters with the Gibbs sampler.

Data Sets. We conduct experiments on three publicly available data sets which are widely used for benchmark. All data sets can be downloaded online. Table 1 summarizes the statistics of the data sets.

Setting of Parameters. The hyper parameters of ECRP were set as follows: $H = \mathbf{Dir}(10, \cdots, 10)$, and $(\alpha = 0.1, \beta = 10^{-4}, \gamma = 0.3, \eta = 0.5)$, and we ran the Gibbs sampler for 2000 iterations in each epoch.

4.2   Results

Method Comparison. Figure 2 compares the perplexity of the clustering result on difference data set. We find that on average ECRP outperforms other methods. The reasons are two folds. First, compared with the parametric Bayesian models, such as DTM, ECRP can learn proper number of topics for each epoch, so data can be more properly clustered. Second, compared with other temporal nonparametric topic model, such as RCRP and dd-CRP, the tree-like evolutionary trace recovered by ECRP is more close to the actual pattern of temporal

---

[1] http://www.cs.nyu.edu/~roweis/data.html
[2] http://www.cs.uiuc.edu/~hbdeng/data/kdd2011.htm

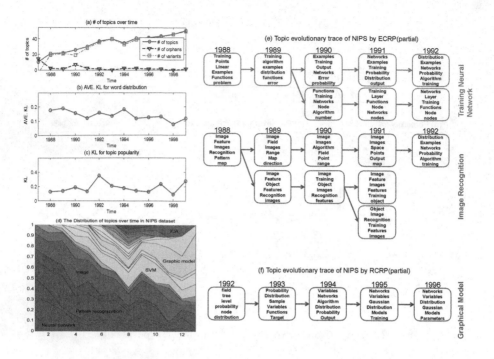

Fig. 3. Analysis results for NIPS data set. (a) the number of topics over time; (b) average KL distance for topics' parameters between parent topic and its variants; (c) KL distance for topics popularity between adjacent epoches; (d) the distribution of topics over time. For clarity, all the variants of a root topic are treated as a whole; (e) the tree-like evolutionary trace recovered by ECRP, the topics are represented by their kernel words; (f) the chain-like evolutionary trace recovered by RCRP.

data. Another benefit of ECRP is that the parameter can be properly smoothed as each cluster is assigned with a proper prior.

NIPS Dataset. We used the ECRP to analyze the NIPS collections. The results are shown in Figure 3. As the entire evolutionary trees of topics are too complex to be contained in a single graph, we plot partial result in 3.(e). We can spot birth, branch and death of topics over time. For example, we can spot that after 1988, the topics of *image* had two branches: *feature extraction* and *MRF*. Compared with the clustering result given by RCRP models where topics are organized as chains (as depicted in Figure 3.(f)), ECRP can discover interesting connections between topics.

With ECRP, we also mined some interesting knowledge about NIPS data set. Figure 3.(a) shows that the number topics grew over time, which confirms with our intuition that originally NIPS focused on the topic of *neural network* and gradually it branched into multiple topics and imported topics from other fields. Figure 3.(b) shows the average KL distance between the parameter of topics between the parent topics and their variants. The value high above the average indicates the lots of topics switch their kernel words. Figure 3.(c) shows the KL distance between the popularity of topics of adjacent epoches. It can be noticed that for the epoch where KL distance high above the average, the number of orphan topics is large. Figure 3.(d) plots evolvement of topics popularity. Note that for clarity, all the variants of a root is treated as whole. We can spot the hot topics of each era and their evolving trend over time. Combine these indexes, we can find important eras for the development of NIPS.

5 Related Work

Nonparametric Temporal Clustering Models. We have mentioned that nonparametric models, such as CRP[15] can clustering data without setting the number of clusters manually. RCRP [2] is an extension of CRP for temporal data, which can recover chain-like evolutionary trace of clusters over time. Besides RCRP, there were also many other nonparametric temporal clustering models [12,16,21], most of these approaches used the stick-breaking construction of the Dirichlet process. In this construction, data are generated with an infinite mixture model, where each mixture component had a weight associated with it. Coupling the weights (some also coupling component parameters) of nearby model state resulted in a form of dependency between them. However, without a proper description of generation of cluster parameters, it was hard to model clusters with tree-like evolutionary pattern. There were also lots parametric Bayesian models [6,17], where the number of cluster/topic was fixed. In this paper, we have demonstrated that parametric models are not suitable for modeling dynamic temporal data.

Hierarchical Clustering Model (HCM). Our work also got inspiration from HCM (short for HDP)[1,5]. However, they were designed for different aspects of clustering problem. HCM aimed to model hierarchical structure within group of data, while ECRP focused on modeling the evolutionary trace of data over time. Blei et al. proposed the nested Chinese restaurant process (nCRP)[5], which assumed that a group of data can be divided into clusters and each cluster can be further divided into smaller clusters recursively. All these clusters were organized as a tree in which the child clusters belonged to their parent clusters. While in ECRP, variant cluster is similar but probabilistically independent to its parent and sibling clusters. nCRP had achieved great success in analyzing text collections and user profiles [1] ect. It inspired us to extend the conjugate prior design for modeling parameter evolution in ECRP.

There were lots of other works [8,9] beyond the Bayesian framework that tried to model the evolutionary pattern of clusters. Due to the limited space, their details are not discussed.

6 Conclusions

In this paper, we studied a new challenging problem of modeling tree-like evolutionary traces of clusters behind temporal data. We presented a new Evolving

Chinese Restaurant Process (ECRP for short) to model tree-like evolutionary traces. ECRP not only can recover evolutionary trees of topics efficiently, but also can capture temporal features such as number of topics, popularity of topics and parameters of topics. Also, we designed an online learning framework for ECRP so as to handle unbounded temporal data. Experimental results on real-world data sets demonstrated the performance of the proposed method.

Acknowledgments. This work was supported by the NSFC (No. 61370025), and the Strategic Leading Science and Technology Projects of Chinese Academy of Sciences (No.XDA06030200), 973 project (No. 2013CB329605) and Australia ARC Discovery Project (DP140102206).

References

1. Ahmed, A., Hong, L., Smola, A.: Nested chinese restaurant franchise process: applications to user tracking and document modeling. In: Proceedings of the 30th International Conference on Machine Learning (ICML-13), pp. 1426–1434 (2013)
2. Ahmed, A., Xing, E.P.: Dynamic non-parametric mixture models and the recurrent chinese restaurant process: with applications to evolutionary clustering. In: SDM, pp. 219–230. SIAM (2008)
3. Ahmed, A., Xing, E.P.: Timeline: A dynamic hierarchical dirichlet process model for recovering birth/death and evolution of topics in text stream (2012). arXiv preprint http://arxiv.org/abs/1203.3463arXiv:1203.3463
4. Blei, D.M., Frazier, P.I.: Distance dependent chinese restaurant processes. The Journal of Machine Learning Research **12**, 2461–2488 (2011)
5. Blei, D.M., Griffiths, T.L., Jordan, M.I., Tenenbaum, J.B.: Hierarchical topic models and the nested chinese restaurant process. In NIPS 16, (2003)
6. Blei, D.M., Lafferty, J.D.: Dynamic topic models. In: Proceedings of the 23rd International Conference on Machine Learning, pp. 113–120. ACM (2006)
7. Blei, D.M., Ng, A.Y., Jordan, M.I.: Latent dirichlet allocation. The Journal of machine Learning research **3**, 993–1022 (2003)
8. Chakrabarti, D., Kumar, R., Tomkins, A.: Evolutionary clustering. In: Proceedings of the 12th ACM SIGKDD International Conference on Knowledge Discovery and Data Mining. KDD 2006, pp. 554–560. ACM, New York (2006)
9. Chi, Y., Song, X., Zhou, D., Hino, K., Tseng, B.L.: Evolutionary spectral clustering by incorporating temporal smoothness. In: Proceedings of the 13th ACM SIGKDD International Conference on Knowledge Discovery and Data Mining, pp. 153–162. ACM (2007)
10. Gao, Z., Song, Y., Liu, S., Wang, H., Wei, H., Chen, Y., Cui, W.: Tracking and connecting topics via incremental hierarchical dirichlet processes. In: 2011 IEEE 11th International Conference on Data Mining (ICDM), pp. 1056–1061, December 2011
11. Gelman, A., Carlin, J.B., Stern, H.S., Dunson, D.B., Vehtari, A., Rubin, D.B.: Bayesian data analysis. CRC Press (2013)
12. Griffin, J.E., Steel, M.J.: Order-based dependent dirichlet processes. Journal of the American Statistical Association **101**(473), 179–194 (2006)

13. Kawamae, N.: Theme chronicle model: Chronicle consists of timestamp and topical words over each theme. In: Proceedings of the 21st ACM International Conference on Information and Knowledge Management, CIKM 2012, pp. 2065–2069, New York. ACM (2012)
14. Neal, R.M.: Markov chain sampling methods for dirichlet process mixture models. Journal of Computational and Graphical Statistics 9(2), 249–265 (2000)
15. Pitman, J.: Exchangeable and partially exchangeable random partitions. Probability Theory and Related Fields 102(2), 145–158 (1995)
16. Ren, L., Dunson, D.B., Carin, L.: The dynamic hierarchical dirichlet process. In: Proceedings of the 25th International Conference on Machine Learning, pp. 824–831. ACM (2008)
17. Wang, X., McCallum, A.: Topics over time: a non-markov continuous-time model of topical trends. In: Proceedings of the 12th ACM SIGKDD International Conference on Knowledge Discovery and Data Mining, pp. 424–433. ACM (2006)
18. Zhang, P., Li, J., Wang, P., Gao, B., Zhu, X., Guo, L.: Enabling fast prediction for ensemble models on data streams. In: KDD (2011)
19. Zhang, P., Zhou, C., Wang, P., Gao, B., Zhu, X., Guo, L.: E-tree: An efficient indexing structure for ensemble models on data streams. IEEE Trans. Knowl. Data Eng. 27(2), 461–474 (2015)
20. Zhang, P., Zhu, X., Shi, Y.: Categorizing and mining concept drifting data streams. In: KDD (2008)
21. Zhu, X., Ghahramani, Z., Lafferty, J.: Time-sensitive dirichlet process mixture models. Technical report, DTIC Document (2005)

Small-Variance Asymptotics for Bayesian Nonparametric Models with Constraints

Cheng Li$^{(\boxtimes)}$, Santu Rana, Dinh Phung, and Svetha Venkatesh

Center for Pattern Recognition and Data Analytics,
Deakin University, Geelong, Australia
{chengl,santu.rana,dinh.phung,svetha.venkatesh}@deakin.edu.au

Abstract. The users often have additional knowledge when Bayesian nonparametric models (BNP) are employed, e.g. for clustering there may be prior knowledge that some of the data instances should be in the same cluster (must-link constraint) or in different clusters (cannot-link constraint), and similarly for topic modeling some words should be grouped together or separately because of an underlying semantic. This can be achieved by imposing appropriate sampling probabilities based on such constraints. However, the traditional inference technique of BNP models via Gibbs sampling is time consuming and is not scalable for large data. Variational approximations are faster but many times they do not offer good solutions. Addressing this we present a small-variance asymptotic analysis of the MAP estimates of BNP models with constraints. We derive the objective function for Dirichlet process mixture model with constraints and devise a simple and efficient K-means type algorithm. We further extend the small-variance analysis to hierarchical BNP models with constraints and devise a similar simple objective function. Experiments on synthetic and real data sets demonstrate the efficiency and effectiveness of our algorithms.

1   Introduction

Bayesian nonparametric (BNP) models have become increasingly popular in many applications since these models can perform automatic model selection based on data, and offer a capacity to increase model complexity as data grows. Dirichlet process mixture models (DPM) [1] and Hierarchical Dirichlet process (HDP) [2] are two of the most popular BNP models. The DPM, where number of clusters is not required to be pre-specified, offers an alternative to the traditional mixture model. The HDP, where number of topics is computed based on data and is not required to be pre-specified, offers an alternative to the traditional topic models such as LDA [3]. Further, in many applications, we may possess additional knowledge about the relationship between instances. For example, in image clustering we know that two images, one of which portrays a stone

---

Electronic supplementary material The online version of this chapter (doi:10.1007/978-3-319-18032-8_8) contains supplementary material, which is available to authorized users.

© Springer International Publishing Switzerland 2015
T. Cao et al. (Eds.): PAKDD 2015, Part II, LNAI 9078, pp. 92–105, 2015.
DOI: 10.1007/978-3-319-18032-8_8

bridge and one showing a wooden bridge, belong to the same semantic cluster of "bridge". In topic modeling, one may prefer that the words "Obama", "United States" and "White Houses" appear in the same topic, because they all describe the symbols of USA. These additional knowledge could be represented in the from of constraints. Typically, there are two types of constraints, *must-link and cannot-link.* They specify that two instances must be in the same cluster (*must-link*) or different clusters (*cannot-link*) [4]. Several works have investigated the use of constraints in partitioning algorithms [5, 6]. However, little work has investigated the use of constraints in BNP models.

We can apply constraints into BNP models by altering the generative probabilities [7]. In particular, the instances with the must-link are generated from the same cluster or topic while the instances with the cannot-link have to be generated from different clusters or topics. More complex approaches, such as [8,9], encode constraints using Markov random field in DPM models. These existing work requires Gibbs sampling or variational techniques for posterior inference.

A traditional way for inference of Bayesian models is to use Gibbs sampling. However, due to the MCMC approach the inference process is painfully slow and this makes most of the Bayesian models including the BNP models unsuitable for large data. Variational techniques offer an alternative recourse, however, many times they do not offer a good approximation. Moreover, including constraints makes it even harder to formulate an inference algorithm for the BNP models [7–9]. Addressing this, in this paper we focus on the efficient inference techniques of BNP models with constraints.

We present a small-variance asymptotic analysis for Dirichlet process (DP) mixtures with constraints and devise scalable hard clustering algorithms with constraints from a Bayesian nonparametric viewpoint. Our analysis is built upon the maximum a posterior probability (MAP) estimate of BNP models when constraints are introduced there. In this process we revisit the exchangeability property of Dirichlet process [10]. We prove that data with must-links or cannot-links are not exchangeable. Therefore, for must-links, we introduce **chunklet** [11] —a set of data points belonging to the same cluster to comply with the must-link constraint sets. Cannot-links are redefined between chunklets. We further prove that data with must-links are exchangeable in chunklets. Subsequently, we derive the optimization function from MAP estimate by using the generic exponential family likelihood, which generalizes varied distributions. Based on the optimization equation, we develop a deterministic clustering algorithm for DPM with must-links. The only one parameter of our algorithm is the tradeoff between the model complexity and the likelihood. Due to the non-exchangability of data with cannot-links, we provide a heuristic and unify must-links and cannot-links into one framework. The optimization equation we derive for must-links is exact and that of cannot-links is approximate. We further extend the analysis to hierarchical models with constraints. Our constrained hard HDP allows users to interactively refine the topics utilizing constraints. We show that our algorithms are the generalization of the algorithms in [12]. Experiment results on synthetic and real data demonstrate the efficiency and effectiveness of our approaches. Our contributions are:

- We prove that data with constraints are not exchangeable and introduce chunklets (transitive closue of must-link constraints) to make it exchangeable. We prove that data with must-links are exchangeable in chunklets;
- We formulate DPM with must-links as an optimization problem using small variance asymptotic analysis. This is derived for general exponential family distribution for mixture model. Cannot-links are handled via an approximation;
- We formulate a similar optimization problem for HDP with constraints;
- We derive efficient clustering algorithms for both DPM and HDP with constraints;
- We validate both the algorithms with extensive experimentation.

2   Background and Preliminaries

2.1   Exponential Family Distributions

An exponential family distribution is defined as [13] $p(\boldsymbol{x}|\boldsymbol{\theta}) = \exp(\langle \boldsymbol{x}, \boldsymbol{\theta} \rangle - \psi(\boldsymbol{\theta}) - h(\boldsymbol{x}))$, where $\boldsymbol{\theta} := \{\theta_j\}_{j=1}^{d}$ is a natural parameter and $\psi(\boldsymbol{\theta}) = \log \int \exp(\langle \boldsymbol{x}, \boldsymbol{\theta} \rangle - h(\boldsymbol{x}))d\boldsymbol{x}$ is the log-partition function. The expected value and covariance are respectively given by $\nabla \psi(\boldsymbol{\theta})$ and $\nabla^2 \psi(\boldsymbol{\theta})$. The conjugate prior of an exponential family distribution is defined as [13] $p(\boldsymbol{\theta} \mid \tau, \eta) = \exp(\langle \boldsymbol{\theta}, \tau \rangle - \eta \psi(\boldsymbol{\theta}) - m(\tau, \eta))$, where τ and η are the hyperparameters. Given the conjugate prior above, the posterior $p(\boldsymbol{\theta} \mid \boldsymbol{x}, \tau, \eta)$ has the same form as the prior.

A bijection between Bregman divergence and exponential families was established in [14]. In particular, given a convex set $S \subseteq \mathbb{R}^d$ and a differentiable, strictly convex function ϕ, the Bregman divergence of any one pair of points $\boldsymbol{x}, \boldsymbol{y} \in S$ is define as $D_\phi(\boldsymbol{x}, \boldsymbol{y}) = \phi(\boldsymbol{x}) - \phi(\boldsymbol{y}) - \langle \boldsymbol{x} - \boldsymbol{y}, \nabla\phi(\boldsymbol{y}) \rangle$. Gaussian and multinomial distributions are two widely used distributions. The corresponding Bregman divergences are Euclidean distance and KL divergence respectively [14]. Further, the bijection allows us to rewrite the likelihood and conjugate prior using the expected value $\boldsymbol{\mu}$ as [14]:

$$p(\boldsymbol{x} \mid \boldsymbol{\theta}) = p(\boldsymbol{x} \mid \boldsymbol{\mu}) = \exp(-D_\phi(\boldsymbol{x}, \boldsymbol{\mu}))f_\phi(\boldsymbol{x}) \tag{1}$$

$$p(\boldsymbol{\theta} \mid \tau, \eta) = p(\boldsymbol{\mu} \mid \tau, \eta) = \exp\left(-\eta D_\phi\left(\frac{\tau}{\eta}, \boldsymbol{\mu}\right)\right) g_\phi(\tau, \eta) \tag{2}$$

where ϕ is the Legendre-conjugate function of ψ, $f_\phi(\boldsymbol{x}) = \exp(\phi(\boldsymbol{x}) - h(\boldsymbol{x}))$ and $g_\phi(\tau, \eta) = \exp(\phi(\tau, \eta) - m(\tau, \eta))$.

2.2   Dirichlet Process Mixture Model

We briefly review the Dirichlet process mixture model (DPM). In DPM, each data point is drawn from one of mixture distributions. The likelihood of one point is expressed by the exponential family characterized by the expected value $\boldsymbol{\mu}_k$ as,

$$p(\boldsymbol{x}) = \sum_{k=1}^{\infty} \pi_k \cdot p(\boldsymbol{x}|\boldsymbol{\mu}_k) = \sum_{k=1}^{\infty} \pi_k \exp(-D_\phi(\boldsymbol{x}, \boldsymbol{\mu}_k))f_\phi(\boldsymbol{x})$$

where $\boldsymbol{\pi} = \{\pi_k\}_{k=1}^{K}$ is the mixture proportion ($1 \geqslant \pi_k \geqslant 0$, for $k = 1, \cdots, K$ and $\sum_{k=1}^{K} \pi_k = 1$). In DPM we do not need to specify the number of clusters upfront. A latent variable $\boldsymbol{z}$ indicates the component from which the data $\boldsymbol{x}$ is generated. The collapsed Gibbs sampling could be used to infer $\boldsymbol{z}$ [15].

2.3 Must-Link and Cannot-Link

Constraint often consists of two types, *must-link (ML) and cannot-link (CL)*. Must-link implies that two data points must belong to the same cluster and cannot-link implies that two data points must not belong to the same cluster. This kind of supervision recently has gained much attention [4–6]. It is easy to find out the transitivity between constraints. For ML constraints, $(u, v) \in \mathcal{M}$ when $(u, w) \in \mathcal{M}$ and $(w, v) \in \mathcal{M}$ are given, where $\mathcal{M}$ is the set of ML constraints. For CL constraints, $(u, v) \in \mathcal{C}$ when $(u, w) \in \mathcal{M}$ and $(w, v) \in \mathcal{C}$ are given, where $\mathcal{C}$ is the set of CL constraints.

Given a series of ML, we construct "**chunklet**"—a set of points that belong to the same cluster [11]. Chunklets are easily obtained through the transitive closure above. Consider a data set $\mathbf{x} = \{\boldsymbol{x}_i\}_{i=1}^{N}$, where N is the number of data points. The transformed data $\mathbf{X} = \{\boldsymbol{X}_j\}_{j=1}^{J}$ ($J \leq N$) denotes the set of J distinct chunklets; $(\boldsymbol{X}_j, \boldsymbol{X}_{j'}) \in \mathcal{C}$ denote that the pair chunklets $(\boldsymbol{X}_j, \boldsymbol{X}_{j'})$ is an instance of the set of CL constraints $\mathcal{C}$. Obviously, the original form of the data is the special case of the chunklet representation when each chunklet just has one point inside. ML constraints now are all complied within chunklets and CL constraints only exist between chunklets.

3 Dirichlet Process Mixture Model with Constraints

We hope to buildup DPM with ML and CL constraints. Vlachos et al. [7] proposed to incorporate constraints into DPM to guide clustering. Must-linked instances are generated by the same component and cannot-linked instances always by different components. However, its Gibbs sampling inference treats each observation separately. Here, based on the concept of chunklet, we develop blocked Gibbs sampling [16] for DPM with constraints (C-DPM). Like Section 2.3, we first construct the chunklet representation of data. Let z_j is the cluster indicator for chunklet $\boldsymbol{X}_j$. $\boldsymbol{z}_{-j}$ is the cluster indicators for chunklets $\boldsymbol{X}_{-j}$. The blocked Gibbs sampling for a chunklet $\boldsymbol{X}_j$ is presented,

$$p(z_j = k \mid \boldsymbol{z}_{-j}, \mathbf{X}, \alpha, G_0) \propto p(z_j = k \mid \boldsymbol{z}_{-j}, \alpha) \cdot p(\boldsymbol{X}_j \mid z_j = k, \boldsymbol{z}_{-j}, \boldsymbol{X}_{-j}, G_0)$$

The equation has the same form with DPM. We expand it,

$$p(z_j = k \mid \boldsymbol{z}_{-j}, \mathbf{X}, \alpha, G_0) = \frac{N_k^{-j}}{Z} \prod_{\boldsymbol{x}_i \in \boldsymbol{X}_j} p(\boldsymbol{x}_i \mid \boldsymbol{\mu}_k)$$

$$p(z_j = k_{new} \mid \boldsymbol{z}_{-j}, \mathbf{X}, \alpha, G_0) = \frac{\alpha}{Z} \int \prod_{\boldsymbol{x}_i \in \boldsymbol{X}_j} p(\boldsymbol{x}_i \mid \boldsymbol{\mu}_{k_{new}}) dG_0$$

where Z is a normalized constant, N_k^{-j} is the number of chunklets (excluding the current chunklet $\boldsymbol{X}_j$) currently assigned to the cluster k. In Markov chain, we update the cluster indicators of chunklets instead of individual data points as DPM does. For $(\boldsymbol{X}_j, \boldsymbol{X}_{j'}) \in \mathcal{C}$, $p(z_j = k \mid \boldsymbol{z}_{-j}, \mathbf{X}, \alpha, G_0) = 0$ if $\boldsymbol{X}_{j'}$ has been assigned to the cluster k. The update of $\boldsymbol{\mu}_k$ depends on the data instances that are assigned to the cluster k.

4   MAP Asymptotic DPM Clustering with Constraints

Due to the slow Gibbs sampling of C-DPM, we derive an efficient inference algorithm for DPM with constraints.

Recent works in [12,17,18] derived scalable Bayesian nonparametric clustering approaches based on the assumption that data are exchangeable. However, with the introduction of constraints the data become non-exchangeable and in that case those derivations cannot directly be applied to our scenario. According to de Finetti's Theorem, an infinite sequence $z_1, z_2, \cdots, z_n, \cdots$ of random variables is exchangeable if,

$$p(z_1 = e_1, z_2 = e_2, \cdots, z_n = e_n, \cdots) = p(z_{\pi(1)} = e_1, z_{\pi(2)} = e_2, \cdots, z_{\pi(n)} = e_n, \cdots),$$

where π is any permutation of $\{1, 2, \cdots, n, \cdots\}$. The traditional Chinese restaurant process (CRP) is exchangeable [10]. Below we discuss the exchangeability property of a sequence with ML and CL constraints.

Lemma 1. Data with must-links are not exchangeable

Proof. We prove it by contradiction. Suppose that there are 4 customers $\boldsymbol{x}_1, \boldsymbol{x}_2,$ $\boldsymbol{x}_3, \boldsymbol{x}_4$ and their table indicators are z_1, z_2, z_3, z_4. $\boldsymbol{x}_3$ and $\boldsymbol{x}_4$ are must-linked. Then, $p(z_1 = 1, z_2 = 2, z_3 = 1, z_4 = 1) = \frac{\alpha}{\alpha+1} \cdot \frac{1}{\alpha+2}$, where α is the concentration parameter of DP. We exchange the sequence of $\boldsymbol{x}_2$ and $\boldsymbol{x}_3$. Then, $p(z_1 = 1, z_3 = 2, z_2 = 1, z_4 = 1) = 0$. Obviously $p(z_1 = 1, z_2 = 2, z_3 = 1, z_4 = 1) \neq p(z_1 = 1, z_3 = 2, z_2 = 1, z_4 = 1)$. Therefore, we have proved Lemma 1.

Addressing this we introduce "chunklet" which is a set of must-linked data points. Must-links now are all compiled within chunklets and no constraint exists between chunklets. Customers are now chunklets. The modified CRP is presented as: the first chunklet sits at the first table, other chunklets sit at one of existing tables with the probability proportional to **the number of chunklets** already in that table or sit at a new table proportional to a constant (α). Therefore, we propose proposition 1 below.

Proposition 1. Data with must-links are exchangeable in chunklets

The proof of proposition 1 is simple since the modified CRP is obviously exchangeable.

4.1 Small-Variance Asymptotic DPM with Must-Links

Following Proposition 1, we perform small-variance asymptotic analysis of posterior using chunklets representation. This is independent of any specific inference algorithms [18]. We derive an optimization objective function for data with only must-links. Before that, we introduce the scaled exponential family distributions with natural parameters $\tilde{\theta} = \beta\theta$ as in [12]. With the scaling of $\tilde{\theta} = \beta\theta$ and $\tilde{\phi} = \beta\phi$, Eq.(1) and (2) become,

$$p(\boldsymbol{x} \mid \tilde{\boldsymbol{\theta}}(\boldsymbol{\mu})) = \exp(-\beta D_\phi(\boldsymbol{x}, \boldsymbol{\mu})) f_{\tilde{\phi}}(\boldsymbol{x}) \tag{3}$$

$$p(\tilde{\boldsymbol{\theta}}(\boldsymbol{\mu}) \mid \beta, \tau, \eta) = \exp\left(-\eta D_\phi\left(\frac{\tau}{\eta}, \boldsymbol{\mu}\right)\right) g_{\tilde{\phi}}\left(\frac{\tau}{\beta}, \frac{\eta}{\beta}\right) \tag{4}$$

Let z_{jk} equal to one if the chunklet $\boldsymbol{X}_j$ belongs to cluster k and 0, otherwise. Let K be the total number of clusters. According to Proposition 1, the joint probability of the clustering is obtained by multiplying the seating probability of J chunklets together:

$$p(z_{1:J,1:K} \mid \alpha) = \alpha^{K-1} \frac{\Gamma(\alpha+1)}{\Gamma(\alpha+J)} \prod_{k=1}^{K} (S_{Jk} - 1)! \propto \alpha^{K-1} \varphi(S) \tag{5}$$

where $S_{Jk} = \sum_{j=1}^{J} z_{jk}$ is the number of chunklets belonging to cluster k. Note that $\frac{\Gamma(\alpha+1)}{\Gamma(\alpha+J)}$ is only dependent on data and the given constraints. We set $\varphi(S) := \prod_{k=1}^{K}(S_{Jk} - 1)!$.

The next step is calculating the joint likelihood of the chunklet set given the clustering. Normally, we assume that data points in one cluster are independently generated from an exponential family distribution with the mean parameter $\boldsymbol{\mu}_k$,

$$p(\boldsymbol{\mu} \mid \tau, \eta) = \prod_{k=1}^{K} p(\boldsymbol{\mu}_k \mid \tau, \eta) \tag{6}$$

The likelihood of a chunklet set $\mathbf{X} = \{\boldsymbol{X}_j\}_{j=1}^{J}$ given $\boldsymbol{z}$ and $\boldsymbol{\mu}$ is,

$$p(\mathbf{X}|\boldsymbol{z}, \boldsymbol{\mu}) = \prod_{k=1}^{K} \prod_{j, z_{jk}=1} p(\boldsymbol{X}_j \mid \boldsymbol{\mu}_k) \tag{7}$$

The MAP point estimate for $\boldsymbol{z}$ and $\boldsymbol{\mu}$ is equivalent to maximizing the posterior: $\arg\max_{K,\boldsymbol{z},\boldsymbol{\mu}} p(\boldsymbol{z}, \boldsymbol{\mu}|\mathbf{X})$. Further, as $p(\mathbf{X}, \boldsymbol{z}, \boldsymbol{\mu}) \propto p(\boldsymbol{z}, \boldsymbol{\mu}|\mathbf{X})$, this can be reformulated as maximizing the joint likelihood: $\arg\max_{K,\boldsymbol{z},\boldsymbol{\mu}} p(\mathbf{X}, \boldsymbol{z}, \boldsymbol{\mu})$.

Combing Eq.(5) (6) (7) and scaling $\tilde{\boldsymbol{\theta}} = \beta\boldsymbol{\theta}$ (See Eq.(3) and (4)), the joint likelihood $p(\mathbf{X}, \boldsymbol{z}, \boldsymbol{\mu})$ is expanded as follows,

$$
p(\mathbf{X}, \boldsymbol{z}, \boldsymbol{\mu}) = p(\mathbf{X} \mid \boldsymbol{z}, \boldsymbol{\mu})p(z_{1:J,1:K} \mid \alpha)p(\boldsymbol{\mu} \mid \tau, \eta)
$$

$$
\propto \prod_{k=1}^{K} \prod_{j:z_{jk}=1} \left(\prod_{\boldsymbol{x}_i \in \boldsymbol{X}_j} f_{\tilde{\phi}}(\boldsymbol{x}_i) \cdot \exp\left(- \sum_{\boldsymbol{x}_i \in \boldsymbol{X}_j} \beta D_\phi(\boldsymbol{x}_i, \boldsymbol{\mu}_k) \right) \right) \cdot \alpha^{K-1} \varphi(S)
$$

$$
\cdot \exp\left(-\sum_{k=1}^{K} \eta D_\phi\left(\frac{\tau}{\eta}, \boldsymbol{\mu}_k \right) \right) \left(g_{\tilde{\phi}}\left(\frac{\tau}{\beta}, \frac{\eta}{\beta} \right) \right)^K \tag{8}
$$

The prior α is set as the function of β, τ, η as [12]: $\alpha = g_{\tilde{\phi}}\left(\frac{\tau}{\beta}, \frac{\eta}{\beta} \right)^{-1} \cdot \exp(-\beta\lambda)$. We then substitute the prior α into Eq.(8). It becomes [1]:

$$
p(\mathbf{X}, \boldsymbol{z}, \boldsymbol{\mu}) \propto \exp\left(-\beta\mathcal{J}(\boldsymbol{z}, \boldsymbol{\mu}, \lambda) - \sum_{k=1}^{K} \eta D_\phi\left(\frac{\tau}{\eta}, \boldsymbol{\mu}_k \right) \right) \cdot \prod_{j=1}^{J} \left(\prod_{\boldsymbol{x}_i \in \boldsymbol{X}_j} f_{\tilde{\phi}}(\boldsymbol{x}_i) \right) \varphi(S)
$$

where $\mathcal{J}(\boldsymbol{z}, \boldsymbol{\mu}, \lambda) = \sum_{k=1}^{K} \sum_{j:z_{jk}=1} \sum_{\boldsymbol{x}_i \in \boldsymbol{X}_j} D_\phi(\boldsymbol{x}_i, \boldsymbol{\mu}_k) + K\lambda$. Note also that the fact $f_{\tilde{\phi}}(\boldsymbol{x})$ and $\varphi(S)$ only depends on $\boldsymbol{x}$. They can be canceled out by applying log partition function in the Equation above. $p(\mathbf{X}, \boldsymbol{z}, \boldsymbol{\mu})$ then involves $\boldsymbol{z}$, $\boldsymbol{\mu}$ and K which minimize $\mathcal{J}(\boldsymbol{z}, \boldsymbol{\mu}, \lambda)$ when $\beta \to \infty$ [2].

Therefore, the MAP estimate of DPM with must-links is asymptotically equivalent to the following optimization problem:

$$
\underset{K, \boldsymbol{z}, \boldsymbol{\mu}}{\arg\min} \sum_{k=1}^{K} \sum_{j:z_{jk}=1} \left(\sum_{\boldsymbol{x}_i \in \boldsymbol{X}_j} D_\phi(\boldsymbol{x}_i, \boldsymbol{\mu}_k) \right) + K\lambda \tag{9}
$$

where λ is the tradeoff between the likelihood and the model complexity. We see that Eq.(9) reduces to the objective function of DP-means [17] when there is no constraint (each chunklet just has one data point inside). The objective function of the DP-means with ML constraints (also termed "C-DP-means") can be expressed as the sum of Bregman divergence between the chunklet to its cluster center plus the penalty of λ for the clusters. Based on the objective function, we propose a deterministic clustering solution for C-DP-means. The algorithm is illustrated in Alg. 1.

Lemma 2. The Alg. 1decreases the objective given in Eq.(9) to local convergence

The proof of Lemma 2 is similar with [17]. We also provide one generalized heuristic of setting up λ. The process is similar with the DP-means in [17], except that we replace distances of points to elements of T with $\sum_{\boldsymbol{x}_i \in \boldsymbol{X}_j} D_\phi(\boldsymbol{x}_i, \boldsymbol{\mu}_T) - \sum_{\boldsymbol{x}_i \in \boldsymbol{X}_j} D_\phi(\boldsymbol{x}_i, \overline{X}_j)$. Next, we discuss the data with cannot-links.

---

[1] We first multiply the prior α into Eq.(8). Since α is independent on the data, it does not change the joint likelihood.

[2] $\sum_{k=1}^{K} \eta D_\phi\left(\frac{\tau}{\eta}, \boldsymbol{\mu}_k \right)$ becomes negligible compared to $\beta\mathcal{J}(\boldsymbol{z}, \boldsymbol{\mu}, \lambda)$ when $\beta \to \infty$.

Algorithm 1. Constrained DP-means with ML constraints

Input: data $\mathbf{x} = \{\boldsymbol{x}_i\}_{i=1}^N$, the cluster penalty parameter λ, ML constraints $\mathcal{M} = \{(\boldsymbol{x}_i, \boldsymbol{x}_j)\}$;
Output: Disjoint partitions $\{l_k\}_{k=1}^K$, the number of clusters K and the cluster indicators $z_{1:J,1:K}$;
Method:
1. **Construct** the chunklet set $\mathbf{X} = \{\boldsymbol{X}_j\}_{j=1}^J$.
2. **Initialize** $K = 1$, $\boldsymbol{\mu}_1$ is the global mean.
3. **Repeat** until convergence

  - **for** $j = 1 : J$
    - **compute** the distance $d_{jk} = \sum_{\boldsymbol{x}_i \in \boldsymbol{X}_j} D_\phi(\boldsymbol{x}_i, \boldsymbol{\mu}_k) - \sum_{\boldsymbol{x}_i \in \boldsymbol{X}_j} D_\phi(\boldsymbol{x}_i, \overline{\boldsymbol{X}}_j)$ for $k = 1, 2, \cdots, K$, where $\overline{\boldsymbol{X}}_j$ is the mean of $\boldsymbol{X}_j$;
    - **if** $\min_k(d_{jk}) > \lambda$, set $K = K + 1$, $z_{jK} = 1$ and $\mu_K = \overline{\boldsymbol{X}}_j$;
    - **else** $k^* = \arg\min_k d_{jk}$, $z_{jk^*} = 1$;
  - **generate** clusters $\{l_k\}_{k=1}^K : l_k = \{\boldsymbol{X}_j \mid z_{jk} = 1\}$;
  - **re-estimate** cluster mean $\boldsymbol{\mu}_k = \frac{1}{|i \in l_k|} \sum_{i \in l_k} \|\boldsymbol{x}_i\|$.

Lemma 3. Data with cannot-links are not exchangeable

Proof. We assume cannot-links exist between $\boldsymbol{x}_1$ and $\boldsymbol{x}_4$, $\boldsymbol{x}_3$ and $\boldsymbol{x}_4$. The probability of the first permutation $p(z_1 = 1, z_2 = 1, z_3 = 2, z_4 = 3) = \frac{1}{\alpha+1} \cdot \frac{\alpha}{\alpha+2}$. We exchange $\boldsymbol{x}_2$ and $\boldsymbol{x}_3$. The probability of this permutation $p(z_1 = 1, z_3 = 1, z_2 = 2, z_4 = 3) = \frac{1}{\alpha+1} \cdot \frac{\alpha}{\alpha+2} \cdot \frac{\alpha}{\alpha+1}$. Obviously $p(z_1 = 1, z_2 = 1, z_3 = 2, z_4 = 3) \neq p(z_1 = 1, z_3 = 1, z_2 = 2, z_4 = 3)$. Therefore, we have proved Lemma 3.

4.2 One Heuristic for DPM with Must-Links and Cannot-Links

Due to the un-exchangeability of cannot-links, we provide a heuristic asymptotic for constrained DPM with cannot-links. We can unify must-links and cannot-links in one optimization function as follows,

$$\arg\min_{K,\boldsymbol{z},\boldsymbol{\mu}} \sum_{k=1}^K \sum_{j:z_{jk}=1} \left(\sum_{\boldsymbol{x}_i \in \boldsymbol{X}_j} D_\phi(\boldsymbol{x}_i, \boldsymbol{\mu}_k) \right) + K\lambda$$

$$s.t. \ \delta_{z_{jk}, z_{j'k}} = 0, \text{ for } \forall(\boldsymbol{X}_j, \boldsymbol{X}_{j'}) \in \mathcal{C} \tag{10}$$

We introduce the binary indicator δ to incorporate constraints into the likelihood. The indicator $\delta_{z_{ik}, z_{i'k}} = 1$ if and only if $z_{ik} = 1$ and $z_{i'k} = 1$, and $(\boldsymbol{X}_i, \boldsymbol{X}_{i'}) \in \mathcal{C}$. In the condition that no constraint exists, the objective function is the same with DP-means [12]. We use the similar algorithm with Alg (1), except that we search legal clusters to comply CL constraints. For data with CL constraints, in the re-assignment step, if the closest cluster is not available, search the next closest cluster until a legal cluster is found (the constraints are not violated). If no legal cluster is found, create a new cluster (The detail algorithm is provided in supplementary material). Experiment results show that this heuristic helps to improve the clustering performance.

5   Extension to Hierarchies with Constraints

Given a text corpus, hierarchical topic modeling [2,3] allows topics to be shared between documents. A fundamental assumption underlying hierarchical topic modeling is that words among one document are conditionally independent. However, in many applications, a user may have additional knowledge about the composition of words that should have high probability in the same or different topics. For example, one prefer that the words "protein", "saccharide" and "axunge" appear in the same topic, since they are three main nutrition matters. We would like such domain knowledge to guide the recovery of latent topics for large scale data sets. Ke et al. [12] derived the exponential family hard HDP (unconstrained hard HDP), which solves the scalability of Bayesian HDP. However, unconstrained Hard HDP still lacks a mechanism for merging such domain knowledge. It motives us to develop hard HDP with constraints.

Here we briefly extend the the small-variance asymptotic analysis to HDP with constraints. Similar with C-DP-means above, we first transform the words into the chunklets. Then local clusters and global clusters are both generated based on chunklets.

Assume that we have Y data sets, $1, 2 \cdots, y, \cdots Y$. Data point x_{yi} refers to data point i from the set y. X_{yj} refers to the chunklet X_j in set y. Chunklets X_{yj} are associated with local cluster indicators z_{yj}. The $z_{yj} = c$ means that the chunklet j in data set y is assigned to local cluster S_{jc}. Local clusters are associated to global clusters $\{l_1, \cdots, l_k, \cdots l_K\}$. The global cluster mean is μ_k. Similar with Eq.(10), we derive the following optimization problem for hierarchical models with constraints,

$$\arg\min_{K, z, \mu} \sum_{k=1}^{K} \sum_{X_{yj} \in l_k} \left(\sum_{x_{yi} \in X_{yj}} D_\phi(x_{yi}, \mu_k) \right) + \lambda_{bottom} t + \lambda_{top} K$$

$$s.t.\ \delta_{z_{yj}, z_{yj'}} = 0, \text{ for } \forall (X_{yj}, X_{yj'}) \in \mathcal{C} \qquad (11)$$

where λ_{top} is the penalty of creating a new global cluster. t and K are the number of local clusters and global clusters respectively. $\delta_{z_{yj}, z_{yj'}} = 1$ if and only if $z_{yj} = z_{yj'}$ and $(X_{yj}, X_{yj'}) \in \mathcal{C}$. λ_{bottom} is the penalty of creating a new local cluster. λ_{top} is the penalty of creating a new global cluster. Our algorithm is the generalization of unconstrained hard HDP [12]. The main difference is that our algorithm assigns linked words in the bottom level while the unconstrained approach assigns individual words. We present the applications of our algorithms to interactive hierarchical modeling.

6   Experiments

6.1   DP-means with Constraints

We compare our algorithm with Gibbs sampling algorithm of C-DPM and COP-Kmeans [4]. We use NMI to measure the performance of algorithms.

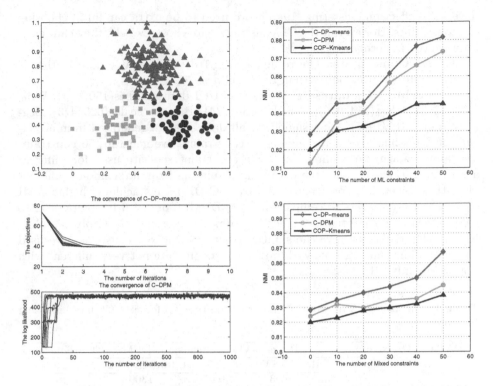

Fig. 1. From left to right, (a) The synthetic data of 3 Gaussians; (b) The NMI over the number of ML constraints; (c) The convergence comparison of two algorithms when 50 ML constraints are given; (d) The NMI over the number of mixed constraints;

Clustering Synthetic Data. We construct a data set with 300 samples from 3 Gaussians (See Fig.1(a)). For C-DPM, we sample the hyperparameter α from a Gamma prior $\Gamma(1,1)$. We average NMI over the final 100 iterations of the 1000 total iterations. Constraint sets are generated randomly for 100 times. We report the average NMI.

Must-Links (ML): C-DP-means (Alg. (1)) can accurately return 3 clusters, whereas C-DPM produces more than 3 clusters. In Fig. 1 (b), we can see that C-DP-means performs better than C-DPM in NMI score. COP-Kmeans just improves a little. We also show the decrease of the objective for C-DP-means and the loglikehood of C-DPM in Fig.1(c). Evidently, C-DP-means converges faster than C-DPM.

Cannot-Links (CL): We randomly select 100 sets of 20 constraints. We find that for 25% constraint sets C-DP-means performs better than the case with no constraint. The best NMI is 0.8924, significantly better than 0.8283 of the case with no constraint. However, most of constraint sets have no improvement for the C-DP-means and some even make the performance lower. The similar trend also is found in C-DPM. A possible reason is that instance-level CL constraints have large effect on clustering [19] and we need to be extra careful in choosing CL constraints. However, if the CL constraints are chosen wisely, they can really improve the performance [20].

Mixed Constraints: The results in Fig.1(d) show that our algorithm is effective and performs better than the baselines.

Clustering UCI datasets. We use six UCI datasets: Iris(150/4/3), Wine (178/13/3), Glass(214/9/6), Vehicle(846/18/4), Balance scale (625/4/3), Segmentation(2310/19/7). We assume that all data sets are generated from a mixture of Gaussians. For a fix number of constraints, we generate 50 constraint sets and average the results. The same constraint sets are used for different algorithms. Table 1 shows the clustering results and running time of different algorithms when ML constraints are given. C-DP-means achieves higher NMI on 5 out of 6 data sets in comparison to C-DPM, 6 out of 6 in comparison to COP-Kmeans, while it requires much less running time than the Gibbs sampling based C-DPM. We observe the similar result when mixed constrains are used (Table 2). The running time with mixed constraints are not very different with that using only ML constraints.

Table 1. The average NMI score (%) and running time (s) for six UCI data sets using ML constraints

DATASETS	No.of ML		DP-means	C-DP-Means	C-DPM	COP-Kmeans
Iris	50	NMI	75.82	83.22±3.90	**90.31±6.91**	74.71±3.86
		time	0.03	0.05	1300	0.08
Wine	50	NMI	39.76	**48.11±2.97**	43.25±5.23	43.05±3.90
		time	0.05	0.08	1500	0.19
Glass	100	NMI	46.36	**51.18±2.74**	42.81±2.91	41.82±4.48
		time	0.06	0.13	520	1.70
Vehicle	200	NMI	18.50	**21.12±1.96**	19.45±2.34	13.51±1.06
		time	1.90	3.51	1700	4.37
Balance scale	200	NMI	27.05	**42.86±6.30**	29.78±1.80	29.11±6.71
		time	1.56	2.28	2000	2.50
Segmentation	400	NMI	20.25	**23.44±1.84**	20.98±1.32	20.18±1.12
		time	6.21	7.50	5200	23.91

Table 2. The average NMI score (%) for six UCI data sets using mixed constraints

DATASETS	No.of Mixed Constraints	DP-means	C-DP-Means	C-DPM	COP-Kmeans
Iris	50	75.8	80.1±3.3	**87.8±5.2**	76.8±3.0
Wine	50	39.7	**45.1±2.6**	40.4±5.2	40.5±2.7
Glass	100	46.3	**49.1±1.8**	38.7±3.1	37.8±1.9
Vehicle	200	18.5	**20.5±2.3**	19.0±1.1	15.42±2.2
Balance scale	200	17.0	**36.8±7.0**	29.2±2.6	28.3±4.5
Segmentation	400	20.2	**22.4±2.8**	19.6±0.9	19.1±2.1

Table 3. Topic changes using interactive hard HDP

Topics	Top Terms
food	**stock**, water, hour, meat, egg, bird, part, chinese, cream, sauce, skin, french, found, protein, light
company	company, zzz_george_bush, zzz_enron, companies, million, firm, business, billion, deal, executives, executive, chief, financial, board, analyst
Merge ("company", "million", "stock")	
food	water, hour, meat, chinese, bird, egg, cream, sauce, french, skin, protein, seed, zzz_roasted_chicken, light, product
company	company, **stock**, zzz_george_bush, million, companies, zzz_enron, tax, billion, firm, business, deal, analyst, president, cut, chief
Split ("company", "zzz_george_bush"), ("company","president")	
food	water, hour, meat, chinese, bird, egg, cream, sauce, french, skin, protein, seed, zzz_roasted_chicken, light, product
company	company, stock, million, companies, zzz_enron, tax, billion, business, firm, deal, analyst, executives, chief, cut, executive

6.2 Hard HDP with Constraints

We now consider *interactive hard HDP*. We guide our topic modeling by providing must-links and cannot-links among a small set of words. We randomly
sample 5611 documents from NYTimes [21] data set. We also eliminate the low-
frequency words with less than 10 occurrences, which finally produce 17,937
words in vocabulary, 1,728,956 words in total. We first run hard HDP with no
constraint between words. We obtain 8 topics, containing topics "food", "company", "sports", "count-terrorist" etc.. We find that the word "stock" is assigned
to food topic and "zzz_george_bush" is assigned to company topic. We apply
Merge ("company", "million", "stock") ("company" and "million" are the two
key words for the company topic), which is compiled into must-links between
these words. After running interactive hard HDP, the number of topics remains
same. "stock" appears in company topic and food topic becomes pure. Other topics have minor changes. We further use split ("company", "zzz_george_bush") and
("company","president"), which are compiled into cannot-links between these
words. Food and company topics both become more consistent than before.
Table 3 shows the topics change within our merge and split steps. Other topics
hardly change and have not been listed. Hard-HDP with constraints converges
within 12 iterations with each iterations costing only 30 sec, making it highly
scalable for large data. In comparison, it takes almost 2000 Gibbs samples to
estimate HDP model, each costing 200 seconds.

7 Conclusion

In this paper, we present a small-variance analysis for Bayesian nonparametric
models with constraints.To handle non-exchangeability arising from the intro-

duction of constraints, we introduce chunklets for must-link constraints. We derive an objective function of Dirichlet process mixture model with must-links from MAP directly. One deterministic K-means type algorithm is derived. We also provide an appropriate heuristic for DPM with cannot-links. We further extend the derivation to hierarchical models with constraints. The experiments of synthetic data and real datasets show the effectiveness and scalability of our proposed algorithms.

References

1. Antoniak, C.E.: Mixtures of Dirichlet Processes with Applications to Bayesian Nonparametric Problems. The Annals of Statistics **2**(6), 1152–1174 (1974)
2. Teh, Y.W., Jordan, M.I., Beal, M.J., Blei, D.M.: Hierarchical dirichlet processes. JASA **101**, 1566–1581 (2006)
3. Blei, D.M., Ng, A.Y., Jordan, M.I.: Latent dirichlet allocation. J. Mach. Learn. Res. **3**, 993–1022 (2003)
4. Wagsta, K., Cardie, C., Schroedl, S.: Constrained k-means clustering with background knowledge. In: Proc. 18th International Conf. on Machine Learning, pp. 577–584 (2001)
5. Wagstaff, K., Cardie, C.: Clustering with instance-level constraints. In: Proceedings of the Seventeenth International Conference on Machine Learning, pp. 1103–1110 (2000)
6. Basu, S., Davidson, I., Wagstaff, K.L. (eds.) Constrained Clustering: Advances in Algorithms, Theory, and Applications, 1st ed. Chapman and Hall/CRC, August 2008
7. Vlachos, A., Korhonen, A., Ghahramani, Z.: Unsupervised and constrained Dirichlet process mixture models for verb clustering. In: GEMS 2009 (2009)
8. Orbanz, P., Buhmann, J.M.: Nonparametric bayesian image segmentation. Int. J. Comput. Vision **77**(1–3), 25–45 (2008)
9. Ross, J., Dy, J.: Nonparametric mixture of gaussian processes with constraints. In: ICML, JMLR Workshop and Conference Proceedings. vol. 28, no. 3, pp. 1346–1354 (2013)
10. Aldous, D.: Exchangeability and related topics. In: Ecole d'Ete de Probabilities de Saint-Flour XIII 1983, pp. 1–198. Springer (1985)
11. Shental, N., Bar-hillel, A., Hertz, T., Weinshall, D.: Computing gaussian mixture models with em using equivalence constraints. In: NIPS. MIT Press (2003)
12. Jiang, K., Kulis, B., Jordan, M.I.: Small-variance asymptotics for exponential family dirichlet process mixture models. In: NIPS, pp. 3167–3175 (2012)
13. Agarwal, A., Daumé III, H.: A geometric view of conjugate priors. Mach. Learn. **81**(1), 99–113 (2010)
14. Banerjee, A., Merugu, S., Dhillon, I.S., Ghosh, J.: Clustering with bregman divergences. J. Mach. Learn. Res. **6**, 1705–1749 (2005)
15. Neal, R.M.: Markov chain sampling methods for dirichlet process mixture models. JCGS **9**(2), 249–265 (2000)
16. Geman, S., Geman, D.: Stochastic relaxation, gibbs distributions, and the bayesian restoration of images, TPAMI, vol. PAMI-6, no. 6, pp. 721–741 (1984)
17. Kulis, B., Jordan, M.I.: Revisiting k-means: New algorithms via bayesian nonparametrics. In: ICML (2012)

18. Broderick, T., Kulis, B., Jordan, M.I.: Mad-bayes: Map-based asymptotic derivations from bayes. In: ICML (2013)
19. Klein, D., Kamvar, S.D., Manning, C.D.: From instance-level constraints to space-level constraints: making the most of prior knowledge in data clustering. In: ICML (2002)
20. Davidson, I., Wagstaff, K.L., Basu, S.: Measuring constraint-set utility for partitional clustering algorithms. In: ECML/PKDD (2006)
21. Bache, K., Lichman, M.: UCI machine learning repository (2013). http://archive.ics.uci.edu/ml

Spectral Clustering for Large-Scale Social Networks via a Pre-Coarsening Sampling Based NystrÖm Method

Ying Kang[1,2(✉)], Bo Yu[1], Weiping Wang[1], and Dan Meng[1]

[1] Institute of Information Engineering, Chinese Academy of Sciences, Beijing, China
{kangying,yubo,wangweiping,mengdan}@iie.ac.cn
[2] University of Chinese Academy of Sciences,
No.89 Minzhuang Road, Haidian District, Beijing, China

Abstract. Spectral clustering has exhibited a superior performance in analyzing the cluster structure of network. However, the exponentially computational complexity limits its application in analyzing large-scale social networks. To tackle this problem, many low-rank matrix approximating algorithms are proposed, of which the NystrÖm method is an approach with proved lower approximate errors. Currently, most existing sampling techniques for NystrÖm method are designed on affinity matrices, which are time-consuming to compute by some similarity metrics. Moreover, the social networks are often built on link relations, in which there is no information to construct an affinity matrix for the approximate computing of NystrÖm method for spectral clustering except for the degrees of nodes. This paper proposes a spectral clustering algorithm for large-scale social networks via a pre-coarsening sampling based NystrÖm method. By virtue of a novel triangle-based coarsening policy, the proposed algorithm first shrinks the social network into a smaller weighted network, and then does an efficient sampling for NystrÖm method to approximate the eigen-decomposition of matrix of spectral clustering. Experimental results on real large-scale social networks demonstrate that the proposed algorithm outperforms the state-of-the-art spectral clustering algorithms, which are realized by the existing sampling techniques based NystrÖm methods.

Keywords: Spectral clustering · NystrÖm method · Pre-coarsening sampling

1 Introduction

Spectral clustering is one of the most popular methods for analyzing the cluster structure of networks. In comparison with other classical clustering algorithms such as k-means or linkage algorithm, spectral clustering often yields superior performances. However, spectral clustering is unable to extend its application in large-scale networks [1]. The key reason is that, the exponential increment of time consumption and space occupation restricts the scalability of spectral clustering, especially facing up to the daily exploding social networks. To break the limitation, a number of low-rank matrix approximating algorithms for spectral clustering have been proposed.

The NystrÖm method is a widely used and efficient low-rank matrix approximating technique to speed up spectral clustering, which is able to reduce the computing

© Springer International Publishing Switzerland 2015
T. Cao et al. (Eds.): PAKDD 2015, Part II, LNAI 9078, pp. 106–118, 2015.
DOI: 10.1007/978-3-319-18032-8_9

and memory burdens enormously [2]. Lots of sampling techniques such as uniform sampling [3], weighted sampling [4], random walking sampling [5], k-means sampling [6] etc. have been designed to lower the error of approximation for spectral clustering as much as possible. In common, these techniques select the interpolation points for NystrÖm method with some probabilities, which are computed in view of the degrees of nodes or the weights of edges. Generally, the weights come from the similarity between pairwise nodes in the networks, and the similarity computation by a metric is time-consuming. Even though based on node degrees, there is no valuable information for the judicious selection of columns and rows for a low-rank approximate matrix. In addition, the social networks are often built on the link structure simply, as well lack of enough useful information for the construction of affinity matrix (or similarity matrix). Hence, there are many obstacles for the NystrÖm method to generate a low-rank matrix, which is used for the approximation of spectral clustering efficiently to mine the cluster property of social networks with large scale nodes.

Apart from link relations, if there is no other resource such as the attribute or feature of nodes in the social networks to construct an affinity matrix, can we extract available information to motivate the approximate computing of NystrÖm method for spectral clustering? As is known to us, the triangle has a strong cluster property due to its any two vertices' connection [7]. If we traverse the triangles in social networks and shrink the encountered triangle into a single node (multi-node), the accumulated edge weights appear, which reflect the local aggregating property and can be extracted to cut down the blindness of further study; moreover, along with the networks become smaller and smaller, the basic cluster structure of original networks has still been kept up. Therefore, In light of what is discussed, we propose a spectral clustering algorithm for large-scale social networks via a pre-coarsening sampling based NystrÖm method, which embeds a new triangle-based coarsening process to make the low-rank matrix approximation of spectral clustering much more efficient and targeted.

2   Related Work

Due to the outstanding capability of identifying the objective clusters in the sampling space with arbitrary shape efficiently and the convergence to the global optimal results [8], spectral clustering has been widely applied to many research fields such as machine learning, computer vision, and bioinformatics etc. In general, spectral clustering makes use of the eigenvectors of an affinity matrix which is derived from the data points, to analyze the aggregating characteristic of the original data implicitly and group the similar points into one cluster [9]. Both the procedures of the construction of affinity matrix and the eigen-decomposition of matrix to obtain eigenvectors are complex, which induces spectral clustering unsuitable for the problems with large data sets. Thus, a number of heuristic methods and approximation algorithms are designed to alleviate the computational burdens of spectral clustering.

Yan et al. [10] utilized a distortion-minimizing local transformation of data to speed up the approximating process of spectral clustering. However, the clustering results are prone to the local optimum and sensitive to the original selection because of k-means, or not steady because of RP tree. Mall et al. [11] employed a primal-dual framework to infer the cluster affiliation for out-of-sample extensions, accelerating spectral clustering in the process of searching for eigenspace. Even if the similarity metric is simplified to the angular cosine between pairwise data points, too much time is cost to compute the affinity matrix.

Inspired by the original application in numerical approximate solution of integral equations [12], the NystrÖm method is introduced to generate a low-rank matrix approximation for the eigen-decomposition of spectral clustering, making spectral clustering capable of studying large-scale networks. Fowlkes et al. [13] simplified the spectral clustering problem by a small random sampling subset of data and then extend the computing results to the whole data set. However, the adopted sampling technique has not improved the computational accuracy. Zhang et al. [6] employed a k-means based sampling scheme to reduce the approximation error of NystrÖm method, but the constraint conditions that the kernel functions comply to make this scheme time-consuming. Belabbas et al. [14] proposed that the probability of choosing interpolation points for NystrÖm method was in proportion to the determinant of similarity matrix, and the Schur complement was used to analyze the NystrÖm reconstruction error. While the bigger the determinant is, the smaller the error is.

Unfortunately, the existing spectral clustering algorithms based on the NystrÖm method cannot be applied to large-scale social networks for the following two main reasons. Firstly, no matter what sampling scheme is adopted for the NystrÖm method, most low-rank matrix approximation algorithms for spectral clustering are based on an affinity matrix, which is too time-consuming to construct due to the high complexity of similarity computing. Secondly, the social networks are often built on link relations, while the node similarity information is insufficient and difficult to obtain. In order to solve the above problems, this paper proposes a spectral clustering algorithm for large-scale social networks via a pre-coarsening sampling based NystrÖm method, which avoids the time-consuming pairwise similarity computation and improves the performance of spectral clustering.

3   Our Approach

3.1   Background

The NystrÖm method is originally used to find the numerical approximations to eigenfunction problems, which are expressed by integral equations of the form [15] as:

$$\int K(x,y)\phi(y)p(y)dy = \lambda\phi(x) \tag{1}$$

where $p(*)$ denotes the probability density function, $K(*,*)$ denotes a kernel function, λ and $\phi(*)$ denote the eigenvalue and eigenvector of the kernel K based on the integral equation respectively. To approximate the integral on the left of Equation

(1), sample q interpolation points $\{x_1, x_2, ..., x_q\}$ drawn from $p(*)$, the approximate result by the empirical average is as follows:

$$\frac{1}{q}\sum_{j=1}^{q} K(x, x_j)\tilde{\phi}(x_j) \simeq \lambda\,\tilde{\phi}(x) \tag{2}$$

where $\tilde{\phi}(x)$ approximates $\phi(x)$ in Equation (1). In addition, choose x in Equation (2) from $\{x_1, x_2, ..., x_q\}$ as well to generate an eigen-decomposition $\bar{K}\bar{U} = q\bar{\Lambda}\bar{U}$, $\bar{K}$ denotes the positive semi-definite matrix with elements $\{K(x_i, x_j) | i, j = 1, 2, ..., q\}$, $\bar{U}$ denotes the eigenvector matrix of $\bar{K}$ and $\bar{U} \in R^{q \times q}$ has orthonormal columns, $\bar{\Lambda} \in R^{q \times q}$ is a diagonal matrix whose non-zero elements are the eigenvalue of $\bar{K}$. Any eigenvector $\phi_i(x)$ and eigenvalue λ_i in Equation (1) can be estimated by $\bar{U}$ and $\bar{\Lambda}$.

$$\phi_i(x) \simeq \sqrt{q}\bar{U}_{ij}, \qquad \lambda_i \simeq \bar{\lambda}_{ii}/q \tag{3}$$

The eigenvector of any point x can be approximated by the eigenvectors of interpolation points in $\{x_1, x_2, ..., x_q\}$, because the k-th eigenvector at an unsampled point x can be computed by:

$$\tilde{\phi}_k(x) \simeq \frac{1}{q\bar{\lambda}_i}\sum_{j=1}^{q} K(x, x_j)\tilde{\phi}_k(x_j) \tag{4}$$

As is shown above, different interpolation points to be selected will lead to different approximating results.

How to extend the NystrÖm method to spectral clustering? Consider a $m \times n$ matrix A, which is partitioned as follows:

$$A = \begin{bmatrix} A_{11} & A_{12} \\ A_{21} & A_{22} \end{bmatrix} \tag{5}$$

without loss of generality, let $A_{11} \in R^{p \times q}$ denotes the submatrix which is generated by the intersection of q columns and p rows sampled in some manner, A_{12} and A_{21} denotes the submatrix consisting of elements with a sampled column label (exclusive) or sampled row label, respectively, and $A_{22} \in R^{(m-p) \times (n-q)}$ denotes the submatrix consisting of the remaining elements of A.

Suppose the eigen-decomposition of A_{11} is $A_{11} = \tilde{U}\tilde{\Lambda}\tilde{U}^T$, the eigenvector matrix of A_{21} can be approximated by $A_{21}\tilde{U}\tilde{\Lambda}^{-1}$, thus the spectral analysis on the submatrix can be extended to the original matrix A, and the estimation of A is as follows:

$$\begin{aligned} \bar{A} &= [\tilde{U}; A_{21}\tilde{U}\tilde{\Lambda}^{-1}]\tilde{\Lambda}[\tilde{U}; A_{21}\tilde{U}\tilde{\Lambda}^{-1}]^T \\ &= \begin{bmatrix} A_{11} & A_{12} \\ A_{21} & A_{21}A_{11}^{-1}A_{12} \end{bmatrix} \end{aligned} \tag{6}$$

the corresponding approximate eigenvectors are $\bar{U} = [\tilde{U}; A_{21}\tilde{U}\tilde{\Lambda}^{-1}]$.

In addition, some details on the normalization of eigenvectors, the transformation of matrix based on whether positive semi-definite or not etc. are not discussed here, which can be referred to [16] for further studies.

3.2   A Pre-Coarsening Sampling Based Nyström Method

From the viewpoint of topology, the social networks are built on link relations ubiquitously, which is the most direct resource for studying social networks. Without any help of other information like attribute or feature of nodes to serve for similarity computing in social networks, is it possible that the task of low-rank matrix approximation based on the Nyström method for spectral clustering can proceed? At least impossible for weighted sampling, random walking sampling or k-means sampling based Nyström methods, because there is no appropriate affinity matrix to match with them. Even though an endeavor is made to search for such a matrix, it is time-consuming to compute the matrix by some similarity metrics like Cosine [17].

How to mine some available information from the link structure of social networks for the approximate computing of Nyström method for spectral clustering is a challenge to our work. Let's start from an example in Figure 1. Figure 1(a) shows a link relation based network which contains many triangles. If we execute operations as follows: traverse the triangles in the network, and when encountering a triangle as labeled by one circle in Figure 1(a), shrink it into one single node (multi-node), the network will be transformed into a smaller weighted network gradually as is shown in Figure 1(b). During the transforming process, the edges adjacent to the shrunk vertices in a triangle are accumulated, thus the unit edge-weight values of original network become numerical values after transformation, which can be utilized as the foundation of further sampling of Nyström method. In addition, it is obvious that the generated network keeps the basic cluster topology structure of the original network.

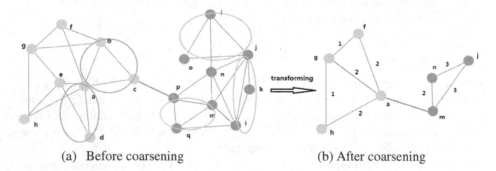

(a) Before coarsening   (b) After coarsening

Fig. 1. The transforming process of the link relation based network by shrinking triangles

As is familiar to us, the three vertices of a triangle are bound to belong to the same cluster, owing to the strong connection between any two of its inner vertices. Therefore, in virtue of the strong cluster property of triangles, the generated smaller weighted network by shrinking the traversed triangles into multi-nodes will not break down the original cluster structure. On the contrary, the weights of edges in the network appear after transformation, which reflect the joint strength of any two nodes. In addition, some connections between nodes in the transformed network are the indirect relations constructed over the shrunk vertices, which is beneficial for mining deeper associations and is an advantage that other methods lack of. So in summary, lots of information about edge weight, cluster and connection etc. has been mined from the transformation of link relation based network, which is able to be utilized for the further sampling of NystrÖm method explicitly.

Extending the principle to the social networks, where triangles are the basic structural elements, we propose a pre-coarsening sampling based Nyström method. We first give the formulized definition of the novel triangle-based coarsening problem:

Definition 1. (Triangle-based Coarsening Problem)
Input: *undirected, link relation based social network* $N = (V, E, W)$, $v \in V$ *denote nodes of N, e $\in$ E denote edges, Δ denote triangles, edge weight $\omega \in W$ are unit.*
Output: *undirected, coarsened weighted network* $N' = (V', E', W')$, *what* $v' \in V'$ *and* $e' \in E'$ *of N' denote is similar to N, $\omega' \in W'$ are numerical.*

Process: *traverse Δ of N in an order (ascending or descending) of node degrees, {v has not been shrunk or to be a multi-node just only one time}; when encounter a Δ, shrink it into a single multi-node $v'_{multi-node}$; accumulate the edges adjacent to shrunk nodes v_{shrunk} and reweight the corresponding remained edges $\omega'_{e-remaining}$.*

The certain order of node degrees in Definition 1 is a necessary condition, which is used to avoid the indeterminacy of generated network by coarsening the random traversed triangles. And if traversing and shrinking triangles in the network without any constraint, the coarsening will make the original cluster structure disappear sometimes, thus to prevent this phenomenon from happening, another condition as '*to be a multi-node just only one time*' must be added. Figure 2 shows an example of the generation of a smaller weighted network by triangle-based coarsening.

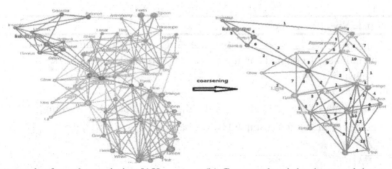

(a) A network of word association [18]   (b) Generated weighted network by coarsening

Fig. 2. The generation of a smaller weighted network

We put the triangle-based coarsening as the preprocessing for sampling of Nyström Method, so the proposed pre-coarsening sampling based Nyström method is:

Definition 2. (Pre-Coarsening Sampling based Nyström Method)
Input: *weight matrix $\tilde{A}$ constructed on the weighted network which is generated by the triangle-based coarsening.*
Output: *low-rank approximating matrix $\bar{A}$.*

Process: *in view of* $\tilde{A}$ *and the probability distribution* $p_i = |A^{(i)}|^2/\|A\|_F^2$, *we sample* p *rows and* q *columns to generate a low-rank approximating matrix* $\bar{A} \in R^{p \times q}$.

3.3   Spectral Clustering for Large-Scale Social Networks

Subsequently, the key problem is how to estimate the cluster affiliation of out-of-sample nodes. Depending on the obtained low-rank approximating matrix $\bar{A}$, the eigen-decomposition of $\bar{A}$ is able to extended to the out-of-samples in $\tilde{A}$, making the approximation of spectral clustering more efficient for large-scale social networks built on link relations. We define the out-of-sample extension problem as follows:

Definition 3. (Out-of-Sample Extension Problem)
Input: *low-rank approximating matrix* $\bar{A}$.
Output: *approximate eigenvector matrix* $\tilde{U}$.

Process: *compute the eigenvector matrix* $\bar{U}$ *by* $\bar{A} = \bar{U}\bar{A}\bar{U}^T$, *and according to Eq.(4)*, *approximate* $\tilde{U}$ *by* $\tilde{U} = [\bar{U}; A_{out-of-sample}\bar{U}\bar{A}^{-1}]$.

In contrast to the process of pre-coarsening sampling of Nyström method, except for the extension of eigenvectors in $\bar{U}$, it is necessary to use k-means to group all of the approximate eigenvectors in $\tilde{U}$ into k clusters, and then unfold the multi-nodes and classify the unfolded nodes into the cluster which the corresponding multi-node belongs to. By now, we have obtained the ultimate spectral clustering results from large-scale social networks. The implementation of integral process of coarsening, sampling and clustering of our spectral clustering is depicted in Algorithm 1.

Algorithm 1. Spectral clustering

Input: Adjacency matrix of social network $A \in R^{n \times n}$, n is the number of nodes, m is the number of sampled columns, r-rank approximation, m and $r \ll n$.
Output: k clusters of A.
1 Begin
2    $\tilde{A}$= weight matrix generated from A by the triangle-based coarsening;
3   S =indices of m columns sampled by probability $p_i = |A^{(i)}|^2/\|A\|_F^2$;
4    $\bar{A} = \tilde{A}(:S)$;
5    $\bar{A} = \bar{U}\bar{A}\bar{U}^T$;
6    $U_r = SmallestEigenVectors(\bar{U}, r)$;
7    $U_{os} = \sqrt{m/n}CU_r\Sigma_r^{-1}$; // U_{os} is the approximating eigenvectors of out-of-sample
                         nodes, Σ_r is the diagonal eigenvalue matrix
8    $Y = NormalizeRows[U_r, U_{os}]$;
9    $\bar{K} = ClusterRows(Y)$;   // group the approximate eigenvector matrix into k
                         clusters by k-means
10 $K = Classify\left(Unfold(\tilde{A})\right)$ by reference to $\bar{K}$;
11 End

3.4   Performance Analysis

By virtue of the triangle-based coarsening, valuable prior information about cluster is extracted from the link structure of social networks, which is useful for the further sampling of NystrÖm method. What's more, due to the intrinsic strong cluster property of triangle, the basic cluster structure of original network is maintained after coarsening preprocessing. Therefore, via the pre-coarsening sampling based NystrÖm method, not only the complexity of spectral clustering can be reduced, but also the computational accuracy of the low-rank matrix approximation of spectral clustering can be promoted greatly.

Lemma 1 (Running Time). *The worst case time complexity of our algorithm is less than* $\Theta(e^{3/2}) + \Theta(m^3) + \Theta(nm^2) + \Theta(nkt)$, *and the worst case space complexity is* $\Theta(mn)$, *where e denotes the number of edges in network A, n denotes the number of columns in matrix* $\tilde{A}$, *m denotes the number of sampled columns from* $\tilde{A}$, *k denotes the number of clusters, t denotes the iteration times of k-means.*

Proof. The process of traversing all of the triangles in the network takes $\Theta(e^{3/2})$ time (refer to compact-forward algorithm), but for the triangle-based coarsening it is unnecessary to do the traversing, because the triangles adjacent to the shrunk nodes become unavailable, so the time complexity of generating matrix $\tilde{A}$ is much less than $\Theta(e^{3/2})$; from the eigen-decomposition of matrix $\tilde{A}$ to the extension to out-of-sample nodes in Algorithm 1, the consuming time is $\Theta(m^3) + \Theta(nm^2)$; and the execution of k-means on Y only needs $\Theta(nkt)$ operations (neglect the time of normalizing Y). Therefore, the *time complexity of our algorithm is less than* $\Theta(e^{3/2}) + \Theta(m^3) + \Theta(nm^2) + \Theta(nkt)$ in the worst case. In the process of approximate computing of spectral clustering based on NystrÖm method, the maximum scale of the matrices which need to be stored is $m \times n$, so the memory usage of our algorithm is $\Theta(mn)$ in the worst case.

4   Experiments

4.1   Dataset and Experiment Setup

Our experiments are designed on the dataset of large-scale social networks which are collected from Stanford University's SNAP networks [19]. The details of each social network are listed in Table 1. Our spectral clustering algorithm for large-scale social networks is realized by a pre-coarsening sampling based NystrÖm method. To test our algorithm's performance, we compare the pre-coarsening sampling with uniform sampling [20], weighted sampling [21], k-means sampling [6] and incremental sampling [22]. Our pre-coarsening sampling can proceed explicitly based on the link relations of social networks, but the weighted sampling, k-means sampling and incremental sampling need to search for some available information by other approaches to compute the similarity matrices firstly. Therefore, we design two sub-experiments for different testing tasks. One is to use all the sampling techniques to analyze the social networks explicitly, while the other is to add the similarity matrix computations for the latter three sampling techniques.

Table 1. Large-scale social networks

Dataset	No. of nodes	No. of edges	No. of clusters	No. of triangles
Youtube	1,134,890	2,987,624	8,385	3,056,386
Orkut	3,072,441	117,185,08	6,288,363	627,584,181
LiveJournal	3,997,962	34,681,189	287,512	177,820,130
Twitter	12,309,718	91,765,139	355,179	519,402,625
Friendster	65,608,366	1,806,067,135	957,154	4,173,724,142

All social networks in Table 1 contain ground-truth clusters, so we can utilize the normalized mutual information (NMI) [23] to evaluate the clustering performance of different clustering algorithms, which are based on different sampling techniques of NystrÖm methods. In general, the larger NMI is, the better the clustering results are.

We perform all the experiments on a Linux machine with 4Core 2.6GHz CPU and 8G main memory. The implementations of all algorithms are in Java. Moreover, we repeat to run each algorithm 30 times to obtain an average result of NMI, making the analytical results more accurate.

4.2   Experimental Results and Analysis

The spectral clustering results corresponding to different social networks are displayed in Figure 3, along with the running time of different algorithms in Figure 4.

From Figure 3(a) we can observe that, our spectral clustering algorithm outperforms other algorithms in analyzing the social networks, in which there is no useful information except for the link structure. Meanwhile, the algorithm adopting a uniform sampling technique has done a little better job than the algorithms which are based on the weighted sampling, k-means sampling, and incremental sampling.

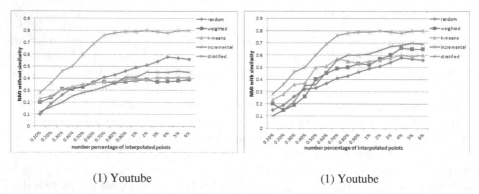

(1) Youtube   (1) Youtube

Fig. 3. The comparison of computing accuracy on social networks

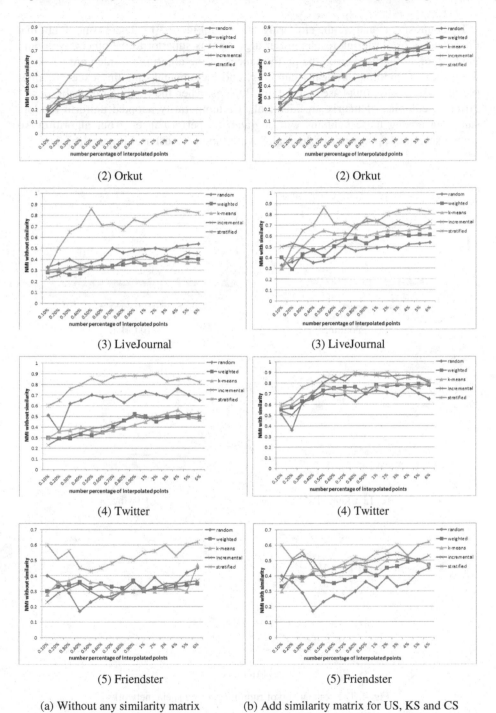

(2) Orkut   (2) Orkut

(3) LiveJournal   (3) LiveJournal

(4) Twitter   (4) Twitter

(5) Friendster   (5) Friendster

(a) Without any similarity matrix   (b) Add similarity matrix for US, KS and CS

Fig. 3. (*continued*)

By contrast, we analyze the clustering results in Figure 3(b). There is no doubt that the accuracy of clustering results of our algorithm is superior to the others. And with the help of similarity matrix, the performances of the algorithms which are based on the weighted sampling, k-means sampling, and incremental sampling excel the ones of the uniform sampling based algorithm. Besides, when the sampling probabilities of different columns of matrix are identical, the temporarily adopted uniform random sampling among these columns will degrade the performance of our algorithm, so some exceptions appear in LiveJournal, Twitter and Friendster of Figure 3(b).

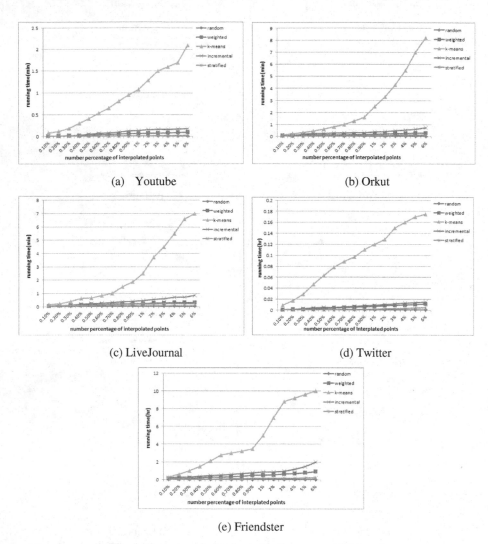

(a) Youtube

(b) Orkut

(c) LiveJournal

(d) Twitter

(e) Friendster

Fig. 4. The comparison of running time on social networks

Subsequently, let us compare the running time between different algorithms (note that we just do comparison in the case of adding similarity matrices). It is obvious that in Figure 4, the algorithm based on the k-means sampling technique spends much more time to tackle the clustering problems of large-scale social networks. The key reason consists in the inherent iterative computing complexity of k-means as an unsupervised method. Because of the relative easier sampling technique, the time consuming of the other algorithms is so small to be neglected in contrast to the k-means sampling based algorithm. As is shown in Figure 4, although our algorithm experiences a coarsening preprocessing before sampling, the running time of our algorithm is less than other algorithms except for the uniform sampling based algorithm. This is because that much more time needs to be cost to compute the similarity matrix for the algorithms which are based on the weighted sampling and incremental sampling.

5   Conclusion

This paper proposes a spectral clustering algorithm for large-scale social networks via a pre-coarsening sampling based NystrÖm method. By virtue of a new triangle-based coarsening policy, this algorithm first extracts a smaller weighted network from the link relation based social network, which reveals some useful prior information about cluster, and then executes an efficient sampling for the NystrÖm method to generate a low-rank matrix approximation for the eigen-decomposition of spectral clustering. Due to the cluster property of triangle, the process of coarsening maintains the original cluster topology structure. Moreover, the pre-coarsening sampling based NystrÖm method makes spectral clustering capable to analyze the social networks explicitly without any other available information except for link relations, promoting the computing accuracy of spectral clustering. Experimental results on real social networks demonstrate that our algorithm outperforms the state-of-the-art spectral clustering algorithms, which are based on other sampling techniques for the NystrÖm method.

Acknowledgement. This work is supported by the National Science Foundation of China under grant number 61402473, KeJiZhiCheng Project under grant number 2012BAH46B03, National HeGaoJi Key Project under grant number 2013ZX01039-002-001 -001, and "Strategic Priority Research Program" of the Chinese Academy of Sciences under grant number XDA06030200.

References

1. Ng, A.Y., Jordan, M.I., Weiss, Y.: On Spectral Clustering: Analysis and an Algorithm. Advances in Neural Information Processing Systems (2002)
2. Bellman, R., Bellman, R.E., et al.: Introduction to Matrix Analysis (1970)
3. Williams, C., Seeger, M.: Using the nyström method to speed up kernel machines. In: Proceedings of the 14th Annual Conference on Neural Information Processing Systems (2001)
4. Drineas, P., Mahoney, M.W.: On the Nyström Method for Approximating a Gram Matrix for Improved Kernel-based Learning. The Journal of Machine Learning Research (2005)

5. Wang, X., Hamilton, H.J.: DBRS: A Density-based Spatial Clustering Method with Random Sampling. Springer, Heidelberg (2003)
6. Zhang, K., Tsang, I.W., Kwok, J.T.: Improved nyström low-rank approximation and error analysis. In: Proceedings of the International Conference on Machine Learning (2008)
7. Kumpula, J.M., Kivelä, M., Kaski, K., Saramäki, J.: Sequential Algorithm for Fast Clique Percolation. Phys. Rev. (2008)
8. Xiang, T., et al.: Spectral Clustering with Eigenvector Selection. Pattern Recognition(2008)
9. Von Luxburg, U.: A Tutorial on Spectral Clustering. Statistics and Computing (2007)
10. Yan, D., Huang, L., Jordan, M.I.: Fast approximate spectral clustering. In: Proceedings of the ACM SIGKDD (2009)
11. Mall, R., Langone, R., Suykens, J.A.K.: Kernel Spectral Clustering for Big Data Networks. Entropy (2013)
12. Sun, Z.H., Wei, X.H., Zhou, W.H.: A Nyström-based Subtractive Clustering Method. Wavelet Active Media Technology and Information Processing (ICWAMTIP) (2012)
13. Fowlkes, C., Belongie, S., Chung, F., et al.: Spectral Grouping Using the Nyström Method. IEEE Transactions on Pattern Analysis and Machine Intelligence (2004)
14. Belabbas, M.A., Wolfe, P.J.: Spectral methods in machine learning and new strategies for very large datasets. In: Proceedings of the National Academy of Sciences (2009)
15. Dick, J., Kritzer, P., et al.: Lattice-Nyström Method for Fredholm Integral Equations of the Second Kind with Convolution Type kernels. Journal of Complexity (2007)
16. Belongie, S., Fowlkes, C., Chung, F., et al.: Spectral Partitioning with Indefinite Kernels Using the Nyström Extension. European Conference on Computer Vision (2002)
17. Adamic, L.A., Adar, E.: You are What You Link. In: The 10th Annual International World Wide Web Conference, Hong Kong (2001)
18. Palla, G., Derényi, I., Farkas, I., Vicsek, T.: Uncovering the Overlapping Community Structure of Complex Networks in Nature and Society. Nature (2005)
19. Yang, J., Leskovec, J.: Defining and evaluating network communities based on ground-truth. In: Proceedings of ACM SIGKDD (2012)
20. Bengio, Y., Paiement, J.F., et al.: Out-of-Sample Extensions for Lle, Isomap, Mds, Eigenmaps and Spectral Clustering. Advances in Neural Information Processing Systems (2004)
21. Choromanska, A., Jebara, T., Kim, H., et al.: Fast Spectral Clustering via the Nyström Method. Algorithmic Learning Theory (2013)
22. Xianchao, Z., Quanzeng, Y.: Clusterability analysis and incremental sampling for nyström extension based spectral clustering. In: International Conference on Data Mining (2011)
23. Danon, L., Duch, J., Diaz-Guilera, A., Arenas, A.: Comparing Community Structure Identification. Stat. Mech. (2005)

pcStream: A Stream Clustering Algorithm for Dynamically Detecting and Managing Temporal Contexts

Yisroel Mirsky$^{(\boxtimes)}$, Bracha Shapira, Lior Rokach, and Yuval Elovici

Department of Information Systems Engineering, Ben Gurion University, Be'er Sheva, Israel
{yisroel,liorrk}@post.bgu.ac.il, {bshapira,yuval}@bgu.ac.il

Abstract. The clustering of unbounded data-streams is a difficult problem since the observed instances cannot be stored for future clustering decisions. Moreover, the probability distribution of streams tends to change over time, making it challenging to differentiate between a concept-drift and an anomaly. Although many excellent data-stream clustering algorithms have been proposed in the past, they are not suitable for capturing the temporal contexts of an entity.

In this paper, we propose pcStream; a novel data-stream clustering algorithm for dynamically detecting and managing sequential temporal contexts. pcStream takes into account the properties of sensor-fused data-streams in order to accurately infer the present concept, and dynamically detect new contexts as they occur. Moreover, the algorithm is capable of detecting point anomalies and can operate with high velocity data-streams. Lastly, we show in our evaluation that pcStream outperforms state-of-the-art stream clustering algorithms in detecting real world contexts from sensor-fused datasets. We also show how pcStream can be used as an analysis tool for contextual sensor streams.

Keywords: Stream clustering · Concept detection · Concept drift · Context-awareness

1 Introduction

Context, in the scope of machine learning, can be described as any information that helps explain an entity's behavior [9]. Context-awareness is the idea of constantly tracking an entity's context over time for some application [18]. For example, an application of context-awareness is the task of data-leakage prevention for smartphones. In this instance, the tracked context is the locomotion of the user (e.g. walking or running) and the behavior of interest is the outgoing emails. By tracking the context, a machine learning algorithm can infer that it is unlikely for an email to be sent while the user is running.

Modern technology can generate vast amounts of sensor data continuously. Even a singular entity, such as a smartphone, can generate a potentially endless amount of data from its sensors. A sensor-stream can be viewed as a sequence of attribute vectors in geometric space [24]. From these sensor-streams it is possible to obtain a context-awareness of the entity [5, 20]. One method is to define an ontology, or a rule set, for each known context—as was done in [22]. However, defining contexts for a sensor-stream is impractical because the definition of contexts may change over time and previously unseen contexts may appear later on. For instance, the definition of a

© Springer International Publishing Switzerland 2015
T. Cao et al. (Eds.): PAKDD 2015, Part II, LNAI 9078, pp. 119–133, 2015.
DOI: 10.1007/978-3-319-18032-8_10

person's home may change when he/she moves, and the sensory definition of a user as he/she is walking will change as he/she gets older.

A concept is an underlying distribution observed from a data-stream which is stable for a period of time [11]. Concepts may reoccur, or evolve over time [17, 28]. In many cases, an entity's context is linked to underlying concepts found in its stream. Therefore, these "hidden contexts" of the entity can be implicitly extracted from the stream itself [13, 14, 26]. Since the hidden contexts have distinct distributions, stream clustering can be performed to detect them in an unsupervised manner.

Clustering a stream is challenging since memory is limited and the stream is potentially boundless. Although many data-stream clustering algorithms exist, they are not suitable for clustering contextual sensor-streams because overlapping clusters cannot be detected. This is because: **1)** *The temporal relation of the arriving points in the clustering decision is not considered (the data-flow is clustered as a sporadic mixture of classes),* **2)** *The clusters' correlated distributions are generally not considered.*

To exemplify the importance of these aspects, consider the case of performing clustering on a smartphone's accelerometer for the application of activity recognition. The objective is to capture and distinguish the underlying contexts found in the stream. Illustrated in Fig. 1 is a possible sequence of points captured from an arbitrary sensor. Here, the captured points form three distinct overlapping distributions (contexts) in a sequential manner.

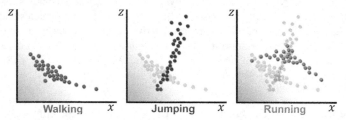

Fig. 1. An illustration of possible sensor values captured as a smartphone user walks (*left*) then repeatedly jumps (*middle*) and then runs (*right*). Here the clusters form distinct correlated distributions which overlap in geometric space.

Since other stream clustering algorithms cannot cope with these types of streams, we propose a different approach to this type of clustering problem. Concretely, when a stream exhibits a certain context, all instances during that time period should be assigned (i.e. clustered) to that context. In other words, the data-stream should be partitioned ad hoc according to the inherent contexts, while detecting reoccurring contexts and accounting for concept drift.

In this paper we present pcStream; a stream clustering algorithm for dynamically detecting and managing temporal contexts. The name "**pc**Stream" is attributed to the **p**rincipal **c**omponents of the distributions in the data-stream which are used to dynamically detect and compare contexts (discussed later in further detail).

The paper's theoretical contributions are: **1)** *a novel method for partitioning data streams considering both temporal and spatial domains during the clustering decision process,* **2)** *a novel method for summarizing (modeling) clusters found in streams (as correlated distributions).*

The algorithm's practical contribution is: *an effective method for detecting and analyzing hidden contexts in a stream, while accounting for context drift.*

The remainder of the paper is organized as follows: In Section 2, we review related work. In Section 3, the notations and problem definition are presented. In Section 4, the core pcStream algorithm and its components are presented. In Section 5, the pcStream algorithm is evaluated as an unsupervised context detection algorithm in comparison to state-of-the-art stream clustering algorithms, and in Section 6 we present our conclusion.

2   Related Work

As opposed to regular clustering algorithms, data-stream clustering algorithms must summarize the data seen in order to preserve memory. CluStream [1] accomplishes this by summarizing the observations into micro-clusters using a tuple of three components (called CF) which describes the micro-cluster's centroid, radius and diameter, which can be updated incrementally. DenStream [8] is a density-based stream clustering algorithm. It uses the CF form to determine whether a group of micro-clusters are a legitimate cluster or a collection of outliers. D-Stream [10] also performs density-based stream clustering, but across a grid.

In contrast to the aforementioned algorithms, pcStream summarizes its clusters with the mean and principal components (vectors of highest variance) of the cluster's last observations (discussed in Section 4). Moreover, none of these algorithms consider the temporal relation between arriving points while making clustering decisions. This makes it difficult to discern between two overlapping concepts and a concept drift. Lastly, they do not cluster a stream as if it were an entity transitioning between concepts. Tracking the stream from this perspective assists in the detection of outliers and new contexts.

In order to assign points to clusters, pcStream uses the Soft Independent Modelling by Class Analogy method (SIMCA) [27] to calculate similarity scores. SIMCA, popular in the domain of chemometrics, is a statistical method for the supervised classification of instances. The classification is "soft" in that it offers fuzzy classifications. Concretely, new instances may be classified as members of one or more classes, or even an outlier, based on their Mahalanobis distance from each of the class's distributions. Only the subspace which describes most of the distribution's variance is retained for this calculation. SIMCA performs well on classes which have distinctly different correlated distributions in multidimensional space [19].

As far as we know, SIMCA has not been used on unbounded streams, nor has it been used as an unsupervised clustering method. Moreover, we have not seen any work where SIMCA has been used to dynamically detect new classes (in our case contexts). Lastly, in contrast to SIMCA, we leave the statistical threshold open to help detect contexts of different categories (discussed later in Section 4).

3   Notation and Problem Definition

In this section we define the notation and basic concepts used in this paper. We also provide a formal problem definition. A full summary of this paper's notations can be found in Table 1.

Definition 1. Let a *context space* be defined as the geometrical space $\mathbb{R}^n$, where n is the number of attributes which define the stream. For instance, one dimension may be the y-axis readings of a smartphone's accelerometer, while another may be the beats per second (bps) of the smartphone's user. This definition is similar to Context Space Theory (CST), formally proposed in [21].

Definition 2. Let a *stream* S be defined as an unbounded sequence of data objects having the form of points in $\mathbb{R}^n$, and let $\vec{x}_i \equiv \left[x_{1,i}, x_{2,i}, \ldots, x_{n,i} \right]$ be the i-th point in the sequence. S can also be viewed as a matrix having n columns and an unbounded number of rows, where row i represents the values sampled at time tick i. We use the notation t to denote the current time tick and the notation $\vec{x}_t$ to refer to the most recent point received from S. Let $f_{a,S}$ be the arrival rate of the row vectors in S measured in Hz.

Definition 3. We define a *high velocity stream* as a *stream* which has an arrival rate that is faster than the stream clustering algorithm's processing rate of new arrivals (f_p). More formally, when $f_{a,S} > f_p$ then S is called a *high velocity stream*.

Definition 4. Let c be a *context* (i.e. concept) defined as a cluster of sequential points having a correlated distribution in $\mathbb{R}^n$, in which S exists within for at least t_{min} time ticks at a time. The distribution of c is generally stationary, but may change gradually over time as it is subjected to concept drift [11]. For instance, with the accelerometer data of a user's smartphone, the *context* which captures the action of jumping may change as the user gets older or sicker. We use the notation c_t to refer to the current context of S.

Definition 5. We define a *contextual stream* to be a *stream* that captures the temporal contexts of a real-world entity. More formally, S is a *contextual stream* if S travels among a finite number of distinct contexts, staying at each for at least d time ticks per visit. The property of revisiting certain distributions is known as a reoccurring drift or reoccurring concepts [17, 28].

Let C be the finite collection of known contexts in which S has been found, such that $c_i \in C$ is the i-th discovered *context*. Let $|C|$ denote the number of known contexts.

It is important to note that C does not necessarily form a distinct partition of $\mathbb{R}^n$. As mentioned earlier, *contexts* are fuzzy by nature and it is possible that two identical points $\vec{x}_a$ and $\vec{x}_b$ belong to two distinctly different contexts c_i and c_j.

For the duration of this paper we will only consider *contextual streams*.

Definition 6. We define a *context category* as all contexts from a *contextual stream* that have the same t_{min}, rate of concept drift, and distinction between their distributions.

Problem Definition. Given the *contextual stream* S, a target *context category* and a limited memory space, dynamically detect the finite number of *contexts* exhibited by S, determine the *current context* (c_t) to some degree of certainty, and provide a fuzzy membership score for $\vec{x}_t$ at any time.

4   Principal Component Stream Clustering

4.1   The Context Model

Since we define *contexts* as correlated distributions in $\mathbb{R}^n$, we model the *contexts* using principal component analysis (PCA) [16]. PCA captures the relationship of the correlation between the dimensions of a collection of observations stored in the $m \times n$

matrix X, where m is the number of observations. The result of performing PCA on X are two $n \times n$ matrices; the diagonal matrix V (the Eigen-values) and the orthonormal matrix P (the Eigen-vectors, a.k.a. *principal components*). The Eigen-vectors $\vec{p}_1, \vec{p}_2, \ldots, \vec{p}_n$ form a basis in $\mathbb{R}^n$ centered on X and oriented according to the correlation of X (see Fig. 2). The Eigen-values $\sigma_1^2, \sigma_2^2, \ldots, \sigma_n^2$ are the variances of the data in the direction of their respective Eigen-vectors. The Eigen-values of V are sorted from highest to lowest variance and the respective Eigen-vectors in P are ordered according-ly. In other words, from the mean of the collection X, $\vec{p}_1$ is the direction of highest variance in the data (with σ_1^2).

We define the contribution of component $\vec{p}_i$ as the percent of total variance it de-scribes for the collection X. More formally,

$$cont_X(\vec{p}_i) = \frac{\sigma_i^2}{\sum_{j=1}^n \sigma_j^2} \tag{1}$$

PCA has been widely used to reduce the dimensionality of a dataset while preserv-ing the information it holds [23]. This is accomplished by projecting observations onto the top *principal components* (PCs). Typically, most of a collection's variance is captured by just a few PCs. By retaining only these PCs, we effectively summarize the distribution and focus our future calculations on the dimensions of interest.

Let ρ be the target percent of variance to be retained. We define $k \in \mathbb{N}$ to be the few-est, most influential PCs in which their cumulative sum of contributions surpasses ρ. Stated otherwise, $\arg\min_k\{\sum_{i=1}^k cont_X(\vec{p}_i) \geq \rho\}$. We denote the k associated with *context* c_i as k_{c_i}.

We model the i-th discoverd *context* as the tuple $c_i \equiv \langle M_i, \mu_i, A_i \rangle$, where M_i is a $m \times n$ for the last m observations assigned to c_i, μ_i is the mean of the observations in M_i, and $A_i = \begin{bmatrix} \dfrac{\vec{p}_1}{\sigma_1}, \dfrac{\vec{p}_2}{\sigma_2}, \ldots, \dfrac{\vec{p}_{k_{c_i}}}{\sigma_{k_{c_i}}} \end{bmatrix}$ is a $n \times k_{c_i}$ transformation matrix.

A_i is essentially a truncated version of P with its Eigen-vectors normalized to their stan-dard deviations (SD). A_i can be calculated by first performing PCA on M_i, to get P_i and V_i, and then by calculating $A_i = Q_i \Lambda_i$ where Λ_i is a diagonal matrix of the top k_{c_i} largest SDs (obtained from V_i), and Q_i is the column-wise truncation of P_i so that it only includes the first k_{c_i} columns. The significance of the transformation matrix A_i will be detailed later in the paper.

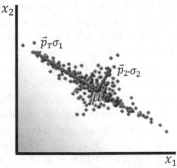

Fig. 2. A visualization of the principal components $(\vec{p}_1, \vec{p}_2)$ scaled to their respective standard deviations (σ_1, σ_2) and centered on the distribution's mean (μ)

Matrix M_i acts as a windowed memory for c_i by discarding the mth oldest observation when a new one is added. Windowing over a stream is an implicit method for dealing with concept drift [3, 24].

4.2   Merging Models

There are cases where the parameters of pcStream will create too many models for the memory space of the host system. Concretely, in the worst case scenario ($k_{c_i} = n$) the maximum memory required for context model c_i is $\theta(n^2)$ for matrix A_i and $\theta(mn)$ for matrix M_i. In total, the memory space of pcStream would be $\theta(|C|(n^2 + mn))$.

To enforce a memory limit, we propose that if a new *context* is discovered though the memory limit has been reached, then two models will be merged into one to make space for the new model. To select which models to merge, we will follow the method used in CluStream where the cluster with the oldest average timestamp (among its observations) is merged with its closest neighbor [1]. Let the merge operation be defined as

$$merge(c_i, c_j) = c_l \tag{2}$$

where c_l is a context model generated from merging the models c_i and c_j. In essence, c_l is created by generating M_l from M_i and M_j, and then calculating μ_l and A_l from M_l. Since we store the last m observations of each model's history, (2) should take into account the temporal history of the observations.

There are at least two ways of accomplishing this. One way is to evenly interleave the first m observations of M_i and M_j into M_l. This will ensure that at least half of each context's information will be preserved. Another way is to merge the memories in order of timestamp. Doing so will place emphasis on retaining the most recent knowledge. The complexity of performing either method is $O(m)$. The selection of which method to use is dependent on the application of pcStream. For example, anomaly detection versus situational awareness.

4.3   Similarity Scores

When a new observation $\vec{x}_t$ arrives, we must determine to what degree it belongs to each known context $c_i \in C$. As mentioned earlier, contexts are fuzzy in their membership, and the point $\vec{x}_t$ can belong to multiple contexts at once. Therefore, we must compute the similarity score of the point in question to each known context.

We calculate the similarity score the same way the SIMCA method does. Therefore the point's statistical similarity to a distribution is calculated as the Mahalanobis distance using only the top k PCs of that distribution. Equivocally, they produce this score by first zero-meaning the point to that distribution, then transforming it onto the distribution's top k PCs, and then by computing the resulting point's magnitude [19]. More formally, the similarity of point $\vec{x}$ to the distribution of c_i is

$$d_{c_i}(\vec{x}) = \|(\vec{x} - \mu_i)A_i\| \tag{3}$$

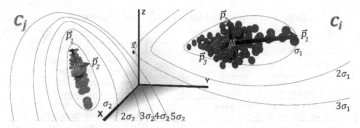

Fig. 3. An illustration of how the ellipsoids of the Mahalanobis distance affect the score of a point ($\vec{x}$) calculated in the perspective of the contexts c_i and c_j. Here $d_{c_i}(\vec{x}) < d_{c_j}(\vec{x})$.

Intuitively, the contours of performing Mahalanobis distance can be seen as ellipsoid shapes extending from the distribution, according to the variance in each direction. To strengthen the concept further, Fig. 3 illustrates an example where two contexts (c_i and c_j) are compared in similarity to point $\vec{x}$. From the perspective of the top k_{c_i} and k_{c_j} normalized PCs of each respective context, their similarity scores differ (even though the Euclidean distances between $\vec{x}$ and both μ_i and μ_j are equivalent). Furthermore, it is possible that the $k_{c_i} > k_{c_j}$ depending on ρ and the correlation of the distribution.

Let $\varphi \in [0, \infty)$ be defined as the similarity threshold. We say that a point $\vec{x}$ is not similar to *context* c_i if $d_{c_i}(\vec{x}) > \varphi$. Let $d_C(\vec{x}) = \left[d_{c_1}(\vec{x}), d_{c_2}(\vec{x}), ..., d_{c_{|C|}}(\vec{x})\right]$ be defined as the score vector (fuzzy membership) of $\vec{x}$ to each of the models in C. We say that $\vec{x}$ does not fit any known context if all elements in $d_C(\vec{x})$ are greater φ. Moreover, we say that $\vec{x}$ is most similar to *context* c_i if the smallest element in $d_C(\vec{x})$ is $d_{c_i}(\vec{x})$.

4.4 Detection of New Contexts

A critical part of the pcStream algorithm is to detect when a previously unseen context has appeared. From Definition 4, contexts are assumed to have rather stationary distributions. Therefore, a new context is detected when the data distribution of S no longer fits the contexts in C for a consistent t_{min} time ticks. At this point, we say that these t_{min} observations constitute a new context, and are then modeled for future use.

To track this behavior, we introduce a new concept called a "drift buffer." Let the drift buffer be called D and have a length of t_{min}. Should D ever be filled without interruption (i.e. should $D = \{\vec{x}_{t-t_{min}-1}, ..., \vec{x}_t\}$) then the content of D is emptied to create a new context model, and is set as current context c_t. However, in the case of a partial drift (i.e. D did not get filled, yet $\vec{x}_t$ fits some context in C) we assume that S is experiencing a wider boundary of c_t, therefore we empty D into c_t.

Table 1. Summary of theory and algorithm notations

Theory Notations		Algorithm Notations	
Nota.	Description	Nota.	Description
n	The number of attributes in the stream	$\vec{p}_i, \sigma_i$	The i-th principal component of collection X and its corresponding variance
$\mathbb{R}^n$	Context Space: The geometric space in which the stream exists, defined by the number of sensors	$cont_x(\vec{p}_i)$	The percentage of total variance component $\vec{p}_i$ describes in collection X
S	Stream: an $\infty \times n$ matrix representing an unbounded sequence of row vectors	ρ	The target percent of total variance to preserve from each distribution in C
t	The row index to an arbitrary time tick in S	k_{c_i}	The number of the fewest top principal components of c_i needed to obtain at least ρ of the variance in the distribution of c_i
$\vec{x}_t$	The samples receives by S at time tick t, represented as the t-th row in R: $[x_{1,t}, x_{2,t}, \ldots, x_{n,t}]$	$d_{c_i}(\vec{x})$	The distance in standard deviations from point $\vec{x}$ to the distribution of c_i measured using the top k principal components of that distribution
$\vec{x}_c$	The current (last received) vector from S	m	Model Memory Parameter: the maximum number of observations a model c_i may store
$f_{a,S}$	The arrival rate of data objects to S, measured in Hz	M_i	Model Memory: a $m \times n$ matrix of the last m observations assigned to context c_i
f_p	The respective algorithm's processing rate, measured in Hz	μ_i	The Mean of a Context: the mean of the observations in M_i
c_i	Context: The i-th discovered context from S. c_i is a correlated distribution in $\mathbb{R}^n$ modeled as the tuple $c_i \equiv \langle M_i, \mu_i, A_i \rangle$	A_i	The Context Transformation Matrix: a $n \times k_{c_i}$ matrix which translates zero-meaned points to context c_i's distribution by using the top k_{c_i} std.-normalized principal components of M_i
C	The ordered collection of all known contexts (c_i) exhibited by S. The size of C is denoted as $|C|$	φ	The Sensitivity Threshold: the distance $d_{c_i}(\vec{x}_t)$ after which $\vec{x}_t$ is not considered to be similar to c_i
t_{min}	The minimum context drift size: the number of time ticks S must stay in a new data distribution (distinct from all those in C) to be considered a new context	D	Drift Buffer: a buffer with at most t_{min} consecutive points that do not fit the contexts in C

4.5   The Core pcStream Algorithm

The basic approach of the core pcStream algorithm is to follow the *stream*'s data distribution. Fuzzy membership scores are available anytime by calculating the statistical similarities between a point and each known context. As long as the arriving points stay within a distribution of a known context, we assign them to that context. The moment the *stream*'s distribution does not fit any known context, we define a new one. Each of the concepts has a window of memory to allow for concept drifts. Should the allocated memory space be filled, then one of two methods for merging context models is performed. Point anomalies are detected as short-term drifts away from all known contexts. Lastly, different context categories are to be detected by adjusting the algorithm's parameters accordingly. The parameters for pcStream are: the sensitivity threshold φ, the context drift size t_{min}, the model memory size m, and the percent of variance to retain in projections ρ. The pseudo-code for pcStream can be found in Algorithm 1.

On lines 1-3, pcStream is initialized by creating the initial collection C with context c_1, and then by setting the current context (c_t) accordingly. The function $init(S, t_{min}, m, \rho)$ runs the function $CreateModel(X, m, \rho)$ on the first t_{min} points of S. The function $CreateModel(X, m, \rho)$ returns a new *context model* c by using the collection of observations X and target total variance retention percentage ρ. Remember that the memory of a *context model* M is a window (FIFO buffer) with a maximum length of m (forgetting the oldest observations). Optionally, an initial set of models for C can be made from a set of observations pre-classified as known contexts of S (e.g. a collection of points that capture running and another that captures walking). From this point on, pcStream enters its running state (lines 4-5).

On lines 4.1-4.3, point $\vec{x}_c$ arrives and $\vec{x}_c$'s similarity score is calculated for all known contexts in C. Stored in i is the index to the model in C to whom $\vec{x}_c$ is most similar. Reminder, the index of C is chronological by order of discovery.

On line 4.4 we determine whether $\vec{x}_c$ fits any of the contexts in C. If it does, then we proceed to lines 4.4.1-4.4.3 where we update the model of best fit (c_i) with instance $\vec{x}_c$, and update c_t accordingly. Since this breaks any consistent drift (between contexts) we empty the drift buffer D into c_i as well (line 4.4.1). The function $UpdateModel(c_i, X)$ re-computes the tuple c_i from C after adding the observation(s) X to the FIFO memory M_i.

If the check on line 4.4 indicates that $\vec{x}_t$ does not any context in C, then we add $\vec{x}_t$ to the drift buffer D, and subsequently check if D is full. If D has reached capacity (t_{min}) then unseen *context* has been discovered. In which case, D is then emptied and formed into a new *context model* (c), which is added to C and set as c_t. The function $AddModel(c, C)$ adds c to C as $c_{|C|+1}$. If the additional model is too much for the memory space allocated to pcStream, then the function $merge(c_i, c_j)$ is used to free one space for c (in C) by merging the average oldest context model c_i with its nearest context model c_j.

Online Algorithm 1: pcStream $\{S\}$
Input Parameters $\{\varphi, t_{min}, m, \rho\}$
Anytime Outputs: $\{c_t, d_C(\vec{x}_t)\}$
1. $C \leftarrow init(S, t_{min}, m, \rho)$
2. $c_t \leftarrow c_1$
3. $D \leftarrow \emptyset$
4.   *loop*
    4.1. $\vec{x}_c \leftarrow next(S)$
    4.2. $scores \leftarrow d_C(\vec{x}_c)$
    4.3. $i \leftarrow IndexOfMin(scores)$
    4.4.   if $scores(i) < \varphi$
        4.4.1. $UpdateModel(c_i, Dump(D))$
        4.4.2. $UpdateModel(c_i, \vec{x}_c)$
        4.4.3. $c_t \leftarrow c_i$
    4.5.   *else*
        4.5.1. $Insert(\vec{x}_c, D)$
        4.5.2.   if $length(D) == t_{min}$
            4.5.2.1. $c \leftarrow CreateModel(Dump(D), m, \rho)$
            4.5.2.2. $AddModel(c, C)$
            4.5.2.3. $c_t \leftarrow c$
5.   *end loop*

4.6   The Algorithm's Computational Cost

In dealing with a streaming algorithm, it is important to understand pcStream's processing rate (f_p) as it relates to the *stream*'s arrival rate (f_a). From a complexity standpoint, there are two calculations that occur on the arrival of $\vec{x}_c$. The first is the calculation of all similarity scores $d_C(\vec{x}_c)$ (line 4.2) and the second is a PCA calculation (performed in either $CreateModel$ or $UpdateModel$). The complexity of calculating the similarity scores is $|C|O(nk)$. Typically $k \ll n$ since most of a model's variance is captured on a few PCs (we found k to range from 1 to 3 per context in our evaluations) making the complexity $O(n)$.

The complexity of performing the PCA calculation using a naïve method is $O(mn^2)$ [12]. It may seem that such a high exponential complexity is unsuitable for a streaming environment since f_p must keep up with f_a. However, for many applications (such as context awareness on a smartphone), n will be in the order of tens and m will typically be in the scale of about a thousand. Therefore, the time it takes to perform the PCA is acceptable in the sense of practicality.

Together, the two calculations that occur at each time tick have a rather linear complexity. Using a dataset with an n of 3, the average f_p for **pcStream**, D-Stream, Den-Stream, and CluStream was **0.4**, 2.9, 44.6, 240.1 milliseconds respectively (R-Studio on a single core of an Intel i5 processor). Moreover, with an n of 561, pcStream's f_p was 2ms. Therefore, pcStream is a practical stream clustering algorithm.

For applications where n and / or m are very large, or where S is a *high velocity stream*, we offer an addition to the core pcStream algorithm (described above) to help f_p maintain a speed at least as fast as f_a. The addition entails decoupling the online calculations (similarity scoring) from the offline calculations (model upkeep) by placing model *update*, *create*, and *merge* procedures into a priority queue (the operation with the highest priority is *merge*, followed by *create* and then *update*). Whenever a point is assigned to context c_i, it is added to the waiting list for c_i's update operation in the queue (if not already in the queue then the operation is added). Lastly, when it is c_i's update operation's turn, the operation's entire waiting list is added chronologically to M_i and $UpdateModel(c_i, X)$ is performed. This operation essentially performs a mini-batch update on the context model.

This addition makes f_p dependent on the score calculations (negligible in respect to the PCA operations) and therefore enables pcStream to be scaled to many more applications. Furthermore, it is possible to parallelize the offline computations over a multiple threads or networked clusters.

4.7   The Detection of Different Context Categories

Each selection of the parameters φ, t_{min}, and m changes pcStream's perspective on S. This essentially causes pcStream to focus on all contexts belonging to a single *context category*, where φ is the degree of distinction between contexts, and m is the rate of concept drift (see Definition 6).

Concretely, a *small* φ will give pcStream the perspective for indistinct contexts (i.e. small nuances) while a *large* φ will give the perspective for ones that are more unique. Similarly, a *small* t_{min} gives the perspective for short term-contexts as opposed to more long-term ones, and a *small* m, gives the perspective for sudden concept drifts as opposed to more gradual. Table 2 summarizes the impact that the pcStream's parameters have on the perspective *context category*.

Intuitively, multiple *context categories* are in a sensor stream at any given time. For example, at a given moment, a smartphone's accelerometer can capture the context whether a user is "*awake or asleep*", "*running or walking*", and "*running to catch the bus, or running for sports*". Therefore, should one be interested in different context categories at the same time, multiple instances of pcStream should be run in parallel with the respective settings.

5   Evaluation

5.1   Datasets and Test Platform

The pcStream algorithm was run in MATLAB, and the evaluation of the various other data-stream clustering algorithms were run in R (a software environment for statistical computing). The packages we used in R were stream and streamMOA [6, 7]. The streamMOA package provided an interface to algorithms implemented for the MOA (Massive Online Analysis) framework for data stream mining [7].

We used three datasets; each one for evaluating a different aspect. The first was the human activity recognition dataset (HAR) [2] for evaluating pcStream in large dimensional spaces. The second was KDD'99 network intrusion dataset [4], selected for testing the effect that mini-batch updates have on *high velocity streams*. Lastly, the third was the HearO smartphone sensor dataset [25] selected for evaluating pcStream's ability to detect different *context categories*. The HearO dataset is a sensor-fused dataset obtained from smartphones. What makes it unique is its explicit context labels provided by the smartphone user at various times over several months. A summarized description of these datasets can be found in Table 3.

Table 2. The effect pcStream's parameter selection has on detecting contexts

		Buffer Size (t_{min})	
		Small	Large
Threshold	Small	*Short-term* indistinct contexts	*Long-term* indistinct contexts
(φ)	Large	*Short-term* distinct contexts	*Long-term* distinct contexts

Table 3. A description of the three datasets used for pcStream's evaluation

Dataset	n	Examples of n	# of rows	Context groups	# labels	Examples of labels
HAR	561	Accl. (x,y,z), FFT...	347	Motion Activity	6	sitting, walking, going upstairs
HearO	5	Accl. correlation (xy,xz,yz), device temperature, battery level.	1764	High-level	4	at home, at work, on break
				Low-level	9	hungry, interested, shopping
				Phone Plugged In	2	yes, no
KDD'99	38	src_bytes, dst_bytes, serror_rate	494,020	Network Attacks	23	buffer overflow, rootkit, teardrop

5.2   Clustering Performance

In the domain of stream clustering, it is common to evaluate the clustering quality by measuring the sum of square distances (SSQ) of every point in a cluster to its cluster's median. However, this metric assumes that clusters do not overlap, and is therefore not suitable for our clustering problem. Therefore, we use the Adjusted Rand Index (ARI) [15].

The ARI is a measure of similarity between two data clustering assignments regardless of their spatial qualities. In our case, we measure an algorithm's performance (of detecting contexts in a dataset) by calculating the ARI between the algorithm's clustering assignment and the dataset's context labels.

The following was performed to evaluate an algorithm's clustering performance with regard to detecting the hidden contexts of streams. First, the dataset was clustered as a stream. Afterward, the ARI of the resulting cluster assignments was calculated using the context labels provided by the dataset. Finally, the clustering performance was recorded as the highest ARI achieved across all possible input parameters of the respective algorithm (achieved with a moderate brute-force search). This was done for each algorithm on both HAR and HearO datasets (note that the HearO dataset has three separate categories of context labels). The algorithm's clustering performances on hidden contexts can be found in Fig. 4. The "Window" is a k-median window algorithm outlined in [3].

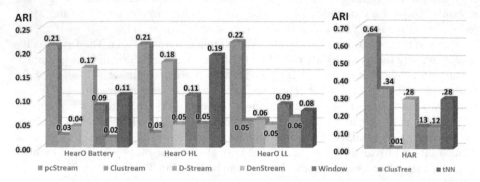

Fig. 4. A comparison of various stream clustering algorithms' best performances against pcStream

Lastly, we tested pcStream's clustering performance in the case of a *high velocity stream*. To measure the impact, we simulated various arrival-rates (f_a) over the KDD'99 dataset. In Fig. 5, there is a performance plot which shows how the clustering is affected as the stream's velocity increases. Noisy spikes are caused because we set a small model memory $(m=500)$, and the duration of most attacks in the dataset last in the order of thousands.

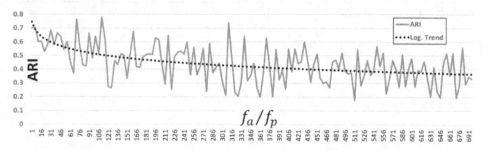

Fig. 5. The performance of pcStream when using mini-batch model updates. Each point is a full pass over the KDD dataset where the on the x-axis indicates the arrival rate of instances per mini-batch update.

5.3   Context Categories

One application of pcStream is to use it as a way to analyze a *contextual stream*. In particular, pcStream can be used to see what hidden contexts are in a stream for some *context category*. Therefore, the following was done in order to analyze how well pcStream detects different *context categories*: First pcStream was run many times on

the HearO dataset, each time targeting a different *context category* by using a different φ and t_{min} (for simplicity we set m to be very large for all trials). Then, for each
of the three context groups (Plugged-In, High-level, and Low-Level), a plot was made
where each coordinate's color in the plot indicates the ARI achieved when using the
respective parameters (see Fig. 6).

It can be seen in Fig. 6 that each plot describes the context categories of the contexts
in each respective group. For instance, these high-level contexts are generally long-term
and semi-distinctive, while the low-level contexts are briefer and even more indistinct.
Moreover, the context of the phone being plugged in or not is a distinct and short-term
context. Note that there are multiple "hot" regions in each plot of Fig. 6. This is because
each group has multiple contexts, each of which belong to a different context category.

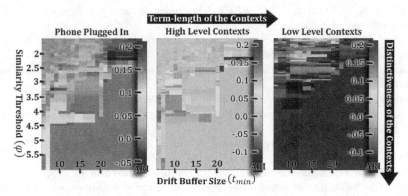

Fig. 6. ARI plots of pcStream for each HearO context group. A coordinate is a parameter selection and its color indicates the level of ARI achieved. The black arrows indicate the general
trend of *context categories* based on Table 2. The labels of each context group form regions of
"best parameter selections" correlated to the label's *context category*.

6 Conclusion

In this paper we have presented a stream clustering algorithm that effectively and
dynamically detects temporal contexts in sensor streams. In addition, we have provided mechanisms which account for gradual concept drifts, reoccurring concepts,
and clusters that overlap in geometrical space. Moreover, we have provided a mechanism
for dealing with *high velocity streams*.

In our evaluations, we have determined that pcStream performs much better than
other data-stream clustering algorithms on sensor-fused datasets. We have demonstrated that pcStream is capable of detecting different *context categories* from the
same dataset, and that pcStream is a useful tool for context analysis of sensor-streams.
Although the focus of this paper was on sensor-fused data-streams and the application
of context-awareness, pcStream is applicable to any data-stream with sequential temporal clusters that have unique correlated distributions.

Acknowledgments. This research was supported by the Ministry of Science and Technology,
Israel.

References

1. Aggarwal, C.C., et al.: A framework for clustering evolving data streams. In: Proceedings of the 29th International Conference on Very Large Data Bases vol. 29, pp. 81–92. VLDB Endowment (2003)
2. Anguita, D., Ghio, A., Oneto, L., Parra, X., Reyes-Ortiz, J.L.: Human Activity Recognition on Smartphones Using a Multiclass Hardware-Friendly Support Vector Machine. In: Bravo, J., Hervás, R., Rodríguez, M. (eds.) IWAAL 2012. LNCS, vol. 7657, pp. 216–223. Springer, Heidelberg (2012)
3. Babcock, B., et al.: Maintaining variance and k-medians over data stream windows. In: Proceedings of the Twenty-Second ACM Symposium On Principles Of Database Systems, pp. 234–243. ACM (2003)
4. Bache, K., Lichman, M.: UCI Machine Learning Repository (2013). http://archive.ics.uci.edu/ml
5. Baldauf, M., et al.: A survey on context-aware systems. International Journal of Ad Hoc and Ubiquitous Computing. 2(4), 263–277 (2007)
6. Bolanos, M., et al.: Introduction to stream: An extensible Framework for Data Stream Clustering Research with R
7. Bolanos, M., et al.: streamMOA: Interface to Algorithms from MOA for stream
8. Cao, F., et al.: Density-based clustering over an evolving data stream with noise. In: SDM, pp. 326–337 SIAM (2006)
9. Chandola, V., et al.: Anomaly Detection: A Survey. ACM Comput. Surv. 41(3), 1–58 (2009)
10. Chen, Y., Tu, L.: Density-based clustering for real-time stream data. In: Proceedings of the 13th ACM SIGKDD International Conference On Knowledge Discovery And Data Mining, pp. 133–142. ACM (2007)
11. Gama, J., Medas, P., Castillo, G., Rodrigues, P.: Learning with Drift Detection. In: Bazzan, A.L., Labidi, S. (eds.) SBIA 2004. LNCS (LNAI), vol. 3171, pp. 286–295. Springer, Heidelberg (2004)
12. Ge, Z., Song, Z.: Multivariate Statistical Process Control: Process Monitoring Methods and Applications. Springer (2012)
13. Gomes, J.B., et al.: CALDS: context-aware learning from data streams. In: Proceedings of the First International Workshop on Novel Data Stream Pattern Mining Techniques, pp. 16–24. ACM, Washington, D.C. (2010)
14. Harries, M.B., et al.: Extracting hidden context. Machine learning. 32(2), 101–126 (1998)
15. Hubert, L., Arabie, P.: Comparing partitions. Journal of classification. 2(1), 193–218 (1985)
16. Jolliffe, I.: Principal Component Analysis. Encyclopedia of Statistics in Behavioral Science. John Wiley & Sons, Ltd (2005)
17. Katakis, I., et al.: Tracking recurring contexts using ensemble classifiers: an application to email filtering. Knowledge and Information Systems. 22(3), 371–391 (2010)
18. Liu, W., et al.: A survey on context awareness. In: 2011 International Conference on Computer Science and Service System (CSSS), pp. 144–147. IEEE (2011)
19. Maesschalck, R.D., et al.: The Mahalanobis distance. Chemometrics and Intelligent Laboratory Systems. 50(1), 1–18 (2000)
20. Makris, P., et al.: A Survey on Context-Aware Mobile and Wireless Networking: On Networking and Computing Environments' Integration. Communications Surveys & Tutorials, IEEE. 15(1), 362–386 (2013)

21. Padovitz, A., et al.: Towards a theory of context spaces. In: 2004, Proceedings of the Second IEEE Annual Conference on Pervasive Computing and Communications Workshops, pp. 38–42. IEEE (2004)
22. Riboni, D., Bettini, C.: COSAR: hybrid reasoning for context-aware activity recognition. Personal and Ubiquitous Computing. **15**(3), 271–289 (2011)
23. Shlens, J.: A tutorial on principal component analysis. arXiv preprint arXiv:1404.1100 (2014)
24. Silva, J.A., et al.: Data Stream Clustering: A Survey. ACM Comput. Surv. **46**(1), 1–31 (2013)
25. Unger, M., et al.: Contexto: lessons learned from mobile context inference. In: ACM 2014 International Joint Conference on Pervasive and Ubiquitous Computing: Adjunct Publication, pp. 175–178. ACM (2014)
26. Widmer, G.: Tracking context changes through meta-learning. Machine Learning. **27**(3), 259–286 (1997)
27. Wold, S., Sjostrom, M.: SIMCA: a method for analyzing chemical data in terms of similarity and analogy. Presented at the (1977)
28. Yang, Y., et al.: Mining in Anticipation for Concept Change: Proactive-Reactive Prediction in Data Streams. Data Mining and Knowledge Discovery. **13**(3), 261–289 (2006)

Clustering Over Data Streams Based on Growing Neural Gas

Mohammed Ghesmoune[✉], Mustapha Lebbah, and Hanene Azzag

University of Paris 13, Sorbonne Paris City LIPN-UMR 7030 - CNRS,
99, av. J-B Clément, 93430 Villetaneuse, France
{mohammed.ghesmoune,mustapha.lebbah,hanene.azzag}@lipn.univ-paris13.fr

Abstract. Clustering data streams requires a process capable of partitioning observations continuously with restrictions of memory and time. In this paper we present a new algorithm, called G-Stream, for clustering data streams by making one pass over the data. G-Stream is based on growing neural gas, that allows us to discover clusters of arbitrary shape without any assumptions on the number of clusters. By using a reservoir, and applying a fading function, the quality of clustering is improved. The performance of the proposed algorithm is evaluated on public data sets.

Keywords: Data stream clustering · Topological structure · GNG

1 Introduction

Clustering is the problem of partitioning a set of observations into clusters such that observations assigned in the same cluster are similar (or close) and the inter-cluster observations are dissimilar (or distant). The other objective of clustering is to quantify the data by replacing a group of observations (cluster) with one representative observation (or prototype). A data stream is a sequence of potentially infinite, non-stationary (i.e., the probability distribution of the unknown data generation process may change over time) data arriving continuously (which requires a single pass through the data) where random access to data is not feasible and storing all arriving data is impractical. The stream model is motivated by emerging applications involving massive data sets; for example, customer click streams, financial transactions, search queries, Twitter updates, telephone records, and observational science data are better modeled as data streams [9]. Mining data streams can be defined as the process of finding complex structures in large data. Clustering data streams requires a process capable of partitioning observations continuously with restrictions of memory and time. In the literature, many data stream algorithms have been adapted from clustering algorithms, e.g., the density-based method DBScan [7,10], the partitioning method k-means [1], or the message passing-based method AP [18]. In this paper, we propose G-Stream, a novel algorithm for discovering clusters of

© Springer International Publishing Switzerland 2015
T. Cao et al. (Eds.): PAKDD 2015, Part II, LNAI 9078, pp. 134–145, 2015.
DOI: 10.1007/978-3-319-18032-8_11

arbitrary shape in an evolving data stream, whose main features and advantages are described as follows: (a) The topological structure is represented by a graph wherein each node represents a cluster, which is a set of "close" data points and neighboring nodes (clusters) are connected by edges. The graph size is not fixed but may evolve; (b) We use an exponential fading function to reduce the impact of old data whose relevance diminishes over time. For the same reason, links between nodes are also weighted by an exponential function; (c) Unlike many other data stream algorithms that start by taking a significant number of data points for initializing the model (these data points can be seen several times), G-Stream starts with only two nodes. Several nodes (clusters) are created in each iteration, unlike the traditional Growing Neural Gas (GNG) [8] algorithm; (d) All aspects of G-Stream (including creation, deletion and fading of nodes, edges management, and reservoir management) are performed online; (e) A reservoir is used to hold, temporarily, the very distant data points, compared to the current prototypes. The remainder of this paper is organized as follows: Section 2 is dedicated to related works. Section 3 describes the G-Stream algorithm. Section 4 reports the experimental evaluation on both synthetic and real-world data sets. Section 5 concludes this paper.

2 Related Works

This section discusses previous works on data stream clustering problems, and highlights the most relevant algorithms proposed in the literature to deal with this problem. Most of the existing algorithms (e.g. *StreamKM++* [1], *CluStream* [2], *DenStream* [7], or *ClusTree* [12]) divide the clustering process in two phases: (a) *Online*, the data will be summarized; (b) *Offline*, the final clusters will be generated. Both *CluStream* [2] and *DenStream* [7] use a temporal extension of the *Clustering Feature vector* [17] (called *micro-clusters*) to maintain statistical summaries about data locality and timestamps during the online phase. By creating two kinds of micro-clusters (*potential* and *outlier micro-clusters*), *DenStream* overcomes one of the drawbacks of *CluStream*, its sensitivity to noise. In the offline phase, the micro-clusters found during the online phase are considered as *pseudo-points* and will be passed to a variant of *k*-means in the *CluStream* algorithm (resp. to a variant of DBScan in the *DenStream* algorithm) in order to determine the final clusters. *StreamKM++* [1] maintains a small outline of the input data using the *merge-and-reduce* technique. The merge step is performed by a means of a data structure, named the *bucket set*. The reduce step is performed by a significantly different summary data structure, the *coreset tree*. *ClusTree* [12] is an anytime algorithm that organizes micro-clusters in a tree structure for faster access and automatically adapts micro-cluster sizes based on the variance of the assigned data points. Any clustering algorithm, e.g. *k*-means or DBScan, can be used in its offline phase. *SOStream* [10] is a density-based clustering algorithm inspired by both the principle of the DBScan algorithm and that of self-organizing maps (SOM) [11]. *E-Stream* [15] classifies the evolution of data into five categories: appearance, disappearance, self evolution, merge, and

Table 1. Comparison between algorithms (WL: weighted links, 2 phases : online+offline)

Algorithms	based on	topology	WL	phases	remove	merge	split	fade
G-Stream	NGas	✓	✓	online	✓	✗	✗	✓
AING	NGas	✓	✗	online	✗	✓	✗	✗
CluStream	k-means	✗	✗	2 phases	✓	offline	✗	✗
DenStream	DBScan	✗	✗	2 phases	✓	offline	✗	✓
SOStream	DBScan, SOM	✗	✗	online	✓	✓	✗	✓
E-Stream	k-means	✗	✗	2 phases	✓	✓	✓	✓
StreamKM++	k-means	✗	✗	2 phases	✓	✓	✓	✓
StrAP	AP	✗	✗	2 phases	✓	✗	✗	✓
SVStream	SVC, SVDD	✗	✗	online	✓	✓	✓	✓

split. It uses another data structure for saving summary statistics, named α-bin histogram. *StrAP* [18], an extension of the Affinity Propagation algorithm for data streams, uses a reservoir for saving potential outliers. In *SVStream* [16], the data elements of a stream are mapped into a kernel space, and the support vectors are used as the summary information of the historical elements to construct cluster boundaries of arbitrary shape. *SVStream* is based on support vector clustering (SVC) and support vector domain description (SVDD) [16]. *AING* [6], an incremental GNG that learns automatically the distance thresholds of nodes based on its neighbors and data points assigned to the node of interest. It merges nodes when their number reaches a given *upper-bound*. Table 1 summarizes the main features offered by each algorithm in terms of: the basic clustering algorithm, whether the algorithm identifies a topological structure or not, whether the links (if they exist) between clusters (nodes) are weighted, how many phases it adopts (online and offline), the types of operations for updating clusters (remove, merge, and split cluster), and whether a *fading* function is used.

3   Growing Neural Gas Over Data Stream

In this section we introduce Growing Neural Gas over Data Stream (G-Stream) and highlight some of its novel features. G-Stream is based on Growing Neural Gas (GNG), which is an incremental self-organizing approach that belongs to the family of topological maps such as Self-Organizing Maps (SOM) [11] or Neural Gas (NG) [13]. It is an unsupervised algorithm capable of representing a high dimensional input space in a low dimensional feature map. Typically, it is used for finding topological structures that closely reflect the structure of the input distribution. We assume that the data stream consists of a sequence $\mathcal{DS} = \{\mathbf{x_1}, \mathbf{x_2}, ..., \mathbf{x_n}\}$ of n (potentially infinite) elements of a data stream arriving at times $T_1, T_2, ..., T_n$, where $\mathbf{x_i} = (x_i^1, x_i^2, ..., x_i^d)$ is a vector in $\Re^d$. The notations used in this paper are presented in Table 2. At each time, G-Stream

Table 2. Notations used in the algorithm

Notation	Description
$\mathcal{DS} = \{\mathbf{x}_1, \mathbf{x}_2, ..., \mathbf{x}_n\}$	set of n (potentially infinite) data streams
$\mathbf{x}_i = (x_i^1, x_i^2, ..., x_i^d)$	d-dimensional data point
t_i	time-stamp of data point $\mathbf{x}_i$
$\mathbf{w}_c$	prototype $\mathbf{w}_c = (w_c^1, w_c^2, ..., w_c^d)$ of node c
δ_c	threshold distance of node c
$error(c)$	local accumulated error variable
$weight(c)$	local weight variable
bmu	best matching unit (the nearest node)
α_1	winning node (the nearest node) adaptation factor
α_2	winning node, neighbor adaptation factor
β	cycle interval between node insertions
age_{max}	oldest age allowed for an edge
λ_1	decay factor in the fading function
λ_2	strength factor in weighting edges

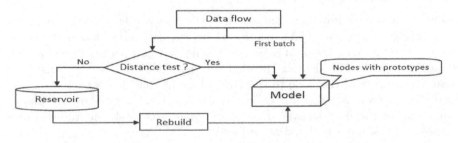

Fig. 1. Diagram of G-Stream algorithm

is represented by a graph $\mathcal{C}$ where each node represents a cluster. Each node $c \in \mathcal{C}$ has a prototype $\mathbf{w}_c = (w_c^1, w_c^2, ..., w_c^d)$ (resp. a distance threshold δ_c) representing its position (resp. the distance from the node to the farthest data point assigned to it). Starting with two nodes, and as a new data point is reached, the nearest and the second-nearest nodes are identified, linked by an edge, and the nearest node with its topological neighbors are moved toward the data point. Each node has an accumulated error variable and a weight, which varies over time using fading function. Using edge management, one, two or three nodes are inserted into the graph between the nodes with the largest error values. Nodes can also be removed if they are identified as being superfluous. Figure 1 represents a schematic diagram of the algorithm.

Fading function: In most data stream scenarios, more recent data can reflect the emergence of new trends or changes in the data distribution [3]. There are three window models commonly studied in data streams: landmark, sliding and damped. We consider, like many others, the damped window model, in which the weight of each data point decreases exponentially with time t via a fading function $f(t) = 2^{-\lambda_1(t-t_0)}$, where $\lambda_1 > 0$, defines the rate of decay of the weight over time, t denotes the current time and t_0 is the timestamp of the data point. The weight of a node is based on data points associated therewith: $weight(c) = \sum_{i=1}^{m} 2^{-\lambda_1(t-t_{i_0})}$, where m is the number of points assigned to the node c at the current time t. If the weight of a node is less than a threshold value then this node is considered as outdated and then deleted (with its links).

Edge management: The edge management procedure performs operations related to updating graph edges, as illustrated in steps 13-16 of Algorithm 1. The way to increase the age of edges is inspired by the fading function in the sense that the creation time of a link is taken into account. Contrary to the *fading* function, the age of the links will be strengthened by the exponential function $2^{\lambda_2(t-t_0)}$, where $\lambda_2 > 0$, defines the rate of growth of the age over time, t denotes the current time and t_0 is the creation time of the edge. The next step is to add a new edge that connects the two closest nodes. The last step is to remove each link exceeding a maximum age, since these links are no longer useful because they were replaced by younger and shorter edges that were created during the graph refinement in steps 18-22.

Reservoir management: The aim of using the reservoir is to hold, temporarily, the distant data points. As mentioned before, each node has a threshold distance. The first batch of data is assigned to nearest nodes without comparing distance thresholds. The distance threshold of each node is learned by taking the maximum distance of the node to the farthest point that it has been assigned. When the reservoir is full, its data is re-passed for learning. They are placed in the heap of the data stream, $\mathcal{DS}$, to be dealt with first and the distance thresholds of nodes are updated accordingly.

Computational complexity: It is obvious that the most consuming operations, in Algorithm 1, are steps 4, 18-22, 23, and 24 with $O(k)$ time complexity each, where k is the number of nodes in the graph. The node insertion phase (step 22) is repeated $\frac{3.n}{\beta}$ times. Seeking the nearest node (step 4), fading function (step 22), and adjusting the error variable (step 24) phases are repeated whenever a new data point is available, i.e. n times. The other steps have a constant time complexity. Therefore, G-Stream has a complexity given by $n.(3.O(k)) + \frac{3.n}{\beta}.O(k) = n.(3 + \frac{3}{\beta}).O(k) = O(nk)$.

4   Experimental Evaluations

In this section, we present an experimental evaluation of the G-Stream algorithm. We compared our algorithm with the GNG algorithm and several well-known and relevant data stream clustering algorithms, including StreamKM++, DenStream, and ClusTree. Our experiments were performed on MATLAB platform using real-world and synthetic data sets. All the experiments are conducted on a

---

Algorithm 1. G-Stream

Data: $\mathcal{DS} = \{x_1, x_2, ..., x_n\}$
Result: set of nodes $\mathcal{C} = \{c_1, c_2, ...\}$ and their prototypes $\mathbf{W} = \{\mathbf{w}_{c_1}, \mathbf{w}_{c_2}, ...\}$

1 Initialize the node set $\mathcal{C}$ to contain two nodes, c_1 and c_2: $\mathcal{C} = \{c_1, c_2\}$;
2 **while** *there is a data point to proceed* **do**
3   Get the next data point in the data stream, x_i;
4   Find the nearest node $bmu_1 \in \mathcal{C}$ and the second nearest node $bmu_2 \in \mathcal{C}$;
5   **if** $\|x_i - \mathbf{w}_{bmu_1}\| > \delta_{bmu_1}$ **then**
6   put x_i in the reservoir;
7   **if** *the reservoir is full* **then** Reservoir management ;
8   **else**
9   Increment the number of points assigned to bmu_1;
10         $error(bmu_1) = error(bmu_1) + \|x_i - \mathbf{w}_{bmu_1}\|^2$;
11   Move bmu_1 and its topological neighbors towards x_i:
        $\mathbf{w}_{bmu_1} = \mathbf{w}_{bmu_1} + \alpha_1.(x_i - \mathbf{w}_{bmu_1})$;
12         $\mathbf{w}_c = \mathbf{w}_c + \alpha_2.(x_i - \mathbf{w}_c)$ for all direct neighbors c of node bmu_1;
13   Increment the age of all edges emanating from bmu_1 and weight them;
14   **if** bmu_1 *and* bmu_2 *are connected by an edge* **then** set the age of this edge to zero ;
15   **else** create an edge between bmu_1 and bmu_2, and mark its time stamp;
16   Remove the edges whose age is greater than age_{max};
17   **if** *the number of points passed is a multiple of a parameter* β **then**
18   **for** *i=1 to 3* **do**
19   Find the node q with the maximum accumulated error;
20   Find the neighbor f of q with the largest accumulated error;
21   Add the new node, r, half-way between nodes q and f;
22   Insert the edges connecting the new node r with nodes q and f, and Remove the original edge between q and f;
23   Apply *fading*, delete outdated and isolated nodes;
24   Finally, decrease the error of all units;

PC with Core(TM)i7-4800MQ with two 2.70 GHz processors, and 8GB of RAM, which runs Windows 7 professional operating system.

4.1   Data Sets and Quality Criteria

To evaluate the clustering quality and scalability of the G-Stream algorithm both real and synthetic data sets are used. The two synthetic data sets used are DS1 and letter4. All the others are real-world publicly available data sets. Table 3 overviews all the data sets used. DS1 is generated by http://impca. curtin.edu.au/local/software/synthetic-data-sets.tar.bz2. The letter4 data set is generated by a Java code https://github.com/feldob/Token-Cluster-Generator. The Sea data set was taken from http://www.liaad.up.pt/kdus/products/ datasets-for-concept-drift. The HyperPlan data set was taken from [19]. The real-world databases were taken from the UCI repository [4], which are the

Table 3. Overview of all data sets

Datasets	#records	#features	#classes
DS1	9,153	2	14
letter4	9,344	2	7
Sea	60,000	3	2
HyperPlan	100,000	10	5
KddCup99	494,021	41	23
CoverType	581,012	54	7

KDD-CUP'99 Network Intrusion Detection stream data set (KddCup99) and the Forest CoverType data set (CoverType) respectively.

The algorithms are evaluated using three performance measures: Accuracy (Purity), Normalized Mutual Information (NMI) and Rand index [14]. The value of each measure lies between 0 and 1. A higher value indicates better clustering results. The Accuracy (Purity) averages the fraction of items belonging to the majority class of in each cluster. $Acc = \frac{\sum_{i=1}^{K} \frac{|N_i^d|}{|N_i|}}{K} \times 100\%$, where K denotes the number of clusters, N_i^d denotes the number of points with the dominant class label in cluster i, and N_i denotes the number of points in cluster i. Intuitively, the accuracy (purity) measures the purity of the clusters with respect to the true cluster (class) labels that are known for our data sets [7]. Normalized mutual information provides a measure that is independent of the number of clusters as compared to purity. It reaches its maximum value of 1 only when the two sets of labels have a perfect one-to-one correspondence [14]. The Rand index measures how accurately a clusterer can classify data elements by comparing cluster labels with the underlying class labels. Given N data points, there are a total of $\binom{N}{2}$ distinct pairs of data points which can be categorized into four categories: (a) pairs having the same cluster label and the same class label (their number denoted as N^{11}); (b) pairs having different cluster labels and different class labels (their number denoted as N^{00}); (c) pairs having the same cluster label but different class labels (their number denoted as N^{10}); (d) pairs having different cluster labels but the same class label (their number denoted as N^{01}). The Rand index is defined as: $Rand = (N^{11} + N^{00})/\binom{N}{2}$.

4.2 Evaluation and Performance Comparison

This section aims to evaluate the clustering quality of the G-Stream and compare it to well-known data stream clustering algorithms, as well as the GNG algorithm. As explained in section 3, the GNG and G-Stream algorithms start with two nodes. We used an online version of GNG but without the parameters that we added expressly to show the interest and contribution of these parameters in G-Stream. Therefore, we carried out experiments by initializing two nodes randomly among the first 20 points and we repeated this 10 times. For comparison purposes, we used DenStream [7] and ClusTree [12] from the **stream** R package [5]. Comparison is also performed with StreamKM++ [1] (this latter algorithm was coded in the C language). StreamKM++ was evaluated by choosing randomly the seed node (please refer to [1] for details) among the first 20 points. DenStream was evaluated by performing a variant of the DBScan algorithm in the offline step. ClusTree was evaluated by performing the k-means algorithm in the offline step by setting the k parameter to 10. All experiments were repeated 10 times and the results (the average value with its standard deviation) are reported in Table 4. In this Table, it is noticeable that G-Stream's Accuracies (Acc) are higher for all data sets as compared to StreamKM++, DenStream and CluStree, except for DenStream for the HyperPlan data set. Its NMI values are higher than the other algorithms except for DenStream for the Sea and HyperPlan data sets. Its Rand index values are higher than the other algorithms except for StreamKM++ for the Sea data set. We recall that G-Stream proceeds in one single phase whereas StreamKM++, DenStream and ClusTree proceed in two phases (online and offline phase).

Figure 2a (resp. Figure 2b) compares G-Stream (red line with circle) with GNG (blue line with cross) with respect to accuracy (resp. RMS error, number of nodes) for the letter4 data set. For almost all cases, the accuracy value (resp. RMS error) of G-Stream is higher (resp. is less) than the one of GNG. Figure 2c compares the two algorithms in terms of the number of nodes creating the graph. Despite that we create several nodes at each iteration (against a single node for GNG), the number of nodes created by G-Stream becomes steady (against a continuous increase for GNG) due to the application of the fading function. The same result can be seen for the remaining data sets.

Table 4. Comparing G-Stream with different algorithms

Datasets		**G-Stream**	StreamKM++	DenStream	ClusTree
DS1	Acc	**0.9809±0.0061**	0.6754±0.0183	0.7740±0.0000	0.6864±0.0275
	NMI	**0.7289±0.0113**	0.7021±0.0209	0.6973±0.0000	0.7064±0.0168
	Rand	**0.8530±0.0024**	0.8443±0.0048	0.8491±0.0000	0.8442±0.0066
letter4	Acc	**0.9832±0.0050**	0.6871±0.0263	0.8110±0.0000	0.8110±0.0000
	NMI	**0.6265±0.0064**	0.5532±0.0219	0.1637±0.0000	0.2425±0.0000
	Rand	**0.8156±0.0015**	0.7941±0.0145	0.5019±0.0000	0.5514±0.0000
Sea	Acc	**0.8386± 0.0021**	0.7886±0.0091	0.8240±0.0001	0.8224±0.0065
	NMI	0.1380±0.0009	0.1463±0.0042	**0.1646±0.0000**	0.1583±0.0095
	Rand	0.4707±0.0001	**0.5072±0.0016**	0.4700±0.006	0.4917±0.0034
HyperPlan	Acc	0.4238±0.0021	0.3966±0.0055	0.4250±0.0000	**0.4380±0.0089**
	NMI	0.0186±0.0009	0.0103±0.0023	**0.0208±0.0000**	0.0170±0.0042
	Rand	**0.7042±0.0008**	0.6674±0.0004	0.6038±0.0000	0.6529±0.0016
KddCup99	Acc	**0.9805±0.0050**	0.6922±0.1140	0.9544±0.0031	0.8182±0.1304
	NMI	**0.6670±0.0089**	0.3926±0.2815	0.6290±0.0300	0.5724±0.2974
	Rand	**0.8380±0.0036**	0.6339±0.2316	0.8164±0.0106	0.8289±0.1798
CoverType	Acc	**0.6085±0.0087**	0.5266±0.0074	0.5850±0.0011	0.5850±0.0000
	NMI	**0.1403±0.0029**	0.0874±0.0086	0.0475±0.0201	0.0362±0.0042
	Rand	**0.6231±0.0008**	0.6106±0.0018	0.4604±0.0070	0.5080±0.0005

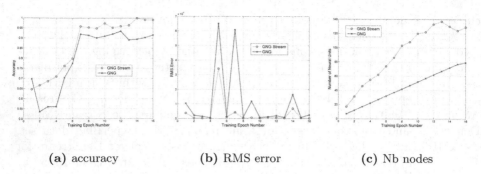

(a) accuracy   (b) RMS error   (c) Nb nodes

Fig. 2. Accuracy, RMS error, and number of nodes for G-Stream and GNG on letter4

4.3 Visual Validation

Figure 3 shows the evolution of the node creation by applying G-Stream on the letter4 data set (green points represent data points of the data stream and blue points are nodes of the graph with edges in blue lines). It illustrates that G-Stream manages to recognize the structures of the data stream and can separate these structures with the best visualization. Figure 4 compares G-Stream with GNG-online on 2-dimensional data sets (DS1 and letter4), in terms of visual results i.e., the final graph found by GNG-online/G-Stream for each data set. As illustrated on these figures, the G-Stream algorithm is superior to the GNG-online with respect to visual structures found.

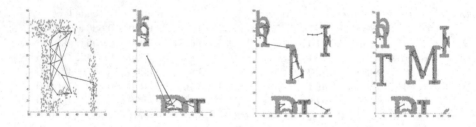

Fig. 3. Evolution of graph creation of G-Stream on letter4 (data set and topological result)

4.4 Evolving Data Streams

In this subsection, we perform G-Stream on different data streams ordered by class labels to demonstrate its effectiveness in clustering evolving data streams (i.e., data points of the first class arrive in first, then the ones of the second, third, etc. class). In this case, old concepts (class labels) disappear due to the use of fading function. In the same time, new concepts (class labels) appear as

(a) G-Stream on **(b)** GNG-online on **(c)** G-Stream on let- **(d)** GNG-online on
DS1   DS1   ter4   letter4

Fig. 4. Visual result comparison of G-Stream with GNG-online (dataset and topological result)

new data points arrive. We use the same experimental protocol as described in section 4.2, i.e., we did experiments by initializing two nodes randomly among the first 20 points, we repeated this 10 times, and we report the average value with its standard deviation in Figure 5a. Figure 5a shows that G-Stream can find clusters with performance measures as comparable to those without ordering classes.

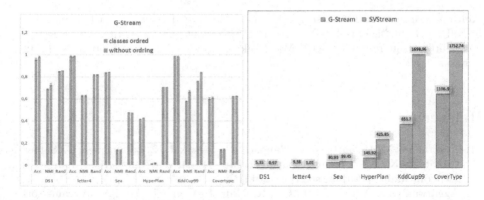

(a) G-Stream with and without ordering   **(b)** Execution time (in seconds)
of classes

4.5   Execution Time

The efficiency of algorithms is measured by the execution time. Referring to the computational complexity we calculated in Section 3, the execution time strongly depends to the number of nodes creating the graph and the size of the data stream. We recall that G-Stream is implemented in MATLAB and SVStream is the only MATLAB program that we have (the other algorithms are implemented in Java, R, or C languages). Figure 5b shows the execution time of G-Stream and that of SVStream. We can see that both the execution time of G-Stream and SVStream grow as the size of the data stream grows, and G-Stream is more efficient than SVStream.

5   Conclusion

In this paper, we have proposed G-Stream, an efficient method for topological clustering an evolving data stream in an online manner. In G-Stream, the nodes are weighted by a fading function and the edges by an exponential function. Starting with two nodes, G-Stream confronts the arriving data points to the current prototypes, storing the very distant ones in a reservoir, learns the threshold distances automatically, and many nodes are created in each iteration. Experimental evaluation over a number of real and synthetic data sets demonstrates the effectiveness and efficiency of G-Stream in discovering clusters of arbitrary shape. Our experiments show that G-Stream outperformed the GNG algorithm in terms of visual results and quantitative criteria such as accuracy, the Rand index and NMI. Its performance, in terms of clustering quality as compared to three relevant data stream algorithms are promising. We plan in the future to implement adaptive windows, make our algorithm as autonomous as possible and develop it in Spark Streaming.

Acknowledgments. This research has been supported by the French Foundation FSN, PIA Grant Big data-Investissements d'Avenir. The project is titled "Square Predict" (http://square-predict.net/). We thank anonymous reviewers for their insightful remarks.

References

1. Ackermann, M.R., Märtens, M., Raupach, C., Swierkot, K., Lammersen, C., Sohler, C.: StreamKM++: A clustering algorithm for data streams. ACM Journal of Experimental Algorithmics, **17**(1) (2012)
2. Aggarwal, C.C., Watson, T.J., Ctr, R., Han, J., Wang, J., Yu, P.S.: A framework for clustering evolving data streams. In: VLDB, pp. 81–92 (2003)
3. de Andrade Silva, J., Faria, E.R., Barros, R.C., Hruschka, E.R., de Carvalho, A.C.P.L.F., Gama, J.: Data stream clustering: A survey. ACM Comput. Surv. **46**(1), 13 (2013)
4. Bache, K., Lichman, M.: UCI machine learning repository (2013). http://archive.ics.uci.edu/ml
5. Bolanos, M., Forrest, J., Hahsler, M.: Stream: Infrastructure for Data Stream Mining (2014). http://CRAN.R-project.org/package=stream, r package version 0.2-0
6. Bouguelia, M.R., Belaïd, Y., Belaïd, A.: An adaptive incremental clustering method based on the growing neural gas algorithm. In: ICPRAM, pp. 42–49 (2013)
7. Cao, F., Ester, M., Qian, W., Zhou, A.: Density-based clustering over an evolving data stream with noise. In: SDM, pp. 328–339 (2006)
8. Fritzke, B.: A growing neural gas network learns topologies. In: NIPS, pp. 625–632 (1994)

9. Guha, S., Meyerson, A., Mishra, N., Motwani, R., O'Callaghan, L.: Clustering data streams: Theory and practice. IEEE Transactions on Knowledge and Data Engineering **15**(3), 515–528 (2003)
10. Isaksson, C., Dunham, M.H., Hahsler, M.: SOStream: Self Organizing Density-Based Clustering over Data Stream. In: Perner, P. (ed.) MLDM 2012. LNCS, vol. 7376, pp. 264–278. Springer, Heidelberg (2012)
11. Kohonen, T., Schroeder, M.R., Huang, T.S. (eds.): Self-Organizing Maps, 3rd edn. Springer, Secaucus (2001)
12. Kranen, P., Assent, I., Baldauf, C., Seidl, T.: The ClusTree: indexing micro-clusters for anytime stream mining. Knowledge and Information Systems **29**(2), 249–272 (2011)
13. Martinetz, T., Schulten, K.: A "Neural-Gas" Network Learns Topologies. Artificial Neural Networks **I**, 397–402 (1991)
14. Strehl, A., Ghosh, J.: Cluster ensembles — a knowledge reuse framework for combining multiple partitions. Journal of Machine Learning Research **3**, 583–617 (2002)
15. Udommanetanakit, K., Rakthanmanon, T., Waiyamai, K.: E-Stream: Evolution-Based Technique for Stream Clustering. In: Alhajj, R., Gao, H., Li, X., Li, J., Zaïane, O.R. (eds.) ADMA 2007. LNCS (LNAI), vol. 4632, pp. 605–615. Springer, Heidelberg (2007)
16. Wang, C., Lai, J., Huang, D., Zheng, W.: SVStream: A support vector-based algorithm for clustering data streams. IEEE Trans. Knowl. Data Eng. **25**(6), 1410–1424 (2013). http://doi.ieeecomputersociety.org/10.1109/TKDE.2011.263
17. Zhang, T., Ramakrishnan, R., Livny, M.: Birch: An efficient data clustering method for very large databases. In: SIGMOD Conference, pp. 103–114 (1996)
18. Zhang, X., Furtlehner, C., Sebag, M.: Data streaming with affinity propagation. In: Daelemans, W., Goethals, B., Morik, K. (eds.) ECML PKDD 2008, Part II. LNCS (LNAI), vol. 5212, pp. 628–643. Springer, Heidelberg (2008)
19. Zhu, X.H.: Stream data mining repository (web site) (2010). http://www.cse.fau.edu/xqzhu/stream.html

Computing and Mining ClustCube Cubes Efficiently

Alfredo Cuzzocrea[✉]

ICAR-CNR and University of Calabria, Cosenza, Italy
cuzzocrea@si.dimes.unical.it

Abstract. A *novel computational paradigm for clustering complex database objects extracted from distributed database settings via well-understood OLAP technology* is proposed and experimentally assessed in this paper. This paradigm conveys in the so-called ClustCube cubes, which define *a novel multidimensional data cube model* according to which (data) cubes store *clustered complex database objects* rather than conventional SQL-based aggregations. A major contribution of this research is represented by effective and efficient algorithms for computing ClustCube cubes that, surprisingly, are capable of reducing computational efforts significantly with respect to traditional approaches. Our analytical contribution is completed by a comprehensive assessment of proposed algorithms against both benchmark and real-life data sets, which clearly confirms the benefits deriving from our proposal.

1   Introduction

While lot of proposals on *mining traditional data sets* exist, Data Mining researchers have devoted poor attention to the problem of *effectively and efficiently mining complex objects*, for instance extracted from *distributed database settings* [12]. Contrary to this actual trend, mining complex objects is indeed relevant in practical application scenarios, as *modern database systems are more and more immersed in object-oriented scenarios rather than tuple-oriented scenarios*. In such applicative settings, it is important to exploit all the potentialities offered by *object-oriented paradigms* like *inheritance*, *information hiding*, *polymorphism*, and so forth, as, in turn, this allows us to achieve more powerful functionalities.

Note that, due to specialized *business processes* defined within the target applications, *complex objects extracted from the underlying distributed database could represent derived information that is significantly different from original information stored in the database itself*. For instance, this could be the case of complex database objects extracted by means of (complex) SQL statements involving multiple JOIN queries across distributed relational tables. As a consequence, *derived knowledge* kept in the application layer as *a collection of complex database objects* could result to be much more interesting to be mined rather than the *primary knowledge* kept in the data layer in the vest of *a collection of tuples*. From this breaking evidence, it clearly follows that taking advantages from a specialized Data Mining component devoted to mine complex objects extracted from data-intensive applications would be very beneficial. In fact, this would allow us to magnify the effectiveness and the expressive power of the *whole* knowledge discovery phase over complex data-intensive software environments dictated by modern scenarios.

© Springer International Publishing Switzerland 2015
T. Cao et al. (Eds.): PAKDD 2015, Part II, LNAI 9078, pp. 146–161, 2015.
DOI: 10.1007/978-3-319-18032-8_12

Among the wide family of Data Mining techniques available in the active literature, since objects essentially aggregate *low-level fields* (which, in turn, are extracted from attribute values of the underlying distributed database) into *complex classes*, it is natural to think of *clustering* (e.g., [5]) as the most suitable technique to mine such so-derived structures. Also, the combined action of clustering techniques and well-consolidated methodologies developed in the context of *OnLine Analytical Processing* (OLAP) [3] clearly offers powerful tools to mine *clustered objects* according to a multidimensional and multi-resolution vision of the underlying *object domain*.

Inspired by motivations above, in this paper we propose and experimentally assess *an innovative OLAP-based framework for clustering and mining complex database objects* extracted from *distributed database settings*, called ClustCube, which encompasses a number of research innovations beyond the capabilities of actual Data Mining methodologies over large and complex-in-nature databases (e.g., [12]). To this end, ClustCube combines the power of clustering techniques over complex database objects and the power of OLAP in supporting multidimensional analysis and knowledge fruition of (clustered) complex database objects, with mining opportunities and expressive power infeasible for traditional methodologies. So-obtained ClustCube cubes store clustered complex database objects within cube cells, rather than conventional SQL-based aggregations like in standard *Business-Intelligence*-oriented OLAP data cubes. According to our vision, the ClustCube proposal realizes a significant improvement towards achieving the *integration of OLAP and Data Mining*, which is a promising paradigm within the broad umbrella of the *OnLine Analytical Mining* (OLAM) discipline [4].

Figure 1 shows the "big picture" of the research we propose, i.e. the ClustCube overview. Basically, ClustCube defines a multiple-layer reference architecture that encompasses the following well-separated layers: (*i*) *Distributed DataBase Layer* (DDBL), where the target distributed database from which complex objects are extracted is located; (*ii*) *Complex Object Definition Layer* (CODL), which supports primitives and functionalities for building and managing complex objects extracted from the DDBL layer; (*iii*) the *Object Layer* (OL), where complex objects are located, along with a suitable *object schema*; (*iv*) the ClustCube *Definition and Management Layer* (CCDML), which supports primitives and functionalities for defining and managing ClustCube cubes; (*v*) the ClustCube *Layer* (CCL), which stores the final ClustCube cube(s).

This paper represents a consistent extension of contribution [14], which has been accepted as short paper. Most part of the novel contribution focuses on a comprehensive and wide experimental analysis addressing performance of algorithms embedded in the proposed framework (see Section 5).

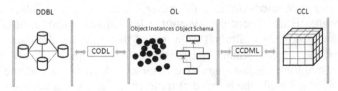

Fig. 1. ClustCube overview

2   From Tuple-Oriented Entities to Complex Database Objects

Building and managing complex database objects extracted from the target distribute database located at the DDBL layer (see Figure 1) plays a critical role within the Clust-Cube framework. Hence, in this Section we focus the attention on this aspect, which is based on consolidated *software methodologies over data-intensive applications*, but targeted to the specific requirements delineated by ClustCube (see Section 1).

From the DDBL layer of the ClustCube framework (see Figure 1), the CODL layer meaningfully extracts complex database objects used to populate the OL layer via a set of *Complex Object Definition Queries* (*CODQ*), denoted by $QS_{CODL} = \{Q_0, Q_1, ..., Q_{|QS_{CODL}|-1}\}$. *CODQ* queries are conceptually located at the CODL layer and defined by the administrator editing the whole analysis/mining process, on the basis of her/his analysis/mining tasks. *CODQ* queries in QS_{CODL} are *combined* by the CODL layer into a singleton *global CODQ query*, denoted by Q_{CODL}, which involves in a *singleton class/object-schema*, denoted by OS_{CODL}, to which the *set of object instances*, denoted by OI_{CODL}, adheres. Both schema OS_{CODL} and set OI_{CODL} are located at the OL layer (see Figure 1). Global *CODQ* query Q_{CODL} can be simply obtained from *CODQ* queries in QS_{CODL} according to several straightforward alternatives, such as *union*, i.e. $Q_{CODL} = \bigcup_{k=0}^{|QS_{CODL}|-1} Q_k$, or *join*, i.e. $Q_{CODL} = \bowtie_{k=0}^{|QS_{CODL}|-1} Q_k$, but, without any loss of generality, in the ClustCube framework the administrator is allowed to edit global *CODQ* queries Q_{CODL} of *arbitrary* nature, even based on *complex user-defined SQL expressions*. Looking into practice, the *CODQ* query Q_{CODL} can be simply thought-of as a canonical SQL query involving multiple JOIN statements over distributed relational tables located at the DDBL layer. We denote as $Sel(Q_{CODL})$ the set of selection attributes of Q_{CODL}. As highlighted in Section 1, this query-based extraction mechanism allows us to define complex objects having no correspondence with singleton tuples stored in singleton local databases of the DDBL layer. As a consequence, ClustCube framework makes it possible to define mining tasks which cannot be originally defined on singleton local databases of the DDBL layer, hence involving in powerful mining opportunities infeasible to traditional mining methodologies. It should be clear enough that this is one of the advantages of ClustCube methodology over conventional mining paradigms.

To give an example of complex objects, consider Figure 2 (*a*), where the E/R schema of the hypothetical database *DataBank* storing bank data and located at the DDBL layer is depicted (up-side), along with complex objects extracted from it (down-side). *DataBank* is characterized by the following relational tables, whose semantics is obvious, hence not detailed for space reasons: (*i*) Employee(EmpId, Name, Surname, Address, Dept), (*ii*) Customer(CustId, Name, Surname, Address, Group), (*iii*) SavingAccount(SavingId, CustId, OpenDate, Balance, InterestRate), (*iv*) LoanAccount(LoanNumber, CustId, OpenDate, TermDate, Amount, InterestRate, MonthlyPayment), (*v*) CheckingAccount(CheckingId, CustId, OpenDate, OverdraftLimit). On top of the database *DataBank*, the *CODQ* query Q_{CODL} of the running example retrieves the set of employees who are also customers and that, during the last year, have obtained a loan whose amount is between 70,000\$ and 80,000\$ while the balance of their saving account is between 5,000\$ and 10,000\$. Q_{CODL} can be reasonably intended as a *mining query* that expresses a derived

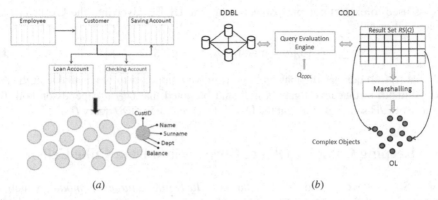

(a)   (b)

Fig. 2. Complex database objects extracted from the example database *DataBank* via the *CODQ* query Q_{CODL} (a) and the overall process to extract complex database objects from the DDBL layer into the OL layer (b)

knowledge that (*i*) is not directly kept within *DataBank* and (*ii*) defines a collection of complex database objects that are capable to support powerful mining tasks over *DataBank* according to the ClustCube guidelines (see Section 1).

The CODL layer takes as input the *CODQ* query Q_{CODL}, and returns as output a pair of kind: $\langle OS_{CODL}, OI_{CODL} \rangle$, which is conceptually located at the OL layer, representing OS_{CODL} the *intensional part* of the OL layer and OI_{CODL} the *extensional part* of the OL layer, respectively. In particular, OI_{CODL} is built in terms of a collection of complex objects generated via *marshaling* relational-tuple bags obtained as answers to the *CODQ* query Q_{CODL}.

The structure of objects of class OS_{CODL} is naturally defined by the structure of the SQL syntax of Q_{CODL}, as follows (assume that Q_{CODL} is characterized by N selection attributes, i.e. $|Sel(Q_{CODL})| = N$): (*i*) $N = |OS_{CODL}|$ fields are introduced in OS_{CODL}, denoted as a_h, one for each selection attribute A_h in $Sel(Q_{CODL})$; (*ii*) type of the field a_h depends on the type of the corresponding selection attribute A_h, and the final in-memory-representation of a_h is determined by the software platform on top of which complex objects are implemented in the AS layer (see Figure 1), e.g. SQL type *Int* is converted into 32-bit platform-dependent *Integer* type; (*iii*) for each attribute a_h in OS_{CODL}, two *accessory methods* are introduced, namely getA_h and setA_h, respectively, which allow us to retrieve and set the actual value of the field a_h in the corresponding object, respectively.

Looking at practical implementations, given the *CODQ* query Q_{CODL}, object instances of class OS_{CODL} populating the set OI_{CODL} generated by the CODL layer are obtained according to the following procedure: (*i*) retrieve the answer to query Q_{CODL} against the DDBL layer, denoted as $RS(Q_{CODL})$; (*ii*) for each tuple t_{o_i} in $RS(Q_{CODL})$, create an object instance o of class OS_{CODL} by means of marshalling tuple t_{o_i} into object o_i, being the baseline marshalling primitive supposed to be available in the AS layer (see Figure 1). This paradigm is reminiscent of similar operations performed to extract EJB objects from relational data sources. Figure 2 (*b*) sketches the overall process used to extract complex objects from the DDBL layer into the OL layer (see Figure 1). Formally:

$$OI_{CODL} = \left\{ o_i | o_i = \text{marsh}(t_{o_i}) \wedge t_{o_i} \in RS(Q_{CODL}) \right\} \qquad (1)$$

such that marsh denotes the marshalling primitive (supposed to be) available at the AS layer. Without further investigation, it should be noted that OI_{CODL} depends on both the structure of E/R schemas stored at the DDBL layer and the CODQ query Q_{CODL}.

3   Keeping Complex Object Clusters into Data Cube Cells

In this Section, we provide details on the *ClustCube data cube model*, which is directly inspired from the traditional OLAP data cube model [3]. As mentioned in Section 1, ClustCube cubes store clustered complex objects within cube cells, contrary to traditional OLAP data cube that store SQL-based aggregations. Focus the attention on the structure of ClustCube cubes in a greater detail. Given a ClustCube cube C characterized by the *set of dimensions* $\mathcal{D} = \{d_0, d_1, \ldots, d_{N-1}\}$, such that $N = |OS_{CODL}|$, being dimensions in $\mathcal{D}$ corresponding to features in $\mathcal{F}(OS_{CODL})$, each ClustCube cube cell $C[i_0][i_1]\ldots[i_{N-1}] \equiv C[\mathbb{I}]$ in C, such that $0 \le i_0 \le |d_0|$, $0 \le i_1 \le |d_1|, \ldots, 0 \le i_{N-1} \le |d_{N-1}|$, denoting $\mathbb{I} = \langle i_0, i_1, \ldots, i_{N-1} \rangle$ an *N-dimensional entry* in the *N-dimensional space* of C, stores a set of clustered objects, denoted by $OI_{CODL}(C[\mathbb{I}])$, that *are obtained by simultaneously clustering objects in* OI_{CODL} *with respect to the dimensions/features $d_0, d_1, \ldots, d_{N-1}$ in* $\mathcal{D}/\mathcal{F}(OS_{CODL})$. Hence, it is trivial to observe that, for each ClustCube cube cell $C[\mathbb{I}]$ in C, *multidimensional boundaries* of $C[\mathbb{I}]$ along the dimension d_i, denoted by $B_{d_i}^{low}$ and $B_{d_i}^{up}$, with $B_{d_i}^{low} < B_{d_i}^{up}$, respectively, are determined by the $\mathcal{A}$-based clustering of objects in OI_{CODL} with respect to the feature d_i in $\mathcal{D}$. In other words, while in traditional OLAP data cubes [3] the multidimensional boundaries of data cube cells along dimensions are determined by the input *OLAP aggregation scheme*, in ClustCube cubes multidimensional boundaries of ClustCube cube cells along dimensions are the *result* of the clustering algorithm $\mathcal{A}$ itself.

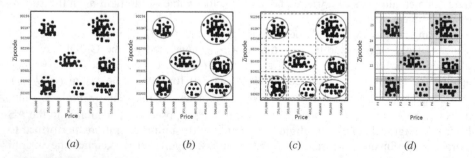

(a)   (b)   (c)   (d)

Fig. 3. A two-dimensional house sale data set (*a*) and its ClustCube clustering-guided aggregation (*d*) ((*b*) and (*c*) are intermediate steps)

This novelty has deep consequences on the way ClustCube cube aggregations are computed. To become convinced of this, consider a simple case study focused on an house sale data set that logically defines a two-dimensional space characterized by the following dimensions (features, respectively – see Figure 3 (*a*)): (*i*) *Price*, which represents the price at which a certain house is sold; (*ii*) *Zipcode*, which represents the zipcode of the city where the house is located. Figure 3 (*b*) shows a possible clustering of such a data set with respect to the features (dimensions, respectively) *Price* and *Zipcode*. Figure 3 (*c*) shows instead the *projection* of so-obtained clusters along both the dimensions, and, lastly, Figure 3 (*d*) shows the final (logical-representation) of the two-dimensional ClustCube cube, where blue lines denote cube cells boundaries. It should be noted that, contrary to traditional **OLAP** data cubes where tuples are aggregated according to *regular* and *somewhat natural* groups along dimensional hierarchies defined by the input **OLAP** aggregation scheme, thus determining *regular partitions* of the target data domain, in novel ClustCube data cubes groups of objects stored in (ClustCube) cube cells correspond to clusters computed by input clustering algorithm $\mathcal{A}$, thus determining *irregular partitions* of the target object domain following sort of a *clustering-guided ClustCube aggregation scheme* (Figure 3 (*d*)). As shown in Figure 3 (*d*), given a ClustCube data cube C, every cell $C[i_0][i_1]\dots[i_{N-1}]$ in C may alternatively contain either a whole cluster generated from $\mathcal{A}$ or a sub-cluster of it. While it is obvious that other representation models exist (e.g., the one constrained to having each cluster stored within one ClustCube cube cell only), the approach we propose embeds several amenities, among which we recall: (*i*) achieving a low-granular representation of clusters across ClustCube cube cells – this would expose us to a wide spectrum of advantages such as supporting meaningful *multi-resolution* [9] and *sub-space* [1] *cluster analysis*; (*ii*) achieving a better support for users during the multidimensional exploration of clustered complex objects by means of the nice property stating that *adjacent ClustCube cube cells store sub-clusters of the same original cluster*, thus supporting meaningful *similarity-based cluster analysis* [12].

One of the most relevant contribution of our research consists in equipping the final ClustCube cube generated by the **CCDML** layer (see Figure 1) with the canonical cubod lattice [3]. In traditional **OLAP**, given an N-dimensional data cube C having $\mathcal{D} = \{d_0, d_1, \dots, d_{N-1}\}$ as dimension set, the cuboid lattice associated to C, denoted by $\mathcal{L}$, is a *hierarchical structure* composed by $2^N - 1$ cuboids, denoted by C_i, i.e. data (sub-)cubes that aggregate original relational data according to arbitrary combinations of dimensions in $\mathcal{D}$, each one at different cardinality. In other words, an n-dimensional cuboid C_i of a data cube C represents a particular n-dimensional view of C, such that $0 \leq n \leq N$. For $n = N$, the base cuboid is defined, which corresponds to the original data cube C. For $n = 0$, the *apex cuboid* is defined, which corresponds to the empty data cube $\varnothing$. Cuboids of a certain cuboid lattice $\mathcal{L}$ are naturally ordered by means of the *precedence relation* $\prec$, such that, for each pair of cuboids C_i and C_j in $\mathcal{L}$, $C_i \prec C_j$ holds iff $\mathcal{D}_i \subset \mathcal{D}_j$, such that $\mathcal{D}_i$ denotes the set of dimensions of C_i and $\mathcal{D}_j$ denotes the set of dimensions of C_j, respectively. This finally determines a *cuboid hierarchy* (associated to C), denoted by $\mathcal{H}(\mathcal{L})$. For instance, Figure 4 (*a*) shows in the left side the cuboid lattice $\mathcal{L}$ for a four-dimensional ClustCube cube C having $\mathcal{D} = \{A, B, C, D\}$ as dimension set. Here, for instance, the property $CD \prec BCD$ holds in $\mathcal{H}(\mathcal{L})$, being the cuboids CD and BCD detailed greatly in Figure 4 (*a*) (right side). Also, from Figure 4 (*a*) (left side), it should be clear enough that in the cuboid lattice $\mathcal{L}$ of an N-dimensional data cube C, cuboids are structurally organized according to $N + 1$ levels such that l-dimensional cuboids located at level l of $\mathcal{L}$ are hierarchically linked to cuboids at level $l - 1$ and $l + 1$ of $\mathcal{L}$, respectively. This realizes the precedence relation $\prec$ in practice.

Now, focus the attention on some important properties of the ClustCube data cube model with respect to the proper clustering. Figure 4 (*a*) (right side) shows clustered objects stored by the cuboids CD and BCD, respectively. Here, it should be clear enough that cuboid BCD stores clusters (of objects) that are *conceptually* obtained from clusters of cuboid CD (which *precedes BCD* in $\mathcal{H}(\mathcal{L})$, i.e. $CD \prec BCD$) *by distributing objects in CD with respect to the newly-added dimension/feature B*. In the ClustCube framework, we *fully* take advantages from such a distributed nature of clustering across hierarchical cuboids, being this nice amenity the key property that allows us to significantly reduce computational needs due to computing ClustCube cube cuboid lattices. In fact, while clusters at level $l = 1$ are retrieved by means of the core algorithm $\mathcal{A}$, thanks to this amenity, *we do not need to compute remaining cuboids from the scratch* but every cuboid C_j at level l of $\mathcal{L}$, such that $2 \le l \le N$, can be obtained from the cuboids at level $l - 1$, denoted by $\left\{ C_{k_0}, C_{k_1}, ..., C_{k_{|l-1|-1}} \right\}$, such that $|l - 1|$ denotes the cardinality of the level $l - 1$ of $\mathcal{L}$, simply by *simultaneously distributing* objects in C_i with respect to the features $\left\{ D_{k_0}, D_{k_1}, ..., D_{k_{|l-1|-1}} \right\}$ of $\left\{ C_{k_0}, C_{k_1}, ..., C_{k_{|l-1|-1}} \right\}$, respectively. Starting from the cuboids/clusters at level $l = 1$, by iterating the procedure above from cuboids at level $l = 2$ towards cuboids at level $l = N$, the cuboid lattice $\mathcal{L}$ of the ClustCube cube C can be finally obtained in a *progressive manner*.

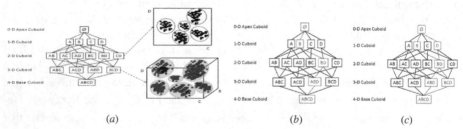

(a)   (b)   (c)

Fig. 4. ClustCube building technique hierarchy (*a*); a regular (*b*) and an irregular (*c*) traversal path for an example four-dimensional ClustCube cube

4   How to Efficiently Compute ClustCube Cubes?

Computing ClustCube cubes enriched by the respective cuboid lattices is the most probing research challenge to face-off within the whole ClustCube proposal, according to guidelines provided in Section 1. In this Section, we provide a collection of algorithms that are capable of fulfilling this goal effectively and efficiently, as demonstrated by our comprehensive experimental evaluation and analysis of ClustCube performance (see Section 5).

Given the input class OS_{CODL}, the collection of complex objects OI_{CODL} and the input clustering algorithm $\mathcal{A}$, the CCDML layer computes the final ClustCube cube to be stored at the CCL layer (see Figure 1). To this end, ClustCube framework comprises several kinds of techniques for efficiently building the ClustCube cube C plus its cuboid lattice $\mathcal{L}$ (each one codified by a respective ClustCube cube building algorithm), which are distinctively characterized by two *orthogonal* strategies, namely *materialization strategy* and *building strategy*, both referring ways C is finally obtained. Materialization strategies specify *which* cuboids, among the $2^N - 1$ cuboids of $\mathcal{L}$, must be *materialized*, i.e. *computed* and *stored* (in secondary memory). On the other hand, building strategies specify *how* cuboids are *computed* actually. With respect to the materialization strategy, the following two alternatives are introduced in the ClustCube framework: (*i*) *full*, denoted by FUL, according to which *all* cuboids of $\mathcal{L}$ are materialized; (*ii*) *partial*, denoted by PAR, according to which a *sub-set* of the $2^N - 1$ cuboids of $\mathcal{L}$ is materialized. As regards the building strategy, ClustCube framework exposes the following two different approaches: (*i*) *baseline*, denoted by BAS, according to which, for each cuboid C_i in $\mathcal{L}$, clusters are *re-computed from the scratch* (i.e., directly from input objects in OI_{CODL}); (*ii*) *drill-down*, denoted by DRI, according to which cuboids at level l of $\mathcal{L}$, for $2 \le l \le N$, are computed from cuboids at level $l - 1$ of $\mathcal{L}$ by means of a meaningfully distributive method. It is critical to notice here that, in the ClustCube framework, the target ClustCube cube C is obtained via computing the whole cuboid lattice $\mathcal{L}$ in terms of the base cuboid [3]. Also, it should be clear enough that the availability of multiple solutions for computing ClustCube cubes fully adheres to the principle of making this critical task *adaptive*, hence customizable to *different* application scenarios where ClustCube cubes can be applied, ranging from conventional centralized scenarios to more probing distributive and stream settings.

Since materialization strategies and building strategies are orthogonal, four techniques for computing the target ClustCube cube are finally obtained in terms of combinations of both materialization and building strategies, namely: ⟨FUL,BAS⟩, ⟨FUL,DRI⟩, ⟨PAR,BAS⟩, ⟨PAR,DRI⟩. In the remainder of this Section, we thoroughly discuss each one of the ClustCube building techniques above, along with the respective ClustCube building algorithm.

Looking at the ClustCube building technique hierarchy, it is obvious that ⟨FUL,BAS⟩ is the most-straightforward technique among all the defined ones. According to ⟨FUL,BAS⟩, the whole cuboid lattice $\mathcal{L}$ is computed and materialized, and each cuboid C_i in $\mathcal{L}$ is computed by applying clustering algorithm $\mathcal{A}$ directly to the input objects in OI_{CODL} with respect to the features in $\mathcal{D}_i$, i.e. the dimension set of C_i (see Section 3). Algorithm BuildClustCube_FUL_BAS implements technique ⟨FUL,BAS⟩. BuildClustCube_FUL_BAS simply computes $\mathcal{L}$ by *visiting* each level of $\mathcal{L}$ and, for each cuboid C_i in $\mathcal{L}$, performing the following steps: (*i*) from cube metadata, determine $\mathcal{D}_i$; (*ii*) execute algorithm $\mathcal{A}$ on OI_{CODL} with respect to features in $\mathcal{D}_i$, thus obtaining the set of clusters $H_i = \{\Lambda_0, \Lambda_1,..., \Lambda_{K-1}\}$; (*iii*) for each cluster Λ_k in H_i and for each dimension d_m in $\mathcal{D}_i$, project Λ_k along d_m thus selecting the set

$U_k = \{\rho_0, \rho_1, ..., \rho_{M-1}\}$ of *dimensional members* of d_m – this finally originates a *multi-dimensional partition* of the underlying object domain, denoted by Φ; (*iv*) materialize clusters Λ_k in H_i within (ClustCube) data cube cells of C_i based on Φ. Building technique $\langle$FUL,DRI$\rangle$ materializes all cuboids in $\mathcal{L}$, but it avoids the computational costs of technique $\langle$FUL,BAS$\rangle$ due to running clustering algorithm $\mathcal{A}$ on *each* cuboid C_i of $\mathcal{L}$. Contrary to technique $\langle$FUL,BAS$\rangle$, technique $\langle$FUL,DRI$\rangle$ instead employs clustering algorithm $\mathcal{A}$ to compute one-dimensional cuboids of $\mathcal{L}$ at level $l = 1$ and then it exploits the clustering results obtained for one-dimensional cuboids to compute *all* the remaining cuboids of $\mathcal{L}$, by fully exploiting the distributed nature of clustering across hierarchical cuboids. Algorithm `BuildClustCube_FUL_DRI` implements technique $\langle$FUL,DRI$\rangle$ by performing the following steps: (*i*) execute clustering algorithm $\mathcal{A}$ on OI_{CODL} via considering the *singleton* feature corresponding to the *singleton* dimension of one-dimensional cuboids in $\mathcal{L}$ (level $l = 1$), thus obtaining clusters $H_i = \{\Lambda_0, \Lambda_1, ..., \Lambda_{K-1}\}$ (it should be noted that this dimension set corresponds to the whole dimension set $\mathcal{D}$ of the target ClustCube data cube model); (*ii*) for each cluster Λ_k in H_i, project Λ_k along the *singleton* dimension d_m in $\mathcal{D}$ thus selecting the set $U_k = \{\rho_0, \rho_1, ..., \rho_{M-1}\}$ of dimensional members of d_m – this finally originates a *one-dimensional partition* of the underlying object domain Φ; (*iii*) for each of the remaining n-dimensional cuboid C_i in $\mathcal{L}$ at level l, such that $2 \leq l \leq N$, perform the following steps: (*iii*.1) from cube metadata, determine $\mathcal{D}_i$; (*iii*.2) compute clusters $H_i = \{\Lambda_0, \Lambda_1, ..., \Lambda_{K-1}\}$ associated to C_i by *distributing* objects in OI_{CODL} with respect to one-dimensional partitions Φ associated to dimensions in $\mathcal{D}_i$ (note that here further clustering steps are *not* required); (*iii*.3) materialize clusters Λ_k in H_i within (ClustCube) data cube cells of C_i based on Φ.

ClustCube partial materialization approaches introduce so-called *complete traversal paths* over cuboid lattices, denoted by $\mathcal{T}$, as yet-another powerful mining capability of the proposed framework. Specifically, this interesting mining capability is inspired from [11], which focuses on the problem of *computing iceberg cubes efficiently*.

The traversal path model is consistent for both traditional **OLAP** and actual ClustCube data cubes. Given an N-dimensional data cube C with cuboid lattice $\mathcal{L}$ having $N + 1$ levels, a complete traversal path $\mathcal{T}$ over $\mathcal{L}$ is defined in terms of a *list of cuboids* C_i in $\mathcal{L}$ of kind: $\mathcal{T} = \langle C_0, C_1, ..., C_{|\mathcal{T}|-1}\rangle$, such that: (*i*) $|\mathcal{T}| \leq 2^N - 1$; (*ii*) $C_i \subseteq C$ for each i in $\{0, 1, ..., |\mathcal{T}| - 1\}$; (*iii*) for each pair $\langle i, j\rangle$ in $\{0, 1, ..., |\mathcal{T}| - 1\}$ such that $i < j$, $C_i \subseteq C_j$; (*iv*) for each pair $\langle i, j\rangle$ in $\{0, 1, ..., |\mathcal{T}| - 1\}$ such that $j = i + 1$, $C_i \prec C_j$ in $\mathcal{H}(\mathcal{L})$; (*v*) C_0 in $\mathcal{T}$ corresponds to the apex cuboid [3] $\varnothing$ in the **OLAP** schema of C (i.e., $C_0 \equiv \varnothing$); (*vi*) $C_{|\mathcal{T}|-1}$ in $\mathcal{T}$ corresponds to the base cuboid (i.e., the cube) C (i.e., $C_{|\mathcal{T}|-1} \equiv C$).

The idea behind the ClustCube partial materialization approach relies on observing that, very often, the administrator is interested in *analyzing/mining specialized portions of the whole cuboid lattice hierarchy* rather than exploring all the available cuboids, which tend to an exponential number, on the basis of the particular application scenario and the particular analysis/mining goals [37G,36G]. While this gracefully involves in ClustCube cubes having a *lower* occupancy in secondary memory, non-materialized cuboids can be still analyzed and mined, thanks to on-the-fly materialization primitives made available by most OLAP server platforms. It should be clear enough that ClustCube partial materialization approach is effective and efficient if the target traversal path T is carefully defined as representing sort of *preferred mining patterns*, i.e. T models a *heavily-mined portion* of the whole cuboid lattice. ClustCube framework supports both *regular*, i.e. the ones defined before, and *irregular traversal paths*, i.e. traversal path such that there exists *at least* one cuboid Ci that precedes *more* than one cudoid in the cuboid hierarchy $\mathcal{H}(\mathcal{L})$, i.e. $\exists\ Ci : \exists\ \langle Cj, Ck \rangle$ in T, $Ci \neq Cj \wedge Ci \neq Ck : Ci < Cj$ in $\mathcal{H}(\mathcal{L}) \wedge Ci < Ck$ in $\mathcal{H}(\mathcal{L})$. As meaningful examples, Figure 4 (*b*) shows a regular traversal path over the cuboid lattice $\mathcal{L}$ of a four-dimensional ClustCube cube C having $\mathcal{D} = \{A, B, C, D\}$ as dimension set, whereas Figure 4 (*c*) shows an irregular (*b*) traversal path over the same cuboid lattice.

Algorithm `BuildClustCube_PAR_BAS` and algorithm `BuildClust-Cube_PAR_DRI` implement the techniques ⟨PAR,BAS⟩ and ⟨PAR,DRI⟩, respectively. These algorithms are straightforward extensions of algorithms `BuildClust-Cube_FUL_BAS` and `BuildClustCube_FUL_DRI`, respectively, with the sole difference that, here, cuboids to be (partially) materialized are taken from the input complete traversal path T rather than the OLAP schema of C (i.e., the whole cuboid lattice structure $\mathcal{L}$) like in the previous solutions.

5   Experimental Results

In order to probe the effectiveness of the proposed ClustCube framework, we performed an extensive experimental evaluation and analysis of the ClustCube performance in computing (ClustCube) cubes on top of two well-known-in-literature data sets, namely the benchmark data set TCP-H [10] and the real-life data set Movie available at the *UCI KDD Archive* [6]. TCP-H stores data about an hypothetical supply company selling parts in several regions and throughout several classes of customers. Movie stores data on movies plus related information such as (movie) categories, casts, actors, studios, and so forth. In particular, we conducted two kinds of experiment aimed at probing the *efficiency* and the *effectiveness* of ClustCube framework, respectively. Each kind of experiment has been performed by adopting as core clustering algorithm $\mathcal{A}$ the following well-known-in-literature clustering algorithms: BIRCH [13], CLARANS [8] and DBSCAN [2]. Briefly, recall that these algorithms adhere to different-in-nature clustering approaches: BIRCH is a *hierarchical clustering algorithm*, CLARANS a *partition-based clustering algorithm*, and DBSCAN a *density-based clustering algorithm*. The variety of core algorithms employed in our experimental analysis conveys much more reliability to our experimental assessment.

As regards the data layer of our experimental framework, we considered several ClustCube cubes built from the two reference data sets TPC-H and Movie at different dimensionalities (ranging from two to ten dimensions). On top so-obtained ClustCube cubes, we finally performed our comprehensive experimental campaign. For what concerns ClustCube partial materialization approaches, we generated synthetic irregular complete traversal paths $\mathcal{T}$ by means of the following approach. Given the target cuboid lattice $\mathcal{L}$ of an N-dimensional data cube, traversal paths $\mathcal{T}$ have been *incrementally* generated from $\mathcal{L}$ by *randomly selecting*, for each level l of $\mathcal{L}$, a sub-set of cuboids of size $max\left\{1 \; ; \; \left\lfloor \binom{N}{l} \cdot \gamma \right\rfloor\right\}$, such that γ denotes an input parameter ranging over the interval [0:1]. γ finally determines how many cuboids per-level are added to $\mathcal{T}$, i.e. what we call as the *selectivity* of $\mathcal{T}$, denoted by σ. In our experimental framework, σ is measured in terms of a percentage value with respect to the total number of cuboids of $\mathcal{L}$, $2^N - 1$. We preferred irregular complete traversal paths over regular complete traversal paths as, clearly (see Section 4), the former represent an applicative setting that is more difficult than the latter, hence its testing provides us with a much more reliable assessment with respect to the case of handling regular complete traversal paths, still being capable to provide us with a yet trustworthy way of assessing the performance of the latter case. The latter claim is supported by the clear evidence stating that computational overheads due to computing partially-materialized ClustCube cubes with irregular complete traversal paths are (computational-complexity) upper bounds for computational overheads due to computing partially-materialized ClustCube cubes with regular complete traversal paths.

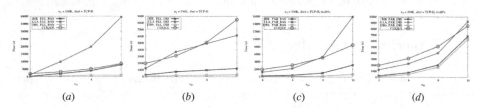

(a)   (b)   (c)   (d)

Fig. 5. ClustCube algorithms' performance in computing ClustCube cubes over the benchmark data set TPC-H with building techniques ⟨FUL,BAS⟩ (a), ⟨FUL,DRI⟩ (b), ⟨PAR,BAS⟩ (c) and ⟨PAR,DRI⟩ (d) versus the number of dimensions N, in comparison with CLIQUE

In our experimental assessment, we compared our proposed algorithms for computing ClustCube cubes against the case of using CLIQUE [1], which is a state-of-the-art approach for subspace clustering of high-dimensional data, as baseline algorithm. Is should be noted that CLIQUE keeps several points of similarity with our approach for computing ClustCube cubes. Indeed, during the subspace clustering phase, CLIQUE searches across the lattice of *all* possible subspaces of a given high-dimensional space in order to find subspaces that allow "*better*" clustering of actual data points rather than the original one (i.e., the clustering of the high-dimensional space). Thanks for this similarity, in our experimental campaign we used CLIQUE as an alternative method to compute ClustCube cubes along with the associated cuboid lattices, thanks to the meaningfully amenity of *associating subspaces of the target multidimensional data set with cuboids of ClustCube cubes*.

For what concerns the experimental parameters of clustering algorithms used as ClustCube core algorithms (i.e., BIRCH, CLARANS and DBSCAN) as well as CLIQUE, we tried our best to set the combinations of those possible values that provide us with the best performance, according to what suggested by the respective authors, or manually chosen based on a try-and-error procedure. This allowed us to obtain the best possible result from each clustering technique, hence achieving a fair experimental assessment.

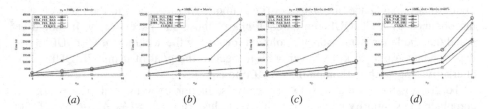

(a)   (b)   (c)   (d)

Fig. 6. ClustCube algorithms' performance in computing ClustCube cubes over the real-life data set Movie with building techniques ⟨FUL,BAS⟩ (a), ⟨FUL,DRI⟩ (b), ⟨PAR,BAS⟩ (c) and ⟨PAR,DRI⟩ (d) versus the number of dimensions N, in comparison with CLIQUE

As mentioned earlier, in the first class of experiments we tested the efficiency of ClustCube in computing (ClustCube) cubes having different number of dimensions. Figure 5 shows the variation of the *execution time* of ClustCube algorithms on the benchmark data set TCP-H for the different building techniques ⟨FUL,BAS⟩ (Figure 5 (a)), ⟨FUL,DRI⟩ (Figure 5 (b)), ⟨PAR,BAS⟩ (Figure 5 (c)) and ⟨PAR,DRI⟩ (Figure 5 (d)), when adopting as core algorithms BIRCH, CLARANS and DBSCAN, respectively, and ranging the number of dimensions N over the range [2:5], in comparison with the performance due to CLIQUE. Figure 6 shows instead the same experimental pattern of Figure 5 retrieved on the real-life data set Movie.

From the experimental results shown in Figure 5 and Figure 6, it follows that the ClustCube framework performs *consistently better* than CLIQUE *in most occurrences*, especially when BIRCH and DBSCAN are employed as core algorithms. CLIQUE exposes instead a better performance in some occurrences when CLARANS is used as core algorithm, an in particular with BAS building techniques. This because, in the latter case, all cuboids must be computed from the scratch (see Section 4), hence the inefficiency of CLARANS (e.g., [2]) decreases, in turn, the performance of ClustCube in computing the whole cuboid lattice. However, as it will be clearer in the following, BAS ClustCube building techniques have a smaller spectrum of applicative settings rather than DRI ClustCube building techniques, hence this drawback does not limit the effectiveness and the efficiency of ClustCube in popular applicative settings [12].

As regards the second class of experiments, they are intended to assess the effectiveness of ClustCube in terms of the *accuracy* of DRI ClustCube building techniques, which are clearly the most significant to be stressed due to fact BAS ClustCube building techniques compute all the cuboids of the lattice from the scratch (see Section 4). Indeed, as stressed before, due to their lower computational requirements, DRI ClustCube building techniques are the most popular to be exploited in a wide range of applicative settings [12]. Again, accuracy due to ClustCube computing algorithms has been compared with the one of CLIQUE.

In particular, accuracy of ClustCube has been measured by computing the *similarity* of the *approximate* clustering produced by DRI ClustCube building techniques with respect to the *exact* clustering produced by BAS ClustCube building techniques. Accuracy of CLIQUE has been instead measured by considering the similarity of the *approximate* clustering obtained by means of a CLIQUE run over the *entire* target high-dimensional space on *subspaces* (i.e., cuboids) with respect to the *exact* clustering obtained by a CLIQUE run on *each one* of those subspaces solely. It should be noted that the latter is a reasonable solution for testing the accuracy of CLIQUE in our experimental framework, and even with respect to the goal of using CLIQUE as an alternative approach for computing ClustCube cubes.

To convey more reliability to our analysis, in our experimental framework we computed the similarity between two different clustering $H_i = \{\Lambda_0, \Lambda_1, \cdots, \Lambda_{K-1}\}$ and $H_i' = \{\Lambda_0, \Lambda_1, \cdots, \Lambda_{K'-1}\}$, such that $K \neq K'$, according to two different approaches, namely *Clustering Similarity (Sim)* [12] and *Adjusted Mutual Information (AMI)* [7]. In more detail, *Sim* [12] is defined in terms of the so-called *degree of subsumption* of clustering H_i with respect to clustering H_i', denoted as $deg(H_i \subset H_i')$, as follows:

$$deg(H_i \subset H_i') = \frac{\sum_{i=1}^{K} max_{1 \leq j \leq K'}(\Lambda_i \cap \Lambda_j)}{\sum_{i=1}^{K} |\Lambda_i|} \tag{2}$$

Based on (2), *Sim* between clustering H_i and clustering H_i', denoted as $Sim(H_i, H_i')$, is defined as follows:

$$Sim(H_i, H_i') = \frac{deg(H_i \subset H_i') + deg(H_i' \subset H_i)}{2} \tag{3}$$

In our experimental evaluation, fixed an N-dimensional ClustCube cube C with the cuboid lattice $\mathcal{L}$, we measured the *average Sim*, denoted by $Sim_{AVG}(\mathcal{A}, l, N)$, of different clustering induced by groups of cuboids of $\mathcal{L}$, such that these cuboids are obtained by means of core algorithm $\mathcal{A}$ according to *different* ClustCube building techniques. In more detail, for each level l of $\mathcal{L}$, we computed the average similarity $Sim_{AVG}(\mathcal{A}, l, N)$ between the i-th cuboid at level l obtained by means of $\mathcal{A}$ and according to the DRI approach, denoted by $C_{l,i}(DRI, \mathcal{A})$, and the analogous cuboid obtained by means of $\mathcal{A}$ and according to the BAS approach, denoted by $C_{l,i}(BAS, \mathcal{A})$. $Sim_{AVG}(\mathcal{A}, l, N)$ is formally defined as follows:

$$Sim_{AVG}(\mathcal{A}, l, N) = \frac{\sum_{i=0}^{\binom{N}{l}} Sim\left(H\left(C_{l,i}(DRI, \mathcal{A})\right), H\left(C_{l,i}(BAS, \mathcal{A})\right)\right)}{\binom{N}{l}} \tag{4}$$

such that $H\left(C_{l,i}(DRI, \mathcal{A})\right)$ and $H\left(C_{l,i}(BAS, \mathcal{A})\right)$ denote the clustering associated to cuboids $C_{l,i}(DRI, \mathcal{A})$ and $C_{l,i}(BAS, \mathcal{A})$, respectively.

AMI [7] is instead defined on the basis of the *mutual information (MI)* between two clustering H_i and clustering H_i', denoted as $MI(H_i, H_i')$, which is defined as follows:

$$MI(H_i, H'_i) = \sum_{i=0}^{k-1} \sum_{j=0}^{k'-1} P(i,j) \cdot \log \frac{P(i,j)}{P(i)P'(j)} \tag{5}$$

such that: (i) $P(i,j) = \frac{|\Lambda_i \cap \Lambda'_j|}{n_T}$; (ii) $P(i) = \frac{|\Lambda_i|}{n_T}$; (iii) $P'(j) = \frac{|\Lambda'_j|}{n_T}$, being n_T the number of objects in the input data set. Based on (5), AMI between clustering H_i and clustering H'_i, denoted as $AMI(H_i, H'_i)$, is defined as follows:

$$AMI(H_i, H'_i) = \frac{MI(H_i, H'_i) - E\{MI(H_i, H'_i)\}}{\max\{\Gamma(H_i), \Gamma(H'_i)\} - E\{MI(H_i, H'_i)\}} \qquad (6)$$

such that: (i) $E\{MI(H_i, H'_i)\}$ denotes the *expected value* of MI; (ii) $\Gamma(H)$ ($\Gamma(H')$, respectively) denotes the *entropy* of clustering H (H', respectively), which is defined as follows:

$$\Gamma(H) = -\sum_{i=0}^{k-1} P(i) \cdot \log P(i) \qquad (7)$$

Finally, similarly to Sim_{AVG} (3), in our experimental framework we introduce the *average AMI*, denoted by $AMI_{AVG}(\mathcal{A}, l, N)$, which is defined as follows:

$$AMI_{AVG}(\mathcal{A}, l, N) = \frac{\sum_{i=0}^{\binom{N}{l}} AMI\Big(H\big(C_{l,i}(\text{DRI}, \mathcal{A})\big), H\big(C_{l,i}(\text{BAS}, \mathcal{A})\big)\Big)}{\binom{N}{l}} \qquad (8)$$

such that such that $H\big(C_{l,i}(\text{DRI}, \mathcal{A})\big)$ and $H\big(C_{l,i}(\text{BAS}, \mathcal{A})\big)$ denote the clustering associated to cuboids $C_{l,i}(\text{DRI}, \mathcal{A})$ and $C_{l,i}(\text{BAS}, \mathcal{A})$, respectively.

Figure 7 shows the variation of metrics Sim_{AVG} and AMI_{AVG} of DRI ClustCube algorithms on the benchmark data set **TPC-H** for the different building techniques $\langle$FUL,DRI$\rangle$ (Figure 7 (a)) and $\langle$PAR,DRI$\rangle$ (Figure 7 (b)), when adopting as core algorithms BIRCH, CLARANS and DBSCAN, respectively, and ranging the level l of the cuboid lattice $\mathcal{L}$ of a cube characterized by $N = 8$ dimensions, still in comparison with the metric results due to CLIQUE. In particular, Figure 7 (a) and Figure 7 (b) show the variation of Sim_{AVG} for the ClustCube building techniques $\langle$FUL,DRI$\rangle$ and $\langle$PAR,DRI$\rangle$, respectively, whereas Figure 7 (c) and Figure 7 (d) show the same results for the case of AMI_{AVG}. To give more details, synthetic irregular complete traversal paths exposing a selectivity equal to $\sigma = 20\%$ have been considered in these experiments. Finally, Figure 8 shows the same experimental pattern of Figure 7 retrieved on the real-life data set **Movie**.

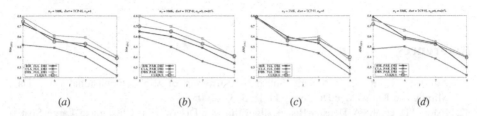

(a)   (b)   (c)   (d)

Fig. 7. ClustCube algorithms' accuracy in computing ClustCube cubes over the benchmark data set **TPC-H** with building techniques $\langle$FUL,DRI$\rangle$ (a) and $\langle$PAR,DRI$\rangle$ (b) according to Sim_{AVG}, and $\langle$FUL,DRI$\rangle$ (c) and $\langle$PAR,DRI$\rangle$ (d) according to AMI_{AVG}, versus the cuboid level l, in comparison with CLIQUE

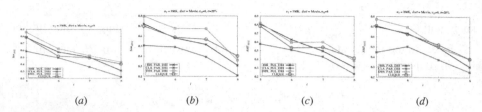

(*a*)	(*b*)	(*c*)	(*d*)

Fig. 8. ClustCube algorithms' accuracy in computing ClustCube cubes over the real-life data set Movie with building techniques ⟨FUL,DRI⟩ (*a*) and ⟨PAR,DRI⟩ (*b*) according to Sim_{AVG}, and ⟨FUL,DRI⟩ (*c*) and ⟨PAR,DRI⟩ (*d*) according to AMI_{AVG}, versus the cuboid level *l*, in comparison with CLIQUE

From the experimental results above, it clearly follows that the accuracy of Clust-Cube is indeed comparable to the one of CLIQUE, and, in some occurrences, it even outperforms CLIQUE's accuracy over both the data sets TPC-H and Movie. Summarizing, by inspecting both efficiency- and effectiveness-oriented experimental results, it should be clear enough that our proposed ClustCube framework allows us to obtain the same (or better) accuracy than CLIQUE, a state-of-the-art subspace clustering method, but with a *substantial improvement* over the performance of CLIQUE in terms of computational costs in most occurrences, still furnishing meaningfully OLAP capabilities in exploring and mining complex database objects. This is a remarkable contribution over state-of-the-art research.

6   Conclusions and Future Work

A complete OLAP-based framework for clustering and mining complex objects extracted from distributed database settings, called ClustCube, has been presented and experimentally assessed in this paper. ClustCube encompasses a spectrum of research innovations towards the seamless integration of consolidated clustering techniques over large databases and well-understood OLAP methodologies for accessing and mining (complex) objects. Future work is mainly oriented towards extending the proposed framework in order to make it capable of dealing with *classification issues over complex database objects*, beyond clustering issues like those investigated in this research.

References

1. Agrawal, R., et al.: Automatic Subspace Clustering of High Dimensional Data. Data Mining and Knowledge Discovery **11**(1), 5–33 (2005)
2. Ester, M., et al.: A Density-Based Algorithm for Discovering Clusters in Large Spatial Databases with Noise. Proc. of KDD **1996**, 226–231 (1996)
3. Gray, J., et al.: Data Cube: A Relational Aggregation Operator Generalizing Group-by, Cross-Tab, and Sub Totals. Data Min. and Know. Disc. **1**(1), 29–53 (1997)
4. Han, J.: Towards On-line Analytical Mining in Large Databases. ACM SIGMOD Record **27**(1), 97–107 (1998)

5. Hinneburg, A., Keim, D.A.: Clustering methods for large databases: from the past to the future. In: Proc. of ACM SIGMOD 1999, p. 509 (1999)
6. University of California, Irvine, UCI KDD Archive. http://kdd.ics.uci.edu/
7. Meila, M.: Comparing Clusterings—an Information Based Distance. Journal of Multivariate Analysis 98(5), 873–895 (2007)
8. Ng, R.T., Han, J.: CLARANS: A Method for Clustering Objects for Spatial Data Mining. IEEE Transactions on Knowledge and Data Engineering 14(5), 1003–1016 (2002)
9. Sheikholeslami, G., et al.: WaveCluster: A Wavelet Based Clustering Approach for Spatial Data in Very Large Databases. VLDB Journal 8(3–4), 289–304 (2000)
10. Transaction Processing Council, TPC Benchmark H. http://www.tpc.org/tpch/
11. Xin, D., et al.: Computing Iceberg Cubes by Top-Down and Bottom-Up Integration. IEEE Trans. on Know. and Data Eng. 19(1), pp. 111—126 (2007)
12. Yin, X., et al.: CrossClus: User-Guided Multi-Relational Clustering. Data Mining and Knowledge Discovery 15(3), 321–348 (2007)
13. Zhang, T., et al.: BIRCH: A New Data Clustering Algorithm and Its Applications. Data Mining and Knowledge Discovery 1(2), 141–182 (1997)
14. Cuzzocrea, A., Serafino, P.: ClustCube: an olap-based framework for clustering and mining complex database objects. In: Proc. of ACM SAC 2011, pp. 976–982 (2011)

Outlier and Anomaly Detection

Contextual Anomaly Detection
Using Log-Linear Tensor Factorization

Alpa Jayesh Shah[1]([✉]), Christian Desrosiers[1], and Robert Sabourin[2]

[1] Department of Software and IT Engineering, Ecole de technologie supérieure,
Montreal, Canada
`AShah@livia.etsmtl.ca`, `Christian.Desrosiers@etsmtl.ca`
[2] Department of Automated Production Engineering,
Ecole de technologie supérieure, Montreal, Canada
`Robert.Sabourin@etsmtl.ca`

Abstract. This paper presents a novel approach for the detection of contextual anomalies. This approach, based on log-linear tensor factorization,
considers a stream of discrete events, each representing the co-occurence
of contextual elements, and detects events with low-probability. A parametric model is used to learn the joint probability of contextual elements,
in which the parameters are the factors of the event tensor. An efficient
method, based on Nesterov's accelerated gradient ascent, is proposed to
learn these parameters. The proposed approach is evaluated on the low-
rank approximation of tensors, the prediction of future of events and the
detection of events representing abnormal behaviors. Results show our
method to outperform state of the art approaches for these problems.

Keywords: Contextual anomaly detection · Tensor factorization · Low-
rank approximation · Future event prediction

1 Introduction

The recent commercialization of technologies for the real-time identification,
location, and tracking of people and objects has opened the door to various
new applications in domains such as safety, logistics and retail. Among these
applications, the real-time detection of malicious or abnormal behaviors is of
critical importance to the safety of the population.

The approaches proposed for this problem over the years can be roughly
divided in two categories: the ones based on probabilistic generative models, and
those using trajectory patterns. Approaches in the first category use a generative
model, for instance based on Markov [7] or Hierarchical Dirichlet [9] processes, to
determine the likelihood of a sequence of observed events/actions, and consider
as abnormal behaviors the ones with a low probability. On the other hand, the
second category of methods represent behaviors as trajectories through space,
and considers as anomalies the trajectories that are significantly different from
commonly observed ones. Trajectories can be encoded in various ways, such
as sequences of points [16] or cubic splines [18]. Moreover, several approaches

© Springer International Publishing Switzerland 2015
T. Cao et al. (Eds.): PAKDD 2015, Part II, LNAI 9078, pp. 165–176, 2015.
DOI: 10.1007/978-3-319-18032-8_13

have been proposed to model the class of normal trajectories, for instance one-class SVM (OCSVM) [16], Gaussian Mixture Model [18], sparse coding [14] and frequent sub-sequence mining [10].

While these solutions are adequate in small and controlled environments, in which well defined activities occur, they usually perform poorly in large and dynamic environments, where the same sequence of events is almost never observed twice. In such complex environments, the context of events (such as location, duration, person ID, type of job, etc.) is often more important than their sequence. Thus, a person might take a slightly different route to go to the office, so analyzing the exact trajectory would likely result in many false positives. Moreover, even though the usual route is taken, this person's behavior can be abnormal if he/she goes to the office at an odd time (e.g., after 9 pm) or on a odd day (e.g., Sunday). This behavior might however be considered normal for other employees, such as security agents working on the evening or weekend shifts.

Although several approaches have been proposed to include contextual information in generative models, for example [3] and [5], these approaches are limited to specific contextual dimensions, such as the duration of events, and are unsuitable for complex environments. On the other hand, tensors have been recognized as a powerful and efficient method to model complex contextual information, and have been used in diverse applications like item recommendation [17] and analyzing email exchanges [1]. Recently, tensor factorization [4,12,21] has been explored as a novel way to detect anomalies, for instance, by tracking the reconstruction error over time [19], detecting outliers in the factor subspace [8,20], or decomposing a tensor as the sum of a low-rank component and a sparse residual representing the anomaly [11].

In this paper, we propose a novel method based on log-linear tensor factorization to detect contextual anomalies in large and complex environments. The advantages of this method are as follows:

1. Unlike existing tensor factorization approaches, which focus on detecting global anomalies [11,19] or distance-based outliers [8,20], our method learns the joint distribution of contextual dimensions. This allows it to evaluate the true probability of incoming events and mark low probability ones as abnormal. Our method also has the ability to detect specific types of anomalies efficiently, by using the probability of a given dimension conditioned over the other ones.

2. While most tensor factorization approaches are based on a linear model, our method uses a log-linear model, which can learn more complex relations in the data. Moreover, the proposed model implicitly enforces non-negativity in the tensor, a useful property when dealing with count data. In comparison, state of the art factorization techniques like Non-negative Tensor Factorization (NTF) [21] and Alternating Poisson Regression (APR) [4] impose non-negativity by constraining the factors, making the inference process more complex.

3. The proposed method uses an efficient inference strategy, based on Nesterov's accelerated gradient ascent, which has a complexity comparable to the state of the art Alternating Least Square (ALS) method [17], but offers more flexibility (e.g., ALS is limited to linear models and does not impose non-negativity).

The rest of this paper is divided as follows. In Section 2, we describe our proposed model, its inference strategy, and the method used to detect anomalies. Section 3 then evaluates our model on the tasks of approximating tensors using a small number of parameters, predicting the occurrence of future events and detecting abnormal events from their context. Finally, we summarize our contributions and results in Section 4.

2   The Proposed Method

2.1   Model Description

We model the multi-dimensional context of an event using a set of discrete random variables $\{X_1, \ldots, X_D\}$, representing the identifier (ID) of contextual elements like person, zone, time of day, etc. Each variable X_j has domain $\Omega_j = \{1, \ldots, N_j\}$, and we denote as $X_j = i_j$ the observation of element $i_j \in \Omega_j$ for dimension j. To simplify the notation, we use x_{i_j} as shorthand for this observation. For example, if the first dimension represents people, then x_{i_1} means the observation of person i_1, from a group of N_1 people, in the event.

We suppose that the observation of events depend on a set of latent factors $\mathcal{Z} = \{Z_1, \ldots, Z_D\}$, providing high-level information about the contextual elements. We define as $\boldsymbol{z}_{i_j} \in \mathbb{R}^K$ the latent factor vector corresponding to the i_j-th element of dimension j, where K is a user-supplied parameter. Using the previous example, $\boldsymbol{z}_{i_1}$ would be the latent factor vector of person i_1.

To model the joint probability of contextual elements, we use the following log-linear model:

$$p(x_{i_1}, \ldots, x_{i_D} \mid \mathcal{Z}) = \frac{\exp\left(\langle \boldsymbol{z}_{i_1}, \ldots, \boldsymbol{z}_{i_D} \rangle\right)}{\displaystyle\sum_{i_1'=1}^{N_1} \cdots \sum_{i_D'=1}^{N_D} \exp\left(\langle \boldsymbol{z}_{i_1'}, \ldots, \boldsymbol{z}_{i_D'} \rangle\right)}, \tag{1}$$

where $\langle \boldsymbol{z}_{i_1}, \ldots, \boldsymbol{z}_{i_D} \rangle$ is the inner product between D vectors of size K:

$$\langle \boldsymbol{z}_{i_1}, \ldots, \boldsymbol{z}_{i_D} \rangle = \sum_{k=1}^{K} z_{i_1,k} \cdot z_{i_2,k} \cdot \cdots \cdot z_{i_{D-1},k} \cdot z_{i_D,k}. \tag{2}$$

To learn the model parameters, we suppose that a set $\mathcal{X}$ of M observed events $(x_{i_1}, \ldots, x_{i_D})$ is available. This set can also be represented as a D-dimension tensor, in which element $(i_1, \ldots, i_D)$ contains the number of events of $\mathcal{X}$ having context $(i_1, \ldots, i_D)$. We call this structure the *event tensor*.

Considering the events as i.i.d., the observation likelihood of the events in $\mathcal{X}$ corresponds to

$$p(\mathcal{X} \mid \mathcal{Z}) = \prod_{(x_{i_1}, \ldots, x_{i_D}) \in \mathcal{X}} p(x_{i_1}, \ldots, x_{i_D} \mid \mathcal{Z})$$

$$= \prod_{i_1=1}^{N_1} \cdots \prod_{i_D=1}^{N_D} p(x_{i_1}, \ldots, x_{i_D} \mid \mathcal{Z})^{M_{i_1, \ldots, i_D}}, \tag{3}$$

where $M_{i_1,\ldots,i_D}$ is the number of events of $\mathcal{X}$ with context $(i_1,\ldots,i_D)$. Since the number of events is small compared to the size of the multi-dimensional event space (i.e., $N_1 \times \ldots \times N_j$), only a few of these values are expected to be non-zero. In other words, the event tensor should be very sparse. To regularize the solution, we suppose the factor vectors as independent and following a zero-mean normal distribution with uniform variance:

$$p(\mathcal{Z}) = \prod_{j=1}^{D} \prod_{i_j=1}^{N_j} \mathcal{N}(\boldsymbol{z}_{i_j}; \boldsymbol{0}, \sigma^{-1}I). \tag{4}$$

The latent factors $\mathcal{Z}$ are found using the *maximum a posteriori* (MAP) estimate, which corresponds to maximizing the following cost function:

$$\begin{aligned}
f(\mathcal{Z}) &= \log p(\mathcal{X} \,|\, \mathcal{Z}) + \log p(\mathcal{Z}) \\
&= \sum_{i_1=1}^{N_1} \cdots \sum_{i_D=1}^{N_D} M_{i_1,\ldots,i_D} \log p(x_{i_1}, \ldots, x_{i_D} \,|\, \mathcal{Z}) - \frac{\sigma}{2} \sum_{j=1}^{D} \sum_{i_j=1}^{N_j} ||\boldsymbol{z}_{i_j}||^2.
\end{aligned} \tag{5}$$

Note that this is equivalent to minimizing the KL divergence between the model and empirical distribution of events, as done in [4].

Since the cost function of Eq. (5) is both non-linear and non-concave, obtaining globally optimum parameters is an intractable problem. Therefore, we must optimize it using an iterative approach like the gradient ascent method, which has a linear convergence rate. However, because this function is concave with respect to *each* factor vector, we can instead use Nesterov's accelerated gradient method [15] for which the convergence rate is quadratic.

Unlike gradient ascent, Nesterov's method performs two different steps at each iteration. The first step is a simple gradient ascent step of size η from the current solution $\boldsymbol{z}_{i_j}^{(t)}$ to a intermediate solution $\boldsymbol{y}_{i_j}^{(t+1)}$:

$$\boldsymbol{y}_{i_j}^{(t+1)} = \boldsymbol{z}_{i_j}^{(t)} + \eta \frac{\partial f}{\partial \boldsymbol{z}_{i_j}}(\mathcal{Z}^{(t)}). \tag{6}$$

Let $R_{i_1,\ldots,i_D}(\mathcal{Z})$ be the difference between the observed number of events in context $(i_1,\ldots,i_D)$ and the expected one according to parameters $\mathcal{Z}$:

$$R_{i_1,\ldots,i_D}(\mathcal{Z}) = M_{i_1,\ldots,i_D} - M \cdot p(x_{i_1}, \ldots, x_{i_D} \,|\, \mathcal{Z}). \tag{7}$$

The gradient with respect to $\boldsymbol{z}_{i_j}$ is given by

$$\frac{\partial f}{\partial \boldsymbol{z}_{i_j}}(\mathcal{Z}) = \sum_{i_1=1}^{N_1} \cdots \sum_{i_{j-1}=1}^{N_{j-1}} \sum_{i_{j+1}=1}^{N_{j+1}} \cdots \sum_{i_D=1}^{N_D} R_{i_1,\ldots,i_D}(\mathcal{Z}) \cdot \hat{\boldsymbol{z}}_{i_j} - \sigma \boldsymbol{z}_{i_j}, \tag{8}$$

where $\hat{\boldsymbol{z}}_{i_j}$ is defined as the Hadamard (i.e., element-wise) product of all factor vectors expect the one of dimension j:

$$\hat{\boldsymbol{z}}_{i_j} = \left(\boldsymbol{z}_{i_1} \circ \ldots \circ \boldsymbol{z}_{i_{j-1}} \circ \boldsymbol{z}_{i_{j+1}} \circ \cdots \circ \boldsymbol{z}_{i_D}\right), \tag{9}$$

Algorithm 1. Parameter inference using Nesterov's method

Input: The event tensor $\mathcal{X}$ and latent factor size K;
Input: The regularization parameter σ and initial gradient step size η;
Output: The factor matrices $\mathbf{Z}_1, \ldots, \mathbf{Z}_D$;

1 **for** $j = 1, \ldots, D$ **do**
2 $\quad\mid\quad$ Initialize the rows of $\mathbf{Z}_j^{(0)}$ following $\mathcal{N}(\mathbf{0}, \sigma^{-1}I)$;
3 $\quad\mid\quad$ $\mathbf{Y}^{(0)} := \mathbf{Z}_j^{(0)}$;

4 Set $t := 0$ and $a := 0$;

5 **while** $f(\mathcal{Z}^{(t)})$ *not converged* **do**

6 $\quad\mid\quad$ Reconstruct estimated tensor $\hat{\mathcal{X}}$ (unfolded along dim. 1):

$$\hat{\mathbf{X}}_{(1)} := \tfrac{1}{T} \exp\left(\mathbf{Y}_1^{(t)}\left(\mathbf{Y}_D^{(t)} \odot \ldots \odot \mathbf{Y}_2^{(t)}\right)\right),$$

$\quad\mid\quad$ where T is such that sum of elements in $\hat{\mathbf{X}}_{(1)}$ is $M = |\mathcal{X}|$;

7 $\quad\mid\quad$ **for** $j = 1, \ldots, D$ **do**
8 $\quad\mid\quad\mid\quad$ StepOK := *false* ;
9 $\quad\mid\quad\mid\quad$ **while** StepOK $= false$ **do**
10 $\quad\mid\quad\mid\quad\mid\quad$ $G_j := \left(\mathbf{X}_{(j)} - \hat{\mathbf{X}}_{(j)}\right)\left(\mathbf{Y}_D^{(t)} \odot \ldots \odot \mathbf{Y}_{j+1}^{(t)} \odot \mathbf{Y}_{j-1}^{(t)} \odot \ldots \odot \mathbf{Y}_1^{(t)}\right) - \sigma \mathbf{Y}_j^{(t)}$;
11 $\quad\mid\quad\mid\quad\mid\quad$ $\mathbf{Z}_j^{(t+1)} := \mathbf{Y}_j^{(t)} + \eta\, G_j$;
12 $\quad\mid\quad\mid\quad\mid\quad$ **if** $f(\mathbf{Z}_j^{(t+1)}) \leq f(\mathbf{Y}_j^t) + \mathrm{Tr}\left(G_j^\top\left(\mathbf{Z}_j^{(t+1)} - \mathbf{Y}_j^t\right)\right) + \frac{1}{2\eta}\|\mathbf{Z}_j^{(t+1)} - \mathbf{Y}_j^t\|_F^2$
$\quad\mid\quad\mid\quad\mid\quad$ **then** $\eta := 0.5\,\eta$;
13 $\quad\mid\quad\mid\quad\mid\quad$ **else** StepOK := *true* ;

14 $\quad\mid\quad$ $a_{t+1} := \frac{1}{2} + \frac{1}{2}\sqrt{1 + 4\,a_t^2}$;
15 $\quad\mid\quad$ **for** $j = 1, \ldots, D$ **do**
16 $\quad\mid\quad\mid\quad$ $\mathbf{Y}_j^{(t+1)} := \frac{a_{t+1} + a_t - 1}{a_{t+1}} \mathbf{Z}_j^{(t+1)} - \frac{a_t - 1}{a_{t+1}} \mathbf{Z}_j^{(t)}$;

17 $\quad\mid\quad$ $t := t + 1$;

18 **return** $\mathcal{Z} = \{\mathbf{Z}_1^{(t)}, \ldots, \mathbf{Z}_D^{(t)}\}$;

The second step then finds the next solution $z_{i_j}^{(t+1)}$ as a convex combination of the two last intermediate solutions:

$$z_{i_j}^{(t+1)} = (1 - \gamma_t)\, y_{i_j}^{(t+1)} + \gamma_t\, y_{i_j}^{(t)}, \tag{10}$$

where γ_t are constants controlling the search momentum (e.g., see Algorithm 1).

2.2 Algorithm Summary and Complexity

The complete inference process is summarized in Algorithm 1. For a greater efficiency, we group latent factor vectors of each dimension j in a single matrix $\mathbf{Z}_j$, and use standard matrix operations. We start by initializing the factor matrices randomly following the prior distribution of parameter σ (lines 1-3). Then, at each iteration of Nesterov's method, the tensor $\hat{\mathcal{X}}$ of expected counts is reconstructed using the current intermediate factors $\mathbf{Y}_j$ (line 6). Operator $\odot$ corresponds to the

Khatri-Rao product and $\mathbf{X}_{(j)}$ denotes the unfolding of tensor $\mathcal{X}$ along dimension j (see [12] for more information). For each dimension j, the gradient is then computed using the residual between the observed and expected counts (line 10), and used to update factor matrix $\mathbf{Z}_j$ (line 11). If the step size η is too large, the solution may diverge. A strategy is thus added to detect such problem and adjust η automatically (lines 12-13), thereby eliminating the need to tune η manually. This strategy is known as *backtracking line search*. The momentum constant and intermediate factors are finally updated as per Nesterov's method (lines 14-16). The process is repeated until converge is attained, or a maximum number of iterations is exhausted.

The computational complexity of this algorithm is as follows. For each iteration, reconstructing the expected tensor $\hat{\mathcal{X}}$ take $O(K \cdot \prod_{j=1}^{D} N_j)$ operations. Likewise, updating the latent factors for each dimension can be done in $O(K \cdot \prod_{j=1}^{D} N_j)$. Therefore, the total complexity is $O(T_{\max} \cdot K \cdot D \cdot \prod_{j=1}^{D} N_j)$, where $T_{\max}$ is the maximum number of iterations.

2.3   Abnormal Event Detection

We use the latent factors learned during training to evaluate the joint probability of new events in real-time, and mark as abnormal those that have a low probability. Let θ be a given probability threshold, an event $(x_{i_1}, \ldots, x_{i_D})$ will be marked as abnormal if $p(x_{i_1}, \ldots, x_{i_D} \mid \mathcal{Z}) < \theta$. By pre-computing the denominator of Eq. (1), evaluating the joint probability of an incoming event requires only $O(D \cdot K)$ operations.

To detect specific types of anomalies, we can instead evaluate the probability of a single dimensional value, conditioned on all other dimensional values. For example, we could evaluate the probability that the event occurs in a certain zone, given the person, day, and time of day corresponding to that event. If the conditional probability of the observed zone is much lower than that of other zones, the event would then be marked as abnormal. Suppose, without loss of generality that the query dimension is $j = 1$. The probability of x_{i_1}, conditioned on all other dimensions, is given by

$$p(x_{i_1} \mid x_{i_2}, \ldots, x_{i_D}, \mathcal{Z}) \;=\; \frac{\exp\left(\langle \mathbf{z}_{i_1}, \ldots, \mathbf{z}_{i_D} \rangle\right)}{\sum\limits_{i_1=1}^{N_1} \exp\left(\langle \mathbf{z}_{i_1}, \ldots, \mathbf{z}_{i_D} \rangle\right)}. \tag{11}$$

To evaluate the computational complexity of this query, we note that the inner product can be decomposed as $\langle \mathbf{z}_{i_1}, \ldots, \mathbf{z}_{i_D} \rangle = \langle \mathbf{z}_{i_1}, \hat{\mathbf{z}}_{i_1} \rangle$. Thus, if we pre-compute $\hat{\mathbf{z}}_{i_1}$, each inner product computation has a time complexity in $O(K)$, where K is the size of the latent subspace. Since we have to compute N_1 of these inner products, the total cost of evaluating the query is in $O(N_1 \cdot K)$.

3   Experiments

We evaluated the performance of our method by conducting three sets of experiments, related to the low-rank approximation of tensors, the prediction of future events, and the detection of contextual anomalies.

3.1   Low-Rank Approximation

The goal of this first experiment is to measure the ability of our method to fit the event tensor using a small number of parameters.

Experimental Design. We generated two synthetic datasets, each containing 11 sparse 3D tensors of size 100×100×100 drawn from two types of distributions: log-linear and Poisson. Each tensor was generated using three sets of 100 latent factor vectors of size $K = 20$, drawn from a Gaussian prior in the case of log-linear tensors and a Gamma prior for Poisson tensors. The parameters of these priors were selected to have a sparsity level between 70% and 90%, and the resulting tensors normalized to have a total number of events equal to $M = 10^6$.

We used the first tensor to tune the regularization parameter σ of our method, and the remaining 10 to evaluate its average performance in terms of Mean Absolute Error (MAE). Factor sizes of $K = 5, 10, 15, 20$ were tested. We compared our Log-Linear Tensor Factorization (LLTF) method to two state-of-the-art factorization approaches: Alternating Poisson Regression (APR) [4] and Non-Negative Tensor Factorization (NTF) [21]. The Matlab Tensor Toolbox v2.5 [2] implementation of these methods was used.

Results and Discussion. Figure 1 (left) gives the average MAE obtained by the three tested approaches on the Poisson (blue curves) and log-linear (green curves) tensors. As expected, the reconstruction error decreases with higher values of K. Moreover, our LLTF method outperforms APR and NTF for log-linear tensors, especially for $K = 20$ where LLTF obtains an average MAE 2.26 times smaller than APR and 2.41 smaller than NTF. For Poisson data, LLTF performs as well as APR, even though this data is tailored to APR's Poisson model not LLTF's log-linear one. Since it is not designed for sparse count data, NTF obtains a lower performance than LLTF and APR for both log-linear and Poisson data.

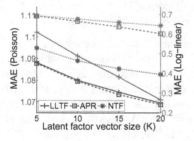 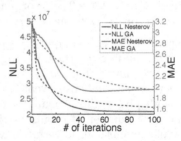

Fig. 1. (left) Average MAE obtained by LLTF, APR and NTF on 10 Poisson (blue curves) and log-linear (green curves) tensors, for latent factor sizes of $K = 5, 10, 15, 20$. (**right**) Convergence of our Nesterov-based method on a sample tensor, compared to simple gradient ascent (GA).

Figure 1 (right) illustrates the convergence rate of our Nesterov-based method on a sample tensor, compared to simple gradient ascent (GA). We see that the convergence in terms of Negative Log-Likelihood (NLL) (blue curves) is attained within 60 iterations, with an average time of 0.3 seconds per iteration, whereas GA has not converged after 100 iterations. In contrast to our method, APR takes on average 1.3 seconds per iteration using the same hardware, and requires over 1000 iterations to converge.

3.2   Future Event Prediction

The second experiment evaluates how well our method can predict the number of future events occurring in a given context. As the event tensor is sparse, this experiment measures the ability of the model to predict events that were not observed in the training data.

Experimental Design. Two real-life datasets were used for this experiment.

- **Reality Mining** [6]: Contains the tracking information of 106 students and faculty members from the MIT Media Laboratory and Sloan Business school, collected through their cell-phones in 2004-2005. From the 106 participants, we picked 87 students as the remaining ones had either less than 7 days of data or no data at all. For the locations, we used the 1027 unique cell-tower IDs, corresponding to the attribute *areaID.cellID* in the data. The timestamps of the tracking events were encoded using 24 discrete values, one for each hour of the day. Combining these three dimensions, we obtained a $87 \times 1027 \times 24$ tensor, each cell containing the number of times a person was recorded as being near a given cell tower, at a given time of the day.
- **Geo-Life Taxi Trajectories** [22]: Contains the GPS trajectories of 10,357 taxis during the period of Feb. 2 to Feb. 8, 2008 within the city of Beijing. From the 10,357 taxis in the data, we selected 259 taxis as the remaining ones had either less than 5000 records of temporal locations or no records at all. Since most records are located near the center of Beijing, we converted the Cartesian coordinates (longitude and latitude) to log polar ones (log radius, angle θ) using the city's center as origin. We divided θ into 12 bins of 30° each, and the log radius into 10 bins, giving a total of 120 zones. Similar to the Reality Mining dataset, we encoded the timestamps using 24 discrete values, one for each hour of the day. Combining these three dimensions, we obtained a $259 \times 120 \times 24$ tensor, each cell containing the number of times a taxi was recorded as being in a specific zone, at a given hour of the day.

We split the datasets temporally, putting the first 60% of each person or taxi's events in the training set, the following 20% in the validation set, which was used to tune the regularization parameter σ, and the remaining 20% in the test set. We predicted the number of events in the test set for each context (i.e., tensor cell) by multiplying the probability obtained for this context during training with the total number of events in the test set. We evaluated the prediction accuracy of our method, in terms of MAE, RMSE (Root Mean Squared Error) and NLL, and compared it once again to with APR and NTF. Note that the Poisson distribution used in APR is specifically tailored to model count data. A latent factor size of $K = 10$ was used for all three methods.

Results and Discussion. The prediction accuracy of the three tested methods on the Reality Mining and Taxi datasets is detailed in Table 1. We see that LLTF outperforms APR and NTF, on both datasets and all three performance metrics. Thus, LLTF obtains a MAE 2.32 times lower than APR in the Reality Mining dataset, and 2.10 times lower in the Taxi dataset. We also note that APR obtained infinite NLL values. This is because it gave a zero probability to the events in the test set that were not observed in the training set. By regularizing the factors, our model can better predict such unobserved events.

Table 1. Prediction error obtained by our LLTF method, as well as the NTF and APR approaches, on the Reality Mining (with $\sigma = 6$) and Taxi (with $\sigma = 4$) datasets. NLL values have been scaled for convenience.

(a) **Reality Mining**

Metric	LLTF	APR [4]	NTF [21]
MAE	0.384	0.892	0.894
RMSE	8.906	19.134	19.743
NLL $\times 10^6$	7.270	∞	7.814

(b) **GeoLife Taxi**

Metric	LLTF	APR [4]	NTF [21]
MAE	1.691	3.549	3.084
RMSE	15.319	19.657	24.220
NLL $\times 10^6$	9.158	∞	9.719

3.3   Abnormal Event Detection

In this last experiment, we assess the usefulness of our method to detect abnormal events from their context.

Experimental Design. Once again, the Reality Mining and GeoLife Taxi Trajectory datasets were considered for this experiment. Three types of synthetic anomalies were generated.

- **Swap People:** This type of anomalies simulates a person (or taxi) behaving like someone different. To generate such anomalies, we first computed the KL divergence between the event distribution (i.e., number of events for each time and zone) of all pairs of persons. We then used weighted sampling to pick random person pairs, those with a higher KL divergence having a greater chance of being selected, and swapped the *personID* of all events involving the corresponding two persons.
- **Swap Times:** This type of anomalies corresponds to a person going to the same places, but at odd times (hours of the day). To generate these anomalies, we randomly picked a person and computed the KL divergence between the event distribution (number of events for each zone) of all time pairs, for this person. Once more, the time pairs were picked using the KL divergence as sampling weight, and the *timeID* of these pairs were swapped in all the events of the selected person.
- **Swap Zones:** This last type anomaly corresponds to a person being active at the same times, but in unusual zones. These anomalies were generated using the same strategy as in *Swap Times*, except that the KL divergence between the time distribution of events of each zone pair was considered.

We split the dataset as in the previous experiment, and used the training data as is to learn the distribution of normal events. For both the validation and testing sets, we generated 10 different sets of random anomalies, using the following procedure. For *Swap People* anomalies, we swapped 5 pairs of people/taxis, while for *Swap Times* and *Swap Zones* anomalies, we randomly picked 3 persons and, for each of them, swapped 3 pairs of *timeID* or *zoneID*. The parameters of the tested methods were tuned using the average Area Under the ROC Curve (AUC) obtained over the 10 validation anomaly sets. Note that this tuning step was necessary to have a fair comparison between the tested methods, but such validation anomalies may not be available in real-life applications. Finally, the performance of the tuned methods was evaluated as the mean AUC obtained over the 10 test anomaly sets.

We tested four variations of our proposed approach. In the first one, called LLTF-Joint, the ROC curves are generated by evaluating the joint probability of test examples, as defined in Eq. (1), and then computing the precision/recall for increasing probability thresholds. The other three methods, denoted by LLTF-D, where $D = \{\text{Person}, \text{Time}, \text{Zone}\}$, instead use the conditional probability of a single dimension D given the other two dimensions, as described in Eq. (11).

We compared the performance of these methods with two well-known unsupervised anomaly detection approaches: One-Class Support Vector Machines (OCSVM) [16] and Kernel Density Estimation (KDE) [13]. For both of these methods, the discrete dimensional values (e.g., *personID*) were first converted to binary features using an indicator function, giving a total of 1138 binary features for Reality Mining and 403 binary features for Taxi. PCA was then applied to these binary features, using a percentage of variance value of 95%, and the resulting components were normalized to have uniform variance. For OCSVM, two parameters required tuning: ν, which controls the fraction of training examples allowed outside the learned region, and the RBF kernel parameter γ. The signed distance to the hyperplane was used to evaluate the normality of test examples, while computing the ROC curves. For KDE, we evaluated the probability of a test example $\boldsymbol{x}$ (projected in PCA space) as

$$p(\boldsymbol{x}) \propto \frac{1}{N} \sum_{n=1}^{N} \exp\left\{ -\frac{1}{h} ||\boldsymbol{x} - \boldsymbol{x}_n||^2 \right\}, \tag{12}$$

where $\boldsymbol{x}_n$ are the training examples (in PCA space) and h is the kernel bandwidth parameter, tuned on the validation data.

Results and Discussion. The mean ROC curves (and corresponding AUC values), computed over the 10 test sets of each anomaly type, are shown in Figure 2. Except for the *Swap Time* anomalies, our LLTF-Joint method obtained a higher mean AUC than OCSVM and KDE. In particular, the AUC of LLTF-Joint is 9% to 35% higher than OCSVM, and 10% to 37% higher than KDE, for *Swap People* anomalies. Furthermore, the conditional probability model can improve the detection of specific types of anomalies. For instance, LLTF-Person obtained mean AUC values of 0.97 and 0.93 on the Swap People anomalies, compared to 0.95 and 0.92 for LLTF-Joint. Similarly, LLTF-Time obtained a mean AUC of 0.85 on the Swap Time anomalies of the Reality Mining dataset, whereas this value was only 0.63 for LLTF-Joint.

The proposed method is also faster and more robust than OCSVM and KDE. Thus, training LLTF for the Taxi dataset took less than 10 minutes on a Quad-Core AMD 2.3 GHz processor with 8 GB of RAM, whereas training OCSVM on this dataset required over 7 hours using the same hardware (no training is necessary for KDE). Likewise, predicting anomalies in the test set took on average 0.03 *ms* for LLTF, compared to 61 *ms* for OCSVM and 7 *ms* for KDE. Moreover, while the best parameters for each type of anomaly (as selected in validation) varied greatly in OCSVM and KDE, our method was more robust to the choice of parameters: $K = 25$, $\sigma = 8$ was used for *all* anomalies in the RM dataset, and $K = 15$, $\sigma = 6$ for *all* anomalies in Taxi.

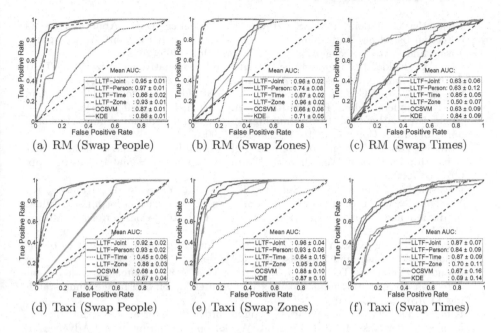

(a) RM (Swap People) (b) RM (Swap Zones) (c) RM (Swap Times)

(d) Taxi (Swap People) (e) Taxi (Swap Zones) (f) Taxi (Swap Times)

Fig. 2. Average ROC curves and AUC obtained by our LLTF, as well as the OCVSM and KDE approaches, on 10 sets of synthetic anomalies generated from the Reality Mining (RM) and Taxi datasets. Three types of anomalies are considered: swapping the events of two people/taxis, and swapping the time/zone of events corresponding to a person/taxi.

4   Conclusion

We presented a new approach, based on log-linear tensor factorization, for the detection of contextual anomalies. A parametric model was proposed to estimate the joint probability of dimensional values, in which the parameters are the factors of an event count tensor. To learn the factors, an efficient technique based on Nesterov's accelerated gradient ascent was presented. The proposed approach was evaluated on three problems: the low-rank approximation of synthetic tensors, the prediction of future of events in real-life data and the detection of events representing abnormal behaviors. Results show our method to outperform state of the art approaches for these problems, while being faster and more robust than these approaches. As future work, we will investigate the use of additional dimensions in the tensor, for instance to model the duration and sequence of events, and extend the method to perform online learning.

References

1. Bader, B.W., Berry, M.W., Browne, M.: Discussion tracking in enron email using parafac. In: Survey of Text Mining II, pp. 147–163. Springer (2008)
2. Bader, B.W., Kolda, T.G., et al.: Matlab tensor toolbox version 2.5 (2012). http://www.sandia.gov/~tgkolda/TensorToolbox/

3. Benezeth, Y., Jodoin, P.M., Saligrama, V., Rosenberger, C.: Abnormal events detection based on spatio-temporal co-occurences. In: IEEE Conf. on Computer Vision and Pattern Recognition (CVPR), pp. 2458–2465 (2009)
4. Chi, E.C., Kolda, T.G.: On tensors, sparsity, and nonnegative factorizations. SIAM Journal on Matrix Analysis and Applications 33(4), 1272–1299 (2012)
5. Chung, P.C., Liu, C.D.: A daily behavior enabled hidden markov model for human behavior understanding. Pattern Recognition 41(5), 1572–1580 (2008)
6. Eagle, N., Pentland, A.: Reality mining: sensing complex social systems. Personal and Ubiquitous Computing 10(4), 255–268 (2006)
7. Hara, K., Omori, T., Ueno, R.: Detection of unusual human behavior in intelligent house. In: 12th IEEE Workshop on Neural Networks for Signal Processing, pp. 697–706 (2002)
8. Hayashi, K., Takenouchi, T., Shibata, T., Kamiya, Y., Kato, D., Kunieda, K., Yamada, K., Ikeda, K.: Exponential family tensor factorization for missing-values prediction and anomaly detection. In: 10th IEEE Int. Conf. on Data Mining (ICDM), pp. 216–225 (2010)
9. Hu, D.H., Zhang, X.X., Yin, J., Zheng, V.W., Yang, Q.: Abnormal activity recognition based on hdp-hmm models. In: IJCAI, pp. 1715–1720 (2009)
10. Jiang, F., Yuan, J., Tsaftaris, S.A., Katsaggelos, A.K.: Anomalous video event detection using spatiotemporal context. Computer Vision and Image Understanding 115(3), 323–333 (2011)
11. Kim, H., Lee, S., Ma, X., Wang, C.: Higher-order PCA for anomaly detection in large-scale networks. In: 3rd IEEE Int. Workshop on Computational Advances in Multi-Sensor Adaptive Processing (CAMSAP), pp. 85–88 (2009)
12. Kolda, T., Bader, B.: Tensor decompositions and applications. SIAM Review 51(3), 455–500 (2009)
13. Latecki, L.J., Lazarevic, A., Pokrajac, D.: Outlier detection with kernel density functions. In: Perner, P. (ed.) MLDM 2007. LNCS (LNAI), vol. 4571, pp. 61–75. Springer, Heidelberg (2007)
14. Li, C., Han, Z., Ye, Q., Jiao, J.: Abnormal behavior detection via sparse reconstruction analysis of trajectory. In: 6th IEEE Int. Conf. on Image and Graphics (ICIG), pp. 807–810 (2011)
15. Nesterov, Y.: Gradient methods for minimizing composite functions. Mathematical Programming 140(1), 125–161 (2013)
16. Piciarelli, C., Micheloni, C., Foresti, G.L.: Trajectory-based anomalous event detection. Circuits and Systems for Video Technology 18(11), 1544–1554 (2008)
17. Rendle, S., Gantner, Z., Freudenthaler, C., Schmidt-Thieme, L.: Fast context-aware recommendations with factorization machines. In: Proceedings of the 34th International ACM SIGIR Conference on Research and Development in Information Retrieval, pp. 635–644. ACM (2011)
18. Sillito, R.R., Fisher, R.B.: Semi-supervised learning for anomalous trajectory detection. In: BMVC, pp. 1–10 (2008)
19. Sun, J., Tao, D., Faloutsos, C.: Beyond streams and graphs: dynamic tensor analysis. In: 12th ACM SIGKDD Int. Conf. on Knowledge Discovery and Data Mining, pp. 374–383 (2006)
20. Tork, H.F., Oliveira, M., Gama, J., Malinowski, S., Morla, R.: Event and anomaly detection using tucker3 decomposition. In: Workshop on Ubiquitous Data Mining, p. 8 (2012)
21. Welling, M., Weber, M.: Positive tensor factorization. Pattern Recognition Letters 22(12), 1255–1261 (2001)
22. Yuan, J., Zheng, Y., Xie, X., Sun, G.: Driving with knowledge from the physical world. In: 17th ACM SIGKDD Int. Conf. on Knowledge Discovery and Data Mining, pp. 316–324 (2011)

A Semi-Supervised Framework for Social Spammer Detection

Zhaoxing Li, Xianchao Zhang$^{(\boxtimes)}$, Hua Shen, Wenxin Liang, and Zengyou He

School of Software Technology, Dalian University of Technology,
116621 Dalian, China
lzhx171@gmail.com, {xczhang,wxliang,zyhe}@dlut.edu.cn,
shenhua_as@126.com

Abstract. Spammers create large number of compromised or fake accounts to disseminate harmful information in social networks like Twitter. Identifying social spammers has become a challenging problem. Most of existing algorithms for social spammer detection are based on supervised learning, which needs a large amount of labeled data for training. However, labeling sufficient training set costs too much resources, which makes supervised learning impractical for social spammer detection. In this paper, we propose a semi-supervised framework for social spammer detection(SSSD), which combines the supervised classification model with a ranking scheme on the social graph. First, we train an original classifier with a small number of labeled data. Second, we propose a ranking model to propagate trust and distrust on the social graph. Third, we select confident users that are judged by the classifier and ranking scores as new training data and retrain the classifier. We repeat the all steps above until the classifier cannot be refined any more. Experimental results show that our framework can effectively detect social spammers in the condition of lacking sufficient labeled data.

Keywords: Semi-supervised Learning · Social Spam · Social Graph

1 Introduction

Social networks, like Twitter and Sina Weibo, are novel web services for online communication and information dissemination. People in social networks can share interested topics via sending short messages which contains plain text and URLs. This kind of web services which combine both micro-blogging and social relationship has attracted more and more users. At the same time, social network have become the main target web platform for spammers to spread unwanted information. It was reported that there had been a 335% growth of social spam during the first half of 2013[1].

---

[1] http://www.nexgate.com.

© Springer International Publishing Switzerland 2015
T. Cao et al. (Eds.): PAKDD 2015, Part II, LNAI 9078, pp. 177–188, 2015.
DOI: 10.1007/978-3-319-18032-8_14

Social spammers create large number of compromised or fake accounts to post phishing URLs, malwares, pornography information and lots of advertisements. In order to spread these spam messages to more users, they pretend to be legitimate that post messages mixed by normal and spam information and follow a lot of legitimate users expecting to acquire more in-links. Moreover, similar to link farming on the Web, social spammers often exchange reciprocal links with other spammers[5] so that they look like more 'reputable', which makes social spammer detection more and more challenging.

Most of existing works adopt supervised machine learning, which extract different features and train a classifier using labeled data to detect spam [11]. However, since a social networks like Twitter has huge amount of users, it is impossible to label abundant training data. This makes supervised learning algorithms impractical. Meanwhile, some researchers proposed unsupervised strategies such as ranking models [5,18] and community methods [4,14,19] that can automatically filter spam messages or accounts based on social relationship. However, this kind of methods often have high false positive rate [14] and are not robust compared with supervised methods.

In this paper, we propose a novel semi-supervised framework for detecting social spammers(SSSD), which take advantages of both supervised learning and unsupervised learning algorithm, and overcomes their difficulties. The framework iteratively trains a classifier by exploring social graph. First, we extract features inspired by previous studies [11,17] and learn an original classifier f with a small number of initial labeled users. Second, we use these labeled data as seeds to propagate trust and distrust on the social graph according to users' social relationships. Third, we use the classifier f to label top ranked users and select most confident users as new training data. Last we retrain the classifier. We repeat the above steps until the classifier cannot be refined any more. Experimental results show that our framework can be effectively detect social spammers in the condition of lacking sufficient labeled data.

The rest of this paper is organized as follows. Section 2 states related work. The whole framework is presented in Section 3. Section 4 shows the performance of our framework and we conclude this paper in Section 5.

2   Related Work

Social spam detection has been actively studied in recent years. Heymann et al. firstly made a survey of approaches and future challenges about defeating spam in social networks [7] and then different detection methods were proposed later.

Some researchers identify spam features and build supervised classification model to detect spammers. Benevenuto et.al [3] discussed video spammers by extracting some features in YouTube. Lee et.al [11] statistically analyzed features of spam profiles on social networks and apply machine learning methods to detect spammers. Other researches [1,2,13,15] similarly extracted different features for spam and trained a supervised model in different social network systems. Yang et.al [17] proposed more comprehensive features and analyzed the robust of different features. Recently, Zhu et.al [21] and Hu et.al [8,9] used optimization

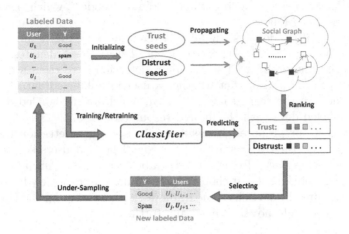

Fig. 1. Social spam detection framework

models combining different social information to detect social spammers. However, this kind of supervised learning models need adequate labeled data, which seems impractical in real social networks due to their huge number of users.

Some other researchers proposed some unsupervised methods. Gao et al. [4] presented a solution to firstly build similarity graph based on the URLs and description texts, then quantified and characterized spam campaigns launched using accounts on online social networks. Later, Zhang et al. [19] promoted the campaigns detection by a novel similarity information entropy. Yang et al. [18] analyzed spammers' inner social relationships, characterized the spam supporters and proposed a ranking model to find more criminal accounts. Ghosh et al. [5] investigated link farming in Twitter and proposed a ranking scheme that penalized users creating link farms. Tan et al. [14] designed an unsupervised spam detection scheme that clustered white users and detected suspected users who frequently posted harmful links. Most of these methods mined the social information on social graph.

Different from previous studies, we propose a semi-supervised framework, which focuses on the problem of building social spammer detection model based on only a few labeled data.

3   The Proposed Framework

3.1   Framework Overview

In social networks, users can be formulated as a set of features. Let $U = \{\mathbf{u_i}\}^n$ be the user account feature set, where $\mathbf{u_i}$ is the feature vector of user i and n is the number of users. From another perspective, users form a social graph

$G = (V, E)$, where V is a set of vertices for users and E is a set of relationships between users. $e(v_i, v_j) \in E$ denotes node i points to node j, which specially like Twitter means user i follows user j.

Let $L = \{(\mathbf{u_i}, y_i)\}^{\ell}$ denotes the labeled data set, where ℓ is the number of labeled data, and $y_i \in \{+1, -1\}$ is the label of user i ($+1$ denotes the spam user). Generally, the problem of social spam detection is to train a classifier f using labeled data L, and predict whether a user in unlabeled data $U - L$ is a spammer or not. In the real situation, $\ell \ll n$, which means labeled data is far less than unlabeled data with the limited resources.

The sketch of our proposed semi-supervised spammer detection framework is shown in Fig. 1. Firstly, it starts with a small labeled users set and trains an original classifier; Secondly, it uses the labeled seeds to rank all unlabeled users in social graph; Thirdly, it chooses the confident users as new labeled data; Fourthly, it retrains the classifier. The framework iteratively run the above steps until the classifier could not be refined any further.

3.2 Identifying Spammer Features

Features are very important to build a classification model, which have been well studied to identify spammers. Inspired by [11,17], we consider four categories of features: user profiles, user contents, user graph features and user neighbors, defined as followings:

- **Profile features.** These features can be extracted from user profiles. The number of followers, followings, all tweets and the account age are straightforward features. F-F ratio is the ratio of followers and followings. Following rate is $\frac{|followings|}{account\ age}$, which indicates whether the account creates much links.
- **Content features.** Content features are extracted from tweets. We calculate the number of tweets in last three months, the average interval of posting tweets and the average similarity of posting tweets of a user. In addition, the count/rate of tweets with urls,@mentions rate, *hashtags* and *retweets* are used as content features.
- **Graph features.** Graph features are extracted from the social graph G. First we simply exploit the bi-directional links with numbers of bi-links and bi-links ratio in graph G. Some other metric of graph are used, including *clustering coefficient,eccentricity, closeness*, and *betweenness*.
- **Neighbor features.** We also collect information of their following neighbors, who can reflect the confidence from its friends. We extract the average number of followers, tweets of their following neighbors, and the ratio of followings and the median number of followers (FMNF) of their neighbors.

Finally, we collect 28 features for each user, and then preprocess these features with log filter and normalization, which is represented by a vector $\mathbf{u_i}$. All detailed features are listed in Table 1. With the features and labeled data, we can train a classification model f using some learning algorithms like SVM, Naive Bayes and so on. Later, we can apply the model f to predict the label y_i of user i, $y_i = f(\mathbf{u_i})$.

Table 1. User Features

Category	Features in Our Study	Used/Proposed in Study
profile	number of followers	[2, 11, 15]
	number of followings	[2, 11, 15, 17]
	number of tweets	[17]
	account age	[11, 17]
	F-F ratio	[1, 2, 11, 17]
	following rate	[17]
content	number of recent tweets	None
	tweets rate	[11, 17]
	tweets similarity	[1, 2, 11, 17]
	url count/rate/unique	[1, 2, 11, 15]
	@*mention* count/rate/unique	[1, 2, 11]
	hashtag count/rate	[2, 15]
	re-tweet count/rate	[2, 11, 15]
graph	bi-links	[11]
	bi-links ratio	[17]
	Clustering Coefficient	[17]
	Eccentricity	None
	Closeness	None
	Betweenness	[17]
neighbor	average neighbors' followers	[17]
	average neighbors' tweets	[17]
	FMNF	[17]

3.3 Trust and Distrust Rank on Social Graph

There are a large number of links between users in social networks, which have built a huge social graph. In some social networks like Twitter and Sina Weibo, the social graph is directed. Inspired by TrustRank and Anti-Trust Rank algorithms on the Web, in this subsection, we introduce a trust ranking scheme and a distrust ranking scheme in social networks, which propagates both trust[6] and distrust[10] respectively from labeled seeds.

In trust propagation, if a trust node links to node v, it will propagate its trust score to v, because legitimate users always follow the legitimates who are also trustful. Then the trust score of a user $\mathbf{t}(v)$ in social graph is recursively defined as follow:

$$\mathbf{t}(v) = \alpha \sum_{p:p \to v} \frac{\mathbf{t}(p)}{outdegree(p)} + (1 - \alpha)s(v) \qquad (1)$$

where α is a decay factor[2], p represents all followers accounts of user v, *outdegree* (p) is the number of followings of user p, and $s(v)$ is the normalized indicator function for legitimate seeds set S^+ that $s(v) = 1/|S^+|$ if $v \in S^+$ and $s(v) = 0$ if $v \notin S^+$.

In distrust propagation, if a node v links to a distrust node, the distrust node will propagate the distrust to v and degrade the confidence of v, based on the observation that in general a legitimate user wouldn't follow a spammer but spammers often create link farms between them to make them more reputable. The distrust score of a user $\mathbf{d}(v)$ is recursively defined as follow:

$$\mathbf{d}(v) = \alpha' \sum_{q:v \to q} \frac{\mathbf{d}(q)}{indegree(q)} + (1 - \alpha')s'(v) \qquad (2)$$

---

[2] Like PageRank and TrustRank, α is set 0.85

where α' is a decay factor[3], q represents all followings accounts of user v, $indegree(q)$ is the number of followers of user q, and $s'(v)$ is the normalized indicator function for spam seeds set S^- that $s'(v) = 1/|S^-|$ if $v \in S^-$ and $s'(v) = 0$ if $v \notin S^-$. To separate spammers more significantly, we define the final distrust score[16] as $\mathbf{d}' = \mathbf{d} - \mathbf{t}$.

3.4   Sampling Initial Labeled Data

Note that the selection of seeds is an important issue because ranking models are sensitive to these initial seeds and may be biased if run with unreasonable seeds. In general ,we would like to select the most useful users to label as the initial data, who can spread more information in graph and identify other users.

In our framework, *Inverse PageRank* [6] is adopted to select legitimate seeds to label, and *PageRank* is applied for spammers seeds to label. Intuitively legitimate users with high *Inverse PageRank* ranking scores have more out-links and can propagate much trust to its neighbors, while spammers with high *PageRank* ranking scores can propagate much distrust. We select the initial labeled data set L from these users to train the original classifier and also use them as the initial seeds of the ranking schemes.

3.5   Semi-Supervised Social Spammer Detection

In this subsection, we describe our Semi-Supervised Spammer Detection (SSSD) framework in details. The main purpose of the framework is to iteratively learn an optimized classifier using enlarged training data set starting from a small labeled data set.

In the real situation, we could get only a small number of labeled users because of limited time and labor resources, while there are abundant unlabeled data with a lot of social relationship information, which is the motivation of the framework. The proposed framework exploits the large number of social relationships of unlabeled data with the ranking model and select the most confident users as new labeled data to retrain the classifier iteratively. The details of the detection framework are presented in Algorithm 1.

First, we initialize variables. Given social graph G, user feature set U and labeled data set L as input. ΔL is the training data set in the current iteration and $\Delta L'$ is the new training set in the next iteration. e and $e\prime$ are the upper bound of the classification error rate in the current iteration and the next iteration, respectively. Then, we learn the initial classification model f using learning algorithm c and labeled data L.

Second, we propagate trust and distrust on the graph G. $InitialSeeds$ is the function that initializes seeds set S^+ and S^- with users in labeled data set L and new labeled set ΔL. After assigning S^+ and S^-, we run the ranking scheme $SocialTDRank$ which calculates trust and distrust scores of each user, t and $d\prime$.

---

[3] Like Anti-Trust Rank, α' is set 0.85.

Algorithm 1. Semi-Supervised Spammer Detection

Input:
      Social network, G;
      User feature set U and Labeled data set L;
      Classification algorithm c;
Output: Classification Model , f;
      $\Delta L \leftarrow \emptyset, \Delta L' \leftarrow L$;
      $e \leftarrow 0, e' \leftarrow 0.5$;
      $f \leftarrow c.learn(L \cup \Delta L)$;
      while True **do**
         $S^+, S^- \leftarrow InitialSeeds(L \cup \Delta L)$;
         $\mathbf{t}, \mathbf{d'} \leftarrow SocialTDRank(G, S^+, S^-)$;
         Rank unlabeled users with $\mathbf{t}$, and select top k as set T;
         Rank unlabeled users with $\mathbf{d'}$, and select top ηk as set D;
         for user u in $T \cup D$ **do**
            if $f(u)$ is $+1$ and $u \in T$ **then**
               $\Delta L \leftarrow \Delta L \cup \{(u, +1)\}$;
            end if
            if $f(u)$ is -1 and $u \in D$ **then**
               $\Delta L \leftarrow \Delta L \cup \{(u, -1)\}$;
            end if
         end for
         $\Delta L \leftarrow UnderSample(\Delta L)$;
         $e \leftarrow TestError(f)$;
         if $e|\Delta L| \le e'|\Delta L'|$ **then**
            $e' \leftarrow e, \Delta L' \leftarrow \Delta L$;
            $f \leftarrow c.learn(L \cup \Delta L)$;
         else
            break
         end if
      end while
      return f;

Third, we select confident users as new labeled data set ΔL. As shown in Algorithm 1, T and D are made up by unlabeled users who have high t score and $d\prime$ score. η is the ratio of spammers and legitimate users, which is used to maintain the class distribution in training. We use the model f to predict label of users in T and D. Users with high t score and predicted $+1$ (legitimate) or users with high $d\prime$ score and predicted -1 (spam) are considered as confident, and are put into the new labeled data set ΔL. To avoiding class imbalance, we use under-sampling to sample the ΔL, which maintains the ratio of spammers and legitimate users greater than η.

Last, we estimate the classification error rate of f and retrain the classifier with the new training data. $TestError(f)$ attempts to estimate the error rate of f. Since it is difficult to estimate the error on the unlabeled data, here we use only the original labeled examples to test the error rate[20]. Whether the new labeled data can be used for training a classifier is determined by:

$$0 < \frac{e}{e'} < \frac{|\Delta L'|}{|\Delta L|} < 1. \tag{3}$$

If Eq.3 is satisfied, we retrain model f with $L \cup \Delta L$. If not, stop the iteration and return the final model f as output.

Note that our framework is an extension of self-training. The difference is that self-training usually select new data according confidence computed by itself, while our framework consider both the social relationships and classifier, introducing the ranking model to jointly select new training set from unlabeled data.

4   Experiments

4.1   Dataset

To empirically study the social spammer detection framework, we use the UDI-Twitter dataset [12], a subset of Twitter containing 50 million tweets for 140 thousand users and 284 million following relationships collected at May 2011. Then we extract all features listed in Table 1 of 140 thousand users based on their profiles and tweets content.

Next, we label the sampled users as spammers as long as satisfied one of the following rules: i)Posting at least one malicious or phishing URLs, ii)Posting much pornographic information or many URLs directed to pornographic websites, iii)Posting a lot of advertisements or URLs directed to online shopping websites for promotion. For the first rule, we use *Google Safe Browsing*[4], a widely used blacklist, to detect the malicious or phishing URLs. For next two rules, we manually scan the tweets content of all sampled accounts and click the URLs to judge whether they are pornographic information or advertisements. We invite 50 volunteers to identify the spammers by majority vote. At last, we have extracted 2527 spammers from 140,000 users and sampled 11282 legitimate users as our dataset, shown as Table 2

Table 2. Dataset

Total users	Relationships	Tweets	Features
13809	860955	50,0000	28
Max indegree	Max outdegree	Spammers	Legitimate users
675	649	2527	11282

4.2   Experiment Setup

In this subsection, we introduce the details about the experimental settings. To simulate the problem of lacking sufficient labeled data, we use just only a few data as the original labeled data. First, we discuss how the initial data affect the $SSSD$ model with different size of initial labeled set and different seeds selection strategy. Second, we compare the performance of $SSSD$ with supervised models. In supervised models, we use 5-fold cross validation, while in $SSSD$, we select 2400 users as its original labeled data and randomly test 20% unlabeled data. Precision, Recall, F1-score and Accuracy are used as the performance metrics. All the metrics are mainly measured for spammers.

---

[4] http://code.google.com/apis/sfebrowsing/

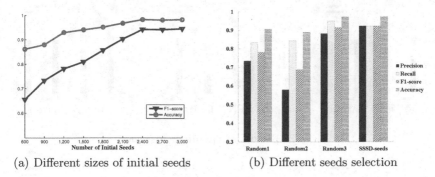

(a) Different sizes of initial seeds   (b) Different seeds selection

Fig. 2. Performance of SSSD with different seeds

We choose most popular learning models including SVM, Logistic Regression (LR), Decision Tree (J48) and Naive Bayes in the experiments, which are implemented by *scikit-learn*[5]. Note that the classes of spammers and legitimate users are extremely imbalanced, where the proportion of spammers is nearly 18% of our data set, we apply under-sampling to solve class imbalance by setting $\eta = 0.5$.

4.3   Experiments on Different Initial Labeled Data

We apply the scheme introduced in 3.4 to select the initial labeled data, and the performance of $SSSD$ with different seeds set is shown in Fig. 2(a). With the size of seeds increasing, F1-measure and Accuracy are rising, which is because that more initial labeled data can provide more precise training information for $SSSD$. Note that two measure metrics tend to be stable when the size is larger than 2400, we suppose that with our seeds selecting scheme it can provide sufficient information for training when it reaches a threshold.

To illustrate the essential of seeds selecting scheme, we compare the performance with random selecting strategy in Fig. 2(b), where the size of seeds is set 2400. As shown in Fig. 2(b), we implement 3 different random seed sets to run our framework and the results show that the random selecting performs sometimes well but it is not robust compared with our proposed strategy. The reason is that propagating methods in graph are a little sensitive to the initial seed set, which would affect the selection of new labeled users in $SSSD$. Our selecting scheme is more robust to build the detection model.

4.4   Experimental Results

In this subsection, we systematically show the robustness of $SSSD$ via comparing with supervised learning using different feature sets and different learning models. For comparison of the different feature sets, we concluded some previous

---

[5] http://scikit-learn.org/

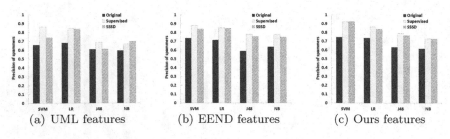

Fig. 3. Results of Spammer Precision

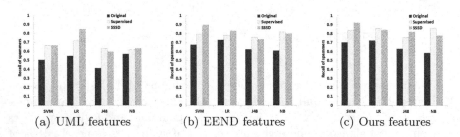

Fig. 4. Results of Spammer Recall

studies, **UML** features[11], **EEND** features[17] and our selected features shown in Table 1. The results of Precision and Recall are presented in Fig. 3 and Fig. 4, where the item **Original** means that the classifier only is trained by the original data without $SSSD$. In general, high precision indicates that most of predicted spammers are correct and only a few legitimate users are wrongly classified as spammers. Table 3 and Table 4 show the F1-score and Accuracy.

For different feature sets, we find that no matter which feature set we choose, the results are definitely optimized by running $SSSD$ compared with the original model. These results prove that $SSSD$ have exploited the large number of social network information of unlabeled data, and selected useful new labeled data for refining the model. From the results, $SSSD$ performs almost as same as supervised models and even better using some classifies and features with less labeled data. For example, when we use SVM on our feature set, we get **0.942** F1-score and **0.982** Accuracy, which is higher than the supervised original model. Furthermore, we notice that our features is better than other feature schemes.

For different learning models, $SSSD$ is effective to improve all classifies from original data with mining the social graph. SVM performs the best among different classifiers in our framework, So we suggest SVM as the final classification model for our data set.

Overall, $SSSD$ is a robust semi-supervised framework and it enhance the performance of model from less labeled data, which also can be generalized on different classifies and more new features.

Table 3. Results of Detection F1-score

	UML features			EEND features			Our features		
	Original	Supervised	SSSD	Original	Supervised	SSSD	Original	Supervised	SSSD
SVM	0.569	0.752	0.701	0.704	0.835	**0.866**	0.723	0.878	**0.922**
LR	0.607	0.777	**0.843**	0.720	0.814	**0.838**	0.730	0.861	0.840
J48	0.491	0.660	0.604	0.606	0.769	0.745	0.632	0.775	**0.794**
NB	0.582	0.644	**0.664**	0.622	0.796	0.772	0.599	0.789	0.753

Table 4. Results of Detection Accuracy

	UML features			EEND features			Our features		
	Original	Supervised	SSSD	Original	Supervised	SSSD	Original	Supervised	SSSD
SVM	0.894	0.917	0.901	0.880	0.941	**0.959**	0.899	0.956	**0.972**
LR	0.849	0.922	**0.952**	0.853	0.927	**0.950**	0.865	0.939	**0.952**
J48	0.833	0.877	0.793	0.836	0.904	0.887	0.834	0.905	0.884
NB	0.806	0.840	0.822	0.818	0.913	0.893	0.801	0.917	0.900

5   Conclusion

Social networks have been the main targets for spammers to disseminate malicious information and promote their advertisements. With the massive users, only a very small fraction of the users could be labeled with limited resources of labor. In this paper, we present a semi-supervised social spammer detection framework to filter the spammers with a small labeled data set. We use ranking models to rank unlabeled users with trust and distrust scores in social graph. Integrating the ranking model into the classification model, the framework automatically labels the data with confidence judged by both of them and retrains the classifier with most confident data iteratively, and finally gets a refined classification model. The experimental results show that our framework is effective to find most of social spammers with only a small labeled seed set. In the future, we will investigate more complicated schemes to select confident users for training such as active learning.

References

1. Amleshwaram, A.A., Reddy, N., Yadav, S., Gu, G., Yang, C.: Cats: characterizing automation of twitter spammers. In: 2013 Fifth International Conference on Communication Systems and Networks (COMSNETS), pp. 1–10. IEEE (2013)
2. Benevenuto, F., Magno, G., Rodrigues, T., Almeida, V.: Detecting spammers on twitter. In: Collaboration, Electronic messaging, Anti-Abuse and Spam Conference (CEAS) (2010)
3. Benevenuto, F., Rodrigues, T., Almeida, V., Almeida, J., Gonçalves, M.: Detecting spammers and content promoters in online video social networks. In: Proceedings of the 32nd International ACM SIGIR Conference on Research and Development in Information Retrieval(SIGIR), pp. 620–627. ACM (2009)
4. Gao, H., Hu, J., Wilson, C., Li, Z., Chen, Y., Zhao, B.: Detecting and characterizing social spam campaigns. In: Proceedings of the 10th Annual Conference on Internet Measurement(IMC), pp. 35–47. ACM (2010)

5. Ghosh, S., Viswanath, B., Kooti, F., Sharma, N.K., Korlam, G., Benevenuto, F., Ganguly, N., Gummadi, K.P.: Understanding and combating link farming in the twitter social network. In: Proceedings of the 21st International Conference on World Wide Web, pp. 61–70. ACM (2012)
6. Gyöngyi, Z., Garcia-Molina, H., Pedersen, J.: Combating web spam with trustrank. In: Proceedings of the Thirtieth International Conference on Very Large Data Bases-vol. 30, pp. 576–587. VLDB Endowment (2004)
7. Heymann, P., Koutrika, G., Garcia-Molina, H.: Fighting spam on social web sites: A survey of approaches and future challenges. IEEE Internet Computing 11(6), 36–45 (2007)
8. Hu, X., Tang, J., Liu, H.: Online social spammer detection. In: Twenty-Eighth AAAI Conference on Artificial Intelligence (2014)
9. Hu, X., Tang, J., Zhang, Y., Liu, H.: Social spammer detection in microblogging. In: Proceedings of the Twenty-Third International Joint Conference on Artificial Intelligence, pp. 2633–2639. AAAI Press (2013)
10. Krishnan, V., Raj, R.: Web spam detection with anti-trust rank. AIRWeb. 6, 37–40 (2006)
11. Lee, K., Caverlee, J., Webb, S.: Uncovering social spammers: social honeypots+ machine learning. In: Proceeding of the 33rd International ACM (SIGIR) Conference on Research and Development in Information Retrieval, pp. 435–442. ACM (2010)
12. Li, R., Wang, S., Deng, H., Wang, R., Chang, K.C.C.: Towards social user profiling: unified and discriminative influence model for inferring home locations. In: KDD, pp. 1023–1031 (2012)
13. Prasetyo, P.K., Lo, D., Achananuparp, P., Tian, Y., Lim, E.P.: Automatic classification of software related microblogs. In: 2012 28th IEEE International Conference on Software Maintenance (ICSM), pp. 596–599. IEEE (2012)
14. Tan, E., Guo, L., Chen, S., Zhang, X., Zhao, Y.: Unik: unsupervised social network spam detection. In: Proceedings of the 22nd ACM International Conference on Conference on Information & Knowledge Management, pp. 479–488. ACM (2013)
15. Wang, A.: Don't follow me: Spam detection in twitter. In: Proceedings of the 2010 International Conference on Security and Cryptography (SECRYPT), pp. 1–10. IEEE (2010)
16. Wu, B., Goel, V., Davison, B.D.: Propagating trust and distrust to demote web spam. MTW **190** (2006)
17. Yang, C., Harkreader, R., Gu, G.: Empirical evaluation and new design for fighting evolving twitter spammers. IEEE Transactions on Information Forensics and Security 8(8), 1280–1293 (2013)
18. Yang, C., Harkreader, R., Zhang, J., Shin, S., Gu, G.: Analyzing spammers' social networks for fun and profit: a case study of cyber criminal ecosystem on twitter. In: Proceedings of the 21st International Conference on World Wide Web, pp. 71–80. ACM (2012)
19. Zhang, X., Zhu, S., Liang, W.: Detecting spam and promoting campaigns in the twitter social network. In: Proceedings of the 2012 IEEE 12th International Conference on Data Mining, pp. 1194–1199. IEEE Computer Society (2012)
20. Zhou, Z.H., Li, M.: Tri-training: Exploiting unlabeled data using three classifiers. IEEE Transactions on Knowledge and Data Engineering 17(11), 1529–1541 (2005)
21. Zhu, Y., Wang, X., Zhong, E., Liu, N.N., Li, H., Yang, Q.: Discovering spammers in social networks. In: AAAI (2012)

Fast One-Class Support Vector Machine
for Novelty Detection

Trung Le[1]([⊠]), Dinh Phung[2], Khanh Nguyen[1], and Svetha Venkatesh[2]

[1] Department of Information Technology, HCMc University of Pedagogy,
Ho Chi Minh City, Vietnam
trunglm@hcmup.edu.vn
[2] Center for Pattern Recognition and Data Analytics,
Deakin University, Geelong, Australia

Abstract. Novelty detection arises as an important learning task in several applications. Kernel-based approach to novelty detection has been widely used due to its theoretical rigor and elegance of geometric interpretation. However, computational complexity is a major obstacle in this approach. In this paper, leveraging on the cutting-plane framework with the well-known One-Class Support Vector Machine, we present a new solution that can scale up seamlessly with data. The first solution is exact and linear when viewed through the cutting-plane; the second employed a sampling strategy that remarkably has a constant computational complexity defined relatively to the probability of approximation accuracy. Several datasets are benchmarked to demonstrate the credibility of our framework.

Keywords: One-class Support Vector Machine · Novelty detection · Large-scale dataset

1 Introduction

Real data rarely conform to regular patterns. Finding subtle deviation from the norm, or *novelty detection* (ND), is an important research topics in many data analytics and machine learning tasks ranging from video security surveillance, network abnormality detection to detection of abnormal gene expression sequence. Novelty detection refers to the task of of finding patterns in data that do not conform to expected behaviors [4]. These anomaly patterns are interesting because they reveal actionable information, the known unknowns and unknown unknowns. Novelty detection methods have been used in many real-world applications, e.g., intrusion detection [8], fraud detection (credit card fraud detection, mobile phone fraud detection, and insurance claim fraud detection) [7].

There are various existing approaches applied to novelty detection, including neural network based approach [2], Bayesian network based approach [1], rule based approach [6], kernel based approach [12,14]. However, these approaches are either ad hoc (rule-based) or lacking a principled approach to scale up with data.

© Springer International Publishing Switzerland 2015
T. Cao et al. (Eds.): PAKDD 2015, Part II, LNAI 9078, pp. 189–200, 2015.
DOI: 10.1007/978-3-319-18032-8_15

Moreover, none of these methods has been experimented with large datasets. In this paper, we depart from a kernel-based approach, which has a established theoretical foundation, and propose a new novelty detection machine that scales up seamlessly with data.

The idea of using kernel-based methods for novelty detection is certainly not new. At its crux, geometry of the normal data is learned from data to define the domain of novelty. One-Class Support Vector Machine (OCSVM) [12] aims at constructing an optimal hyperplane that separates the origin and the data samples such that the margin, the distance from the origin to the hyperplane, is maximized. OCSVM is used in case data of only one class is available and we wish to learn the description of this class. In another approach, Support Vector Data Description (SVDD) [14], the novelty domain of the normal data is defined as an optimal hypersphere in the feature space, which becomes as a set of contours tightly covering the normal data when mapped back the input space.

There are also noticeable research efforts in scaling up kernel-based methods to linear complexity including Pegasos [13] (kernel case), SVM^{Perf} [10]. However, to our best of knowledge, none has attempted at novelty detection problem.

Starting with the formulation of cutting plane method [11], we propose in this paper a new approach for novelty detection. Our approach is a combination of cutting plane method applied to the well-known One-Class Support Vector Machine [12] – we term our model Fast One-Class SVM (FOSVM). We propose two complementary solutions for training and detection. The first is an exact solution that has linear complexity. The number of iterations to reach a very close accuracy, termed as $\theta-$precision (defined later in the paper) is $O\left(\frac{1}{\theta}\right)$, making it attractive in dealing with large datasets. The second solution employed a sampling strategy that remarkably has constant computational complexity defined relatively to probability of approximation accuracy. In all cases, rigorous proofs are given to the convergence analysis as well as complexity analysis.

2   Related Work

Most closely to our present work is the Core Vector Machine (CVM) [18] and its simplified version Ball Vector Machine (BVM) [17]. CVM is based on the achievement in computational geometry [3] to reformulate a variation of L2-SVM as a problem of finding minimal enclosing ball (MEB). CVM shares with ours in using the principle of extremity in developing the algorithm. Nonetheless, CVM does not solve directly its optimization problem and does not offer the insight view of cutting plane method as ours.

Cutting plane method [11] has been applied to building solvers for kernel methods in the work [9,10,15,16]. In [10], the cutting plane method is employed to train SVM^{struct} ; the $N -$ slack optimization problem with N constrains is converted to an equivalent $1 -$ slack optimization problem with 2^N constrains (binary classification); the constrains are subsequently added to the optimization problem. In [15,16], the bundle method is used for solving the regularized risk minimization problem. Though those work have been proven that bundle method is very efficient for the regularized risk minimization problem and the number of iterations to gain $\epsilon-$precision solution are $O\left(\frac{1}{\epsilon}\right)$ for non-smooth problem and $O\left(\log\left(\frac{1}{\epsilon}\right)\right)$ for smooth problem, they are implausible to extend to a non-linear kernel case.

3   Proposed Fast One-Class Support Vector Machine (FOSVM1)

3.1   One-Class Support Vector Machine

We depart from the One-Class Support Vector Machine proposed by [12]. Given a training set with only positive samples $\mathcal{D} = \{x_1, \ldots, x_N\}$, One-Class Support Vector Machine (OCSVM) learns the optimal hyperplane that can separate the origin and all samples such that the margin, the distance from the origin to the hyperplane, is maximized. The optimization problem of soft model of OCSVM is as follows:

$$\min_{w,\rho} \left(\frac{1}{2} \|w\|^2 - \rho + \frac{1}{vN} \sum_{i=1}^{N} \xi_i \right) \tag{1}$$
$$\text{s.t.} : \forall_{i=1}^{N} : w^T \Phi(x_i) \geq \rho - \xi_i$$
$$\forall_{i=1}^{N} : \xi_i \geq 0$$

If the slacks variables are not used, i.e., all samples in the training set must be completely stayed on the positive side of the optimal hyperplane, the soft model becomes the hard model associated with the following optimization problem:

$$\min_{w,\rho} \left(\frac{1}{2} \|w\|^2 - \rho \right) \tag{2}$$
$$\text{s.t.} : \forall_{i=1}^{N} : w^T \Phi(x_i) \geq \rho$$

3.2   Cutting Plane Method for One-Class SVM

To apply the cutting plane method for the optimization problem of OCSVM, we rewrite the OCSVM optimization problem as follows:

$$\min_{w,\rho} \left(\frac{1}{2} \|w\|^2 - \rho \right)$$
$$\text{s.t.} : \forall_{i=1}^{N} : \begin{bmatrix} \Phi(x_i) \\ -1 \end{bmatrix}^T \begin{bmatrix} w \\ \rho \end{bmatrix} \geq 0 \,(\text{constrainst } \mathscr{C}_i) \tag{3}$$

The feasible set of the OCSVM optimization problem is composed by intersection of N half-hyperplanes corresponding to the constrains $\mathscr{C}_i\,(1 \leq i \leq N)$ (see Figure 1). Inspired by the cutting plane method, the following algorithm is proposed:

Algorithm 1. Cutting method to find the θ-approximate solution.

---

for $n = 2, 3, 4...$

$\quad (w_n, \rho_n) = solve(n);$

$\quad n + 1 = i_{n+1} = \underset{i>n}{\arg\min} \begin{bmatrix} \Phi(x_i) \\ -1 \end{bmatrix}^T \begin{bmatrix} w_n \\ \rho_n \end{bmatrix}$

$\quad = \underset{i>n}{\arg\min} \left(w_n^T \Phi(x_i) - \rho_n \right);$

$\quad o_{n+1} = w_n^T \Phi(x_{n+1}) - \rho_n;$

$\quad$ **if** $(o_{n+1} \geq -\theta\rho_n)$ **return** $(w_n, \rho_n);$

endfor

The procedure $solve(n)$ stands for solving the optimization problem in Eq. (3) where first n constrains $\mathscr{C}_i$ $(1 \leq i \leq n)$ are activated. By convenience, we assume that the chosen index $i_{n+1} = n + 1$. The satisfaction of condition $o_{n+1} \geq -\theta\rho_n$ where $\theta \in (0,1)$ means that the current solution (w_n, ρ_n) is that of the relaxation of the optimization problem in Eq. (3) while the constrains $\mathscr{C}_i$ $(i > n)$ are replaced by its relaxation $\mathscr{C}'_i$ $(i > n)$ as follows:

$$\min_{w,\rho} \left(\frac{1}{2} \|w\|^2 - \rho \right)$$

$$\text{s.t.} : \forall_{i=1}^n : \begin{bmatrix} \Phi(x_i) \\ -1 \end{bmatrix}^T \begin{bmatrix} w \\ \rho \end{bmatrix} \geq 0 \, (\text{constrainst } \mathscr{C}_i)$$

$$\forall_{i=n+1}^N : \begin{bmatrix} \Phi(x_i) \\ -1 \end{bmatrix}^T \begin{bmatrix} w \\ \rho \end{bmatrix} \geq -\theta\rho_n \, (\text{constrainst } \mathscr{C}'_i)$$

Otherwise, the hyperplane corresponding with the constrains $\mathscr{C}_{n+1}$ separates the current solution $s_n = (w_n, \rho_n)$ and the feasible set of the full optimization problem in Eq. (1) $\left(\text{since } \begin{bmatrix} \Phi(x_{n+1}) \\ -1 \end{bmatrix}^T \begin{bmatrix} w_n \\ \rho_n \end{bmatrix} < -\rho_n\theta < 0 \right)$; the constrains $\mathscr{C}_{n+1}$ is added to the active constrains set (see Figure 1). Furthermore, the distance from the current solution as shown in Eq. (4) to the chosen hyperplane is maximized so that the current feasible set rapidly approaches the destined one (see Figure 1).

$$d_{n+1} = d(s_n, \mathscr{C}_{n+1}) = \frac{\rho_n - w_n^T \Phi(x_{n+1})}{\|w_n\|} \tag{4}$$

3.3 Approximate Hyperplane

We present an interpretation of the proposed *Algorithm 1* under the framework of approximate hyperplane. Let us start with clarifying the approximate hyperplane notion. Let $A \subseteq [N]$ be a subset of the set including first N positive integer numbers and $\mathcal{D}_A$ be a subset of the training set $\mathcal{D}$ including samples whose indices are in A.

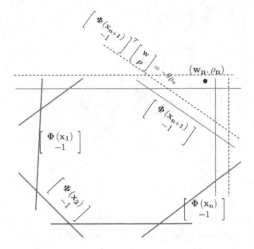

Fig. 1. The constrains $\mathscr{C}_{n+1}$ is chosen to add to the current active constrains set (red is active and green is inactive)

Denote the optimal hyperplane induced by $\mathcal{D}_A$ by $(\mathcal{H}_A): w_A^T \Phi(x) - \rho_A = 0$. Given a sample x, let us define the membership of x with regard to the positive side of the hyperplane $(\mathcal{H}_A)$ by :

$$m_A(x) = \begin{cases} 1 - \frac{d(\Phi(x), \mathcal{H}_A)}{d(0, \mathcal{H}_A)} = 1 + \frac{w_A^T \Phi(x) - \rho_A}{\rho_A} = \frac{w_A^T \Phi(x)}{\rho_A} & \text{if } w_A^T \Phi(x) - \rho_A < 0 \\ 1 & \text{otherwise} \end{cases}$$

Intuitively, the membership of x is exactly 1 if $\Phi(x)$ lies on the positive side of $\mathcal{H}_A$; otherwise this membership decreases when $\Phi(x)$ is moved further the hyperplane on the negative side.

We say that hyperplane $\mathcal{H}_A$ is an θ – approximate hyperplane if it is the optimal hard-margin hyperplane induced by $\mathcal{D}_A$ and the memberships with respect to $\mathcal{H}_A$ of data samples in $\mathcal{D} \setminus \mathcal{D}_A$ are all greater than or equal $1 - \theta$, i.e., $\forall x \in \mathcal{D} \setminus \mathcal{D}_A : m_A(x) \geq 1 - \theta$. The visualization of θ – approximate hyperplane is shown in Figure 2.

Using the θ – approximate hyperplane, *Algorithm 1* can now be rewritten as:

Algorithm 2. Algorithm to find θ – approximate hyperplane.

$A = \{1, 2\}$;
$n = 2$;
do
    $(w_n, \rho_n) = \textbf{OCSVM}(\{x_i : i \in A\})$;
    $n + 1 = i_{n+1} = \underset{i \in A^c}{\text{argmin}} (w_n^T \Phi(x_i) - \rho_n)$;
    $o_{n+1} = w_n^T \Phi(x_{n+1}) - \rho_n$;
    if $(o_{n+1} \geq -\theta \rho_n)$ **return** (w_n, ρ_n);
    else $A = [n + 1]$;
while (true)

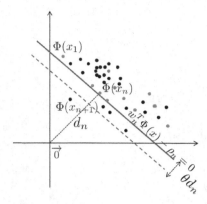

Fig. 2. θ – approximate hyperplane (green is active, black is inactive), the red line stands for the optimal hard-margin hyperplane induced by $\mathcal{D}_A$

Algorithm 2 aims at finding a θ – approximate hyperplane in which θ – approximate notion can be interpreted as slack variables of soft model. It is worthwhile to note that the stopping criterion in *Algorithm* 2 means that $\forall_{i=n+1}^{N}$: $m_A(x_i) \geq 1 - \theta$.

In the next section, we prove that *Algorithm* 2 terminates after a finite number of iterations independent of the training size N. To simplify the derivation, we assume that isometric kernel, e.g., Gaussian kernel, is used which means that $\|\Phi(x)\| = K(x,x)^{1/2} = 1, \forall x$. Furthermore, let us define $w^* = w_\infty = \lim_{n\to\infty} w_n$ and $d^* = d_\infty = \lim_{n\to\infty} d_n$ where $d_n = d(\mathbf{0}, \mathcal{H}_n) = \frac{\rho_n}{\|w_n\|}$ is the margin of the current active set $\mathcal{D}_{[n]}$ and d^* is the margin of the general dataset.

Theorem 1. *Algorithm 2 terminates after at most n_0 iterations where n_0 depends only on θ and the margin of the general data set $d^* = d_\infty$.*

Proof. We sketch the proof as follows. The main task is to prove that:

$$\forall_{j=1}^{N} : m_n(x_j) \geq 1 - \frac{\left(\sqrt{\frac{2}{1+n/2}} + \frac{1}{1+n/2}\right)\left(1 - \|d^*\|^2\right)}{\|d^*\|^2} = 1 - g(n)$$

It is easy to see that $\lim_{n\to\infty} g(n) = 0$. Therefore, there exists n_0 where n_0 is a constant which is independent with N such that $m_n(x_j) > 1 - \theta$ for all $n \geq n_0$.

Theorem 2. *The number of iterations in Algorithm 2 is $O(1/\theta)$.*

Proof. Let us denote $\theta' = \frac{\theta d^{*2}}{1 - d^{*2}}$. We have:

$$1 - g(n) \geq 1 - \theta \Leftrightarrow \sqrt{\frac{2}{1+n/2} + \frac{1}{1+n/2}} \leq \theta' \iff \frac{n}{2} + 1 \geq \frac{\sqrt{2}\left(\sqrt{2\theta'+1}+1\right)}{2\theta'}$$

It is obvious that:

$$\frac{\sqrt{2}\left(\sqrt{2\theta'+1}+1\right)}{2\theta'} \leq \frac{const}{2\theta'} \sim O\left(1/\theta'\right) = O\left(1/\theta\right)$$

Therefore, we gain the conclusion.

From *Theorem* 2, we also gain the number of iterations in *Algorithm* 2 is $O\left(1/\theta\right)$. *Theorem* 1 reveals that the complexity of *Algorithm* 2 is $\mathbf{O}(N)$ since the complexity of each iteration is $\mathbf{O}(N)$.

4   Sampling Fast One-Class Support Vector Machine (FOSVM2)

In this section, we develop a second solution to train the proposed model using sampling strategy. We now show how to use sampling technique to predict this furthest vector with the accuracy can be made as close as we like to the exact solution specified via a bounded probability similar to concept of p-value.

Let $0 < \epsilon < 1$ and m vectors are sampled from N vectors of the training set $\mathcal{D}$. We denote the set of ϵ-top furthest vectors in $\mathcal{D}$ by B. We estimate the probability in order that at least one of m sampled vectors belonging to B. We call the probability of this event as $\Pr(\epsilon\text{-top})$. First, we note from our definition that $n = |B| = \lceil \epsilon N \rceil$. Hence, the probability of interest becomes $\Pr(\epsilon\text{-top}) = 1 - \binom{N-n}{m} / \binom{N}{m}$. With some effort of manipulation, this reduces to:

$$\Pr(\epsilon\text{-top}) = 1 - \left[(N-n)!(N-m)!\right]\left[N!(N-n-m)!\right]^{-1}$$

$$= 1 - \left[\prod_{i=N-n-m+1}^{N-m} i\right]\left[\prod_{i=N-n+1}^{N} i\right]^{-1}$$

Taking the logarithm yields

$$\ln\left[1 - \Pr(\epsilon\text{-top})\right] = \sum_{i=N-n-m+1}^{N-m} \ln\left(\frac{i}{i+m}\right) = \sum_{i=N-n-m+1}^{N-m} \ln\left(1 - \frac{m}{i+m}\right)$$

Algorithm 3. Algorithm to find $\theta -$ approximate hyperplane.

$A = \{1, 2\}$;
$n = 2$;
do
    $(w_n, \rho_n) = \mathbf{OCSVM}\left(\{x_i : i \in A\}\right)$;
    $B = \text{Sampling}(A^c, m)$;
    $n + 1 = i_{n+1} = \underset{i \in B}{\operatorname{argmin}}\left(w_n^T \Phi\left(x_i\right) - \rho_n\right)$;
    $o_{n+1} = w_n^T \Phi\left(x_{i_{n+1}}\right) - \rho_n$;
    if $\left(o_{n+1} \geq -\theta\rho_n\right)$ **return** (w_n, ρ_n);
    else $A = [n+1]$;
while (true)

Now applying the inequality $\ln(1 - x) \leq -x, \forall x \in [0, 1)$, we have:

$$\ln\left[1 - \Pr(\epsilon\text{-top})\right] \leq - \sum_{i=N-n-m+1}^{N-m} \frac{m}{i+m} \leq - \sum_{i=N-n+1}^{N} \frac{m}{i} \leq -\frac{mn}{N}$$

Therefore, $\Pr(\epsilon\text{-top}) \geq 1 - e^{\frac{-mn}{N}} = 1 - e^{-\epsilon m}$. If we wish to have at least 95% probability in approximate accuracy (to the exact solution) with $\epsilon = 5\%$, we can calculate the number of sampling points m from this bound: $\Pr(\epsilon\text{-top}) \geq 1 - e^{-\epsilon m} > 0.95$. With $\epsilon = 0.05$ we have $m \geq 20 \ln 20 \approx 59$. Remarkably, this is a *constant* complexity w.r.t. to N. We note that the disappearance of N is due to the fact that the accuracy of our approximation to the exact solution is defined in terms of the probability. What remarkable is that with a very tight bound required on a accuracy (e.g., within 95%), the number of points required to be sampled is remarkably small (e.g., 59 points regardless of the data size N).

Hence the complexity of *Algorithm 3* is $\mathbf{O}(\text{const})$.

5   Experimental Results

In the first set of experiments, we demonstrate that our proposed FOSVM is comparable with both *OCSVM* and *SVDD* in terms of the learning capacity but are much faster. *OCSVM* and *SVDD* are implemented by using the state-of-the-art *LIBSVM solver* [5]. All comparing methods are coded in C# and run on a computer with 4 GB of memory.

The datasets in the experiment were constructed by choosing one class as the normal class, appointing the remaining classes as the abnormal class, and randomly removing the samples from the abnormal class such that the ratio of data in the normal and abnormal classes are 50 : 1. The details of the experimental datasets are given in Table 1.

We used three-fold cross validation. For the training folds, we only used the positive samples to train the model and we used all samples of the testing folds for testing. Gaussian kernel, which is given by $K(x, x') = e^{-\gamma \|x - x'\|^2}$, was employed. The width of kernel γ was searched in the grid $\{2^{-15}, 2^{-13}, \ldots, 2^5\}$. The trade-off parameter C was varied in the grid $\{2^{-15}, 2^{-13}, \ldots, 2^5\}$. For *FOSVM2*, we set the parameter θ to 0.05 and $m = 100$ corresponding to the confidence of 99.33% for obtaining the top $\epsilon = 5\%$ furthest samples. To measure the accuracy, we applied the measurement given by $acc = \frac{acc^+ + acc^-}{2}$ where acc^+, acc^- are the accuracies on the positive and negative classes, respectively. This measurement is appropriate for ND because it encourages the high accuracies for both two classes.

Table 2 shows the performance of our proposed models versus baselines on a diverse set of datasets. As can be seen from Table 2, *FOSVM1* and *FOSVM2* are comparable with others in terms of the accuracies but much faster. Particularly, *FOSVM2* is superior in its time complexity as expected from our theoretical result. Surprisingly, *FOSVM2* also gains the highest accuracies on 9 out of 13

Table 1. The details of the experimental datasets

Datasets	#Features	#Pos	#Neg
Satimage	36	1,072	21
Usps	256	1,194	23
News20	1,355,191	3,986	79
a9a	123	7,841	150
Porker	10	10,599	211
Acoustic	407	18,261	366
Seismic	60	18,320	366
Real-sim	20,958	22,213	444
Shuttle	9	34,108	682
Connect-4	126	44,473	889
CodRna	8	90,539	1,810
IJCNN	22	91,701	1,834
Covertype	54	297,711	5,954
Rvc1	47,236	355,460	7,190

datasets. This may partially be explained from the fact that the furthest sample found by *FOSVM1* may be noise or outlier. Hence, *FOSVM1* and others are more sensitive to noises and outliers than *FOSVM2*. In Table 2, we boldfaced the datasets at which the training times are sped up more than 10 times. As seen from Table 2, for the large-scale datasets including CodRna, Covertype, and Rvc1 whose sizes amount greater than $90,000$, the training time speed-up ratios are all greater than 10 and are approximately equal to 40, 12, 28, respectively. Especially, for the dataset Seismic, our proposed method *FOSVM2* are 808 times faster than the baseline.

In the second experiment, we aim to investigate the behaviors of our proposed methods. For each dataset, we randomly chose $10\%, 20\%, \ldots, 100\%$ of data and

Table 2. The accuracies on the experimental datasets

Datasets	SVDD		OCSVM		FOSVM1		FOSVM2	
	Acc	Time	Acc	Time	Acc	Time	Acc	Time
Satimage	94%	63s	94%	45s	94%	35s	94%	20s
Usps	96%	113s	96%	85s	91%	117s	91%	85s
News20	60%	6,970s	60%	7,354s	59%	4,673s	61%	3,236s
a9a	65%	821s	58%	548s	65%	822s	69%	122s
Porker	50%	551s	50%	562s	50%	554s	53%	82s
Acoustic	66%	6,211s	67%	6,138s	60%	1,066s	70%	59s
Seismic	71%	54,177s	71%	42,244s	68%	924s	73%	67s
Real-sim	70%	8,690s	69%	8,885s	61%	3,566s	64%	529s
Shuttle	92%	723s	91%	726s	94%	167s	94%	43s
Connect-4	53%	1,772s	53%	1,022s	55%	460s	55%	157s
CodRna	62%	13,224s	62%	13,468s	60%	655s	64%	315s
Covertype	55%	105,157s	54%	107,760s	52%	78,942s	54%	8,151s
Rvc1	63%	204,331s	63%	204,298s	56%	9,236s	54%	7,439s

evaluated both the training accuracies and times for each sub-dataset. To ensure the stability of the proposed methods, we ran each case ten times. To explicitly observe the complexities of *FOSVM1* and *FOSVM2*, we took average of the training times of ten times for each percentage and then plotted them. As seen in Figure 3 (left), the training times of *FOSVM2* is increased at first and eventually does slightly change when the training size is sufficiently large. As shown in Figure 3 (right), the training time of *FOSVM1* is approximately linear.

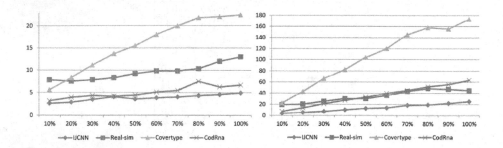

Fig. 3. The averages of the training times of FOSVM1 (right) and FOSVM2 (left)

6   Conclusion

In this paper, we integrate the formulation of cutting-plane method and the well-known One-Class Support Vector Machine to propose Fast One-Class Support Vector Machine (FOSCM) for novelty detection. We actually propose two complementary solutions FOSVM1 and FOSVM2 for training and detection. The first is an exact solution thas has linear complexity. The number of iterations to reach a very close accuracy, termed as θ-precision is $O\left(\frac{1}{\theta}\right)$, making it attractive in dealing with large datasets. The second solution employed a sampling strategy that remarkably has constant computational complexity defined relative to probability of approximation accuracy. The experiment results indicate that our proposed methods are comparable with OCSVM and SVDD in terms of accuracy but much faster than them. The speed-up in training time could reach 808 times for some dataset.

A   Appendix

Because of the limit space, in this appendix, we only sketch out the theoretical results used in the paper.

Lemma 1. *The following holds:*

1. $\rho_n = \|w_n\|^2, d_n = \|w_n\|$
2. $1 > \|w_1\| \geq \|w_2\| \geq \ldots \geq \|w_n\| \geq \ldots \geq \|w^*\|$ where $w^* = w_\infty = \lim\limits_{n \to \infty} w_n$.

Lemma 2. *There exists* x_i $(1 \le i \le n)$ *such that* $\|\Phi(x_i) - w_n\|^2 = 1 - \|w_n\|^2$ *and* $(\Phi(x_i) - w_n)^T u \le 0$ *for any vector* $u \ne \mathbf{0}$.

Lemma 3. *The following holds:*

1. $\|w_j - \Phi(x_i)\|^2 \le 1 - \|w_j\|^2$ $(1 \le i \le j)$
2. $\|w_n\|^2 - \|w_j\|^2 \ge \|w_n - w_j\|^2$ $(j \ge n+1)$
3. $\|x_{n+1} - w_n\|^2 \ge 1 - \|w^*\|^2$

Lemma 4. *Let us denote* $\lambda_n = \sqrt{\frac{1-\|w_n\|^2}{1-\|w^*\|^2}}$. *We have* $\lambda_n \ge 1 - \frac{1}{1+n/2}$.

References

1. Abbey, A.S., Temitope, O., Lionel, S.: Active platform security through intrusion detection using naive baycsian network for anomaly detection. In: Proceedings of London Communications Symposium (2002)
2. Augusteijn, M.F., Folkert, B.A.: Neural network classification and novelty detection. International Journal of Remote Sensing **23**, 2891–2902 (2002)
3. Badoiu, M., Clarkson, K.L.: Optimal core-sets for balls. In: Proc. of DIMACS Workshop on Computational Geometry (2002)
4. Chandola, V., Banerjee, A., Kumar, V.: Anomaly detection: A survey. ACM Comput. Surv. **41**(3), 1–58 (2009)
5. Chang, C.-C., Lin, C.-J.: Libsvm: A library for support vector machines. ACM Trans. Intell. Syst. Technol. **2**(3), 27:1–27:27 (2011)
6. Fan, W., Miller, M., Stolfo, S.J., Lee, W.: Using artificial anomalies to detect unknown and known network intrusions. In: Proceedings of the first IEEE International Conference on Data Mining, pp. 123–130. IEEE Computer Society (2001)
7. Fawcett, T., Provost, F.: Activity monitoring: noticing interesting changes in behavior. In: Proceedings of the Fifth ACM SIGKDD International Conference on Knowledge Discovery and Data Mining, pp. 53–62 (1999)
8. Gwadera, R., Atallah, M.J., Szpankowski, W.: Reliable detection of episodes in event sequences. Knowl. Inf. Syst. **7**(4), 415–437 (2005)
9. Joachims, T., Finley, T., Yu, C.-N.J.: Cutting-plane training of structural svms. Machine Learning **77**(1), 27–59 (2009)
10. Joachims, T., Yu, C.-N.J.: Sparse kernel svms via cutting-plane training. Machine Learning **76**(2–3), 179–193 (2009)
11. Kelley, J.E.: The cutting plane method for solving convex programs. Journal of the SIAM **8**, 703–712 (1960)
12. Schölkopf, B., Platt, J.C., Shawe-Taylor, J.C., Smola, A.J., Williamson, R.C.: Estimating the support of a high-dimensional distribution. Neural Comput. **13**(7), 1443–1471 (2001)
13. Shalev-Shwartz, S., Singer, Y., Srebro, N., Cotter, A.: Pegasos: Primal estimated sub-gradient solver for svm. Mathematical Programming **127**(1), 3–30 (2011)
14. Tax, D.M.J., Duin, R.P.W.: Support vector data description. Journal of Machine Learning Research **54**(1), 45–66 (2004)

15. Teo, C.H., Smola, A., Vishwanathan, SVN, Le, Q.V.: A scalable modular convex solver for regularized risk minimization. In: Proceedings of the 13th ACM SIGKDD International Conference on Knowledge Discovery and Data Mining, pp. 727–736. ACM (2007)
16. Teo, C.H., Vishwanthan, S.V.N., Smola, A.J., Le, Q.V.: Bundle methods for regularized risk minimization. The Journal of Machine Learning Research **11**, 311–365 (2010)
17. Tsang, I.W., Kocsor, A., Kwok, J.T.: Simpler core vector machines with enclosing balls. In: Proceedings of the 24th International Conference on Machine Learning, ICML 2007, pp. 911–918 (2007)
18. Tsang, I.W., Kwok, J.T., Cheung, P., Cristianini, N.: Core vector machines: Fast svm training on very large data sets. Journal of Machine Learning Research **6**, 363–392 (2005)

ND-Sync: Detecting Synchronized Fraud Activities

Maria Giatsoglou[1]($\boxtimes$), Despoina Chatzakou[1], Neil Shah[2], Alex Beutel[2],
Christos Faloutsos[2], and Athena Vakali[1]

[1] Informatics Department, Aristotle University of Thessaloniki, Thessaloniki, Greece
{mgiatsog,deppych,avakali}@csd.auth.gr
[2] School of Computer Science, Carnegie Mellon University, Pittsburgh, USA
{neilshah,abeutel,christos}@cs.cmu.edu

Abstract. Given the retweeting activity for the posts of several Twitter users, how can we distinguish organic activity from spammy retweets by paid followers to boost a post's appearance of popularity? More generally, given groups of observations, can we spot strange groups? Our main intuition is that organic behavior has more variability, while fraudulent behavior, like retweets by botnet members, is more synchronized. We refer to the detection of such *synchronized* observations as the *Synchonization Fraud* problem, and we study a specific instance of it, *Retweet Fraud Detection*, manifested in Twitter. Here, we propose: (A) ND-Sync, an efficient method for detecting *group fraud*, and (B) a set of carefully designed features for characterizing retweet threads. ND-Sync is *effective* in spotting retweet fraudsters, *robust* to different types of abnormal activity, and *adaptable* as it can easily incorporate additional features. Our method achieves a 97% accuracy on a real dataset of 12 million retweets crawled from Twitter.

1 Introduction

Suppose that Twitter user "John" posts a tweet, and in 10 minutes it is retweeted by 3,000 people. Is this suspicious? Not necessarily. Suppose that this happens for the next 5 tweets of John: roughly the same 3,000 people all retweet his post in a few minutes.

Now, *that* is suspicious because such synchronized behavior among so many users is not natural. This is exactly the main intuition behind our work: events (like retweet threads) that belong to the same group are suspicious if they are highly synchronized; that is, if all of the retweet threads for John's tweets are synchronized. The challenge therefore is, given many events belonging to many groups (retweet threads for user "Jane", etc.), find the suspicious groups. We assume that organic behavior is the norm, and deviations from it constitute suspicious (anomalous, fraudulent) behavior.

Anomaly and outlier detection has attracted a lot of interest, both at the individual level (e.g. a single retweet thread in our example) [3,12,20] as well as *group* [25] or *collective anomalies* [6]. While most anomaly detection focused

© Springer International Publishing Switzerland 2015
T. Cao et al. (Eds.): PAKDD 2015, Part II, LNAI 9078, pp. 201–214, 2015.
DOI: 10.1007/978-3-319-18032-8_16

on a cloud of p-dimensional points (entities), extensions to more complex data have been proposed [6], such as for rare subsequences, subgraphs, and image subregions. Here, we propose a novel, general approach to collective anomaly detection, informally defined as follows:

Informal Problem 1 (Synchronicity Fraud Detection)

Given: *N groups of entities; a representation for each entity in a p-dimensional space;*

Identify *groups of entities abnormally synchronized in some feature subspaces.*

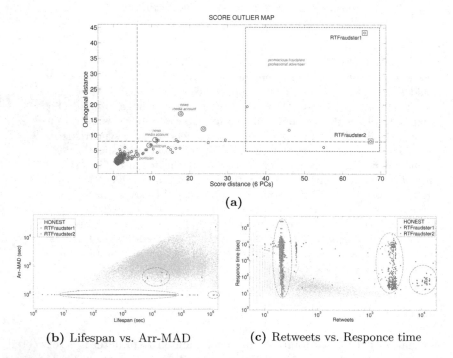

(b) Lifespan vs. Arr-MAD   (c) Retweets vs. Responce time

Fig. 1. ND-SYNC detects outliers clearly separated from the majority of users. In (1a), points right of the vertical dashed line are spotted as RT fraudsters. (1b), (1c) reveal high synchronicity in the retweet threads of users RTFraudster1 and RTFraudster2, for different feature combinations.

At a high level, our proposed ND-SYNC methodology is as follows:

1. Extract p features from each entity (i.e. retweet thread), such as the average inter-arrival time of retweets, variance of it, and number of retweets.
2. Analyze the collective behavior of each group (the set of retweet threads for posts from a given user), and compare it to the behavior of the rest of the threads. Using the concepts of "intra-synchronicity" and "inter-synchronicity" (see Section 4.2), assign a suspiciousness-score to each user in all 2^p available subspaces.

3. Combine the scores for each user and report the most suspicious such groups.

Figure 1 illustrates the effectiveness of ND-Sync in detecting Twitter accounts whose posts trigger fake retweet threads. In Figure 1a each point is a user plotted in terms of two dimensions that reflect deviation from normal users' behavior. The vertical dashed line clearly separates fraudulent from normal users. The most "anomalous" users (at the rightmost part of the figure) correspond to bot accounts acting as professional promoters of content or other users. Figures 1b and 1c depict the retweet threads of normal users (grey points) along with the retweet threads of the "significantly outlying" caught users RTFraudster1 and RTFraudster2 projected in 2-D subspaces with respect to two pairs of our proposed features. Compared to honest retweet threads, we can clearly observe fraudulent users' retweet threads are abnormaly clustered together.

The contributions of this work are the following:

- **Methodology**: ND-Sync is a general, effective pipeline that automatically detects group anomalies.
- **Feature engineering**: we customize ND-Sync for the case of retweet fraud, using a carefully selected set of features.

Reproducibility: We share our (anonymized) data at: http://oswinds.csd.auth.gr/project/NDSYNC.

2   Related Work

We first discuss approaches addressing fraud detection in Twitter, and then review efforts on collective anomaly detection. Table 1 compares ND-Sync to the methods that are most relevant to our problem setting. ND-Sync does not require textual content, graph structure, or user attributes (such as account creation date) to detect fraudsters.

Table 1. ND-Sync comparison against alternatives. (*: in multiple feature subspaces, **: searches the best, ***: in a single 2D feature space)

		Textual content-agnostic?	Graph structure-agnostic?	User attributes-agnostic?	Parameter free?	Unsupervised?	Designed for RTFRAUD?	Detects synchronicity fraud?
ND-Sync		✓	✓	✓	✓	✓	✓	✓*
Twitter Fraud Detection	[23]	✓	✓	✓	✓			
	[27]			✓		✓		
	[7]				✓			
	[11]	✓	✓	✓	✓**	✓	✓	
Collective Anomaly Detection	[25]	✓	✓	✓				
	[28]	✓		✓				
	[16]	✓		✓	✓	✓		✓***

Fraud on Twitter: *Retweet Fraud Detection* identifies Twitter users that obtain fake retweets for their posted content (RT fraudsters). Fraud is a serious problem on Twitter [24], with fraudsters exhibiting several classes of strange behaviors [8], and varying degrees of automation. [9] lists several such attempts of fraudsters to mimick organic behavior, for example by re-broadcasting others' posts. Earlier work focuses on account tweeting activity and/or social connectivity. [27] analyzes the relationships of fraudsters inside and outside of criminal communities to infer types of accounts serving as criminal supporters. [7,8] leverage tweeting behavior, tweet content and account properties to compute the likelihood of an unknown user being a human, bot or cyborg. In general, such feature-based methods (e.g. username pattern, age) have been shown to fail to catch more sophisticated fraud schemes that exploit real users' accounts, such as the *pyramid follower markets* [22] and *account compromization* [1]. [23] shows the effectiveness of temporal features for distinguishing between account types. [11] addresses a problem similar to ours but uses the URLs found in tweets instead of retweet threads in conjunction with a time and user-based entropy to classify posting activity and content.

Collective anomaly detection: The goal of collective anomaly detection is to find groups of entities that jointly exhibit abnormal behavior. Variants exist for sequential [5], spatial [13] and graph [19] data, while other approaches are more general, simply assuming p-dimensional points [25,26,28]. Synchronized ("lock-step") behavior is often an indication of fraud, like e.g. users Liking the same Facebook Pages at the same time [2] or following the same accounts [16,17,21].

3   Background

This section provides the necessary background for ND-SYNC: a formula to measure the suspiciousness of a group of 2-D points, given a large set of 2-D points (Section 3.1) and a robust multivariate outlier detection method (Section 3.2).

3.1   Measuring Group Strangeness

Given a large cloud of 2-D points $\mathcal{E}$, and a set of points $\mathcal{E}' \subset \mathcal{E}$, how unusual is $\mathcal{E}'$ with respect to $\mathcal{E}$? [16] gave such a score, namely, the *residual score* (rs_score, see Eq. 4 below). The definition needs some auxiliary concepts: *synchronicity* (how coherent/lockstep is the subset $\mathcal{E}'$) and *normality* (how similar it is to the presumed-normal cloud $\mathcal{E}$). In the next 3 paragraphs, we give their mathematical definitions, as well as the equation for a crucial lower-bound.

Synchronicity: For a group $\mathcal{E}'$, it is the average closeness between all pairs of its members

$$sync(\mathcal{E}') = \bar{c}(e, e') \qquad e, e' \in \mathcal{E}' \tag{1}$$

The closeness $c(e, e')$ of two entities e and e', is a similarity function - in [16], it was binary: after dividing the address space into grid-cells, the closeness is 1, if the elements are in the same grid-cell, and zero otherwise.

Normality: The *normality*, of a group $\mathcal{E}'$ with respect to a (superset) group $\mathcal{E}$, is the average closeness of the members of $\mathcal{E}'$ to the members of $\mathcal{E}$.

$$norm(\mathcal{E}') = \bar{c}(e, e') \qquad e \in \mathcal{E}, e' \in \mathcal{E}' \tag{2}$$

Lower bound: Given a group $\mathcal{E}'$, with normality value n with respect to a (superset) group $\mathcal{E}$, the synchronicity $sync(\mathcal{E}')$ is lower-bounded by $sync_{min}(n)$:

$$sync_{min}(n) = (-Mn^2 + 2n - s_b)/(1 - Ms_b) \tag{3}$$

where M is the count of non-empty grid-cells for $\mathcal{E}$, and s_b is the synchronicity of $\mathcal{E}$.

Residual score: For group $\mathcal{E}'$ with synchronicity $sync(\mathcal{E}')$ and normality $norm(\mathcal{E}')$ wrt $\mathcal{E}$, the **residual score** is given by:

$$rs\_score(\mathcal{E}') = sync(\mathcal{E}') - sync_{min}(norm(\mathcal{E}')) \tag{4}$$

3.2 Robust Outlier Detection

ROBPCA-AO [15] is a robust *Principal Component Analysis* (PCA) and outlier detection method that is suitable for multivariate, high-dimensional data and independent of their features' distribution. Proposed as a robust alternative to classic PCA, ROBPCA-AO identifies Principal Components (PCs) that best describe the uncontaminated data, while at the process, it detects outliers.

Initially, the method applies SVD on the set of observations to project them into the (restricted) space they span. Then, the dimensionality of data is reduced by keeping only the first k PCs of the data's covariance matrix. Taking into account the possibility of skewed data, ROBPCA-AO computes a robust k-subspace V_r that fits the majority of observations and projects them on it. Outliers are detected on subspace V_r based on two distance-based scores: (a) the *orthogonal distance* (*od*) of each observation from its projection on V_r, and (b) *robust score distance* (*sd*), which is taken as the *adjusted outlyingness* of the observation. *Adjusted outlyingness* is a measure suitable for multivariate, asymmetric data that estimates the distance of a given observation from the bulk of observations as its maximum robust distance (outlyingness) over B directions. Each such direction is perpendicular to the subspace spanned by k randomly sampled observations. Observations whose *od* or *sd* score surpasses a data-dependent cutoff threshold (defined as the upper whisker of the *adjusted boxplot* [15]) are characterized as outliers.

4 Proposed Method: ND-SYNC

This section first outlines *Retweet Fraud Detection* as a special case of the *Synchronicity Fraud* problem, and then presents ND-SYNC, an effective and robust solution.

4.1   Problem Defintition

Retweet fraud detection (RTFRAUD) is a problem of various dimensions since: (a) it can be practiced by different types of user accounts (automated or bot orchestrated, semi-automated, human managed), (b) the inflation of content's popularity can be the sole purpose of the suspected user account or an occasional tactic hidden (*camouflaged*) in organic activities, (c) the promoted content can attract both fake and honest retweets. Thus, it is important to find features able to separate such diverse fraudulent activities from honest user behavior, and to design a method that can effectively leverage their variety. To study the retweeting activity in terms of time and retweeting users, given a user u_m (*author*) we represent the ith tweet posted by u_m with $tw_{m,i}$ as a tuple $(u_m, t_{m,i})$, where $t_{m,i}$ is the tweet's creation time. A retweet thread is defined as follows:

Definition 1 (Retweet thread). *Given an author u_m and a tweet $tw_{m,i}$, a retweet thread $R_{m,i}$ is defined as the set of all tweets that retweeted $tw_{m,i}$.*

Here, we formulate RTFRAUD as an instance of the *Synchronicity fraud* (SYNCFRAUD) problem, which is defined at two levels (group and entity) below:

Problem 1 (SYNCFRAUD).

Given: N groups of entities G, where each group $g_m \in G$ comprises a variable number of entities $e_{m,i} \in E_m$, and a set of p features for the entities' representation,
Extract: a set of features at the group-level, and
Identify: suspicious groups S that exhibit highly *synchronized* characteristics.

Even though essentially, the RTFRAUD problem involves three levels instead of two – lower (individual retweets), middle (retweet threads) and upper (users) – we simplify it by collapsing a post's retweets into a single retweet thread and by defining features for its characterization. Then, RTFRAUD can be directly mapped to the SYNCFRAUD problem where each user u_m has a group g_m containing all retweet threads $R_{m,i}$ for that user, and the suspicious groups S are the detected fraudsters (RT fraudsters).

4.2   Proposed Approach

In this section we provide the pipeline of ND-SYNC, our approach to the SYNCFRAUD problem, and describe its steps. Then, we propose a 7-dimensional feature space for representing the retweet threads in the RTFRAUD problem.

ND-SYNC Pipeline. ND-SYNC comprises three main steps: (1) *Feature subspace sweeping*, which generates and bins all entity-wise feature subspaces and projects entities in them; (2) *User scoring*, which calculates the group-based suspiciousness score vectors; (3) *Multivariate outlier detection*, which identifies suspicious groups based on their deviation from normal behavior. ND-SYNC's pipeline is outlined in Algorithm 1.

Data: $E = \{e_i\}$: Set of p-dimensional entities, $G = \{g_m\}$: Set of N groups
     where $g_m = \{e_{m,i}\}$, I : number of iterations
Result: S : suspicious groups
generate all 2^p subspaces F and project e_i in them ;
logarithmically segment all subspaces and assign e_i to bins ;
for *each group g_m in G* **do**
    for *each feature subspace f_i in F* **do**
        calculate *suspiciousness* $susp(g_m, f_i)$ (Def. 2);
    combine $susp$ over all subspaces in vector $SU(g_m)$ (Def. 3);
robustly scale and center $SU(g_i)$ vectors, $i \in [1, N]$;
for $iter = 1 : I$ **do**
    extract set of outliers S_{iter} from $SU(g_i)$ applying multivariate outlier
    detection ;
extract final S set by applying majority voting on all S_{iter} ;

Algorithm 1. Pseudocode for ND-SYNC

Feature subspace sweeping. How can we detect microclusters in a p-dimensional space? Given N groups of entities represented in a p–feature space, ND-SYNC first: (a) projects all entities into the desired feature subspaces, and then (b) reduces the statistical noise in each subspace to prepare the data for synchronicity detection. Given a p-feature space, we take all possible q-feature combinations for $q \in [1, p]$; this produces 2^p possible subspaces to analyze. In anomaly detection, it is difficult to estimate apriori the most effective feature combinations for discriminating suspicious groups from normal ones. Thus, a straightforward approach is to generate and apply ND-SYNC's next steps on all 2^p feature subspaces. In cases when p is too high, though, practitioners can select a subset of subspaces for extra efficiency. As we show in Section 5.2, considering only the 3-D and 2-D subspaces is relatively effective for RTFRAUD.

To compute the suspiciousness score, we need to bin the feature space (see subsection 3.1); we choose logarithmic binning, in powers of 2, for each dimension/feature.

User scoring. At this point, we need to combine the entity-level features in a way that reflects the synchronicity of the group. These group features should allow us to identify the suspicious set S from the normal groups. Here, on each entity-feature subspace f_i, where $i \in [1, 2^p]$, we calculate the *intra-synchronicity* $intra\_sync(g_m, f_i)$ and *inter-synchronicity* $inter\_sync(g_m, f_i)$ measures for each group of entities g_m, based on Eq. 1 and Eq. 2, respectively.

We expect that suspicious groups will have significantly higher intra-synchronicity compared to normal groups. However, depending on the feature subspace, the deviation between the intra-synchronicity of suspicious and normal groups may vary (due to differences in the features' discriminative power and distribution). Thus, given $F = 2^p$ feature subspaces, we generate an F-dimensional feature vector for each group g_m, i.e. the *suspiciousness score* vector, SV, where each dimension represents the group's *suspiciousness*, *susp*, in the corresponding subspace (given in Definition 2). The *susp* scores correspond to our user-level features.

Definition 2 (Suspiciousness). *For subspace f_l and group g_m with intra-synchronicity $intra\_sync(g_m, f_l)$ and inter-synchronicity $inter\_sync(g_m, f_l)$, the group's suspiciousness is given by* $\mathbf{susp(g_m, f_l)} = rs\_score(projection(g_m, f_l))$.

Definition 3 (Suspiciousness score vector). *For group g_m and $i \in [1, F]$, the suspiciousness vector is given by* $\mathbf{SV(g_m)} = [susp(g_m, f_i)]_{i=1}^{F}$.

Unlike previous approaches, here we do not assume that suspicious groups are characterized by low inter-synchronicity. We claim that inter-synchronicity is more difficult to interpret as an indication of normality vs. suspiciousness, since its value depends on the selected features and their distribution over all entities in normal and suspicious groups. For example, our experiments on RTFRAUD indicated that normal users are approximately at the same scale of inter-synchronicity, whereas suspicious users can have either very low values (retweet threads have rare feature values) or very high (retweet threads of several suspicious users have the same feature values, e.g. zero inter-arrival time of retweets; for normal users the corresponding values are more diverse).

Multivariate outlier detection. The last step of ND-SYNC takes as input the set of groups G with their F-dimensional suspiciousness score vectors, and proceeds to the identification of the suspicious groups S. First, we standardize the vectors using their *median* and *mean average deviation* (MAD), which are considered as robust estimators of center and scale. ND-SYNC then spots as suspicious the groups that largely deviate from the majority of groups in G, considering their standardized scores in all feature subspaces, based on the outlier detection approach described in Section 3.2.

To address the non-deterministic nature of this outlier detection method, mainly in terms of entities positioned close to the distance cutoffs, ND-SYNC applies it iteratively, maintains a list of the identified outliers in each iteration, and classifies a group as suspicious based on the majority vote over all runs. Our experiments on RTFRAUD revealed that even a small number of iterations (e.g. 10) is enough to estimate the suspicious groups. Moreover, to eliminate the need for selecting parameters in ND-SYNC:

1. We propose automatic selection of the k principal components via a heuristic technique such as the *95% cumulative variance explained* criterion [18]. According to this criterion, during dimentionality reduction, the first k components that together explain more than 95% of the data's variance are maintained;

2. We use all entities, instead of a subset of them (as in the approach of Section 3.2), to estimate the robust feature subspaces. This is reasonable since typically the percentage of outliers in a dataset is small, and difficult to estimate apriori.

Feature Engineering for Retweet Threads. Earlier works have associated bot activity with temporal activity anomalies, such as low entropy in the time intervals between posts. Here, we also expect that the retweet threads of RT fraudsters will exhibit different inter-synchronicity and high intra-synchronicity compared to honest users, with respect to their temporal characteristics. In addition, due to automation tools, as well as due to the way retweet markets operate, we expect RT fraudsters to be synchronized in terms of the number of retweets their posts receive. Based on the above, after experimenting with several features which are omitted here for brevity, we ended up with the following features for a retweet thread's representation:

- **Retweets**: number of retweets
- **Response time**: time elapsed between the tweet's posting time and its first retweet
- **Lifespan**: time elapsed between the first and last (observed) retweet, constrained to 3 weeks to remove bias with respect to later tweets
- **RT-Q3 response time**: time elapsed after the tweet's posting time to generate the first 3 quarters of the (observed) retweets
- **RT-Q2 response time**: time elapsed after the tweet's posting time to generate the first half of the (observed) retweets
- **Arr-MAD**: mean absolute deviation of inter-arrival times for retweets
- **Arr-IQR**: inter-quantile range of inter-arrival times for retweets

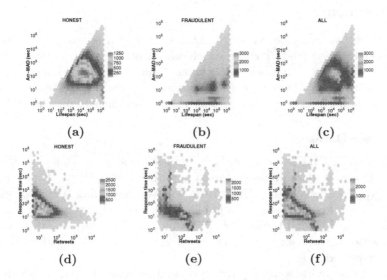

Fig. 2. Synchronicity patterns are revealed as microclusters in RT fraudsters behavior. a, b, c (d, e, f) correspond to the Lifespan vs. Arr-MAD (Retweets vs. Response time) scatter plots for "honest", "fraudulent", and both types of users. The majority of organic retweet threads are clustered around a limited range of values, clearly separated from the fraudsters' microclusters.

To estimate the suitability of the proposed features for revealing retweet fraud, we examined the projections of the retweet threads of the Twitter dataset described in Section 5.1 in all 2-D feature subspaces derived by the proposed feature set. To assist visualization, we binned all feature subspaces and generated 2-D heatmaps in logarithmic scales.

Figure 2 indicatively depicts the scatter plots of *Lifespan vs. Arr-MAD* and *Retweets vs. Response time* for all dataset's users (Fig. 2c and 2f), and only for those annotated as "honest" (Fig. 2a and 2d) and "fraudulent" (Fig. 2b and 2e).

The *Lifespan vs. Arr-MAD* plots reveal microclusters of fraudulent retweet threads (at very low values of Arr-MAD and at high Lifespan values), whereas the majority of honest users' retweet threads seem to be concentrated around a certain area of the feature subspace, clearly separated from RT fraudsters. Similar observations are made from the *Retweets vs. Response time* plots where we observe a microcluster at abnormaly low values for both features, and another one for high Response time. Some of these results were anticipated based on our intuition, e.g. bots of the same network may retweet all at once, having on average a zero Arr-MAD, but it seems that our proposed features can reveal more complex retweet fraud practices. For example, promoted posts may continue to receive retweets for a prolonged period of time, which explains the microcluster of long Lifespan, whereas certain RT fraudsters may wait for some time, after posting their tweets, before applying to some retweet market for their promotion.

5   Experiments

In this section, we evaluate ND-SYNC by conducting a series of experiments on a dataset crawled from Twitter which is comprised of over *130K retweet threads* characterizing more than *11M retweets* to posts of several hundred active Twitter users. We detail our data collection approach, describe the settings of our numerous experiments, and finally present the performance of our ND-SYNC.

5.1   Twitter Dataset

The evaluation of ND-SYNC requires a dataset of several, *complete* retweet threads of honest and fraudulent users, i.e. with no gaps in the tuples representing a given post's retweets. Due to the Twitter Streaming API's constraint of allowing access to only a sample of published tweets, our requirement for complete retweet threads, and the lack of a relevant (labeled) dataset, we manually selected a set of target users for whom we could track all tweets and retweets in a given time period.

Target user accounts were selected in several fashions. Firstly, we examined a recent, 2-day sample of the global Twitter timeline and identified the users who posted the most retweeted tweets and those who posted tweets containing keywords heavily used in spam campaigns (*casino, followback,* etc). Our next approach involved selecting users based on "Twitter Counter"[1], a web application that publishes lists ranking Twitter users based on criteria including follower

---

[1] http://twittercounter.com/

count and number of tweets. We chose users based on their posting frequency and influence – specifically, we kept only users who tweeted several times per week and received more than 100 retweets on their recent posts. Lastly, we collected users active in specific topics (European affairs and Automobile), given that they were added in such topic-related lists by other Twitter users.

We manually labeled target users as "fraudulent" in cases where (a) inspection of their tweets' content led to the discovery of spammy links to external web pages, spam-related keywords, and multiple posts with similar promotions or vacuous content (e.g. quotes), and (b) profile information was clearly fabricated. The rest of the target users were labeled as "honest". We monitored the set of target users for time periods spanning from 2 to 6 months and eliminated those who had less than 20 retweet threads or a maximum-length retweet thread of less than 50 retweets. This process left us with a total of 298 users in the dataset, of which 270 were labeled honest and 28 labeled fraudulent. For each user, we extracted the retweet threads and mapped them to our proposed 7-D feature space. The dataset includes **134,022 retweet threads** (83,587 with respect to honest and 50,435 of fraudulent users) which in total comprise **11,727,258 retweets** (2,939,455 to posts of honest and 8,787,803 to posts of fraudulent users).

5.2 Results

Next, we present the experimental results of ND-SYNC's application on the Twitter dataset described above. We discuss its effectivess when ND-SYNC was applied on (a) all available 2^7 feature subspaces, (b) a restricted subset of the feature subspaces.

Preliminary Observations on the Data. Before applying the final outlier detection step of ND-SYNC, we examine the distribution of standardized user-level scores for honest users with respect to each feature subspace. All score variables were found to be significantly asymmetric, having a *medcouple*[2] [4] between 0.16 and 0.46, and the corresponding p-values rejected the normality hypothesis at the 1% significance level. If honest users' scores had symmetric distributions, we could apply typical thresholded outlier detection techniques that assume normal distribution [10,14]. However, in our case, the skewness of the "uncontaminated" data would likely result in many false positives. Conversely, ND-SYNC's outlier detection approach is rather suitable for the RTFRAUD problem.

Detection Effectiveness on Real Data. Figure 3 shows ND-SYNC's performance on detecting RT fraudsters in terms of *F1-score* and *accuracy*. To examine robustness of ND-SYNC to the number of dimensions maintained (k)

---

[2] Medcouple measures the skewness of a distribution in [0, 1] range. Right and left skewed distributions have positive and negative medcouple respectively; symmetric distributions have zero medcouple.

in the beginning of the outlier detection step, we provide perfomance measures for k from 1 to 10 and all feature subspaces considered for the users' suspiciousness estimation. We observe that ND-SYNC is relatively robust with respect to k: we attain $[95\% - 97\%]$ accuracy and $[0.73 - 0.82]$ F1-score. For $k = 6$, based on the *95% cumulative variance explained criterion*, ND-SYNC has the best performance with respect to precision-recall balance and accuracy. Our experiments with ND-SYNC considering only the 3-D and the 2-D feature subpaces showed effectiveness in catching several cases of fraud. Specifically, the accuracy (F1-score) with respect to the best performing ND-SYNC run in all feature subspaces was reduced only by 4.5% (0.4%) and 10.6% (1%) for 3-D and 2-D subspaces, respectively.

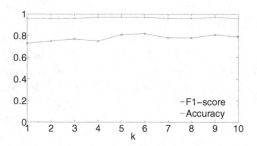

Fig. 3. ND-SYNC **is highly accurate and robust to the selection of** k **(number of dimensions maintained).** The best performance is at $k = 6$ which explain approximately 95% of the data's cumulative variance.

Observations on the Outlier Map. Figure 1a illustrates the outlier map of *sd* and *od* scores generated for the best run of ND-SYNC, where red lines correspond to adaptive cutoff values for *sd* and *od* – the plot clearly discerns the outliers from the majority of users which lie in the bottom-left region. All discovered outliers have an abnormal *sd* score, whereas 36% also have outlying *od* scores (in the top-right quartile of the figure).

A closer examination of RT fraudsters that were caught by ND-SYNC reveals that the ones who scored high in *sd* and *od* were exemplary bot accounts that are typically hired for promotion or advertisement. For example, RTFraudster1, enclosed within a rectangle in the top-right quartile of Figure 1a, is an example of such a *promiscuous* fraudster with 800 followers that had 65 retweet threads in a 4-month time period – 80% (60%) of these were comprised of more than 1k (10k) retweets and had almost 0 Arr-IQR[3]. The remainder of the "caught" RT fraudsters have a more *subtle* profile, resembling cyborg behavior: the accounts often create vacuous posts, but occasionaly interact genuinely with other users, thus indicating a human operator. We found that the five false positives detected by ND-SYNC (enclosed by a red circle in Figure 1a) belonged to media accounts and politicians. Three of these accounts have significantly abnormal *sd* and *od*

---

[3] This account was later suspended by Twitter.

scores, while the others are situated very close to RT fraudsters, suggesting that they may have tampered with the organic behavior of their retweet threads.

6   Conclusions

In this work, we broach the problem of discerning fraudulent, group-based activity from organic online behavior. The contributions of this work are the following:
- **Methodology**: we present ND-SYNC, a general, effective pipeline, which automatically detects group anomalies
- **Feature engineering**: a carefully designed set of features, that customize ND-SYNC for the case of retweet fraud.

We present experiments on real data from Twitter consisting of almost *12 million retweets*, where the proposed ND-SYNC achieved excellent classification accuracy of 97% in distinguishing fraudulent from honest users.

Reproducibility: For the reproducibility of our results we share an (anonymized) version of our data at: http://oswinds.csd.auth.gr/project/NDSYNC.

Acknowledgments. The authors would like to thank Dr Stephan Günnemann for his insightful comments and valuable discussions about this work.

References

1. Almaatouq, A., et al.: Twitter: who gets caught? observed trends in social micro-blogging spam. In: WebSci, pp. 33–41. ACM (2014)
2. Beutel, A., et al.: CopyCatch: stopping group attacks by spotting lockstep behavior in social networks. In: WWW, pp. 119–130. ACM (2013)
3. Breunig, M., et al.: LOF: identifying density-based local outliers. In: Proc. ACM SIGMOD Conf. 2000, pp. 93–104 (2000)
4. Brys, G., et al.: A Robust Measure of Skewness. Journal of Computational and Graphical Statistics **13**, 996–1017 (2004)
5. Chan, P. K., et al.:Modeling multiple time series for anomaly detection. In: ICDM, pp. 90–97. IEEE Computer Society (2005)
6. Chandola, V., et al.: Anomaly Detection: A Survey. ACM Comput. Surv. **41**(3), 15:1–15:58 (2009)
7. Chu, Z., et al.: Detecting Automation of Twitter Accounts: Are You a Human, Bot, or Cyborg? IEEE Trans. Dependable Secur. Comput. **9**(6), 811–824 (2012)
8. Cook, D., et al.: Twitter Deception and Influence: Issues of Identity, Slacktivism, and Puppetry. Journal of Information Warfare **13**(1), 58–71 (2014)
9. Freitas, C.A., et al.: Reverse Engineering Socialbot Infiltration Strategies in Twitter. ArXiv e-prints (2014)
10. Garrett, R.G.: The Chi-square Plot: a Tool for Multivariate Outlier Recognition. Journal of Geochemical Exploration **32**, 319–341 (1989)
11. Ghosh, R., et al.: Entropy-based classification of 'Retweeting' activity on twitter. In: KDD Workshop on Social Network Analysis (SNA-KDD) (2011)
12. Ghoting, A., et al.: Fast mining of distance-based outliers in high-dimensional datasets. Data Mining and Knowledge Discovery **16**(3), 349–364 (2008)

13. Hazel, G.: Multivariate Gaussian MRF for Multispectral Scene Segmentation and Anomaly Detection. IEEE Transactions on Geoscience and Remote Sensing **38**(3), 1199–1211 (2000)

14. Hubert, M., et al.: ROBPCA: A New Approach to Robust Principal Component Analysis. Technometrics **47**, 64–79 (2005)

15. Hubert, M., et al.: Robust PCA for Skewed Data and its Outlier Map. Computational Statistics & Data Analysis **53**(6), 2264–2274 (2009)

16. Jiang, M., et al.: CatchSync: catching synchronized behavior in large directed graphs. In: KDD, pp. 941–950. ACM (2014)

17. Jiang, M., Cui, P., Beutel, A., Faloutsos, C., Yang, S.: Inferring strange behavior from connectivity pattern in social networks. In: Tseng, V.S., Ho, T.B., Zhou, Z.-H., Chen, A.L.P., Kao, H.-Y. (eds.) PAKDD 2014, Part I. LNCS, vol. 8443, pp. 126–138. Springer, Heidelberg (2014)

18. Jolliffe, I.T.: Discarding Variables in a Principal Component Analysis. II: Real Data. Journal of the Royal Statistical Society. Series C (Applied Statistics) **22**(1), 21–31 (1973)

19. Noble, C.C., et al.: Graph-based anomaly detection. In: KDD (2003)

20. Papadimitriou, S., et al.: LOCI: Fast outlier detection using the local correlation integral. In: ICDE 2003 (2003)

21. Shah, N., et al.: Spotting suspicious link behavior with fBox: an adversarial perspective. In: ICDM (2014)

22. Stringhini, G., et al.: Follow the green: growth and dynamics in twitter follower markets. In: IMC, pp. 163–176. ACM (2013)

23. Tavares, G., et al.: Scaling-Laws of Human Broadcast Communication Enable Distinction between Human, Corporate and Robot Twitter Users. PLoS ONE **8**(7), e65774 (2013)

24. Twitter Inc. S-1 Filing, US Securities and Exchange Commission (2013). http:// www.sec.gov/Archives/edgar/data/1418091/000119312513390321/d564001ds1. htm

25. Xiong, L., et al.: Group Anomaly Detection using Flexible Genre Models. Advances in Neural Information Processing Systems **24**, 1071–1079 (2011)

26. Xiong, L., et al.: Efficient learning on point sets. In: ICDM, pp. 847–856 (2013)

27. Yang, C., et al.: Analyzing spammers' social networks for fun and profit: a case study of cyber criminal ecosystem on twitter. In: WWW, pp. 71–80 (2012)

28. Yu, R., et al.: GLAD: group anomaly detection in social media analysis. In: KDD, pp. 372–381. ACM (2014)

An Embedding Scheme for Detecting Anomalous Block Structured Graphs

Lida Rashidi[✉], Sutharshan Rajasegarar, and Christopher Leckie

NICTA Victoria, Department of Computing and Information Systems,
The University of Melbourne, Victoria, Australia
lrashidi@student.unimelb.edu.au, {sraja,caleckie}@unimelb.edu.au

Abstract. Graph-based anomaly detection plays a vital role in various application domains such as network intrusion detection, social network analysis and road traffic monitoring. Although these evolving networks impose a curse of dimensionality on the learning models, they usually contain structural properties that anomaly detection schemes can exploit. The major challenge is finding a feature extraction technique that preserves graph structure while balancing the accuracy of the model against its scalability. We propose the use of a scalable technique known as random projection as a method for structure aware embedding, which extracts relational properties of the network, and present an analytical proof of this claim. We also analyze the effect of embedding on the accuracy of one-class support vector machines for anomaly detection on real and synthetic datasets. We demonstrate that the embedding can be effective in terms of scalability without detrimental influence on the accuracy of the learned model.

Keywords: Anomaly detection · Block structured graph · One-class SVM · Random projection · Embedding

1 Introduction

Anomaly detection or outlier detection refers to the study of a system's normal state and the detection of unusual patterns based on the learned normal model. Mining abnormal, i.e., anomalous, patterns is a significant component of many data mining tasks. Numerous methodologies have been developed for detecting anomalous data objects under the assumption that there is no relational information between these objects [1]. However, in many scenarios such as biology, social sciences and information systems, the data points cannot be considered as independent entities. Data objects may demonstrate relationships or dependencies that must be considered in the process of detecting abnormal behavior. Graphs are known as a means of representing these relationships and network structures in real world datasets.

Anomalies in graphs can be determined within one graph [2,3], i.e., static, or over a sequence of graphs [4], i.e., dynamic. In this paper, we focus on the

© Springer International Publishing Switzerland 2015
T. Cao et al. (Eds.): PAKDD 2015, Part II, LNAI 9078, pp. 215–227, 2015.
DOI: 10.1007/978-3-319-18032-8_17

latter case where dynamic graphs can represent snapshots of evolving networks. Our objective is to determine an anomalous graph by constructing a normal model of the observed graphs. A major challenge in this task is to handle the complexity of the relational data structure, and find a technique that can exploit the relational properties of a graph data structure and represent the graph in a simple abstraction. We also need to find a way to handle possible noisy datasets. Therefore, we need a robust and scalable algorithm to summarize the graph dataset and detect abnormal graphs instances among a set of graphs.

Graph embedding is a common approach for simplifying graph structure and can be considered as a feature extraction process. Graph embedding methods assume that the data in a high dimensional space, i.e., a graph, usually lies near a non-linear manifold with lower complexity [5]. Therefore, a required pre-processing step in graph anomaly detection is graph embedding. Graph embedding techniques can help us in devising more efficient and interpretable anomaly detection techniques. Moreover, they provide a method of visualization for analyzing graph data [5,6]. However we need to extract features that preserve structural information of a graph and help us in detecting abnormal patterns.

To address this challenge, we propose a structure aware graph embedding scheme. Our embedding approach is based on random projection and exploits the Johnson and Lindenstrauss lemma [7] to provide a theoretical proof of its performance. Although random projection has been proven to preserve pairwise distances in Euclidean space, its suitability for non-relational datasets has received little attention. As opposed to traditional object or vectorial datasets, graphs are relational data structures known for their ability to capture topological proximity and structural properties.

In order to confirm that random projection has structure preserving properties for relational datasets, we need to prove that the Euclidean distance between entities in a graph can be representative of node proximity. Therefore, we consider the case of community structured graphs and assert that the Johnson and Lindenstrauss theorem is also applicable in this case [7]. Moreover, we can infer that the lower bound properties on the accuracy of random projection still holds for the case of community structured graphs.

We apply these embedding techniques on community structured datasets and analyze their influence on one-class Support Vector Machines (OCSVM) [8] as the dimension of the embedding decreases. We demonstrate that for dimensions much less than the Johnson and Lindestrauss lower bound, we can still achieve high levels of accuracy. Moreover, we introduce perturbations to graphs in the form of background noise and discuss their effect on the performance of our anomaly detection approach, i.e., OCSVM.

The main contributions of this paper are as follows: (i) We propose a structure aware embedding of community structured graphs based on random projections such that the proximity of the nodes comprising a community is maintained during embedding. (ii) We provide an analytical proof on the structure preserving property of our embedding approach as well as stating that the lower bound on the dimension of the random projection technique can also be applicable in our embedding approach. (iii) We demonstrate the effect of random projection on the scalability and accuracy of OCSVM. (iv) We analyze the effect of background noise on the performance of OCSVM in addition to random projection.

2 Related Work

Detecting abnormal events in graphs has received considerable attention in various disciplines [9,10]. If we are presented with a set of evolving graphs, an important form of actionable information that we may extract is the normal and abnormal patterns in such a set of graphs. Due to the large volume of data and lack of labels, this task is usually performed in an unsupervised manner.

We discuss anomaly detection in graph data only for plain graphs where there are only nodes and edges representing the data without any associated feature. However these techniques can be extended to attributed graphs as well. Nodes and/or edges in an attributed graph represent various features. For instance, a node in a social network may have various education levels or interests, and links may have different strengths.

Several approaches to pattern mining in graphs stem from distance based techniques, which utilize a distance measure in order to detect abnormal vs. normal structures. An example of such an approach is the k-medians algorithm [11], which employs graph edit distance as a measure of graph similarity. Other approaches take advantage of graph kernels [12], where kernel-based algorithms are applied to graphs. They compare graphs based on common sequences of nodes, or subgraphs. However, the computational complexity of these kernels can become a limitation when applied to large graphs.

Graph centric features are common forms of information to extract from a graph. These features can be computed from the combination of two, three or more nodes, i.e., dyads, triads and communities. They can also be extracted from the combination of all nodes in a more general manner [3]. Many intrusion detection approaches [2] have utilized graph centric features in their process of anomaly detection.

The main task in many graph-based anomaly detection schemes is to utilize graph structure in the process of detecting anomalies. Many techniques try to transform the problem of anomaly detection in graphs to the well-known problem of spotting outliers in an n-dimensional space. This step, known as graph embedding, is considered as a necessary pre-processing phase in many domains. Therefore we have considered a brief overview of embedding techniques. After performing graph embedding, standard unsupervised anomaly detection schemes such as OCSVM can be employed [13,14]. A thorough survey of such techniques can be found in [1,15].

2.1 Large Scale Graph Embedding

Graphs represent relational information between various entities in a dataset. However the structure of the graph is complex and is not readily suited to traditional classification and anomaly detection techniques that assume the availability of input data in a d-dimensional space. The input of such techniques is the adjacency or distance matrix of the graph, and the outcome is the equivalent point coordinates for each vetrex. One of the techniques to transform a graph into its corresponding point coordinates is spectral embedding [16]. This approach employs singular value decomposition to the adjacency matrix of a graph. The result is a set of eigenvalues and eigenvectors. The largest eigenvalues and their corresponding eigenvectors correspond to the dimensions that capture the variability in the input data. Therefore, spectral embedding preserves the eigenvectors corresponding to the largest eigenvalues [17].

Another technique that has been developed for graph embedding based on eigendecomposition is proposed in [17] where they use the Laplacian instead of the adjacency matrix. The Laplacian matrix represents the connectedness of a graph and can be computed from the adjacency matrix. In addition to eigendecomposition based approaches, techniques such as spring embedding have been developed. The intuition behind spring embedding is to simulate nodes as mass particles and edges as springs. The optimum state for such a system is the state where the energy is minimized. Note that such an objective function is non-convex, and due to random initialization the results may be highly suboptimal.

Both of the mentioned approaches ignore the topology of the graph. Therefore the outcome of embedding is not reversible for these techniques. The authors in [5] try to learn a positive semi-definite kernel matrix from the adjacency matrix and apply eigendecomposition to the learned kernel matrix. This method makes the following assumption: the data in a high dimensional space lies near a low dimensional nonlinear manifold. The kernel matrix aims to preserve the local pairwise distances between neighboring nodes in a graph and therefore simulates the distances on the manifold, as opposed to spectral techniques where the outcomes, i.e., eigenvectors, were arbitrary directions. The embedding results may be reversible to the original input by using algorithms such as nearest neighbor and maximum weight spanning tree.

However, these techniques do not consider the distances between non-neighbor nodes. Therefore the outcome of the reversed embedding will not have the same connectivity as the original graph. In order to handle this problem, another technique called structure preserving embedding (SPE) has been developed [6]. The learned kernel matrix in this approach considers the distances between neighbor and non-neighbor nodes. SPE uses semidefinite programming to learn a kernel matrix, and then applies eigen-decomposition on this matrix in order to find the embedding coordinates.

In summary, graphs are complicated data structures and in order to detect anomalies in these datasets, we need to begin by extracting the structural information in the graphs. Many approaches try to embed these graphs into lower dimensions without studying the structural proprieties that graphs offer. Moreover, the features extracted by these techniques may not be useful in constructing an anomaly detection model. Therefore, we have analyzed a specific graph structure before embedding and proven its suitability for a dimensionality reduction technique known as random projection. This graph embedding approach preserves the structure of the graphs and makes the anomaly detection scheme more scalable.

3   Problem Statement and Proposed Approach

In this section, we present our hybrid scheme that comprises random projection with a OCSVM for the purpose of anomaly detection in block structured graphs.

3.1   Preliminaries

We begin by formally defining the problem of anomaly detection in graphs. A graph, $G = (V, E)$, is characterized as a set of vertices V and edges E. In this paper, we consider the case of observing a set of graphs evolving over time with consistent node labeling, $\vartheta^{1..t} = \{G_n^1, G_n^2, ..., G_n^t\}$. Note that the number of vertices n does not change over time, but edges can be removed or added.

We assume the graphs are plain and directed, but the method can also be applied to undirected graphs. The adjacency matrix A of such graphs is an $n \times n$ matrix where each $A_{ij} \in \{0, 1\}$. We also assume that the majority of the observed graphs are normal and have a specific community structured model. Block structured graphs are discussed in the following section.

Our aim is to detect abnormal graphs, i.e., graphs with different structures, by learning a normal model of the observed graphs. However, learning from several hundreds of graphs that are each presented in an $n \times n$ matrix can be computationally inefficient. The problem we address is to make the learning process more scalable without losing accuracy by extracting structure aware features from the graphs. This approach can also be viewed as a graph embedding scheme.

The key intuition behind our approach is the fact that normal graphs share common topological features. However, one of the main challenges is to find a balance between the number of extracted features and the model accuracy. The feature extraction phase results in $A^{1...t} = \{A_{n \times d}^1, A_{n \times d}^2, ..., A_{n \times d}^t\}$ where d is the number of extracted features from each node. Thereafter, we can determine the abnormality of new graphs using the learned model of normal inputs.

3.2   Block Structured Graphs

We now define the properties of block structured graphs [18], specifically community structured graph models. In this paper, we mainly focus on unweighted directed graphs. The edges in such graphs demonstrate the existence of links between vertices and can be represented in an adjacency matrix, $A_{n \times n}$, where n is the number vertices.

Block structured graphs are abundant in real world application such as social networks [19]. A simple approach to generating such graphs is applying a stochastic block model, i.e., a generative model for creating blocks in graphs [18, 19].

Such models can build realistic network structures such as community, core-periphery and hierarchical network structures [18].

A stochastic block model generates graphs with the following characteristics:

- They fall into the category of random graph models.
- They can be decomposed to a set of $k, 1 \leq k \leq n$, smaller blocks.
- The membership of each vertex to these blocks is demonstrated through a membership matrix $M \in [0,1]_{n \times k}$.
- The probability of a link between blocks is defined in a matrix $\omega_{k \times k}$.

The overall process of generating a block structured graph is formulated in Equation 1:

$$\omega_{ij} = \lambda \omega_{ij}^{planted} + (1 - \lambda) \omega_{ij}^{random} . \tag{1}$$

The variable $\omega_{ij}^{planted}$ creates the underlying blocks in the network based on the block model that we choose. An example of a community structured graph with four blocks is shown as follows:

$$\begin{bmatrix} b_{11} & 0 & 0 & 0 \\ 0 & b_{22} & 0 & 0 \\ 0 & 0 & b_{33} & 0 \\ 0 & 0 & 0 & b_{44} \end{bmatrix} \tag{2}$$

However ω_{ij}^{random} generates a random graph without any block model. The parameter λ modifies the form of the graph from the extreme cases of fully random, i.e., $\lambda = 0$, to fully structured, i.e., $\lambda = 1$. A detailed description of this method can be found in [18].

3.3 Proposed Graph-Based Anomaly Detection Scheme

We now describe the main phases of our graph-based anomaly detection scheme. We first define our structure aware embedding approach followed by an unsupervised classifier to learn the normal graph model.

Graph Embedding Scheme. Graph embedding approaches assign point coordinates to each vertex of a graph by optimizing a specific objective. For instance, a possible objective of graph embedding can be minimizing edge crossings between nodes. These approaches can also aim to preserve properties like node proximity in order to capture the topology of a graph [6].

Our aim is to preserve node proximity in our graph embedding scheme, in order to help identify community structures present in the input graphs. We propose to use a dimensionality reduction technique known as random projections for this purpose. Random projection approaches are based on the Johnson and Lindenstrauss lemma [7]. This lemma asserts that a set of points in Euclidean space, $P^{1 \cdots n} \in \mathbb{R}^{n \times m}$, can be embedded into a d-dimensional Euclidean space, $P'^{1 \cdots n} \in R^{n \times d}$ while preserving all pairwise distances to within a small factor ϵ. The Johnson and Lindenstrauss lemma is presented in Lemma 1.

Lemma 1. *Given an integer n and $\epsilon > 0$, let d be a positive integer such that $d \geq d_0 = O(\epsilon^{-2} \log n)$. For every set P of n points in $\mathbb{R}^m$, there exists $f : \mathbb{R}^m \to \mathbb{R}^d$ such that for all $u, v \in P$*

$$(1 - \epsilon)||u - v||^2 \leq ||f(u) - f(v)||^2 \leq (1 + \epsilon)||u - v||^2 \tag{3}$$

We discuss three random projection matrices that have been shown to preserve pairwise distances [20]. Since we are dealing with graphs, we consider each node, v_i, to be an instance and its associated row in the adjacency matrix, A_i, as its $m = n$ features. We denote n as the number of features in the original space in the rest of the paper. The construction of the random projection matrix, $R_{n \times d} = r_{ij}^{\{i=1...n, j=1...d\}}$, can be based on the following structures formulated in Equations 4 ,5, 6:

$$r_{ij} = \begin{cases} +1 & with\,probability\,1/2 \\ -1 & with\,probability\,1/2 \end{cases} \tag{4}$$

$$r_{ij} \sim \sqrt{2}\mathcal{N}(0,1) \tag{5}$$

$$r_{ij} = \sqrt{3} \begin{cases} +1 & with\,probability\,1/6 \\ 0 & with\,probability\,2/3 \\ -1 & with\,probability\,1/6 \end{cases} \tag{6}$$

The embedded graph A' is computed as $A'_{n \times d} = \frac{1}{\sqrt{d}} AR$. In order to prove that random projection extracts structure aware features from the graph, we propose Lemma 2 which asserts that the expected Euclidean distance between the vertices within the same block is close to zero while nodes belonging to different blocks result in a larger expected Euclidean distance. Therefore, vertices in a community structured graph can be treated as points in a Euclidean space where the Euclidean distance reflects the nodes' memberships to each block.

Lemma 2. *Given a graph with a community block structure generated by ω, the expected Euclidean distance between any two vertices u and v belonging to the same block b_{ii} is close to zero.*

$$E[||v - u||^2] \simeq 0 \tag{7}$$

Proof. The density of a single block b_{ij} in the adjacency matrix is generated by a binomial distribution, $B(n, p)$, where n is the number of trials and p is the probability of success. The number of trials and probability of success are determined by the size of the block β_{ij} and $p_{ij} \sim \mathcal{N}(\mu_{ij}, \sigma_{ij}^2)$ respectively. Assigning 0 or 1 to a cell in a block is determined according to a uniform distribution where every element in a block has the same probability of being 1, $\frac{B(n,p)}{\beta_{ij}}$. We assume that β_{ij} is large enough and $0 \leq p_{ij} \leq 1$, so that we can approximate $B(n, p)$ with a normal distribution $Bi(n, p) \sim \mathcal{N}(np, np(1 - p))$.

According to the properties of the normal distribution, the sum or difference of two normal distributions with the same mean μ and variance σ^2 is another normal distribution $\mathcal{N}(0, 2\sigma^2)$. The expected Euclidean distance of two vertices in the same block b_{ii} can be determined by the sum of squared differences of normal distributions within all blocks:

$$E[||u - v||^2] = E[||D < b_{ii} > ||^2 + \sum_{i \neq j} ||D < b_{ij} > ||^2]$$

$$\text{where} \quad D < b_{ij} >= X_i - X_j \quad (8)$$

$$X_i, Y_j \sim \mathcal{N}(\beta_{ij} p_{ij}, \beta_{ij} p_{ij}(1 - p_{ij}))$$

$$D < b_{ij} > \sim \mathcal{N}(0, 2\beta_{ij} p_{ij}(1 - p_{ij}))$$

Note that the parameter λ in Equation 1 controls the amount of background noise, therefore we assume that $(1 - \lambda) \leq \varepsilon$ and as a result $p_{ij}, i \neq j$ is a small non-zero value. The expected value of Equation 8 is summarized in Equation 9 according to the rule $E[X^2] = d\sigma^2 + \mu^2$ for any random variable $X \in \mathbb{R}^d$.

$$\sum_{i,j} E[||D < b_{ij} > ||^2] = \sum_{i,j} 2d_{ij} \beta_{ij} p_{ij}(1 - p_{ij}) \quad (9)$$

The same approach can be followed for determining the expected Euclidean distance of vertices coming from different blocks. It can be shown that given a reasonable level of noise in the graphs, λ, the Euclidean distance can be considered as a proximity measure of community structured graphs. Therefore by preserving pairwise distances, we are also maintaining the structural information of the adjacency matrix.

The embedded adjacency matrix $A'_{n \times d}$ makes the learning model more scalable by reducing the number of inputs form n^2 to $n \times d$. The trade-off between accuracy and scalability can be determined using the Johnson-Lindenstrauss Lemma. However, our empirical results demonstrate that even for dimensions d much less than the lower bound $\epsilon^{-2} \log n$, we achieve high levels of accuracy.

One-Class Support Vector Machine. We briefly describe the use of the OCSVM algorithm for the purpose of anomaly detection [8]. A OCSVM maps the reshaped embedded graphs $A'' = [A'_{11}...A'_{1d}A'_{21}...A'_{2d}...A'_{n1}...A'_{nd}]$ into a high dimensional feature space Φ by using a kernel $k(x, y) = (\Phi(x).\Phi(y))$ [8].

The dot product of the images in $\Phi(.)$ can be determined using a kernel such as the radial basis function. OCSVM tries to find the maximum margin hyperplane that separates the majority of the observed data, assuming mostly normal samples, from the origin. Let $f(x) = \langle w, x \rangle + b$ to denote the resulting hyperplane where the terms w and b are the normal vector and bias term of the hyperplane respectively. When a test graph arrives, we use the reshaped embedded representation of the graph and determine its label, i.e., normal or anomalous, using $f(x)$ as a measure of how anomalous is the graph.

4 Empirical Results

In this section we evaluate the quality of our anomaly detection scheme with graph embedding as a pre-processing phase. The main objective of this empirical study is to determine the effect of dimensionality reduction on the computation time, scalability and accuracy of our approach.

Lemma 2 provides the theoretical proof of the suitability of random projection as a graph embedding approach. In the beginning, the anomaly detection schemes have been provided with the reshaped adjacency matrix of the graphs, i.e., high dimensional data. Thereafter, we apply random projection on the input graphs and evaluate the accuracy and scalability of the anomaly detection scheme. We also study the influence of noise on the embedding technique.

We assume that the training data mainly consists of normal samples. However, in order to make the problem more realistic, we insert 5% of anomalous instances into the unlabeled train dataset along with the normal instances.

4.1   Datasets

We generate the synthetic datasets with the aim of evaluating the structure awareness of our embedding scheme and its robustness to the level of background noise. In order to generate the datasets based on the stochastic block model, we begin by defining the number of communities in the normal and anomalous networks, the distribution of the node-to-community assignment, the underlying density of each community and the background noise level.

The distribution of node-to-community assignments is uniform. Therefore, we can make sure that the graphs have a number of dominant communities. We determine the node membership to blocks by drawing random values from a hypergeometric distribution. The density of each block is determined by a Gaussian distribution with $\mu = 0.6, \sigma^2 = 0.1$. There are 1000 normal graphs in addition to 100 anomalous ones where the number of nodes in each graph is 200.

In order to generate multiple normal and anomalous graphs, we preserve the node-to-community assignments but modify the density of blocks. The levels of introduced noise can be adjusted using the parameter λ in Equation 1. We have varied the noise level from 1% to 19%. The number of communities in the normal and anomalous graphs are 3 and 5, respectively. All other parameters remain the same for the normal and abnormal graphs.

In addition to this synthetic dataset, we have used the network of American football games, Karate club social network and 1997 US Air flights graph as base datasets [21]. We added 1% noise by changing the values of 1% of the edges in the adjacency matrix and created the anomalous dataset by introducing 10% noise to the original dataset.

4.2   Results and Discussion

In order to evaluate the results, we have applied random projection with three different methods according to Equations 4, 5, 6. The embedded graphs are then used as training instances for OCSVM. We have used the hyperbolic tangent kernel and polynomial kernel for our synthetic and real datasets. These kernel functions are formulated as $k(u,v) = tanh(\gamma \times u' \times v + coef)$ and $k(u,v) = (\gamma \times u' \times v + coef)^3$ respectively. The settings used for the real and synthetic datasets are summarized in Table 1. The results on the synthetic dataset demonstrate

Table 1. Dataset Description and OCSVM parameter settings

Dataset	Number of Nodes	OCSVM Kernel	OCSVM kernel parameters
Synthetic	200	Hyperbolic Tangent	$\gamma = 0.00000001, coef = 0$
Football	115	Polynomial	$\gamma = 1.0 \times 10^{-8}, coef = 1$
Karate	34	Hyperbolic Tangent	$\gamma = 1.0 \times 10^{-8}, coef = 0$
US Air	332	Hyperbolic Tangent	$\gamma = 1.0 \times 10^{-11}, coef = 0$

that the outcome of OCSVM varies when the embedding dimension is well below the lower bound defined in Equation 3. We consider $\epsilon = 0.25$, therefore the lower bound on this dataset is 85. Fig. 1, 2 and Table. 2 depict the average training and test accuracy over various levels of noise against the number of dimensions in the projected space. As can be seen using Method 1 (Equation 4), we can achieve high

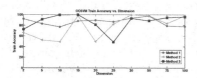

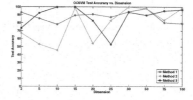

Fig. 1. Synthetic graphs: Average training accuracy vs. the dimension of embedding, where randomization methods 1, 2 and 3 are defined according to Equations 4, 5, 6.

Fig. 2. Synthetic graphs: Average test accuracy vs. the dimension of embedding, where randomization methods 1, 2 and 3 are defined according to Equations 4, 5, 6.

levels of accuracy from approximately 80% up to 100% given a graph embedding with dimensions as low as $d = 2$ in most cases. The fluctuations in the diagrams demonstrate the appropriate dimension and accuracy trade-off.

It is worth mentioning that the computation time of OCSVM as shown in Table 3 on the original dataset without the embedding, i.e., $d = 200$, is dramatically higher than when random projection is used, i.e., $d < 20$. Therefore we achieve the scalability without losing high levels of accuracy.

Table 2. Test accuracy of OCSVM using various dimensions of random projection on Football and Karate datasets. Note that $d = 115$ and $d = 34$ in the top and bottom tables corresponds to no embedding.

Football	Method 1	Method 2	Method 3
d=2	93.43 ± 0.78	95.13 ± 0.086	13.2 ± 3.76
d=50	91.12 ± 0.24	98.73 ± 0.65	91.45 ± 6.04
d=115	100 ± 0.0	100 ± 0.0	100 ± 0.0

Karate	Method 1	Method 2	Method 3
d=2	90.93 ± 0.52	91.13 ± 3.25	91.87 ± 0.31
d=7	90.13 ± 0.09	91.13 ± 0.26	$93.07 + 0.74$
d=34	96.33 ± 0.47	96.33 ± 0.47	96.33 ± 0.47

Table 3. Processing time in seconds of OCSVM using various dimensions of random projection on the synthetic dataset. Note that $d = 200$ corresponds to no embedding.

Random Projection	$d = 2$	$d = 10$	$d = 20$	$d = 100$	$d = 200$
Method 1	1.1×10^{-6}	1.2×10^{-6}	1.0×10^{-6}	1.1×10^{-6}	24.627
Method 2	1.1×10^{-6}	1.0×10^{-6}	1.3×10^{-6}	1.1×10^{-6}	26.551
Method 3	1.0×10^{-6}	1.1×10^{-6}	1.1×10^{-6}	1.0×10^{-6}	24.566

5   Conclusion and Future Work

In this paper, we have presented an approach for graph embedding and provided an analytical proof as well as empirical evidence that this embedding technique can preserve the underlying structure of communities in graph databases such as social networks. This graph embedding technique has been used as a pre-processing step for anomaly detection, i.e., using a OCSVM. We achieved high accuracy after performing the graph embedding, therefore this technique can provide a balance in terms of anomaly detection precision as well as scalability.

After applying different levels of perturbation to the real and synthetic datasets, we observed that OCSVM still performs well after the embedding. Therefore, we can infer that random projection is a robust technique for graph embedding. According to the experimental studies, the combination of embedding and OCSVM achieves high accuracy for dimensions much less that the lower bound of Johnson and Lindenstrauss.

As a follow-up to this preliminary work, we are investigating the use of matrix re-ordering techniques in order to pre-process other types of block structured graphs such as core-periphery and hierarchy for random projection embedding.

Acknowledgments. The authors would like to thank National ICT Australia (NICTA) for providing funds and support.

References

1. Chandola, V., Banerjee, A., Kumar, V.: Anomaly Detection: A Survey. ACM Computing Surveys **4**(3), 1–58 (2009)
2. Ding, Q., Katenka, N., Barford, P., Kolaczyk, E.D., Crovella, M.: Intrusion as (anti)social communication: characterization and detection. In: Proceedings of the 18th ACM International Conference on Knowledge Discovery and Data Mining (SIGKDD), pp. 886–894 (2012)
3. Henderson, K., Eliassi-Rad, T., Faloutsos, C., Akoglu, L., Li, L., Maruhashi, K., Prakash, B.A., Tong, H.: Metricforensics: a multi-level approach for mining volatile graphs. In: Proceedings of the 16th ACM International Conference on Knowledge Discovery and Data Mining (SIGKDD), pp. 163–172 (2010)
4. Aggarwal, C.C., Zhao, Y., Yu, and P.S.: Outlier detection in graph streams. In: Proceedings of the 27th International Conference on Data Engineering (ICDE), pp. 399–409 (2011)
5. Shaw, B., Jebara, T.: Minimum volume embedding. In: Proceedings of the 11th International Conference on Artificial Intelligence and Statistics, pp. 460–467 (2007)
6. Shaw, B., Jebara, T.: Structure preserving embedding. In: Proceedings of the 26th International Conference on Machine Learning, pp. 937–944 (2009)
7. Johnson, W.B., Lindenstrauss, J.: Extensions of Lipschitz mappings into a Hilbert space. In: Conference in Modern Analysis and Probability, pp. 189–206 (1984)
8. Schlkopf, B., Williamson, R.C., Smola, A., Shawe-Taylor, J.: Support vector method for novelty detection. In: NIPS, vol. 12, pp. 582–588 (1999)
9. Cook, D., Holder, L.: Mining Graph Data. Wiley (2007)
10. Kang, U., Faloutsos, C.: Big graph mining: algorithms and discoveries. ACM SIGKDD Explorations Newsletter **14**(2), 29–36 (2012)
11. Riesen, K., Bunke, H.: Classification and clustering of vector space embedded graphs. In: Emerging Topics in Computer Vision and Its Applications, pp. 49–70. World Scientific (2012)
12. Neuhaus, M., Bunke, H.: Bridging the Gap Between Graph Edit Distance and Kernel Machines. World Scientific (2007)
13. Moshtaghi, M., Leckie, C., Karunasekera, S., Bezdek, J.C., Rajasegarar, S., Palaniswami, M.: Incremental elliptical boundary estimation for anomaly detection in wireless sensor networks. In: Proceedings of the 11th IEEE International Conference on Data Mining, pp. 467–476 (2011)
14. Rajasegarar, S., Leckie, C., Bezdek, J.C., Palaniswami, M.: Centered Hyperspherical and Hyperellipsoidal One-Class Support Vector Machines for Anomaly Detection in Sensor Networks. IEEE Transactions on Information Forensics and Security **5**(3), 518–533 (2010)
15. Akoglu, L., Tong, H., Koutra, D.: Graph-based Anomaly Detection and Description: A Survey. Data Mining and Knowledge Discovery, pp. 1–63 (2014)
16. Chung, F.R.K.: Spectral Graph Theory. American Mathematical Society (1997)
17. Belkin, M., Niyogi, P.: Laplacian Eigenmaps for Dimensionality Reduction and Data Representation. Neural Computation, pp. 1373–1396 (2002)
18. Newman, M.: Modularity and community structure in networks. In: Proceedings of the National Academy of Sciences, vol. 23, pp. 8577–8582 (2006)

19. Chan, J., Liu, W., Kan, A., Leckie, C., Bailey, J., Ramamohanarao, K.: Discovering latent blockmodels in sparse and noisy graphs using non-negative matrix factorisation. In: Proceedings of the 22nd ACM International Conference on Information & Knowledge Management, pp. 811–816 (2013)
20. Achlioptas, D.: Database-friendly Random Projections: Johnson-Lindenstrauss with Binary Coins. Journal of Computer and System Sciences **66**(4), 671–687 (2003)
21. http://www-personal.umich.edu/~mejn/netdata/

A Core-Attach Based Method for Identifying Protein Complexes in Dynamic PPI Networks

Jiawei Luo[✉], Chengchen Liu, and Hoang Tu Nguyen

School of Information Science and Engineering, Hunan University,
Changsha 410082, China
luojiawei@hnu.edu.cn

Abstract. Indentifying protein complexes is essential to understanding the principles of cellular systems. Many computational methods have been developed to identify protein complexes in static protein-protein interaction (PPI) network. However, PPI network changes over time, the important dynamics within PPI network is overlooked by these methods. Therefore, discovering complexes in dynamic PPI networks (DPN) is important. DPN contains a series of time-sequenced subnetworks which represent PPI at different time points of cell cycle. In this paper, we propose a dynamic core-attachment algorithm (DCA) to discover protein complexes in DPN. Based on core-attachment assumption, we first detect cores which are small, dense subgraphs and frequently active in the DPN, and then we form complexes by adding short-lived attachments to cores. We apply our DCA to the data of S.cerevisiae and the experimental result shows that DCA outperforms six other complex discovery algorithms, moreover, it reveals that our DCA not only provides dynamic information but also discovers more accurate protein complexes.

Keywords: Clustering · Protein complexes · Dynamic PPI networks · Core-attachment

1 Introduction

Detecting protein complexes in available PPI networks is an important and challenging task in the post-genomic era. Protein complexes are molecular aggregations of proteins assembled by multiple PPIs. They are key molecular entities to perform cellular functions. For example, complex "RNA polymerase II" transcribes genetic information into messages for ribosomes to produce proteins [1].

Up to now, many computational methods have been proposed to detect complexes in static PPI networks. Bader et al. [2] presented an algorithm MCODE, its a local-searched method which relies on the topological structure of the PPI network. Altaf-UI-Amin et al. [3] proposed a complex discovery method called DPClus which based on the combination of density and peripheral proteins to mine densely connected subgraphs. Moreover, the core-attachment concept has been proposed to identify complexes. Gavin et al. [4] illustrated the protein

© Springer International Publishing Switzerland 2015
T. Cao et al. (Eds.): PAKDD 2015, Part II, LNAI 9078, pp. 228–239, 2015.
DOI: 10.1007/978-3-319-18032-8_18

complex generally contain a core and attachments. The core is a small group of proteins with high degree of functional similarity, its the heart of a complex. While attachments are several peripheral neighbors of a core that assist their core to perform subordinate functions, which are often short-lived. Wu et al. [1] proposed the COACH algorithm which defined the core vertices among the neighborhood graphs, this method added attachments into the cores to form protein complexes.

However, all these methods discussed above only consider PPI networks as a static graph and overlook the dynamics inherent within them. In fact, PPI networks are varying with time and space. Therefore, understanding the dynamics of PPI networks is important to further understand molecular systems. Tang et al. [5] used gene expression data construct DPN by splitting the static PPI network into a series of time-sequenced subnetworks. In the framework of DPN, Li et al. [6] proposed a new framework to identify protein complexes and functional modules in DPN. Li et al. [7] discovered a novel method to identify dynamic complexes that integrate PPI network and gene expression data. All these efforts have made significant progress in protein complex discovery. However, only a few of these algorithms can both achieve high accuracy and capture the dynamic topology structure of DPN.

The protein complex consists of two parts in this paper: frequently active core and almost short-lived attachments. So our DCA operates in two phases: it first detects protein-complex cores and then identifies protein complexes by including attachment into cores. We compare our DCA with six competing complex discovery algorithm: DFM-CIN [6], COACH [1], ClusterOne [8], MCL [9], MCODE [2] and SPICI [10], which including the clustering method on the same DPN (DFM_CIN) and core-attachment method (COACH). Experiment results based on core analysis, F-measure, Coverage rate and functional enrichment show that our DCA performs better than these algorithms and can efficiently acquire the dynamic features of complexes.

2   Method

The static PPI network is generally considered as an undirected graph G(V, E), where a vertex in vertex set V represents a protein and an edge in edge set E represents an interaction between two proteins.

The dynamic PPI network (DPN) is constructed from static PPI network, which containing n time-sequenced subnetworks denoted as $\{D_1, D_2, \ldots, D_n\}$, as reported by Tang et al [5]. In each subnetwork of DPN, all the proteins and interactions activate at the same time, eg., subnetwork D_i is modeled as (V_i, E_i) where V_i represents the protein set and E_i represents the interaction set in the ith subnetwork.

Our DCA is differing from the previous core-attachment method. We redefine the proteins in one core are not only highly connected with each other but also simultaneously occur at multiple subnetworks. Figure1 shows an example to illustrate the cores in DPN. While, attachments have much interact with cores and often short-lived. We first introduce some related definitions below.

For a protein v, its active sub-network set can be abstracted into $Protein\_actives(v) = \{i, j, \ldots, k, \ldots\}$, where $i, j, \ldots, k$ denote the corresponding subnetworks that v appeared in. For a core S, $Core\_actives(S) = \{i, j, \ldots, h, \ldots\}$ represents the subnetwork set that S active in. Where $i, j, \ldots, h$ denote the corresponding subnetworks that the whole vertices of S are completely appeared in, it can be acquired by computing the intersection of all the proteins' $Protein\_actives()$, that is,

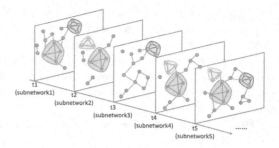

Fig. 1. Protein-complex cores in dynamic PPI networks: The nodes in pink, yellow, and blue color represent different cores in DPN. The pink core appears 4 times in DPN; the yellow and blue cores active 3 times in DPN; the purple nodes denote the remainder proteins in DPN.

$$Core\_actives(S) = \cap_{v \in S} Protein\_actives(v) \tag{1}$$

2.1   Complex Cores Mining

Based on the definition of DPN and complex cores, we assume n is the total number of subnetworks and $m(1 < m \leq n)$ is the least number of subnetworks where a core must be appeared in. In this paper, we define complex core in DPN should satisfy following four constrains with considering its topological structure and dynamic properties: (1) it's a dense subgraph of the PPI; (2) the core should active in no less than m subnetworks; (3) a core should include at least k proteins; (4) every two cores have no common proteins.

Algorithm 1 illustrates the detailed procedure on detecting cores in DPN. Before clustering, we should initialize the proteins in line 1-3. For each protein $v \in V$, we first calculate $Protein\_actives(v)$ in line 1-2, which is the set of subnetworks that v appeared in, eg., if $v \in V_i$, we then put i into the set $Protein\_actives(v)$, we iteratively traverse all the subnetworks in DPN to obtain $Protein\_actives()$ of each vertices. Next, we calculate the local clustering coefficient (CC) of v in graph G in line 3, CC(v) quantifies how close the v's neighbors are to being a clique or complete graph. The CC(v) is defined as the number of edges between v's neighbors, divided by the maximum number of edges that might potentially include in v's neighbors [11]:

$$CC(v) = \frac{\sum_{u,w \in N_v} e(u, w) | e(u, w) \in E}{d_v \times (d_v - 1)/2} \tag{2}$$

Where N_v is the set of all v's neighbors. It is obvious that protein with high CC value more tends to be involved in the core, and has a higher priority to be considered as a seed.

After initialization, all the vertices are queued into Q in descending order in terms of their CC in line 4. The first unused vertex v in Q is selected as a seed to expand a new probable complex core S in line 5-15. When we expand the seed v, we should first collect the New_N_v of v in line 7, New_N_v are the core's candidate proteins and consist of v's direct neighborhoods that are still in Q. Then we will calculate the closeness(cl) between S and each vertex $u \in New\_N_v$ in line 9, the closeness function [3] is given as follows:

$$cl(u, S) = \frac{E_{uS}}{d_S \times |V_S|} \tag{3}$$

Where E_{uS} is the number of edges that connect vertex u to core S; $|V_S|$ is the number of vertices in core S; d_S is the density of core S which is formed in the equation (4):

$$d_S = \frac{2 \times |E_S|}{|V_S| \times (|V_S| - 1)} \tag{4}$$

Algorithm 1. Cores Mining

Input:

$G = (V, E)$: static PPI network

$DPN = \{D_1, D_2, \ldots, D_n\}$: n time-sequenced dynamic PPI subnetwork

α : closeness threshold for expanding cores

m: the threshold of subnetworks number that a core active in

k : the threshold of vertices number in each core

Output:

CS: the set of cores in DPN

$Protein\_actives(v_1), Protein\_actives(v_2), \ldots, Protein\_actives(v_p)$: subnetwork set of each protein active in

1. for $i = 1 : n$ do
2.   for each vertex v in V_i, $Protein\_actives(v) = Protein\_actives(v) \cup \{i\}$
3. for each protein $v \in V$, compute $CC(v)$
4. sort proteins into queue Q in descending order by CC
5. for $v \leftarrow Q$ //the first vertex v in Q is selected as a seed to expand core S
6. $S = \{v\}$ // initialize v as a singleton core S
7. $New\_N_v = \{u \mid u \in N_v \wedge u \in Q\}$
8.   while $New\_N_v \neq \phi$
9.   for all the $u_i \in New\_N_v$, compute $cl(u_i, S)$
10.   if $max_{u_i \in New\_N_v} cl(u_i, S) \geq \alpha$
11.   if $comNetwork(u_i, S) \geq m$, then add u_i into S
12.   remove u_i from New_N_v
13.   else break   //stop expanding core S
14.   if number of proteins in $S \geq k$
15.   remove all proteins of S from Q, and put S into CS

A higher cl value of a neighbor indicates that it is part of the core while a lower cl indicates that it's part of the periphery. Next we choose the neighbor u_i with maximum cl, if $cl(u_i, S)$ is smaller than a prefixed threshold α, we will stop expanding S. Otherwise, we should determine whether u_i has no less than m common subnetworks with proteins in S, as $comNetwork(u_i, S) \geq m$, and

$$comNetwork(u_i, S) = \cap_{w \in S} Protein\_actives(w) \cap Protein\_actives(u_i) \quad (5)$$

If u_i shares m common subnetworks with core S, add u_i into core S. Next, u_i is removed from New_N_v to prevent recalculated in line 10-13. We will repeatedly add neighbor to S until all the vertices in New_N_v is removed in line 8-13.

Once the preliminary core S is formed, we need to judge whether S includes at least k proteins, if it is, put S into final core set CS, and remove the whole vertices of S from Q to avoid being included into any other cores in line 14-15. Another round of expanding is performed until Q is empty in line 5-15, and output CS.

2.2 Protein Complexes Formation

Considering that cores in DPN have been generated, we should select attachments to cores to construct complexes. As attachments are often short-lived, we will detect attachments on each subnetwork of DPN respectively. The description of forming protein-complexes is shown in Algorithm 2.

Algorithm 2. Complexes Formation

Input:
$DPN = \{D_1, D_2, \ldots, D_n\}$: n time-sequenced dynamic PPI subnetwork
CS: the set of cores in DPN
$Protein\_actives(v_1), Protein\_actives(v_2), \ldots, Protein\_actives(v_p)$: subnetwork set of each protein active in
Output:
DC: the set of complexes in DPN

1. for each $S \in CS$
2.   computer $Core\_active(S)$
3.   for each $i \in Core\_active(S)$ //form a complex C_i of S in D_i
4. $C_i \leftarrow S$ //initialize core S as a complex C_i in the ith subnetwork
5.   compute neighborhood proteins N_{S_i} of S in D_i
6.   for each $u \in N_{S\_i}$
7.   if $E_{uS} \geq 0.5 \times |V_S|$ then
8.   put u into C_i //select u as an attachment of core S
9. $C = \cup_{i \in Core\_actives(S)} C_i$, and put C into DC
10. filtering DC

For each core $S \in CS$. First, we need to calculate $Core\_actives(S)$ in line 2, which is the subnetwork set that S appeared in. And then, for every sub-network D_i with $i \in Core\_Active(S)$, we add attachments to S to construct temp_complex C_i in line 3-8. The final complex C is made up of all the temp_complexes in corresponding subnetworks in line 9. When we choose attachments of core S, we based on the idea of majority rule that neighbor vertices interacting with no less than half of the proteins in the core S will be selected as attachments [1] in line 6-8. Although our cores are non-overlapped, the complexes detected by DCA may be overlapped as they could have common attachments. So we need to filter completely overlapped complexes, and output the filtered complexes of DPN in line 10. Its obvious that attachments may be active in one or several subnetworks and usually short-lived.

The time complexity is $O(cV^3)$ of algorithm 1 and $O(ncV^2)$ of algorithm 2 in the worst case. Where V is the number of whole vertices in PPI, c is the number of vertices in a core and n is the number of subnetworks. As $n, c \ll V$, the time complexity of our DCA is approximating $O(V^3)$.

3 Experiments and Results

3.1 Datasets and DPN Construction

We performed our method on two different yeast PPI networks, including DIP [12] and Krogan [13] data. The DIP data consist of 4930 proteins and 17201 interactions, while Krogan contain 3581 proteins and 14076 interactions. The gene-expressing profiles of S.cerevisiae were retrieved from Ref. [14] with the accession number GSE3431, there are 4851 genes involved in DIP and 3509 genes in Krogan. GO data was downloaded from Ref.[15]. For evaluating our identified complexes, the benchmark set consists of 428 complexes[16], from three source: (I)MIPS, (II)Aloy et al. and (III)SGD database based on GeneOntology(GO) annotations.

Previous studies [5,6] have shown that integrating gene expression profiles with the PPI networks can acquire efficient DPN. So we construct DPN as Tang et al have done in Ref. [5]. Considering that GSE3431 covers three successive cell cycles and each cycle includes 12 time points, the average expression value of gene at the same time point for three cycles is used as its expression value at the given time point. We normalize the expression values of each gene to make the values of genes range from 0 to 1, and use a proper threshold value 0.3 to screen gene products at each time point. Finally, we create the DPN based on these filtered expression values and obtain 12 time-sequenced subnetworks.

For the sake of evaluating our algorithm DCA, we compared it with six competing clustering algorithms: DFM-CIN[6], COACH[1], ClusterOne[8], MCL[9], MCODE[2] and SPICI[10]. DFM-CIN is a functional module detecting algorithm which performs on the same DPN. The others are all well-known complex discovery algorithms. Their values of the parameters are selected from those recommended by the authors.

3.2   Core Analysis

As core is the key functional unit of protein complex, we will analyze the biological similarity of our cores on DIP data. The GO annotations are used to evaluate the GO functional similarities of cores, complexes, DPN and PPI network. Two interacting proteins can have a similarity score based on their GO terms. Here, functional similarity between two proteins is calculated by the method in [17]. We sum up the similarity of all interactions in each component using three sub-ontologies (BP, CC, MF) of GO, and then average the overall similarity. Table 1 shows the average similarity of each component detected by our DCA and COACH algorithms respectively, as COACH is developed based on core-attachments and achieves an excellent performance among current algorithms.

Table 1. Average similarity of interactions involved in Cores, Complexes and PPI data on DIP

(a)Extracted by DCA

Interactions	BP	CC	MF
In DCA cores	0.335	0.451	0.233
In DCA complexes	0.243	0.447	0.193
In DPN	0.117	0.244	0.112
In PPI network	0.115	0.242	0.111

(b)Extracted by COACH

Interactions	BP	CC	MF
In COACH cores	0.241	0.436	0.185
In COACH complexes	0.190	0.440	0.155
In PPI network	0.115	0.242	0.111

Table 1(a) shows the interactions within cores in DCA achieve the highest similarity on DIP dataset, no matter which GO domain (BP,CC,MF) they are. The GO similarity scores are declined orderly in these four components, which suggests the cores have higher degree of functional similarities, and can be seen the biological hearts of protein complexes. From Table 1(b) we can see that, although cores identified by COACH is of highly biological similarities, their GO similarity score is much less than that of our DCA, which also indicates our DCA is better than COACH for producing high biological significance cores.

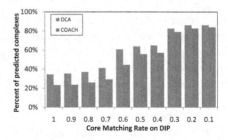

Fig. 2. The percentage of cores discovered by DCA and COACH with the CMR threshold from 1.0 to 0.1 on DIP

To further evaluate the quality of cores detected by our DCA, we quote the Core Matching Rate(CMR) [7] to measure the quality of complexes cores, which is defined as:

$$CMR(C) = max\Big(\frac{|C \cap K_i|}{|C|}\Big), \quad K_i \in K \tag{6}$$

Where K is the known benchmark complexes set [16]. $|C \cap K_i|$ denotes the number of proteins of core C included in one known complex; When a core C is completely included in a known complex K_i, $CMR(C) = 1$. Figure 2 shows the comparison results with respect to different CMR ranging from 1.0 to 0.1 on DIP. From Figure 2 we can see that as the threshold of CMR changing from 1.0 to 0.1, the percentage of DCA cores remains higher than that of COACH cores, which indicates that our DCA can produce more accurate cores than COACH method.

3.3 Functional Enrichment Analysis

To evaluate the biological enrichment and functional relevance of identified complexes, the functional homogeneity P-value [3] is applied. Accordingly, a predicted complex with a low P-value indicates it achieves a high statistical significance. The complex with corrected P-value of less than 0.01 [1] is considered significant. The proportion of significant complexes over the predictions can be used as an evaluation for assess the overall performance of various algorithms. Table 2 shows the comparison results obtained from six algorithms on DIP and Krogan datasets respectively.

Table 2. Functional enrichments of the identified complexes detected by DCA and other algorithms on DIP and Krogan datasets

Dataset	Algorithms	Identified complexes	% of significant complexes	Average P-values	Significant complexes (P)			
					< E-15	E-15 to E-10	E-10 to E-5	E-5 to E-0.01
	DCA	381	94.23%	1.63E-04	30.45%	21.26%	25.46%	17.06%
	DFM_CIN	395	74.94%	3.56E-04	20.3%	9.1%	25.3%	20.3%
	COACH	746	87.67%	3.02E-04	19.6%	14.7%	27.6%	25.7%
DIP	ClusterOne	343	67.64%	2.25E-04	10.8%	11.1%	26.5%	19.2%
	MCL	1246	30.74%	8.21E-04	2.6%	2.6%	8.5%	17%
	MCODE	59	89.83%	1.36E-04	18.7%	18.7%	37.3%	15.3%
	SPICI	583	53.52%	9.11E-04	5.8%	5.5%	15.6%	26.6%
	DCA	240	94.17%	1.55E-04	43.33%	16.67%	17.5%	16.67%
	DFM_CIN	358	75.42%	2.58E-04	21.23%	10.61%	26.26%	17.32%
	COACH	570	87.89%	3.85E-04	22.46%	10.53%	28.07%	26.84%
Krogan	ClusterOne	225	78.67%	4.42E-04	14.67%	12.89%	28.44%	22.67%
	MCL	834	37.89%	9.17E-04	3.12%	19.06%	12.47%	19.06%
	MCODE	50	94%	1.04E-04	24%	20%	36%	14%
	SPICI	383	59.53%	8.27E-04	7.83%	6.01%	18.02%	27.68%

From Table 2, it is easy to see that the percentage of significant complexes predicted by our DCA achieves the highest in the fourth column on both DIP and Krogan datasets. Moreover, the average P-values of DCA is much smaller than other algorithms except MCODE, the percentage of complexes produced by DCA with P-values less than E-15 is much higher than that of other algorithms, especially higher than DFM_CIN which is perform on DPN as well. This indicates that complexes predicted by DCA are quite accurately and have good functional enrichments.

3.4 F-measure and Coverage Rate

In order to estimate the performance of protein complexes discovered by DCA, two comprehensive evaluation methods called F-measure and Coverage rate (CR)[1] are used. F-measure is the harmonic mean of Precision and Recall. Precision measures how many correct predictions that matched real complex, Recall measures how many real complexes that matched predicted complex. CR evaluates the amount of proteins in the real complexes that can be covered by the predicted complexes. Generally, high F-measure and CR values indicate that the prediction has good efficiency.

Table 3. The precision and recall results of various algorithms on DIP and Krogan datasets

	DIP			Krogan	
Algorithms	Precision	Recall		Precision	Recall
DCA	0.546	0.437		0.704	0.348
DFM_CIN	0.387	0.4		0.492	0.418
COACH	0.382	0.582		0.421	0.449
ClusterOne	0.347	0.367		0.453	0.322
MCL	0.17	0.598		0.176	0.46
MCODE	0.525	0.143		0.56	0.105
SPICI	0.226	0.488		0.272	0.416

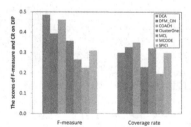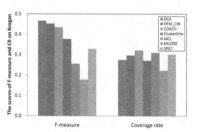

Fig. 3. The performance comparison for DCA and other algorithms on F-measure and Coverage rate on DIP(left) and Krogan(right) datasets

The basic informations for prediction by various algorithms on DIP and Krogan datasests are presented in Table 3, the precision of our DCA achieves the highest on two datasets showing DCA can identify precise complexes. In Figure 3 we can see that DCA achieves the best performance on F-measure. In detail, on DIP dataset, the F-measure of SCAIA is 48.5%, which is 2.4% higher than the second one COACH. For the number of proteins in DPN is less than that in static PPI network [6], the CR of complexes detected by DCA couldn't achieve a rather high value, but its better than ClusterOne and MCODE. All above results demonstrate that our algorithm can obtain good performance, and complexes detected by DCA match quite well with benchmark complexes.

3.5 An Example of Protein-Complex

To further reveal the results obtained by our algorithm, we display one of our protein complex that generated by DCA on DIP dataset. Figure 4 shows an example of *complex#238*. The biological process of *complex#238* is "nuclear pore organization and biogenesis"(annotated in GO:0006999) with the lowest P-value=8.23E-26 which is carried out at the cellular level that results in the assembly, arrangement of constituent parts, or disassembly of the nuclear pore [18]. It contains 10 proteins and all of them are participating in the mechanism of nuclear pore organization.

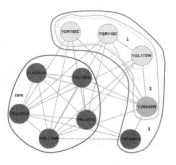

As shown in Figure 4, the core of *complex#238* contains 5 proteins that simultaneously active in the $2^{th}, 3^{th}, 7^{th}, 8^{th}, 9^{th}, 10^{th}, 11^{th}$ and 12th subnetworks. It is a complete graph in which every pair of distinct proteins is connected by a unique edge. The core is perfectly recalled by the benchmark *complex#50*. However, three groups of attachments active in different subnetworks, eg., attachment YKL068W actives in the $8^{th}, 9^{th}, 11^{th}$ and 12th subnetworks, while YJR042W is expressed in the $2^{th}, 3^{th}, 8^{th}, 9^{th}, 10^{th}, 11^{th}$ and 12th subnetworks. It illustrates the dynamic properties of our complexes detected in DPN.

Fig. 4. An Example of protein complexes generated by DCA on DIP: 1,2,3 represent three sets of attachments in different timestamps; the red nodes represent the Core's proteins; the yellow nodes in yellow cycle denotes attachment set1; the nodes in blue cycle belong to attachment set2; the nodes in purple cycle belong to attachment set3

3.6 Effects of Parameters

We now discuss the effects of parameters of DCA in F-measure and CR, on DIP data. The parameters are the closeness threshold α for expanding cores, the number of subnetworks m that a core active in, the least vertices number k of each core.

As we can see from Figure 5(a), when α is small enough, the nodes with a very low value of *cl* are allowed to be included in the core, the core's size is much bigger than the real one, so the F-measure increases while α increases and achieves highest when α =0.8. As CR is not very sensitive to α , the optimum

α value is 0.8. Figure 5(b) reveals that as subnetwork number m increases, the number of cores decreases, thus leading to decreases in CR. Here, we set m=5, which obtains the best F-measure and comparable CR. Figure 5(c) shows that our algorithm can achieve a good balance between F-measure and CR while k=3 in our experiments.

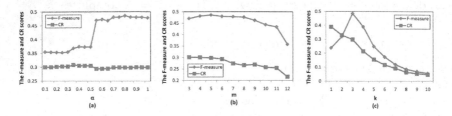

Fig. 5. The effects of α , m and k: (a)the variation of α affects the F-measure and CR; (b)the plot of the F-measure and CR for different values m; (c)the relation between k and F-measure and CR

4   Conclusion

Exploring biologically significant protein complexes are important challenges in post-genomic era. However, current complexes discovering algorithms have mainly focused on the static PPI network and failed to consider the inherent dynamics within them. Hence, we proposed a DCA algorithm to identify protein complexes in DPN. The DPN is constructed according to Tang et al' study[5]. We first detect cores which active frequently in DPN, and then extract the short-lived attachment to form complexes.

We tested our DCA on two yeast PPI data. First, we use GO similarity and CMR to analyze our core by comparing with another core-attachment method COACH. It demonstrates that proteins in our cores have high functional similarity and are highly recalled by known complexes. Second, we employ the P-value to evaluate the functional enrichment of predict complexes, the proportion of significant complexes over the predicted ones by DCA is much greater than other algorithms. The last, a lot of comparison based on F-measure and CR reveals our DCA outperforms other six algorithms, as we achieve the highest F-measure and get a comparable CR values. In conclusion, all the experiments show our DCA can not only provide a new way of detecting complexes in DPN but also identify more accurate protein complexes.

Acknowledgments. This work is supported by National Natural Science Foundation of China (Grant no.61240046) and Hunan Provincial Natural Science Foundation of China (Grant no.13JJ2017).

References

1. Wu, M., Li, X., Kwoh, C.K., Ng, S.K.: A core-attachment based method to detect protein complexes in PPI networks. J. BMC bioinformatics. **10**(1), 169 (2009)
2. Bader, G.D., Hogue, C.W.: An automated method for finding molecular complexes in large protein interaction networks. J. BMC bioinformatics. **4**(1), 2 (2003)
3. Altaf-Ul-Amin, M., Shinbo, Y., Mihara, K., Kurokawa, K.: Development and implementation of an algorithm for detection of protein complexes in large interaction networks. J. BMC bioinformatics. **7**(1), 207 (2006)
4. Gavin, A.C., Aloy, P., Grandi, P., Krause, R., Boesche, M., Marzioch, M., Superti-Furga, G.: Proteome survey reveals modularity of the yeast cell machinery. J. Nature. **440**(7084), 631–636 (2006)
5. Tang, X., Wang, J., Liu, B., Li, M., Chen, G., Pan, Y.: A comparison of the functional modules identified from time course and static PPI network data. J. BMC bioinformatics. **12**(1), 339 (2011)
6. Li, M., Wu, X., Wang, J., Pan, Y.: Towards the identification of protein complexes and functional modules by integrating PPI network and gene expression data. J. BMC bioinformatics. **13**(1), 109 (2012)
7. Li, M., Chen, W., Wang, J.: Identifying Dynamic Protein Complexes Based on Gene Expression Profiles and PPI Networks. J. BioMed Research International **2014** (2014)
8. Nepusz, T., Yu, H., Paccanaro, A.: Detecting overlapping protein complexes in protein-protein interaction networks. J. Nature methods. **9**(5), 471–472 (2012)
9. Enright, A.J., Van, D.S., Ouzounis, C.A.: An efficient algorithm for large-scale detection of protein families. J. Nucleic acids research. **30**(7), 1575–1584 (2002)
10. Jiang, P., Singh, M.: SPICi: a fast clustering algorithm for large biological networks. J. Bioinformatics. **26**(8), 1105–1111 (2010)
11. Watts, D.J., Strogatz, S.H.: Collective dynamics of 'small-world' networks. J. nature. **393**(6684), 440–442 (1998)
12. Xenarios, I., Salwinski, L., Duan, X.J., Higney, P., Kim, S.M.: Eisenberg D. DIP, the Database of Interacting Proteins: a research tool for studying cellular networks of protein interactions. J Nucleic acids research. **30**(1), 303–305 (2002)
13. Krogan, N.J., Cagney, G., Yu, H., Zhong, G., Guo, X., Ignatchenko, A., et al.: Global landscape of protein complexes in the yeast Saccharomyces cerevisiae. Nature **440**, 637–643 (2006)
14. Gene Expression Omnibus database. http://www.ncbi.nlm.nih.gov/geo/query/acc.cgi?acc=GSE3431
15. Gene Ontology Database. http://www.geneontology.org/GO.database.shtml
16. Friedel, C.C., Krumsiek, J., Zimmer, R.: Bootstrapping the interactome: unsupervised identification of protein complexes in yeast. In: Vingron, M., Wong, L. (eds.) RECOMB 2008. LNCS (LNBI), vol. 4955, pp. 3–16. Springer, Heidelberg (2008)
17. Wu, H., Su, Z., Mao, F., Olman, V., Xu, Y.: Prediction of functional modules based on comparative genome analysis and Gene Ontology application. J. Nucleic acids research. **33**(9), 2822–2837 (2005)
18. Huntley, R.P., Sawford, T., Mutowo-Meullenet P.: The GOA database: gene ontology annotation updates for 2015. J. Nucleic acids research. gku1113 (2014)

Mining Uncertain and Imprecise Data

Mining Uncertain Sequential Patterns in Iterative MapReduce

Jiaqi Ge[1(✉)], Yuni Xia[1], and Jian Wang[2]

[1] Department of Computer and Information Science, Indiana University Purdue University Indianapolis, Indianapolis, IN 46202, USA
{jiaqige,yxia}@cs.iupui.edu
[2] School of Electronic Science and Engineering,
Nanjing University, Nanjing, Jiangsu 210023, China
wangjnju@nju.edu.cn

Abstract. This paper proposes a sequential pattern mining (SPM) algorithm in large scale uncertain databases. Uncertain sequence databases are widely used to model inaccurate or imprecise timestamped data in many real applications, where traditional SPM algorithms are inapplicable because of data uncertainty and scalability. In this paper, we develop an efficient approach to manage data uncertainty in SPM and design an iterative MapReduce framework to execute the uncertain SPM algorithm in parallel. We conduct extensive experiments in both synthetic and real uncertain datasets. And the experimental results prove that our algorithm is efficient and scalable.

Keywords: Uncertain databases · Sequential pattern mining

1 Introduction

Sequential pattern mining (SPM) is an important data mining application. It provides inter-transactional analysis for timestamped data which are modeled by sequence databases. In real applications, uncertainty is almost everywhere and it may cause probabilistic event existence in sequence databases. For example, in an employee tracking RFID network, the tag read by sensors are modeled by a relation $see(t, aId, tId)$, which denotes that the RFID tag tId is detected by an antenna aId at time t. Since an RFID sensor can only identify a tag with a certain probability within its working range, the PEEX system [10] outputs an uncertain event such as $meet(100, \text{Alice}, \text{Bob}, 0.4)$, which indicates that the event that Alice and Bob meet at time 100 happens with probability 0.4.

Possible world semantics is widely used to interpret uncertain databases[5, 17]; however, it also brings efficiency and scalability challenges to uncertain SPM problems. Meanwhile, applications in the areas of biology, Internet and business informatics encounter limitations due to large scale datasets. While MapReduce is a widely used programming framework for processing big data in parallel, its basic framework can not directly be used in SPM because it does not support the iterative computing model which is required by most SPM algorithms.

© Springer International Publishing Switzerland 2015
T. Cao et al. (Eds.): PAKDD 2015, Part II, LNAI 9078, pp. 243–254, 2015.
DOI: 10.1007/978-3-319-18032-8_19

In this paper, we propose a sequential pattern mining algorithm in iterative MapReduce for large scale uncertain databases. Our main contributions are summarized as follows:

(1) We use possible world semantics to interpret uncertain sequence databases and analyze the naturally correlated possible worlds.
(2) We design a vertical format of uncertain sequence databases in which we save and reuse intermediate computational results to significantly reduces the time complexity.
(3) We design an iterative MapReduce framework to execute our uncertain algorithm in parallel.
(4) Extensive experiments are conducted in both synthetic and real uncertain datasets, which prove the efficiency and scalability of our algorithm.

2   Related Works

A lot of traditional database and data mining techniques have been extended to be applied to uncertain data [2]. Muzammal and Raman propose the SPM algorithm in probabilistic database using expected support to measure pattern frequentness, which has weakness in mining high quality sequential patterns[14,15]. Zhao et al. define probabilistic frequent sequential patterns using possible world semantics and propose their complimentary uncertain SPM algorithm UPrefixSpan [17,18]; however, it uses the depth-first strategy to search frequent patterns and cannot be directly extended to MapReduce framework. A dynamic programming approach of mining probabilistic spatial-temporal frequent sequential patterns is introduced in [11]; Wan et al. [16] propose a dynamic programming algorithm of mining frequent serial episodes within an uncertain sequence. However, dynamic programming also cannot be directly extended to MapReduce.

Jeong et al. propose a MapReduce framework for mining sequential patterns in DNA sequences with only four distinct items [8], in contrast to this paper where unlimited number of items are allowed; Chen et al. extend the classic SPAM algorithm to its MapReduce version SPAMC [7]. However, SPAMC relies on a global bitmap and it is still not scalable enough for mining extremely large databases. Miliaraki et al. propose a gap-constraint frequent sequence mining algorithm in MapReduce [13]. However, all these algorithms are applied in the context of deterministic data, while our work aims to solve large scale uncertain SPM problems.

3   Problem Statement

3.1   Uncertain Model

An uncertain database contains a collection of uncertain sequences. An uncertain sequence is an ordered list of uncertain events. An uncertain event is represented by $e = \langle sid, eid, I, p_e \rangle$. Here sid is the sequence id and eid is the event id.

sid	eid	I	Pe
1	1	{A,B}	0.8
1	2	{C}	0.8
2	1	{B}	1
2	2	{C}	0.8
2	3	{C}	0.4

wid	Possible world
1	$<e_{11}><e_{21}, e_{22}, e_{23}>$
2	$<e_{11}, e_{12}><e_{21}, e_{22}>$
3	$<e_{11}, e_{12}><e_{21}, e_{23}>$
4	$<e_{11}, e_{12}><e_{21}, e_{22}, e_{23}>$
...	...

Fig. 1. An example of uncertain database   **Fig. 2.** possible worlds table

$\langle sid, eid \rangle$ identifies a unique event. I is an itemset that describes event e, and p_e is the existential probability of event e. Figure 1 shows an example of an uncertain sequence database. Here, for instance, the uncertain event $e_{11} = \langle 1, 1, \{AB\}, 0.8 \rangle$ indicates that the itemset $\{AB\}$ occurs in e_{11} with probability 0.8.

We use possible world semantics to interpret uncertain sequence databases. A possible world is instantiated by generating every event according to its existential probability. The number of possible worlds grows exponentially to the number of sequences and events. It is widely assumed that uncertain sequences in the sequence database are mutually independent, which is known as the *tuple-level independence* [2,9] in probabilistic databases. Events are also assumed to be independent of each other [5,17], which can be justified by the assumption that events are often observed independently in real world applications. Therefore, we can compute the existential probability of a possible world w in Equation (1).

$$P_e(w) = \prod_{\forall d_i \in w} \{ \prod_{\forall e_{ij} \in d_i} P(e_{ij}) * \prod_{e_{ij} \notin d_i} (1 - P(e_{ij})) \} \tag{1}$$

Where $d_i \in w$ is a sequence in w and $e_{ij} \in d_i$ is an event in d_i. Here e_{ij} is instantiated from the original database and $P(e_{ij})$ is its existential probability. Figure 2 is a table which contains four possible worlds of the uncertain sequence database in Figure 1. Then, for example, we can compute the existential probability of possible world w_1 by $P(w_1) = (0.8 * 0.2) * (1 * 0.8 * 0.4) = 0.0512$.

When one or more uncertain event occurs multiple times in the sequence such as C in e_{22} and e_{23}, some possible worlds are correlated even under the independent assumptions. For example, w_1 and w_3 in Figure 2 are correlated in supporting a pattern, because each event in w_1 is also present in w_3.

3.2   Uncertain SPM Problem

A sequential pattern $\alpha = \langle X_1 \cdots X_n \rangle$ is *supported* by a sequence $\beta = \langle Y_1 \cdots Y_m \rangle$, denoted by $\alpha \sqsubseteq \beta$, if and only if there exists integers $1 \leq k_1 < \cdots < k_n \leq m$ so that $X_i.I \subseteq Y_{k_i}.I, \forall i \in [1, n]$. In deterministic databases, a sequential pattern s is frequent if and only if it satisfies $sup(s) \geq t_s$, where $sup(s)$ is the total number of sequences that support s and t_s is the user-defined minimal threshold. In an uncertain database D, the frequentness of s is probabilistic and it can be computed by Equation (2).

$$P(sup(s) \geq t_s) = \sum_{\forall w, sup(s|w) \geq t_s} P(w) \qquad (2)$$

Where w is a possible world in which s is frequent and $P(w)$ is the existential probability of w.

Then the uncertain sequential pattern mining problem is defined as follows. *Given an uncertain sequence database D, a minimal support threshold t_s and a minimal frequentness probability threshold t_p, find every probabilistic frequent sequential pattern s in D which has $P(sup(s) \geq t_s) \geq t_p$.*

4   Solution

4.1   Approximation of Frequentness Probability

Suppose $D = \{d_1, \ldots, d_n\}$ is an uncertain database and s is a sequential pattern. Because $d_1, \ldots, d_n$ in D are mutually independent, the probabilistic support of s in D, denoted by $sup(s)$, can be computed by Equation (3).

$$sup(s) = \sum_{i=1}^{n} sup(s|d_i) \qquad (3)$$

Where $sup(s|d_i)$ $(i = 1, \ldots, n)$ are Bernoulli random variables, whose success probabilities are $P(sup(s|d_i) = 1) = P(s \sqsubseteq d_i)$. And we will discuss the computation of $P(s \sqsubseteq d_i)$ in section 4.2.

We find that $sup(s)$ is a Poisson-Binomial random variable, because it is the sum of n independent but non-identical Bernoulli random variables. And $sup(s)$ can be modeled by its probability mass function (pmf), denoted by $sup(s) = \{sup(s)|0 : p_0, 1 : p_1, \ldots, n : p_n\}$. Here $n = |D|$ is the number of sequences in D.

According to *central limit theorem*, $sup(s)$ converges to the Gaussian distribution when n goes to infinity. Therefore, in the large scale database D, we can approximate the distribution of $sup(s)$ by Equation (4).

$$sup(s) = \sum_{i=1}^{n} X_i \longrightarrow N(\sum_{i=1}^{n} p_i, \sum_{i=1}^{n} p_i * (1 - p_i)) \qquad (4)$$

Here we approximate $sup(s)$ by the Gaussian distribution $\mathcal{N}(\mu, \sigma^2)$, and then the approximated frequentness probability $P(sup(s) \geq t_s)$ can be computed in linear time.

4.2   Support Probability

The support probability $P(s \sqsubseteq d)$ is the probability that a sequential pattern s is supported by an uncertain sequence d and it can be computed in (5) according to possible world semantics.

$$P(s \sqsubseteq d) = \sum_{\forall w, s \sqsubseteq w} P(w) \tag{5}$$

Where w is a possible world of d which supports s and $p(w)$ is its existential probability. However, suppose each item in a k-length pattern s has m multiple occurrences in d in average, there are $O(k^m)$ possible worlds that may support s in the worst case. And directly enumerating all of them is usually too complex in practice.

Therefore, we design an incremental approach to compute support probability efficiently. Let l be the last item of sequential pattern s. In uncertain sequence d, suppose there are q possible occurrences of l in events $e_{k_1}, \ldots, e_{k_q}$, then all the possible worlds that may support s can be divided into q disjoint subsets $(g_1, \ldots, g_q)$ by the most recent occurrence of item l.

Let $P(g_i)$ be the probability that the latest occurrence of item l (the last item of s) is in e_{k_i}, then it can be computed by Equation (6).

$$P(g_i) = P(l \in e_{k_i}) * \prod_{t=i+1}^{q} P(l \notin e_{k_t}) \tag{6}$$

The amortized cost of Equation (6) is $O(1)$, when events are pre-sorted by their $eids$. And the support probability $P(s \sqsubseteq d)$ can be computed in (7).

$$P(s \sqsubseteq d) = \sum_{i=1}^{q} P(s \sqsubseteq d|g_i) * P(g_i) = P(s \sqsubseteq d \cap g_i) \tag{7}$$

For example, given $d = \langle (B:0.5)(C^1:0.4)(C^2:0.4) \rangle$ and $s = \langle BC \rangle$, according to possible world semantics, there are three possible worlds of d that may support s: $w_1 = \{BC^1\}$, $w_2 = \{BC^2\}$ and $w_3 = \{BC^1C^2\}$, and we divide them into two disjoint groups by the latest occurrence of item C in the possible worlds as $g_1 = \{w_1\}$ and $g_2 = \{w_2, w_3\}$. We first compute $P(g_1) = 0.4 * 0.6 = 0.24$ and $P(g_2) = 0.4$, then we have $P(s \sqsubseteq d) = 0.5 * 0.24 + 0.5 * 0.4 = 0.22$.

Suppose l is the last item of s, then $s' = s - \{l\}$ is a $(k-1)$-length sequential pattern. $P(s \sqsubseteq d|g_i)$ in (7) can be computed by (8).

$$P(s \sqsubseteq d|g_i) = \sum_{j=1}^{p} P(s' \sqsubseteq d|g_j) * P(g_j|g_i) = \sum_{j=1}^{p} P(s' \sqsubseteq d \cap g_j) * \delta(g_i, g_j) \tag{8}$$

Where g_j ($\forall j \in [1, p]$) are p disjoint subsets of possible worlds in which the latest occurrence of the last item of s' in the event e_{k_j}. And $\delta(g_j, g_i) = 1$, if the last item of s' occurs before the last item of s; otherwise, $\delta(g_j, g_i) = 0$.

By substituting (8) into (7), we can compute the support probability in (9).

$$P(s \sqsubseteq d) = \sum_{i=1}^{q} \sum_{j=1}^{p} P(s' \sqsubseteq d \cap g_j) * P(g_i) * \delta(g_i, g_j) \tag{9}$$

Therefore, if we save and reuse the values of $P(s' \sqsubseteq d \cap g_j)$, we can avoid repeated computation which reduces the time complexity of support probability computation from exponential to $O(p * q)$.

D		
sid	tid	I
1	1	A:0.3
1	2	A:0.5
1	3	B:0.4
2	1	A:0.4
2	2	B:0.8
2	3	B:0.7

D1

sid	1			2		
c	\<A>	\<B>	\<A>		\<B>	
tid	1	2	3	1	2	3
Pc	0.15	0.5	0.4	0.4	0.24	0.7
Pi	0.3	0.5	0.4	0.4	0.8	0.7

D2

sid	1	2	
c	\<AB>	\<AB>	
tid	2	2	3
Pc	0.26	0.096	0.28
Pi	0.4	0.8	0.7

Fig. 3. An example of constructing the vertical data structure

4.3 Vertical Data Structure

We develop a vertical data format D_k to save occurrences of k-length candidate patterns. The schema of D_k is $\langle sid, c, tid, P_c, P_i \rangle$, where sid identifies an uncertain sequence d, c is a candidate pattern and $\langle tid, P_c, P_i \rangle$ records an occurrence of c in d. Suppose i is the last item of c and e is the event identified by (sid, tid), then we have $P_c = P(c \sqsubseteq d \cap g_i)$, where g_i is a subset of possible worlds in which the latest occurrence of item i locates in event e. And $P_i = P(i \in e)$ is the existential probability of i in e.

We transform the original sequence database into its vertical format which is a set of candidate occurrences. Figure 3 shows an example of constructing the vertical data format D_k. Here D is the original database, and D_1 is transformed from D. For example, let $s = \langle A \rangle$, then we have two groups g_1 and g_2 of occurrences of s in sequence d_1. We compute $P_{c1}(s) = 1 * P(g_1) = 0.3 * 0.5 = 0.15$ and $P_{c2}(s) = 0.5$ and save the results in D_1. Thereafter, we can compute the support probabilities $P(s \sqsubseteq d_1) = 0.65$ and $P(s \sqsubseteq d_2) = 0.4$ from D_1, which are used to calculate the frequentness probability. In this example, if we set $minsup = 1$ and $minprob = 0.5$, then $\langle A \rangle$ and $\langle B \rangle$ are two frequent patterns. 2-length candidates are generated by self-joining 1-length frequent patterns, and their occurrences are saved in D_2. For example, let $s' = \langle AB \rangle$, then $P(s' \sqsubseteq d_1) = 0.65 * 0.4 = 0.26$. Since there are two occurrences of item B in d_2, we first compute $P(g_1) = 0.8 * 0.3 = 0.24$ and $P(g_2) = 0.7$, then we have $P_{c1}(s') = 0.4 * 0.24 = 0.096$ and $P_{c2}(s') = 0.4 * 0.7 = 0.28$. Thereafter, the support probability $P(s' \sqsubseteq d_2) = 0.376$.

In our approach, we only refer to D_k in searching k-length frequent patterns. And D_k is usually in a much smaller size than the original database because it only contains occurrences of potential frequent candidate patterns.

4.4 Uncertain SPM in Iterative MapReduce

Our iterative MapReduce framework helps to traverse a huge sequence tree[4] in searching frequent patterns in parallel. In each iteration, we start a MapReduce job to search k-length frequent patterns on a cluster of computers.

Fig. 4 shows our iterative MapReduce framework for uncertain SPM. In the first iteration, the original database is split and input to mappers; in the

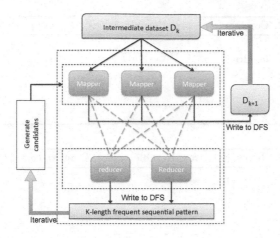

Fig. 4. Iterative MapReduce framework for uncertain sequential pattern mining

---

ALGORITHM 1. Map(Key *key*, Value *value*, Context *context*)

$d_{k-1} \leftarrow$ pase(*value*)   /* $d_{k-1} \in D_{k-1}$ parsed from *value* */
$C_k \leftarrow$ DistributedCache.file
$d_k \leftarrow$ *construct from C_k and d_{k-1}*
foreach $c \in C_k$ **do**
    $p \leftarrow P(c \sqsubseteq d_k)$ /* computed by summing $P_c(c)$ in d_k */
    $key \leftarrow c$;
    $value \leftarrow \langle p, p * (1 - p) \rangle$ /* composited value */
    context.collect(*key*, *value*)
end
DFS.file f = new DFSFile("D_k");
f.append(d);

---

k^{th} $(k > 1)$ iteration, the input data of a mapper is a chunk of D_{k-1}. We modify the data split function in MapReduce to make sure that all occurrences in one sequence are input to the same map function. A set of k-length candidate patterns are distributed to mappers, which is denote by C_k.

(1) **Mapper Function:** The mapper function is shown in Algorithm 1. It first constructs d_k from d_{k-1} and C_k, where $d_{k-1} \in D_{k-1}$ contains occurrences of $(k-1)$-length frequent patterns in one uncertain sequence. Given a candidate pattern c, the mapper computes the support probability $p = P(c \sqsubseteq d_k)$ using the newly updated data structure and outputs a key-value pair $\langle c, \langle \mu, \sigma^2 \rangle \rangle$ if $p = P(c \sqsubseteq d) > 0$. Here $\mu = p$ and $\sigma^2 = p * (1 - p)$ are the mean and variance of the Bernoulli random variable $sup(c|d_k)$. Thereafter, d_k is written to distributed file system (DFS) to be used in the next iteration.

ALGORITHM 2. Combine(Key *key*, Iterable *values*, Context *context*)

$\mu \leftarrow 0,\ \sigma^2 \leftarrow 0$
foreach *value* $\in$ *values* **do**
    $\mu = \mu + value.\mu$
    $\sigma^2 = \sigma^2 + value.\sigma^2$
end
context.collect(*key*, $\langle \mu, \sigma^2 \rangle$)

ALGORITHM 3. Reduce(Key *key*, Iterable *values*, Context *context*)

$c \leftarrow key$
$\mu \leftarrow 0,\ \sigma^2 \leftarrow 0$
foreach *value* $\in$ *values* **do**
    $\mu = \mu + value.\mu$
    $\sigma^2 = \sigma^2 + value.\sigma^2$
end
$sup(c) \sim N(\mu, \sigma^2)$
$t_s \leftarrow$ *context.minsup*, $t_p \leftarrow$ *context.minprob*
if $P(sup(c) \geq t_s) \geq t_p$ **then**
    DFS.file f = new DFSFile("frequent pattern");
    f.append(c);
end

---

(2) **Combiner Function:** We design a combiner function in Algorithm 2 to help improve the performance. Suppose a mapper function emits n key-value pairs $\langle c, \langle \mu_i, \sigma_i^2 \rangle \rangle$ ($i = 1, \ldots, n$) which are associated with the identical pattern c. As the value filed of the mapper output is associative and commutative, they can be condensed to a single pair $\langle c, \langle \sum_1^n u_i, \sum_1^n \sigma_i^2 \rangle \rangle$. Then each mapper sends only one key-value pair to the reducer for each candidate pattern, which dramatically reduce the total bandwidth cost of data shuffling.

(3) **Reducer function**: Algorithm 3 shows the reducer function. The input key-value pair of the reducer is in the form of $\langle c, \langle \mu_i, \sigma_i^2 \rangle \rangle$, where $\mu_i = \sum p$ and $\sigma_i^2 = \sum p * (1 - p)$ are the partially aggregated mean and variance of the probabilistic support of candidate c. The reducer function accumulates the overall mean and variance of c in the entire uncertain database and uses the Gaussian distribution to approximate the distribution of overall support $sup(c)$. Given $minsup = t_s$ and $minprob = t_p$, the reducer outputs the probabilistic frequent sequential patterns to the file, if $P(sup(c) \geq t_s) \geq t_p$; otherwise, c is not probabilistic frequent and is discarded by the reducer.

A MapReduce iteration is finished after all k-length probabilistic frequent sequential patterns are discovered and written to DFS files. After that, we self-join k-length frequent patterns to generate $(k + 1)$-length candidate patterns for the next iteration. This process continues until all frequent patterns are discovered.

5   Evaluation

In this section, we implement our uncertain SPM algorithm in iterative MapReduce, denoted by *IMRSPM*, and evaluate its performance using both synthetic and real world datasets in a 10-node Hadoop cluster.

A naive method directly enumerates possible worlds table without reusing previous computational results. We implement this naive approach in Iterative MapReduce as *baseline*, which is denoted by *BL* here. We also compare our algorithm with the single-machine uncertain sequential pattern mining algorithm, denoted by *UPrefix* [17,18], to show the benefit from parallel computing.

5.1   Synthetic Dataset Generation

The IBM market-basket data generator [3] uses the following parameters to generate sequence datasets in various scales: (1) C: number of customers; (2) T: average number of transactions per sequence; (3) L: average number of items per transaction per sequence; (4) I: number of different items.

We assume that an event existential probability follows normal distribution $t \sim N(\mu, \sigma^2)$, where μ is randomly drawn from range $[0.7, 0.9]$ and σ is randomly drawn from range $[1/21, 1/12]$. Then we draw a value from t and assign it to an event in the original synthetic datasets as its existential probability. This approach has been used in previous work [1] to generate synthetic uncertain datasets. We name a synthetic uncertain dataset by its parameters. For example, a dataset T4L10I10C10 indicates $T = 4$, $L = 10$, $I = 10*1000$ and $C = 10*1000$.

5.2   Scalability

In Figure 5, we evaluate the scalability of IMRSPM on synthetic datasets generated by different parameters. Here we set $minsup = 0.2\%$ and $minprob = 0.7$. Fig. 5(a) shows the running time variations of IMRSPM when C varies from $10\,000$ to $10\,000\,000$, where $T = 4$, $L = 4$, $I = 10\,000$. Fig. 5(b) shows the running time variations of IMRSPM when T varies from 5 to 25, where $C = 100\,000$, $L = 4$, $I = 10\,000$. Fig. 5(c) shows the running time variations of IMRSPM when L varies from 2 to 32, where $C = 100\,000$, $T = 4$, $I = 10\,000$. Fig. 5(d) shows the running time variations of IMRSPM when I varies from $2\,000$ to $32\,000$, where $C = 100\,000$, $T = 4$, $L = 4$.

In Figure 5, we observe the following phenomenons:
(1) IMRSPM outperforms BL under every setting of the parameters, which proves the effectiveness of our incremental temporal uncertainty management approach; meanwhile, IMRSPM is much more scalable than UPrefix, which demonstrates the advantage of using iterative MapReduce framework.
(2) The running time increase with the increment of C, T, L, as increasing these parameters generates larger scale datasets. Furthermore, when T or L are set to larger values, there are more repeated items in uncertain sequences. And our incremental uncertainty management approach shows its effectiveness in improving the efficiency especially in such cases.

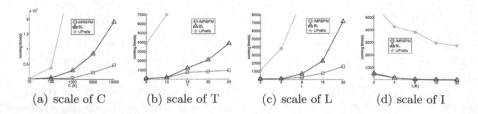

Fig. 5. Scalability of IMRSPM-A algorithm

(3) The running time slightly drops with the increment of I. When the value of I grows, the number of repeated item in one sequence become less because items are randomly selected from a fixed set of items.

5.3   Mining Customer Behavior Patterns from Amazon Reviews

We apply our IMRSPM algorithm in Amazon review dataset[12] to discover customer behavior patterns. The Amazon review dataset includes 34 686 770 reviews of 2 441 053 products from 6 643 669 customers between June 1995 to March 2013. Each review is scored by an integer between 1 to 5, which indicates a user opinion toward a product. However, this score is a lose measurement of subjective satisfaction. Suppose a customer gives a score t to a product, then we believe that the probability that this customer likes this product is $p = t/5$. An ordered list of user reviews is regarded as an uncertain sequence. A probabilistic frequent sequential pattern $\langle A, B \rangle$ mined from this database can be explained as: if a customer likes product A, then it is very likely that he/she will like product B in the future.

For example, given $minsup = 0.005\%$ and $minprob = 0.7$, we have discovered the sequential pattern $\langle$B000TZ19TC $\rightarrow$ B000GL8UMI$\rangle$. Here B000TZ19TC is the Amazon Standard Identification Number (ASIN) of the book *Fahrenheit 451* published in 1953. And this pattern reveals that users who now like product B000TZ19TC may also like B000GL8UMI in the future, which is a newer edition of the same book published in 1963. We also discover other non-trivial patterns as $\langle$B000MZWXNA $\rightarrow$ B000PBZH6Q$\rangle$, where B000MZWXNA is associated with the book *The Martian Way* and ASIN B000PBZH6Q identifies the book *Foundation*.

Figure 6 and Figure 7 show the effect of user-defined parameters $minsup$ and $minprob$ in Amazon dataset. We initially set $minprob = 0.7$ and $minsup = 0.04\%$. In Figure 6(a) and 7(a), we vary the value of $minsup$ from 0.02% to 0.04%; while $minprob$ is varied from 0.5 to 0.8 in Figure 6(b) and 7(b). From Figure 6 and Figure 7 , we observe that:

(1) In Figure 6(a), the running time of IMRSPM decreases with the increment of $minsup$; meanwhile, the effect of $minsup$ to the computing time is more significantly than that to the shuffling time. The reason is that fewer frequent patterns are mined when $minsup$ is larger, which can be proved by Figure 7(a).
(2) The performance remains relatively stable to the variation of $minprob$.

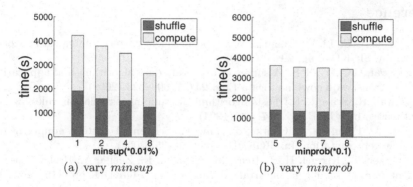

(a) vary *minsup*   (b) vary *minprob*

Fig. 6. Effect of user-define parameters in efficiency

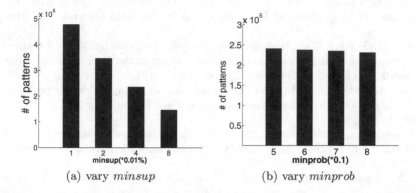

(a) vary *minsup*   (b) vary *minprob*

Fig. 7. Effect of user-define parameters in number of patterns

The probabilistic support of a sequential pattern is bounded to its expected value (*Chernoff bound*). Thus, the frequentness of a large number of candidate patterns becomes deterministic, and this explains why the running time and the number of frequent patterns do not significantly fluctuate in Figure 6(b) and 7(b).

6   Conclusions

In this paper, we propose a SPM algorithm in an iterative MapReduce framework for large scale uncertain databases to discover customer behavior patterns in Amazon review dataset. In the future, we will continue to explore the facilitation of other distributed platforms in solving uncertain SPM problems.

References

1. Aggarwal, C.C., Li, Y., Wang, J., Wang, J.: Frequent pattern mining with uncertain data. In: SIGKDD, pp. 29–38 (2009)
2. Aggarwal, C.C., Yu, P.S.: A survey of uncertain data algorithms and applications. IEEE Trans. on Knowl. and Data Eng. **21**(5), 609–623 (2009)
3. Agrawal, R. Srikant, R.: Fast algorithms for mining association rules in large databases. In: VLDB, pp. 487–499 (1994)
4. Ayres, J., Flannick, J., Gehrke, J., Yiu, T.: Sequential pattern mining using a bitmap representation. In: SIGKDD, pp. 429–435 (2002)
5. Bernecker, T., Kriegel, H.-P., Renz, M., Verhein, F., Zuefle, A.: Probabilistic frequent itemset mining in uncertain databases. In: SIGKDD, pp. 119–128. ACM (2009)
6. Chernoff, H.: A measure of asymptotic efficiency for tests of a hypothesis based on the sum of observations. Annals of Mathematical Statistics **23**, 493–507 (1952)
7. Chun-Chieh Chen, M.-S.C., Tseng, C.-Y.: Highly scalable sequential pattern mining based on mapreduce model on the cloud. In: BigData Congress, pp. 310–317 (2013)
8. Jeong, B.-S., Choi, H.-J., Hossain, M.A., Rashid, M.M., Karim, M.R.: A mapreduce framework for mining maximal contiguous frequent patterns in large dna sequence datasets. IETE Technical Review **29**, 162–168 (2012)
9. Jestes, J., Cormode, G., Li, F., Yi, K.: Semantics of ranking queries for probabilistic data. IEEE Transactions on Knowledge and Data Engineering **23**(12), 1903–1917 (2011)
10. Khoussainova, N., Balazinska, M., Suciu, D.: Probabilistic event extraction from rfid data. In: Proceedings of the 24th IEEE International Conference on Data Engineering, pp. 1480–1482 (2008)
11. Li, Y., Bailey, J., Kulik, L., Pei, J.: Mining probabilistic frequent spatio-temporal sequential patterns with gap constraints from uncertain databases. In: IEEE International Conference on Data Mining, pp. 448–457 (2013)
12. McAuley, J., Leskovec, J.: Hidden factors and hidden topics: understanding rating dimensions with review text. In: RecSys (2013)
13. Miliaraki, I., Berberich, K., Gemulla, R., Zoupanos, S.: Mind the gap: large-scale frequent sequence mining. In: SIGKDD, pp. 797–808 (2013)
14. Muzammal, M., Raman, R.: Mining sequential patterns from probabilistic databases. In: Huang, J.Z., Cao, L., Srivastava, J. (eds.) PAKDD 2011, Part II. LNCS, vol. 6635, pp. 210–221. Springer, Heidelberg (2011)
15. Tong, Y., Chen, L., Cheng, Y., Yu, P.S.: Mining frequent itemsets over uncertain databases. Proceeding of the VLDB Endowment **5**, 1650–1661 (2012)
16. Wan, L., Chen, L., Zhang, C.: Mining frequent serial episodes over uncertain sequence data. In: EDBT, pp. 215–226 (2013)
17. Zhao, Z., Yan, D., Ng, W.: Mining probabilistically frequent sequential patterns in uncertain databases. In: EDBT, pp. 74–85 (2012)
18. Zhao, Z., Yan, D., Ng, W.: Mining probabilistically frequent sequential patterns in large uncertain databases. IEEE Transactions on Knowledge and Data Engineering **26**, 1171–1184 (2013)

Quality Control for Crowdsourced POI Collection

Shunsuke Kajimura[1]([⊠]), Yukino Baba[2,3],
Hiroshi Kajino[1], and Hisashi Kashima[4]

[1] The University of Tokyo, Tokyo, Japan
{shunsuke_kajimura,hiroshi_kajino}@mist.i.u-tokyo.ac.jp
[2] National Institute of Informatics, Tokyo, Japan
ybaba@nii.ac.jp
[3] JST, ERATO, Kawarabayashi Large Graph Project, Tokyo, Japan
[4] Kyoto University, Kyoto, Japan
kashima@i.kyoto-u.ac.jp

Abstract. Crowdsourcing allows human intelligence tasks to be outsourced to a large number of unspecified people at low costs. However, because of the uneven ability and diligence of crowd workers, the quality of their submitted work is also uneven and sometimes quite low. Therefore, quality control is one of the central issues in crowdsourcing research. In this paper, we consider a quality control problem of POI (points of interest) collection tasks, in which workers are asked to enumerate location information of POIs. Since workers neither necessarily provide correct answers nor provide exactly the same answers even if the answers indicate the same place, we propose a two-stage quality control method consisting of an answer clustering stage and a reliability estimation stage. Implemented with a new constrained exemplar clustering and a modified HITS algorithm, the effectiveness of our method is demonstrated as compared to baseline methods on several real crowdsourcing datasets.

Keywords: Quality control · POI · Constrained exemplar clustering · HITS

1 Introduction

The idea of crowdsourcing is to outsource human intelligence tasks to a large number of unspecified people via the Internet. Stimulated by the successes of varioius crowdsourcing projects, and the emergence of crowdsourcing marketplaces, such as Amazon Mechanical Turk, crowdsourcing is rapidly expanding in various areas. Since crowd workers comprise people with different skill levels and diligence, the quality of their work is quite uneven, and sometimes untrustworthy workers called spam workers produce significantly low-quality work. Therefore, quality control of crowdsourcing is one of the central issues in crowdsourcing research [8]. A common strategy for controlling crowdsourcing quality is to introduce redundancy, that is, to assign a single task to multiple workers and

© Springer International Publishing Switzerland 2015
T. Cao et al. (Eds.): PAKDD 2015, Part II, LNAI 9078, pp. 255–267, 2015.
DOI: 10.1007/978-3-319-18032-8_20

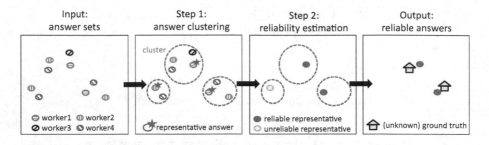

Fig. 1. Illustration of our problem setting and quality control method in POI collection

aggregate their answers to obtain a more reliable answer. Not only simple aggregation methods, such as majority voting and averaging have been proposed but also various sophisticated statistical models [2,4,12,17,18]. Recently, beyond multiple-choice tasks, tasks with more complex structured responses, such as orders [3,11,20], sequences [19], spatial data [15], and general tasks [1,9], have been addressed.

Although a number of efforts have been made to develop quality control methods for respective tasks, there has been no quality control method for POI (point of interest) collection tasks, which require workers to enumerate POIs, in spite of its great significance. One of the important applications is citizen science, which calls for voluntary participation of amateur people in collecting scientific information to cover wide areas of interests. There are several citizen science projects collecting location information on animals, such as birds' nests[1] and diseased or bleached points of the coral reefs of Hawaii[2]. POI collection tasks also appear in crowdsensing, which asks people to submit current location information together with additional information such as pictures of POIs. Collecting location information such as road-side parkings [10] can be considered as POI collection tasks.

However, the quality control issue of a POI task has not been addressed because of its task-specific uncertainty. The intrinsic difficulty of quality control in POI collection tasks is the lack of correspondence between each POI and each of workers' answers. In other words, we cannot distinguish between two given answers corresponding to two distinct POIs and those corresponding to the identical POI. Therefore, the ideas in existing studies assuming the correspondences between questions and answers cannot be directly applied to the POI collection task.

We consider that the difficulty involved in the quality control issue of POI tasks is caused by the fact that there are two types of uncertainties in POI collection tasks: workers neither necessarily provide correct answers nor provide exactly the same answers for a specific POI because of the variations in numerical values. Considering these two types of uncertainty enables us to propose a

---

[1] http://nestwatch.org/learn/how-to-nestwatch/understanding-nestwatch-data/
[2] http://eorhawaii.org/

quality control method for POI collection tasks. Our quality control method has two-stages: *answer clustering* followed by *reliability estimation*. As illustrated in Figure 1, the first stage is designed to select the representative answers that indicate distinct POIs, and the second to find the correct answers from among them.

The technical contributions of our quality control method are twofold. First, we construct an answer clustering method by incorporating *cannot-link* constraints [16] into the *exemplar clustering* formula proposed by Elhamifar et al. [5]. Our new clustering formula enables us to divide answers into clusters and choose the most representative answer from each cluster, while the answers given by the identical worker are separated from each other by cannot-link constraints. Second, we develop a new reliability estimation method by modifying the HITS algorithm [7]. The modified HITS algorithm using a normalized iterative matrix handles "streakers," who give many more answers in crowdsourced enumeration [6,14], while the original HITS algorithm suffers from these workers.

We conducted experiments using four real-world datasets on POI collection tasks and evaluated the performance of our two-stage quality control method. Our method appropriately handles the two types of uncertainties pertaining to the answers and provides high quality answer lists. In fact, it makes 6.4% to 19.4% improvements of the F-measures over the baseline methods in four experimental tasks. In addition, we synthetically added spam workers to the original datasets and confirmed that our method is robust against the presence of spam workers. Our method curbed the decreases of the F-measures by 8.2% to 72.5% of those of NONE in four tasks.

In summary, this paper makes three main contributions:

1. We indicate there are two types of uncertainties in crowdsourced POI collection.
2. We introduce a two-stage quality control method that deals with the two types of uncertainties in the enumeration tasks. Our quality control method yields a constrained exemplar clustering formulation.
3. We describe experiments using four POI collection tasks and investigate worker characteristics that affect the performance of our quality control method.

2   Two-stage Quality Control Method for POI Collection Tasks

We propose a two-stage quality control method to obtain the set of correct answers from the answers enumerated by workers. As shown in Figure 1, all the answers are clustered so that those in each cluster indicate the same POI in the first stage. Each cluster is represented by a single answer, which we call a *representative answer*. The objective of this answer clustering stage is to reduce the uncertainty due to noises in answers, and remove redundant answers indicating the same POIs. Subsequently, the second stage estimates the reliability of each representative answer, and then, we remove representative answers with low reliability.

2.1   Problem Setting

We first define the quality control problem in crowdsourced POI collection. A POI collection task requires crowdworkers to enumerate location information (i.e., longitude and latitude) of POIs. The workers are usually expected to give one or more answers (as many as they can provide) that indicate distinct POIs according to explicit instructions.

Let us assume that there are W workers. Let $\mathcal{T}_i$ denote the set of answers given by worker i, N the total number of the answers collected from all the workers, and $\mathcal{T} = \mathcal{T}_1 \cup \mathcal{T}_2 \cup \cdots \cup \mathcal{T}_W$ the set of all the answers. In addition, we assume that we have a distance d_{uv} between every pair of answers u and v in $\mathcal{T}$.

Given the answer set $\{\mathcal{T}_i\}_{i=1,\ldots,W}$ and the distance matrix $D = (d_{uv})_{u,v=1,\ldots,N}$, our goal is to obtain a subset of given answers, $\mathcal{P} \subseteq \mathcal{T}$, such that each answer in the answer subset correctly satisfies the task requirement and indicates a distinct POI.

2.2   Answer Clustering

To select representative answers from noisy and duplicate answers, we apply a clustering method and consider that the answers in each cluster indicate the identical POI. Standard clustering methods usually adopt the cluster centers as the representatives of the obtained clusters; however, this often results in undesirable representatives because the middle point of several data points, such as locations of restaurants, frequently falls in unreasonable points, such as the middle of a road. This observation leads us to resort to *exemplar clustering*, which chooses representatives from among data points.

We assume that the answers given by a single worker indicate distinct POIs even if some of the answers are close to each other. This assumption motivates us to introduce cannot-link constraints in the exemplar clustering. The cannot-link constraints assign the answers given by a single worker to distinct clusters. We believe this is a reasonable assumption, because we can prevent workers from providing multiple answers indicating the identical POI by giving appropriate instructions, and simple methods may be effective for detecting untrustworthy workers who provide the same answer many times to earn easy money.

Although several constrained clustering methods such as COP-Kmeans [16] have been proposed, no constrained extensions of exemplar clustering methods exist; therefore, we formalize a constrained exemplar clustering by incorporating cannot-link constraints into an exemplar clustering method.

Exemplar Clustering. We first review the convex exemplar clustering proposed by Elhamifar et al. [5]. Let $z_{uv} \in [0, 1]$ denote the probability that answer v belongs to the cluster represented by answer u and $Z = (z_{uv})_{u,v=1,\ldots,N}$ denote the representative matrix. Given the distance matrix D, the exemplar clustering problem is defined as

$$\min_{Z} \quad \sum_{v=1}^{N}\sum_{u=1}^{N} d_{uv}z_{uv} + \lambda \sum_{u=1}^{N} ||z_{u,:}||_q \tag{1}$$

$$\text{s.t.} \quad \sum_{u=1}^{N} z_{uv} = 1, \ \forall v, \quad z_{uv} \geq 0, \ \forall u, v,$$

where $z_{u,:}$ denotes the u-th row of Z and $\lambda > 0$ is a regularization parameter that indirectly controls the number of clusters. When q is chosen appropriately, the regularization term $||z_{u,:}||_q$ forces $z_{u,:} = 0$ for some u; $q \in \{2, \infty\}$ is chosen to ensure the convexity of the object function (1).

Constrained Exemplar Clustering. We create a constrained exemplar clustering method by appending cannot-link constraints to the exemplar clustering problem (1). We first assume that $z_{uv} \in \{0, 1\}$. We can ensure that none of two answers of worker i belongs to representative answer u by adding the constraint, $\sum_{v \in T_i} z_{u,v} \leq 1$. In order to force any pair of the answers given by a single worker to belong to distinct clusters, we add the constraints

$$\max_{u} \sum_{v \in T_i} z_{u,v} = \left\| \sum_{v \in T_i} z_{:,v} \right\|_{\infty} \leq 1, \ \forall i,$$

where $z_{:,v}$ denotes the v-th column of Z. We relax $z_{uv} \in \{0, 1\}$ into the interval $z_{uv} \in [0, 1]$, and then, our constrained exemplar clustering is formulated as

$$\min_{Z} \quad \sum_{v=1}^{N}\sum_{v=1}^{N} d_{uv}z_{uv} + \lambda \sum_{u=1}^{N} ||z_{u,:}||_q$$

$$\text{s.t.} \quad \sum_{u=1}^{N} z_{uv} = 1, \ \forall v, \quad z_{uv} \geq 0, \ \forall u, v, \quad \left\| \sum_{v \in T_i} z_{:,v} \right\|_{\infty} \leq 1, \ \forall i.$$

We then adopt answer u such that $||z_{u,:}|| > 0$ as a representative answer, which represents the cluster. Let $\mathcal{U} = \{u \in \mathcal{T} \mid ||z_{u,:}|| \neq 0\}$ denote a set of representative answers.

2.3 Reliability Estimation

As discussed in the Introduction, some of the representative answers given by the first answer clustering stage may not be correct; hence, we next estimate the reliability of each representative answer, as shown in Step 2 of Figure 1. We assume that an answer is reliable if it is supported by multiple reliable workers; a worker is reliable if s/he provides many reliable answers. This notion is similar to that employed in the HITS algorithm [7], a link analysis algorithm for estimating the importance of Web pages. In HITS, a Web page is considered highly *authoritative* if it has links from multiple good *hub* pages, and a Web page is regarded as a reliable hub if it has links to multiple authorities. The HITS algorithm estimates both the authority and hub values of each Web page using the link structure between the pages. We modify the HITS algorithm by using a normalized iterative matrix and apply it to the answer reliability estimation with analogies between authorities and answers and between hubs and workers. The modified HITS using the normalized iterative matrix performs well even when we have some streakers.

Modified HITS. We first introduce how to apply the HITS algorithm to reliability estimation. Let $a = (a_1, \ldots, a_M)^\top$ denote a vector of authority values of representative answers in $\mathcal{U}$, where M is the size of $\mathcal{U}$, and $h = (h_1, \ldots, h_W)^\top$ denote a vector of hub values of workers. Instead of the hyperlink structure used in the HITS, in reliability estimation we exploit the connection weights between workers and answers. Let us denote by $(u_1, u_2, \ldots, u_M)$ the indices of representative answers in $\mathcal{T}$. We define the connection weight matrix $L = (\ell_{ik})_{i,k}$ as $L_{ik} = \sum_{v \in T_i} z_{v,u_k}$, which represents the total amount of supports for the representative answer u_k of worker i.

Given L and the initial values of authority a and of hub h, the HITS algorithm updates a and h iteratively as $a \leftarrow \alpha L^\top h$, $h \leftarrow \beta L a$, where $\alpha > 0$ and $\beta > 0$ are normalization constants to keep the norms of $\mathbf{a}$ and $\mathbf{h}$ constant. The update operations can be combined into $a \leftarrow c L^\top L a$, where $c = \alpha\beta$. The solution of a that we seek is obtained by finding the eigenvector corresponding to the largest eigenvalue of $L^\top L$. We finally output a set of answers $\mathcal{P}$ whose authority values are greater than some threshold, ϵ.

In order to estimate the reliability of workers accurately in the presence of streakers, we updated the authority scores by a normalized matrix instead of $L^\top L$ as $a \leftarrow c N^{-1/2} L^\top L N^{-1/2} a$, where $N = \mathrm{diag}(n_1, n_2, \ldots, n_W)$ and n_i denotes the number of answers given by the i-th worker.

3 Experiments

We evaluated the effectiveness of the proposed two-stage quality control method, using four real-world datasets, on POI collection tasks. A comparison with two baselines demonstrates that our method appropriately handles the two types of uncertainties of the answers in these tasks and provides high quality answer lists. In addition, we confirm the robustness of the proposed method against the participation of spam workers by synthetically adding spam workers to the original datasets.

3.1 Datasets

We conducted experiments on several POI collection tasks posted to the Lancers crowdsourcing marketplace (http://www.lancers.jp). A POI collection task asks

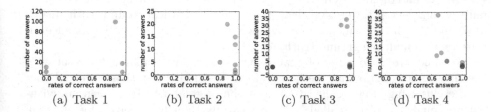

(a) Task 1   (b) Task 2   (c) Task 3   (d) Task 4

Fig. 2. Distribution of the accuracy and the number of answers provided by each worker. Spam workers are observed in Tasks 1 and 3, and unreliable streakers are observed in Tasks 2 and 4.

workers to enumerate the longitude and latitude of points that satisfy a given requirement of the task. We ordered four collection tasks for POIs in Japan asking for locations of telephone booths around Shimbashi station (Task 1), noodle restaurants around Takamatsu station (Task 2), mail boxes around Ueno station (Task 3), and public toilets around Shinjuku station (Task 4). Workers were shown a Google Maps-based interface where a target area was specified, and asked to extract the latitude and longitude of a point that they thought satisfied the given requirement. The workers were instructed to enumerate as many points as possible but not to provide multiple points indicating the same POI. Their rewards were propotional to the number of their answers.

We collected ground truth data from authoritative Web pages that list the spots satisfying the requirement of each task. We obtained the list of noodle restaurants near Takamatsu station from an online restaurant review portal (http://tabelog.com/) as the ground truth for Task 2. The other online portals we used to create the ground truths for Tasks 1, 3, and 4 were http://www.telmap.net/, http://postmap.org/ and http://toilet.blog.shinobi.jp/. The number of answers, workers, and ground truth answers in Tasks 1, 2, 3, and 4 were (133,4,96), (63,7,58), (122,14,84), and (82, 11, 150), respectively. Figure 2 shows the distribution of the accuracy and the number of answers provided by each worker.

3.2   Evaluation Methodology

We used precision, recall, the F-measure, and AUCs (areas under the receiver operation characteristic curve) as our evaluation metrics. We now explain the method of judging whether each answer is correct by using coordinate data. Let v denote an answer represented as a two-dimensional vector, (longitude, latitude). In POI tasks, we considered that each obtained answer $v \in \mathcal{P}$ was correct if it was close to one of the answers appearing in the ground truth dataset, namely, $\exists g \in \mathcal{G}, \|v - g\|_2 < d$, where $\mathcal{G}$ denotes a set of ground truth answers for a given task. d denotes a distance threshold value; we set $d = 0.0003$ degrees[3]. Since our

---

[3] Varying the value of d resulted in no significant change in the overall trends.

goal was to find distinct POIs, we judged that only a single answer was correct if multiple answers in $\mathcal{P}$ were sufficiently close to the identical ground truth answer in $\mathcal{G}$, and that a single answer was tied with a single ground truth answer even if there were multiple ground truth answers near it.

We conducted two types of experiments. First, we compared the quality of the answers selected by our method (called CL+modHITS), those selected only by the clustering stage (called CL), and the original set of the answers without any quality control (called NONE) to verify the effectiveness of each stage of our two-stage method, by calculating precisions, recalls, and the F-measures as the evaluation metrics. We employ these metrics because the baseline methods have no threshold hyperparameter. Note that we only used the subset of the ground truth answers that were enumerated by at least one of the workers, because the rest of them were not useful for comparing the methods. Second, we compared the performance of our method with one consisting of the clustering and HITS (called CL+HITS) and one consisting of the clustering and a majority vote, which regards the number of workers indicating a cluster as the cluster's reliability (called CL+MAJ) by calculating AUCs as the evaluation metrics. We employ the AUC because all the tested methods have threshold hyperparameters. It should be noted that the set of answers obtained by NONE corresponds to $\mathcal{T}$, CL corresponds to $\mathcal{U}$, and CL+modHITS, CL+HITS, and CL+MAJ corresponds to $\mathcal{P}$.

We calculated the Euclidean distance between each pair of answers u and v as the distance matrix D. In the answer clustering stage, we set $q = \infty$ for solving the problem as a linear program. The regularization parameter λ was set to $\lambda = 0.2$.

3.3 Evaluation with Original Data

We first conducted experiments with the original datasets, (i.e., without synthetic spam workers). Figure 3 shows the comparison of the F-measures and Figure 4, and Table 1 show the comparisons of precision-recall curves and AUCs.

In all the tasks, our two-stage method achieved higher F-measures than any other method. The improvements of the F-measure of CL+modHITS from that of NONE for Tasks 1, 2, 3, and 4 were 8.1%, 18.1%,19.4%, and 6.4%, respectively. We observed the effectiveness of both the answer clustering and the reliability estimation by comparing NONE with CL, and CL with CL+modHITS, respectively. The F-measures were improved by the answer clustering in Tasks 2, 3, and 4, while not in Task 1 because the number of workers and that of multiple answers indicating the identical POI were so small that the decrease of recall affected the F-measure more than the increase of precision. Nonetheless, the reliability estimation was effective in all the tasks, which indicates the answer clustering appropriately clustered the answers given by workers. Moreover, we observed our method still allows us to control a balance between precision and recall or improve either of them to attain the objectives of tasks by adjusting the threshold in the reliability estimation, ϵ, as shown in Figure 4.

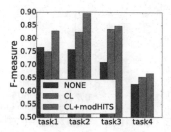

Fig. 3. Comparison of the F-measures in the experiments with original data. In all the tasks, our two-stage control method improved the F-measures of answers.

Table 1. Comparison of the AUCs with the majority vote, the HITS algorithm, and the modified HITS algorithm. The AUCs of the modified HITS were approximately the best for all tasks.

Task	Majority vote	HITS	Modified HITS
1	0.68	0.76	0.77
2	0.91	0.78	0.89
3	0.77	0.83	0.82
4	0.65	0.62	0.66

Then, we verified the effectiveness of the modified HITS algorithm by comparing the AUCs with those of the majority vote and the HITS algorithm as shown in Table 1. In all the tasks, the performance of the modified HITS algorithm was the best, or at least, comparable to the best one. The HITS algorithm also performed well in some tasks, but not in Task 2 or 4 because in these tasks, streakers provided relatively unreliable answers as compared to the other workers, as shown in Figure 2. When a part of the answers given by an unreliable streaker accidentally coincide with those by trustworthy workers, the streaker acquires a high reliability value in the reliability estimation stage; therefore, the answers of the unreliable streaker are considered reliable. In order to prevent the reliabilities of streakers from rising up too much, the modified HITS normalized the iterative matrix by the number of answers given by each worker, and thereby, it successfully handled this problem.

3.4   Evaluation with Synthetic Spam Workers

Although we observed only a few apparent spam workers in our experiments as shown in Figure 2, the number of workers may easily increase if we pay them more money or we use a different crowdsourcing service. In order to investigate the robustness of our proposed method against the presence of spam workers, we synthetically added spam workers to our datasets and investigated the effects. Specifically, we generated artificial datasets by the following steps. In each task, we first fixed the number of spam workers, and sampled the number of answers of each spam worker from a Poisson distribution $x_i \sim Po(\nu)$, where ν is the

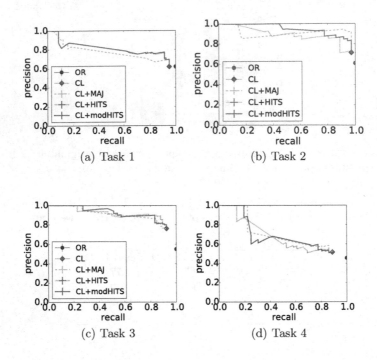

Fig. 4. Comparison of the precision-recall curves for the experiments with original data. Our reliability estimation stage improved the precision of answers and enabled us to control a balance between precision and recall by adjusting ϵ.

average number of answers per worker in the original dataset. Next, we sampled pairs of longitude and latitude from inside the specified area from a uniform distribution and counted the pairs as answers of spam workers. We added the synthetic data to the original datasets. We repeated the experimental procedure 20 times for each setting.

Figure 5 shows the F-measures for different numbers of spam workers. In all the tasks, our two-stage quality control method is quite robust to the presence of spam workers; the CL+modHITS method more effectively curbed the performance reduction as compared with NONE and CL.

4   Related Work

A number of unsupervised methods have been proposed for quality control in crowdsourcing. A groundbreaking study was conducted by Dawid and Skene [4], who modeled the differences in reliabilities of workers (doctors in their context) to estimate true answers (diagnoses). Extensions of the Dawid-Skene model have recently been studied in depth by, for example, introducing task difficulty [18] or incorporating the affinity between workers and tasks [17].

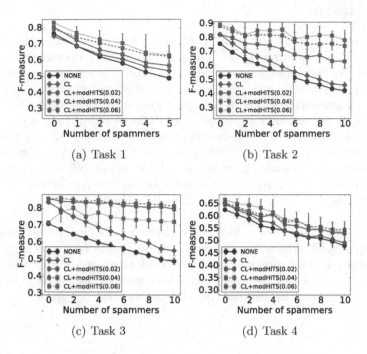

(a) Task 1   (b) Task 2

(c) Task 3   (d) Task 4

Fig. 5. Comparison of the F-measures together with the number of synthetic spam workers with the three values of ϵ, 0.02, 0,04, and 0.06. Because the reliability estimation step successfully removed spam workers, our two-stage quality control method curbed the decreases of the F-measures by, at least, 13.4%, 43.0%, 72.5%, and 8.2% of those of NONE in Tasks 1, 2, 3, and 4, respectively.

While most of the existing approaches focused on binary question tasks (e.g., yes-or-no questions) or multiple-choice questions, several studies addressed the quality control problem in more complex tasks. Chen et al. [3], Yi et al. [20], and Matsui et al. [11] proposed quality control methods for item ordering tasks; Wu et al. [19] focused on tasks with annotation tasks on sequential data, such as a named entity recognition task in natural language texts; Venanzi et al. [15] showed community-based aggregation can learn accurate estimations of worker accuracies from a limited amount of interactions; Lin et al. [9] addressed tasks with unstructured output formats, and Baba and Kashima [1] focused on more general tasks where workers do not necessarily agree on one answer.

Although Trushkowsky et al. [14] stated the necessity of quality control in enumeration tasks, no quality control method was provided for them. Tian and Zhu [13] proposed a quality control method for tasks with multiple correct answers; however, their method cannot be applied to the POI collection tasks where candidates of answers are not explicitly given to the workers. To the best of our knowledge, our work is the first attempt to address quality control for crowdsourced POI collection tasks.

5   Conclusion

We addressed the quality control problem for POI collection tasks that ask workers to enumerate correct location information of POIs. In order to resolve the uncertainties of enumerated answers and obtain reliable answers, we proposed a two-stage quality control method consisting of an answer clustering stage and a reliability estimation stage. The answer clustering stage was designed for filtering out redundant answers indicating the same POIs. We formalized a constrained exemplar clustering to utilize the assumption that multiple answers given by a single worker indicate distinct POIs. The reliability estimation stage used the HITS algorithm to estimate the reliability of answers as well as that of workers in order to select the correct answers from among the representative answers obtained in the previous stage, and modified it to handle the problem caused by streakers.

The experimental results on four POI collection tasks showed that our method successfully produced reliable answer sets and was quite robust to the presence of spam workers. Our future work includes developing a model that can usefully be applied to general enumeration tasks where workers can be asked to provide character strings and the distances between answers are not given.

References

1. Baba, Y., Kashima, H.: Statistical quality estimation for general crowdsourcing tasks. In: Proceedings of the 19th ACM SIGKDD International Conference on Knowledge Discovery and Data Mining (2013)
2. Bachrach, Y., Graepel, T., Minka, T., Guiver, J.: How to grade a test without knowing the answers–a Bayesian graphical model for adaptive crowdsourcing and aptitude testing. In: Proceedings of the 29th International Conference on Machine Learning (2012)
3. Chen, X., Bennett, P.N., Collins-Thompson, K., Horvitz, E.: Pairwise ranking aggregation in a crowdsourced setting. In: Proceedings of the 6th ACM International Conference on Web Search and Data Mining (2013)
4. Dawid, A.P., Skene, A.M.: Maximum likelihood estimation of observer error-rates using the EM algorithm. Journal of the Royal Statistical Society **28**(1), 20–28 (1979). Series C (Applied Statics)
5. Elhamifar, E., Sapiro, G., Vidal, R.: Finding exemplars from pairwise dissimilarities via simultaneous sparse recovery. In: Advances in Neural Information Processing Systems 25 (2012)
6. Heer, J., Bostock, M.: Crowdsourcing graphical perception: using mechanical turk to assess visualization design. In: Proceedings of the 28th SIGCHI Conference on Human Factors in Computing Systems (2010)
7. Kleinberg, J.: Authoritative sources in a hyperlinked environment. In: Proceedings of the 9th Annual ACM-SIAM Symposium on Discrete Algorithms (1998)
8. Lease, M.: On quality control and machine learning in crowdsourcing. In: Proceedings of the 3rd Human Computation Workshop (2011)
9. Lin, C., Mausam, M., Weld, D.: Crowdsourcing control: moving beyond multiple choice. In: Proceedings of the 28th Conference on Uncertainty in Artificial Intelligence (2012)

10. Mathur, S., Jin, T., Kasturirangan, N., Chandrasekaran, J., Xue, W., Gruteser, M., Trappe, W.: Parknet: Drive-by sensing of road-side parking statistics. In: Proceedings of the 8th International Conference on Mobile Systems, Applications, and Services (2010)
11. Matsui, T., Baba, Y., Kamishima, T., Kashima, H.: Crowdordering. In: Tseng, V.S., Ho, T.B., Zhou, Z.-H., Chen, A.L.P., Kao, H.-Y. (eds.) PAKDD 2014, Part II. LNCS, vol. 8444, pp. 336–347. Springer, Heidelberg (2014)
12. Raykar, V.C., Yu, S.: Ranking annotators for crowdsourced labeling tasks. In: Advances in Neural Information Processing 24 (2011)
13. Tian, Y., Zhu, J.: Learning from crowds in the presence of schools of thought. In: Proceedings of the 18th ACM SIGKDD Conference on Knowledge Discovery and Data Mining (2012)
14. Trushkowsky, B., Kraska, T., Franklin, M.J., Sarkar, P.: Crowdsourced enumeration queries. In: Proceedings of the 52th IEEE International Conference on Data Engineering (2013)
15. Venanzi, M., Rogers, A., Jennings, N.R.: Crowdsourcing spatial phenomena using trust-based heteroskedastic gaussian processes. In: Proceedings of the 1st AAAI Conference on Human Computation and Crowdsourcing (2013)
16. Wagstaff, K., Rogers, S., Schroedl, S.: Constrained k-means clustering with background knowledge. In: Proceedings of the 8th International Conference on Machine Learning (2001)
17. Welinder, P., Branson, S., Belongie, S., Perona, P.: The multidimensional wisdom of crowds. In: Advances in Neural Information Processing Systems 23 (2010)
18. Whitehill, J., Ruvolo, P., Wu, T., Bergsma, J., Movellan, J.: Whose vote should count more: optimal integration of labels from labelers of unknown expertise. In: Advances in Neural Information Processing Systems 22 (2009)
19. Wu, X., Fan, W., Yu, Y.: Sembler: ensembling crowd sequential labeling for improved quality. In: Proceedings of the 26th AAAI Conference on Artificial Intelligence (2012)
20. Yi, J., Jin, R., Jain, S., Jain, A.: Inferring users' preferences from crowdsourced pairwise comparisons: A matrix completion approach. In: Proceedings of the 1st AAAI Conference on Human Computation and Crowdsourcing (2013)

Towards Efficient Sequential Pattern Mining in Temporal Uncertain Databases

Jiaqi Ge[1](✉), Yuni Xia[1], and Jian Wang[2]

[1] Department of Computer and Information Science, Indiana University Purdue University Indianapolis, Indianapolis, IN 46202, USA
{jiaqige,yxia}@cs.iupui.edu
[2] School of Electronic Science and Engineering, Nanjing University, Jiangsu 210023, China
wangjnju@nju.edu.cn

Abstract. Uncertain sequence databases are widely used to model data with inaccurate or imprecise timestamps in many real world applications. In this paper, we use uniform distributions to model uncertain timestamps and adopt possible world semantics to interpret temporal uncertain database. We design an incremental approach to manage temporal uncertainty efficiently, which is integrated into the classic pattern-growth SPM algorithm to mine uncertain sequential patterns. Extensive experiments prove that our algorithm performs well in both efficiency and scalability.

Keywords: Temporal uncertainty · Sequential pattern mining

1 Introduction

Sequential pattern mining (SPM) provides inter-transactional analysis for times-tamped data and mines frequent patterns in sequence databases. However, it is very common that timestamps of events might be inaccurate or imprecise in real applications. And temporal uncertainty is usually caused by the following reasons:

- The exact time of an event is often unavailable. For example, in temperature monitoring sensor networks, temperatures are measured periodically. The exact time of a sudden temperature change is unknown, and it can only be inferred from raw data probabilistically.
- Temporal uncertainty arises when data are collected in different temporal scales. For example, a handhold GPS device may update the position every 10 minutes; while a GPS on a fast-moving vehicle may report every 5 seconds. And the temporal relationship is uncertain between two events within different granularities.
- Temporal uncertainty can also be caused by aggregation operations on temporal scales. For example, an economic indicator may be aggregated from weekly or monthly data to represent high level abstracted information in this time period.

© Springer International Publishing Switzerland 2015
T. Cao et al. (Eds.): PAKDD 2015, Part II, LNAI 9078, pp. 268–279, 2015.
DOI: 10.1007/978-3-319-18032-8_21

– Temporal uncertainty is also used to protect privacy and confidentiality. Precise time information in monitoring data usually is not released if there is a potential to identify individuals. Therefore, uncertainty is introduced to original time points, which is unquantifiable and unknown by the data user.

A time series $T = \{t, (t + 1), \ldots, (t + n)\}$ that bounds a set of consecutive timestamps is used to model an uncertain event time in probabilistic temporal databases, where it assumes that all events are defined within the same discrete time domain. However, this model becomes inaccurate and inconvenient when data are actually collected in different time scales. Instead, we use uniform distributions to represent uncertain timestamps in our model, which do no rely on any discrete time domain.

It is very important to carefully manage temporal uncertainty in SPM problems; otherwise, the mined patterns might be inaccurate. Possible world semantics is widely used to interpret probabilistic databases; however, it also brings efficiency and scalability challenges to uncertain SPM problems. Therefore, in this paper, we propose an efficient SPM algorithm in temporal uncertain sequence databases. Our main contributions are listed as follows:

(1) We model uncertain timestamps by uniform distributions. And we use possible world semantics to interpret this type of temporal uncertainty.
(2) We develop a novel approach to manage temporal uncertainty in the process of mining uncertain sequential patterns by a pattern-growth algorithm.
(3) We conduct extensive experiments to demonstrate the efficiency and scalability of our algorithm.

2   Related Works

Data mining in uncertain databases has been an active area of research recently. A lot of traditional database and data mining techniques have been extended to be applied to uncertain data [1]. Particularly, Muzammal and Raman proposal the SPM algorithm in probabilistic database using the expected support as the measurement of pattern frequentness [10]; however, expected support has inherent weakness in mining high-quality sequential patterns[12]. Zhao et al. measure pattern frequentness using possible world semantics and propose a pattern-growth uncertain SPM algorithm [14,15]. Miliaraki et al use approximation with probabilistic guarantee to improve the efficiency of uncertain SPM problem [9]. A dynamic programming approach is used to mine probabilistic spatial-temporal frequent sequential patterns [8]. However, these methods are all designed for sequence databases with accurate timestamps.

Dyreson and Snodgrass introduced probabilistic temporal databases which models uncertain timestamp by a set of time points with equal probabilities [4]. Zhang et al. proposed a pattern recognition algorithm in temporal uncertain streams[13]; while pattern queries in temporal uncertain sequences is studied in [16]. However, our work distinguishes from the above in that we use uniform distributions to represent uncertain timestamps, which is more flexible in modeling data collected from different scales. Meanwhile, the above works focused on

sid	eid	T	I
1	1	[100,103]	{A,C}
1	2	[102,105]	B
2	1	[160,163]	A
2	2	[162,164]	B
2	3	[163,166]	B
2	4	[167,168]	C

sid	eid	t	I
1	1	102.5	{A,C}
1	2	103.9	B
2	1	163	A
2	2	162	B
2	3	165	B
2	4	166	C

Fig. 1. An example of uncertain database   **Fig. 2.** An example of a possible world

matching patterns in one sequence, while the SPM problem is more complicated because it mines patterns from a large number of uncertain sequences so that their techniques cannot be directly employed.

3   Problem Statement

3.1   Temporal Uncertain Sequence Database

A temporal uncertain sequence database contains a collection of uncertain sequences, and an uncertain sequence is a set of temporal uncertain events. A temporal uncertain event is represented by $e = \langle sid, eid, T, I \rangle$. Here sid is the sequence-id, eid is the event-id and $\langle sid, eid \rangle$ identifies a unique event. Note that events are not guaranteed to be ordered by their $eids$. An uncertain timestamp T is modeled by a uniform distribution $T \sim U(t^-, t^+)$, where $[t^-, t^+]$ is the range of T. I is an itemset that describes the content of event e.

Fig. 1 shows an example of temporal uncertain database. A sequence is a list of events that are associated with the same sid and an event identified by $sid = i$ and $eid = j$ is denoted by e_{ij}. For example, $e_{11} = \langle 1, 1, \{[100, 103]\}, \{A, C\} \rangle$ indicates that event $\{A, C\}$ occurs at time T, where $T \sim U(100, 103)$ is uniformly distributed within 100 and 103.

3.2   Temporal Possible Worlds

We use possible world semantics to interpret temporal uncertain databases. Temporal possible worlds of an uncertain database D are generated by instantiating all possible values of each uncertain timestamp. Fig. 2 shows an example a temporal possible worlds that are instantiated from the uncertain database in Fig. 1, in which the time point of an event is randomly drawn from the corresponding uncertain timestamps.

It is widely assumed that uncertain sequences in D are mutually independent, which is known as the *tuple-level independence* [1,7] in probabilistic databases. Meanwhile, event time are also assumed to be independent of each other

[3,5,6,14], which can be justified by the fact that events are often observed independently in real applications. Thus, the probability density function (pdf) of the possible words can be computed by Equation (1).

$$f(w) = \prod_{i=1}^{m} f(d_i) = \prod_{i=1}^{m} \prod_{j=1}^{n_i} f(T_{ij} = t_{ij}) \tag{1}$$

Where d_i is a sequence in the database D and e_{ij} is an event in d_i. $m = |D|$ is the number of sequences in D and $n_i = |d_i|$ is the number of events in d_i. Let $T_{ij} \sim U(t_{ij}^-, t_{ij}^+)$ be the uncertain time of event e_{ij}, then its pdf $f(T_{ij} = t_{ij})$ is shown in Equation (2).

$$f(T_{ij} = t_{ij}) = \begin{cases} \frac{1}{t_{ij}^+ - t_{ij}^-} & , t \in [t_{ij}^-, t_{ij}^+] \\ 0 & , \text{otherwise} \end{cases} \tag{2}$$

3.3 Uncertain Sequential Pattern Mining Problem

A sequential pattern $\alpha = \langle X_1 \cdots X_n \rangle$ is *supported* by a sequence $\beta = \langle Y_1 \cdots Y_m \rangle$, denoted by $\alpha \preceq \beta$, if and only if there exists integers $\{k_1, \ldots, k_n\}$ so that we have $X_i.I \subseteq Y_{k_i}.I$ $(\forall i \in [1, n])$ and $l \le Y_{k_{i+1}}.t - Y_{k_i}.t \le h$ $(\forall i \in [1, n-1])$. Here $l = mingap$ is the minimal time gap constraint between two adjacent events of α and $h = maxgap$ is the maximal time gap constraint.

In deterministic databases, a sequential pattern s is frequent if and only if it satisfies $sup(s) \ge t_s$, where $sup(s)$ is the total number of sequences that support s and t_s is the user-defined minimal threshold. However, In an uncertain database D, the frequentness of s is probabilistic and it can be computed by Equation (3).

$$P(sup(s) \ge t_s) = \int_{sup(s|w) \ge t_s} f(w) \mathrm{d}w \tag{3}$$

Where w is a possible world in which s is frequent and $f(w)$ is the pdf of w.

The SPM problem in temporal uncertain databases can be defined as follows. *Given a minimal support t_s, a minimal frequentness probability threshold t_p, a minimal time gap l and a maximal time gap h, find every probabilistic frequent sequential pattern s in a temporal uncertain database, which has $P(sup(s) \ge t_s) \ge t_p$.*

4 Solution

4.1 Frequentness Probability

Suppose $D = \{d_1, \ldots, d_n\}$ is a temporal uncertain database and s is a sequential pattern. Because $d_1, \ldots, d_n$ in D are mutually independent, the probabilistic support of s in D, denoted by $sup(s)$, can be computed by Equation (4).

$$sup(s) = \sum_{i=1}^{n} sup(s|d_i) \tag{4}$$

Where $sup(s|d_i)$ ($\forall i \in [1, n]$) is a Bernoulli random variable, whose success probability is $P(sup(s|d_i) = 1) = P(s \preceq d_i)$.

$sup(s)$ is a Poisson-Binomial random variable, since it is the sum of n independent but non-identical Bernoulli random variables. And the probability mass function (pmf) of $sup(s)$ is $P(sup(s) = i) = p_i$, where p_i is the probability that the support of s in D equals to i ($i \in [1, |D|]$). Here we adopt the Fast Fourier Transform (FFT) technology in [14] to compute the pmf of $sup(s)$ in $O(nlogn)$ time . Thereafter, the *frequentness probability* of s is computed by Equation (5).

$$P(sup(s) \geq t_s) = \sum_{i \geq t_s} p_i = 1 - \sum_{i < t_s} p_i \tag{5}$$

Where, t_s is the minimal support threshold and $p_i = P(sup(s) = i)$ ($i \in [1, n]$) is the probability that the support of s in D equals to i. Given the minimal frequentness probability threshold t_p, s is probabilistically frequent if and only if $P(sup(s) \geq t_s) \geq t_p$.

4.2   Support Probability

We first define the *minimal possible occurrence* of a sequential pattern s in an uncertain sequence d.

Definition 1. *Given a sequential pattern s and an uncertain sequence d, a subset d' of d (e.g. $d' \subseteq d$) is called a minimal possible occurrence of s if and only if (1) $P(s \preceq d') > 0$; (2) $\forall d'' \subset d', P(s \preceq d'') = 0$.*

For example, in Fig. 1, $\{e_{21}, e_{22}\}$ and $\{e_{21}, e_{23}\}$ are two minimal possible occurrences of the sequential pattern $\langle A, B \rangle$ in the sequence s_2; while $\{e_{21}, e_{22}, e_{23}\}$ is not a minimal occurrence of $\langle A, B \rangle$. Then the *support probability* $P(s \preceq d)$ can be computed by Equation (6), since event timestamps are independent.

$$P(s \preceq d) = \sum_{i=1}^{N} P(s \preceq o_{s_i}) \tag{6}$$

Here o_{s_i} ($i = 1, \ldots, N$) are N minimal possible occurrences of s, and the computation of $P(s \preceq o_{s_i})$ is discussed in section 4.3.

4.3   Probability of Satisfying Time Constraints

Let $o_s = \{e_{k_1}, \ldots, e_{k_n}\}$ be a minimal possible occurrence of sequential pattern $s = \langle s_1, \ldots, s_n \rangle$. Suppose T_i is the uncertain time of the event e_{k_i}, then $P(s \preceq o_s)$, denoted by $P(\langle T_1 \cdots T_n \rangle)$, is the probability that $T_1, \cdots, T_n$ satisfy time constraints $l \leq T_{i+1} - T_i \leq h$, $\forall i \in [1, n)$. Here l is the minimal time gap between two adjacent timestamps and h is the maximal time gap.

A naive approach of computing $P(\langle T_1 \cdots T_n \rangle)$ is to use the *chain rule*, which is shown in Equation (7).

$$
\begin{aligned}
P(\langle T_1 \cdots T_n \rangle) &= \underset{l \le t_i - t_{i-1} \le h}{\int \cdots \int} f(T_{k_1} = t_1, \ldots, T_{k_n} = t_n) dt_1 \cdots dt_n \\
&= \underset{l \le t_i - t_{i-1} \le h}{\int \cdots \int} f(t_n | t_1 \cdots t_{n-1}) \cdots f(t_2 | t_1) f(t_1) dt_1 \cdots dt_n
\end{aligned}
\tag{7}
$$

However, this method is usually too complex in practice. Therefore, we design a new approach to compute $P(\langle T_1 \cdots T_n \rangle)$ efficiently.

Basic Case. We first consider the basic case of two uncertain timestamps $X \sim U(x^-, x^+)$ and $Y \sim U(y^-, y^+)$. Given time constraints $mingap = l$, $maxgap = h$, $P(\langle XY \rangle)$ can be computed in Equation (8).

$$
P(\langle XY \rangle) = \int_{max(y^-, x^- + l)}^{min(y^+, x^+ + h)} \int_{max(x^-, y - h)}^{min(x^+, y - l)} \frac{1}{(x^+ - x^-)(y^+ - y^-)} dx dy \tag{8}
$$

Equation (8) is decomposed into p deterministic cases, if $[y^-, y^+]$ is divided into p disjoint subintervals by the endpoints $\{x^+ + l, x^- + l, x^+ + h, x^- + h\}$ as $[y^-, y^+] = \bigcup[y_i^-, y_i^+], \forall i \in [1, p]$. Here $Y_k \sim U[y_k^-, y_k^+]$ is a uniformly distributed random variable, and $P(\langle XY \rangle)$ can be computed by Equation (9).

$$
P(\langle XY \rangle) = \sum_{k=1}^{p} P(\langle XY_k \rangle) P(Y = Y_k) \tag{9}
$$

Where $P(Y = Y_k) = (y_k^+ - y_k^-)/(y^+ - y^-)$. We use a geographic method to compute $P(\langle XY_k \rangle)$ in $O(1)$ time, which is shown in Equation (10).

$$
P(\langle XY_k \rangle) = \frac{S_k}{A_k} = \frac{(1/2) * (L_1 + L_2) * H}{(y_k^+ - y_k^l)(x^+ - x^-)} \tag{10}
$$

Where, A_k is the area of the rectangle defined by the 2-dimensional uniform distribution of X and Y_k, and S_k is the area within A_k which satisfies the time constraints. Here $H = y_k^+ - y_k^-$, L_1 and L_2 are computed as follows.

$$
L_1 = max(0, L_1'), L_1' = min(y_k^- - l, x^+) - max(y_k^- - h, x^-)
$$
$$
L_2 = max(0, L_2'), L_2' = min(y_k^+ - l, x^+) - max(y_k^+ - h, x^-)
$$

Fig. 3(a) shows an example of computing $P(\langle XY \rangle)$ with $l = 0$ and $h = 5$, where $X \sim U[60, 63]$ and $Y \sim U[62, 68]$. There two endpoints $\{63, 65\}$ within the range of Y, which divide $[62, 68]$ into three disjoint subintervals as $[62, 68] = [62, 63] \cup [63, 65] \cup [65, 68]$.

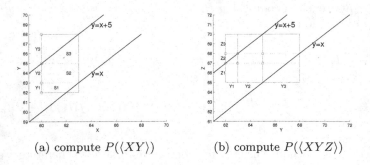

(a) compute $P(\langle XY \rangle)$ (b) compute $P(\langle XYZ \rangle)$

Fig. 3. An example of compute the probability of satisfying time constraints

Let $Y_1 \sim U[62, 63]$, $Y_2 \sim U[63, 65]$ and $Y_3 \sim U[65, 68]$, then we have

$$P(Y = Y_1) = \frac{1}{6} \qquad P(\langle XY_1 \rangle) = \frac{S_1}{A_1} = \frac{2.5}{3} \qquad P(\langle XY_1 \rangle \cap Y_1) = \frac{2.5}{18}$$

$$P(Y = Y_2) = \frac{2}{6} \qquad P(\langle XY_2 \rangle) = \frac{S_2}{A_2} = \frac{6}{6} \qquad P(\langle XY_2 \rangle \cap Y_2) = \frac{1}{3} \qquad (11)$$

$$P(Y = Y_1) = \frac{3}{6} \qquad P(\langle XY_3 \rangle) = \frac{S_3}{A_3} = \frac{4.5}{9} \qquad P(\langle XY_3 \rangle \cap Y_3) = \frac{1}{4}$$

Thereafter, $P(\langle XY \rangle) = \sum_{i=1}^{3} P(\langle XY_i \rangle \cap Y_i) = 0.72$.

General Case. Given uniformly distributed uncertain timestamps $T_1, \ldots, T_n$, suppose the range of T_n is divided into p sub-partitions as $[t_n^-, t_n^+] = \bigcup_{i=1}^{p} [t_{n_i}^-, t_{n_i}^+]$, and $T_{n_i} \sim U(t_{n_i}^-, t_{n_i}^+)$ is a uniform distributed random variable, then we can compute $P(\langle T_1 \cdots T_n \rangle)$ by Equation (12).

$$
\begin{aligned}
P(\langle T_1, \ldots, T_n \rangle) &= \sum_{i=1}^{p} P(\langle T_1, \ldots, T_{n_i} \rangle) * P(T_n = T_{n_i}) \\
&= \sum_{i=1}^{p} P(\langle T_1, \ldots, T_{n_i} \rangle \cap T_{n_i})
\end{aligned}
\qquad (12)
$$

Where $P(\langle T_1 \cdots T_{n_i} \rangle)$ can be computed by Equation (13).

$$
\begin{aligned}
P(\langle T_1, \ldots, T_{n_i} \rangle) &= \sum_{j=1}^{q} P(\langle T_1, \ldots, T_{(n-1)_j} \rangle) P(\langle T_{(n-1)_j} T_{n_i} \rangle) P(T_{(n-1)_j}) \\
&= \sum_{j=1}^{q} P(\langle T_{(n-1)_j} T_{n_i} \rangle) P(\langle T_1, \ldots, T_{(n-1)_j} \rangle \cap T_{(n-1)_j})
\end{aligned}
\qquad (13)
$$

Let $s' = \langle s_1, \ldots, s_{n-1} \rangle$ be a sequential pattern by removing the last element of $s = \langle s_1, \ldots, s_n \rangle$. In SPM process, we have already computed

$P(\langle T_1, \ldots, T_{(n-1)_j} \rangle \cap T_{(n-1)_j})$ in searching s'. Thus, we can save and reuse these values when we search pattern s, in order to avoid repeated computation.

Given another uncertain time $Z \sim U[65, 70]$, Fig. 3(b) shows the process of computing $P(\langle XYZ \rangle)$ by reusing previous computational results. First, we compute potential end points by the ranges of Y_1, Y_2 and Y_3 as follows.

$$z_{11} = y_1^- + l = 62, z_{12} = y_1^+ + l = 63, z_{13} = y_1^- + r = 67, z_{14} = y_1^+ + h = 68$$

$$z_{21} = y_2^- + l = 63, z_{22} = y_2^+ + l = 65, z_{23} = y_2^- + r = 68, z_{24} = y_2^+ + h = 70$$

$$z_{31} = y_3^- + l = 65, z_{32} = y_3^+ + l = 68, z_{33} = y_3^- + r = 70, z_{34} = y_3^+ + h = 73$$

Therefore, the range of Z is divided into three disjoint sub-partitions as $[65, 70] = [65, 67] \cup [67, 68] \cup [68, 70]$. Let $Z_1 \sim U[65, 67]$, $Z_2 \sim U[67, 68]$ and $Z_3 \sim [68, 70]$. Here we take the computation of $P(\langle XYZ_1 \rangle)$ in Equation (14) as an example.

$$P(\langle XYZ_1 \rangle) = \sum_{i=1}^{3} P(\langle XY_i \rangle \cap Y_i) P(\langle Y_i Z_1 \rangle) \tag{14}$$

Where $P(\langle XY_i \rangle \cap Y_i)$ is already computed in Equation (11). Referring to Equation (10), we can compute $P(\langle Y_1 Z_1 \rangle) = 1$, $P(\langle Y_2 Z_1 \rangle) = 1$, $P(\langle Y_3 Z_1 \rangle) = 1/3$. Thereafter, we have $P(\langle XYZ_1 \rangle) = 1 * \frac{1}{6} * \frac{2.5}{3} + 1 * \frac{2}{6} * \frac{6}{6} + \frac{1}{3} * \frac{3}{6} * \frac{4.5}{9} = 0.5555$.

Similarly, we can compute $P(\langle XYZ_2 \rangle) = 0.6111$ and $P(\langle XYZ_3 \rangle) = 0.3333$. Therefore, we arrive to the final result $P(\langle XYZ \rangle) = 0.4 * 0.5555 + 0.2 * 0.6111 + 0.4 * 0.3333 = 0.4777$.

4.4   Uncertain SPM Algorithm

We integrate our uncertain management approach into the classic SPM algorithm PrefixSPan[11]. There are two major modifications to the original PrefixSpan in our uncertain SPM algorithm.

(1) We project the database by minimal possible occurrences. Suppose $o_s = \{e_{k_1}, \ldots, e_{k_n}\}$ is a minimal possible occurrence of s in sequence d. The projection of d w.r.t. o_s, denoted by $d|_{o_s}$, eliminates any event e_i in d if $P(e_i.T \geq e_{k_n}.T + mingap) = 0$. A projected database $D|_s = \{d_1|_{o_1, \ldots, o_p}, \ldots, d_t|_{o_1, \ldots, o_q}\}$ is a collection of projected sequences, where $d_i|_{o_1, \ldots, o_p} = \{d_i|_{o_1}, \ldots, d_i|_{o_p}\}$ is a set of p projected sequences of d_i w.r.t. to the minimal occurrences $o_1, \ldots, o_p$ of s in d_i. For example, in Fig. 1, if we set $mingap = 1$ and let $s = \langle AB \rangle$, then $D|_s = \{d_2|_{o_1, o_2}\}$, where $d_2|_{o_1} = \{e_{23}, e_{24}\}$ and $d_2|_{o_2} = \{e_{24}\}$.

(2) We save intermediate computational results for each minimal possible occurrence. Let $o_s = \{e_{k_1}, \ldots, e_{k_n}\}$ be a minimal possible occurrence of s in d and $T_i = e_{k_i}.T \; \forall i \in [1, n]$. Suppose the range $[t_n^-, t_n^+]$ of T_n is divided into k subintervals $[t_n^-, t_n^+] = \bigcup [t_{n_i}^-, t_{n_i}^+] \; (\forall i \in [1, k])$, then we compute $p_i = P(\langle T_1, \ldots, T_{n_i} \rangle)$ by Equation (12) and save the results in the form as $T(o_s) = \{[t_{n_1}^-, t_{n_1}^+] : p_1, \ldots, [t_{n_k}^-, t_{n_k}^+] : p_k\}$. Therefore, we can reuse $T(o_s)$ in searching longer sequences.

We adopt the pattern-growth approach to search new patterns in Algorithm 1. We first mine frequent items in $D|_s$, denoted by $I = \{i_1, i_2, \ldots, i_n\}$.

ALGORITHM 1. USPM$(s, D|_s)$

Input: sequential pattern s, uncertain projected database $D|_s$,
        $minsup = t_s$, $minprob = t_p$
Output: L: a set of frequent sequential patterns
Find all frequent items $I = \{i_1, i_2, ..., i_n\}$ in $D|_s$
if $D|_s = \phi$ *or* $I = \phi$ **then**
    return L
end
foreach *item* $i \in I$ **do**
    $s' = s + \{i\}$
    for $d_i|_{o_1,...,o_n} \in D|_s$ **do**
        for $d_i|_{o_j} \in d_i|_{o_1,...,o_n}$ **do**
            construct a projected sequence $d_i|_{o_{s'}}$ from $d_i|_{o_j}$
            compute $T(o_{s'})$ from $T(o_{s_j})$ in d_i by Equation (12) and Equation (13)
        end
        compute the support probability $P(s' \preceq d_i)$ by Equation (6).
    end
    use FFT to compute the Poisson Binomial distribution of $sup(s')$
    if $P(sup(s') \geq t_s) \geq t_p$ **then**
        $L = L \cup \{s'\}$;
        USPM$(D|_{s'}, s')$;
    end
end

This process is straightforward because it does not need to consider temporal uncertainty. A candidate pattern $s' = s + \{i\}$ is generated for each $i \in I$. Then, we extract all minimal possible occurrences of s' and construct their projected databases. For each minimal possible occurrence $o_{s'}$ of s', we compute and save its probability of satisfying time constraints by Equation (12) and Equation (13).

We compute the support probability of s' in each uncertain sequence by Equation (6). Thereafter, we adopt the FFT technique in [14] to compute the Poisson Binomial distribution of the overall support $sup(s')$. The frequentness probability is computed in Equation (5), by which we can determine if s' is a probabilistic frequent sequential pattern. The searching process stops until no frequent patterns are mined.

5   Evaluation

5.1   Synthetic Data Generation

We use the IBM market-basket data generator [2] to generate synthetic sequence datasets in different scales with the following parameters: (1)C: number of sequences; (2)T: average number of transactions/itemsets per data-sequence; (3)L: average number of items per transaction/itemset per data-sequence; (4)I: number of different items.

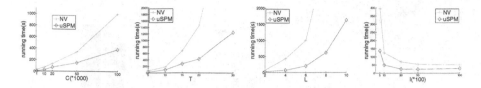

Fig. 4. Scalability of uSPM in synthetic uncertain datasets

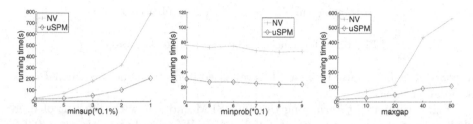

Fig. 5. Effect of parameters in synthetic uncertain datasets

To add temporal uncertainty, we replace a point-value timestamp t in the original synthetic datasets by a uniform distribution in $[(1 - r) * t, (1 + r) * t]$, where r is randomly drawn from the uniform distribution $U(0, 1)$. We name the generated synthetic dataset by parameters. For example, the dataset named $T4L10I1C10$ indicates that $T = 4$, $L = 10$, $I = 1 * 1000$ and $C = 10 * 1000$.

Our uncertain sequential pattern mining algorithm is called *uSPM* for short. Recall from Section 4.3 that a naive method to compute the probability of an occurrence satisfying time constraints is to directly evaluate Equation (7) using chain rule. This naive method is implemented and abbreviate as *NV*. We compare uSPM with NV to evaluate the performance of our algorithm. All the experiments were done in the desktop with Intel(R) Core (TM) Duo CPU @ 2.33GHz and 4GB memory.

5.2 Scalability and Efficiency

In Fig. 4, we compare the running time of uSPM and NV on synthetic datasets with different scales, where we set $minsup = 0.5\%$, $minprob = 0.7$, $mingap = 1$, and $maxgap = 10$. We initially have $C = 10\,000$, $T = 4$, $I = 10\,000$ and $L = 2$. In Figigure 4(a), C varies from $1\,000$ to $100\,000$; In Fig. 4(b), T varies from 5 to 30; In Fig. 4(c), L varies from 2 to 10; and In Fig. 4(c) I varies from 500 to $10\,000$.

In Fig. 4, we observe the following phenomenons: (1) uSPM is significantly faster than NV under every setting of the parameters, which proves the effectiveness of our temporal uncertainty management approach. (2) The running time increases with the increment of C, T, L, as the increment of these parameters generates larger synthetic datasets. (3) The running time drops slightly with the

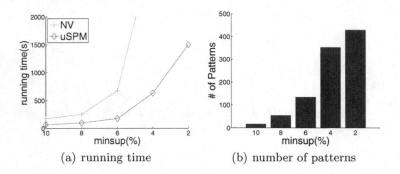

(a) running time   (b) number of patterns

Fig. 6. Performance of uSPM in real stock dataset

increment of I, because there are less repeated items in sequences when I is set to a larger value.

Fig. 5 compares the running time of uSPM and NV with different of user-defined parameters in the dataset T4L2I10C10. We initially set $minsup = 0.2\%$, $minprob = 0.7$, $mingap = 1$, and $maxgap = 10$. In Fig. 5(a), $minsup$ decreases from 0.8% to 0.1%; in Fig. 5(b), $minprob$ varies from 0.4 to 0.9; and $maxgap$ varies from 5 to 80 in Fig. 5.

In Fig. 5, we observe that: (1) The running time of uSPM increase with the decrement of $minsup$; however, the performance is relatively stable to the variations of $minprob$. The probabilistic support of a sequential pattern is bounded to its expected value (*Chernoff bound*) so that the frequentness of a large number of patterns become deterministic. This explains why the running time of uSPM does not fluctuate significantly in Fig. 5(b). (3) The running time of uSPM increases when we set a larger value to $maxgap$. This is intuitive because a larger $maxgap$ indicates a less strict constraint of sequential patterns.

We also apply uSPM to a real world stock market dataset. The prices for 882 stocks are extracted from Shanghai Stock Exchange Center in 16 weeks from 12-03-2012 to 03-24-2013. Each stock corresponds to a sequence. We define three events such as price going up $(+)$, going down $(-)$ and no change (0). An uncertain event is aggregated from consecutive events. For example, if price goes up at time 1, 2 and 3, then we aggregate them to form an uncertain event $([1,3], +)$.

Here we set $minprob = 0.7$, $mingap = 1$ and $maxgap = 5$. Fig. 6(a) shows that the running time of uSPM in the stock dataset increases with the decrement of $minsup$. As we only define three distinct events, there are many repeated items in sequences; however, uSPM still significantly outperforms NV in this dataset. In Fig. 6(b), we can see that the number of frequent sequential patterns in the stock dataset increases significantly when we decrease the value of $minsup$ from 10% to 2%. And a mined pattern $\langle +, -, +, - \rangle$ from this dataset reveals that stock prices are fluctuated in general during the time when data are collected, which is consistent with intuitive observations.

6   Conclusion

In this paper, we study the problem of mining probabilistic frequent sequential patterns in databases with temporal uncertainty. We design an incremental approach to manage temporal uncertainty efficiently and integrate it into classic pattern-growth SPM algorithm. The experimental results prove that our algorithm is efficient and scalable.

References

1. Aggarwal, C.C., Yu, P.S.: A survey of uncertain data algorithms and applications. IEEE Trans. on Knowl. and Data Eng. **21**(5), 609–623 (2009)
2. Agrawal, R., Srikant, R.: Fast algorithms for mining association rules in large databases. In: VLDB, pp. 487–499 (1994)
3. Bernecker, T., Kriegel, H.-P., Renz, M., Verhein, F., Zuefle, A.: Probabilistic frequent itemset mining in uncertain databases. In: SIGKDD, pp. 119–128 (2009)
4. Dyreson, C.E., Snodgrass, R.T.: Supporting valid-time indeterminacy. In: TODS (1998)
5. Chui, C.-K., Kao, B.: A decremental approach for mining frequent itemsets from uncertain data. In: Washio, T., Suzuki, E., Ting, K.M., Inokuchi, A. (eds.) PAKDD 2008. LNCS (LNAI), vol. 5012, pp. 64–75. Springer, Heidelberg (2008)
6. Chui, C.-K., Kao, B., Hung, E.: Mining frequent itemsets from uncertain data. In: Zhou, Z.-H., Li, H., Yang, Q. (eds.) PAKDD 2007. LNCS (LNAI), vol. 4426, pp. 47–58. Springer, Heidelberg (2007)
7. Jestes, J., Cormode, G., Li, F., Yi, K.: Semantics of ranking queries for probabilistic data. IEEE Transactions on Knowledge and Data Engineering **23**(12), 1903–1917 (2011)
8. Li, Y., Bailey, J., Kulik, L., Pei, J.: Mining probabilistic frequent spatio-temporal sequential patterns with gap constraints from uncertain databases. In: ICDM, pp. 448–457 (2013)
9. Miliaraki, I., Berberich, K., Gemulla, R., Zoupanos, S.: Mind the gap: Large-scale frequent sequence mining. In: SIGKDD, pp. 797–808 (2013)
10. Muzammal, M., Raman, R.: Mining sequential patterns from probabilistic databases. In: Huang, J.Z., Cao, L., Srivastava, J. (eds.) PAKDD 2011, Part II. LNCS, vol. 6635, pp. 210–221. Springer, Heidelberg (2011)
11. Pei, J., Han, J., Mortazavi-asl, B., Pinto, H., Chen, Q., Dayal, U., Chun Hsu, M.: Prefixspan: mining sequential patterns efficiently by prefix-projected pattern growth. In: ICDE, pp. 215–224 (2001)
12. Tong, Y., Chen, L., Cheng, Y., Yu, P.S.: Mining frequent itemsets over uncertain databases. Proceeding of the VLDB Endowment **5**, 1650–1661 (2012)
13. Zhang, H., Diao, Y., Immerman, N.: Recognizing patterns in streams with imprecise timestamps. Proc. VLDB Endow. **3**(1–2), 244–255 (2010)
14. Zhao, Z., Yan, D., Ng, W.: Mining probabilistically frequent sequential patterns in uncertain databases. In: EDBT, pp. 74–85 (2012)
15. Zhao, Z., Yan, D., Ng, W.: Mining probabilistically frequent sequential patterns in large uncertain databases. IEEE Transactions on Knowledge and Data Engineering **26**, 1171–1184 (2013)
16. Zhou, Y., Ma, C., Guo, Q., Shou, L., Chen, G.: Sequence pattern matching over time-series data with temporal uncertainty. In: EDBT, pp. 205–216 (2014)

Preference-Based Top-k Representative Skyline Queries on Uncertain Databases

Ha Thanh Huynh Nguyen[✉] and Jinli Cao

Department of Computer Science and Information Technology,
La Trobe University, 10 Macarthur Court, Mill Park, VIC 3082, Australia
ht34nguyen@students.latrobe.edu.au, j.cao@latrobe.edu.au

Abstract. Top-k representative skyline queries are important for multi-criteria decision making applications since they provide an intuitive way to identify the k most significant objects for data analysts. Despite their importance, top-k representative skyline queries have not received adequate attention from the research community. Existing work addressing the problem focuses only on certain data models. For this reason, in this paper, we present the first study on processing top-k representative skyline queries in uncertain databases, based on user-defined references, regarding the priority of individual dimensions. We also apply the odds ratio to restrict the cardinality of the result set, instead of using a threshold which might be difficult for an end-user to define. We then develop two novel algorithms for answering top-k representative skyline queries on uncertain data. In addition, several pruning conditions are proposed to enhance the efficiency of our proposed algorithms. Performance evaluations are conducted on both real-life and synthetic datasets to demonstrate the efficiency, effectiveness and scalability of our proposed approaches.

1 Introduction

Extracting a few interesting data objects from a potentially huge database to support multi-criteria decision making is challenging for the database community. There is an increasing number of real-world applications where the end-users need to make their selections by ranking multi-dimensional objects. If user-defined scoring functions over the objects' dimensions are available, a subset of the most interesting objects can be returned based on their corresponding scores. However, in some scenarios, the users may find it difficult to formulate a proper scoring function due to conflicting multiple ranking criteria. Consider the following example. A tourist wants to book a hotel for a holiday on the hotel reservation system with conditions of cheapest in price, closest distance to CBD and good facilities in room. It is clear that the room price and the good facilities are contradictory criteria. Driven by these multi-criteria decision-making applications, ranking queries which retrieve the most important answers to help users make better decisions have received considerable attention in recent years.

In the literature, there are a number of ranking tools that have been studied extensively in order to assist users in ranking processes. Top-k and skyline queries are two

© Springer International Publishing Switzerland 2015
T. Cao et al. (Eds.): PAKDD 2015, Part II, LNAI 9078, pp. 280–292, 2015.
DOI: 10.1007/978-3-319-18032-8_22

classic types of these ranking tools. A top-k query retrieves k objects having the highest scores that are based on some user-defined scoring functions. However, in multi-criteria decision making applications, it might be rather difficult for an end-user to formulate a proper scoring function to best describe the trade-offs offered by different ranking criteria, e.g. room price and quality of hotels [1]. A skyline query is an alternative ranking approach which relies on the dominance relationship. Specifically, a skyline query aims at efficiently retrieving a subset of points, called "skyline points", which are not dominated by any other point when all dimensions are considered together. However, the size of the skyline answer set is arbitrarily large even if there are only a few criteria involved.

To overcome the deficiencies of these two ranking techniques, a number of variants of ranking approaches have been proposed recently to identify the truly interesting objects. Yiu and Mamouslis [2,3] propose top-k dominating queries which return k data objects which dominate the highest number of other objects. However, this ranking approach is unstable since the final results can be affected by the addition of unimportant data objects (non-skyline objects) [3]. Another limitation is that this approach does not always return the skyline objects which present all best possible trade-offs among different ranking dimensions [4]. Top-k dominating queries have been explored further in the context of uncertain databases [1,5-7] and continuous processing [8].

Motivation and Challenges
In recent years, uncertain data has emerged in many real-world application domains including sensor networks, moving object tracking, data integration, data cleaning, information extraction and others. The uncertainty is inherent in such application domains due to various factors, including data randomness and incompleteness, limitations of measuring equipment, delay or loss of data updates, etc. Since the number of such application domains is rapidly increasing, developing advanced data analysis tools over uncertain data has become more crucial.

As aforementioned, skyline, top-k and top-k dominating queries have been widely studied in uncertain environments. However, these ranking approaches have their own deficiencies as analyzed above. By contrast, top-k representative queries play an important role as they provide a novel ranking mechanism for applications where multi-criteria are involved. This approach has been studied widely in traditional databases. However, to the best of our knowledge, no previous work has been undertaken on top-k representative queries in the context of uncertain data. Motivated by this, we develop novel algorithms to efficiently compute *top-k representative skyline queries* in uncertain databases. Dealing with imprecise data in the ranking processes is much more complex and challenging since there are many factors that need to be taken into account, such as the existent probabilities of objects, the skyline probability calculated for each instance, the uncertainty of the results, and the exponential number of possible instance combinations, and so on.

Our Principal Contributions Can be Summarized as Follows:
- We introduce, for the first time, a new formalization of probabilistic top-k representative skyline queries in uncertain databases based on user-specified preferences regarding the priority of individual dimensions.

- We apply odds ratios to control the size of the answer set. We also prove that, by using odds ratios, our methods are not sensitive to the rescaling of any dimension or the insertion of the non-skyline instances.
- We propose two efficient algorithms based on the R-Tree indexing structure to answer preference-based top-k representative queries on uncertain data.
- We present a detailed performance evaluation of our proposed algorithms based on real-life and synthetic datasets to demonstrate the efficiency, effectiveness and the scalability of our proposed algorithms.

Organization of the Paper: The rest of this paper is organized as follows. Section 2 describes the preliminary concepts related to the proposed topic and clearly presents the data model to be used and the problem definition. In Section 3, we develop two novels algorithms for answering top-k representative queries in uncertain databases. A number of pruning conditions to enhance the efficiency of our methods are also proposed in this section. Section 4 briefly analyses related work in this area. Section 5 demonstrates the performance of the proposed algorithms through extensive experiments. Finally, section 6 concludes this research and briefly discusses future work.

2   Preliminaries

2.1   Data Model

In this section, we present the data model to be used for formulating our framework and related concepts regarding ranking queries on uncertain databases.

Let P_n be a set of n uncertain data objects in an m-dimensional space $D_m = \{d_1, d_2, ..., d_m\}$, where each dimension has a domain $dom(d_i)$ of a non-negative rational number. In our data model, each uncertain object $O_l \in P_n$ may have a set of mutually exclusive alternatives (or instances) $O_l = \{o_1, o_2, ..., o_j\}$. For $1 \leq i \leq j$, an instance o_i may appear with non-zero probability $Pr(o_i) > 0$, and $\sum_{i=1}^{j} Pr(o_i) \leq 1$. As a result, the existing probability of each uncertain object is presented by the probability distribution over its instance set.

Generally, we consider each instance of an uncertain object as an m-dimensional point. The j-th dimensional coordinate value of an instance o_i is denoted by $o_i[d_j]$. Without loss of generality, we assume that $\forall d_j$: $o_i[d_j] \geq 0$.

Given a set of uncertain objects P_n, a possible world W is a combination of instances where each instance represents an uncertain object uniquely. The probability of the appearance of possible world W, denoted by $Pr(W)$, is the product of the existing probabilities of all instances in the world W and the absence of all other objects in the dataset. Let $\mathcal{W}$ be the set of all possible worlds, and then $\sum_{W \in \mathcal{W}} Pr(W) = 1$. We use $SKY(W)$ to denote the skyline set of possible world W.

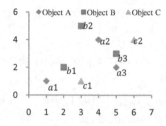

Object	A			B			C	
Instance	a1	a2	a3	b1	b2	b3	c1	c2
Value	(1,1)	(4,4)	(5,2)	(2,2)	(3,5)	(5,3)	(3,1)	(6,4)
Existing probability	0.2	0.3	0.5	0.4	0.2	0.2	0.2	0.8
Personalized skyline prob.	0.2	0.096	0.16	0.32	0.128	0	0.096	0

Fig. 1. An example of uncertain dataset containing 3 objects and 8 instances

Consider the example given in Figure 1, there are 3 uncertain objects (A, B and C). Each uncertain object has a number of its instances and each instance is associated with an existing probability. It is noted that object B does not exist with probability 0.2 as the total probability of its three instances is $0.8 < 1$.

2.2   Problem Definition

The notion of user preference has been widely exploited in a number of important research areas, such as Information Retrieval Systems [9,10] and Recommender Systems [11]. The authors in [12] have proved that applying user preferences to reduce the volume of data presented to the user is an efficient strategy when explicit scoring functions are not available.

The main limitation of the previously proposed ranking methods is that they assume all dimensions of objects are equally important and totally ignore user preferences on the priority of the individual dimensions. In fact, a user may be interested in only a subset of dimensions or may prefer one specific dimension which is more important than the others. Motivated by this, we complement existing work by exploiting user-specified preferences in identifying top-k representative skyline objects in uncertain databases.

User Preference
The formal definition of the user-specified preferences is defined as an ordered subset of dimensions in D_m.

Definition 1 (User Preference). Given a set of all dimensions D_m, a user preference $U_r = \{u_1, ..., u_r\}$ $(r > 1)$ is an ordered subset of D_m $(U_r \subseteq D_m)$, such that $\forall i, j\ (1 \leq i < j \leq r): u_i, u_j \in U_r$, then $u_i \vdash u_j$ (i.e. users prefer u_i over u_j).

Definition 2 (Personalized Dominance Relationship). Given two instances $p, q \in P_n$ and a user preference $U_r = \{u_1, ..., u_r\}$ $(r > 1, U_r \subseteq D_m)$, we say p dominates q based on U_r, denoted by $p \prec_{U_r} q$, if and only if $\exists\, u_i \in U_r, p[u_i] < q[u_i]$ and for all $j < i$, we have $p[u_j] = q[u_j]$.

Based on the personalized dominance relationship against two instances, we now can compute the personalized skyline probability of an instance as follows.

Definition 3 (Personalized Skyline Probability). Given a set of uncertain jects $P_n = \{O_1, O_2, \dots, O_n\}$ and a user preference $U_r = \{u_1, \dots, u_r\}$ $(r > 1, U_r \subseteq D_m)$, the personalized skyline probability of an instance $o_i \in O_l$ $(1 \le l \le n)$, denoted by $Pr_{SKY_{U_r}}(o_i)$, is the probability that o_i exists multiplied by the probability that no instance of other uncertain objects dominating o_i exists.

$$Pr_{SKY_{U_r}}(o_i) = Pr(o_i) \cdot \prod_{j=1, j \ne l}^{n} \left(1 - \sum_{q \in O_j, q \prec_{U_r} o_i} Pr(q) \right) \tag{1}$$

The personalized skyline probability of an object $O \in P$ is defined as:

$$Pr_{SKY_{U_r}}(O_l) = \sum_{o_i \in O_l} Pr_{SKY_{U_r}}(o_i) \tag{2}$$

Theorem 1. Given a possible world W $(W \in \mathcal{W}$ and $|W| > 0)$ and a user preference $U_r = \{u_1, \dots, u_r\}$ $(r > 1, U_r \subseteq D_m)$, there is an unique order among instances in W. Let $SKY_{U_r}(W)$ be a personalized skyline set of the possible world W, there only one instance exists in $SKY_{U_r}(W)$, $|SKY_{U_r}(W)| = 1$.

Lemma 1. Given a set of uncertain objects $P_n = \{O_1, O_2, \dots, O_n\}$ and a user preference $U_r = \{u_1, \dots, u_r\}$ $(r > 1, U_r \subseteq D_m)$, let $|O_i|$ be the number of instances of O_i and I_s be the set of all instances in P_n where $s = \sum_{i=1}^{n}|O_i|$, the total of the personalized skyline probabilities of all instances in P_n equals to 1.

$$\sum_{o_i \in I_s} Pr_{SKY_{U_r}}(o_i) = 1 \tag{3}$$

Due to the space limitation, the proof of the correctness of Theorem 1 and Lemma 1 have been omitted. The next stage of our proposed method is to efficiently extract a small subset (say k) of truly interesting objects from the others in the skyline set. The method of using a threshold to control the size of the answer set has its own limitations which may cause difficulty for the user in defining an appropriate threshold. To alleviate this shortcoming, in this paper, we apply the odds ratio instead of using a probability threshold to narrow down the whole skyline to a subset of a controllable size. The odds ratio is one of the important statistics which is commonly used in information retrieval [13] and data mining [14]. Specifically, we define the odds of an instance as follows.

Definition 4 (Odds Ratio of an Instance): The odds of an instance o, denoted by $\phi(o)$, is the ratio of the probability that the instance will be in the personalized skyline set to its existing probability.

$$\phi(o) = \frac{Pr_{SKY_{U_r}}(o)}{Pr(o)} \tag{4}$$

Definition 5 (Preference-Based Top-k Representative Skyline Query): Given a set of uncertain objects $P_n = \{O_1, O_2, \dots, O_n\}$, a user preference $U_r = \{u_1, \dots, u_r\}$ $(r > 1, U_r \subseteq D_m)$, and an integer k $(k \ge 1)$, the preference-based top-k representative skyline query retrieves k instances with the highest odds ratio such that k instances belonging to k different objects.

We define the odds ratio of a skyline instance based on intuition: the larger the odds ratio of an instance, the more interesting the instance.

Significance of Using Odds Ratio

In the context of top-k representative skyline queries, our approach using the odds ratio to limit the size of the result set satisfies two basic properties of the top-k representative skyline query, namely as *scale invariance* and *stability*, which are proposed in [15-17]. *Scale invariance* means that the result is not sensitive to the scaling of dimensions. That is, rescaling the values of any dimension should not alter the final result (only positive rescaling factors are used). *Stability* means that the addition of non-skyline instances should not affect the final result. Satisfying this property is important as it can avoid the scenario where malicious users manipulate databases to reach their objective by simply adding non-skyline (unimportant) instance.

3   Computing Preference-Based Top-k Representative Skyline Queries in Uncertain Databases

The Naïve approach for the problem of top-k representative skyline queries is to first calculate the skyline probabilities for each instance by conducting dominance checks against instances of other objects. Then, we compute the odds ratio of each instance and return k instances (belonging to k objects) having the highest odds. However, this method is inefficient as the total number of instances may be too large and performing pair-wise dominance checks between instances is too costly. Motivated by this limitation, we now present two algorithms, namely Sweep-Based Algorithm and Bounding Probabilities Algorithm, which are efficient and effective in terms of the response CPU time and number of objects pruned, especially with high dimensional datasets.

3.1   Pruning Techniques

The objective of this section is to define a number of sufficient pruning conditions which allow us to reduce the number of dominance checks by eliminating redundant instances or objects. Hence, the computational cost will be significantly reduced.

Eliminating Redundant Instances and Objects

The instances/objects having personalized skyline probabilities equal to 0 are defined as redundant instances/objects. We exploit an important technique using the minimum bounding box of each uncertain object to eliminate redundant instances/objects. Let O be an uncertain object having ℓ instances, $O = \{o_1, \dots, o_\ell\}$ and $U_r = \{u_1, \dots, u_r\}$ be a user-defined preference. We denote O_{max} and O_{min} for the upper-right corner and the lower-left corner of the minimum bounding box (MBB for short) of the uncertain object O respectively.

Pruning Rule 1 (Eliminate Redundant Instances): Let O be an uncertain object that has l intances and q is an instance such that $q \notin O$

$$\text{If } O_{max} \prec_{U_r} q \text{ and } \sum_{i=1}^{l} Pr(o_i) = 1 \text{ then } Pr_{SKY_{U_r}}(q) = 0$$

Pruning Rule 2 (Eliminate Redundant Objects): Let O be an uncertain object that has l intances and Y is another uncertain object such that $O \neq Y$

$$\text{If } O_{max} \prec_{U_r} Y_{min} \text{ and } \sum_{i=1}^{l} Pr(o_i) = 1 \text{ then } Pr_{SKY_{U_r}}(Y) = 0$$

3.2   Sweep-Based Algorithm

The main idea of this algorithm is that we use lines sweeping across dimensions specified in the user-defined preference U_r to efficiently identify the most k representative skyline points. To facilitate the sweep-based algorithm, we assume that all instances are lexicographically sorted based on U_r and it is named as the input list L. We also maintain a SKY list which is a maximal heap based on the odds ratio to store the processed instances having non-zero odds ratios.

Suppose we are processing instance $t_i \in T$ in L and we denote $t_{j_1}, t_{j_2}, ..., t_{j_{i-1}}$ to be instances of object T with position before t_i. Then, we define the *accum* value of an instance as follows:

$$accum(t_i) = 1 - Pr(t_{j_1}) - Pr(t_{j_2}) - \cdots - Pr(t_{j_{i-1}}) - Pr(t_i)$$

The parameter *accum* acts as a stopping condition for our algorithm as it enables us to terminate the searching process as early as possible since it guarantees that the rest of the instances in the list L are redundant. In our example given in Figure 1, we can terminate the iteration after processing instance a_3 as all instances with positions after a_3 in L have zero personalized skyline probabilities. Algorithm 1 illustrates the pseudo code of the Sweep Based Algorithm. We assume that the first k instances in L have existing probabilities smaller than 1 to guarantee that at least k instances are returned.

Algorithm 1. Sweep-Based Algorithm

Input: An input list L of uncertain instances which are lexicographically ordered based on the user-specified preference U_r
Output: k representative instances which belong to k different objects
Method:
For each instance o_i in list L do
    Compute $Pr_{SKY_{U_r}}(o_i)$ and $\phi(o_i)$
    If there exists an instance o_j in the SKY list such that $o_i \neq o_j$ and $o_i, o_j \in O_l$ then
        If $\phi(o_i) > \phi(o_j)$ then replace o_j by o_i in SKY
        Else
            Update $o_j.accum$
    Else
        Compute $o_i.accum$ and insert o_i to SKY
If there exists an instance v in SKY where $v.accum = 0$ and $|SKY| \geq k$ then terminate the search and return k instances having the highest values of ϕ in SKY;

3.3   Bounding Probabilities of Instances

The limitation of the sweep-based algorithm is that it requires the instances in the input list to be lexicographically sorted, based on U_r and the cost of sorting is considerably expensive when the number of dimensionalities increases. We develop an efficient algorithm to answer the top-k representative queries based on R-tree indexing structure. Specifically, this algorithm exploits an upper bound for the existing probabilities to check whether an instance needs further consideration or should be pruned.

Pruning Rule 3: Let o_i be the current processing instance and Pr^+ be the upper bound probability. If $Pr(o_i) > Pr^+$ then o_i is pruned where Pr^+ can be obtained as follows:

$$Pr^+ = \frac{1 - PrSKY_{sum}}{\phi_{min}} \tag{5}$$

Proof: In formula (5), we refer ϕ_{min} to the current smallest odds ratio of instances in SKY and $PrSKY_{sum}$ to the current sum of personalized skyline probabilities of all instances that have been processed so far. The numerator of the above equation acts as an upper bound of the personalized skyline probabilities of all unseen instances (i.e. the maximum personalized skyline probability of any unseen instance). If instance o_i wants to be inserted into SKY, its odds value has to be greater than ϕ_{min}. This means $\frac{1 - PrSKY_{sum}}{Pr(o_i)} > \phi_{min}$.

Algorithm 2 illustrates the pseudo code of the Bounding Probabilities Algorithm. Specifically, the minimum corners of MBBs of all uncertain objects in P_n are organized into a global R-tree. As the personalized skyline probability of an object depends on only the objects or instances that dominate it (based on formula (1)), the calculation of the personalized skyline probability of an object O is performed by a window query on the R-tree with the origin and O_{max} as the opposite corners. Specifically, the minimum corners of MBBs of all uncertain objects in P_n are organized into a heap. The top instance of the heap is iteratively processed. We also create a minimal heap SKY to store k instances having the highest odds ratio so far. To support efficient processing, the pruning rules are firstly checked to reduce the number of instances that need to be processed. When the heap top cannot be pruned, its personalized skyline probability and odds ratio are computed. If there is another instance of the same object that is already in the SKY, the odds ratios of both are compared and the instance having the lower odds ratio is discarded. The variables ϕ_{min}, $PrSKY_{sum}$ and Pr^+ are updated when there is an addition or replacement performed in the SKY heap accordingly. Once the instance of an object is processed, the existing probability of the next instance of the object is compared with the upper bound before it is inserted into the heap. As shown in our experimental results, in our proposed algorithms, only a small portion of instances are involved in the searching processes. Therefore, it is clear that our proposed algorithms have good scalability on large data sets.

Algorithm 2. Bounding Probability Algorithm

Input: $\mathcal{R}_P$: the R-Tree of P_n which is built based on O_{min} of all object $O \in P_n$ and an integer k
Output: k representative instances which belong to k different objects
Method:
    $SKY = \emptyset$; $PrSKY_{sum} = \phi_{min} = Pr^+ = 0$;
    Build a heap $\mathcal{H}$ on O_{min} for all $O \in P_n$
    While $\mathcal{H} \neq \emptyset$
        Let $o \in O$ be the top instance in $\mathcal{H}$
        If o is redundant then NEXT; (Pruning rule 1 and 2)
        If $Pr^+ \neq 0$
            If $Pr(o) > Pr^+$ then NEXT; (Pruning rule 3)
        Compute $Pr_{SKY_{U_r}}(o)$ and $\phi(o)$;
        If $|SKY| < k$ then

$\qquad$ Add o to SKY and update $\phi_{min}, PrSKY_{sum}, Pr^+$;

Else $\quad$ // $|SKY| \geq k$

$\qquad$ If $\phi(o) > \phi_{min}$ then

$\qquad\qquad$ If there exists another instance $o_x \in O$ and $\phi(o) > \phi(o_x)$ then

$\qquad\qquad\qquad$ Replace o_x by o;

$\qquad\qquad\qquad$ Update $\phi_{min}, PrSKY_{sum}$ and Pr^+;

$\qquad\qquad$ Replace instance having ϕ_{min} by o;

$\qquad\qquad$ Update $\phi_{min}, PrSKY_{sum}$ and Pr^+;

$\qquad$ If $Pr(next\ instance\ of\ O) < Pr^+$

$\qquad\qquad$ Insert $next\ instance\ of\ O$ into $\mathcal{H}$;

4   Related Work

Skyline computation has received considerable attention in the database community in recent years. While the skyline processing on certain data is straightforward and well defined for the domination relationship, the skyline computation on uncertain data (also known as probabilistic skyline queries) is much more complicated and challenging since each uncertain object takes a probability of not being dominated by any other objects. In addition, returning the full skyline set would be impractical and may cause serious confusion for users in the selection processes. Therefore, significant research efforts have been devoted to restrict the size of the skyline set. The authors in [18] and [19] aimed to return k uncertain objects that have the highest skyline probabilities while other researchers assigned the quantity of subsets of uncertain objects based on their dominant power (i.e. the number of objects dominated by the subset) [5,7,1]. Pei et al. [20] proposed the p-skyline query which retrieves those objects with the skyline probabilities above a given threshold p. However, these aforementioned methods have one of the following limitations: (1) they totally ignore the user preferences on the priority of dimensions; (2) it might be difficult for end-users to specify an appropriate threshold value when the end-users have limited knowledge about the dataset; (3) the final result sets are unstable since they can be controlled by the addition of unimportant (non-skyline) points.

Motivated by these identified weaknesses, *top-k representative skyline queries* are developed as a novel ranking mechanism for applications where multi-criteria are involved. The work in [4] is the first to study top-k representative skyline queries and aims to return k skyline objects so that the number of objects that are not dominated by any of k representative objects is minimized. Paper [21] proposed an alternative approach for top-k representative skyline queries by retrieving k skyline points that minimize the Euclidean distance between a non-representative skyline point and its nearest representative skyline point. Paper [22] explored the queries under the setting of distributed data. However, these aforementioned methods focus on traditional certain data and cannot be applied directly to uncertain datasets. Our study extends top-k representative skyline queries to uncertain data. To the best of our knowledge, this is the first study of top-k representative skyline queries in uncertain databases.

5 Experimental results

In this section, we present the experimental results of comprehensive performance studies to evaluate the efficiency, effectiveness and scalability of our proposed algorithms on both real-life and synthetic datasets. We have implemented all the algorithms presented in this paper using Java programming language. The experiments were conducted on a PC computer with a Core-2 Duo CPU running Windows 7 with 4 GB of main memory. We evaluate the algorithms proposed in this paper for the issues of efficiency, effectiveness and scalability.

In our experiments, we use the NBA data set which is downloaded from www.basketball-reference.com as the real data set. The real data set contains 260,599 records of 1,092 players from the 1999 to 2008 season. We treat each player as an uncertain object and the records of the players in each game as instances with three dimensions: number of points, number of assists, and number of rebounds. Existing probabilities are randomly assigned to instances of the same object such that the total value of existing probabilities of an object is up to 1.We also employ two synthetic datasets, Anti-correlated (**ANT**) and Independent (**IND**) which are produced by following the same data generator in [23] regarding the parameters: dimensionality **d,** the number **n** of uncertain objects (i.e. cardinality), the value **k** and the edge size **h** of the hyper-rectangle region of each object where the instances of the object appear.

5.1 Evaluating Efficiency and Scalability

In this subsection, we evaluate the efficiency and scalability of our proposed algorithms by conducting extensive experiments regarding to the response time over varying parameters.

We evaluate the efficiency of our proposed algorithms over varying dimensionality **d**, cardinality **n**, retrieval size **k** and the average edge length **h** of the objects' minimum bounding rectangle, as reported in Figure 2, 3, 4 and 5. It is observed that our algorithms are faster than the Naive approach by orders of magnitude. We also observe that the response time on the anti-correlated dataset increases more sharply than the relative performance on the independent dataset. This is because when the dimensions are anti-correlated, the number of skylines exponentially increases with the growth of dimensions. The results also consistently show that our proposed algorithms achieve high scalability, low response time and are much more efficient than the Naive approach. It is clear that the growth in the size of the dataset leads to a slower response time since the cost of the pruning processes becomes more expensive. In addition, the average edge length of the objects' minimum bounding rectangle is another factor that can directly affect the response time. The larger the average edge length of the objects' minimum bounding rectangles, the higher the cost of the pruning processes (due to the fact that there are more overlapping regions between sibling nodes in R-tree). The smaller the average edge length of the objects' minimum bounding rectangles, the more effective the pruning processes.

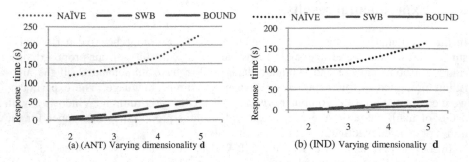

Fig. 2. a(ANT), b(IND) on varying dimensionality

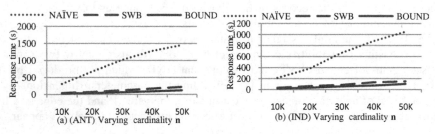

Fig. 3. a(ANT), b(IND) on varying cardinality (number of objects)

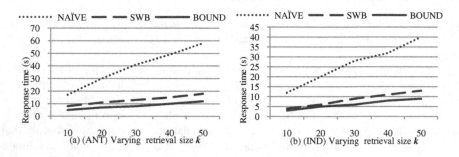

Fig. 4. a(ANT), b(IND) on varying retrieval size

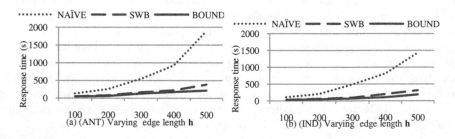

Fig. 5. a(ANT), b(IND) on varying edge length

5.2   Evaluating Effectiveness

We also evaluate the effectiveness of our proposed algorithms by investigating the percentage of uncertain objects pruned by the pruning rules proposed in this paper. We conducted experiments on both real-life and synthetic data sets. Our results in Figure 6 show that our proposed pruning rules are highly effective in filtering out unqualified instances and that BOUND is more effective than SWB since it prunes not only redundant objects/instances but also non-promising objects/instances by using the upper bound on the probability.

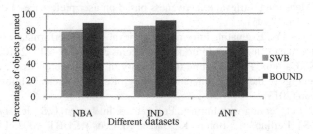

Fig. 6. Effectiveness of pruning rules by measuring the percentage of objects pruned

6   Conclusion

In this paper, we are the first to study the problem of top-*k* representative queries on uncertain databases. One of the main drawbacks of previous work is that their ranking approaches totally ignore user-defined preferences on the priority of individual dimension or a subset of dimensions. Moreover, applying the odds ratio to restrict the cardinality of the answer set is more effective than those techniques using a given threshold as it may be inconvenient for an end-user. We also investigated the efficiency, effectiveness and scalability of our proposed algorithms by conducting extensive experiments on both real and synthetic datasets. An interesting direction for future work is to consider the problem in a streamed or distributed setting, offering additional scalability, especially for massive big data.

References

1. Zhan, L., Zhang, Y., Zhang, W., Lin, X.: Identifying top *k* dominating objects over uncertain data. In: Bhowmick, S.S., Dyreson, C.E., Jensen, C.S., Lee, M.L., Muliantara, A., Thalheim, B. (eds.) DASFAA 2014, Part I. LNCS 8421, vol. 8421, pp. 388–405. Springer, Heidelberg (2012)
2. Yiu, M.L., Mamoulis, N.: Multi-dimensional top-k dominating queries. The VLDB Journal **18**(3), 695–718 (2009)
3. Yiu, M.L., Mamoulis, N.: Efficient processing of top-k dominating queries on multi-dimensional data. In: Proceedings of the 33rd International Conference on Very Large data Bases, pp. 483-494. VLDB Endowment (2007)
4. Lin, X., Yuan, Y., Zhang, Q., Zhang, Y.: Selecting stars: The k most representative skyline operator. In: IEEE 23rd International Conference on Data Engineering, ICDE 2007, pp. 86-95. IEEE (2007)

5. Lian, X., Chen, L.: Probabilistic top-k dominating queries in uncertain databases. Information Sciences **226**, 23–46 (2013)
6. Lian, X., Chen, L.: Top-k dominating queries in uncertain databases. In: Proceedings of the 12th International Conference on Extending Database Technology: Advances in Database Technology, pp. 660-671. ACM (2009)
7. Zhang, W., Lin, X., Zhang, Y., Pei, J., Wang, W.: Threshold-based probabilistic top-k dominating queries. The VLDB Journal **19**(2), 283–305 (2010)
8. Kontaki, M., Papadopoulos, A.N., Manolopoulos, Y.: Continuous top-k dominating queries in subspaces. In: Panhellenic Conference on Informatics, PCI 2008, pp. 31-35. IEEE (2008)
9. Yao, Y.: Measuring retrieval effectiveness based on user preference of documents. JASIS **46**(2), 133–145 (1995)
10. Zhou, B., Yao, Y.: Evaluating information retrieval system performance based on user preference. Journal of Intelligent Information Systems **34**(3), 227–248 (2010)
11. Vargas, S., Castells, P.: Exploiting the diversity of user preferences for recommendation. In: Proceedings of the 10th Conference on Open Research Areas in Information Retrieval, pp. 129-136 (2013)
12. Chomicki, J.: Querying with Intrinsic Preferences. In: Jensen, C.S., Jeffery, K., Pokorný, J., Šaltenis, S., Bertino, E., Böhm, K., Jarke, M. (eds.) EDBT 2002. LNCS, vol. 2287, pp. 34–51. Springer, Heidelberg (2002)
13. Manning, C.D., Raghavan, P., Schütze, H.: Introduction to information retrieval, vol. 1. Cambridge university press Cambridge, (2008)
14. Li, H., Li, J., Wong, L., Feng, M., Tan, Y.-P.: Relative risk and odds ratio: a data mining perspective. In: Proceedings of the Twenty-Fourth ACM SIGMOD-SIGACT-SIGART Symposium on Principles of Database Systems, pp. 368-377. ACM (2005)
15. Nanongkai, D., Sarma, A.D., Lall, A., Lipton, R.J., Xu, J.: Regret-minimizing representative databases. Proceedings of the VLDB Endowment **3**(1–2), 1114–1124 (2010)
16. Das Sarma, A., Lall, A., Nanongkai, D., Lipton, R.J., Xu, J.: Representative skylines using threshold-based preference distributions. In: 2011 IEEE 27th International Conference on Data Engineering (ICDE), pp. 387-398. IEEE (2011)
17. Magnani, M., Assent, I., Mortensen, M.L.: Taking the Big Picture: representative skylines based on significance and diversity. The VLDB Journal 1-21 (2014)
18. Zhang, Y., Zhang, W., Lin, X., Jiang, B., Pei, J.: Ranking uncertain sky: The probabilistic top-k skyline operator. Information Systems **36**(5), 898–915 (2011)
19. Yong, H., Lee, J., Kim, J., Hwang, S.-W.: Skyline ranking for uncertain databases. Information Sciences **273**, 247–262 (2014)
20. Pei, J., Jiang, B., Lin, X., Yuan, Y.: Probabilistic skylines on uncertain data. In: Proceedings of the 33rd International Conference on Very large data bases, pp. 15-26. VLDB Endowment (2007)
21. Tao, Y., Ding, L., Lin, X., Pei, J.: Distance-based representative skyline. In: IEEE 25th International Conference on Data Engineering, ICDE 2009, pp. 892-903. IEEE (2009)
22. Vlachou, A., Doulkeridis, C., Halkidi, M.: Discovering representative skyline points over distributed data. In: Ailamaki, A., Bowers, S. (eds.) SSDBM 2012. LNCS, vol. 7338, pp. 141–158. Springer, Heidelberg (2012)
23. Borzsony, S., Kossmann, D., Stocker, K.: The skyline operator. In: Proceedings. 17th International Conference on Data Engineering, 2001, pp. 421-430. IEEE (2001)

Cluster Sequence Mining: Causal Inference with Time and Space Proximity Under Uncertainty

Yoshiyuki Okada, Ken-ichi Fukui[✉], Koichi Moriyama, and Masayuki Numao

The Institute of Scientific and Industrial Research,
Osaka University, 8-1 Mihogaoka, Osaka, Ibaraki, Japan
fukui@ai.sanken.osaka-u.ac.jp
http://www.ai.sanken.osaka-u.ac.jp

Abstract. We propose a pattern mining algorithm for numerical multidimensional event sequences, called cluster sequence mining (CSM). CSM extracts patterns with a pair of clusters that satisfies space proximity of the individual clusters and time proximity in time intervals between events from different clusters. CSM is an extension of a unique algorithm (co-occurrence cluster mining (CCM)), considering the order of events and the distribution of time intervals. The probability density of the time intervals is inferred by utilizing Bayesian inference for robustness against uncertainty. In an experiment using synthetic data, we confirmed that CSM is capable of extracting clusters with high F-measure and low estimation error of the time interval distribution even under uncertainty. CSM was applied to an earthquake event sequence in Japan after the 2011 Tohoku Earthquake to infer causality of earthquake occurrences. The results demonstrate that CSM suggests some high affecting/affected areas in the subduction zone farther away from the main shock of the Tohoku Earthquake.

Keywords: Hierarchical clustering · Pattern mining · Bayesian learning · Earthquake

1 Introduction

In this study, we focus on a task that combines well-known clustering techniques and the concept of frequent pattern mining. Clustering[3,12] attempts to produce groups of similar objects within a so-called data space or feature space, which is typically represented by a multidimensional numerical vector. In contrast, frequent pattern mining[1,5] attempts to extract and list frequently appearing item sets, wherein an item is typically identified via nominal variables.

In this study, given a sequence of events, where each event is represented by a multidimensional numerical vector (e.g., sequence of signal events, image events, and position events), the goal is *to find and list pairs of clusters that frequently appear in a sequence.* This task may induce novel applications, e.g.,

© Springer International Publishing Switzerland 2015
T. Cao et al. (Eds.): PAKDD 2015, Part II, LNAI 9078, pp. 293–304, 2015.
DOI: 10.1007/978-3-319-18032-8_23

identification of weather change patterns from a sequence of satellite images, inference of health change patterns from a sequence of medical inspection data, and inference of mechanical interaction patterns between components from a sequence of sounds caused by damage.

Various studies have been conducted in the field of spatio-temporal data mining. Spacetime scan statistics[7] detects outbreak regions (clusters) in certain periods on the basis of statistical tests, e.g., detecting region(s) and period(s) of infectious disease. However, spacetime scan statistical methods do not extract causality between different regions; the purpose is to detect a single region and its period.

We have proposed a unique algorithm called co-occurring cluster mining (CCM)[6]. CCM extracts and lists pairs of clusters with both high intra-cluster density in the data space and high inter-cluster co-occurrence in the sequence of events. They applied CCM to extract co-occurrence patterns from acoustic emission (signal) events in a fuel cell to infer mechanical interactions among components. In addition, they also applied CCM to extract earthquake co-occurrence patterns wherein different areas may affect each other[4].

However, to accommodate model simplicity, CCM does not consider the order and time intervals of events when evaluating co-occurrence. Obviously, order and time intervals play important roles because causal events must precede an observable effect. In this work, we extend CCM by introducing the order and time intervals of events into co-occurrence. We refer to this algorithm as *cluster sequence mining (CSM)* because co-occurrence does not imply an order of events. In physical phenomena such as earthquakes, there is no typical time interval, i.e., power law. Therefore, we utilize a Bayesian framework to infer a probability density function (pdf) of time intervals. Bayesian inference is robust for a relatively small number of observations and noise because it combines a prior distribution and observations, i.e., Bayesian learning. We then introduce the inferred pdf of the time intervals into the evaluation of candidate patterns in CSM. To the best of our knowledge, no other method can extract such a sequence of clusters from multidimensional event sequence with order and time intervals.

In our experiment, we validated CSM using synthetic data. CSM is capable of extracting more precise cluster patterns than CCM under uncertainty of time intervals in terms of the F-measure between an embedded true pattern and an extracted pattern. We then applied CSM to earthquake event sequence data to infer the causality of earthquake occurrences. The patterns extracted from the hypocenter catalog data from the 2011 Tohoku Earthquake indicated that high affecting/affected areas were identifiable in the subduction zone farther away from the main shock.

2　Proposed Method: Cluster Sequence Mining

2.1　Problem Definition

This section introduces the characteristics of data addressed by this work. We then define the requirements of cluster sequence patterns.

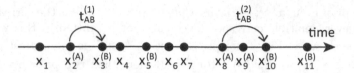

Fig. 1. Example of time intervals for a candidate cluster sequence pattern

Definition 1 (*event sequence with time*). Assume a data set $\mathcal{D}$ with N numerical event data points $\boldsymbol{x}_i = (x_{i,1}, \cdots, x_{i,v})$, $(i = 1, \cdots, N)$ in v-dimensional space obtained in order $\boldsymbol{x}_1 \prec \cdots \prec \boldsymbol{x}_N$ with occurrence time $t(\boldsymbol{x}_i)$.

Given the above definition, the extracted patterns must meet the following requirements:

Requirement 1 (*time proximity*). Given two clusters $\mathbf{A}, \mathbf{B} \subset \mathcal{D}$ ($\mathbf{A} \cap \mathbf{B} = \emptyset$), and assuming that causality depends on time proximity, corresponding events in different clusters $\boldsymbol{x}^{(A)} \in \mathbf{A}$ and $\boldsymbol{x}^{(B)} \in \mathbf{B}$ must occur near in time with order $\boldsymbol{x}^{(A)} \prec \boldsymbol{x}^{(B)}$, i.e., $0 < t_{AB} \equiv t(\boldsymbol{x}^{(B)}) - t(\boldsymbol{x}^{(A)}) < \epsilon$.
Requirement 2 (*frequency*). Assuming that the confidence of a pattern depends on occurrence frequency, the number of times in which the events from $\mathbf{A}$ and $\mathbf{B}$ occur near in time must be high, i.e., $\#\{t_{AB} | 0 < t_{AB} < \epsilon\} > Supp_{min}$, where $Supp_{min}$ is the minimum support threshold.
Requirement 3 (*space proximity*). For the given two clusters $\mathbf{A}$ and $\mathbf{B}$, events in each cluster must be similar, e.g., the sum of squares within (SSW) clusters, SSW($\mathbf{A}$) and SSW($\mathbf{B}$) must be small. Note that clusters $\mathbf{A}$ and $\mathbf{B}$ should be evaluated independently.

The correspondence of events to calculate t_{AB} in requirement 1 is introduced in the next section. Given the above requirements, we define the cluster sequence pattern as follows:

Definition 2 (*cluster sequence pattern*). Let a set of time intervals between corresponding events in clusters $\mathbf{A}$ and $\mathbf{B}$ be $\mathbf{T}_{AB} = \{t_{AB}\}$. Assuming the time intervals follow a probability distribution function $\psi(\theta)$, i.e., $t_{AB} \sim \psi(\theta)$, where θ is a parameter, if $\mathbf{A}$ and $\mathbf{B}$ meet the above three requirements, then $P_{A \rightarrow B} =< \mathbf{A}, \mathbf{B}, \psi(\theta) >$ is called a *cluster sequence pattern*. Here, $\mathbf{A}$ refers to a prior cluster and $\mathbf{B}$ is a posterior cluster.

2.2 Calculation of Time Intervals

The CSM algorithm employs candidate generation and a testing procedure. An example in this section shows calculation of time intervals for a given candidate cluster pair. Assume that the candidate clusters are given as shown in Fig. 1; events $\boldsymbol{x}_2$, $\boldsymbol{x}_8$ and $\boldsymbol{x}_9$ are assigned to cluster $\mathbf{A}$, and $\boldsymbol{x}_3$, $\boldsymbol{x}_5$, $\boldsymbol{x}_{10}$, and $\boldsymbol{x}_{11}$ are assigned to cluster $\mathbf{B}$. Note that other events are not assigned to any cluster. Then, the time intervals are evaluated in the time line. A problem here is how to define corresponding events between two clusters in order to calculate time intervals.

For the prior event $\boldsymbol{x}^{(A)}$, the nearest posterior event $\boldsymbol{x}^{(B)}$ is the corresponding event. However, multiple correspondences are not permitted in this work. For the example shown in Fig. 1, for prior event $\boldsymbol{x}_2^{(A)}$, $\boldsymbol{x}_3^{(B)}$ is the nearest posterior event; therefore, $t_{AB}^{(1)} = t(\boldsymbol{x}_3^{(B)}) - t(\boldsymbol{x}_2^{(A)})$. If more than two prior events occur successively as $\boldsymbol{x}_8^{(A)}$ and $\boldsymbol{x}_9^{(A)}$, the first event $\boldsymbol{x}_8^{(A)}$ is selected, i.e., $t_{AB}^{(2)} = t(\boldsymbol{x}_{10}^{(B)}) - t(\boldsymbol{x}_8^{(A)})$ because the time interval at the most is $t_{AB}^{(2)}$. $\boldsymbol{x}_5^{(B)}$ remains because there is no prior corresponding event. Note that calculation of time intervals has several options; however, in this work we employ the simplest definition, i.e., the nearest posterior event for a prior event with a single correspondence.

2.3   Inference of Probability Density Function of Time Intervals

Observed time intervals have uncertainty caused by noise, the number of observations, and the definition of time interval calculation. Therefore, we introduce a Bayesian framework based on a probabilistic model. Assume that time intervals are distributed by a pdf, e.g., a Gaussian, gamma, or exponential distribution.

For an earthquake event application, we assume an exponential distribution, i.e., $\psi(t_{AB}) = \lambda \exp(-\lambda t_{AB})$, where λ is a slope parameter. All other parts use an exponential distribution for pdf in this paper. Let $\pi(\lambda)$ be a prior distribution of parameter λ, $\psi(t_{AB}|\lambda)$ be a likelihood function for time intervals given λ, and $\pi(\lambda|t_{AB})$ be a posterior distribution of λ given time intervals. From Bayes' rule,

$$\pi(\lambda|t_{AB}) \propto \psi(t_{AB}|\lambda) \times \pi(\lambda). \tag{1}$$

Here, because we assume exponential distribution for the likelihood, $\pi(\lambda)$ and $\pi(\lambda|t_{AB})$ are both gamma distributions; $\mathrm{Ga}(\alpha_{prior}, \beta_{prior})$ and $\mathrm{Ga}(\alpha_{post}, \beta_{post})$ as gamma distributions are conjugate distributions of the exponential distribution. α_{prior} and β_{prior} are pre-defined parameters given by the user. In this work, $\mathrm{Ga}(1,1)$ was used for all experiments. By using Bayes' rule (Eq. (1)), the update rule of the posterior is given by

$$\alpha_{post} = \alpha_{prior} + n, \quad \beta_{post} = \beta_{prior} + n\bar{t}_{AB}, \tag{2}$$

where $\bar{t}_{AB}$ denotes the average of the time intervals $\{t_{AB}\}$, and n is its number $\#\{t_{AB}\}$. From Eq. (2), when the number of observations is small, the prior distribution has greater effect. From the definition of gamma distribution, we obtain the average of the posterior as $\mu_{post} = \alpha_{post}/\beta_{post}$. We then estimate an optimal parameter of the pdf for a given pattern $\mathrm{P}_{A \to B}$ as $\hat{\lambda}_{AB} = \mu_{post}$.

2.4   Evaluation Function

Here, we define an evaluation function to search for cluster sequence patterns. We search pairs of clusters $\mathbf{A}, \mathbf{B} \subset \mathcal{D}$ using the following evaluation function:

$$\mathcal{L}(\mathrm{P}_{A \to B}) = \mathcal{F}(\hat{\lambda}_{AB})^\gamma \cdot \mathcal{G}(\mathbf{A}, \mathbf{B})^{(1-\gamma)}, \tag{3}$$

$$\mathcal{F}(\hat{\lambda}_{AB}) = \frac{1}{1 + \exp(-\tau \hat{\lambda}_{AB})}, \tag{4}$$

$$\mathcal{G}(\mathbf{A}, \mathbf{B}) = \exp\left(-\frac{\mathrm{SSW}(\mathbf{A})^2 + \mathrm{SSW}(\mathbf{B})^2}{2\sigma^2}\right). \tag{5}$$

Function $\mathcal{F}(\hat{\lambda}_{AB})$ evaluates the time proximity for requirement 1. The higher the $\mathcal{F}(\hat{\lambda}_{AB})$ value, the higher the time proximity. $\hat{\lambda}_{AB}$ is the estimated parameter of the pdf of time intervals described in Section 2.3. We assume an exponential distribution for the pdf; therefore, greater $\hat{\lambda}_{AB}$ gives steeper distribution. Thus, time proximity can be interpreted as greater $\hat{\lambda}_{AB}$. Then, a sigmoid function is wrapped for normalization, where $\tau > 0$ is an adjusting parameter.

Function $\mathcal{G}(\mathbf{A}, \mathbf{B})$ denotes the space proximity for requirement 3. Note that greater $\mathcal{G}(\mathbf{A}, \mathbf{B})$ values result in denser clusters. SSW indicates the (SSW) clusters. Here, a Gaussian function is used to adjust the bias of SSW in a dataset, where a radius σ is a parameter to control correction of the bias. Note that function $\mathcal{G}$ is the same as CCM[6]; however, we replace $\mathcal{F}$ for our purpose.

Evaluation function $\mathcal{L}$ is defined as the product of $\mathcal{F}$ and $\mathcal{G}$ to simultaneously meet the requirements of time and space proximity. γ is a hyper parameter to weight $\mathcal{F}$ or $\mathcal{G}$. In this work, $\gamma = 0.5$ is used for simplicity. In addition, requirement 2 for occurrence frequency can be met using minimum support $Supp_{min}$ as a threshold.

2.5   The CSM Algorithm

Most of the CSM algorithm is similar to that of CCM, with the exception of the Bayesian inference. To generate candidate clusters $\mathbf{A}$ and $\mathbf{B}$, CSM utilizes aggregative hierarchical clustering (AHC), similar to CCM. In AHC, once the clustering merge process is obtained, cluster sequence patterns can be searched within the dendrogram of the merge process. The other benefit of using AHC is reduction of the search space, although the degree of freedom for the cluster shape decreases.

The pseudo-code of the CSM algorithm is presented in Algorithm 1. The algorithm first generates possible sub-clusters from the dendrogram obtained by AHC in the data space. All combinations of sub-clusters can be candidate patterns (Step 1; Algorithm 1 $l.$ 1-2, Fig. 2(a)). Second, the algorithm evaluates each candidate pattern via function $\mathcal{L}$. If the evaluation score exceeds the minimum

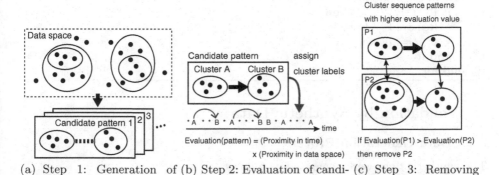

(a) Step 1: Generation of candidate patterns

(b) Step 2: Evaluation of candidate patterns

(c) Step 3: Removing overlapped patterns

Fig. 2. Cluster sequence mining algorithm

thresholds $\mathcal{L}_{min}$ and $Supp_{min}$, then these patterns are added to the output pattern list $\mathcal{P}$ (Step 2; Algorithm 1 l. 3-10, Fig. 2(b)). Here, $Supp(\mathrm{P}_{\mathbf{H}_i \to \mathbf{H}_j})$ denotes the number of time intervals between $\mathbf{H}_i$ and $\mathbf{H}_j$ described in Section 2.2. Third, to eliminate similar patterns, the algorithm checks the inclusion between the patterns and removes patterns from $\mathcal{P}$ that have lower evaluation score (Step 3; Algorithm 1 l. 11-15, Fig. 2(c)). Here, in the algorithm description, $\mathrm{P}^{(l)} \cap \mathrm{P}^{(m)}$ (Algorithm 1 l. 12) means $\mathbf{A}_l \cap \mathbf{A}_m$ and $\mathbf{B}_l \cap \mathbf{B}_m$, i.e., $\mathrm{P}^{(l)} \cap \mathrm{P}^{(m)} \neq \emptyset$ indicates, due to a property of AHC, that both prior and posterior clusters from two patterns are an inclusion relation.

Algorithm 1. CLUSTER SEQUENCE MINING

Input: event sequence with occurred time
$\quad \mathcal{D} = \{< \boldsymbol{x}_k, t(k) > | \boldsymbol{x}_1 \prec \cdots \prec \boldsymbol{x}_N\}_{k=1}^{N}$
$\quad$ dendrogram by hierarchical clustering ($\mathcal{HC}$) from $\mathcal{D}' = \{\boldsymbol{x}_k\}_{k=1}^{N}$
$\quad$ minimum evaluation score $\mathcal{L}_{min}$
$\quad$ minimum support score $Supp_{min}$
$\quad$ parameters $\tau, \sigma, \gamma, \alpha_{prior}, \beta_{prior}$
Output: cluster sequence patterns $\mathcal{P} = \{\mathrm{P}_{\mathbf{A} \to \mathbf{B}}^{(k)} =< \mathbf{A}, \mathbf{B}, \psi(\hat{\lambda}_{AB}) > | \mathbf{A} \cap \mathbf{B} = \emptyset, \mathbf{A}, \mathbf{B} \subseteq \mathcal{D}\}$
$\quad$ **Step 1**: Generate candidate patterns
1: Generate possible sub-clusters
$\quad \mathbf{H}_1, \mathbf{H}_2, \cdots, \mathbf{H}_{N-2} \subset \mathcal{HC}$;
2: $\forall i, j (i \neq j)\ \mathrm{P}_{\mathbf{H}_i \to \mathbf{H}_j} =< \mathbf{H}_i, \mathbf{H}_j, * >$;
$\quad$ **Step 2**: Evaluate candidate patterns
3: Initialize $k \gets 0$;
4: **for all** combinations of $\mathbf{H}$ **do**
5: $\quad$ Infer $\hat{\lambda}_{\mathbf{H}_i \mathbf{H}_j}$ by Bayesian inference (section 2.3)
6: $\quad$ **if** $\mathbf{H}_i \cap \mathbf{H}_j = \emptyset$ and $\mathcal{L}(\mathrm{P}_{\mathbf{H}_i \to \mathbf{H}_j}) > \mathcal{L}_{min}$ and $Supp(\mathrm{P}_{\mathbf{H}_i \to \mathbf{H}_j}) > Supp_{min}$ **then**
7: $\quad\quad \mathrm{P}_{\mathbf{A} \to \mathbf{B}}^{(k)} \gets \mathrm{P}_{\mathbf{H}_i \to \mathbf{H}_j} =< \mathbf{H}_i, \mathbf{H}_j, \psi(\hat{\lambda}_{\mathbf{H}_i \mathbf{H}_j}) >$
8: $\quad\quad k \gets k + 1$;
9: $\quad$ **end if**
10: **end for**
$\quad$ **Step 3**: Eliminate patterns with inclusion relation
11: **for all** combinations of P **do**
12: $\quad$ **if** $\mathrm{P}^{(l)} \cap \mathrm{P}^{(m)} \neq \emptyset$ **then**
13: $\quad\quad$ Remove $\mathrm{P}^{(i)}$ from $\mathcal{P}$
$\quad\quad\quad$ s.t. $i = \arg\min\{\mathcal{L}(\mathrm{P}^{(l)}), \mathcal{L}(\mathrm{P}^{(m)})\}$;
14: $\quad$ **end if**
15: **end for**

---

3 Validation by Synthetic Data

3.1 Data Generation Process

The synthetic data were generated as follows:

1. Generate N data points in two-dimensional space from a normal distribution for two classes; $\boldsymbol{x} \sim \mathrm{N}(m, \Sigma)$, where the center of class 1 $m_1 = (-2, 0)$ and of class 2 $m_2 = (2, 0)$ with a standard deviation 1.0. (Fig. 3(a))

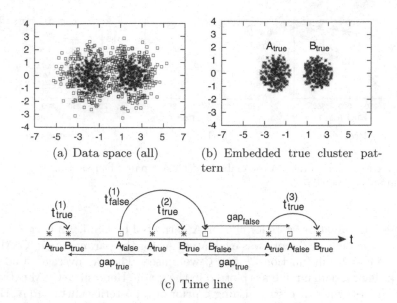

(a) Data space (all)

(b) Embedded true cluster pattern

(c) Time line

Fig. 3. Examples of synthetic data with $T = 300$ embedded true pairs, where * are samples of true cluster patterns and □ are false samples

2. Select T pairs of data points from classes 1 and 2 in order of closer points from the class center. These data points are embedded as a true cluster pair $\mathbf{A}_{true}$ and $\mathbf{B}_{true}$ (Fig. 3(b)). The rest of the data points are false points that exhibit noise in the data space.
3. For the selected T pairs, the time intervals of each pair are generated from an exponential distribution; $t_{true} \sim$ exponential(λ_{true}), where λ_{true} is a parameter of an exponential distribution. Then, true pairs are allocated in the time line with a sufficiently large constant gap between pairs; $\text{gap}_{true} = 10 \cdot \max_i\{t_{true}^{(i)}\}$. (Fig. 3(c))
4. ime intervals for the remaining false pairs are generated from a uniform distribution that exhibits noise in the time line; $t_{false} \sim \mathrm{U}(0, t_{total}/(N - T)]$, where $t_{total} = \sum_i t_{true}^{(i)} + (T - 1)\text{gap}_{true}$. Then, false pairs are allocated with a constant gap given by $\text{gap}_{false} = (t_{total} - \sum_i t_{false}^{(i)})/(N - T - 1)$. (Fig. 3(c))

In this experiment, we generated $N = 500$ samples for each cluster. Parameter T controls the balance between the amount of a true cluster sequence pattern and noise, and λ_{true} controls the intensity of causality of the pattern in the time line. These parameters were varied in the experiment, where $\lambda_{true} = 0.05$ was used as a default setting.

3.2   Extraction Accuracy

Figure 4 shows the precision, recall, and F-measure of clusters for an extracted pattern $P_{A \rightarrow B}$ when varying the number of true pairs T. For example, CSM(A)

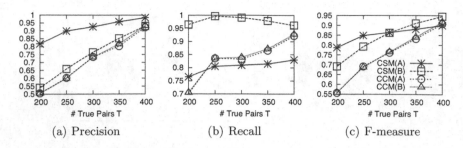

(a) Precision   (b) Recall   (c) F-measure

Fig. 4. Extraction accuracy of the embedded cluster pair when varying the number of true pairs ($\lambda_{true} = 0.05$)

indicates an evaluation score for cluster **A** extracted by CSM. An extracted pattern with the highest $\mathcal{L}$ score was evaluated. As for the parameters in CSM, $\tau = 100.0$ and $\sigma = 1.0$ in functions $\mathcal{F}$ and $\mathcal{G}$ were used. Then an average of 30 trials with varying random numbers is plotted in the graph. Here, since CCM extracts a pair of clusters without distinguishing a prior or a posterior cluster $P(\mathbf{A}, \mathbf{B})$, we employed F-measure($P_{\mathbf{A} \rightarrow \mathbf{B}}^{\text{CCM}}$) $= \max\{\text{F-measure}(P(\mathbf{A}, \mathbf{B}), \text{F-measure}(P(\mathbf{B}, \mathbf{A})\}$ to compare with CSM.

Note that since CSM uses an assumption on a pdf that is also used in the generation process of the synthetic data, it is obvious that CSM outperforms CCM, however we refer CCM as a baseline. In addition, CSM can infer a prior and a posterior clusters and time intervals between them.

From the figures, CSM obtained high precision in cluster **A** (greater than 0.8) and high recall in cluster **B** (greater than 0.95) for all true pairs. Especially, CSM maintains a high F-measure with a lower number of true pairs, i.e., with significant amounts of noise. When $T = 400$, CCM is competitive with CSM; however, the standard deviation of the scores in 30 trials was lower than that of CSM (0.015 (CSM) and 0.14 (CCM)), which means that CSM is more stable than CCM.

We also defined $E_\lambda = |\hat{\lambda}_{\mathbf{AB}} - \lambda_{true}|$ as error for the estimation of the parameter of time interval distribution of a pattern. Even when varying the number of true pairs, CSM can stably estimate λ_{true} as shown in Table 1. However, the estimated $\hat{\lambda}_{\mathbf{AB}}$ were always lower than λ_{true}, due to existence of the false pairs which have longer intervals than the true pairs.

Table 1. Average of estimated parameter of the time interval distribution $\hat{\lambda}_{\mathbf{AB}}$ and its error E_λ ($\lambda_{true} = 0.05$); standard deviation is indicated in the bracket

# true pairs T	$\hat{\lambda}_{\mathbf{AB}}$	E_λ
200	0.0324 (0.0133)	0.0185 (0.0120)
300	0.0282 (0.0120)	0.0222 (0.0109)
400	0.0312 (0.0186)	0.0221 (0.0143)

Table 2. Average of F-measure for clusters A and B, the estimated time interval parameter $\hat{\lambda}_{AB}$, and its estimation error E_λ when varying λ_{true} ($T = 400$); standard deviation is indicated in the bracket

	F-measure for CSM			
λ_{true}	Cluster A	Cluster B	$\hat{\lambda}_{AB}^{CSM}$	$\hat{\lambda}_{AB}^{CCM}$
0.01	0.897 (0.015)	0.937 (0.018)	0.0048 (0.0032)	0.0017 (0.0007)
0.05	0.900 (0.015)	0.945 (0.016)	0.0312 (0.0186)	0.0097 (0.0040)
0.10	0.892 (0.014)	0.944 (0.013)	0.0526 (0.0369)	0.0172 (0.0074)

Next, when we varied λ_{true}, i.e., with different slope degrees, CSM can obtain the clusters with a high F-measure robustly (approximately 0.9 and 0.94) as shown in Table 2. Regading estimation of λ_{AB}, CSM can estimate more accurate than CCM, where we apply CCM and then estimate λ_{AB} by the same Beyesian inference as the CSM. The result shows that two-step of the time interval estimation cannot provide accurate estimation at all. Nevertheless, $\hat{\lambda}_{AB}^{CSM}$ is still far from λ_{true}, which can be improved if we fine tune the parameters as described in the next section.

3.3 Parameter Study

Here, we examine the effects of control parameters τ and σ in the evaluation functions $\mathcal{F}$ and $\mathcal{G}$. Figure 5 shows clear stable regions for both parameters because the F-measure suddenly drops after a certain value. In addition, note that values lower than those illustrated in the graph are impractical regions because $\mathcal{L}$ scores are nearly equal; thus, it is difficult to find a distinguishing pattern. The results shown in the graphs indicate that these parameters are not sensitive to the results and are easy to adjust.

Moreover, E_λ has a trend wherein low error can be obtained when decreasing the parameter in $\mathcal{F}$ or $\mathcal{G}$. The parameter study revealed that high F-measure in clusters and a low E_λ in time interval estimation is possible to obtain by adjusting the main parameters of CSM.

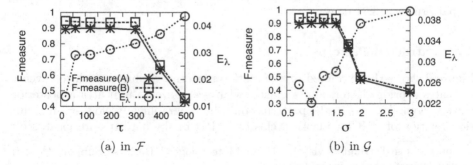

(a) in $\mathcal{F}$ (b) in $\mathcal{G}$

Fig. 5. Effects of parameters in $\mathcal{F}$ and $\mathcal{G}$ ($T = 400$, $\lambda_{true} = 0.05$)

4   Application to Causal Inference of Earthquake Occurrences

We applied the proposed CSM to earthquake event data to identify seismic causality between different areas afflicted by the 2011 Tohoku Earthquake. The purpose of this study is not earthquake prediction but to advance the understanding of the mechanical interactions of seismic induced activity.

4.1   Earthquake Event Data and Settings

We applied CSM to the hypocenter catalog data recorded from Jan. 1st, 2011 to Dec. 31st, 2012 released by the Japan Meteorological Agency[1]. Each event has an origin time (JST), hypocenter (latitude, longitude, and depth), and magnitude. We omitted events distant from Japan, i.e., only events between 23 °N and 50 °N and between 129 °E and 156 °E were used. The events within this region with a magnitude greater than four were used for analysis; 5954 seismic events were recorded in that period, and these events are plotted in Fig. 6(a).

Regarding AHC in CSM, we used latitude and longitude as attributes for merging clusters, and Euclidean distance was used for the distance metric. We did not include depth because there were only minor differences in depth in the same areas. We employed Ward's method empirically by checking the dendrogram, e.g., a chaining effect and monotonicity.

In seismology, Bayesian learning has been widely utilized for long-term prediction [8] or aftershock activity forecasting [10]; however, it has not been applied to causal inference. In these methods, earthquake occurrence is modeled by a Poisson process; thus, an exponential distribution is also assumed.

We used the following parameters in CSM. The thresholds were set to $\mathcal{L}_{\min} = 0.8$ and $Supp_{\min} = 8$, the balancing parameter in $\mathcal{L}$ was set to $\gamma = 0.5$, which is equal weights in $\mathcal{F}$ and $\mathcal{G}$, and the sharpness parameters in $\mathcal{F}$ and $\mathcal{G}$ were set to $\tau = 1.0$, $\sigma = 0.03$ in order to obtain wide ranging scores. Although we have used several parameters, most of the parameters are not difficult to adjust if a user knows the meaning and effects of each parameter. Note that we performed approximately 10 trials to adjust these parameters. With these parameters, 37 patterns were extracted. The average of estimated time intervals λ_{AB} among the 37 patterns was 3.71 days.

4.2   Extracted Patterns

Figure 6 shows the extracted representative patterns. A cluster with red events indicates a prior cluster, and a cluster with green events indicates a posterior cluster. Some clusters in the patterns could be geographically close or an inclusion relation of AHC or identical clusters. Most of the patterns are located off

---

[1] The data is distributed via the Japan Meteorological Business Support Center: http://www.jmbsc.or.jp (in Japanese).

 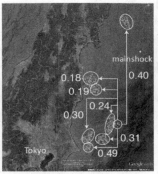 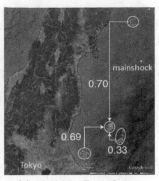

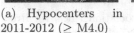

(a) Hypocenters in 2011-2012 ($\geq$ M4.0)   (b) Highly affecting area   (c) Highly affected area

Fig. 6. Hypocenters and representative patterns extracted by CSM. Average of time intervals (days) of each pattern are indicated together.

the coast of the Tohoku area, even though earthquakes also occurred in the mid-island, as is shown in Fig. 6(a). Moreover, both neighboring prior and posterior clusters were obtained, which may indicate interesting causality.

The shared prior clusters shown in Fig. 6(b) indicate *affecting areas*, while the shared posterior clusters shown Fig. 6(c) indicate *affected areas*, which may indicate high causality areas. Here, CCM[6] and KeyGraph[9] cannot identify the direction of such effects. Note that this is the greatest advantage of the proposed method.

A possible cause of these patterns could be interaction between asperities[11]. An asperity is an area in a normally stable subduction zone that suddenly slips and produces an earthquake when it reaches its limit. In seismology, a hypothesis about interactions between asperities exists [2]. As far as we have surveyed, no method exists for determining the actual interactions among asperities, whereas the proposed method is promising in that it can suggest such hidden interactions.

5   Conclusion

We have proposed a pattern mining algorithm called cluster sequence mining (CSM) that can extract pairs of clusters that satisfy time and space proximity for a multidimensional event sequence. In calculating time proximity, a probability distribution function of time intervals of events in different clusters is inferred by utilizing Bayesian inference.

In our experiment, we validated the performance of the proposed CSM using synthetic data. The results revealed that CSM was able to extract a prior and a posterior clusters with high F-measure and low estimation error of the time interval distribution, even under uncertainty. We then applied CSM to an earthquake event sequence to infer causality between different areas. The results demonstrated that CSM extracts highly affecting/affected earthquake occurrence patterns. Note that the current calculation of time intervals in CSM uses only a single correspondence between events; thus, in future DP (dynamic programming) matching could be employed to allow multiple correspondences properly.

Acknowledgments. This study was supported by JSPS KAKENHI Grant Number 24650068.

References

1. Agrawal, R., Srikant, R.: Fast algorithms for mining association rules. In: Proc. 20th Int. Conf. on Very Large Databases (ICVLD), pp. 487–499 (1994)
2. Ariyoshi, K., Hori, T., Ampuero, J.P., Kaneda, Y., Matsuzawa, T., Hino, R., Hasegawa, A.: Influence of interaction between small asperities on various types of slow earthquakes in a 3-d simulation for a subduction plate boundary. Gondwana Research 16, 534–544 (2009)
3. Everitt, B.S., Landau, S., Leese, M., Stahl, D.: Cluster Analysis, 5th edn. Wiley (2011)
4. Fukui, K., Inaba, D., Numao, M.: Discovering seismic interactions after the 2011 tohoku earthquake by co-occurring cluster mining. Transactions of Japanese Society for Artificial Intelligence 29(6), 493–502 (2014)
5. Han, J., Pei, J., Yin, Y.: Mining frequent patterns without candidate generation. In: Proc. ACM Conf. on Management of Data (SIGMOD), pp. 1–12 (2000)
6. Inaba, D., Fukui, K., Sato, K., Mizusaki, J., Numao, M.: Co-occurring Cluster Mining for Damage Patterns Analysis of a Fuel Cell. In: Tan, P.-N., Chawla, S., Ho, C.K., Bailey, J. (eds.) PAKDD 2012, Part II. LNCS, vol. 7302, pp. 49–60. Springer, Heidelberg (2012)
7. Kulldorff, M.: Prospective time periodic geographical disease surveillance using a scan statistic. Journal of the Royal Statistical Society, Series A **164**, 61–72 (2001)
8. Nomura, S., Ogata, Y., Komaki, F., Toda, S.: Bayesian forecasting of recurrent earthquakes and predictive performance for a small sample size. Journal of Geophysical Research **116**, B04315 (2011)
9. Ohsawa, Y.: Keygraph as risk explorer in earthquake-sequence. Journal of Contingencies and Crisis Management **10**(3), 119–128 (2002)
10. Omi, T., Ogata, Y., Hirata, Y., Aihara, K.: Estimating the etas model from an early aftershock sequence. Geophysical Research Letters **41**, 850–857 (2014)
11. Ruff, L.J.: Asperity distributions and large earthquake occurrence in subduction zones. Tectonophysics **211**, 61–83 (1992)
12. Xu, R., Wunsch-II, D.C.: CLUSTERING. IEEE Press Series on Computational Intelligence (2008)

Achieving Accuracy Guarantee for Answering Batch Queries with Differential Privacy

Dong Huang$^{(\boxtimes)}$, Shuguo Han, and Xiaoli Li

Institute for Infocomm Research, Singapore, Singapore
{huangd,shan,xlli}@i2r.a-star.edu.sg

Abstract. In this paper, we develop a novel strategy for the privacy budget allocation on answering a batch of queries for statistical databases under differential privacy framework. Under such a strategy, the noisy results are more meaningful and achieve better utility of the dataset. In particular, we first formulate the privacy allocation as an optimization problem. Then derive explicit approximation of the relationships among privacy budget, dataset size and confidence interval. Based on the derived formulas, one can automatically determine optimal privacy budget allocation for batch queries with the given accuracy requirements. Extensive experiments across a synthetic dataset and a real dataset are conducted to demonstrate the effectiveness of the proposed approach.

1 Introduction

Differential privacy (DP), is a promising strategy for providing privacy for data publishing and data queries [6,8]. A simple but feasible method to achieve differential privacy is to insert noises to the query outputs [4]. Currently, most of the related work focus on privacy protection but don't further analyze how useful of the noisy results. If these noise results show what level of accuracy can be achieved, they will help data analysts further investigate and improve the effectiveness of them. Moreover, the privacy allocation is very important to how useful of the noisy results. To answer multiple queries, a simple way is to allocate the privacy budget to these queries *equally* [7]. However, such a strategy may cause some noisy results to be unmeaningful due to large noise magnitude relative to the original results.

We use the following intuitive example to illustrate the problem. The *Adult* dataset, extracted from UCI machine learning repository [2], has $32,561$ individuals. Suppose we want to use two queries $q = [q_1, q_2]$ on the *Adult* dataset to infer the real values of the whole population, where q_1 is the proportion of individuals with Sex="Male", Race="Black" and Income=">50K" over the number of individuals with Sex="Male" and Income=">50K", which is equal to 0.0446; q_2 is the proportion of individuals with Sex="Male", Race="White" and Income=">50K" over the number of individuals with Sex="Male" and Income=">50K", which is equal to 0.9140.

We observe that the returned real value from q_1 (i.e. 0.0446) is much smaller than that from q_2 (i.e. 0.9140). Under the aforementioned "uniformly split" strategy, equal budgets will be assigned to q_1 and q_2, e.g. each gets a budget 0.5.

© Springer International Publishing Switzerland 2015
T. Cao et al. (Eds.): PAKDD 2015, Part II, LNAI 9078, pp. 305–316, 2015.
DOI: 10.1007/978-3-319-18032-8_24

Table 1. Statistics of the *Adult* dataset

		Income	
Sex	Race	$> 50K$	$<= 50K$
Male	Amer-Indian-Eskimo	24	168
Male	Black	297	1272
Female	Other	6	103
Male	White	6089	13085
...	...	...	...

Correspondingly, the noise magnitude for q_1 will be larger compared with its smaller value, making its noisy result less useful, although the q_2's noisy result could be relatively close to its true value. Ideally, we should assign a higher budget for q_1 and lower budget for q_2, such that the lower noise magnitude will be added to q_1, making both of them are useful. As such, how to reasonably allocate the limited privacy budget to multiple queries is a crucial problem to ensure the overall accuracy guarantee for all the queries.

For the impact of noise on the noisy results, while most existing work provide the analysis of the upper error bound [3,17] in implementing differential privacy, this is not sufficient for the utility evaluation of the noisy results. From data mining perspective, it will be helpful for data analysts to understand the utility of the noisy results if they can visualize the level of accuracy achieved after adding noise.

The objective of this research is thus to design a framework to allocate privacy budget among the queries with differential privacy and further provides analysis of how useful of the noisy results. We have further investigated the framework proposed by A. Smith [14]. We consider the problem of multiple queries with ϵ-differential privacy under this framework, where the queries studied in the paper are the ratios of multiple subsets to the given dataset. The contributions of this paper can be summarized as follows:

- We formulate the optimization problem with accuracy guarantee in terms of confidence interval (CI). This enables data analysts to better understand what are the accuracy guarantee of the noisy statistical results.
- We formulate the noisy results with normal-Laplace distribution. This property enables us to derive its cumulative distribution function (i.e. cdf).
- We further approximate the minimum privacy budget required for given level of accuracy with explicit formulas.

The remaining parts of the paper are organized as follows: First, section 2 provides a brief discussion about the related work. Then, we describe the background information in Section 3. Next, section 4 presents the differential privacy framework and discusses the normal-Laplace distribution. Section 5 introduces our novel approximation formulas for accuracy guarantee. Finally, we evaluate the proposed approach and conclude the paper in section 6 and section 7 respectively.

2   The Related Work

Currently, most of existing work studied the noise reduction in terms of sensitivity. For example, a data publishing technique, *Privelet*, based on wavelet transforms, was proposed in [16]. *Privelet* not only ensures ϵ- differential privacy, but also guarantees that the variance of the noisy results is polylogarithmic in terms of m where m denotes the number of queries. There are some database/ data mining applications, where the given dataset is a correlated time-series data or the dataset is distributively collected. For such types of the applications, a differential privacy framework, PASTE, was proposed in [12]. To provide differential privacy for time-series data, PASTE developed a Fourier perturbation algorithm. For the case of absence of a trusted central server, PASTE used a distributed Laplace perturbation algorithm to guarantee differential privacy. In order to publish cuboids for data cubes with small noise, an efficient method was proposed in [5]. The proposed method ensures that the maximal noise in all published cuboids will be within a factor $(\ln |\mathcal{L}| + 1)^2$ of the optimal, where $|\mathcal{L}|$ is the number of cuboids to be published. To handle the problem of differential private data release for a class of counting queries, a new computationally efficient method based on learning thresholds was proposed in [9].

We notice that privacy budget has important impact on the noise magnitude. A few related work on this topic have been investigated. For example, K. Nissim *et al.* have proposed a framework, subsample and aggregate, to reduce the noise magnitude [11,14]. In such a framework, the dataset is first divided into k groups. Then it estimates the parameters based on the k results. Compared with traditional Laplace mechanism in [8], the framework reduces the error dramatically, where the errors decrease with the increasing number of data. The GUPT was proposed in [10] to allocate the privacy budget by ensuring the same noise magnitude for each query.

3   Background

Definition 1 ([15]). *Two databases* x, $x' \subseteq D^n$ *are neighbouring databases if they differ on exactly one record, i.e.,*

$$x = \{x_1, \ldots, x_i, \ldots, x_n\} \quad \text{and} \quad x' = \{x_1, \ldots, x_i', \ldots, x_n\}$$

From the definition, note that two neighbouring databases differ only one record while they have the same cardinality. Before discussing the Laplace mechanism, we first give the definition of ϵ-differential privacy proposed by Dwork [8].

Definition 2 ([8]). *With* $\epsilon > 0$, *a randomized algorithm* $\mathcal{K} : D^n \rightarrow \mathbb{R}^l$ *is said to satisfy* ϵ-*differential privacy, if for any two neighbouring databases* x, $x' \subseteq D^n$ *and for any subset of outputs* $S \subseteq Range(\mathcal{K})$, *the following condition holds:*

$$\frac{Pr(\mathcal{K}(x) \in S)}{Pr(\mathcal{K}(x') \in S)} \leq \exp(\epsilon) \tag{1}$$

where the probability is taken over the randomness of $\mathcal{K}$.

A simple way to achieve ϵ-differential privacy is to insert noise to the true value. For instance, the Laplace mechanism (LM) generates the noise by using a random variable of Laplace distribution with mean equal to 0 and scale parameter equal to S_T/ϵ, where S_T is the sensitivity of $T(X)$.

Definition 3 ([11]). *The sensitivity of a function $T : D^n \rightarrow \mathbb{R}^d$ is*

$$S_T = \max_{x,x':d(x,x')=1} \|T(x) - T(x')\|. \tag{2}$$

Lemma 1 (Laplace Mechanism [8]). *Given $\epsilon > 0$, a statistic $T(X) \in \mathbb{R}^l$ and its sensitivity S_T, the noisy result*

$$T^*(X) = T(X) + Lap(S_T/\epsilon)) \tag{3}$$

satisfies ϵ-differential privacy.

For the convenience of discussion, we define the Laplace mechanism with accuracy guarantee as follows.

Definition 4 (Accuracy Guarantee). *Let $\theta \in \mathbb{R}^l$ be the true value of the statistic $T(X)$. Suppose $\Phi(W)$ is the cumulative distribution function of a random variable of Laplace distribution with mean equal to 0 and scale parameter equal to S_T/ϵ. If there exist $d = (d_1, \ldots, d_l) > 0$ and $\alpha = (\alpha_1, \ldots, \alpha_l) \in (0,1)$ such that*

$$\Phi(d_i) \geq 1 - \alpha_i/2, \forall i = 1, 2, \ldots, l,$$

then the noisy results obtained from Eq. (3) is at least $100(1 - \alpha)\%$ to be $\pm d$ of the true value θ.

Note that the accuracy achieved by LM may be smaller than the required accuracy since there exists estimation error between the statistic $T(X)$ and the true value θ but the error is not considered for the accuracy estimation (i.e., only the variance of the Laplace noise is considered here). We need to find a more accurate formula to describe the relationships among the accuracy achieved and other related parameters in order to ensure the accuracy achieved is close to the required accuracy. Specifically, we focus on the maximum likelihood estimator (MLE), which is obtained by maximizing the likelihood function.

Theorem 1 (Asymptotic Distribution). *Let $x_1, x_2, \ldots, x_n$ be independently identically distributed with density $f(x|\theta)$, $\theta \in \Theta$ and let θ_0 denote the true value of θ. Suppose the MLE estimator of θ_0 is $T(x)$. Then the probability distribution of*

$$\sqrt{nI(\theta_0)}(T(x) - \theta_0)$$

tends to be a standard normal distribution, i.e.,

$$\sqrt{nI(\theta_0)}(T(x)) - \theta_0) \xrightarrow{D} N(0,1)^l, \tag{4}$$

where $I(\theta_0)$ is Fisher information.

4   Differential Privacy Framework

In this section, we first formulate the optimization problem and then show the relationship between the problem and normal-Laplace distribution.

4.1   Problem Definition

The scenarios we consider in the paper is illustrated in Figure 1. Users first send queries with accuracy requirements to the database server, then the database server passes the requirements to the computing server to optimize the parameters such as privacy budget and confidence interval in order to minimize the expectation of the errors in Eq. (5). Finally, the database server executes queries based on the optimized parameters and returns users the noisy results with accuracy description.

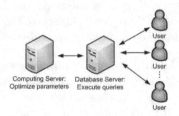

Fig. 1. The query execution model

Given a dataset $x = (x_1, \ldots, x_n) \in D^n$, the value of x_i in D is a real number, where D is a value space of x_i. Here, we use the capital letter $X = (X_1, \ldots, X_n)$ to denote a random vector variable and lower case $x = (x_1, \ldots, x_n) \in D^n$ to denote a specific value in D^n. Suppose $X = (X_1, \ldots, X_n)$ is drawn according to the distribution $f(x|\theta)$, where $\theta \in \mathbb{R}^l$ is unknown parameter vector. Let

$$T(X) = [T_1(X), \ldots, T_l(X)]$$

be the estimator of θ. In this paper, we study the problem of how to ensure that the parameter estimations under ϵ-differential privacy satisfy the given level of precision. In other words, we wish to estimate θ using an estimator based on the given dataset $x = (x_1, \ldots, x_n) \in D^n$ with $\boldsymbol{\alpha} = [\alpha_1, \ldots, \alpha_l]$ confidence interval to be $\pm\mathbf{d}$, where $\mathbf{d} = [d_1, \ldots, d_l]$, of the true value θ_0. Here, we want to minimize the expected squared deviation from the real parameter θ. Specifically, we wish to minimize the following objective function:

$$\min \quad J_\theta(T^*(X)) = \mathrm{E}\{\|T^*(X) - \theta\|^2\} \tag{5}$$

$$\text{s.t} \quad \Pr(|T_i^*(X) - \theta_i| \leq d_i) \geq 1 - \alpha_i, \forall i \in \{1, 2, \ldots, l\} \tag{6}$$

$$\sum \epsilon_i = \epsilon_{total}, \forall i \in \{1, 2, \ldots, l\} \tag{7}$$

$$0 < k \leq n \tag{8}$$

where ϵ_{total} is the total privacy budget and k is the number of blocks. In order to solve this problem, we need to derive explicit formula to characterize the relationships among the privacy budget, the number of blocks, the statistics of the data and accuracy guarantee. Without loss of generality, we assume the solution of the problem in Eqs. (5)-(8) always exists. Specifically, the multiple queries we consider in this paper are the ratios of multiple subsets to the given dataset.

Specifically, suppose users send queries $Q_1(\mathbf{q}, \boldsymbol{\chi})$, where $\mathbf{q} = [q_1, \ldots, q_l]$ denotes the query vector while $\boldsymbol{\chi}$ represents the corresponding accuracy constraints. The database server then passes the queries with privacy budget, $Q_2(\mathbf{q}, \boldsymbol{\chi}, \boldsymbol{\vartheta}, \epsilon_{total})$, where $\boldsymbol{\vartheta}$ denotes the required statistics of related dataset for parameter optimization, to the computing server. It optimizes the privacy budget among the queries by solving the problem in Eqs. (5)-(8) and returns the execution queries with optimized parameters, $Q_3(\mathbf{q}, \boldsymbol{\chi}, \boldsymbol{\epsilon}, \varsigma)$, where $\boldsymbol{\epsilon} = [\epsilon_1, \ldots, \epsilon_l]$ denotes the privacy budget allocation for the query vector $\mathbf{q}$ and ς is the number of blocks, to the execution server. Finally, the execution server executes the queries according to given optimized parameters and returns the noisy results, $R(\mathbf{q}, \boldsymbol{\chi}')$, where $\boldsymbol{\chi}'$ is the accuracy obtained, to users. In order to illustrate the model clearly, we use an example to show how is the process of our proposed model.

Example 1. Consider the Adult dataset. Suppose users are interested in two queries $\mathbf{q} = [q_1, q_2]$ where q_1 is the ratio of individuals with race="black", sex="female" and income="> 50K" to those with sex="female" and income="> 50K" and q_2 is the ratio of individuals with race="white", sex="female" and income="> 50K to those with sex="female" and income="> 50K. The corresponding accuracy requirement for the two queries is $\boldsymbol{\chi} = [\chi_1, \chi_2]$, where χ_1 is that the noisy result should be $\pm d_1$ with $d_1 = 0.05$ of the true value with $\alpha_1 = 95\%$ and χ_2 denotes that the noisy result should be $\pm d_2$ with $d_2 = 0.1$ of the true value with $\alpha_2 = 90\%$. Suppose $\epsilon_{total} = 1$. The database server will pass $Q_2(\mathbf{q}, \boldsymbol{\chi}, \boldsymbol{\vartheta}, 1)$ to the computing server after it receives the queries $Q_1(\mathbf{q}, \boldsymbol{\chi})$. Here $\boldsymbol{\vartheta}$ may include the mean and variance of an estimator, sample size and the sensitivity of a query. The computing server then optimizes the privacy budget between the two queries by solving the problem in Eqs. (5)-(8). Finally, the database server returns users the query results, $R(\mathbf{q}, \boldsymbol{\chi}')$.

4.2   Differential Privacy Framework

In this paper, we apply the differential privacy framework proposed by [14], called "sample and aggregate" [11]. It is an effective method to decrease the noise magnitude, where it randomly divides the data set into k blocks with size roughly equal to n/k. Then the estimation is applied in each block and finally the estimates are aggregated by using a differentially private function. Especially, the MLE estimator developed by Algorithm 1 [15] can asymptotically approach the true value θ_0.

Lemma 2 ([15]). *Algorithm 1 satisfies ϵ-differential privacy.*

Algorithm 1. An ϵ-Differential Privacy Algorithm

Input: $x = (x_1, \ldots, x_n) \in D^n$, $\epsilon > 0$
Output: $T_i^*(x)$, $i = 1, \ldots, m$
1: Let Γ be the range of $T_i(x)$ or diameter of the parameter space
2: Suppose $T_1(x), \ldots, T_m$ are the sufficient statistics for a set of parameters $\theta_1, \ldots, \theta_m$.
3: Calculate T_i, $i = 1, \ldots, m$ based on the input data x
4: **for** $i = 1$ to m **do**
5:    Draw a random observation R_i from a laplace distribution with mean 0 and standard deviation $\sqrt{2}\Gamma/(n\epsilon)$
6: **end for**
7: Output $T_i^*(x) = T_i(x) + R_i$

4.3   The Normal-Laplace Distribution

Suppose an MLE estimator is used to estimate the ratios of multiple subsets to a given dataset in Algorithm 1. Then the output T^* is the summation of two independent random variables Z and Y, where Z is drawn from the normal distribution with $N(E_\theta(T(X)), \mathrm{Var}(T(X)))$ and Y is drawn from the Laplace distribution with $\mathrm{Lap}(\lambda)$. The distribution of T^* is called normal-Laplace distribution [13]. In general, let $W = Z + Y$, where Z and Y are independent random variables with $Z \sim N(\mu, \sigma^2)$ and Y with following an asymmetric Laplace distribution with pdf

$$f_Y(y) = \begin{cases} \frac{\eta}{2} e^{\eta y}, & \text{for } y \leq 0 \\ \frac{\eta}{2} e^{-\eta y}, & \text{for } y > 0 \end{cases}$$

The distribution of W is called normal-Laplace distribution. We use

$$W \sim \mathrm{NL}(\mu, \sigma^2, \eta, \eta)$$

to denote such a distribution.

From the properties of characteristic function [1], we can derive the mean and variance of W as

$$E\{W\} = \mu, \quad \text{and} \quad \mathrm{Var}(W) = \sigma^2 + 2/\eta^2.$$

A closed-form expression for the cumulative distribution function of the normal-Laplace distribution can be expressed as [13]

$$F(W) = \Phi\left(\frac{W - \mu}{\sigma}\right) - \phi\left(\frac{W - \mu}{\sigma}\right)\frac{R(\varphi_1) - R(\varphi_2)}{2} \tag{9}$$

with $\varphi_1 = \eta\sigma - (W - \mu)/\sigma$ and $\varphi_2 = \eta\sigma + (W - \mu)/\sigma$, where Φ and ϕ are the cdf and the pdf of a standard normal random variable, respectively. R is *Mill's ratio*.

5   Accuracy Guarantee

Suppose μ and σ are the mean and standard deviation of variable X. Let $T_1(X) = \frac{1}{n}\sum_{i=1}^{n} X_i$ be the estimator of μ. The noisy result is derived from Algorithm 1. We approximate the minimum privacy budget required for given level of precision requirement according to Eq. (9). In general, constructing an exact confidence interval requires complete information about the distribution of the variable. However, this information is not available in practice. Note that it is not easy to derive $W_{\alpha/2}$ such that $F(W_{\alpha/2}) = 1 - \alpha/2$ in Eq. (9). A feasible way is to construct confidence interval based on the large sample theory. Suppose $\sqrt{nI(\hat{\theta})}(\hat{\theta} - \theta_0)$ is approximately the standard normal distribution, then we get

$$\Pr(-y_{\alpha/2} \le \sqrt{nI(\hat{\theta})}(\hat{\theta} - \theta_0) \le y_{\alpha/2}) \approx 1 - \alpha.$$

That is, we can get an approximate $100(1 - \alpha)\%$ confidence interval such that

$$\hat{\theta} - y_{\alpha/2}\frac{1}{\sqrt{nI(\hat{\theta})}} \le \theta_0 \le \hat{\theta} + y_{\alpha/2}\frac{1}{\sqrt{nI(\hat{\theta})}}.$$

Infinite Case. Consider the population is infinite. Let X be a variable. Assume X has a normal, bell-shaped frequency distribution. We wish to estimate the mean of the population subject to the following constraint

$$\Pr(\hat{\theta} - y_{\alpha/2}s_{\hat{\theta}} < \theta < \hat{\theta} + y_{\alpha/2}s_{\hat{\theta}}) = 1 - \alpha,$$

where $s_{\hat{\theta}} = \frac{1}{\sqrt{nI(\hat{\theta})}}$ is the estimated standard deviation. We can determine the sample size by

$$n = (\frac{y_{\alpha/2}s}{d})^2,$$

where d is the desired absolute error and s is the standard deviation. Suppose $X \sim \mathcal{NL}(\mu, \frac{\sigma^2}{n}, \frac{k\epsilon}{\Gamma}, \frac{k\epsilon}{\Gamma})$. Then we get $Y \sim \mathcal{NL}(0, 1, \frac{k\epsilon\sigma}{\Gamma\sqrt{n}}, \frac{k\epsilon\sigma}{\Gamma\sqrt{n}})$. Here, we can characterize the accuracy guarantee as

$$y_{\alpha/2} \cdot \sqrt{\frac{\sigma^2}{n} + \frac{2\Gamma^2}{k^2\epsilon^2}} \le d.$$

Given accuracy requirement and dataset size, minimum the privacy budget ϵ required is expressed as

$$\epsilon = \phi_1(n, \sigma^2, \Gamma, d, y_{\alpha/2}) = \frac{\Gamma}{k} \cdot \sqrt{\frac{2}{(d/y_{\alpha/2})^2 - \sigma^2/n}}. \tag{10}$$

Finite Case. When the population is finite, the accuracy guarantee is different. Suppose the population is N. We need to derive explicit formula to express the relationships among those parameters discussed above for given level of precision

$1 - \alpha$. Let $x_1, \ldots, x_N$ be the population and $X_1, \ldots, X_n$ be the variables selected for estimation. Let $p_i \in \{0, 1\}$ be the indicator variable. $p_i = 1$ if x_i belongs to a given sample. Then we can see that $\sum_{i=1}^{n} X_i = \sum_{i=1}^{N} p_i x_i$. Therefore,

$$E\{1/n \sum_{i=1}^{n} X_i\} = 1/n \cdot n/N \sum_{i=1}^{N} x_i = m.$$

Let $X \sim \mathcal{NL}(\mu, \frac{\sigma^2}{n} \cdot \frac{N-n}{N-1}, \frac{k\epsilon}{\Gamma}, \frac{k\epsilon}{\Gamma})$. If $Y = (X - \mu)/(\frac{\sigma}{\sqrt{n}} \cdot \sqrt{\frac{N-n}{N-1}})$, then we have $Y \sim \mathcal{NL}(0, 1, \frac{k\epsilon}{\Gamma} \cdot \frac{\sigma}{\sqrt{n}} \cdot \sqrt{\frac{N-n}{N-1}}, \frac{k\epsilon}{\Gamma} \cdot \frac{\sigma}{\sqrt{n}} \cdot \sqrt{\frac{N-n}{N-1}})$. The accuracy guarantee can be expressed as

$$y_{\alpha/2} \cdot \sqrt{\frac{\sigma^2}{n} \cdot \frac{N-n}{N-1} + \frac{2\Gamma^2}{k^2 \epsilon^2}} \leq d.$$

Here, we also derive similar function such that

$$\epsilon = \phi_2(n, \sigma^2, \Gamma, d, y_{\alpha/2}) = \frac{\Gamma}{k} \cdot \sqrt{\frac{2}{(d/y_{\alpha/2})^2 - \sigma^2 \cdot (N - n)/(n \cdot (N - 1))}} \tag{11}$$

for the minimum privacy budget required.

We have derived explicit formulas to describe given accuracy guarantee in terms of the privacy budget, number of blocks and dataset size. These formulas characterize how the parameters affect mutually. Thus, we can solve the optimization problem in Eqs. (5)-(8)) based on Lagrangian method. In the following section, we conduct simulations to demonstrate the effectiveness and feasibility of them.

6   Empirical Evaluations

In this section, we evaluate the performance of the proposed algorithm (denoted as NL) by comparing it with two state-of-the-art mechanisms, including LM and GUPT, which are proposed in [8] and [10] respectively. Particularly, we first evaluate the effectiveness and feasibility of the proposed algorithm based on a synthetic data and a real dataset. Then we further study the privacy budget allocation for the optimization problem in Eqs. (5)-(8).

6.1   Approximation Formulas Evaluation

We evaluate the relationships among accuracy, dataset size and privacy budget for the infinite case through synthetic data. We first generate the synthetic data with binomial distribution and dataset size $n_1 = 200$ and $n_2 = 100$ by Monte Carlo method. Two cases are considered, where $p = 0.4$. We wish to estimate the mean here. Then we test the cumulative accuracy of noisy results derived from Algorithm 1 falling into the given interval with $d = 0.05$, where the number

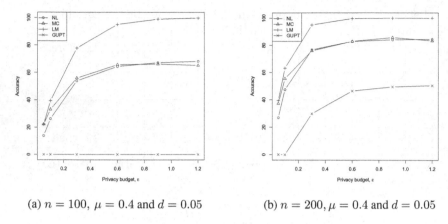

(a) $n = 100$, $\mu = 0.4$ and $d = 0.05$ (b) $n = 200$, $\mu = 0.4$ and $d = 0.05$

Fig. 2. Comparison of privacy budget, ϵ, versus accuracy for the infinite case

of generations is set to 1000. The results are shown in Figure 2, where NL and MC denote the theoretical results obtained by the proposed algorithm and the true results obtained by Monte Carlo method, LM and GUPT are the results obtained by Laplace mechanism and the GUPT algorithm, respectively.

From the above figure, the accuracy increases with the increase of ϵ. The results obtained by NL are the most close to those of MC. Especially in the case of high accuracy. In contrast, the theoretical accuracy obtained by LM is much higher than the true accuracy while the theoretical accuracy obtained by GUPT is much lower. This demonstrates that NL is able to achieve higher accurate estimation than the two state-of-the-art techniques.

Next, we employ a real dataset, (i.e., *Adult* dataset from UCI dataset), to further prove the correctness of the approximation for finite case. Consider the estimation of the proportion of individuals with race="black" and sex="females" with income="> 50K" in terms of race="black" and income="> 50K". The total number of individuals with race="black" and income=">50K" is $N = 387$. We first randomly select $n_1 = 100$ and $n_2 = 200$ samples from the 387 individuals. Then we calculate the theoretical accuracy by using NL, LM and GUPT for different privacy budget.

The results are shown in Figure 3. The results are very similar to the infinite cases, as shown in Figure 2. This means that the proposed NL algorithm accurately characterizes the relationships among the three parameters.

6.2 Privacy Budget Allocation for Multiple Queries

We further investigate the expected squared estimation errors of multiple queries from the optimization problem in Eqs. (5)-(8). We consider two queries $q = [q_1, q_2]$ with $\alpha = (0.05, 0.1)$ and $d = (0.05, 0.1)$. Figure 4(a) shows the comparison of the cases with different dataset size. It can be observed that the expected squared errors obtained from different cases decrease with the increasing ϵ_{total}. Particularly, given a ϵ_{total}, the expected squared errors decrease with the increasing dataset size. Figure 4(b) shows the comparison of the corresponding privacy budget allocation under different datasets for the two queries. It can be seen that the

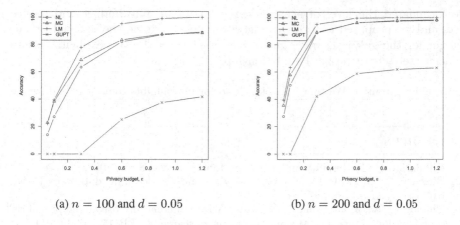

(a) $n = 100$ and $d = 0.05$ (b) $n = 200$ and $d = 0.05$

Fig. 3. Comparison of privacy budget, ϵ, versus accuracy for the finite case

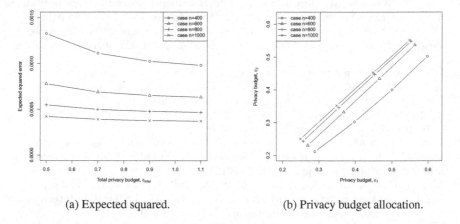

(a) Expected squared.   (b) Privacy budget allocation.

Fig. 4. Performance comparison of privacy budget allocation

privacy budget allocated to q_2 linearly increases with the privacy budget allocated to q_1. Moreover, given a total privacy budget ϵ_{total}, when the dataset size increases, the privacy allocated to it decreases while the privacy allocated to q_2 increases.

In summary, the above simulation results demonstrate that the proposed NL algorithm accurately describes the relationships among the parameters, namely the privacy budget, dataset size, accuracy and confidence interval, as well as how the privacy budget varies with the accuracy requirement.

7 Conclusion

In this paper, we have investigated the problem of how to allocate privacy budget among a batch of queries under the differential privacy framework. Particularly, we formulated the level of accuracy in terms of privacy budget and dataset size, and we proposed a novel NL algorithm to determine the optimal privacy budget for the given accuracy guarantee. We further derived explicit formulas to accurately characterize the relationships among three parameters.

Acknowledgments. We thank Xiaokui Xiao for his valuable comments and suggestions.

References

1. Billingsley, P.: Probability and measure. John Wiley & Sons (2008)
2. Blake, C.L., Merz, C.J.: UCI repository of machine learning databases (1998); Robustness of maximum boxes
3. Chaudhuri, K., Monteleoni, C., Sarwate, A.D.: Differentially private empirical risk minimization. Journal of Machine Learning Research: JMLR **12**, 1069 (2011)
4. Clifton, C., Tassa, T.: On syntactic anonymity and differential privacy. In: First Workshop on Privacy-Preserving Data Publication and Analysis at ICDE, pp. 8–12 (2013)
5. Ding, B., Winslett, M., Han, J., Li, Z.: Differentially private data cubes: optimizing noise sources and consistency. In: SIGMOD Conference, pp. 217–228 (2011)
6. Dwork, C.: Differential Privacy. In: Bugliesi, M., Preneel, B., Sassone, V., Wegener, I. (eds.) ICALP 2006. LNCS, vol. 4052, pp. 1–12. Springer, Heidelberg (2006)
7. Dwork, C.: A firm foundation for private data analysis. Communications of the ACM **54**(1), 86–95 (2011)
8. Dwork, C., McSherry, F., Nissim, K., Smith, A.: Calibrating Noise to Sensitivity in Private Data Analysis. In: Halevi, S., Rabin, T. (eds.) TCC 2006. LNCS, vol. 3876, pp. 265–284. Springer, Heidelberg (2006)
9. Hardt, M., Rothblum, G.N., Servedio, R.A.: Private data release via learning thresholds. In: Proceedings of the Twenty-Third Annual ACM-SIAM Symposium on Discrete Algorithms, pp. 168–187 (2012)
10. Mohan, P., Thakurta, A., Shi, E., Song, D., Culler, D.: Gupt: privacy preserving data analysis made easy. In: Proceedings of the ACM SIGMOD International Conference on Management of Data, pp. 349–360 (2012)
11. Nissim, K., Raskhodnikova, S., Smith, A.: Smooth sensitivity and sampling in private data analysis. In: Proceedings of the 39th Annual ACM Symposium on Theory of Computing, pp. 75–84 (2007)
12. Rastogi, V., Nath, S.: Differentially private aggregation of distributed time-series with transformation and encryption. In: Proceedings of the International Conference on Management of Data, pp. 735–746. ACM (2010)
13. Reed, W.J.: The normal-laplace distribution and its relatives. In: Advances in Distribution Theory, Order Statistics, and Inference, pp. 61–74. Springer (2006)
14. Smith, A.: Efficient, differentially private point estimators. arXiv preprint arXiv:0809.4794 (2008)
15. Vu, D., Slavkovic, A.: Differential privacy for clinical trial data: Preliminary evaluations. In: IEEE International Conference on Data Mining Workshops, pp. 138–143 (2009)
16. Xiao, X., Wang, G., Gehrke, J.: Differential privacy via wavelet transforms. IEEE Transactions on Knowledge and Data Engineering **23**(8), 1200–1214 (2011)
17. Zhang, J., Zhang, Z., Xiao, X., Yang, Y., Winslett, M.: Functional mechanism: regression analysis under differential privacy. Proceedings of the VLDB Endowment **5**(11), 1364–1375 (2012)

Mining Temporal and Spatial Data

Automated Classification of Passing in Football

Michael Horton[1]([✉]), Joachim Gudmundsson[1], Sanjay Chawla[1,2],
and Joël Estephan[1]

[1] School of Information Technologies, The University of Sydney, Sydney, Australia
mhor9676@uni.sydney.edu.au,
{joachim.gudmundsson,sanjay.chawla}@sydney.edu.au,
j.estephan4@gmail.com
[2] Qatar Computing Research Institute, Doha, Qatar

Abstract. A knowledgeable observer of a game of football (soccer) can make a subjective evaluation of the quality of passes made between players during the game. In this paper we consider the problem of producing an automated system to make the same evaluation of passes. We present a model that constructs numerical predictor variables from spatiotemporal match data using feature functions based on methods from computational geometry, and then learns a classification function from labelled examples of the predictor variables. In addition, we show that the predictor variables computed using methods from computational geometry are among the most important to the learned classifiers.

1 Introduction

There is currently considerable research into developing objective methods of analysing sports, including football. Football analysis has practical applications in player evaluation for coaching and scouting; development of game and competition strategies; and also to enhance the viewing experience of televised matches. Currently, football analysis is typically done manually or using simple frequency analysis, and thus applying probabilistic algorithms may offer more sophisticated analysis of player and team performance.

In this paper we use methods from computational geometry to compute complex predictor variables, and then apply machine learning classification algorithms to produce a quantitative measure of passing performance. The results show that this combination is promising, and may be applied to problems in domains other than football.

The application of geometric and stochastic algorithms to facilitate sports analysis is a relatively recent innovation. Initial efforts were focussed was on sports where play was structured where the collection of raw performance data was a more straightforward proposition, such as baseball and cricket. Within the

J. Gudmundsson—Joachim Gudmundsson was supported by the Australian Research Council (project numbers DP150101134 and FT100100755).
S. Chawla—Sanjay Chawla's research was supported by ARC Discovery and Linkage grants.

© Springer International Publishing Switzerland 2015
T. Cao et al. (Eds.): PAKDD 2015, Part II, LNAI 9078, pp. 319–330, 2015.
DOI: 10.1007/978-3-319-18032-8_25

past decade, advances in computer vision and image processing have resulted in systems that can capture accurate positional data about players in less structured games like football, hockey and basketball. However, the analytic capabilities that have been built on these types of data are fairly limited.

Analysing the performance of football players during matches is important for effective coaching. However, experiments have found that the level of recollection of critical events in football matches is as low as 42% [4]. Given this level of recall, the use of a systematic approach to collecting data about events occurring during matches is desirable. The performance evaluation on these data have typically been through frequency analysis of events, such as the number of passes made by a player, or the number of unforced errors [2]. However, the actions taken by players in a match are influenced by the interactions with other players, and simple frequency data may not fully capture the situation.

Event data represents only a small proportion of the data that could be captured from football matches. Collecting positional data is possible using modern image capture and processing technology [11,16]. There are commercial companies with similar systems such as Prozone [14] and TRACAB [3] which are used in professional leagues. State of the art systems are currently able to do this with high precision, for example the Prozone system computes trajectories accurate to 10cm at an frequency of 10Hz [14].

The availability of accurate positional and event data provides the basis for complex analysis of matches and player performance. To date, a range of diverse approaches have been explored.

Pattern matching has been used in a number of applications. Borrie et al. [2] uses T-pattern detection to extract similar passing sequences from matches. Gudmundsson and Wolle [5] analysed sub-trajectories using the Fréchet distance as a metric to cluster sub-trajectories that occurred multiple times.

Taki and Hasegawa [16] defined a geometric subdivision called a *dominant region*, similar to a Voronoi region [1], that subdivides the pitch into cells owned by players such that the player can reach all areas of the cell before any other player. The dominant region considers the direction and velocity of the players. Gudmundsson and Wolle [6] consider a related problem of computing the passing options that the player in possession has, based on the areas of the pitch that are reachable by the other players.

The remainder of the paper is structured as follows. Section 2 defines the problem and the model. The feature functions that compute the predictor variables used for the pass classification are described in Section 3. The learning algorithms used are described in Section 4. The experimental setup and results are reported in Section 5, and an analysis of the results is provided in Section 6.

2   Problem Definition

We will consider the problem of classifying passes made during a football match based on the quality of the pass. A pass is an event where the player in possession of the ball kicks the ball to a teammate with the intention that the receiver will be able to control the ball and thus obtain possession. The approach that we selected was to use supervised machine learning algorithms to learn a classification function. A significant component of our approach was thus to take the raw spatiotemporal signal and transform it to an appropriate format for the learning algorithms.

The input to the framework consists of three datasets for each match: a set of trajectories for each player; a sequence of events that occurred during the match; and a mapping of players to their respective teams. We had data available from four home matches played by Arsenal Football Club in the English Premiership during the 2007/08 season. The data was provided by Prozone [14]. These matches contained 2,932 passes in total. The input data is used to construct a vector of predictor variables for each pass made during the match.

Furthermore, a rating of the quality of each of the passes made in the four matches was made by two human observers. The ratings were combined and used to train and evaluate the classifiers. A detailed description of the rating process is provided in a longer version of this paper [7].

2.1 Preliminaries

For a given match, assume that there is a global clock which increment in steps of uniform size where $S = \{s_i : i \in \mathbb{N}\}$. Here, the step size is 10Hz.

Let $P = \{p_1, p_2, ..., p_n\}$ be the set of players who appeared in the match. Associated with each player p_j is a trajectory $t_{p_j} = \{(x_i, y_i, s_i) : s_i \in TI(p_j)\}$, where $TI(p_j)$ is the time interval when the player p_j is on the field.

Each player belongs to a team and, although there are five teams included in our experimental dataset, it is only necessary to consider the two teams playing in a particular match, and we use the convention of denoting these the home-team and away-team. We thus define a mapping from player to team as $M = \{(p_j, l) : p_j \in P, l \in \{\text{home}, \text{away}\}\}$.

Associated with each match is a set of events E. Each event $e_k \in E$ has a type (from a fixed-set), such as a pass, a shot, a tackle, etc. Formally, an event e_k is a triple $e_k = (s_i, u, R)$. Here s_i is the time step when the event occurred, u is the type of the event and R is the set of players involved in the event. The event that is of particular interest is the pass event, which is characterized as $\{(s_i, u, \{p_j\}) : s_i \in S, u = \text{pass}, p_j \in P\}$, where p_j is the player who makes the pass at time step s_i.

Now, the predictor variables associated with each event are defined as follows. The predictor variables can (in principle) depend upon all the previous and future events. Similarly, a predictor variable can depend on the current location of all players, and possibly their previous and future locations.

Thus, the trajectory sequences, event sequence and player mappings are used to compute the predictor variables for each pass event. Let ϕ_j be the feature function that produces the j-th predictor variable: $\phi_j : E \times Tr \times E \times M \to \mathbb{R}$.

We define vector $x^{(i)}$ of predictor variables for the i-th pass event. In order to train the learning algorithms, example labels are required for each class as response variables. The response variable is drawn from a discrete set $Y = \{1, 0, -1\}$, corresponding to {good, OK, bad}. Combining the response variable vectors with the corresponding predictor variables provides the training set for the machine learning algorithms.

$$
x^{(i)} = \begin{bmatrix} \phi_1(e_i, Tr, E, M) \\ \phi_2(e_i, Tr, E, M) \\ \vdots \\ \phi_n(e_i, Tr, E, M) \end{bmatrix} \qquad X = \begin{bmatrix} -(x^{(1)})^\top - \\ -(x^{(2)})^\top - \\ \vdots \\ -(x^{(m)})^\top - \end{bmatrix} \qquad y = \begin{bmatrix} y^{(1)} \\ y^{(2)} \\ \vdots \\ y^{(m)} \end{bmatrix}
$$

$$
x^{(i)} \in \mathbb{R}^n \qquad\qquad X \in \mathbb{R}^{m \times n} \qquad\qquad y \in \mathbb{R}^m
$$

The classification problem is defined as follows: given the training set of pass events X and labels y, and a cost function $J(X, y, \theta)$, the objective is to learn a parameterisation, θ such that: $\hat{\theta} = \mathrm{argmin}_\theta\, J(X, y, \theta)$ This parameterisation characterizes the classifier function $h_\theta(x^{(i)})$ that will predict the response variable $y^{(i)}$, given the input vector $x^{(i)}$.

3   Predictor Variables

The challenge of the feature engineering task is to extract information from the spatiotemporal match data so that the classifiers are able to make accurate inferences about the quality of the passes made in a match. The objective is to ensure that sufficient information about the match state is inherent in the constructed predictor variables. To motivate this task, consider how an informed observer of a football match would make an assessment about the quality of a pass.

At a basic level, the observer would consider the fundamentals of the pass, such as the distance and speed of the pass and whether the intended recipient of the pass was able to control the ball. These are the basic geometric aspects of the pass, but even at this level, the observer is required to make some inferences, such as who the intended recipient of the pass is. They would also likely consider the context of the match state when the pass was made. For example, was the passer under pressure from opposition players? To make such assessments, the observer would consider the positions of the players and the speed and direction that they were moving in, and the observer would make assumptions about whether the players are physically able to influence the pass by pressuring the passer or intercepting the pass. The observer thus has an idea of the physiological capabilities of the players, and will consider this in their estimation of the quality of a pass made.

At a higher level, passes are not made simply to move the ball from one player to a teammate, but also to improve the tactical or strategic position of their team. Passes can be made to improve the position of the ball, typically by trying to move the ball closer to the opponent's goal in order to have an opportunity to score. Passes may also be made to improve the match state by moving the ball from a congested area of the pitch to an area where the team in possession has a numerical or positional advantage. Meanwhile, the opposition will be actively trying to reduce the options of the player in possession to make passes. Thus, the match observer would need to consider the tactical and strategic objectives of the passer, and thus would have an understanding of the tactics and strategies employed by the player and team, and apply them to their estimate [18]. Likewise, the observer would consider the defensive team and their strategies and tactics.

A football match can also be viewed as a sequence of events occurring at particular times. The event-type that we are concerned with, *pass*, can thus be viewed as part of a sequence. This sequence can be subdivided in various ways, for example by unbroken sub-sequences of events where a single player or team is in control of the ball, or by a sub-sequence of events that occur between stoppages in play such as fouls, goals or injuries. When assessing the quality of a pass, the observer may consider the context of the pass in the sequence of events.

Finally, the observer may also consider the opportunity cost of the pass. By making the pass, the player forsook the other options available, such as passing to other players, dribbling or shooting.

For the trained observer, synthesising all this disparate information and making a prediction is a mental exercise that can be done in a matter of seconds. The problem described in this paper is to replicate this in a computational process.

3.1   Feature Functions

Feature functions are used to compute the predictor variables that are input to the classifier. The feature function $\phi_j(e_k, Tr, E, M)$ outputs the j-th predictor variable for the k-th pass event. The predictor variables are divided into the following categories in a manner consistent with our analysis of the types of information discussed above. The full list of features is provided in [8].

Basic geometric predictor variables are the predictor variables derived from the basic orientation of the players and ball on the football pitch.

Sequential predictor variables are constructed from the event sequence data. Currently three types of sequences are modelled: player possession; team possession; and play possession.

Physiological predictor variables are predictor variables that incorporate some aspect of the physiological capabilities of the players, generally how quickly they can reach a given point, given their current location and velocity.

Strategic predictor variables are predictor variables that are designed to provide some information about the strategic elements of the match.

3.2   Player Motion Model

The physiological and strategic features used in the model are based to some degree on an estimate of how quickly a player currently travelling at a given speed and direction can reach a given point. In particular, if a player can reach a point before any other player, then that player is said to dominate the point.

This notion is the basis for the physiological and strategic predictor variables we have defined in our model. In order to determine the time required to reach a given point, a motion model of the player is required.

The motion model is defined as a function $g(x, t_{p_j}, s_i)$ for a player p_j that takes as arguments: the coordinates of a point $x \in \mathbb{R}^2$, the time step s_i at which to determine the distance, and the trajectory t_{p_j} of the player p_j. The function returns the time $t \in \mathbb{R}^+$ it would take for the player to reach the point. As this time is dependent on the existing direction and velocity of the player, these factors are extrapolated from the players trajectory immediately prior to s_i.

Gudmundsson and Wolle [6] propose three simple motion models to approximate the reachable region for a player, that is, the region a player can in a given time. The models discretize the reachable region by introducing a time-step $\tau \in T$ to approximate t. The motion model would thus map a series of boundary curves that surround the initial starting point of the player. Furthermore, each curve is approximated by an n-sided polygon. The three motion models considered are based on a circular boundary, an elliptical boundary and a boundary constructed by sampling from the trajectory sequences.

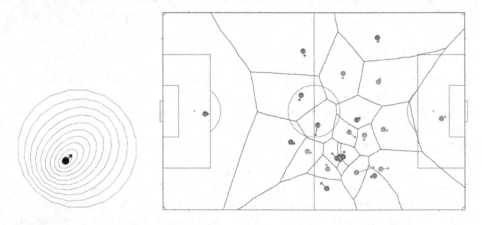

Fig. 1. Reachable region boundaries using ellipse motion model

Fig. 2. Dominant regions constructed using ellipse motion model

The motion models are used in several of the feature functions, in particular those based on the dominant region, below. We evaluated all three motion models in [6], however since the variance between the models is small, particularly for small distances such as on a football field, we consider only the ellipse motion model in this paper, see Figure 1.

3.3   The Dominant Region

Taki and Hasegawa [16] presents the dominant region as a dynamic area of influence that a player in a football match can exert dominance over, where dominance is defined as being the regions of the pitch that the player is able to reach before any other player. We propose to use the dominant region as a measure to approximate the strategic position of a team at a given point in time, and to subsequently construct feature functions based on it. Furthermore, we use the dominant region to also construct predictor variables that model the pressure exerted on the player in possession of the ball by opposition players in close proximity.

Intuitively, these appear to be useful predictors for the task of rating passes. The passing player wants to put the ball at a point where the intended recipient can reach it first, and this is, by definition, in the receiving player's dominant region. Thus, the proportion of the pitch that the team in possession dominates is a factor in the passing options of the passing player. Similarly, the size of the dominant region surrounding the passing and receiving players provide information about the pressure the player is under.

The football pitch can be partitioned into a subdivision of dominant regions, each dominated by a particular player. It is thus conceptually similar to the Voronoi region [1], the difference being the function that determines the region that a particular point belongs to: the function for a Voronoi region is usually the Euclidean distance; and for the dominant region is the time it takes for a player to reach a given point.

The dominant region is defined in [16] by the following equation for a player p_j at time-step s_i, where g is defined in Section 3.2.

$$D(p_j, s_i) = \{x \in \mathbb{R}^2 | g(x, t_{p_j}, s_i) \leq g(x, t_{p_m}, s_i) \text{ for } m \neq j, p_m \in P, s_i \in S\} \quad (1)$$

The subdivision of the dominant regions for all players will thus partition the football pitch into cells, as can be seen in Figure 2. However, $x \in \mathbb{R}^2$ is continuous, and there is currently no algorithm available to efficiently compute this continuous function. In fact, the dominant region of a player may not even be a single connected region [16]. Computing the intersection of surfaces in three dimensions, as required in Equation (1) is non-trivial and time-consuming. As such, we use an approximation algorithm to compute the dominant region. Taki and Hasegawa [16] and Nakanishi et al. [13] both present approximation algorithms where x is approximated by a discrete grid $Y \subset \mathbb{N}^2$, and the dominant regions are thus computed for all points in Y.

We implemented an approximation algorithm to compute the dominant region using the motion models in [6]. The algorithm has three steps. First, for every pairwise combination of players, the intersection points are determined between the reachable region polygons for each time-step. In the second step, the intersection points at each time-step are used to produce a reachable boundary between each pair of players. This is done using a modified version of Kruskal's minimum spanning tree algorithm [10], constrained so that the degree of every vertex is at most two. The third step constructs the smallest enclosing polygon around each player from the boundaries, and this is the player's dominant region. Pseudo-code for the algorithm is provided in Algorithm 1. This algorithm was able to efficiently compute the dominant region arrangement in quadratic time

Algorithm 1. Approximation algorithm to compute the dominant region at a given time

---

$P \leftarrow \{p_o, \cdots, p_n\}$ ▷ The players
$T \leftarrow \{\tau_0, \cdots, \tau_{max}\}$ ▷ The time-steps for boundaries
$V_{MSP} \leftarrow \emptyset, E_{MSP} \leftarrow \emptyset$ ▷ The vertices and edges of candidate boundaries between players' regions
for all $(p_i, p_j) \in \{(p_i, p_j) : p_i, p_j \in P, p_i \neq p_j\}$ **do**
     $V \leftarrow \emptyset, E \leftarrow \emptyset$
     for all $\tau \in T$ **do**   ▷ Step 1
         $V' \leftarrow \text{INTERSECTIONPOINTS}(p_i, p_j, \tau)$
         $E' \leftarrow \{(v_p, v_q) : (\tau(v_q) = \tau(v_p) + 1) \vee$
         $(\tau(v_p) = \tau(v_q) = \min(\{\tau(v_r) : r \in V'\}))\}$
         $V \leftarrow V \cup V', E \leftarrow E \cup E'$
     end for
     $(V'_{MSP}, E'_{MSP}) \leftarrow \text{MINIMUMSPANNINGPATH}(V, E)$ ▷ Step 2
     $V_{MSP} \leftarrow V_{MSP} \cup V'_{MSP}$
     $E_{MSP} \leftarrow E_{MSP} \cup E'_{MSP}$
end for
$D \leftarrow \emptyset$
for all $p \in P$ **do**   ▷ Step 3
     $E_p \leftarrow \{e : e \in E_{MSP} \wedge p(e) = p\}$
     $d \leftarrow \text{SMALLESTENCLOSINGPOLYGON}(E_p)$
     $D \leftarrow D \cup \{d\}$
end for
D is the set of dominant regions for all players

---

on the number of players, albeit with a large constant dominated by the square of the number of line segments in the boundary polygons.

Seven of the feature functions constructed are based on the dominant region and motion model. Examples of these feature functions are the computation of the dominant region when the player in possession passes; the net change in dominant region area between when a pass is made and when it is received; and the pressure the passer is under, defined by the area of the passers dominant region [8]. Our intuition is that the computed predictor variables will be important to the classifier, as they provide domain-specific information about the physiology and strategy of the players; and this information would not otherwise be available to the classifier. We examine the importance applied to features in Section 6.2.

4 Learning Algorithms

The pass rating task is a classification problem and we evaluate several supervised machine learning algorithms for this task. The goal of the learning algorithm is to produce a hypothesis function $h_\theta(x^{(i)})$ that can predict the ground truth variable $y^{(i)}$ for a given input vector of predictor variables, $x^{(i)}$. The algorithm is trained on the labelled example data with the objective of learning a parameterisation θ for the hypothesis function $h_\theta(x^{(i)})$ such that the prediction error is minimised.

The distribution of example data used in our experiments is unevenly distributed amongst the classes. The majority of examples were clustered towards the middle of the scale, see Table 1, and thus learning algorithms designed to handle class imbalance were selected. We examined a support vector machine (SVM) classifier; a RUSBoost classifier; and the multinomial logistic regression (MLR) with three different cost functions. The intention was to perform the experiments using diverse types of classifiers: SVM being a maximum margin classifier [17] and RUSBoost is an ensemble method that utilises sampling and boosting of weak classifiers [15].

For multinomial logistic regression, we evaluated three models learned using different empirical risk functions. The function (2) is the risk function derived from the maximum likelihood estimation of θ. The arithmetic (3) and quadratic (4) risk functions are intended to perform better under class imbalance conditions by computing the per-class risk [12]. The arithmetic risk takes the sum of the per-class values, whereas the quadratic risk uses the root of the sum of the squared values.

$$R^L(X,y,\theta) = -\frac{1}{m}\left[\sum_{i=1}^{m}\sum_{j=1}^{k}1_{\{y^{(i)}=j\}}\log\frac{e^{\theta_j^\top x^{(i)}}}{\sum_{l=1}^{k}e^{\theta_l^\top x^{(i)}}}\right] \tag{2}$$

$$R^A(X,y,\theta) = -\frac{1}{k}\sum_{j=1}^{k}\left[\frac{1}{m_j}\sum_{i=1}^{m_j}1_{\{y^{(i)}=j\}}\log\frac{e^{\theta_j^\top x^{(i)}}}{\sum_{l=1}^{k}e^{\theta_l^\top x^{(i)}}}\right] \tag{3}$$

$$R^Q(X,y,\theta) = -\sqrt{\frac{1}{k}\sum_{j=1}^{k}\left[\frac{1}{m_j}\sum_{i=1}^{m_j}1_{\{y^{(i)}=j\}}\log\frac{e^{\theta_j^\top x^{(i)}}}{\sum_{l=1}^{k}e^{\theta_l^\top x^{(i)}}}\right]^2} \tag{4}$$

$$\hat{\theta} = \underset{\theta}{\operatorname{argmin}}\, R(X,y,\theta) + \lambda\|\theta\|_p \tag{5}$$

Included in the cost function J is a regularization term. We evaluate the ℓ_1- and ℓ_2-norms in our experiments, i.e. $p \in \{1,2\}$ in Equation 5. Moreover, the ℓ_1-norm will induce a sparse parameterisation of θ [9], and we investigate the predictor variables whose corresponding value in θ is non-zero as a measure of the importance of the predictor variable.

5   Experiments

The experiments were intended to answer the following questions:

1. Is it possible to find a classification function and set of predictor variables to accurately predict the quality of a pass?
2. To what extent are the predictor variables computed using algorithms from computational geometry contributors to the performance of the classifier?

5.1 Setup

The objective of the experiments conducted for this research was to learn an optimal set of parameters $\hat{\theta}$ that characterize the hypothesis function $h_\theta(x)$ such that the hypothesis function makes correct predictions on unseen examples. We held out 20% of the labelled data for testing, stratified in proportion to the class frequencies detailed in Table 1. The classifiers were trained using tenfold cross-validation of the training examples, and evaluated comparing the classifier prediction to the ground-truth label on the test examples.

We evaluated the hypothesis functions using the following objective functions: accuracy; precision; recall and $F_{\beta=1}$-score. The metrics for precision, recall and $F_{\beta=1}$-Score are calculated on a per-class basis, and a simple mean of the per-class values was used.

5.2 Results

The results of the experiments are summarised in Table 2. The overall accuracy of the classifiers shows that SVM is the best configuration. The multinomial logistic regression classifiers using the arithmetic and quadratic loss functions showed improved recall, but this was at a cost of precision. This is consistent with expectations, as these cost functions are less sensitive to the class imbalance, and would be more likely to assign examples to the minority classes. However, the improvement in recall was not sufficient to outweigh the reduction in precision, and so the overall accuracy was reduced. Similarly, RUSBoost learned a classifier with good recall, but at the cost of relatively poor precision.

Table 1. Per-class frequencies   **Table 2.** Summary results

Class	Relative Frequency	Count
Good	0.066	193
OK	0.789	2314
Bad	0.145	425

Classifier	Accuracy	Precision	Recall	$F_{\beta=1}$-Score
MLR	0.829	0.666	0.752	0.638
MLR-Arith	0.730	0.580	0.780	0.612
MLR-Quad	0.741	0.581	**0.784**	0.617
SVM	**0.858**	**0.713**	0.734	**0.711**
RUSBoost	0.756	0.600	0.781	0.646

6 Analysis

This section analyses the performance of the learned classifiers and considers the results in the context of football analysis.

6.1 Classifier Performance

The experimental results show that it is possible to learn a classifier that performs better than random, and also better than deterministically selecting the majority class. The error present in these classifiers appears to be the result of bias, as the accuracy of the classifiers on the unseen test data is close to the accuracy on the training data. For example, the SVM classifier has an accuracy of 85.8% on the test data and 85.9% on the training data.

6.2 Predictor Variable Importance

The predictor variables were computed by feature functions of varying complexity. This begs the question whether the effort involved in implementing and computing complex feature functions resulted in improved performance. We selected the features that were computed using the dominant region described in Section 3.3. Seven of the 49 features constructed were based on the dominant region.

In this situation, none of the predictor variables was a strong predictor of the quality label assigned to the pass. A visual examination of scatter-plots of pairs of response variables did not reveal any pairs that clearly separated the labelled data. Moreover, applying principal component analysis produced a coefficient of 0.03 on the first eigenvector. Therefore, the importance of any given feature is likely to be incremental.

We investigated the importance that various algorithms assigned to these seven features based on the dominant region. The first approach was to investigate the weights learned when using sparsity-inducing ℓ_1 regularization [9]. Informally, this means that features that are of low importance will have a zero weight and important features will be non-zero. We examined the weights vector for the MLR classifier using ℓ_1 regularization. Seven variables were assigned non-zero weights and three of these were based on the dominant region. Furthermore, we considered the predictor importances derived from the RUSBoost classifier. In this case, the predictor variable of the dominant region of the passer was ranked the most important variable. Thus, in both these cases, features based on the dominant region were important to the classifiers, suggesting that the effort computing them is worthwhile.

7   Conclusion

In this paper we present a model that is able to learn a classifier to rate the quality of passes made during a football match with an accuracy of up to 85.8%. The model uses feature functions based on methods from computational geometry, in particular the dominant region [16]. This structure is intended to provide information about the strategic and physiological state of the match, however it is costly to compute. We evaluate the importance to the classifier of the predictor variables based on the dominant region, and find them to be important to the classifiers, suggesting that the cost of computation is worthwhile in this case.

References

1. de Berg, M., Cheong, O., van Kreveld, M., Overmars, M.: Computational geometry. Springer (2008)
2. Borrie, A., Jonsson, G.K., Magnusson, M.S.: Temporal pattern analysis and its applicability in sport: an explanation and exemplar data. Journal of Sports Sciences **20**(10), 845–852 (2002)
3. ChyronHego Corporation, : Tracab player tracking system. http://chyronhego.com/sports-data/player-tracking (2014)
4. Franks, I.M., Miller, G.: Eyewitness testimony in sport. Journal of Sport Behavior (1986)
5. Gudmundsson, J., Wolle, T.: Towards automated football analysis: Algorithms and data structures. In: Proc. 10th Australasian Conf. on Mathematics and Computers in Sport, Citeseer (2010)
6. Gudmundsson, J., Wolle, T.: Football analysis using spatio-temporal tools. Computers, Environment and Urban Systems (2013)
7. Horton, M., Gudmundsson, J., Chawla, S., Estephan, J.: Classification of passes in football matches using spatiotemporal data. arXiv preprint arXiv:1407.5093 (2014)
8. Horton, M., Gudmundsson, J., Chawla, S., Estephan, J.: Feature descriptions for pass classifier. http://bit.ly/1ynhNm5 (2014)
9. Jenatton, R., Audibert, J.Y., Bach, F.: Structured variable selection with sparsity-inducing norms. The Journal of Machine Learning Research **12**, 2777–2824 (2011)
10. Kruskal, J.B.: On the shortest spanning subtree of a graph and the traveling salesman problem. Proceedings of the American Mathematical society **7**(1), 48–50 (1956)
11. Leo, M., Mosca, N., Spagnolo, P., Mazzeo, P.L., D'Orazio, T., Distante, A.: Real-time multiview analysis of soccer matches for understanding interactions between ball and players. In: Proceedings of the 2008 International Conference on Content-Based Image and Video Retrieval, pp. 525–534. ACM, 1386419 (2008)
12. Liu, W., Chawla, S.: A quadratic mean based supervised learning model for managing data skewness. In: SDM, pp. 188–198. SIAM (2011)
13. Nakanishi, R., Maeno, J., Murakami, K., Naruse, T.: An Approximate Computation of the Dominant Region Diagram for the Real-Time Analysis of Group Behaviors. In: Baltes, J., Lagoudakis, M.G., Naruse, T., Ghidary, S.S. (eds.) RoboCup 2009. LNCS, vol. 5949, pp. 228–239. Springer, Heidelberg (2010)
14. Prozone Sports Ltd: Prozone: Prozone sports - prozone 3 - sports performance analysis. http://www.prozonesports.com/product/prozone3/ (2013)
15. Seiffert, C., Khoshgoftaar, T.M., Van Hulse, J., Napolitano, A.: Rusboost: A hybrid approach to alleviating class imbalance. IEEE Transactions on Systems, Man and Cybernetics, Part A: Systems and Humans **40**(1), 185–197 (2010)
16. Taki, T., Hasegawa, J.: Visualization of dominant region in team games and its application to teamwork analysis. In: Proceedings of the Computer Graphics International, pp. 227–235 (2000)
17. Vapnik, V.N.: The nature of statistical learning theory. Springer-Verlag New York, Inc. (1995)
18. Wilson, J.: Inverting the pyramid: The history of football tactics. Hachette UK (2010)

Stabilizing Sparse Cox Model Using Statistic and Semantic Structures in Electronic Medical Records

Shivapratap Gopakumar[✉], Tu Dinh Nguyen, Truyen Tran,
Dinh Phung, and Svetha Venkatesh

Center for Pattern Recognition and Data Analytics, Deakin University,
Melbourne 3216, Australia
{sgopakum,ngtu,truyen.tran,dinh.phung,svetha.venkatesh}@deakin.edu.au

Abstract. Stability in clinical prediction models is crucial for transferability between studies, yet has received little attention. The problem is paramount in high dimensional data, which invites sparse models with feature selection capability. We introduce an effective method to stabilize sparse Cox model of time-to-events using statistical and semantic structures inherent in Electronic Medical Records (EMR). Model estimation is stabilized using three feature graphs built from (i) Jaccard similarity among features (ii) aggregation of Jaccard similarity graph and a recently introduced semantic EMR graph (iii) Jaccard similarity among features transferred from a related cohort. Our experiments are conducted on two real world hospital datasets: a heart failure cohort and a diabetes cohort. On two stability measures – the Consistency index and signal-to-noise ratio (SNR) – the use of our proposed methods significantly increased feature stability when compared with the baselines.

1 Introduction

Stability is fundamental to prognosis. Besides good performance, a prognostic model needs to be interpretable and stable to warrant clinical adoption. This translates to a small group of succinct predictors that are consistent in the face of data re-sampling. Hence strong feature selection is key when deriving clinical models.

When data is high dimensional and correlated, automated feature selection causes instability in linear [1] and survival models [2]. These aspects are intrinsic to modern healthcare data. Medical events often co-occur, especially in aged cohorts. Comorbidities or diseases that co-exist with the primary disease in a patient, cause multiple diagnoses that are strongly correlated to each other. For example, Fig. (1a)[1] shows the common complications in a diabetic cohort.

---

[1] Image courtsey: http://www.clker.com/clipart-human-body-anatomy-basics-no-lines.html

© Springer International Publishing Switzerland 2015
T. Cao et al. (Eds.): PAKDD 2015, Part II, LNAI 9078, pp. 331–343, 2015.
DOI: 10.1007/978-3-319-18032-8_26

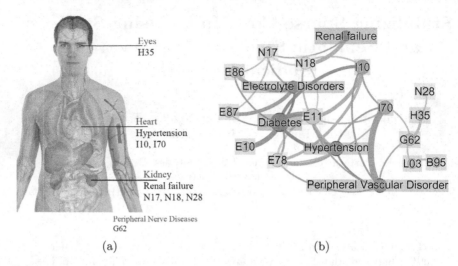

(a)   (b)

Fig. 1. Extracting disease correlations in diabetic cohort. Common comorbidities and diagnosis codes are shown in (a). A portion of the disease graph constructed using Jaccard similarity between EMR features in a diabetic cohort is shown in (b). The nodes represent EMR features, and links represent interaction strength, measured using Jaccard index. Blue nodes are co-occurring diseases, green nodes are diagnosis codes for diabetes, orange nodes for heart diseases, and yellow nodes for urinary diseases.

These correlations can be visualized as a network, as in Fig. (1b). When deriving a prediction model from such data, we have to account for the complex interconnectedness of the features.

Integrating domain knowledge to improve learning has been gaining much attention recently [3]. Biological understanding of gene-disease networks, for example, has enabled discovery of what genes contribute to a disease, and what proteins would bind with a particular chemical compound [4]. However, little has been explored in networks derived from the healthcare processes and their contribution to prediction models. Domain knowledge, represented as networks, should ideally guide the feature selection process in clinical prediction.

In this paper, we address the problem of stabilizing a high dimensional model derived from routinely collected EMR data. We focus on minimizing the variance in feature subsets and model estimation parameters. We construct a feature graph with nodes as EMR features and edges as relationship between the features. We look at three feature relationships: (i) Jaccard score between features (ii) aggregate of Jaccard score and the semantic EMR link used in [5] (iii) Jaccard scores between features transferred from a related cohort. A random walk regularization of the proposed graphs is used to stabilize a sparse Cox model that predicts time to readmission. Our experiments are conducted on 2 real world hospital datasets: a heart failure cohort and a diabetes cohort. We measure feature stability using the Consistency index and model estimation stability using signal-to-noise ratio (SNR). Our proposed method, when compared with elastic net and recently introduced semantic EMR graph regularization [5], confirmed better feature and model stability when validated against both cohorts.

In summary, our main contributions are:

1. Representation of medical domain knowledge as feature graphs that embed (i) statistical correlations between features using Jaccard index (ii) aggregate of statistical and semantic correlation among features (iii) correlations between features transferred from a related cohort.
2. A random walk regularizer based on the proposed feature graphs to stabilize a Cox model as opposed to the traditional Laplacian regularizer. While Laplacian regularizer focuses on pairwise similarity, the random walk regularizer encourages groupwise similarity.
3. Demonstration of improved feature stability as measured by the Consistency index and improved model stability as measured by signal-to-noise ratio (SNR) for model regularization using proposed feature graphs. The stability measures were compared with lasso, elastic net and a recently introduced Laplacian semantic EMR graph regularization [5] on a cohort of $1,784$ index admissions in heart failure patients and $2,370$ index admissions in diabetic patients admitted to a regional hospital in Australia.
4. Demonstration of improved stability, using transfer learning on related cohorts. Related cohorts like heart failure and diabetes share comorbidities and predictors. Hence, the feature graph constructed from one cohort is used to stabilize the model derived from related cohort. Stability is measured using the Consistency index and SNR.

The significance of our study lies in understanding the importance of incorporating underlying feature relationships into model learning. This promotes stable feature selection in a clinical setting.

2   Related Background

Stabilizing clinical prediction has received little attention, partly because most models are built using a small subset of well-defined predictors, chosen either by domain experts or from prior knowledge. In most high dimensional models, the primary regularizer of choice is the lasso because of its convexity and sparsity inducing property [6]. However, when data is correlated, as in the healthcare domain, lasso regularized models are susceptible to data variations resulting in loss of stability [1]. The inconsistency of Lasso to handle data correlation has been demonstrated for linear models [7] and recently for survival models [2]. When prior knowledge about feature relationships are available, Sandler *et al.* proposed additional regularization using graph networks where the nodes are features and edges represent relationships [3]. Graph regularization ensures statistical weight sharing among related features. In such scenarios, the first challenge is to identify useful prior information.

High-dimensional models in bioinformatics domain often resort to online databases like KEGG, Pathway Commons and BioCarta to extract context specific data to build feature graphs [8]. Recent linear classification models have started using such gene-pathway networks and protein interaction networks to improve prediction accuracy and model interpretability [9]. For genomic data, a recent study employed a quadratic Laplacian regularizer into Cox regression, where the Laplacian graph was derived from prior gene regulatory network information [10]. Another study investigated the significance of including domain knowledge as network information into Cox model for identifying biomarkers in breast cancer [11]. Specifically, the study compared eight network based regularizers and three non-network based regularizers for Cox regression. All methods were validated on five public breast cancer datasets. The study observed no significant advantage for network-based approaches over non-network-based approaches in terms of prediction performance or signature stability.

In the healthcare domain, elastic net regularized Cox model showed superior performance over Lasso for prostate cancer dataset [12]. Though elastic net regularization can handle correlated features, it cannot incorporate structural relationships. Vinzamuri *et al.* introduced a modification to the elastic net involving an RBF kernel for handling feature correlations to improve accuracy and reduce redundancy [13]. In contrast, our work focuses on handling correlations to stabilize the model estimation and top predictors. Feature similarity is captured by building a Jaccard similarity graph for additional regularization. Recent studies use semantic graphs constructed from ICD-10 code relations and temporal relations in medical events to stabilize linear readmission models derived from electronic medical records [5]. We compare our approach with this EMR graph regularization. Further, we investigate the effect of aggregating the semantic EMR graph with the statistical Jaccard graph on model stability.

3   Framework

3.1   Sparse Cox Model

We use Cox regression to model risk of readmission (hazard function) at a future time instance, based on data from EMR. Unlike logistic regression where each patient is assigned a nominal label, Cox regression models the readmission time directly [13]. The proportional hazards assumption in Cox regression assumes a constant relationship between readmission time and EMR-derived explanatory variables. Let $\mathcal{D} = \{\boldsymbol{x}_n, y_n\}_{n=1}^{N}$ be the training dataset with N observations, ordered on increasing y_n, where $\boldsymbol{x}_n \in \mathbb{R}^K$ denotes the feature vector for n^{th} index admission and y_n is the time to next unplanned readmission. When a patient withdraws from the hospital or does not encounter readmission in our data during the follow-up period, the observation is treated as right censored. Let M observations be uncensored and $R(t_n)$ be the remaining events at readmission time t_n.

Since the data $\mathcal{D}$ is high dimensional (possibly $K \gg N$), we apply lasso regularization for sparsity induction [14]. The feature weights $\boldsymbol{w} \in \mathbb{R}^K$ are estimated by maximizing the ℓ_1-penalized partial likelihood:

$$\mathcal{L}_{\text{lasso}} = \frac{1}{N}\mathcal{L}\left(\boldsymbol{w}; \mathcal{D}\right) - \alpha \sum_{k} |w_k| \tag{1}$$

where $\|\boldsymbol{w}\|_1 = \sum_k |w_k|$, $\alpha > 0$ is the regularizing constant, and $\mathcal{L}(\mathbf{w}; \mathcal{D})$ is the log partial likelihood [15] computed as:

$$\mathcal{L}(\mathbf{w}; \mathcal{D}) = \sum_{m=1}^{M} \left\{ \boldsymbol{w}^\top \boldsymbol{x}_m - \log \left[\sum_{j \in R(t_m)} \exp\left(\boldsymbol{w}^\top \boldsymbol{x}_j \right) \right] \right\}$$

Lasso induces sparsity by driving the weights of weak features towards zero. However, sparsity induction is known to cause instability in feature selection [16]. Instability occurs because Lasso randomly chooses one in two highly correlated features. Each training run with slightly different data could result in a different feature from the correlated pair. The nature of EMR data further aggravates this problem. The EMR data is, by design, highly correlated and redundant. Also, features in the EMR data maybe weakly predictive for some task, thereby limiting the probability that they are selected. These sum up to lack of reproducibility between model updates or external validations, hindering the method credibility and adoption by clinicians.

A popular solution to instability is elastic net [12]. Elastic net regularization modifies the likelihood function in Eq. (1) as:

$$\mathcal{L}_{\text{elastic.net}} = \frac{1}{N} \mathcal{L}(\boldsymbol{w}; \mathcal{D}) - \alpha \left(\lambda \sum_k |w_k| + (1 - \lambda) \sum_k w_k^2 \right) \qquad (2)$$

Here, the ridge regression term $\sum_k w_k^2$ tends to give equal weights to correlated features, while the lasso term $\sum_k |w_k|$ introduces sparsity. However, this formulation ignores domain knowledge.

3.2   Stabilization Using Feature Graph

Medical events often co-occur, especially in aged cohorts. For example, the presence of comorbidities causes multiple diagnoses at the same time. We capture feature correlation in a knowledge network, with features as nodes and relations between features as edges. Let the adjacency matrix of the feature graph be G, where $G_{ij} = g \in (0, 1)$ represents the weighted similarity score between features i and j. We ensure all features have equal prominence by constraining the out-links of each node to sum to one. The medical events linked together in the feature graph should have similar weights. We introduce a random walk regularizer [3]:

$$\Omega(\boldsymbol{w}; \mathbf{G}) = \sum_k \left(w_k - \sum_i G_{ki} w_i \right)^2$$
$$= \boldsymbol{w}^\top (I - \mathbf{G})^\top (I - \mathbf{G}) \boldsymbol{w} \qquad (3)$$

where I is the identity matrix. The graph stabilized model likelihood can be written as:

$$\mathcal{L}_{\text{graph}} = \mathcal{L}_{\text{lasso}} - \frac{1}{2} \beta \boldsymbol{w}^\top (I - \mathbf{G})^\top (I - \mathbf{G}) \boldsymbol{w} \qquad (4)$$

Here the $\ell_1$regularizer introduces sparsity by pushing weak features towards zero, while the random walk regularizer distributes smoothness equally among correlated features. The gradient of Eq. (4) becomes:

$$\frac{\partial \mathcal{L}_{\text{graph}}}{\partial w} = \sum_{m=1}^{M} \left\{ x_m - \frac{\sum\limits_{j \in R(t_m)} x_j \exp(w^\top x_j)}{\sum\limits_{j \in R(t_m)} \exp(w^\top x_j)} \right\}$$
$$- \alpha \operatorname{sign}(w) - \beta (I - \mathbf{G})^\top (I - \mathbf{G}) w \tag{5}$$

Parameter estimation is done by maximizing the likelihood in Eq. (4) using L-BFGS algorithm [17]. We build and compare different feature graphs to stabilize our model. Each feature graph differs in the construction of its adjacency matrix $\mathbf{G}$. A recent study [5] introduced a semantic EMR graph, where nodes denoted features and edges denoted a temporal relation or ICD-10 structural relation between features (Fig. 2b). Using this as a baseline, we construct $\mathbf{G}$ using the following methods. First, we represent the edges using the Jaccard index between features, as in Fig. 2a. Second, we aggregate the baseline semantic EMR graph and the Jaccard graph. Here, each edge is the maximum of Jaccard and semantic scores between the features (Fig. 2c). Finally, we investigate transferring the adjacency matrix between related cohorts. Specifically, the Jaccard similarity scores between features in one cohort is transferred to a related cohort (Fig. 2d). We detail these methods below.

Jaccard Graph. The *Jaccard index* measures the percentage of agreement between components among feature vectors. Given two feature vectors F_i and F_j, the pairwise Jaccard score reads:

$$J_{ij} = \frac{a}{a + b + c} \tag{6}$$

where a is the number of non-zero components in F_i and F_j, b is the number of non-zero components in F_i but not in F_j and c is number of non-zero components

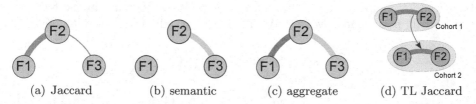

| (a) Jaccard | (b) semantic | (c) aggregate | (d) TL Jaccard |

Fig. 2. Feature correlation captured by constructing feature graph with nodes as features and edges as: (a) statistical correlation measured using Jaccard score (b) semantic relations derived from temporal and ICD-10 structures (c) aggregation of Jaccard and Semantic graphs. (d) transfer of Jaccard similarity between features from a related cohort

Table 1. Characteristics of training and validation cohorts

	Heart failure		Diabetes	
	Training set	Testing set	Training set	Testing set
Checkpoint	Sep 2010		Dec 2008	
Number of admissions	1,415	369	1,341	1,029
Unique patients	1,088	317	951	765
Gender				
Male	541 (49.7%)	155 (48.9%)	501 (52.68%)	407 (53.20%)
Female	547 (50.2%)	162 (51.1%)	450 (47.32%)	358 (46.80%)
Mean age (years)	78.3	79.4	57.8	56.4

in F_j but not in F_i. We construct an undirected graph with nodes as features and edges representing the Jaccard score between features.

Graph Aggregation. We investigate the effect of combining the semantic EMR graph with Jaccard graph on model stability and feature stability. The semantic EMR graph captures the general relationship between diagnostic codes based on the ICD-10 structures, while the Jaccard graph is cohort specific. Here, we use a simple aggregation technique to construct the final $\langle$EMR; Jaccard$\rangle$ graph as:

$$G_{\langle\text{EMR;Jaccard}\rangle} = \max(G_{\text{EMR}}, G_{\text{Jaccard}}) \tag{7}$$

Transfer Learning. Finally, we examine the capability of our proposed method in transfer learning. Knowledge from one domain can be transferred to a related domain when data is scarce or expensive to collect [18]. Getting high quality training data is often difficult, particularly in a medical setting. Cohorts that share comorbidities and diagnoses, as in diabetes and cardiovascular diseases, are likely to have similar correlations among features. Accordingly, we propose to stabilize a Cox model derived from one cohort using the Jaccard similarity graph constructed from a related cohort. We denote the transferred graph as: TL-Jaccard graph. Further, we use TL-Jaccard graph to construct the aggregated graph:

$$G_{\langle\text{EMR;TL-Jaccard}\rangle} = \max(G_{\text{EMR}}, G_{\text{TL-Jaccard}})$$

Here, the temporal and hierarchical feature relations in the cohort are captured by the EMR graph. The statistical relations among features, which can be expensive to calculate, are transferred from the related cohort using TL-Jaccard graph.

4 Experiments

In this section, we evaluate feature and model stability of our framework. The results are reported on two cohorts: heart failure (HF) and diabetes (DB), provided by Barwon Health, a regional health service provider which has been

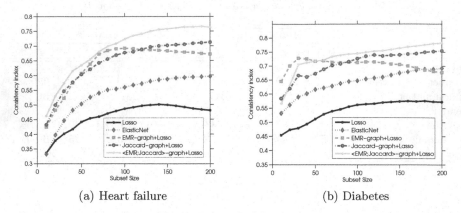

(a) Heart failure   (b) Diabetes

Fig. 3. Stabilization using statistical and semantic structures. Feature stability measured by the Consistency index as functions of the subset size for readmission prediction within 6 months for heart failure (Fig. 3a) and 12 months for diabetes patients (Fig. 3b). Larger indices imply more stability.

serving more than $350,000$ residents in Victoria, Australia[2].We collect retrospective data for heart failure and diabetes patients from the hospital EMR database. The heart failure cohort contains all patients with at least one ICD-10 diagnosis code I50, while the diabetes cohort includes all patients with at least one diagnosis code between E10-E14. This resulted in $1,885$ heart failure admissions and $2,840$ diabetes admissions between January 2007 and December 2011. Patients of all age groups were included whilst inpatient deaths were excluded. We focus our study on emergency attendances and unplanned admissions of patients.

We use the one-sided convolutional filter bank introduced in [19] to extract a large pool of features from EMR databases. The filter bank summarizes event statistics over multiple time periods and granularities. The feature extraction process resulted in $3,338$ features for heart failure cohort and $7,641$ features for diabetes cohort. The extracted features are used to derive a sparse Cox model. Our proposed feature graphs capture correlations between these features to stabilize model learning.

4.1   Evaluation Protocol

The baseline regularization methods for Cox regression are chosen to be (i) lasso (ii) elastic net (iii) semantic EMR graph (as in [5]). Based on the construction of the feature graph, we arrive at four different models: (i) Jaccard graph regularized model: feature graph is the Jaccard similarity graph among features in

---

[2] Ethics approval was obtained from the Hospital and Research Ethics Commitee at Barwon Health (number 12/83) and Deakin University.

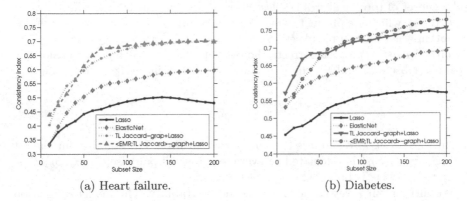

(a) Heart failure.   (b) Diabetes.

Fig. 4. Stabilization using transfer of Jaccard graph (TL Jaccard). Stabilization using statistical and semantic structures. Feature stability measured by the Consistency index as functions of the subset size for readmission prediction within 6 months for heart failure (Fig. 4a) and 12 months for diabetes patients (Fig. 4b). Larger indices imply more stability.

the given cohort (ii) EMRJaccard regularized model: feature graph is the aggregation of Jaccard graph with semantic EMR graph, as in Eq. (7) in the given cohort (iii) TL Jaccard regularized model: feature graph is the Jaccard similarity graph transferred from a related cohort (iv) EMR; TL Jaccard regularized model: feature graph is the aggregation of semantic EMR graph from the given cohort and Jaccard graph transferred from a related cohort.

Temporal Validation. We ensure that the training and testing sets are completely separated in time. This validation strategy is chosen because it better reflects the common practice of training the model in the past and using it in the future. We gather admissions which have discharge dates before September 2010 for heart failure and before 2009 for diabetes patients to form the training set and after that for testing. Next we specify the set of unique patients in the training set. We then remove all admissions of such patients in the testing set to guarantee no overlap between two sets. The statistical characteristics of two cohorts are summarized in Table 1.

Our model is then learned using training data and evaluated on testing data. Model performance is evaluated using measures of AUC (area under the ROC curve) with confidence intervals based on Mann-Whitney statistic. The AUC is computed from the ranking of hazard rates of the patient readmissions.

Measuring Model Stability. We use the Consistency index [20] to measure stability of feature selection process. The Consistency index (CI) supports feature selection in obtaining several desirable properties, i.e., monotonicity, limits and correction for chance. To simulate data variations due to sampling, we create B data bootstraps of original size n. For each bootstrap, a model is trained and a subset of top k features is selected. Features are ranked according to their importance, which are product of feature weight and standard deviation. Finally, we obtain a list of feature subsets $S = \{S_1, S_2, ..., S_B\}$ where $|S_b| = k$.

The *Consistency index* corrects the overlapping due to chance. Considering a pair of subsets S_i and S_j, the pairwise Consistency index I_C is defined as:

$$CI(S_a, S_b) = \frac{rd - k^2}{k(d - k)} \tag{8}$$

in which $|S_a \cap S_b| = r$ and d is the number of features. The stability for the set $S = \{S_1, S_2, ..., S_B\}$ is calculated as average across all pairwise $CI(S_a, S_b)$. The Consistency index is bounded in $[-1, +1]$.

We further our investigations on the model stability. The model estimation stability is defined as variance in parameters. A measure is the signal-to-noise ratio (SNR): $SNR(i) = \bar{w}_i/\sigma_i$ in which $\bar{w}_i$ is the mean feature weight across bootstraps for feature i, and σ_i is its standard deviation. We take the average of the 20 highest SNR values. Higher score indicates better stability.

4.2   Results

Our models are designed using two hyper-parameters: lasso regularization parameter α and graph regularization parameter β. We empirically tune these parameters to improve feature stability without hurting model discrimination.

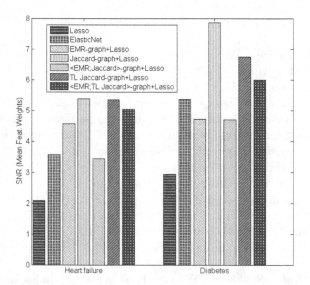

Fig. 5. Model estimation stability measured by signal-to-noise ratios (SNR) of feature weights. High value of SNR indicates more stability. *"TL Jaccard"* means the Jaccard-graph used in transfer learning settings: heart failure Jaccard graph for stabilizing diabetes data and diabetes Jaccard graph for stabilizing heart failure cohort.

Overall, feature stability depended more on graph parameter β, while model discrimination was more sensitive to α. A good tradeoff was achieved at $\alpha = 0.003$ and $\beta = 0.8$. All models are externally validated against (i) heart failure cohort with a 6-month horizon (ii) diabetes cohort with a 12-month horizon. Table 2 reports the AUC scores with confidence intervals for the different models. The predictive performance of our proposed graph stabilization models and transfer learning techniques are comparable with the baselines.

Table 2. AUC scores with confidence intervals for readmission prediction within 6 months for heart failure and 12 months for diabetes patients. Model performance on individual cohorts and on cohorts with Jaccard graph transferred from the other cohort is shown in separate sections.

	Heart failure	Diabetes
Lasso	0.60 [0.55; 0.66]	0.74 [0.70; 0.77]
Elastic net	0.61 [0.55; 0.67]	0.75 [0.72; 0.79]
EMR-graph+Lasso	0.61 [0.56; 0.67]	0.76 [0.72; 0.79]
Jaccard-graph+Lasso	0.62 [0.56; 0.67]	0.76 [0.73; 0.79]
⟨EMR; Jaccard⟩-graph+Lasso	0.62 [0.56; 0.67]	0.76 [0.73; 0.79]
Stabilization using Transfer Learning		
TL_Jaccard-graph+Lasso	0.62 [0.56; 0.67]	—
⟨HF_EMR; TL_Jaccard⟩-graph+Lasso	0.62 [0.57; 0.68]	—
TL_Jaccard-graph+Lasso	—	0.76 [0.73, 0.79]
⟨DB_EMR; TL_Jaccard⟩-graph+Lasso	—	0.75 [0.72, 0.79]

Stabilization Using Statistical and Semantic Graphs. Graph regularized models consistently produced more stable features than lasso and elastic net regularized models. When comparing different graph regularizations, we found the semantic EMR graph to be more effective for small feature subsets (see Fig. 3). For increasing feature subset sizes (>100), Jaccard graphs proved effective. The temporal and structural relations of diagnosis codes have stronger effect for small set of features, while Jaccard index was effectual on larger sets. This behavior suggests aggregating statistical and semantic structures. For the top 100 predictors, EMRJaccard graph stabilization demonstrated the highest feature stability in both cohorts (see Fig. 3).

Next, we compare variance in parameter weights using SNR measures. In Fig. 5, each model is represented by average of its 20 highest SNR values. The Jaccard graph regularized model proved to be most robust in both cohorts. Interestingly, model stability using EMRJaccard graph was similar to elastic net and was not able to improve upon semantic EMR graph or Jaccard graph.

Stabilization Using Transfer Learning. We investigate transfer of feature graphs between related cohorts. For the heart failure cohort, TL Jaccard graph represents the Jaccard scores transferred from diabetes cohort, while EMR;TL

Jaccard graph is the aggregation of the semantic EMR graph of heart failure cohort and Jaccard graph transferred from diabetes cohort. The same technique is applied to stabilize diabetes cohort, where the Jaccard scores are transferred from heart failure cohort. We compare the transferred graph stabilizations with lasso and elastic net. Figs. (4;5) show that the cross-domain graphs also help the stabilities of feature selections and model estimation.

5   Discussion and Conclusion

Novel methods in feature selection often concentrate on model performance and overlook stability [21,22]. Stability facilitates reproducibility between model updates and generalization across medical studies. Stable predictors inspire confidence in prognosis, as they are often subjected to further examinations. In this paper, we utilize statistical and semantic relations in EMR data to stabilize a sparse Cox model for predicting readmission. The model is validated on two different retrospective cohorts. The random walk regularization of the aggregated feature graph promotes group level selection and rare-but-important features. On two stability measures, the proposed method has demonstrated largely improved stability. In related cohorts, when collecting data becomes expensive, transferring domain knowledge using TL-Jaccard graph was also found to improve stability. Also, our proposed model is derived entirely from commonly available data in medical databases. All these factors suggest that our model could be easily integrated into the clinical pathway to serve as a fast and inexpensive screening tool in selecting features and patients for further investigation. Future work includes applying the same technique for a variety of cohorts and investigating other latent correlations in EMR to enhance feature stability.

References

1. Austin, P.C., Tu, J.V.: Automated variable selection methods for logistic regression produced unstable models for predicting acute myocardial infarction mortality. Journal of Clinical Epidemiology **57**, 1138–1146 (2004)
2. Lin, W., Lv, J.: High-dimensional sparse additive hazards regression. Journal of the American Statistical Association **108**, 247–264 (2013)
3. Sandler, T., Blitzer, J., Talukdar, P.P., Ungar, L.H.: Regularized learning with networks of features. In: Advances in Neural Information Processing Systems 21. Curran Associates, Inc., pp. 1401–1408 (2009)
4. Barabási, A.L., Gulbahce, N., Loscalzo, J.: Network medicine: a network-based approach to human disease. Nature Reviews Genetics **12**, 56–68 (2011)
5. Tran, T., Phung, D., Luo, W., Venkatesh, S.: Stabilized sparse ordinal regression for medical risk stratification. Knowledge and Information Systems, 1–28 (2014)
6. Ye, J., Liu, J.: Sparse methods for biomedical data. ACM SIGKDD Explorations Newsletter **14**, 4–15 (2012)
7. Zhao, P., Yu, B.: On model selection consistency of lasso. The Journal of Machine Learning Research **7**, 2541–2563 (2006)

8. Cun, Y., Fröhlich, H.: Biomarker gene signature discovery integrating network knowledge. Biology **1**, 5–17 (2012)
9. Dao, P., Wang, K., Collins, C., Ester, M., Lapuk, A., Sahinalp, S.C.: Optimally discriminative subnetwork markers predict response to chemotherapy. Bioinformatics **27**, i205–i213 (2011)
10. Sun, H., Lin, W., Feng, R., Li, H.: Network-regularized high-dimensional cox regression for analysis of genomic data. Statistica Sinica **24**, 1433–1459 (2014)
11. Fröhlich, H.: Including network knowledge into cox regression models for biomarker signature discovery. Biometrical Journal **56**, 287–306 (2014)
12. Simon, N., Friedman, J., Hastie, T., Tibshirani, R.: Regularization paths for cox's proportional hazards model via coordinate descent. Journal of Statistical Software **39**, 1–13 (2011)
13. Vinzamuri, B., Reddy, C.: Cox regression with correlation based regularization for electronic health records. In: ICDM, pp. 757–766 (2013)
14. Tibshirani, R., et al.: The lasso method for variable selection in the cox model. Statistics in Medicine **16**, 385–395 (1997)
15. Cox, D.R.: Partial likelihood. Biometrika **62**, 269–276 (1975)
16. Xu, H., Caramanis, C., Mannor, S.: Sparse algorithms are not stable: A no-free-lunch theorem. IEEE Transactions on Pattern Analysis and Machine Intelligence **34**, 187–193 (2012)
17. Liu, D.C., Nocedal, J.: On the limited memory bfgs method for large scale optimization. Mathematical Programming **45**, 503–528 (1989)
18. Pan, S.J., Yang, Q.: A survey on transfer learning. IEEE Transactions on Knowledge and Data Engineering **22**, 1345–1359 (2010)
19. Tran, T., Phung, D.Q., Luo, W., Harvey, R., Berk, M., Venkatesh, S.: An integrated framework for suicide risk prediction. In: KDD, 1410–1418 (2013)
20. Kuncheva, L.I.: A stability index for feature selection. In: Artificial Intelligence and Applications, 421–427 (2007)
21. Vinzamuri, B., Li, Y., Reddy, C.K.: Active learning based survival regression for censored data. In: CIKM 2014, 241–250. ACM, New York (2014)
22. Bilal, E., Dutkowski, J., Guinney, J., Jang, I.S., Logsdon, B.A., Pandey, G., Sauerwine, B.A., Shimoni, Y., Vollan, H.K.M., Mecham, B.H., et al.: Improving breast cancer survival analysis through competition-based multidimensional modeling. PLoS Computational Biology **9**, e1003047 (2013)

Predicting Next Locations with Object Clustering and Trajectory Clustering

Meng Chen[1], Yang Liu[1], and Xiaohui Yu[1,2 (✉)]

[1] School of Computer Science and Technology, Shandong University,
Jinan 250101, China
chenmeng114@hotmail.com, {yliu,xyu}@sdu.edu.cn
[2] School of Information Technology, York University,
Toronto, ON M3J 1P3, Canada

Abstract. Next location prediction is of great importance for many location based applications. In many cases, understanding the similarity between objects and the similarity between trajectories may lead to more accurate predictions. In this paper, we propose two novel models exploiting these two types of similarities respectively. The first model, named *object-clustered Markov model (object-MM)*, first clusters similar objects based on their spatial localities, and then builds variable-order Markov models with the trajectories of objects in the same cluster. The second model, named *trajectory-clustered Markov model (tra-MM)*, considers the similarity between trajectories, and clusters the trajectories to form the training set used in building the Markov models. The two models are integrated to produce the final next location predictor *(objectTra-MM)*. Experiments based on a real data set demonstrate significant increase in prediction accuracy of *objectTra-MM* over existing methods.

Keywords: Next location prediction · Object similarity · Trajectory similarity

1   Introduction

With the wide spread use of positioning technology and the increasing deployment of surveillance infrastructures, it is increasingly possible to track the movement of people and other objects (e.g., vehicles). The availability of such trajectories gives rise to a dazzling array of location-based applications, such as route planning [2,7] and location-based recommendations [1,13]. One key operation in such applications is predicting the next location of a moving object. For example, if we could predict the next locations of vehicles given the past trajectories, this knowledge can help us forecast traffic conditions and routes the drivers so as to alleviate traffic jams.

As another example, consider the location-based social network applications (e.g., Foursquare) that allow users to automatically check in at locations they are

© Springer International Publishing Switzerland 2015
T. Cao et al. (Eds.): PAKDD 2015, Part II, LNAI 9078, pp. 344–356, 2015.
DOI: 10.1007/978-3-319-18032-8_27

visiting. A set of check-ins can be regarded as a trajectory because each check-in has a location tag and a time-stamp respectively, corresponding to where and when the check-in is made. For instance, let us assume that Emma has just checked in at a shopping mall in Foursquare. If we could predict the location (the area in the mall) that she will pass by, we can provide some useful information to her, such as the dress on promotions or the most popular restaurant in the area.

The existing approaches to *next location prediction* are based on discovering the individual movement patterns [15], the collective movement patterns of a group [10], or both of them [3]. However, these methods have some serious drawbacks. First, prediction based on individual patterns may suffer from the data sparsity problem, as in some circumstances (e.g., social check-in, and traffic surveillance), an object usually has few past trajectories available, and meaningful movement patterns thus cannot be mined. Second, both individual and collective patterns fail to give proper consideration to the similarity between objects. Since different objects may have different travel preferences, objects with similar preferences tend to have similar moving patterns. Finally, methods based on collective patterns of all available trajectories make predictions at too coarse a granularity, and do not make good use of the inherent similarity between some trajectories.

In this paper, we propose two novel models to predict the next locations of moving objects, given the sequences of locations that they have just passed by. Inspired by user-based collaborative filtering adopted in recommender systems, we first propose a model named *object-clustered Markov model* (*object-MM*). The basic idea of *object-MM* is to predict the next location of an object using the mobility patterns of similar objects. We first compute the spatial distribution of each object, and then cluster the objects with similar localities. Finally, variable-order Markov models are constructed with trajectories in each cluster, which can quantify the movement patterns within that cluster.

In addition to *object-MM*, we propose a model called *trajectory-clustered Markov model* (*tra-MM*) that exploits the similarity between trajectories. In *tra-MM*, we first cluster similar trajectories according to a given similarity metric, and then for each cluster, train a variable-order Markov model using the trajectories contained therein. Since *object-MM* and *tra-MM* focus on different aspects of the movement patterns and thus complement each other, we propose to combine them using logistic regression. This results in a new model termed *objectTra-MM*, which is capable of producing more complete and accurate predictions.

When making predictions, given the sequence of locations the object has just passed by, we first identify the most relevant cluster based on the object (for *object-MM*) or the sequence (for *tra-MM*), and then obtain the probability of reaching each possible next location. Finally, the top ranked ones are returned as answers.

We present experimental results on a real data set consisting of the vehicle passage records over a period of 31 days in a metropolitan area. The quantitative results show that *objectTra-MM* significantly outperforms the other existing methods.

The contributions of this paper can be summarized as follows.

- We propose *object-MM*, an object-clustered Markov model that utilizes the past trajectories of similar objects to predict the next location of a given object.
- We present *tra-MM*, a trajectory-clustered Markov model that predicts the next location using patterns mined from similar trajectories.
- We build *objectTra-MM*, a model that integrates *object-MM* and *tra-MM*. To the best of our knowledge, *objectTra-MM* is the first model that takes a holistic approach and considers the similarity between objects and the similarity between trajectories in making next location predictions.
- We conduct extensive experiments with real-world traffic data to investigate the effectiveness of *objectTra-MM*, showing remarkable improvement as compared with baselines.

Roadmap. The rest of the paper is organized as follows. We review the related work in Section 2, and give the preliminaries of our work in Section 3. In Section 4, we present an object-clustered Markov model. In Section 5, we propose a trajectory-clustered Markov model, and integrate the proposed two models. We present the experimental results and performance analysis in Section 6, and conclude this paper in Section 7.

2   Related Work

2.1   Trajectory Mining

There have appeared a considerable body of studies that use only the individual movements to predict the next locations [9,12,15]. Xue et al. [15] use taxi traces to construct a Probabilistic Suffix Tree and predict short-term routes with Variable-order Markov Models. Simmons et al. [12] build a hidden Markov model (HMM) for every driver, which can predict the future destination and route of the target driver. Liao et al. [9] introduce a hierarchical Markov model that can learn and infer a user's daily movements through an urban community.

The other direction focuses on predicting next locations with the historical movements of all the moving objects [10,11,14]. Monreale et al. [10] build a T-pattern tree with all the trajectories to make future location predictions. Morzy [11] uses all the moving objects' locations to discover frequent trajectories and movement rules with the PrefixSpan algorithm. Xue et al. [14] decompose historical trajectories into sub-trajectories, and connects the sub-trajectories into "synthesised" trajectories, and then predict the destination with a Markov model. In addition, Chen et al. [3] consider both individual and collective movement patterns in predicting next locations.

2.2   Social-Media Mining

In addition to the trajectories of moving objects, there has also appeared work on trajectory mining using social-media data [6,17,19]. Zheng et al. [19] mine interesting locations and classical travel sequences in a given region with multiple users' GPS trajectories. Yin et al. [17] investigate the problem of trajectory pattern ranking and diversification based on geo-tagged social media. Hasan et al. [6] analyze urban human mobility and activity patterns using location-based data collected from social media applications.

Probably more related to the problem addressed in this paper is the work [16,18] on predicting next locations with semantic trajectories, where semantic trajectories are a sequence of visited places tagged with semantic information. Ye et al. [16] first predict the category of user activities with a hidden Markov model and then predict the most likely locations, given the estimated category distribution. Ying et al. [18] improve the accuracy of predicting future locations by integrating the semantic information with the location data.

3 Preliminaries

In this section, we first introduce some concepts which are required for the subsequent discussion, and then give an overview of the problem addressed in this paper.

Definition 1 (location). *An object o passes through a set of locations, where each location l is defined as a point or a region where the position of o is recorded.*

Definition 2 (Trajectory). *The trajectory T is defined as a time-ordered sequence of locations:* $< l_1, l_2, \ldots, l_n >$.

Definition 3 (Prefix Sequence). *For a location l_i and a given trajectory $T =< l_1, l_2, \ldots, l_n >$, its prefix sequence L_i^j refers to a length-j subsequence of T ending with l_i.*

4 Object-Clustered Markov Modeling

A naive approach to predicting next locations is to train a Markov model for each object using his/her past trajectories. However, this can be problematic: the individual trajectories are often spotty, and when the object goes to some locations that he (she) has not visited before, the Markov model has no ground to make a prediction. Indeed, the movement patterns of different objects exhibit similarity, and we can infer some unseen patterns of an object through modeling the trajectories of similar objects. This prompts us to propose an *object-clustered Markov model (object-MM)*.

object-MM predicts a given object's next locations based on the past trajectories of similar objects. The group of similar objects can be identified by analyzing the spatial distributions. For a known object's testing trajectory, only the trajectories of similar objects are used, and the posterior probabilities are computed with variable-order Markov models.

4.1   Computing the Spatial Locality Matrix

For the objects, who are usually localized to the same area are more likely to have similar spatial distributions, and their frequencies of visited locations are usually similar. So the spatial localities of objects may be expressed as the frequency of each location that one object has visited, which can be described with the formalized definition of *global location probability*.

Definition 4 (global location probability). *For an object o, the* global location probability *of location $l_i \in \mathcal{L}_o$ is defined as*

$$p(o, l_i) = \frac{\sharp(l_i)}{\sum_{l \in \mathcal{L}_o} \sharp(l)}.$$

In the above definition, $\mathcal{L}_o$ is the location set of object o and $\sharp(\cdot)$ is the counting number of corresponding location.

We then build an object spatial locality matrix $\mathbf{P}$ with global location probability. The row of $\mathbf{P}$ represents the objects; the column stands for the locations; the element of $\mathbf{P}$ represents the frequency that an object arrives at a location. A row vector reflects the frequency that one object arrives at every location. The matrix $\mathbf{P}$ can be described as follows:

$$\mathbf{P} = \begin{pmatrix} p_{11} & p_{12} & \cdots & p_{1l^*} \\ p_{21} & p_{22} & \cdots & p_{2l^*} \\ \vdots & \vdots & \cdots & \vdots \\ p_{o^*1} & p_{o^*2} & \cdots & p_{o^*l^*} \end{pmatrix}$$

$$p_{ij} = p(i, j), \tag{1}$$

where o^* is the number of unique objects and l^* is the number of unique locations. p_{ij} stands for the global location probability of object i arriving at location j. The matrix $\mathbf{P}$ may contain a few zero probabilities, and we can adopt a standard technique (e.g., smoothing) to obtain non-zero conditional probabilities for such states.

4.2   Clustering Objects

Having computed the spatial locality matrix, we cluster the similar objects to mine some unseen patterns. Here, the similarity between two objects can be measured by calculating the Kullback-Leibler divergence [4] between the corresponding row vectors of $\mathbf{P}$, but the same methodology still applies when other distance metrics (e.g., the cosine value) are used.

For the object i and j, $\mathbf{P}_i$ and $\mathbf{P}_j$ respectively stand for the ith and jth row vector of $\mathbf{P}$, which indicates the object's frequency of arriving at different locations, and the similarity between object i and j is measured by

$$Sim(\mathbf{P}_i, \mathbf{P}_j) = 1 - \frac{1}{2} \sum_{w=1}^{l^*} \left(p_{iw} \log \frac{p_{iw}}{p_{jw}} + p_{jw} \log \frac{p_{jw}}{p_{iw}} \right). \tag{2}$$

With the similarity metric defined, we use a method like K-means[5] to perform clustering for the objects.

4.3   Markov Modeling

The result of the object clustering process is K_c clusters, each containing a set of similar objects. We then train K_c Markov models for each cluster using only the trajectories contained therein. Let T be an object's trajectory of length n (i.e., it contains n locations), and let $p(l_{n+1}|T)$ be the probability that the object will arrive at location l_{n+1} next. The location l_{n+1} is given by

$$l_{n+1} = \underset{l_i \in \mathcal{L}}{\arg\max} \left\{ p\left(l_{n+1} = l_i | l_{n-(m-1)}, l_{n-(m-2)}, \ldots, l_n \right) \right\}, \tag{3}$$

where $\mathcal{L}$ is the set of all the locations. The number m is the order of the Markov model.

In order to use the mth-order Markov model, we learn l_{n+1} for each prefix sequence containing m locations, $L_n^m = \left(l_{n-(m-1)}, \ldots, l_{n-1}, l_n\right)$, by estimating the various conditional probabilities,

$$p\left(l_i | L_n^m\right) = \frac{\sharp(L_n^m, l_i)}{\sharp(L_n^m)}, \tag{4}$$

where $\sharp(L_n^m)$ is the number of times that prefix sequence L_n^m occurs in the training set, and $\sharp(L_n^m, l_i)$ is the number of times that location l_i occurs immediately after L_n^m.

For K_c clusters, we need to train K_c variable-order Markov models from the first-order to the mth-order with trajectories in every cluster. For the cluster c, the location l_{n+1} that the object will arrive at next is given by

$$
\begin{aligned}
l_{n+1} &= \underset{l_i \in \mathcal{L}}{\arg\max} \left\{ p\left(l_{n+1} = l_i | L_n^m, \mathcal{T}_c\right) \right\} \\
&= \underset{l_i \in \mathcal{L}}{\arg\max} \left\{ \frac{\sharp(L_n^m, l_i, \mathcal{T}_c)}{\sharp(L_n^m, \mathcal{T}_c)} \right\},
\end{aligned} \tag{5}
$$

where $\mathcal{T}_c$ is the training set of trajectories in cluster c, and $\sharp(L_n^m, \mathcal{T}_c)$ is the number of times that prefix sequence L_n^m occurs in $\mathcal{T}_c$, and $\sharp(L_n^m, l_i, \mathcal{T}_c)$ is the number of times that location l_i occurs immediately after L_n^m in $\mathcal{T}_c$.

In order to make predictions, for a given sequence of locations that object o has just passed by, we first identify the cluster that o belongs to, and then adopt the principle of longest match to predict next locations with the most relevant model. That is, if we build the first-, second-, and mth-order Markov models, for the given sequence, we first try to make predictions using the mth-order model. If this model does not contain the corresponding state, we then try to predict next locations using the $(m-1)$th-order model, and so on.

5   Trajectory-Clustered Markov Modeling

The movement of people often exhibits strong collective characteristics, and we can build a Markov model using the trajectories of all moving objects. However, the model makes predictions at too coarse a granularity, and does not consider the inherent similarity between different trajectories. Therefore, we present a *trajectory-clustered Markov model* (tra-MM), which predicts next locations with the mobility patterns mined from the historical trajectories similar to the testing trajectory sequence.

5.1   Trajectory Clustering

One critical step in *tra-MM* is to cluster the similar trajectories. A trajectory consists of a time-ordered sequence of locations, and is a kind of time series data. There exist a large volume of distance measures for similarity of time series data, e.g., Euclidean distance and Dynamic Time Warping (DTW). Every distance measure has its own particular superiority, and here we choose to use DTW to measure the distance between two trajectories, but other distance metrics can also be used. The distance between trajectory T_i and T_j is formalized as follows:
$DTW(T_i, T_j) =$

$$
\begin{cases}
0 & \text{if } |T_i| = |T_j| = 0 \\
\infty & \text{if } |T_i| = 0 \text{ or } |T_j| = 0 \\
dist(l_1^i, l_1^j) + min\left\{DTW(Rest(T_i), Rest(T_j)),\right. & \text{otherwise} \\
\left.DTW(Rest(T_i), T_j),\ DTW(T_i, Rest(T_j))\right\},
\end{cases}
\tag{6}
$$

where $|T|$ is the length of trajectory T and $Rest(T)$ is the subsequence of T without the first location. $dist(l_1^i, l_1^j)$ is the distance between a pair of location l_1^i and l_1^j, in which l_1^i and l_1^j represent the first location in the trajectory T_i and T_j respectively. In our experiments, we set $dist(l_1^i, l_1^j)$ at 1 if l_1^i is not the same as l_1^j and 0 otherwise.

With the distance metric defined, we perform clustering on the trajectories. Since huge amounts of trajectories are often received incrementally, clustering should be processed efficiently. However, traditional clustering algorithms are developed for static and small data sets, they are not suitable for large-scale trajectory clustering. Therefore, we propose an *Incremental Trajectory Clustering* method similar to Trajectory Micro-Clustering used in [8], which is detailed in Algorithm 1.

5.2   Next Location Prediction

After clustering trajectories, we train a variable-order Markov model with order ranging from 1 to m for every cluster using only the trajectories contained therein. Specially, we first compute the transition probability of $P(l_{n+1}|L_n^1)$ from prefix sequence L_n^1 to next location l_{n+1}, followed by $P(l_{n+1}|L_n^2)$ from prefix sequence L_n^2 to l_{n+1}, until $P(l_{n+1}|L_n^m)$ is computed, to train a variable-order *tra-MM*. For a given trajectory, we first compute its closest cluster and then choose the corresponding model of the cluster to predict the next locations.

Algorithm 1. Incremental Trajectory Clustering

Input: New trajectories $\{T_1, T_2, \ldots, T_n\}$, existed clusters $C = \{C_1, C_2, \ldots, C_r\}$ and distance threshold δ;
Output: Updated clusters with new trajectories;
  1: **for** every new trajectory T_i **do**
  2:     Find the closest cluster C_j;
  3:     **if** $DTW(T_i, C_j) < \delta$ **then**
  4:         Add T_i into C_j and update C_j accordingly;
  5:     **else**
  6:         Create a new cluster for T_i;
  7:     **end if**
  8: **end for**

5.3 Integration of object-MM and tra-MM

object-MM and *tra-MM* concentrate on different aspects of the movement patterns and thus complement each other, so we integrate them through logistic regression to obtain the final *objectTra-MM*.

For the given i-th trajectory sequence, we can get a vector of probabilities, $\mathbf{p}_i^w = \left(p_1^i, p_2^i, \cdots, p_n^i\right)'$ ($w = 1$ for *object-MM* and $w = 2$ for *tra-MM*), where n is the number of the locations, and p_j^i is the probability of location j being the next location. Let $\mathbf{y}_i = (y_1^i, y_2^i, \cdots, y_n^i)'$ be a prediction vector for the i-th trajectory sequence, and

$$\mathbf{y}_i = \beta_0 \mathbf{1} + \sum_{w=1}^{2} \beta_w \mathbf{p}_i^w, \tag{7}$$

where $\mathbf{1}$ is a unit vector, and β_w is the coefficient for different individual models. We also have a vector of indicators $\mathbf{r}_i = (r_1^i, r_2^i, \cdots, r_n^i)'$ for the i-th trajectory sequence, where $r_j^i = 1$ if the actual next location is j and 0 otherwise. We can predict $\mathbf{r}_i$ through the logistic regression approach:

$$\hat{\mathbf{r}}_i = \frac{1}{1 + \exp(-1 * \mathbf{y}_i)}. \tag{8}$$

The optimal values of β_w can be easily obtained through the logistic regression approach. Then the proposed *objectTra-MM* is trained again using the same parameters on the entire training set. For a testing trajectory sequence, we can predict the next location by identifying the largest elements in the estimator $\hat{\mathbf{r}}$.

6 Performance Evaluation

We present experiments using a real vehicle passage data set to conduct quantitative evaluation for the proposed models. In this section, we first describe the data set and settings adopted in our empirical study, and then present the experimental results of our proposal. Finally, we compare the *objectTra-MM* with other state-of-the-art strategies.

6.1   Datasets and Settings

In our study, we exploit a real vehicle passage data set which is collected over the traffic surveillance system in a major metropolitan city. We accumulate 10,344,058 records from the data center during a period of 31 days. Each record here contains a vehicle ID, a location of the surveillance camera, and a timestamp of passage. We pre-process them to formulate trajectories, and only consider trajectories that contain at least three locations to make the model more robust. According to statistics, one vehicle only has about 28 trajectories in a month on average, which validates the data sparsity problem mentioned above. In the task of predicting next locations, we first choose trajectories in the first 20 days to construct the models, and adopt the data in the next 7 days to fine tune the parameters. Next, 746,790 trajectories in the first 27 days are used to train the final models with the tuned parameters. Finally, we use the remaining 4-day 104,129 trajectories to formulate the test data set.

To compare different Markov model-based methods, we use two evaluation metrics, namely, *accuracy* and *average precision*.

Accuracy is defined as the ratio of the number of trajectories for which the model is able to correctly predict to the total number of trajectories in the test set. That is, $accuracy = \sum p(l) / |T'|$, where $|T'|$ is the number of trajectories in the testing set, and $p(l)$ is 1 if l is the true next location and 0 otherwise.

Average precision is defined as $ap = \sum (p(l_i)/i) / |T'|$, where i denotes the position in the predicted list, and $p(l_i)$ takes the value of 1 if the predicted location at the i-th position in the list is the actual next location and 0 otherwise.

6.2   Evaluation of Models

In this section, we evaluate the performance of our proposed models, namely, *object-MM*, *tra-MM* and their integration *objectTra-MM*. For each experiment, we perform 50 runs and report the average of the results. First, we study the effect of the order of Markov model by varying order from 1 to 5. We respectively set different parameters for the models: *object-MM* (cluster number $K_c = 40$, 80), *tra-MM* (distance threshold $\delta = 6, 8$), and *objectTra-MM* ($K_c = 80$, $\delta = 8$). Since the values of cluster number and distance threshold are estimations, we vary them in a range to obtain more comprehensive performance evaluation. We demonstrate the performances using top-1 accuracy (accuracy@1) and top-5 average precision (ap@5).

As shown in Fig. 1, the accuracy and average precision of all the models have an apparent improvement when the order increases from 1 to 2, and start to decrease as we further continue to increase the order. The higher-order model has a number of limitations associated with high state-space complexity, and gradually deteriorates the overall prediction accuracy. So we set the order of Markov model at 2 in the following experiments.

Next, We study the effect of the cluster number K_c on the performance of *object-MM*. To this end, we vary K_c from 5 to 200 at a step of 5, and demonstrate

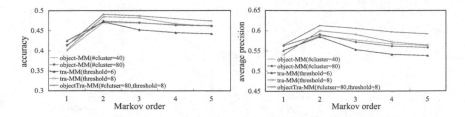

Fig. 1. Effect of Markov order on accuracy and average precision

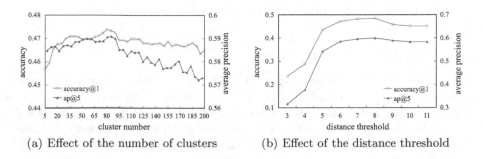

(a) Effect of the number of clusters   (b) Effect of the distance threshold

Fig. 2. Effect of parameters

the results in Fig. 2(a). Clearly, as we increase the cluster number, the *object-MM* achieves its maximum accuracy at the cluster number 80, and the accuracy starts to decrease when we continue to further increase the cluster number. The decrease is because having too many or too few clusters with either hurts the cohesiveness or the separation of the clusters. The average precision has similar change, and comes to the maximum when the cluster number is around 80.

We then vary the distance threshold δ from 3 to 11 to study the effect of it on *tra-MM*. The results are shown in Fig. 2(b). Initially an increase in the threshold, is accompanied by an increase in the accuracy and average precision, and they reach the peak values when the threshold is 8. As we continue to further increase the distance threshold, the accuracy and average precision start to decrease and remain stable. This decrease is because we have to put some dissimilar trajectories in the same cluster, affecting the overall accuracy.

Finally, we predict top k next locations with *object-MM (cluster number $K_c = 80$), tra-MM (distance threshold $\delta = 8$)* and *objectTra-MM ($K_c = 80$, $\delta = 8$)*, and evaluate their performances. A number of interesting observations can be made from Fig. 3. On one hand, for the three models, it is obvious that the accuracy and average precision improve as we increase k. On the other hand, for any k, *tra-MM* performs better than *object-MM*, and *objectTra-MM* obtains the best performance, validating the effectiveness of the integration of two models.

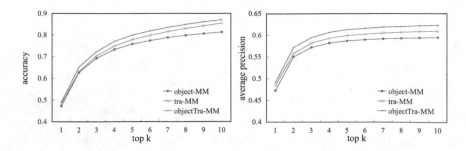

Fig. 3. Performance of object-MM, tra-MM and objectTra-MM

6.3   Comparisons with Baselines

To evaluate the effectiveness of *objectTra-MM*, we compare against some start-of-the-art approaches including *VMM* [15], *WhereNext* [10] and *NLPMM* [3]. *VMM* discovers individual patterns for each object based on its own historical data; *WhereNext* uses the previous movements of all moving objects to make future location predictions; *NLPMM* considers both individual and collective movement patterns in making predictions. We choose the optimal parameters for the start-of-the-art approaches after many experiments. The parameters of *VMM* are set as follows: memory length $N=2$, $\sigma=0.3$, and $N_{min}=1$. For *WhereNext*, the support for constructing T-pattern tree is 20. For the *NLPMM*, the Markov order is 2. For the *objectTra-MM*, the cluster number K_c is 80, and the distance threshold δ is 8.

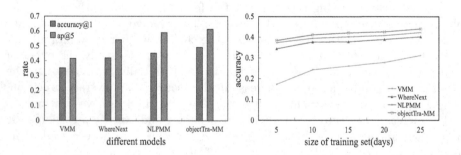

Fig. 4. Performance comparison with baselines

Fig. 4 shows the performance comparison of *objectTra-MM* with baselines in terms of top-1 accuracy and top-5 average precision. *objectTra-MM* performs better than the start-of-the-art approaches, as it considers the spatial localities of different objects and the similarity between trajectories in location prediction. In addition, we also evaluate the performances of all the models with different sizes of training set. It is apparent that *objectTra-MM* has the best top-1 accuracy over other methods for any training set size.

7   Conclusions

In this paper, we have proposed a next location predictor *objectTra-MM* to predict the next location of an object with a given sequence of locations. The proposed *objectTra-MM* consists of two models: *object-clustered Markov model* (object-MM) and *trajectory-clustered Markov model* (tra-MM). *object-MM* analyzes the objects' spatial localities, and clusters the objects with similar localities. For each cluster, *object-MM* trains a variable-order Markov model using the trajectories contained therein. *tra-MM* clusters trajectories using a given similarity metric, and trains a series of Markov models with trajectories in each cluster. We have evaluated the proposed models using a real vehicle passage record data set, and the experiments show that the proposed model significantly outperforms the state-of-the-art methods (*VMM*, *WhereNext* and *NLPMM*).

Acknowledgments. This work was supported in part by the 973 Program (2015CB352 500), the National Natural Science Foundation of China Grant (6127 2092), the Shandong Provincial Natural Science Foundation Grant (ZR2012 FZ004), the Science and Technology Development Program of Shandong Province (2014GGE27178), the Independent Innovation Foun-dation of Shandong University (2012ZD012), the Taishan Scholars Program, and NSERC Discovery Grants.

References

1. Bao, J., Zheng, Y., Mokbel, M.F.: Location-based and preference-aware recommendation using sparse geo-social networking data. In: GIS, pp. 199–208. ACM (2012)
2. Chen, L., Lv, M., Ye, Q., Chen, G., Woodward, J.: A personal route prediction system based on trajectory data mining. Information Sciences **181**(7), 1264–1284 (2011)
3. Chen, M., Liu, Y., Yu, X.: NLPMM: A Next Location Predictor with Markov Modeling. In: Tseng, V.S., Ho, T.B., Zhou, Z.-H., Chen, A.L.P., Kao, H.-Y. (eds.) PAKDD 2014, Part II. LNCS, vol. 8444, pp. 186–197. Springer, Heidelberg (2014)
4. Ertoz, L., Steinbach, M., Kumar, V.: A new shared nearest neighbor clustering algorithm and its applications. In: SDM, pp. 105–115 (2002)
5. Hartigan, J.A., Wong, M.A.: Algorithm as 136: A k-means clustering algorithm. Applied Statistics, 100–108 (1979)
6. Hasan, S., Zhan, X., Ukkusuri, S.V.: Understanding urban human activity and mobility patterns using large-scale location-based data from online social media. In: UrbComp, p. 6. ACM (2013)
7. Kurashima, T., Iwata, T., Irie, G., Fujimura, K.: Travel route recommendation using geotags in photo sharing sites. In: CIKM, pp. 579–588 (2010)
8. Li, Z., Lee, J.-G., Li, X., Han, J.: Incremental Clustering for Trajectories. In: Kitagawa, H., Ishikawa, Y., Li, Q., Watanabe, C. (eds.) DASFAA 2010. LNCS, vol. 5982, pp. 32–46. Springer, Heidelberg (2010)
9. Liao, L., Patterson, D.J., Fox, D., Kautz, H.: Learning and inferring transportation routines. AI **171**(5), 311–331 (2007)

10. Monreale, A., Pinelli, F., Trasarti, R., Giannotti, F.: Wherenext: a location predictor on trajectory pattern mining. In: SIGKDD, pp. 637–646 (2009)
11. Morzy, M.: Mining Frequent Trajectories of Moving Objects for Location Prediction. In: Perner, P. (ed.) MLDM 2007. LNCS (LNAI), vol. 4571, pp. 667–680. Springer, Heidelberg (2007)
12. Simmons, R., Browning, B., Zhang, Y., Sadekar, V.: Learning to predict driver route and destination intent. In: ITSC, pp. 127–132 (2006)
13. Son, J.W., Kim, A., Park, S.B., et al.: A location-based news article recommendation with explicit localized semantic analysis. In: SIGIR, pp. 293–302. ACM (2013)
14. Xue, A.Y., Zhang, R., Zheng, Y., Xie, X., Huang, J., Xu, Z.: Destination prediction by sub-trajectory synthesis and privacy protection against such prediction. In: ICDE, pp. 254–265. IEEE (2013)
15. Xue, G., Li, Z., Zhu, H., Liu, Y.: Traffic-known urban vehicular route prediction based on partial mobility patterns. In: ICPADS, pp. 369–375 (2009)
16. Ye, J., Zhu, Z., Cheng, H.: Whats your next move: User activity prediction in location-based social networks. In: SDM, pp. 171–179 (2013)
17. Yin, Z., Cao, L., Han, J., Luo, J., Huang, T.: Diversified trajectory pattern ranking in geo-tagged social media. In: SDM, pp. 980–991 (2011)
18. Ying, J.J.C., Lee, W.C., Weng, T.C., Tseng, V.S.: Semantic trajectory mining for location prediction. In: GIS, pp. 34–43. ACM (2011)
19. Zheng, Y., Zhang, L., Xie, X., Ma, W.Y.: Mining interesting locations and travel sequences from gps trajectories. In: WWW, pp. 791–800 (2009)

A Plane Moving Average Algorithm for Short-Term Traffic Flow Prediction

Lei Lv[1], Meng Chen[1], Yang Liu[1], and Xiaohui Yu[1,2(✉)]

[1] School of Computer Science and Technology, Shandong University,
Jinan 250101, China
llsdu13@gmail.com, chenmeng114@hotmail.com, {yliu,xyu}@sdu.edu.cn
[2] School of Information Technology, York University, Toronto,
ON M3J 1P3, Canada

Abstract. In this paper, a plane moving average algorithm is proposed for solving the urban road flow forecasting problem. This new approach assembles information from relevant traffic time series and has the following advantages: (1) it integrates both individual and similar flow patterns in making prediction, (2) the training data set does not need to be large, (3) it has more generalization capabilities in predicting unpredictable and much complex urban traffic flow than previously used methods. To assess the new model, we have performed extensive experiments on a real data set, and the results give evidence of its superiority over existing methods.

Keywords: Plane moving average(PMA) · Time series · Flow pattern · Traffic flow prediction

1 Introduction

Traffic flow forecasting is a vital component of transportation planning, traffic control, intelligent transportation systems, and forecasting accurate traffic flow conditions has long been considered as an active approach to regional traffic control [16]. The approach can be broadly classified under: i) short-term and ii) long-term traffic flow forecasting [1]. Long-term forecasting provides monthly or yearly traffic flow forecasting conditions and is commonly used for long-term planning of transportation or construction. Short-term forecasting focuses on making predictions about traffic flow changes in the short-term, typically within one hour. In particular, short-term traffic volume forecasts support proactive dynamic traffic control. As a result, prediction technologies have gotten the attention of traffic engineers and researchers. A wide variety of techniques have been applied in the context of short-term traffic flow forecasting, depending upon the type of data that are available and the potential end use of the forecast. These techniques include moving average methods [17], k-nearest-neighbor methods [8], autoregressive MA (ARIMA) model or seasonal ARIMA (SARIMA) [19,20], neural networks (NNs) [9], and combining technologies such as DA approach [18].

© Springer International Publishing Switzerland 2015
T. Cao et al. (Eds.): PAKDD 2015, Part II, LNAI 9078, pp. 357–369, 2015.
DOI: 10.1007/978-3-319-18032-8_28

Development of traffic flow forecasting models relies mainly on historical and current traffic flow data. The problem of traffic flow forecasting is a standard time series prediction task and the goal is to approximate the function that relates future values of traffic flow to previous and current observations of traffic flow, and the short-term traffic flow predictor represents a multi-input-single-output system, which relates the past traffic flow conditions to the future traffic flow conditions. Recent literatures have been concentrating on three aspects: i). presenting new or enhanced approaches [2,11,15], ii). merging efficient forecasting results [3,7,18], and iii). proposing effective pre-processing technologies [4,5,10].

However, it becomes clear that there are a few major problems with the existing methods. Firstly, those methods use data collected from motorways and freeways, where the change of traffic flow seems quite stable. Secondly, they use data collected from single point sources, which are limited in a more complicated situation. Thirdly, existing forecasting methods do not consider the temporal data as complex interactions in densely populated urban road networks and fresh comprehensive enough as well. Precisely, short-term traffic forecasting at urban arterials forms a more complex problem than freeway predictions due to constraints such as signalization, and urban flow collections contain lots of data points that previous models cannot handle their relations. The irrelevant traffic junctions may get the similar change rule and present the same flow pattern. The traditional methods do not consider the similar flow patterns and mine these similarities as well, they are often narrow in the angle of the data source and some of them are modeled so complicated, these reasons make them not appear to get practical application.

To solve those problems, we propose a Plane Moving Average (PMA) algorithm. The main idea of PMA is that it uses the closest flow data to make predictions through a new perspective. The PMA model is simple but effective and builds upon two models: the Individual Model and the Similarity Model. The first one, focuses on individual traffic junction using its own past vehicle volumes, the second one utilizes all traffic junctions' vehicle flow data based on the customized WWL (Where We Like) method which is a kind of Top-K technology and both of two sub-models use PMA algorithm to make predictions. Finally, We combine these two forecasting results by neural network to produce a more accurate predictor.

The contributions of this paper can be summarized as follows.

- We propose a Plane Moving Average to predict the short-term traffic flow. It is much simple but effective and uses a new angle of flow patterns in making prediction.
- To the best of our knowledge, PMA is the first model that takes similarity flow patterns into account and integrates both individual and similarity outcomes.
- We perform extensive experiments using a real data set and the results demonstrate the effectiveness of PMA.

The rest of this paper is organized as follows. Section 2 reviews related work. Section 3 provides a brief description of preliminary preparation. Section 4 designs and describes the PMA Modeling of the short-term traffic flow predictor. In Section 5, the performance is discussed. Finally, the discussion of the conclusions and the future visions, regarding short-term traffic flow predictor design using the PMA with WWL algorithm, is presented in Section 6.

2 Related Work

For the past two decades, a considerable effort has made to develop efficient traffic prediction methods, which is backed by a huge number of literatures in this field. In what follows, we will describe three categories of studies that are most closely related to us.

Long-term prediction: Long-term prediction is researched in [13,14] and provides monthly or yearly flow forecasting conditions that is commonly used for long-term planning. Papagiannaki explores the properties of the network traffic, and propose a methodology that can be applied to forecast network traffic volume months in the future.

Short-term prediction: Short-term prediction has been widely investigated [2, 3,7,11,15,18–20], which is concerned with the prediction of only the nearly next period of time. Some of these methods make prediction with only the individual models [19,20], while others use the combination models [3,7,18]. Williams uses the well-known time series model ARIMA and SARIMA to make vehicular traffic flow prediction [19,20]. Tan et al. proposes an aggregation model (DA) that is elaborated using different fitting functions: the moving average (MA), exponential smoothing (ES), ARIMA and simple neural network for evaluation of prediction values [18]. Chan et al. provides a neural-network-based models for short-term traffic flow forecasting using a hybrid exponential smoothing and levenberg? Cmarquardt algorithm [3]. Davarynejad describes a multi-phase time series architecture to solve the motorway flow forecasting problem [7]. [2,11,15] incorporate multiple factors more or less, such as travel speed, weather conditions and geographical features of a road. Given the travel speed and traffic volume of a road segment, more accurate predicting model can be built.

Preprocessing techniques: Several studies adopt data preprocessing technique to improve the final forecasting results [4,5,10]. Chan et al. proposes a simple but effective training method to pre-process traffic flow data before training purposes and the pre-processing approach intends to aid the back-propagation algorithm to develop more accurate neural networks [4,5]. Gao et al. decomposes the original data into burst data traffic and non-burst data and be predicted, respectively [10]. These techniques indicate that the flow predictor based on these pre-processed data outperform those that are developed based on original data.

3 Preliminary Preparation

In this section, we will define a few terms that are required for the subsequent discussion and describe our modeling strategy for the vehicle traffic flow forecasting.

Definition 1 (Sampling Flow). *The traffic flow data are collected all the time, each sampling flow refers to a total number of passed vehicles during a certain period of time. Unaffectedly, we choose Δt minutes per period that the traffic flow is collected within the time interval $(t - \Delta t, t]$, t is an integer. $q(t)$ is also an integer and stands for the traffic volume in the No.t periods.*

Definition 2 (Flow Pattern). *For each sampling flow, $q(t)$ is the source time series. By analyzing the observed traffic flow data, it can be found that the traffic flow pattern is almost cyclical every day. Thus, two relevant time series are constructed as the period series $s_a(t)$ and the daily series $s_b(t)$, denoted as follows:*
I. *$s_a(t)$ is a set that includes the previous k_a periods traffic flow data before $q(t)$.*

$$s_a(t) = \{q(t - k_a\Delta t), q(t - (k_a - 1)\Delta t), ..., q(t - \Delta t), \hat{q}(t)\}.$$

II. *$s_b(t)$ is a set that includes the previous traffic flow record within the same time interval on k_b days before $q(t)$.*

$$s_b(t) = \{q(t - 1440k_b), q(t - 1440(k_b - 1)), ..., q(t - 1440), \hat{q}(t)\}.$$

4   PMA Modeling

We use a new PMA model to solve the short-term traffic flow prediction problem. PMA is an improvement algorithm of moving average(MA). Specifically, PMA runs on the plane flow pattern (a 2-dimensional data structure).

$S_b(t-k_a\triangle t)$		$\cdots$	$S_b(t)$	
$q(t-1440k_b-k_a\triangle t)$	$q(t-1440k_b-(k_a-1)\triangle t)$	$\cdots$	$q(t-1440k_b)$	$S_a(t-1440k_b)$
$q(t-1440(k_b-1)-k_a\triangle t)$	$q(t-1440(k_b-1)-(k_a-1)\triangle t)$	$\cdots$	$q(t-1440(k_b-1))$	
$q(t-1440(k_b-2)-k_a\triangle t)$	$q(t-1440(k_b-2)-(k_a-1)\triangle t)$	$\cdots$	$q(t-1440(k_b-2))$	
$\cdots$	$\cdots$	$\cdots$	$\cdots$	$\vdots$
$q(t-1440-k_a\triangle t)$	$q(t-1440-(k_a-1)\triangle t)$	$\cdots$	$q(t-1440)$	
$q(t-k_a\triangle t)$	$q(t-(k_a-1)\triangle t)$	$\cdots$	$\hat{q}(t)$	$S_a(t)$

$$PFP_{individual}(t)$$

Fig. 1. Plane flow pattern of individual model. $\hat{q}(t)$ is the predicted value.

In order to take into account both individual and similar flow patterns to make the prediction, we propose two models, an Individual Model to model the individual flow patterns by using its own past traffic flow data and a Similarity Model to model the similar flow patterns by using alike historical traffic flow data. They are combined by using neural network to generate a predictor. What calls for special attention is that these two models share the PMA algorithm. Next we will present data construction methods about similarity model, individual model, the PMA training approach and the model integration.

4.1   Individual Model

We call $s_a(t)$ the horizontal flow pattern and call $s_b(t)$ vertical flow pattern, and propose a new perspective to build a plane flow pattern (PFP) (see Figure 1). There are $s_a(t)$, $s_b(t)$, k_a periods before $s_b(t - k_a \Delta t)$ and k_b days before $s_a(t - 1440 k_b)$ combined together to build plane flow pattern (PFP) and each two adjacent value q's timestamp differs either of Δt minutes and a day. We may find out it is possible that $k_a \Delta t$ probably equals to 1440 which means that a typical sampling flow will appear more than once in PFP and get reutilization.

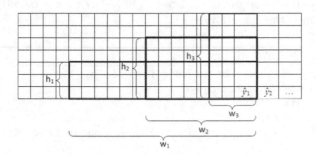

Fig. 2. Different size of PFP during the PMA training process and $\hat{y}$ is the forecasting value

PMA algorithm is used to produce the forecasting results. For convenience, we will use $y(i, j)$ to refer the ith line of the jth column's value of the plane flow pattern. In order to describe PMA algorithm, we must define two factors in the first place.

Pattern Weight. Each one of flow pattern could be benefited to the forecasting accuracy, so we define a kind of pattern weight ρ to describe different pattern's contribution to the target predictions.

$$\rho(m - i) = \frac{1}{L} \sum_{k_a=1}^{L} y(m, \delta) \, / \, y(m - i, \delta - k_a) \quad 0 < i < m \qquad (1)$$

where L is the length of the PFP (see Figure 2) and δ refers to $\hat{q}(t)$'s column, m is the target day number of the data set.

PMA Forecasting Value. The computing method is similar to MA, a kind of simple smoothing technique.

$$PMA(\delta) = \frac{1}{W} \sum_{k_b=1}^{W} y(m - k_b, \delta) \times \rho(m - k_b) \qquad (2)$$

where W is the width of the PFP (see Figure 2). Each PMA forecasting value corresponds to a specific PMA size $L_i \times W_i$.

For example, consider this PFP
$$\begin{array}{|cccc|}
\hline
3 & 4 & 5 & 10 \\
2 & 3 & 4 & 8 \\
1 & 2 & 3 & [4] \\
\hline
\end{array}$$
that consists of three kinds of s_a

patterns and four kinds of s_b patterns, updown adjacent patterns are one day apart, the target value is 4. If we simply use flow pattern $\{1, 2, 3\}$ and $\{10, 8\}$ to make the average forecasting value, we receive value 2 and 9 but none of them is closed to 4. Using pattern weight we will get two weight 0.478 and 0.639, and the final forecasting value would change to $(10 \times 0.478 + 8 \times 0.639)/2 = 4.946$. Obviously, 4.946 is the nearest value.

Algorithm 1. PMA Algorithm

1: Firstly, for the PFP, work out the pattern weight ρ by (1);
2: $\langle X, Y \rangle$ is the size of the training set;
3: **for** $L = 1...X$ **do**
4:     **for** $W = 1...Y$ **do**
5:         To each $\langle L, W \rangle$, compute PMA forecasting value of all training samples by (2);
6:         Calculate the average RMSE between the each pair of forecasting value and the actual value;
7:         Update the optimal PMA size according to the minimum RMSE;
8:     **end for**
9: **end for**
10: Use the optimal PMA size to make predictions.

In designing PMA training approach, the PMA size in the training set is an important feature that needs to be carefully chosen. The PMA algorithm is detailed in Algorithm 1. To obtain a PMA size that is capable of generalizing and performing well with new cases, data samples are usually subdivided into two sets: 1) a training set, 2) a testing set.

4.2   Similarity Model

In most cases, time series in the same category show the similar change rule, especially in a certain field (e.g., same area). For example, a flow that over 1000 vehicles /h must not happen in a small town's main road. Therefore, we put forward a hybrid distance calculation criterion which combined Euclidean Distance and Pearson Correlation Coefficient to evaluate time series similarity, and select top-N similar time series to make predictions. Notice that the comparable time series should be at least **one** day before and if not, no known value could be available.

WWL (Where We Like) Algorithm. Between each two time series(equilong), we come up with a hybrid distance calculation which combined Euclidean Distance and Pearson Correlation Coefficient as the distance metric to define their similarities. Minimizing these two metrics can both ensure a highly similar level about traffic volume and landmark as well. Firstly, filter time series by calculating Pearson Correlation Coefficient $pcc(s^i, s^j) = \frac{\sum_{t=1}^{m}(s^i(t)-\bar{s^i})(s^j(t)-\bar{s^j})}{\sqrt{\sum_{t=1}^{m}(s^i(t)-\bar{s^i})^2\sum_{t=1}^{m}(s^j(t)-\bar{s^j})^2}}$ that pcc value is over 0.8 to obtain the highly closed shape. Secondly, calculate Euclidean Distance by equation $ed(s^i, s^j) = ||s^i - s^j||$ to make sure that the traffic volume is as much as close and order them in the following way.

$$PFP_{similarity}(t) = \{s(t), s^1(t), ..., s^i(t), ..., s^d(t)\}. \tag{3}$$

where $s(t)$ is the target flow pattern, $s^i(t)$ stands for the $No.i$ similar flow pattern to $s(t)$ and the number d is the parameter of WWL algorithm. Remember that each $s^i(t)$ should be at least **one** day before.

Empirically, in terms of flow data, the more similar the shape, the more useful the flow pattern. Hence, for ease of running PMA, we put them bottom-top in a straight line according to the order of similarity by WWL algorithm just the same as $PFP_{individual}(t)$ does in individual model and PMA algorithm is used to produce its forecasting results.

4.3   Integration

Choosing a proper model to merge these two forecasting results is the primary task. There are many popular ways can be applied to do combination, such as Linear Regression, Logistic Regression, Neural Network, etc. For the effectiveness and simplicity, we choose neural network to integrate these two models that we have proposed.

As a result, the similarity model and individual model can get a series of predicting values, $\mathbf{P}^w = (p_1^w, p_2^w, ..., p_m^w)$ ($w = 1$ for similarity model and $w = 2$ for individual model), where m is the number of the time-ordered sampling results, and p_i^w is the forecasting value of time m. Certainly, we have a series of real traffic flow, $\mathbf{R} = (r_1, r_2, ..., r_m)$ as the predicting target. Then we can build a data set like $(\mathbf{P}^1, \mathbf{P}^2, \mathbf{R})$.

In designing NN models, the activation functions in the hidden layer and the output layer are *tansig* and *purelin*, respectively. We use the Levenberg-Marquardt BP algorithm as the learning algorithm because LM algorithm has the quickest convergence and it is the best learning rule in this case.

$$h_i = tansig(\sum_{j=1}^{2} w_{ij}p^j - \theta_1)$$
$$net = purelin(\sum_{i=1}^{n} w_i h_i - \theta_2) \tag{4}$$

where n is the number of neurons in hidden layer and h_i is the hidden layer output and net is the final output. Although there is no precise rule on the optimum size of the training set and testing set, it is recommended that the training set should be larger.

5   Performance Evaluation

We have made extensive experiments to evaluate the performance of the proposed PMA model by using a real traffic flow data set. In this section, we will first explain the dataset and experimental settings, followed by the evaluation metrics to measure the performance and then, show the experimental results.

5.1   Study Area

The dataset used in the experiments consists of real vehicle passage records from April 1, 2013 to May 5, 2013 that collected from the traffic surveillance system of a major metropolitan area [6]. The dataset contains 140,440,933 real vehicle passage records, involves totally 308 camera locations on the main roads and we divide it into two parts, training set and testing set. The training set for the above models covers the data from April 1, 2013 to May 2, 2013 and the testing set is the traffic records from May 3 to May 5, 2013. We average the experimental results of all camera locations. The traffic flow data were aggregated and averaged into Δt minutes per period.

5.2   Pre-processing

We pre-process the dataset to form the flow patterns, counting the number of vehicles in each peroid of time, [18] suggests setting time interval Δt as 60. If a sampling flow is missing (always be 0 in this case) due to the camera's break or any other reason, we fix it by averaging the adjacent values. After the preprocessing, we get a total of 10780 horizontal flow patterns and 7392 vertical flow patterns and we predict traffic flow of future one period in the experiments.

5.3   Goodness-of-Fit Statistics

We use two widely employed evaluation statistics to assess the forecast accuracy of the results.
1) The **Root Mean Squared Error (RMSE)** is a way to measure the average error about the forecasting results and is calculated as

$$RMSE = \sqrt{\frac{1}{N} \sum_{n=1}^{N} (y_n - \hat{y}_n)^2} \tag{5}$$

2) The **Mean Absolute Percentage Error (MPAE)** is a way to measure the proportional error about forecasting results and is calculated as

$$MPAE = \frac{1}{N} \sum_{n=1}^{N} \frac{|y_n - \hat{y}_n|}{y_n} \times 100\% \tag{6}$$

Here, y_n and $\hat{y}_n$ are the observed and the forecast values of observation n, respectively, and N is the total number of observations.

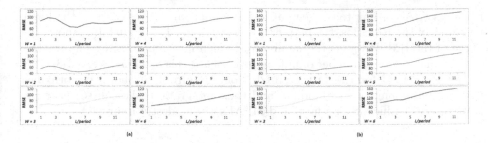

Fig. 3. The training RMSEs of the individual models and similarity models with different values of $\langle L, W \rangle$, respectively

5.4   Evaluation of PMA

We evaluate performance of PMA algorithm in these two sub-models, individual model and similarity model, and their integrations. For each experiment, we make predictions for all camera locations and report the average of the evaluation statistics.

Firstly, We vary the parameters $\langle L, W \rangle$ (mentioned before, is the size of plane series) in (1) and (2) of individual model and similarity model. By reason of the RMSE could reflect the absolute error in traffic volume, we choose it to help training our models. Figure 3(a) and (b) show the RMSEs of the individual models and similarity models with different values of $\langle L, W \rangle$, in which the RMSE is calculated based on 24 observations (60 mins, one day). Based on the results showed in the picture, we choose $L = 6, W = 2$ for individual model and $L = 7, W = 2$ for similarity model as the optimal PMA size, when the RMSE is the smallest, respectively.

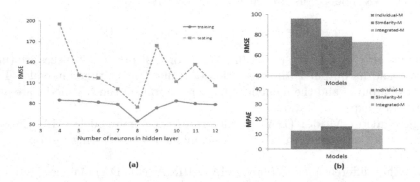

Fig. 4. Hidden-node numbers versus the RMSE on the training set and testing sets, and performance comparison among sub-models and their integration

Secondly, for the NN integration model, a series of NNs with different numbers of neurons in the hidden layer are trained. The number of neurons varies from 4 to 12, and the RMSEs are calculated for both the training set and the testing set. According to its generalization ability on the testing set, the lower the value of the RMSE is, the better the network model is. Figure 4(a) shows the curve of the RMSE versus the number of hidden-layer neurons. In Figure 4(a), we find that the best number of hidden-layer neurons is 8. Therefore, a 2-8-1 NN model is selected for the further predictions. Figure 4(b) shows performance comparison among two sub-models and their integration. We can see that the RMSEs are significantly reduced and MPAEs are modestly improved.

5.5   Comparison of Results

Several single-source models including the naïve, MA, ARIMA, and NN model, and an aggregation model are applied to time series $s_a(t)$ and $s_b(t)$. We compare their predictions on the testing sets with the PMA model.

1) **Naïve (or no-change) model** for traffic flow forecasting has the simplest form

$$\hat{q}_{Na}(t) = q(t-1)$$

where $\hat{q}_{Na}(t)$ is the forecast value at time interval t.

2) **MA Model**: An MA of order is computed by

$$\hat{y}_{t+1} = \frac{y_t + y_{t-1} + t_{t-2} + \dots + t_{t-k+1}}{k}$$

where k is the number of terms in the MA [12]. The MA technique deals only with the latest k periods of known data; the number of data points in each average dose not change as time continues.

3) **ARIMA Model**: A general ARIMA model of order (r, d, s), where d is the order of differencing. and orders r and s are the AR and MA operators.

4) **Artificial NN**: It is a single-source model that used for comparison with the PMA. The NN model is used to fit the nonlinear relationship

$$\hat{q}_{NN}(t) = f_1(q(t-1), q(t-2), \dots, q(t-l))$$

The inputs of NN model are the traffic flow records at previous l successive time intervals, and its output is the prediction of traffic flow at time interval t. The number of inputs l and the number of hidden neurons in the NN model are also optimized by experimentation.

5) **Aggregation Model (DA)**: It integrates three submodels (MA, ES and ARIMA) using simple neural network to combine the forecasting results.

Note that different models (naïve, MA, ARIMA, NN, DA) may need a different length of historical data. Hence, the sample size should be properly chosen for each model. For example, the length of time of the training data for the naïve model is no more than one hour before the forecasting time, while that for the training data for the ARIMA model at least covers two days before. For each model, we choose the parameters by observing the best fitting or forecasting and compare their forecasts on the same test sets.

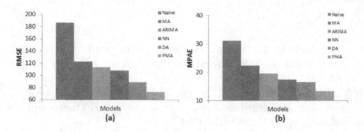

Fig. 5. RMSE and MPAE of different models in comparison

We generate forecasts for the future one period. Figure 5 show the RMSEs and MAPEs of testing set with the different forecasting models. For the Naïve, we simply use the last value as the forecasting value. As to the MA model, we set $k = 3$ which is the most optimal parameter. For the ARIMA model, we set parameter as $(1, 1, 0)$ where RMSE is the smallest. We set $\alpha = 0.1$ and $\gamma = 0.1$ in the ES model to build DA approach. For the last one, we select a 3-12-1 NN model for comparison. Figure 5 (a) and (b) show the predictions that resulted from the PMA model are better than the predictions that resulted from the other five models.

6 Conclusion

In this paper, aiming at the repeatable pattern of the similar traffic flow time series that previously methods did not consider, PFP has been constructed, a PMA algorithm and an aggregation strategy have been proposed to build a short-term traffic flow predictor. The PFP can be easily constructed from the source time series and the forecasting value of the PMA model can be automatically obtained by a computer program once the parameters are specified. Therefore, this new proposed approach is not a time-consuming but, rather a feasible job.

The PFP make full use of the information in the source time series which is collected on individual model as well as similarity model. By analyzing the forecasting performance of the naïve, MA, ARIMA, NN, DA and PMA, we have shown that the PMA can provide more accurate results than those of the other models.

For the further work, with the development of big data, data mining and machine learning, not limited to digital correlation, it is possible that we can consider the geographic correlation and even each flow movements to describe more potential characteristics and relationships of traffic flow. This problem deserves further study.

Acknowledgments. This work was supported in part by the 973 Program (2015CB-352 500), the National Natural Science Foundation of China Grant (6127 2092), the Shandong Provincial Natural Science Foundation Grant (ZR2012 FZ004), the Science and Technology Development Program of Shandong Province (2014GGE27178), the Independent Innovation Foun-dation of Shandong University (2012ZD012), the Taishan Scholars Program, and NSERC Discovery Grants.

References

1. Abdulhai, B., Porwal, H., Recker, W.: Short-term traffic flow prediction using neuro-genetic algorithms. ITS Journal-Intelligent Transportation Systems Journal **7**(1), 3–41 (2002)
2. Chan, K.Y., Dillon, T., Chang, E., Singh, J.: Prediction of short-term traffic variables using intelligent swarm-based neural networks. IEEE Transactions on Control Systems Technology, 21(1), 263–274 (2013)
3. Chan, K.Y., Dillon, T.S., Singh, J., Chang, E.: Neural-network-based models for short-term traffic flow forecasting using a hybrid exponential smoothing and levenberg-marquardt algorithm. IEEE Transactions on Intelligent Transportation Systems 13(2), 644–654 (2012)
4. Chan, K.Y., Dillon, T.S., Singh, J., Chang, E.: Traffic flow forecasting neural networks based on exponential smoothing method. In: 2011 6th IEEE Conference on Industrial Electronics and Applications (ICIEA), pp. 376–381. IEEE (2011)
5. Chan, K.Y., Yiu, C.K.: Development of neural network based traffic flow predictors using pre-processed data. In: Optimization and Control Methods in Industrial Engineering and Construction, pp. 125–138. Springer (2014)
6. Chen, M., Liu, Y., Yu, X.: NLPMM: A Next Location Predictor with Markov Modeling. In: Tseng, V.S., Ho, T.B., Zhou, Z.-H., Chen, A.L.P., Kao, H.-Y. (eds.) PAKDD 2014, Part II. LNCS, vol. 8444, pp. 186–197. Springer, Heidelberg (2014)
7. Davarynejad, M., Wang, Y., Vrancken, J., van den Berg, J.: Multi-phase time series models for motorway flow forecasting. In: 2011 14th International IEEE Conference on Intelligent Transportation Systems (ITSC), pp. 2033–2038. IEEE (2011)
8. Davis, G.A., Nihan, N.L.: Nonparametric regression and short-term freeway traffic forecasting. Journal of Transportation Engineering **117**(2), 178–188 (1991)
9. Dia, H.: An object-oriented neural network approach to short-term traffic forecasting. European Journal of Operational Research **131**(2), 253–261 (2001)
10. Gao, Q., Li, G.: A traffic prediction method based on ann and adaptive template matching (2011)
11. Guo, H., Xiao, X., Tang, Y.: Short-Term Traffic Flow Forecasting Based on Grey Delay Model. In: Lei, J., Wang, F.L., Deng, H., Miao, D. (eds.) AICI 2012. LNCS, vol. 7530, pp. 357–364. Springer, Heidelberg (2012)
12. Hanke, J.E., Reitsch, A.G., Wichern, D.W.: Business forecasting. Prentice Hall Upper Saddle River, NJ (2001)
13. Papagiannaki, K., Taft, N., Zhang, Z.-L., Diot, C.: Long-term forecasting of internet backbone traffic: Observations and initial models. In: INFOCOM 2003, Twenty-Second Annual Joint Conference of the IEEE Computer and Communications. IEEE Societies, vol. 2, pp. 1178–1188. IEEE (2003)
14. Papagiannaki, K., Taft, N., Zhang, Z.-L., Diot, C.: Long-term forecasting of internet backbone traffic. IEEE Transactions on Neural Networks **16**(5), 1110–1124 (2005)
15. Shang, J., Zheng, Y., Tong, W., Chang, E., Yu, Y.: Inferring gas consumption and pollution emission of vehicles throughout a city. In: Proceedings of the 20th ACM SIGKDD International Conference on Knowledge Discovery and Data Mining, pp. 1027–1036. ACM (2014)
16. Smith, B.L., Demetsky, M.J.: Traffic flow forecasting: comparison of modeling approaches. Journal of Transportation Engineering **123**(4), 261–266 (1997)

17. Smith, B.L., Williams, B.M., Oswald, R.K.: Comparison of parametric and non-parametric models for traffic flow forecasting. Transportation Research Part C: Emerging Technologies **10**(4), 303–321 (2002)
18. Tan, M.-C., Wong, S.C., Xu, J.-M., Guan, Z.-R., Zhang, P.: An aggregation approach to short-term traffic flow prediction. IEEE Transactions on Intelligent Transportation Systems 10(1), 60–69 (2009)
19. Williams, B.M.: Multivariate vehicular traffic flow prediction: Evaluation of arimax modeling. Transportation Research Record: Journal of the Transportation Research Board 1776(1), 194–200 (2001)
20. Williams, B.M., Hoel, L.A.: Modeling and forecasting vehicular traffic flow as a seasonal arima process: Theoretical basis and empirical results. Journal of Transportation Engineering **129**(6), 664–672 (2003)

Recommending Profitable Taxi Travel Routes Based on Big Taxi Trajectories Data

Wenxin Yang[1(✉)], Xin Wang[1], Seyyed Mohammadreza Rahimi[1], and Jun Luo[2,3]

[1] Department of Geomatics Engineering, University of Calgary, Calgary, AB, Canada
{weyang,xcwang,smrahimi}@ucalgary.ca
[2] HK Advanced Technology Center, Ecosystem and Cloud Service Group, Lenovo, China
[3] Shenzhen Institutes of Advanced Technology, Chinese Academy of Sciences, China
jun.luo@siat.ac.cn

Abstract. Recommending routes with the shortest cruising distance based on big taxi trajectories is an active research topic. In this paper, we first introduce a temporal probability grid network generated from the taxi trajectories, then a profitable route recommendation algorithm called Adaptive Shortest Expected Cruising Route (ASECR) algorithm is proposed. ASECR recommends profitable routes based on assigned potential profitable grids and updates the profitable route constantly based on taxis' movements as well as utilizing the temporal probability grid network dynamically. To handle the big trajectory data and improve the efficiency of updating route constantly, a data structure kdS-tree is proposed and implemented for ASECR. The experiments on two real taxi trajectory datasets demonstrate the effectiveness and efficiency of the proposed algorithm.

Keywords: GPS-trajectory · Route recommendation · kdS-tree

1 Introduction

Taxi is an important way of transportation as it provides fast, secure and personalized travels for passengers. However, taxis spend 35%-60% of their time cruising along the roads looking for passengers [1]. With the advent of GPS tracking technology on taxis, improving taxis' performance based on taxi GPS trajectory data becomes a promising solution [2,3,4,5,6].

A few challenges exist when using GPS trajectories to improve taxi performance. Firstly, most route recommendations are based on the shortest traveling path to a location that passengers may appear [7]. However, usually the locations of potential passengers are unknown and may vary as time changes. How to summarize and represent the passenger information from the big historical taxi trajectory data and utilize the information to calculate the expected cruising distance of taxis is still challenging. Secondly, most existing works do not consider temporal factor and the route recommendation is not customized for the individual taxi. It may result in sending all cruising taxis to the same location to compete for the same group of passengers. Thirdly, taxi trajectories are big temporal and spatial datasets. For example, the size of daily trajectories of a taxi with 5 second interval could be around 4 GB [8]. Therefore, how to make recommendation based on the big trajectory data is still challenging.

© Springer International Publishing Switzerland 2015
T. Cao et al. (Eds.): PAKDD 2015, Part II, LNAI 9078, pp. 370–382, 2015.
DOI: 10.1007/978-3-319-18032-8_29

To address above problems, we propose a temporal probability grid network generated from historical taxi trajectories. Then a profitable route recommendation algorithm based on map-reduce technology is proposed to reduce the cruising distances. Specifically, the contributions of the paper are summarized as follows.

First, we present a temporal probability grid network by modeling potential profitability of locations based on the taxis historical GPS data. The grid network separates the interested region into small grids. Each grid of the grid network includes two temporal properties, probability of finding a passenger and capability of accommodating the cruising taxis.

Second, a profitable grid assignment algorithm is proposed based on the spatial and temporal factors of taxis. The assignment algorithm distributes the potential profitable grids among taxis according to the probability and capacity of grids. The assignment also balances the workload among a group of taxis.

Third, a concept of the shortest expected cruising distance is proposed. The shortest expected cruising distance indicates the potential cruising distance from a taxi to potential passenger based on the probability of picking up a passenger at the locations. Based on it, an Adaptive Shortest Expected Cruising Route (ASECR) algorithm, which can dynamically recommend the route, is implemented by finding the most potential profitable locations based on the current time and taxi's movement. In addition, the temporal probability grid network will also be updated via different time stamps.

Fourth, to efficiently handle the big trajectory data and support updating k nearest profitable grids, a new data structure kdS-tree is implemented in ASECR.

The remainder of this paper is organized as follows. Section 2 reviews some related works. Section 3 formally defines the problem discussed in the paper. Section 4 presents a temporal probabilistic grid network model. ASECR algorithm is discussed in details in Section 5. Section 6 presents the experimental results and Section 7 concludes the paper.

2   Related Works

In this section, we briefly review some related works on taxi route recommendation. Ge et al. [3] develop a mobile recommender system that recommends a sequence of pick-up points to maximize a taxi's profit. They propose two algorithms, called LCP and Sky-Route, to find a route with minimal potential travel distance before finding a passenger. Specifically, for the length of suggested route L, the LCP algorithm enumerates all the L-length sub-routes from all pick-up points and recommends the best route among these candidate routes to the taxi driver. The SkyRoute algorithm efficiently computes the candidate routes. However, these two methods do not consider temporal factor. Yamamoto et al. [4] and Powell et al. [5] present the probabilistic based methods for routing multiple taxis. In [4], roads are divided into road segments based on the appearance frequency of passengers and vacant taxis are assigned to the locations based on the expectation of potential passengers. In [5], a Spatio-Temporal Profitability Map (STP map) is used to guide cruising taxis to better profitability locations. However, none of the above methods provide personalized route recommendation to individual taxi drivers based on their locations and times. Yuan et al. [6] present a method to calculate the fastest route to destination at a given departure time. Zhang et al. [9] propose a cruising system called PCruise that can find the shortest route for one taxi to pick up a passenger. However, they do not consider other taxis that may lead to the overall imbalance taxi load.

3   Preliminary

Definition 1. A *taxi* t is defined as a 5-tuple, denoted as:
$$t = (g_o, g_d, r, \tau, s),$$
where g_o and g_d are the origin and destination of the taxi. r is a route from the origin to the destination. τ is the starting time of the route r. s is a Boolean value to show whether the taxi is occupied.

Definition 2. A *GPS-reading* γ^t of a taxi t is defined as a 3-tuple, denoted as:
$$\gamma^t = (\tau, l, s),$$
where τ is the time stamp of the GPS-reading. l is the location of the taxi at time τ and s is a Boolean value showing whether taxi is occupied by a passenger at time τ.

The state of a taxi can be determined using its GPS-readings. When a taxi has a passenger, it is in *occupied* state. Otherwise, it is in *cruising* state. However, if the time difference between the two consecutive GPS-readings is greater than a defined threshold θ, then the taxi is in *out-of-service* state. The *out-of-service* state divides trajectories into sub-trajectories, or called "trips".

Definition 3. A *trip* of a taxi t is defined as a set of GPS readings, denoted as:
$$trip^t = (\gamma_1^t, \gamma_2^t, \gamma_3^t, \dots, \gamma_m^t); \text{where } 0 < \gamma_i^t.\tau - \gamma_{i-1}^t.\tau \le \theta, 1 < i \le m.$$
Each trip is a sub-trajectory during which the taxi t is in service.

Since GPS-readings from taxis are generally imprecise, directly matching the GPS-readings on existing road network could be problematic. In addition, the operations on the existing road network are computational expensive due to the large amount of vertices (i.e. junctions) and edges (i.e. road segments) on the road network. In this paper, a temporal probabilistic grid network from historical GPS trajectories is defined to divide the interested region into small grids and summarize the temporal and spatial information from taxis' GPS trajectories.

Definition 4. A *grid* g is a 4-tuple, denoted as:
$$g = (l, p, c, \tau),$$
where l is the center location of the grid g, p and c are the temporal probability and capacity of g at time τ. The temporal probability is the probability that a taxi finds a passenger in the grid g at the time τ. The capacity is the number of the possible passengers of the grid g at the time τ.

Definition 5. A *grid network* GN is a weighted directed graph, denoted as:
$$GN = (G, E, D),$$
where G is a set of grids, and E is a set of edges connecting grids. An edge is built to connect two neighbor grids if consecutive GPS-readings exist between the two grids. D is a set of weights of the edges, representing the physical traveling distances of edges.

Definition 6. Given a set of taxis T and the time interval $[\tau 1, \tau 2]$, the taxis' *distance performance* starting between $[\tau 1, \tau 2]$ is defined as:
$$P(T) = \frac{\sum_{t \in T} \sum_{t.r} dist_{occupied}}{\sum_{t \in T} \sum_{t.r} (dist_{occupied} + dist_{cruising})},$$
where $dist_{cruising}$ is the travelled distance that a taxi is in *cruising* state. $dist_{occupied}$ is the travelled distance that a taxi is in *occupied* state.

The distance performance measures the ratio of distance travelled by all the taxis in T having a passenger on board to the whole distance they travelled. A larger value of distance performance indicates the higher distance efficiency of the set of taxis. In this paper, we assume the distance performance reflects the profits of taxis.

Hence, the problem of finding profitable routes for a set of taxis can be defined as:

Definition 7. Given a set of taxis T, the taxi *profitable routing problem* is to recommend routes to a set of taxis so that the overall distance performance of taxis is maximized. It is formulized as:

$$Arg_{t,r}\max P(T) \ ,$$

where $t \epsilon T$. This definition is to recommend routes to a set of taxis so that the overall profit of all taxis is maximized. This can be achieved by minimizing the cruising distances for all taxis. In other words, the goal is to maximize the average distance performance of the whole group of the taxis.

4   Temporal Probabilistic Grid Network

Building the temporal probabilistic grid network from taxi trajectories includes two steps: primary grid network generation, and probability and capability calculation for grids.

4.1   Primary Grid Network Generation

To generate the primary grid network, the interested region is divided into small grids and GPS-readings are projected onto grids according to their coordinates. The following two steps briefly show how to build the grid network: 1) The GPS-readings are overlaid with gridlines, showing in which grid each GPS-reading is placed. 2) The center of each grid becomes a node of the network, called grid node. An edge exists between two grid nodes if a pair of consecutive GPS-readings satisfies the two conditions: a) Two readings are placed in different grids; b) The two grids are neighbors.

The size of grids is important to build the grid network. If the grid size is too big, more than one GPS-readings would located in the same grid and make the graph less accurate for tracking GPS data sequences. If it is too small, then no or very few pair of consecutive GPS-readings will be placed in the neighboring grids.

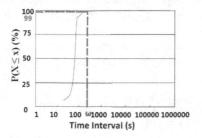

Fig. 1. Distribution of the Time Interval

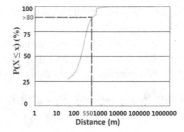

Fig. 2. Distribution of the Distance Interval

To determine the grid size, the first step is to identify the *out-of-service* states from each trajectory because the distance between the two consecutive readings is not meaningful when the taxi is *out-of-service*. The *out-of-service* state can be determined by analyzing the frequency of GPS-readings. To do so, the time intervals between every pair of consecutive GPS-readings are plotted based on the following empirical distribution function:

$$P\left(x_i \le \omega\right) = \frac{|x_i \mid x_i \le \omega|}{n} \quad ; \quad x_i = \gamma_i \cdot \tau - \gamma_{i-1} \cdot \tau,$$

where n is the total number of the time intervals of two consecutive GPS-readings and x_i is the time interval between two consecutive GPS-readings less than ω. The time interval threshold is set to the value when gradient of the curve dramatically dropped.

Fig. 1 shows an example plot of the time interval distribution. In this example, we can assume that taxi is *out-of-service* if the time interval between two GPS-readings is over 420 seconds, i.e. 7 minutes, because the gradient of the curve dramatically descend and almost keep even after it. It means most time intervals between consecutive GPS-readings are less than 420 seconds. After identifying the time interval threshold for out-of-service consecutive GPS-readings, we can use the similar method to plot the distances between two consecutive GPS-readings so as to determine the grid size. Fig. 2 shows a plot of the distribution of distances between consecutive GPS-readings. The gradient of distribution curve approaches to zero at around 550 meters, which represents the consecutive GPS-readings are at most 550 meters apart. Then 550 meters can be a good value for the grid size.

4.2   Calculation of Probability and Capacity of Grids

Probability of finding a passenger in a location and capacity of a grid are two temporal properties of the grid. After generating the primary grid network, the two temporal properties will be calculated. The definitions are given as follows.

Definition 8. The *probability* of finding a passenger in a grid l in a given time interval $[\tau1, \tau2]$ is defined as:

$$p\left(l, \tau1, \tau2\right) = \frac{|P|}{|E|},$$

where $|P|$ is the number of taxis showing pickups in grid l in the time interval $[\tau1, \tau2]$ and $|E|$ is the total number of taxis entering grid l in the time interval $[\tau1, \tau2]$.

However, if all taxis are assigned to the same set of grids, it may result in non-equilibrium of supply and demand. To avoid that, the capacity of a grid is defined to control the number of taxis assigned to it.

Definition 9. The *capacity* of a grid l in a time interval $[\tau1, \tau2]$ is defined as:

$$c(l, \tau) = c(l, \tau1, \tau2) = |P|,$$

where $|P|$ is the number of pickups in the grid in the given time interval.

5   Profitable Route Recommender

The profitable route recommendation algorithm consists of two steps. At the first step, in order to reduce the total cruising distance, potential profitable grids are assigned to each taxi with respect to the probability and capacity of the grids. Then the potential profitable route is recommended for each taxi. The recommended route connects assigned potential profitable grids with potential minimum cruising distance. The capacity of each released potential profitable grid is updated after a taxi picks up a passenger.

5.1   Potential Profitable Grids Assignment

The potential profitable grid assignment aims to assign a set of grids that have the probability to pick up passengers to each taxi. Each potential profitable grid can only be assigned to the limited number of taxis based on its capacity. First, the potential profitable grid assignment algorithm examines all potential profitable grids with the capacity greater than the number of taxis in the grid. If c is the capacity of the grid, it assigns the grid to the c-nearest taxis. Then, all taxis are checked to see whether it is a starving taxi. A starving taxi is a taxi which has no potential profitable grid assigned. For each starving taxi, the algorithm identifies the closest rich taxi that has been assigned to multiple grids, removes one potential profitable grid from it and assigns to the starving taxi.

5.2   Profitable Route Recommendation

Profitable route recommendation is to find a route with the shortest expected cruising distance that connects the assigned potential profitable grids of each taxi. First, the concept of the expected cruising distance is introduced. Then a method to search the route with the shortest expected cruising distance route is proposed.

5.2.1   Expected Cruising Distance

The most profitable route for a taxi is the route with the minimum cruising distance. However, the potential passengers' locations are unknown in advance. To find a profitable route for each taxi, we find the passengers through the expected cruising distance.

The expected cruising distance from the taxi to a potential passenger is calculated based on the grids' probabilities. If a taxi is assigned n potential profitable grids and going to move along a route $r = \{l_1, l_2, \dots, l_n\}$ in time interval $[\tau1, \tau2]$ from grid l_0, it may pick up passengers at any grid of the route. For each grid $l_i (1 \leq i \leq n)$, the probability that the taxi pick up a passenger at the grid can be calculated with the following equation:

$$p(l_i | r, \tau1, \tau2) = \begin{cases} p(l_i, \tau1, \tau2), & i = 1 \\ p(l_i, \tau1, \tau2) \prod_{j=1}^{i-1} (1 - p(l_i, \tau1, \tau2)), & i > 1 \end{cases},$$

where $p(l_i | r, \tau1, \tau2)$ is a conditional probability that taxi does not pick up a passenger at previous grids but in grid l_i.

Note that the cruising distance before picking up a passenger at l_i is $\sum_{j=0}^{i-1} dist(l_j, l_{j+1})$, where $dist(l_j, l_{j+1})$ is the Euclidean distance between grid l_j and grid l_{j+1}. The expected cruising distance could be calculated by integrating the distribution of the conditional probability with the distribution of the cruising distance before picking up a passenger. We define the expected cruising distance as follows:

Definition 10. Given a taxi t located at grid l_0, and a set of potential profitable grids with a route sequence $r = \{l_1, l_2, \ldots, l_n\}$, the *expected cruising distance* in the time interval $[\tau1, \tau2]$ is defined as:

$$dist_{expected}(t.r) = \sum_{i=1}^{n} \left(p(l_i | r, \tau1, \tau2) \sum_{j=0}^{i-1} dist(l_j, l_{j+1}) \right),$$

Definition 11. Given a taxi t and a set of potential profitable grids $G = \{l_1, l_2, \ldots, l_n\}$, *the shortest expected cruising distance* in the time interval $[\tau1, \tau2]$ is the minimal expected cruising distance of all possible routes which connect all grids in G.

5.2.3   Adaptive Shortest Expected Cruising Route (ASECR) Algorithm

To find the shortest expected cruising distance, a naive solution is to check all possible combinations of the assigned profitable grids and find the shortest cruising routes among them. However, the time complexity for this solution is O(n!), where n is the number of assigned grids. In this paper, a heuristic method ASECR is proposed with a threshold K on the number of assigned profitable grids in a recommended route.

Algorithm findAdaptiveShorestExpectedCruisingRoute (T, G, k)

Input: T is a set of empty taxis; G is the grid network.

1. **for** each taxi t in T {
2.   **while** taxi t is empty {
3.   currentGrid = t. getCurrentLocation( );
4.   assignGridstoTaxis (T, G, currentTime, currentGrid);
5.   assignGrids = t.getTopKNearestAssignedGrid(k);
6.   t.findShortestExpectedCruisingRoute(currentGrid, assignGrids); }
7.   G.updateCapacity(t); }

Fig. 3. ASECR Algorithm

As shown in Fig. 3, for all empty taxis, the ASECR algorithm updates the potential profitable grids in line 4 and then finds the k nearest potential profitable grids at each grid along the recommended route (line 5). Next, it calculates the shortest expected cruising route based on taxi's current location and current time in line 6. The process continues until the taxi picks up a passenger. Finally, it updates the capability of the grid network in line 7.

The most time consuming operation of the ASECR algorithm is to generate and update a profitable route to each taxi based on its current time and location. However, the volume of GPS dataset is so large that the response time on a single computer is too long to be practically useful. Furthermore, the data cannot be all loaded into memory and will lead to frequent disk accesses. ASECR is built based on the grid network that partitions the space into regular regions and the simple structure is inherently suitable for parallel. Therefore, we utilize a kdS-tree data structure to tackle the problems.

5.2.4   kdS-Tree

As mentioned, ASECR algorithm constantly updates the K nearest potential profitable grids based on current time and taxi's current location. Finding K nearest neighborhood (kNN) has been discussed in many literatures and can be solved through spatial index structures, such as kd-tree[10] and R-tree[11]. However, these spatial index structures do not perform well for continuous kNN because they are not designed for frequently updating tree structures and efficiently querying profitable potential grids. A new data structure named kdS-tree is proposed to solve the problem.

kdS-tree is extended from kd-tree and ball-tree[12]. It is a binary tree in which each node is associated with a circle. Instead of only having one point on splitting line for 2-dimensional kd-tree, at least two points are on the splitting circles for kdS-tree. kdS-tree is constructed in a top-down manner. Given a set of profitable potential grids, first, a minimum enclosing circle is created to include all grids. Each node of kdS-tree contains five fields: center and radius of the associated splitting circle, profitable grid nodes located on the circle, left and right subtrees. The root node of a kdS-tree has the circle including all potential profitable grids. Then, two grid nodes are identified inside the circle as two seeds, between which is the largest distance among all pairs of grid nodes inside the circle. All other grid nodes inside the circle are assigned to the closer seed to form two clusters, and two minimum enclosing circles are formed for two clusters. Based on the x-coordinate of center of these two clusters, two circles will be added to the left and right subtrees. The procedure repeats until no profitable grid is inside the circle. As shown in Fig. 4, (a) shows 21 profitable potential grids. (b) shows the generated minimum enclosing circles based on the 21 grids, and (c) is the resulting kdS-tree.

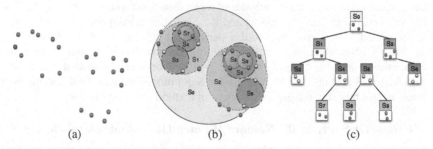

(a)   (b)   (c)

Fig. 4. kdS-tree for 21 grids: (a) the profitable grids; (b) the circles; (c) the kdS-tree

To retrieve k nearest grids for a given taxi, the query starts with traversing the tree from root to locate the lowest level node that contains the location of the requested taxi. Meanwhile, all distances between grids on visited nodes and location of the taxi are computed. The top-k nearest distances from the taxi to the grids are recorded. Then a query circle can be generated with the center as the location of the taxi and the radius as the current k-th nearest distance. Therefore, the upper bound for searching the k nearest grids is the radius of the query circle. Next, for each node whose splitting circle over-laps with the query circle, the distances from profitable grid nodes located on the circle to the taxi location are calculated. If the distance is less than the upper bound, the upper bound is updated. To speed up the query process, two pruning rules are proposed based on the branch-and-bound technique in [13]:

The Downward Pruning rule removes tree nodes whose circle does not intersect the query circle. If a tree node does not intersect the query circle, then it is impossible for any of the grid in the tree node and its subtree nodes inside the query circle. So the entire branch of that tree node can be pruned. The Upward Pruning rule removes tree nodes whose circle completely contains the query circle. If query circle is completely inside the circle of a tree node, the tree node cannot have any grids closer to than any of the current k nearest grids as well as its ancestor nodes. So this tree node and its ancestor tree nodes can be pruned. The two rules can be specifically represented as:

Downward Pruning: If $dist(TC, QC) > r(TC) + r(QC)$, then for each grid g_j on TC, $dist(g_j, QC) > r(QC)$. Therefore, the tree node of TC can be pruned.

Upward Pruning: If $dist(TC, QC) < r(TC) - r(QC)$, then for each grid g_j on the corresponding circle of the node TC, $dist(g_j, QC) > r(QC)$. Therefore, TC can be pruned.

Note that TC presents the circle corresponding to the tree node, $r(TC)$ is the radius of TC, QC represents the query circle of the requested taxi, $r(QC)$ is the radius of QC, g_j is a grid point located on the TC, $dist(TC, QC)$ is the distance between the center of TC and the center of QC (i.e. taxi location), and $dist(g_j, QC)$ is the distance between g_j and the center of QC. Meanwhile, a heap H of size K is used to hold the current candidates for the kNN and allows fast check and insertion.

6   Experiments

In this section, we evaluate the performance of ASECR algorithm on two real taxi datasets. The algorithms are implemented in Java. All the reported values are the average of 20 runs. Two real taxi datasets are from San Francisco, USA and Shenzhen, China. The San Francisco dataset contains GPS trajectories of more than 500 taxis in a 25-day period with the frequency of 10 minutes per reading and the Shenzhen dataset contains GPS trajectories of more than 13,000 taxis in a 30-day period with the frequency of 15 seconds per reading. Each GPS-reading includes fields: unique taxi ID, longitude and latitude, status (occupied/empty) and the time stamp. The grid networks are generated using 2-hour intervals. The grid size is set to 550 meters for San Francisco dataset and 350 meters for Shenzhen dataset by using the grid determination method discussed above.

6.1   Parameter Study on the Number of Potential Profitable Grids K

K is the number of potential profitable grids assigned to each taxi to generate the profitable route. Since the passenger's location cannot be predicted, it is difficult to set the value for the parameter. In this experiment, we investigate the distance performance of taxis changing over the different values of K. The parameter K is set from 2 to 12. Fig. 5 shows a progressive increase of the distance performance until some certain K value, beyond which the distance performance levels off. The reason is when K increases, more assigned profitable potential grids lead to higher pick-up probability. Therefore, the distance performance for taxis increases with decreasing cruising distances. However, after the certain value (e.g. when more than 8 grids for San Francisco dataset) is assigned to each taxi, the distance performance stabilized. The reason is that most taxis would have already picked up passengers after visiting K number of potential profitable grids. The level off value for San Francisco dataset is 8 grids, which indicates assigning more than 8 potential grids does not improve taxis distance performance very much. Similar observations can be found from Shenzhen dataset, which the level off K value is 6.

We also investigate the running time of querying the K potential profitable grids with respect to different K values. As shown in Fig. 6, the running time increases with the increasing of K value. In practice, the distance performance and the computation time needs to be balanced. So, we set K to 8 for San Francisco dataset and 6 for Shenzhen dataset for the rest of experiments.

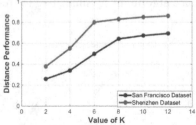

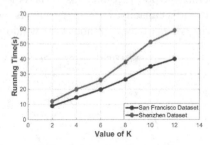

Fig. 5. Distance Performance vs. K   **Fig. 6.** Running time vs. K

6.2   KdS-Tree Performance Pruning Effectiveness and Query Efficiency

In this section, we first compare the query efficiency of the proposed kdS-tree over the kd-tree. The query efficiency is defined as the difference of the query response CPU time between kdS-tree and kd-tree over the query response CPU time of kd-tree. Fig. 7 shows the query efficiency over different K values. The figure shows that kdS-tree outperforms the kd-tree and the query efficiency increases as K increases. The reason is that the radius of query circle extends as K increases, and the larger query circle leads to more nodes which need to be examined by kd-tree than kdS-tree.

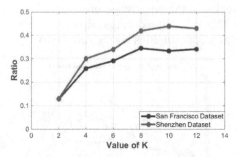

Fig. 7. Query Efficiency vs. K

Table 1. The number of kdS tree nodes visited with and without pruning rules

Value of K	San Francisco Dataset		Shenzhen Dataset	
	Before Pruning	After Pruning	Before Pruning	After Pruning
2	13	11	16	14
4	52	31	89	67
6	108	53	562	388
8	361	197	1498	864
10	776	397	3168	2082
12	1298	695	4398	3602

Then we evaluate the effectiveness of the proposed pruning rules on the kdS-tree. We build one kdS-tree of Shenzhen dataset contains 6612 nodes and the one of San Francisco dataset contains 1862 nodes. Table 1 shows the average number of visited kdS-tree nodes for one taxi to find K nearest profitable grids with and without the two pruning rules. The number of potential profitable grids K varies from 2 to 12. As expected, the pruning rules can reduce the number of visited nodes. For example, when K is set to 8, 197 nodes are visited with pruning rules while 361 nodes are visited without pruning rules. It shows that with two pruning rules, about 45% of the nodes can be eliminated during the search process.

6.3   Evaluation on Recommendation

In this experiment, we compare the ASECR algorithm with two other methods: the LCP method[3] and the baseline method. Here, the length of route for LCP method is set to the same value as the number of potential profitable grids for ASECR. The baseline method recommends all taxis to their nearest profitable grid with respect of each grid's capacity. In the experiments, we evaluate algorithms by simulating the real world events on every 2-hour time window of a day.

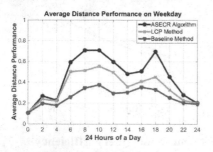

Fig. 8. San Francisco: Distance Performance   **Fig. 9.** Shenzhen: Distance Performance

Fig. 8 and 9 study the overall performance of the algorithms based on San Francisco data and Shenzhen data on weekday. The ASECR algorithm outperforms LCP method and baseline method in term of distance performance of taxis, which show the effectiveness of the algorithm. This is because during rush hours, the ASECR algorithms assigns more taxis to the grids which have more passengers, which leads to the significant improvement for both distance performance. The other observation from Fig. 8 and 9 is that during the rush hours of a weekday, the percentage of distance performance is higher than that of non-rush hours. This is because during rush hours, there are more passengers, which indicates more pick-up opportunities during rush hours.

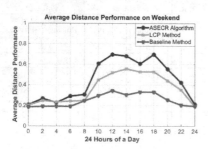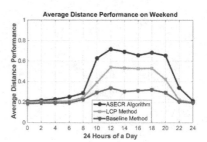

Fig. 10. San Francisco: Distance Performance   **Fig. 11.** Shenzhen: Distance Performance

For further investigations, Fig. 10 and 11 give the average distance performance on weekend in 2-hour time windows. Similar conclusions can be drawn from the two figures. Different than weekdays, the distance performance of ASECR are better than LCP and baseline algorithms from 10:00 to 22:00 for both San Francisco and Shenzhen data. It shows that people take taxis at different times on weekends than weekdays.

7   Conclusions and Future Works

In this paper, a temporal profitable grid network is built by mining taxis historical GPS data. Based on it, we define an expected cruising distance to formulate the potential cruising distance. In order to find the shortest expected cruising distance for taxis, regarding the time and location of a taxi are changing over time we propose the ASECR algorithm, an adaptive route searching method based on the continuous movement of taxis. To improve the efficiency of the algorithm, a new data structure called kds-tree and map-reduce models are implemented in the algorithm. The experiments show that the method can provide better profitable routes to a group of taxis to improve the taxis performance.

Taxi GPS trajectories are big spatial and temporal data. Future research will focus on the evaluating of the proposed algorithm on larger datasets. In addition, more efforts will be taken on exploring new architecture and data structures to efficiently update and query for the big data.

Acknowledgment. The research is supported by the Natural Sciences and Engineering Research Council of Canada IPS scholarship to the first author, Discovery Grant to the second author. National Natural Science Foundation of China (Grant No. 11271351 and No. 41271387), and the Shenzhen New Industry Development Fund under grant No. JCYJ20120617120716224.

References

1. Schaller Consulting: The New York City Taxicab Fact Book, Schaller Consulting, Brooklyn, NY (2006). http://www.schallerconsult.com/taxi/taxifb.pdf
2. Hunter, T., Herring, R., Abbeel, P., Bayen, A.: Path and travel time inference from GPS probe vehicle data. In: Proc. Neural Information Processing Systems (NIPS). New York, USA, pp. 316–324 (2009)
3. Ge, Y., Xiong, H., Alexander, T., Xiao, K., Marco, G., Michael, P.: An energy-efficient mobile recommender system. In: Proc. Knowledge Discovery and Data Mining, KDD 2010. Washington, DC, pp. 899–908 (2010)
4. Yamamoto, K., Uesugi, K., Watanabe, Y.: Adaptive routing of cruising taxis by mutual exchange of pathways. In: Proc. Knowledge-Based Intelligent Information and Engineering Systems. Zagreb, Croatia, pp. 559–566 (2008)
5. Powell, J.W., Huang, Y., Bastani, F., Ji, M,.: Towards reducing taxi time using spatio-temporal profitability maps. In: Proc. Advances in Spatial and Temporal Databases. Minneapolis, MN, USA, pp. 242–260 (2011)
6. Yuan, J., Zhen, Y., Zhang, L., Xie, X., Sun, G.: Where to find my next passenger? In: Proc. Ubiquitous Computing. Beijing, China, pp. 109–118 (2011)
7. Ma, S., Zheng, Y., Wolfson, O.: T-share: A large-scale dynamic taxi ridesharing service. In: Proc. Data Engineering Brisbane, Australia, pp. 410–421 (2013)
8. Long, C., Wong, R.C.W., Jagadish, H.V.: Direction-preserving trajectory simplification. In: Proc. the VLDB Endowment, pp. 949–960 (2013)
9. Zhang, D., He, Y.: pCruise: reducing cruising miles for taxicab networks. In: Proc. Real-Time Systems Symposium. San Juan, PR, USA, pp. 85–94 (2012)
10. Bentley, J.L.: Multidimensional binary search trees used for associative searching. In: Communications of the ACM, pp. 509–517 (1975)

11. Guttman, A.: R-trees: A dynamic index structure for spatial searching. In: Proc. Management of Data, pp. 47–57 (1984). Agrawal, D., Bernstein, P., Bertino, E.: Challenges and opportunities with big data (2012). http://www.cra.org/ccc/files/docs/init/bigdatawhite paper.pdf
12. Omohundro, S.M.: Five balltree construction algorithms. In: ICSI Technical Report (1989)
13. Roussopoulos, N., Kelley, S., Vincent, F.: Nearest neighbor queries. In: Proc. Management of Data, San Jose, CA, pp. 71–79 (1995)

Semi Supervised Adaptive Framework for Classifying Evolving Data Stream

Ahsanul Haque[1]([⊠]), Latifur Khan[1], and Michael Baron[2]

[1] Department of Computer Science, The University of Texas at Dallas,
Richardson, TX, USA
{ahsanul.haque,lkhan}@utdallas.edu
[2] Department of Mathematical Sciences, The University of Texas at Dallas,
Richardson, TX, USA
mbaron@utdallas.edu

Abstract. Most of the approaches for classifying evolving data stream divide the stream into fixed size chunks to address infinite length and concept drift problems. These approaches suffer from trade-off between performance and sensitivity. To address this problem, existing adaptive sliding window techniques determine chunk boundaries dynamically by detecting changes in classifier error rate which requires true labels for all of the data instances. However, true labels are scarce and often delayed in reality. In this paper, we propose an approach which determines dynamic chunk boundaries by detecting significant changes in classifier confidence scores using only limited number of labeled data instances. Moreover, we integrate suitable classification technique with it to propose a complete semi supervised framework which uses dynamic chunk boundaries to address concept drift and concept evolution efficiently. Results from the experiments using benchmark data sets show the effectiveness of our proposed framework in terms of handling both concept drift and concept evolution.

Keywords: Dynamic chunk size · Change detection · Concept drift

1 Introduction

Data streams have inherent properties which make it difficult for the traditional data mining techniques to classify stream data. Some of the most challenging properties of data streams include but not limited to infinite length, concept drift, concept evolution, limited labeled data and delayed labeling. Since data stream is an infinite stream of data, it cannot be stored into any storage for analyzing, e.g., labeling. So, data stream classification is essentially a single pass process. Concept drift occurs when the target class or concept evolves within the feature space such that, the class encroaches or crosses previously defined decision boundaries of the classifier [1]. So, any classification method used in the context of data streams need to be updated to cope up with changing concepts. Concept evolution occurs when a new class emerges in the data stream [2,3].

© Springer International Publishing Switzerland 2015
T. Cao et al. (Eds.): PAKDD 2015, Part II, LNAI 9078, pp. 383–394, 2015.
DOI: 10.1007/978-3-319-18032-8_30

To address infinite length and concept drift problems, most of the approaches in the literature divide the data stream into fixed size chunks [2–4]. As a result, these approaches fail to adapt to the change of concepts immediately. If the chunk size is too small, classification method may end up with frequent training during stable period when there is no concept drift, causing performance drawback due to unnecessary update of the classifier. On the contrary, if the chunk size is too large, the classifier may remain outdated for a long period of time. Some other approaches [5,6] use *gradual forgetting* to address infinite length and concept drift problems. These approaches use various decay functions to assign weight to the instances based on their age. This strategy also suffers from similar trade-off while choosing the decay rate to match unknown rate of change.

To solve the problems due to fixed chunk size or decay rate, a dynamic sliding window is maintained in [7,8] by tracking any major change in error rate of the classifier. These approaches mostly assume that, true labels of the data instances will be available to calculate the error rate as soon as it is tested. However, in the real world, labeled data is scarce since labeling data instances manually is costly and time consuming [9]. In data streams, where data arrives very quickly, it may not be possible to label all the data instances as soon as they arrive. So, a good classifier in streaming context should be able to defer the training until true labels become available yet continuing labeling newly arrived instances using the current classifier. Moreover, it should be able to use partially labeled training data [9].

In this paper, we present a complete framework which addresses all of the above challenges. It uses similar semi supervised approach as [2] for classification and novel class detection. However, unlike [2], our proposed framework divides the data stream into dynamically determined chunks using change detection technique. To avoid the use of true labels of data instances for change detection, our proposed framework calculates a confidence value while predicting label of each data instance. It then uses a change point detection technique to detect any significant change in the classifier confidence scores and determines the chunk size dynamically. If a significant change is detected, the classifier is updated using only recent labeled training data instances. To address the concept drift problem, our framework maintains an ensemble of classifier models, each trained on different dynamically determined chunks. Both of classification and confidence value calculation in our framework are semi supervised. So, the proposed framework can work with delayed labeling and partially labeled training data. To the best of our knowledge, our framework is the first semi supervised approach which addresses both of concept drift and concept evolution using dynamically determined chunk boundaries.

The primary contributions of our work are as follows: 1) We present a technique to estimate classifier confidences in predicting labels of stream data instances. 2) We propose a change detection approach which takes classifier confidence values as input and detects if there is any significant change in the classifier confidence. It detects the chunk boundary dynamically if there is a significant change, which triggers updating of the existing classifier on recent labeled training data. Unlike other adaptive sliding window techniques which detect changes in error rates of the classifier, our approach does not need true labels of all the data instances for determining the chunk boundary dynamically. 3) We present a semi supervised framework by integrating classification and novel class detection technique with the change detection approach to address both of concept drift and concept evolution using dynamic chunk sizes. 4) We implement and evaluate our proposed framework on several benchmark and synthetic data sets. Results from the experiments show that, our framework outperforms other state of the art approaches for data stream classification and novel class detection.

The rest of the paper is organized as follows: In Section 2, we briefly discuss some related works. Section 3 describes our approach in detail. We describe the data sets, evaluation metrics and present experiment results in Section 4. Finally, Section 5 concludes the paper.

2   Related Works

Typically, existing data stream classification approaches address infinite length and concept drift in data stream by dividing it in fixed length chunk sizes [2–4] or using gradual forgetting [5,6]. However, setting a fixed size of the chunks or finding the perfect decay function for gradual forgetting are challenging tasks if the information on time-scale of change is not available [8]. Unlike these approaches, we determine the chunk size dynamically based on a significant change in classifier confidence in predicting labels of test data instances.

There are two types of techniques in terms of change detection, i.e., detecting change in the posterior distribution of the classes given the features $P(y|\mathbf{X})$, another one is detecting change in the generating distribution $P(\mathbf{X})$ [10]. In the literature, several methods[11,12] exist to deal with change of $P(\mathbf{X})$ in multidimensional data. However, detecting change in $P(\mathbf{X})$ is a hard problem especially in case of multi-dimensional data and does not work well to detect change of concept in multi-dimensional multi-class data streaming context [13]. In this paper, we focus on detecting changes in one dimensional classifier confidence values.

Various techniques to detect changes in $P(y|\mathbf{X})$ have been proposed in [7, 8,13], which are mostly based on loss estimation of a predictor performance. These approaches track any significant change in classifier error rate over time which requires the true labels of the data instances. Instead of using the error rate, we calculate classifier confidence in the prediction. Monteith et al. propose a method to estimate classifier confidence in [14], but they use confidence scores only for weighted voting. Unlike this approach, we use confidence scores both for weighted voting and for determining chunk boundaries dynamically. We use two sample *t-test* for one sided right tail hypothesis testing to detect changes in classifier confidence scores. Our proposed framework uses this change detection technique to address both of concept drift and concept evolution problems where [7,8,13] address only the concept drift problem. Unlike most of the above approaches, both of the classification and change detection of our approach are semi supervised in nature.

3   Proposed Approach

As discussed in Section 2, finding the fixed size of chunks or rate of decay is a non trivial task without prior knowledge on time-scale of change [8,10]. Moreover, approaches which use dynamic sliding window using change detection techniques [7,8,13] are based on loss estimation of predictor performance. This estimation needs true labels of the data instances to calculate the predictive performance. However, in the real world data streams, labeled data is scarce and not readily available. The above mentioned approaches might suffer in these scenario.

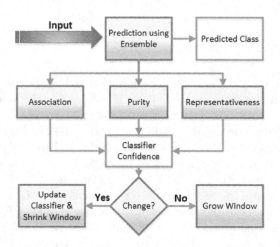

Fig. 1. High level work flow of the framework

In this paper, we present a complete framework *SCDMiner* (Adaptive $\underline{\text{S}}$emi supervised $\underline{\text{C}}$oncept $\underline{\text{D}}$rift $\underline{\text{Miner}}$ with novel class detection and delayed labeling) for classifying evolving data streams with novel class detection. It predicts the label of a data instance along with a confidence behind this prediction. Moreover, we also propose a change detection technique which takes these confidence values as input and detects any significant change in the classifier confidence over the time. If a significant change is detected, chunk boundary is determined and the classifier is updated using only the recent labeled data. In this way, *SCDMiner* addresses different challenges of data stream mining discussed in Section 1.

Figure 1 depicts the high level workflow of *SCDMiner*. It maintains an ensemble of L classification models and a dynamic window W containing classifier confidence scores in predicting labels of data instances in the stream. Let $\{M_1, ..., M_L\}$ be the models in the ensemble. As soon as an instance of the data stream arrives, label for this instance is predicted by the current ensemble along with a confidence score which is inserted into W. Subsequently, a change detection technique is executed on W. If it detects a significant change in the confidence scores, i.e., values stored in W, *SCDMiner* determines the chunk boundary which contains all the instances corresponding to the values stored in W. A new model is trained on the instances which are already labeled in this chunk, and the ensemble is updated by including the newly trained model. On the other hand, if the change detector finds no significant change in the confidence scores, the current ensemble is retained and W keeps growing. Since, *SCDMiner* tracks changes in the confidence values instead of the predictive performance, so it does not need all the true labels immediately after the prediction.

3.1   Classification and Novel Class Detection

SCDMiner uses similar techniques as *ECSMiner* [2] for classification and novel class detection. A k-NN based classifier is trained with the training data. Rather than storing the raw training data, K clusters are built using a semi-supervised K-means clustering, and the cluster summaries (mentioned as *pseudopoints*) of each cluster are saved. These pseudopoints constitute the classification model. The summary contains the *centroid, radius,* and *frequencies* of data points belonging to each class. The radius of a pseudopoint is equal to the distance between the centroid and the farthest data point in the cluster. The raw data points are discarded after creating the summary. Therefore, each model M_i is a collection of K pseudopoints. A test instance x_j is classified using M_i as follows. Let $h \in M_i$ be the pseudopoint whose centroid is nearest from x_j. The predicted class of x_j is the class that has the highest frequency in h. A confidence score between 0 to 1 is calculated based on certain criteria (will be discussed shortly) which is used as the weight of this prediction. The data point x_j is classified using the ensemble M by taking a weighted majority vote among all the classifiers.

Each pseudopoint corresponds to a "hypersphere" in the feature space with a corresponding centroid and radius. The *decision boundary* of a model M_i is therefore the union of the feature spaces encompassed by all pseudopoints $h \in M_i$. The decision boundary of the ensemble M is the union of the decision boundaries of all models $M_i \in M$. If a test instance is outside of the ensemble decision boundary, it is declared as an *F-outlier*, or filtered outlier. These are potential novel class instances, and are temporarily stored in a buffer *buf* to observe whether they are close to each other (*cohesion*) and farther apart from the data points of other classes (*separation*) [2]. A *new class* is declared if there are sufficient number of *F-outliers* fulfilling these conditions.

3.2   Calculation of Confidence Scores

We calculate three different confidence estimator values on each of the test instance to calculate confidence of each individual model. Finally, we combine all these individual model confidences to calculate the overall confidence of the ensemble classifier. Assuming h is the closest pseudopoint from labeled data instance x in model M_i, our proposed confidence estimators are as follows:

- *Association* is calculated by $R_h - D_i(x)$, where R_h is the radius of h and $D_i(x)$ is the distance of x from h.

- *Purity* is calculated by N_m/N_s, where N_s is the sum of all the *frequencies* and N_m is the highest *frequency* in h.
- *Representativeness* is calculated by N_s/N_t, where N_t is the number of labeled training instances used to build M_i and N_s is the sum of all the *frequencies* in h.

Association, *Purity* and *Representativeness* of the model M_i for instance x are denoted by $\mathcal{A}_i^x$, $\mathcal{P}_i^x$ and $\mathcal{R}_i^x$ respectively. Each of the confidence estimators contribute to the final confidence in prediction of a model according to their estimation capability. We measure this capability by calculating the correlation coefficient between confidence estimator values and classification accuracy for each model M_i using the labeled training instances as follows. M_i calculates confidence estimator values for each of the labeled training instances. Let h_{ij}^k be the value of jth confidence estimator in M_i's classification of instance k. Since we use three confidence estimators, $j \in \{1, 2, 3\}$. Let $\hat{y}_i^k$ be the prediction of M_i on instance k and y^k be the true label of that instance. Let v_i be the vector containing v_i^k values indicating whether the classification of instance k by model M_i is correct or not. In other words, $v_i^k = 1$ if $\hat{y}_i^k = y^k$ and $v_i^k = 0$ if $\hat{y}_i^k \neq y^k$. Finally, correlation vector r_i is calculated for model M_i. It contains r_{ij} values which are *pearson*'s correlation coefficients between h_{ij} and v_i for different j.

Correlation coefficients calculated in the training phase are used for classification and confidence estimation during testing phase as follows. First, *SCDMiner* calculates confidence estimator values h_i^x for a test instance x. Let c_i^x be the confidence value of model M_i in predicting test instance x. c_i^x is calculated by taking the dot product of h_i^x and v_i, i.e., $c_i^x = h_i^x . v_i$. Similarly, *SCDMiner* calculates confidence value of each of the models in the ensemble along with the prediction for each test instance. Each confidence value is normalized between 0 and 1. Normalized confidence value is treated as the weight of the prediction $\hat{y}_i$ by model M_i. Finally, to estimate confidence of the entire ensemble denoted by c^x, *SCDMiner* takes the average confidence of the models in the ensemble towards the predicted class.

3.3   Change Detection and Updating the Ensemble

As discussed earlier, *SCDMiner* maintains a variable size window W to monitor confidence scores of the ensemble classifier on recent data instances. The expectation is, the size of W will increase during stable period, and will decrease when there is a concept drift. The basic intuition behind this is, concept drift or concept evolution causes change of class boundaries which worsens performance of the classifier if not updated timely [15]. Confidence estimators are chosen in such a way that estimator values are expected to be decreased if class boundaries are changed. For example, if class boundaries are changed due to concept drift, more recent instances will be nearby the decision boundary or outside of the decision boundary of the ensemble classifier. So, in case of these instances, classifier models will have low *association* values. A change detection algorithm is therefore applied on the confidence values stored in W to detect any significant change in classifier confidence scores. If a change is detected, the base learning algorithm is invoked to build a new model on data instances corresponding to the current window W and subsequently W is shrinked. On the contrary, if no change is detected, W keeps growing indicating a stable period.

Algorithm 1.. Change detection algorithm

1: $W \leftarrow \emptyset$
2: **while** true **do**
3:     $x \leftarrow$ the latest data point in the stream
4:     $[\hat{y}, c^x] \leftarrow$ Classify(x) // c^x is calculated as discussed in Section 3.2
5:     $W \leftarrow W \cup c^x$
6:     **for** $n \leftarrow \triangle$ **to** $N - \triangle$ **do**
7:        $W_b \leftarrow W[1 : n]$
8:        $W_a \leftarrow W[n + 1 : N]$
9:        $t_{obs} \leftarrow$ calcObsStat()
10:       $pVal \leftarrow \Pr(t > t_{obs} \mid H_0)$
11:       **if** $pVal \leq \alpha$ **then**
12:          RetrainClassifier()
13:          ShrinkW()
14:       **end if**
15:     **end for**
16: **end while**

Algorithm 1 sketches our proposed change detection method. The variable size window W is maintained as follows. After inserting each confidence value of the ensemble classifier, our change detection technique divides W into two sub windows. Let W_b and W_a are two sub windows within W, where W_a contains performance values on more recent data instances than W_b. Change detection algorithm detects any significant change of statistical properties between contents of the sub windows for all possible combinations of *sufficiently large* W_a and W_b (Lines 6 to 15). By mentioning *sufficiently large*, we mean that each of the sub windows must have atleast $\triangle$ number of values. We use one tenth of the size of W as $\triangle$ in our experiments.

We use statistical hypothesis testing to detect change of a statistical property θ between elements of W_b and W_a. In this paper, we use the mean of the population as θ. Let μ_a and μ_b be the mean of population of distribution in W_a and W_b respectively. Let D be the difference between the mean of two sub windows, i.e., $D = \mu_b - \mu_a$. Since we want to detect the case where W_b contains greater average confidence value than W_a, i.e., decreasing classifier confidence, we perform a one sided right tail hypothesis testing. In this hypothesis testing, *Null Hypothesis* is $D \leq \delta$; in other words μ_b is at most δ more than μ_a. On the contrary, *Alternative Hypothesis* is $D > \delta$; in other words μ_b is greater than μ_a and difference between them is more than δ. Here, δ is a small real number and an user defined parameter.

Sample average $\bar{X}$ is the estimator of the mean of the population μ. Let $\bar{X}_a$ and $\bar{X}_b$ be the sample averages of the values in W_a and W_b respectively.

According to Central Limit Theorem, if sample size n is large, $\bar{X}$ follows a normal distribution with expectation μ and variance σ^2/n, where σ^2 is the variance of the population. Since we expect each of the sub windows W_a and W_b to contain sufficiently large number of values and true variance of the distribution of these values are unknown, we perform a two-sample t-test for the hypothesis testing. Let n and m be the number of values in W_b and W_a respectively. Test statistic for our case is the following-

$$t = \frac{\bar{X}_b - \bar{X}_a - \delta}{\sqrt{\frac{s_b^2}{n} + \frac{s_a^2}{m}}} \tag{1}$$

Where s_a^2 and s_b^2 are the sample variances of elements in W_a and W_b respectively. Since the samples in W_a and W_b are not paired, i.e., independent samples, the test statistic in Equation 1 follows a Student's t-distribution with degree of freedom γ, which is the Null Distribution in our case. The value γ is approximated using the following Satterthwaite's Formula. To limit the possibility of false positives, we use a small value α as the level of significance so that $Pr[Reject\ H_0 \mid H_0\ is\ true] \leq \alpha$. H_0 is rejected and a change point is detected if $t_{obs} \geq t_{\gamma,\alpha}$ holds. We build a new model on the data instances corresponding to confidence values stored in W and drop the sub window W_b subsequently. Once a new model is trained, it replaces the oldest model in the ensemble classifier. This ensures that we have exactly L models in the ensemble at any given point of time.

4   Experiment Results

We evaluate our proposed approach *SCDMiner* both on several benchmark real world and synthetic data sets. In this section, we present and analyze the experiment results.

4.1   Data Sets

We use three real and four different types of synthetic data sets to test performance of *SCDMiner* along with some other baseline approaches. Table 1 depicts the characteristics of the data sets.

ForestCover [16] contains geospatial descriptions of different types of forests. We normalize the data set, and arrange the data in order to prepare it for novel class detection so that in any chunk at most three and at least two classes co-occur, and new classes appear randomly. In PAMAP [17], nine persons were equipped with sensors that gathered a total of 52 streaming metrics features whilst they performed activities. Electricity [16] data set contains data collected from the Australian New South Wales Electricity Market.

SynCN (Synthetic Data with Concept-Drift and Novel Class) is a synthetic data set generated using the following equation: $\sum_{i=1}^{d} a_i x_i = a_0$ as explained

Table 1. Characteristics of Data Sets

Name of Data set	Num of Instances	Num of Classes	Num of Features
ForestCover	150,000	7	54
PAMAP	150,000	19	52
Electricity	45,312	2	8
SynCN	100,000	20	40
SynRBF@0.001	100,000	7	70
SynRBF@0.002	100,000	7	70
SynRBF@0.003	100,000	7	70

in [2]. SynRBF@X are synthetic data sets generated using *RandomRBFGeneratorDrift* of MOA [18] framework where X is the Speed of change of centroids in the model. We generate three such data sets using different X to check how efficiently different approaches can adapt to a concept drift.

We use ForestCover, PAMAP and SynCN data sets for simulating both concept drift and novel classes. On the contrary, rest of the data sets are used to test only concept drift capturing ability of different approaches.

4.2   Experiment Setup

We implement *SCDMiner* in Java version 1.7.0.51. To evaluate performance, we use a virtual machine which is configured with 8 cores and 16 *GB* of RAM. The clock speed of each virtual core is 2.4 *GHZ*.

We compare classification and novel class detection performance of our approach *SCDMiner* with *ECSMiner* [2]. We choose *ECSMiner* since it is one of the most robust and efficient frameworks available in the literature for classifying data streams having both concept drift and concept evolution. However, *ECSMiner* uses fixed chunk size where our proposed approach uses variable chunk size based on the change in classifier confidence.

Other than that, we compare performance of *SCDMiner* with *OzaBagAdwin (OBA)* and *Adaptive Hoeffding Tree (AHT)* implemented in MOA [18] framework, since these approaches seem to have superior performance than others on the data sets used in the experiments. Both of *OBA* and *AHT* use *ADWIN* [8] as the change detector. These approaches do not have novel class detection feature. So, we compare these approaches with *SCDMiner* only in terms of classification performance.

We evaluate the above classifiers on a stream by first testing and then training. To evaluate *SCDMiner* and *ECSMiner*, we use 50 pseudopoints, ensemble size 6, and 95% of labeled training data as suggested in [2]. On the contrary, we use 100% labeled training data in case of *OzaBagAdwin (OBA)* and *Adaptive Hoeffding Tree (AHT)*, since training and updating of these approaches are fully supervised.

4.3   Performance Metrics

Let FN = total number of novel class instances misclassified as existing class, FP = total number of existing class instances misclassified as novel class, TP = total number of novel class instances correctly classified as novel class, Fe = total number of existing class instances misclassified (other than FP), N_c = total number of novel class instances in the stream, N = total number of instances the stream. We use the following performance metrics to evaluate our technique:

1. ERR: Total misclassification error (percent), i.e., $\frac{(FP+FN+Fe)*100}{N}$.
2. M_{new}: % of novel class instances Misclassified as existing class, i.e., $\frac{FN*100}{N_c}$.
3. F_{new}: % of existing class instances Falsely identified as novel class, i.e., $\frac{FP*100}{N-N_c}$.
4. F_2: F_β score provides the overall performance of a classifier. In this paper, we use $\beta = 2$, which gives us $F_2 = \frac{5*TP}{5*TP+4*FN+FP}$.

Table 2. Summary of classification results

Name of Data set	SCDMiner Error%	ECSMiner Error%	AHT Error%	OBA Error%
ForestCover	**2.62**	4.23	22.89	18.06
PAMAP	**4.96**	35.26	8.76	7.27
Electricity	**0.0**	0.02	27.72	22.26
SynCN	1.20	**0.01**	4.81	4.5
SynRBF@0.001	**9.26**	34.64	18.82	11.31
SynRBF@0.002	**19.72**	63.43	38.75	37.04
SynRBF@0.003	**39.77**	65.39	48.65	46.86

4.4   Classification Performance

As discussed earlier, *SCDMiner* avoids unnecessary training during stable period and frequently updates the classifier when needed using dynamically determined chunks. As an instance, with increasing speed of change of centroids X in SynRBF@X data sets, our change detection technique helps *SCDMiner* to update the ensemble classifier more frequently to cope up with increasing concept drift. *SCDMiner* creates 111 and 160 number of chunks while classifying SynRBF@0.001 and SynRBF@0.003 data sets respectively where *ECSMiner* creates same 69 number of chunks in both of the cases. We do not report the number of chunks for all the data sets due to limited space in this paper.

Table 2 summarizes the classification error of the techniques on each data set described in Section 4.1. In almost all the cases, our proposed approach *SCD-Miner* clearly outperforms all the other approaches by large margin in terms of classification accuracy. For example, in case of ForestCover data set, *SCDMiner*

Table 3. Summary of novel class detection results

Data set	Method	M_{new}	F_{new}	F_2
ForestCover	SCDMiner	0.07	2.39	**0.95**
	ECSMiner	8.42	2.13	0.88
PAMAP	SCDMiner	0.09	13.65	**0.98**
	ECSMiner	0.05	37.53	0.45
SynCN	SCDMiner	0.0	0.01	0.99
	ECSMiner	0.0	0.0	**1.0**

shows around 38%, 89% and 85% better performance than *ECSMiner*, *AHT* and *OBA* respectively. Only in case of *SynCN* data set, *ECSMiner* shows slightly better performance than *SCDMiner*. It can be observed that, in case of *SynCN* data set, all the approaches show comparatively better result than the other data sets which indicates that *SynCN* data set contains less frequent concept drift. Since, *ECSMiner* uses fixed chunk size, it updates the model more frequently during stable time period comparing with *SCDMiner*. From the experiment, we know that *SCDMiner* updates the ensemble classifier only 7 times comparing with 44 number of updates by *ECSMiner*. So, *ECSMiner* gains slightly better accuracy in expense of more frequent training and updating the ensemble.

4.5 Novel Class Detection

Table 3 summarizes novel class detection performance of *SCDMiner* and *ECS-Miner* on different data sets. From the experiment data, it is clear that *SCD-Miner* outperforms *ECSMiner* by a large margin based on F_2 measure on all the data sets except *SynCN*. For example, in case of *ForestCover* data set, *SCDMiner* shows 8% better performance than *ECSMiner* in terms of F_2 measure. In case of *SynCN* data set, *SCDMiner* shows competitive performance. *ECSMiner* gains slightly better performance due to more frequent updates as discussed above.

5 Conclusion

In this paper, we present a framework *SCDMiner* which addresses most of the challenges of classifying evolving data streams. It exploits a change detection technique to determine the chunk boundaries dynamically. As a result, *SCD-Miner* determines number of training based on the frequency and intensity of concept drift. Results from the experiments show that, *SCDMiner* outperforms other approaches in the stream mining domain in terms of both classification and novel class detection accuracy.

Acknowledgments. This material is based upon work supported by NSF award no. CNS-1229652 and DMS-1322353.

References

1. Aggarwal, C.C., Han, J., Wang, J., Yu, P.S.: A framework for on-demand classification of evolving data streams. IEEE Transactions on Knowledge and Data Engineering **18**(5), 577–589 (2006)
2. Masud, M.M., Gao, J., Khan, L., Han, J., Thuraisingham, B.M.: Classification and novel class detection in concept-drifting data streams under time constraints. IEEE Trans. Knowl. Data Eng. **23**(6), 859–874 (2011)
3. Parker, B., Khan, L.: Detecting and tracking concept class drift and emergence in non-stationary fast data streams. In: Twenty-Ninth AAAI Conference on Artificial Intelligence, January 2015
4. Aggarwal, C.C., Yu, P.S.: On classification of high-cardinality data streams. In: SDM, pp. 802–813. SIAM (2010)
5. Koychev, I.: Tracking changing user interests through prior-learning of context. In: De Bra, P., Brusilovsky, P., Conejo, R. (eds.) AH 2002. LNCS, vol. 2347, pp. 223–232. Springer, Heidelberg (2002)
6. Klinkenberg, R.: Learning drifting concepts: Example selection vs. example weighting. Intell. Data Anal. **8**(3), 281–300 (2004)
7. Gama, J., Medas, P., Castillo, G., Rodrigues, P.: Learning with drift detection. In: Bazzan, A.L.C., Labidi, S. (eds.) SBIA 2004. LNCS (LNAI), vol. 3171, pp. 286–295. Springer, Heidelberg (2004)
8. Bifet, A., Gavald, R.: Learning from time-changing data with adaptive windowing. In: SDM. SIAM (2007)
9. Masud, M.M., Gao, J., Khan, L., Han, J., Thuraisingham, B.M.: A practical approach to classify evolving data streams: Training with limited amount of labeled data. In: ICDM, pp. 929–934 (2008)
10. Gama, J.A., Žliobaitė, I., Bifet, A., Pechenizkiy, M., Bouchachia, A.: A survey on concept drift adaptation. ACM Comput. Surv. **46**(4), 44:1–44:37 (2014)
11. Song, X., Wu, M., Jermaine, C., Ranka, S.: Statistical change detection for multi-dimensional data. In: 13th ACM SIGKDD International Conference on Knowledge Discovery and Data Mining, pp. 667–676, New York. ACM (2007)
12. Kuncheva, L.I., Faithfull, W.J.: PCA feature extraction for change detection in multidimensional unlabelled data. IEEE Transactions on Neural Networks and Learning Systems (2013)
13. Harel, M., Mannor, S., El-yaniv, R., Crammer, K.: Concept drift detection through resampling. In: Proceedings of the 31st International Conference on Machine Learning (ICML 2014), JMLR Workshop and Conference Proceedings, pp. 1009–1017 (2014)
14. Monteith, K., Martinez, T.: Using multiple measures to predict confidence in instance classification. In: The 2010 International Joint Conference on Neural Networks (IJCNN), pp. 1–8, July 2010
15. Vapnik, V.N.: Statistical Learning Theory. Wiley-Interscience (1998)
16. MOA: Moa massive online analysis-real time analytics for data streams repository data sets (2015). http://moa.cms.waikato.ac.nz/datasets/
17. Reiss, A., Stricker, D.: Introducing a new benchmarked dataset for activity monitoring. In: ISWC, pp. 108–109. IEEE (2012)
18. Bifet, A., Holmes, G., Pfahringer, B., Kranen, P., Kremer, H., Jansen, T., Seidl, T.: Moa: massive online analysis, a framework for stream classification and clustering. Journal of Machine Learning Research, 44–50 (2010)

Mining Temporal and Spatial Data

Cost-Sensitive Feature Selection
on Heterogeneous Data

Wenbin Qian[1], Wenhao Shu[2]([✉]), Jun Yang[1], and Yinglong Wang[1]

[1] School of Software, Jiangxi Agriculture University, Nanchang, China
qianwenbin1027@126.com
[2] School of Computer and Information Technology, Beijing Jiaotong University,
Beijing, China
11112084@bjtu.edu.cn

Abstract. Evaluation functions, used to measure the quality of features, have great influence on the feature selection algorithms in areas of data mining and knowledge discovery. However, the existing evaluation functions are often inadequately measured candidate features on cost-sensitive heterogeneous data. To address this problem, an entropy-based evaluation function is firstly proposed for measuring the uncertainty for heterogeneous data. To further evaluate the quality of candidate features, we propose a multi-criteria based evaluation function, which attempts to find candidate features with the minimal total costs and the same information as the whole feature set. On this basis, a cost-sensitive feature selection algorithm on heterogeneous data is developed. Compared with the existing feature selection algorithms, the experimental results show that the proposed algorithm is more efficient to find a subset of features without losing the classification performance.

Keywords: Feature selection · Cost-sensitive · Entropy · Heterogeneous data · Rough sets

1 Introduction

Handling high-dimensional data represents one of the most challenging problems in areas of data mining and knowledge discovery. Dimensionality reduction has been shown effective in dealing with high-dimensional data for learning, which refers to the study of methods for reducing the number of dimensions describing data [1]. It can bring many potential benefits: alleviating the curse of dimensionality, speeding up the learning process, and improving the generalization capability of a learning model.

Many dimensionality reduction algorithms have been developed at present. In general, they can be broadly classified into two categories: feature extraction and feature selection [3,5]. Feature extraction constructs new features with a linear or nonlinear transformation by projecting the original feature space to a lower dimensional one. Unlike feature extraction methods, feature selection methods preserve the original meaning of the features after reduction, which

© Springer International Publishing Switzerland 2015
T. Cao et al. (Eds.): PAKDD 2015, Part II, LNAI 9078, pp. 397–408, 2015.
DOI: 10.1007/978-3-319-18032-8_31

can be broadly categorized into wrapper and filter methods [2,4]. The wrapper
method uses the predictive accuracy of a predetermined learning algorithm to
determine the quality of selected features. One drawback of the wrapper method,
however, is that it is very expensive to run for data with numbers of features.
The filter method separates feature selection from classifier learning so that the
bias of a learning algorithm does not interact with feature selection algorithms.
It relies on many feature measures. Much attention has been paid to filter feature selection. Rough set theory offers a formal methodology for filter feature
selection. The main advantage of rough set theory is that no additional information about the data is required for data analysis such as thresholds or expert
knowledge on a particular domain. It provides a mathematical tool to handle
uncertainty in many data analysis tasks [6]. From the perspective of evaluation
functions in rough set theory, the feature selection algorithms encountered in the
literature can be mainly categorized into three representative types: consistency-
based feature selection [7], discernibility matrix-based feature selection [10], and
entropy-based feature selection [9]. The main differences of these algorithms lie
in the metrics that are used to evaluate the quality of candidate features to find
optimal solutions.

It should be noticed that the classical feature selection algorithms established
on the rough set theory are developed for a single type of features. However,
there often coexist some heterogeneous features, such as missing, categorical and
numerical ones, in data sets of most practical problems. For example, in a clinic
data set, some clinic results of a patient cannot obtain due to the expense or difficulty for some unconventional clinical measurements. In addition, most feature
selection techniques assume the data are already stored in data sets and available
without cost. But data are not free in real-world applications. In general, there
are two main types of cost [11]. i.e.,the test cost and misclassification cost, in real
applications, they are intrinsic to data. It is important to consider both types of
costs together. Under this context, the classical feature selection algorithms cannot deal with the heterogeneous data. Thus, it is meaningful and challenging to
effectively perform feature selection for heterogeneous data. In the literature, a
simple method to deal with heterogeneous data is to preprocess them into single-
typed data [14]. For instance, numerical features can be discreted as categorical
ones, and missing features can be filled with known ones. This transformation of
feature types might cause the loss of information. Recently, the methods found
that deal with heterogeneous data [15,16]. However, they have not taken into
the cost-sensitive data consideration. In most real-world applications, a user is
not only interested in removing irrelevant and redundant features, but also in
reducing costs that may be associated to features. The methods found that deal
with cost-sensitive data [11–13]. Min et al. [11] used the covering rough set to
deal with the minimal test cost feature selection problem on complete data with
test cost constraint. Weiss et al. [12] proposed the CASH algorithm cost-sensitive
feature selection using histograms. Bolon-Canedo et al. [13] proposed a new general framework for cost-based feature selection, together with the filter approach.
Based on the above observations, it is challenging to effectively perform the feature selection issue from both of heterogeneous features and costs intrinsic to

data. In this paper, we will focus on the feature selection work for selecting a feature subset with a minimal total cost as well as preserving a particular property of the heterogeneous data.

This paper is organized as follows. In Section 2, we introduce some relevant concepts involved in the paper. In Section 3, a multi-criteria based evaluation function for selecting features is proposed, and a cost-sensitive feature selection algorithm on heterogeneous data is developed. In Section 4, comparison experiments are made to show the efficiency and effectiveness of the proposed algorithm. Finally, the conclusions are presented in Section 5.

2   Preliminaries

In this section, we briefly review the basic concepts of rough set theory [6–8], and introduce a hybrid decision system with costs to represent the cost-sensitive heterogeneous data.

Rough set based data analysis starts from a data table called an information system [4–11]. More formally, an information system is a 4-tuple $IS =< U, A, V, f >$, where U is a set of nonempty and finite objects; A is the set of features characterizing the objects; V is the union of feature domains, i.e., $V = \cup_{a \in A} V_a$, where V_a is the value set of feature a, called the domain of a; and $f : U \times A \rightarrow V$ is an information function, which assigns feature values from domains of feature to objects such as $\forall a \in A$, $x \in U$, and $f(x, a) \in V_a$, where $f(x, a)$ denotes the value of feature a for object x. Each nonempty subset $B \subseteq A$ determines an indiscernibility relation, which is $IND(B) = \{(x, y) \in U \times U : f(x, a) = f(y, a), \forall a \in B\}$, the relation $IND(B)$ partitions U into some equivalence classes given by $U/IND(B) = \{[x]_B, x \in U\}$, where $[x]_B$ denotes the equivalence class determined by x with respect to B, i.e., $[x]_B = \{y \in U : (x, y) \in IND(B)\}$. If the feature set A is divided into condition feature set C and decision feature set D, the system is called a decision system.

In the rough set theory, the heterogeneous data can be expressed as a hybrid decision system. A hybrid decision system can be written as $HDS =< U, C \cup D, V, f >$, the heterogeneous data mainly include three parts: categorical, numerical and missing feature values, where C^c is a categorical feature set and C^n is a numerical feature set, respectively. To simplify this, we denote the i th numeric or categorical feature in C as c_i^n or c_i^c. In addition, if there exist $x \in U$ and $c \in C$ such that $f(x, c)$ is equal to a missing value (a null or unknown value, denoted as $*$), i.e., $* \in V_C$.

In real applications, both test and misclassification costs are taken into account in data sets. The cost-sensitive heterogeneous data can be expressed as a hybrid decision system with test and misclassification costs (HDS-TMC), it is represented as the 6-tuple $S =< U, C \cup D, V, f, tc, mc >$, where U, V and f have the same meanings as HDS, $tc : C \rightarrow \mathbb{R}^+ \cup \{0\}$ is the test cost function of feature set, $\mathbb{R}^+$ is the set of positive real numbers. Test costs are independent of one another, that is, $tc(B) = \sum_{b \in B} tc(b)$ for $B \subseteq C$, where $tc(b)$ is the test cost of single feature b. Central to the test cost is a vector, which can be easily be represented by $tc = [tc(c_1), tc(c_2), \cdots, tc(c_{|C|})]$. In addition, $mc : k \times k \rightarrow \mathbb{R}^+ \cup \{0\}$ is the misclassification cost function, which can be represented by a matrix $MC = \{mc_{k \times k}\}$, where $k = |V_D|$. Fundamental to the misclassification cost is the concept of the cost matrix. The cost matrix can be considered as a numerical representation of the penalty of classifying objects from one class to another.

3 Cost-Sensitive Feature Selection on Heterogeneous Data

In this section, an entropy based evaluation function is firstly proposed to measuring the uncertainty for heterogeneous data. On this basis, we present a multi-criteria based evaluation function for selecting candidate features that achieve the balance between the informative power and computational cost on cost-sensitive heterogeneous data. Finally, based on the proposed multi-criteria evaluation function, we develop a cost-sensitive feature selection algorithm on heterogeneous data.

3.1 Evaluation Function on Heterogeneous Data

Since the indiscernibility relation in the classical rough set is inapplicable to heterogeneous data, we first propose the hybrid relation to handle the heterogeneous data, including categorical, numerical and missing feature values.

Definition 1. Given a hybrid decision system $HDS = < U, C \cup D, V, f >$, for $B = B^c \cup B^n \subseteq C$, let $HR(B)$ denote the hybrid relation between objects that are possibly indiscernible in terms of B, $HR(B) = \{(x,y) \in U \times U | \forall b_1 \in B^c, f(x, b_1) = f(y, b_1) \vee f(x, b_1) = * \vee f(y, b_1) = *\} \wedge \{(x,y) \in U \times U | \forall b_2 \in B^n, |f(x, b_2) - f(y, b_2)| \leq \varepsilon \vee f(x, b_2) = * \vee f(y, b_2) = *\}$, where ε is a threshold.

The hybrid relation can be used to deal with the heterogeneous data, which greatly enhances the application scope of rough sets. In this sense, it naturally extends the definition of indiscernibility relation in rough sets to deal with numerical, categorical and missing feature values without resorting to the discretization process. Obviously, $HR(B)$ is reflexive and symmetric but not transitive. It can be easily shown that $HR(B) = \cap_{b \in B} HR(\{b\})$.

The neighborhood information granule of object x with reference to a hybrid feature set B is denoted as $\delta_B(x) = \{y \in U | (x, y) \in HR(B)\}$. Let $U/HR(B)$ denote the family set $\{\delta_B(x) | x \in U\}$, which is the classification induced by B. For $X \subseteq U$, the lower and upper approximations of X in B are $\underline{B}(X) = \{x \in U | \delta_B(x) \subseteq X\}$ and $\overline{B}(X) = \{x \in U | \delta_B(x) \cap X \neq \varnothing\}$. Furthermore, since $* \notin V_D, * \in V_C, U/IND(D) = \{Y_1, Y_2, \cdots, Y_n\}$ is a partition of U, the lower and upper approximations with respect to the decision D are $\underline{B}(D) = \cup_{i=1}^n \underline{B}(Y_i)$ and $\overline{B}(D) = \{x \in U | \delta_B(x) \cap X \neq \varnothing\}$. To further analyze the essential principle of feature selection with rough sets, we introduce the entropy as feature measure to evaluate the uncertainty of heterogeneous data.

One of the main problems in real life data analysis is uncertainty. In this background, uncertainty measures evaluate the roughness and accuracy of knowledge, which can provide us with principled methodologies to analyze uncertain data and unveil the substantive characteristics of the data sets. As common measures of uncertainty, various entropies as well as their conditional ones, and the extended one have been widely applied to devise feature selection algorithms. The conditional entropy can be used as a reasonable information measure in incomplete decision systems [17], and it is quite representative among other entropies. We extend the entropy to the hybrid data environment.

Definition 2. Let $HDS =< U, C \cup D, V, f >$ be a hybrid decision system, $* \notin V_D$, $* \in V_C$, $U/IND(D) = \{Y_1, Y_2, \cdots, Y_n\}$, for $B \subseteq C$, $\forall x_i \in U$, the entropy of D with respect to B is defined by

$$H(D|B) = -\sum_{i=1}^{|U|} \sum_{j=1}^{n} \frac{|\delta_B(x_i) \cap Y_j|}{|U|} log \frac{|\delta_B(x_i) \cap Y_j|}{|\delta_B(x_i)|}$$

Lemma 1. *Given a hybrid decision system $HDS =< U, C \cup D, V, f >$, for $B_1, B_2 \subseteq C$, and $B_1 \subseteq B_2$, with the same threshold ε for computing neighborhoods, we have the following relation $H(D|B_2) \le H(D|B_1)$.*

Lemma 1 indicates the validity of the proposed entropy. The validity of the entropy is important for constructing a greedy search algorithm. In the following, an evaluation function for measuring the significance of candidate features is defined as follows.

Definition 3. Let $HDS =< U, C \cup D, V, f >$ be a hybrid decision system, for $B \subseteq C$, $\forall b \in B$, the evaluation function for measuring the feature b in B is denoted by $E(b) = H(D|B) - H(D|B \cup \{b\})$.

The higher the change in the entropy, the more significant the feature is. We can sort all the features according to the evaluation values in descending order and select the top feature in the feature selection process.

To further evaluate the significance of candidate features on the cost-sensitive heterogeneous data, a new term is added to the above evaluation function so that the cost is taken into account. The multi-criteria based evaluation function is given as follows.

Definition 4. Given a hybrid decision system with costs (HDS-TMC) $S =< U, C \cup D, V, f, tc, mc >$, for $B \subseteq C$, $\forall b \in B$, the evaluation function for measuring the feature b is defined by $M(b) = E(b) + \lambda.(tc(b) + mc(b))$, where λ is a parameter introduced to weight the influence of the cost in the evaluation function, $tc(b)$ is the test cost of the feature b, and $mc(b)$ is the misclassification cost induced by feature b.

3.2 Cost-Sensitive Feature Selection Algorithm on Heterogeneous Data

Based on the proposed multi-criteria evaluation function, we can develop a feature selection algorithm to find a feature subset with cost minimization and the same information power as the whole feature set. However, to quicken the feature selection process, we will introduce a strategy to speed up this process in this subsection. Given a hybrid decision system with costs (HDS-TMC) $S = < U, C \cup D, V, f, tc, mc >$, for $B \subseteq C$, B is the selected feature subset, the objects in U have been partitioned into two parts with respect to B, i.e., recognized objects $U_R \subseteq U$ and unrecognized objects $U_{\overline{R}} \subseteq U$, where $U_R \cap U_{\overline{R}} = \varnothing$ and $U_R \cup U_{\overline{R}} = U$. For the recognized objects U_R, we have the following property. For the objects $U_R \subseteq U$ recognized by B, if $b \in C - B$ is a candidate feature, then b is redundant to the objects U_R, i.e., $E(B \cup \{b\}) = E(B)$.

From above, it indicates that any candidate feature b has no impact on the objects that have been recognized by the selected feature subset B. Thus in the process of feature selection, the recognized objects U_R can be deleted from the whole object set U, and the remaining objects get fewer as the selection of candidate features goes on. Incorporated into this strategy, we will develop the feature selection algorithm with the forward greedy on the hybrid decision system with costs (HDS-TMC) as follows.

Algorithm 1. Cost-sensitive feature selection algorithm on heterogeneous data (Algorithm **CSFS**)

Input: A hybrid decision system with costs $S = < U, C \cup D, V, f, tc, mc >$, λ;
Output: A feature subset Red.
Begin

1. Initialize $Red \leftarrow \varnothing$; $U_R \leftarrow \varnothing$; $U_{\overline{R}} \leftarrow U$;
2. **For** $\forall c_i \in C$ $(1 \leq i \leq |C|)$ **do**
3.   compute $E(C - \{c_i\})$ and $E(C)$ on $U_{\overline{R}}$;
4.   if $E(C - \{c_i\}) \neq E(C)$ then $Red \leftarrow Red \cup \{c_i\}$;
5. **End for**
6. Compute the recognized objects U_R from $U_{\overline{R}}$ induced by Red, and let $U_{\overline{R}} \leftarrow U_{\overline{R}} - U_R$;
7. Sort the features in $C - Red$ by $M(c)$ in a descending order, and record the results by $\{c'_1, c'_2, \cdots, c'_{|C-Red|}\}$;
8. **While** $(U_{\overline{R}} \neq \varnothing)$ **do**
9.   for $i = 1$ to $|C - Red|$ do
10.   let $Red \leftarrow Red \cup \{c'_i\}$;
11.   compute the recognized objects U_R from $U_{\overline{R}}$ induced by c'_i, and let $U_{\overline{R}} \leftarrow U_{\overline{R}} - U_R$;
12. **End while**
13. **For** $\forall p \in Red$ do
14.   if $E(Red - \{p\}) = E(Red)$ then $Red \leftarrow Red - \{p\}$;
15. **End for**
16. Return Red.

End

As shown above, to select relevant features with low associated costs, Algorithm CSFS uses a greedy forward strategy combined with the multi-criteria evaluation function to estimate the merit of candidate features. Steps 2-5 are to select the indispensable features from the whole feature set. Step 6 is to delete the objects recognized by the selected features. Steps 7-12 are to add the current best feature $c_i'(1 \leq i \leq |C - Red|)$ to Red by the multi-criteria based evaluation function until satisfying the unrecognized object set is empty. In addition, the objects recognized by the selected features are deleted from the object set, this process is the key step of this framework. Steps 13-15 are to delete some redundant features from the selection result Red.

4   Experimental Analysis

In order to test the efficiency and effectiveness of the proposed feature selection algorithm, we conduct some experiments on a PC with Windows 7, Intel (R) Core(TM) Duo CPU 2.93 GHz and 4GB memory. Algorithms are coded in C++ and the software being used is Microsoft Visual 2010. We perform the experiments on six real UCI data sets [18]. The characteristics of six heterogeneous data sets are described in Table 1. The data sets have no intrinsic test and misclassification costs associated. In order to study the performance of the feature selection algorithm, we will generate the random costs for their input features for experiments. For each feature, the costs are generated as a random number between 0 and 1.

Table 1. A description of six data sets

Data sets	No. of Instances	No. of Features	Classes
Adult	48842	14	2
Annealing	798	38	5
Arrhythmia	452	279	16
Dermatology	366	33	6
Hepatitis	155	19	2
Stat_Credit	1000	20	2

In what follows, to illustrate the effectiveness and efficiency of the proposed algorithm, three feature selection algorithms are used for comparisons. One is the positive region-based feature selection algorithm [7], the other is the rough entropy-based feature selection algorithm [9], and another is the discernibility matrix-based feature selection algorithm [10]. For convenience, the three feature selection algorithms are denoted as PRFS, REFS and DMFS, respectively.

4.1   Optimum Value of the Parameter λ

To determine the optimum value of the parameter λ in the multi-criteria evaluation function, extensive experiments are done on the six data sets. In general, different values of λ may lead to different feature subsets selected. The proposed algorithm CSFS runs 10 times with different λ settings on each data set. The values of λ investigated are from 0 to 0.5 with step 0.05. With different λ settings on the data set, there are different test and misclassification costs, we use the minimal total cost to evaluate the quality of a particular feature subset. Table 2 records the optimal value λ^* on the eight data sets for selecting feature subset with minimal costs.

Table 2. The optimal value λ^* on different data sets

Dataset	Adult	Annealing	Arrhythmia	Dermatology	Hepatitis	Stat_Credit
λ^*	0.10	0.15	0.25	0.10	0.35	0.20

From Table 2, we can observe that the optimal value λ^* varies for different data sets. There does not exist a rational setting of λ^* for all data sets, the optimum value is hard to specify for all data sets.

4.2 Feature Subset Size and Total Costs

For the data sets shown in Table 1, to compare the size of the selected feature subsets, Table 3 records the results of the four feature selection algorithms in terms of the feature subset size.

Table 3. Size of feature subsets selected by four algorithms

Dataset	PRFS	REFS	DMFS	CSFS
Adult	10	10	12	10
Annealing	13	13	15	11
Arrhythmia	26	28	31	25
Dermatology	11	11	12	9
Hepatitis	8	8	7	7
Stat_Credit	6	6	8	6

From Table 3, we can observe that the number of the selected features has largely been reduced by the four feature selection algorithms. However, the size of selected features by CSFS is fewer than other three algorithms at most cases. The detailed observations can be seen as follows. Take the data set Annealing as an example, CSFS selects 11 features, while PRFS, REFS and DMFS select 13, 13 and 15 features, respectively. The reason why the number of feature subset selected by the proposed algorithm CSFS is fewer than other three algorithms attributed to a redundancy-removing step, some redundant features are removed from the selected feature subset.

To compare the total costs of the selected feature subsets, Fig.1 records the results of the four feature selection algorithms in terms of the total costs. For this figure, the x-axis represents the six different data sets, for instance, the coordinate value "1" correspond to the Adult data set. And the y-axis represents the total costs of the selected feature subset.   Also, we can see that the

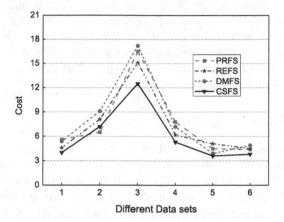

Fig. 1. Costs of features selected by the four algorithms on different data sets

cost values of the selected feature subset by CSFS are lower than that of other three algorithms in most of the data sets. Take the data set Arrhythmia as an example, the cost value of the feature subset selected by CSFS is 12.35, while the cost value of the feature subset selected by PRFS, REFS and DMFS is 15.19, 16.83 and 17.46, respectively. This may happen due to the multi-criteria based evaluation function, the proposed evaluation function considers the test and mis-classification cost of candidate features. The cost value of the selected features in CSFS, however, remains expensive in the Annealing date set. This phenomenon happens attributed to the costs have a greater effect than information power in this data set.

4.3   Computational Time

A summary of computational time for the four feature selection algorithms is given in Table 4. The computational time is expressed in seconds.

Table 4. Comparison of computational time for the four algorithms

Dataset	PRFS	REFS	DMFS	CSFS
Adult	290.583	367.411	541.275	56.201
Annealing	20.062	18.235	26.308	4.9132
Arrhythmia	59.127	67.460	83.159	15.360
Dermatology	3.895	4.521	4.036	1.148
Hepatitis	1.670	1.914	2.210	0.531
Stat_Credit	4.521	4.805	5.169	1.407

The experimental results on Table 4 show that the computational time cost of CSFS is much less than other three algorithms. The advantage of CSFS over other three algorithms is evident for large-scale data sets. Take the Adult data set as another example, the time cost of CSFS is about 1/5, 1/6 and 1/9 than that of PRFS, REFS, and DMFS, respectively. The results can be caused because the objects that have been recognized by the selected features in CSFS are deleted from the object set, such that the selection of features is in a compact search space. In contrast, the objects in other three algorithms always keep the same in the process of selecting candidate features.

4.4   Classification Accuracy

In the process of classification, ten-fold cross validation is applied on the data sets shown in Table 1. We randomly divide the objects into ten parts, and nine of them are used as the training data and the rest one as test data. The process is repeated for 10 times, after ten rounds, the results achieved on each time are recorded and averaged to obtain the final performance. The four feature selection algorithms are performed on each partition. To evaluate the subsets of features selected for different feature selection methods, the redial base function support vector machine (SVM) classifier, is trained on these subsets. The results achieved on each partition in terms of classification accuracy are recorded and averaged to obtain the final results. The recorded accuracy values are presented as a percentage.

Table 5. Comparison of classification accuracy for the SVM classifier

Data sets	Raw	PRFS	REFS	DMFS	CSFS
Adult	83.65±3.17	83.27±3.95	84.01±3.68	82.49±4.22	84.71±3.50
Annealing	99.15±0.41	100.00±0.0	100.00±0.0	100.00±0.0	100.00±0.0
Arrhythmia	71.60±7.53	74.68±8.26	72.15±6.97	72.78±7.02	74.29±6.84
Dermatology	90.35±2.64	88.50±2.91	89.31±3.40	86.53±3.74	89.17±2.15
Hepatitis	82.97±4.21	84.16±3.10	84.27±4.78	85.21±4.39	86.03±3.77
Stat_Credit	62.18±5.95	67.72±6.79	67.03±5.92	66.24±7.58	68.90±5.36
Average	81.65	83.06	82.79	82.21	83.85

The results shown in Table 5 indicate that the four feature selection algorithms produces better classification performances after feature selection, when compared with the raw data sets. Considering the average results in the last low, we see that there are no big differences among these four algorithms. However, there is a difference in the classification results among the four algorithms. CSFS outperforms the other three comparative algorithms slightly as to SVM. To illustrate the variation of classification accuracies versus the number of selected features, we choose two data sets Adult and Hepatitis to make an experimental comparison. Fig.2 shows the plots of the classification accuracies for the four feature selection algorithms.

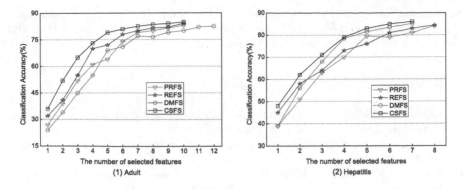

Fig. 2. Variation of classification accuracies versus the number of selected features

From the results presented in Fig.2, it is seen that the classification accuracies of the four algorithms increases rapidly at first, then the classification performance of them is relatively stable as the number of features increases. However, CSFS yields superior classification accuracy comparing with other three algorithms in most cases. Take the data set Adult as an example, when selecting 5 features, the classification accuracy of CSFS is 79.58%, which is higher than that of PRFS, REFS and DMFS 15.46%, 7.13% and 10.29%, respectively. The same phenomenon occurs as to the Hepatitis data set. With different number of selected features, CSFS always gives feature selection that is better than or comparable to that of others. The better performance of the algorithm CSFS is achieved in most cases, due to the fact that the multi-criteria can identify relevant and significant features from the data sets more efficiently than other algorithms. From the experimental results, we confirm that the proposed algorithm can find a feature subset without losing the classification performance.

5   Conclusions

In this paper, we introduce a multi-criteria based evaluation function to develop a feature selection algorithm with forward greedy search on cost-sensitive heterogeneous data. Two main conclusions are drawn as follows. On the one hand, compared with the existing feature selection algorithms, the proposed algorithm can find a feature subset with costs minimization. The main reason attributed to the multi-criteria based evaluation function that not only describes the information power of candidate features, but also considers the test and misclassification cost of candidate features. On the other hand, compared with other algorithms, the proposed algorithm can find a feature subset in a much shorter time without losing the classification performance.The main reason attributed to the multi-criteria based evaluation function implemented in dwindling object set in the feature selection process.Therefore, the proposed algorithm supplies a reasonable solution for feature selection on cost-sensitive heterogeneous data.

Acknowledgments. This work was supported by the Natural Science Foundation of China(61175048) and the Natural Science Foundation of Jiangxi Province(20132BAB2-01045). The corresponding author is Wenhao Shu.

References

1. Guyon, I., Elisseeff, A.: An introduction to variable and feature selection. Journal of Machine Learning Research **3**, 1157–1182 (2003)
2. Kohavi, R., John, G.H.: Wrappers for feature subset selection. Artificial Intelligence **97**, 273–324 (1997)
3. Farahat A.K., Ghodsi A., Kamel M.S.: An efficient greedy method for unsupervised feature selection. In: The 11th IEEE International Conference on Data Mining (ICDM), pp. 161–170 (2011)
4. Xue, B., Cervante, L., et al.: Multi-Objective Evolutionary Algorithms for Filter Based Feature Selection in Classification. International Journal on Artificial Intelligence Tools. **22**(4), 1350024, 1–31 (2013)
5. Xue, B., Zhang, M.J., et al.: Particle Swarm Optimization for Feature Selection in Classification: A Multi-Objective Approach. IEEE Transactions on Cybernetics **43**(6), 1656–1671 (2013)
6. Pawlak, Z., Skowron, A.: Rough sets and Boolean reasoning. Information Sciences **177**(1), 41–73 (2007)
7. Hu, Q., Zhao, H., Xie, Z., Yu, D.: Consistency based attribute reduction. In: Zhou, Z.-H., Li, H., Yang, Q. (eds.) PAKDD 2007. LNCS (LNAI), vol. 4426, pp. 96–107. Springer, Heidelberg (2007)
8. Qian, Y.H., Liang, J.Y., Pedrycz, W.: Positive approximation: an accelerator for attribute reduction in rough set theory. Artificial Intelligence **174**, 597–618 (2010)
9. Sun, L., Xu, J.C.: Feature selection using rough entropy-based uncertainty measures in incomplete decision systems. Knowledge-Based Systems **36**, 206–216 (2012)
10. Yang, M., Yang, P.: A novel condensing tree structure for rough set feature selection. Neurocomputing **71**, 1092–1100 (2008)
11. Min, F., Hu, Q.H., Zhu, W.: Feature selection with test cost constraint. International Journal of Approximate Reasoning **55**, 167–179 (2014)
12. Weiss, Y., Elovici, Y., Rokach, L.: The CASH algorithm cost-sensitive attribute selection using histograms. Information Sciences **222**, 247–268 (2013)
13. Bolon-Canedo, V., Porto-Daz, I., Sanchez-Marono, N.: A framework for cost-based feature selection. Pattern Recognition **47**, 2481–2489 (2014)
14. Yu, L., Liu, H.: Efficient feature selection via analysis of relevance and redundancy. Journal of Machine Learning Research **5**, 1205–1224 (2004)
15. Hu, Q.H., Pedrycz, W., Yu, D.R., Lang, J.: Selecting discrete and continuous features based on neighborhood decision error minimization. IEEE Transactions on Systems, Man, and Cybernetics-Part B: Cybernetics **40**(1), 137–150 (2010)
16. Chen, D.G., Yang, Y.Y.: Attribute reduction for heterogeneous data based on the combination of classical and fuzzy rough set models. IEEE Transactions on Fuzzy Systems **22**(5), 1325–1334 (2014)
17. Dai, J.H., Wang, W.T.: An uncertainty measure for incomplete decision tables and its applications. IEEE Transactions on Cybernetics **43**(4), 1277–1289 (2013)
18. UCI Dataset: http://www.ics.uci.edu/mlearn/MLRepository.html

A Feature Extraction Method for Multivariate Time Series Classification Using Temporal Patterns

Pei-Yuan Zhou[✉] and Keith C.C. Chan

Department of Computing, The Hong Kong Polytechnic University,
Hung Hom, Kowloon, Hong Kong
choupeiyuan@gmail.com, cskcchan@comp.polyu.edu.hk

Abstract. Multiple variables and high dimensions are two main challenges for classification of Multivariate Time Series (MTS) data. In order to overcome these challenges, feature extraction should be performed before performing classification. However, the existing feature extraction methods lose the important correlations among the variables while reducing high dimensions of MTS. Hence, in this paper, we propose a new feature extraction method combined with different classifiers to provide a general classification strategy for MTS data which can be applied for different area problems of MTS data. The proposed algorithm can handle data of high feature dimensions efficiently with unequal length and discover the relationship within the same and between different component univariate time series for MTS data. Hence, the proposed feature extraction method is application-independent and therefore does not depend on domain knowledge of relevant features or assumption about underling data models. We evaluate the algorithm on one synthetic dataset and two real-world datasets. The comparison experimental result shows that the proposed algorithm can achieve higher classification accuracy and F-measure value.

Keywords: Multivariate time series · Time series classification · Intra-temporal patterns · Inter-temporal pattern

1 Introduction

A multivariate time series (MTS) can be considered as made up of a collection of data values taken by a set of temporally interrelated variables monitored over a period of time at successive time instants spaced at uniform time intervals [4]. Effective classifying of such data can be applied into various problem domains. For example, in medicine and healthcare, the values taken by many variables representing different signs and symptoms may be temporally related or interrelated and they have to be monitored for a patient over a period of time for such relationship or interrelationship to be discovered. In financial analysis, as another example, the performance of a stock in terms of such variables as highs and lows, opening and closing prices, trading volumes, may also be temporally related or interrelated and for such relationship or interrelationship to be understood.

© Springer International Publishing Switzerland 2015
T. Cao et al. (Eds.): PAKDD 2015, Part II, LNAI 9078, pp. 409–421, 2015.
DOI: 10.1007/978-3-319-18032-8_32

Time series classification has received a great deal of attention in the past, and it also brings some new challenges to the data mining and machine learning community [1]. A number of different approaches have been proposed for univariate time series classification [9-11], however, few papers are found about multivariate time series classification in the literature [6].

Generally, to classify MTS data, feature extraction need to be performed for the original data, and then the classifiers, such as Support Vector Machine (SVM) or Artificial Neural Network (ANN), is used to classify the feature vectors [14-16]. The challenges of the classification process can be summarized as follows [2,7]: i) The most of the classifiers (e.g. decision trees, neural networks) can only take input data as a vector of features, but there are no explicit features in sequence data. ii) The dimensionality of feature space should be very high and the computation is costly. iii) The important correlations information among variables may be lost if the value of one variable is broken into MTS or each processed separately. Finally, the MTS can be of different lengths that cannot be extracted features by traditional method easily.

Hence, for satisfying the above requirement, we propose a new feature extraction algorithm combined with traditional classifier that provides a general classification strategy, called as Multivariate Time Series Classifier (MTSC). The algorithmic contributions of the proposed algorithm are: i) it can handle MTS data of high feature dimensions efficiently with unequal length; ii) it focuses on discovering the relationship in the same or among different variables; iii) it is a general method that can be applied to different problems on MTS data.

The structure of this paper is arranged as follows. In Section 2, we present a summary of existing work on feature extraction and classification for MTS data. In Section 3, we describe the details of the proposed algorithm. In Section 4, the results of the algorithm for both simulated and real world data sets are performed and presented. In the same section, we discuss results of the various tests carried out for effectiveness of its tasks. In the last section, we present a summary of the paper and the possible directions for future work.

2   Literature Survey

A multivariate time series (MTS) can be defined as a sequence of vectors, which may carry a class label [2]. For example, Electrocardiography (ECG) is a kind of MTS data, which is recorded from several sensors to describe the electrical activity of the heart, and it may come from either a healthy or ill person, labeled as "healthy" and "ill". The classification of MTS is the problem of classifying a set of MTS samples into a pre-defined set of classes [8]. To summary the existing algorithms of classification for MTS data, two main steps are considered: extract features using classical feature extraction method as pre-processing for MTS data, and classify feature vectors using classifier.

Feature extraction greatly affects the design and performance of the classifier and feature extraction is to use the existing feature parameters to comprise a lower-dimensional feature space, map useful information contained by original features to a small number of features, ignoring redundant and irrelevant information [15]. The classical feature extraction methods are based on statistical analysis. As the representative, there are some classical methods such as Principal Component Analysis (PCA), Linear Discriminant Analysis (LDA) [16], Factor Analysis [16], and so on [15]. In addition, for improving the performance of classification MTS data, some new feature extraction methods are proposed. Li et al. [7] proposed a feature extraction method, Singular Value Decomposition (SVD), to reduce the different length of data to feature vectors, and then apply SVM on the feature vectors to classify MTS data. Weng et al. [6] project original MTS into PCA subspace by throwing away the smallest principal components firstly, and then MTS samples in the PCA subspace are projected into a lower-dimensional space by using supervised Locality Preserving Projection (LPP). However, the above existing extraction method may lose the dependency relationship information among different univariate time series.

Hence, to extract the features among the same and different variables can retain more significant information for further classification. For proving the efficiency of the proposed algorithm, we compare the proposed algorithm with classical feature extraction methods, PCA. We focus on discovering the relationship within the same variable (intra-temporal patterns) and between different variables (inter-temporal patterns) at different time points, and combine the degree value of all patterns as feature vector.

After extracting features from original MTS data, a classifier, such as Support Vector Machine (SVM) [18] and Artificial Neural Networks (ANN) [19] can be applied to classify output feature vectors. The SVM transforms original input data into a higher dimensional space using a nonlinear mapping and then searches for a linear separating hyper-plane [20]. Considering the input vector is a $m \times n$ matrix, m is the number of MTS, n is the number of features, in order to classify MTS, it applies a kernel function to the original input data. ANN is composed of interconnecting artificial neurons that can compute values from inputs. Multi-Layer Perceptron (MLP) and Radial Basis Function (RBF) are two of the most widely used neural network architecture in literature for classification or regression problems. RBF is a local type of learning which is responsive only to a limited section of input space, and MLP is a distributed approach [21].

3   Methodology

With the above requirements in mind, we developed a feature extraction method to catch the dependency relationship between different variables, and classify MTS data based on feature vectors. Combining the proposed feature extraction method and classical classifier, the proposed strategy is called as Multivariate Time Series Classifier (MTSC).

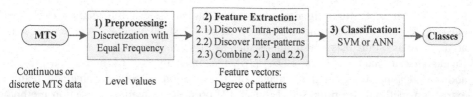

Fig. 1. The process of the proposed algorithm

The task of the proposed algorithm is to uncover the temporal relationship or inter-relationships between different variables. These temporal relationship or interrelation-ships constituent what we call intra- and inter- temporal patters respectively. The proposed algorithm performs its tasks in several steps: 1) discretize MTS data into level value using Equal Frequency; 2) extract features from MTS data which contains three sub-steps: 2.1) discover intra-temporal patterns within each component time series in each MTS; 2.2) discover inter-temporal patterns between different compo-nent time series within each MTS; 2.3) combine the value of degree of patterns disco-vering in 2.1) and 2.2); 3) classify MTSs based on feature vectors outputting in 2.3) using SVM with RBF-kernel function or MLP ANN. The structure of our proposed algorithm is shown in Figure 1. The definitions and notations are given in the Section 3.1, and then Section 3.2 specifies the proposed algorithm in detail.

3.1   The Problem Definition and Notations

Let S represent a set of MTS with the following characteristics:
1. S consists of m MTS represented as $\mathbf{S} = \{\mathbf{S}^1, \mathbf{S}^2, \dots, \mathbf{S}^m\}$.
2. For ith MTS, $\mathbf{S}^i, i = \{1, \dots, m\}$ consists of n components univariate time series (variables) that can be represented as $S_j^i, j = 1, \dots, n$, so that $\mathbf{S}^i = \{S_1^i, S_2^i, \dots, S_n^i\}$, and S_j^i represent the jth univariate time series in ith MTS.
3. The values in the vector $S_j^i, j = 1, \dots, n$, takes on the time instants of $1, \dots, p$ can be represented as $S_j^i = (s_{j,1}^i, s_{j,2}^i, \dots, s_{j,t-\tau}^i, \dots, s_{j,t}^i, \dots, s_{j,p}^i), 1 \le \tau \le p, \tau \in \mathbb{Z}^+$.
4. The domains of the value of variable, $V_j, j = 1, \dots, n,$, are represented as $(V_j) = [L_{V_j}, U_{V_j}], j = 1, \dots, n,$ L_{V_j} represent the lower bound and U_{V_j} represent the upper bound of the values that V_j can take on.

Given a set of MTS, S, with characteristics as described above, they are pre-classified into k classes, where, $C_1 = \{ \mathbf{S}_{C_1}^{(1)}, \mathbf{S}_{C_1}^{(2)}, \dots, \mathbf{S}_{C_1}^{(n_1)} \}$, $C_2 = \{ \mathbf{S}_{C_2}^{(1)}, \mathbf{S}_{C_2}^{(2)}, \dots, \mathbf{S}_{C_2}^{(n_2)} \}$, …, $C_k = \{ \mathbf{S}_{C_k}^{(1)}, \mathbf{S}_{C_k}^{(2)}, \dots, \mathbf{S}_{C_k}^{(n_k)} \}$. To classify these MTS, we need to find their class labels C_1 to C_k.

3.2   The Proposed Algorithm

3.2.1   Discretization
Before discovering patterns from MTS data, the preprocessing is needed. For reduc-ing and simplifying the original data, numerous values of a continuous variable is

always be replaced by a small number of interval labels, which leads to a concise, easy-to-use, knowledge-level representation of mining results [22]. Data discretization is a frequently used technique to partition the value space of a continuous variable into a finite number of intervals and assigning a nominal value of each of them [23, 24]. *Equal Width* and *Equal Frequency* are two simplest discretization methods. However, if uncharacteristic extreme values (outliers) exist in the data set, *Equal Width* can hardly handle this situation. Hence, in our case, we transform original numerical $s_{j,t}^i$ which represents the ith MTS of jth variable at t time point into $D_{j,t}^i$ using *Equal Frequency* [17] algorithm. And we set the number of bins is three, that is to say, the original numerical data is transformed into three levels (such as {high, medium, low}).

3.2.2   Discover Intra-/Inter-Temporal Patterns

After preprocess, discovering intra- and inter-temporal patterns among one MTS is applied. As one MTS contains several variables (univariate time series) which are temporally interrelated, the value that a particular variable take on at any time instant can be related to the variable's previous values or to the previous values of other variables. These interrelationships constitute, respectively, the intra- (within one univariate time series) and inter- (between different univariate time series) temporal patterns in one MTS. We use the proposed *significant discrepancy* measures to evaluate these patterns, and use a set of value of these measures to represent MTS. This process can be treated as feature extraction process from MTS, and the value of degree that describes dependency patterns can be treated as Feature Vectors. Two main sub-steps are specified as following.

Step One: Discovering Intra-Temporal Patterns.
Given a value, say, $S_{j,t}^i$, which represents the value on the time point t in jth time series in ith MTS, the magnitude of the difference between the conditional probability $Pr(s_{j,t}^i|s_{j,t-\tau}^i)$, $1 \leq \tau \leq \tau_{max} < t$, and the a priori probability $Pr(s_{j,t}^i)$ can be estimated as follows:

$$Pr\left(S_{j,t}^i|S_{j,t-\tau}^i\right) = \frac{freq\left(s_{j,t}^i, s_{j,t-\tau}^i\right)}{freq\left(s_{j,t-\tau}^i\right)} \tag{1}$$

where $freq(S_{j,t}^i, S_{j,t-\tau}^i)$ is the number of instants of the value, $S_{j,t}^i$, being preceded at τ instants ahead by $S_{j,t-\tau}^i$ and $freq(S_{j,t-\tau}^i)$ is the number of instants of the value $S_{j,t-\tau}^i$ that appear in S_j^i in S^i. As for $Pr(S_{j,t}^i)$, it can be estimated as follows:

$$Pr\left(S_{j,t}^i\right) = \frac{freq\left(s_{j,t}^i\right)}{p} \quad (p \text{ is the total number of time points}) \tag{2}$$

Given these probability estimations, the magnitude of the difference between the conditional probability $Pr\left(S_{j,t}^i|S_{j,t-\tau}^i\right)$ and the apriori probability $Pr(S_{j,t}^i)$ can be defined simply as: $Pr\left(S_{j,t}^i|S_{j,t-\tau}^i\right) - Pr(S_{j,t}^i)$. And the differences in the two probabilities are normalized using (3) [23].

$$d^i_{j,\tau} = \frac{p*\left(Pr\left(s^i_{j,t}\middle|s^i_{j,t-\tau}\right)-Pr\left(s^i_{j,t}\right)\right)}{\sqrt{p_{ij}\,Pr\left(s^i_{j,t}\right)Pr\left(s^i_{j,t-\tau}\right)\left(1-Pr\left(s^i_{j,t}\right)\right)\left(1-Pr\left(s^i_{j,t-\tau}\right)\right)}} \tag{3}$$

The significance of the temporal relationship depends on the magnitude of normalized difference, $d^i_{j,\tau}$ [23] which can be either ≥ 0 or ≤ 0. If $|d^i_{j,\tau}|$ is large, the presence or absence of $S^i_{j,t-\tau}$ would likely imply that at τ time instants later, the component univariate time series will or will not take on the value $S^i_{j,t}$, respectively. The magnitude of the normalized differences in conditional captures the strength of the temporal relationships [23] and they constitute the intra-temporal patterns for S^i_j. Hence, $d^i_{j,\tau}$ can be defined as the *significant discrepancy measure* to evaluate the relationship of two values within the same variable.

Step Two: Discovering Inter-Channels Temporal Patterns.
Similar to Step One, the inter-temporal patterns can be also discovered. The inter-temporal patterns are defined between two different variables say, S^i_j, and $S^i_{j'}$, both within $\mathbf{S}^i$, to consist of a set of temporal relationships or interrelationships detected between a value of S^i_j at a particular time instant, t, and those that it takes on at an earlier time instant, τ, $1\leq\tau \leq \tau_{max} < t$. These temporal relationships or interrelationships can be determined as follows.

Algorithm 1. The Proposed Feature Extraction Method

```
Input:S = {S¹,S²,...,Sᵐ}(m is the number of MTS)
       Sⁱ = {S₁ⁱ,S₂ⁱ,...,Sₙⁱ} (n is the number of variable)
       τ(the time window 1•τ< t)
Output: finalResult(feature vector for one MTS)

for each MTS Sⁱ in S
        Discretization using Equal Frequency for Sⁱ
        for each univariate time series Sⱼⁱ in Sⁱ
            Calculatedⱼ,τⁱ
            Result + = dⱼ,τⁱ /*degree vector of intra-patterns*/
        end
        for each channels sⱼ',tⁱ ≠ sⱼ,tⁱ  in MTS
            Calculate dⱼⱼ',τⁱ
            Result + = dⱼⱼ',τⁱ /*degree vector of inter-patterns*/
        end
        finalResult += Result
end /*finalResult is a set of degrees of all patterns*/
```

Given a value, say, $S^i_{j,t}$, and another value, say, $S^i_{j',t-\tau}$, The magnitude of the differences can be determined by first estimating the two probabilities $Pr(S^i_{j,t}|S^i_{j',t-\tau})$ and

$Pr(S_{j,t}^i)$, $1 \leq \tau \leq \tau_{max} < t$, based on the data and calculated using (1) and (2). And similarly, the differences in the two probabilities, $d_{jj',\tau}^i$, are normalized using same equation of (3). As the *significant discrepancy measure* can evaluate the relationship between two variables S_j^i and $S_{j'}^i$ in MTS, $d_{jj',\tau}^i$ is used to represent degree of relationship between S_j^i and $S_{j'}^i$. The pseudo code of the whole process of the feature extraction method is shown in Algorithm 1.

3.2.3   Classification Using SVM or ANN

Once all the temporal patterns are discovered, for MTS, we get a set of intra-temporal patterns measure $d_{j,\tau}^i$, ($i = 1, 2, ..., m, j = 1, 2, ..., n$, and $\tau = 1,2,... p$), and a set of inter-temporal patterns measure $d_{jj',\tau}^i$ ($i = 1, 2, ..., m; jj' = 1, 2, ..., n$, j $\neq$ j' and $\tau = 1,2,... p_{ij}$), associated with it. These intra- and inter- temporal patterns forms feature vectors: $\{d_{1,\tau}^i, ... d_{j,\tau}^i, d_{1\ 2,\tau}^i, d_{1\ 3,\tau}^i, ... , d_{jj',\tau}^i, ... d_{m-1\ m,\tau}^i\}$, where n is the total number of variables, $1 \leq \tau \leq \tau_{max}$ and i = 1, 2, ..., m. In addition, each $d_{jj',\tau}^i$ may contain a set of value for when $t_{j,t}^i$ equal to different discrete value. And then the classifier, SVM with RBF or MPL ANN, can be used to classify these feature vectors. The detail of classification algorithm has been specified in Section 2.

4   Experimental Result

To evaluate the performance of MTSC, a number of different experiments were carried out using both synthetic and real world data. In order to prove the proposed algorithm can handle both categorical and numerical data, the synthetic data set is a discrete data set which was generated by embedding different temporal patterns in the data to see if MTSC can discover them for classification. The other two real world data sets included EMG: Physical Action data set and ECG data set. For the purpose of performance evaluation, the test samples are compared with the known class labels using two performance measures: F-measure and classification accuracy. In the following, we describe the data set in section 4.1; and then the experimental results using the proposed algorithm are provided in section 4.2; finally, a comparison result between the proposed algorithm and traditional methods are given in section 4.3.

4.1   Data Set Description

4.1.1   Discrete Data Set: Synthetic Data Set

The synthetic data set is a discrete dataset that consists of 45 MTS generated randomly. Each of these MTS consists, in turn, of 5 variables, $v_1, ... v_5$, and $v_i = (A_i, B_i, C_i)$, i= 1,2,...5. There are total of 500 data values are generated for each variable to make all 5 univariate time series consists of 500 time points. Hence the synthetic data set is a 45 (MTS) × 5 (variables) × 500 (time points) dimensions data set.

Table 1. The Inserted Rules for Synthetic Dataset

Classes	Rules
Class 1	1. v_1 and v_5 are totally random. 2. v_4 takes on "A_4" at every interval of 2 time units and then V_2 at next time point, is generated to be "B_2", others are "A_2" or "C_2" randomly. 3. If v_4 not to be "B_4", v_3 takes on "A_3" at the next time instant 50% or "C_3" at the next time instant 50% of the time.
Class 2	1. v_2 and v_5 are totally random. 2. If v_2 in A_2 then v_1 in B_1, others are random 3. If v_2 in B_2, v_3 takes values in C_3, others are random 4. If v_2 in C_2, then v_4 in A_4 others are random
Class 3	1. v_3 and v_5 are totally random. 2. If v_3 in A_3 then v_2 in C_2, others are random 3. v_1 takes on A_1 at every interval of 3 time units and then at the next time points, v_4 is generated to be within A_4, others are random.

The different rules that belong to three classes are shown in Table 1. For example, in Class 1, we generate v_1 to v_5 randomly that takes on the value of $v_i = (A_i, B_i, C_i)$, $i = 1,2,\ldots5$ firstly. The Rule 2 in Class 1 means, we insert patterns into v_4 to make it take on "A_4" at every interval of 2 time units, and 1hen Variable 4 takes on value of "A_4", v_2 is generated to be "B_2" at next time point. Similarly, Rule 3 means, if value of v_4 is not equal to "B_4", the value of v_3 is generated as "A_3" at the next time instant 50% or "C_3" at the next time instant 50% of the time. Hence, for Class 1 v_1 and v_5 are noise data since they are totally random.

4.1.2   Numerical Real-World Data Set 1: EMG Physical Action Dataset

Physical Action Dataset (EMG) which is a benchmark data set from UCI repository [12]. The subjects are three male and one female (age 25 to 30), who have experienced aggression in scenarios such as physical fighting, took part in the experiment. Throughout 20 individual experiments, each subject had to perform ten normal and ten aggressive activities. 8 skin-surface electrodes correspond 8 input time series, muscle channel 1 to muscle channel 8, placed on the upper arms and upper legs to detect the position of actions of muscle. Each time series contains ~10000 samples (time points) with sampling frequency of 200Hz. Hence the EMG data is a 80 (MTS) × 8 (channels) × ~10000 (time points) dimensions data set.

4.1.3   Numerical Real-World Data Set 2: ECG Dataset

The other real-world dataset is ECG data set [13]. This data set comprises a collection of time-series data sets where each file contains the sequence of measurements recorded with two electrodes by one electrode during one heartbeat. Each heartbeat has an assigned classification of normal or abnormal. It contains 200 data sets where 133 were identified as normal and 67 were identified as abnormal. Hence, in this dataset, there are 200 MTS in total, each MTS contains 2 univariate time series, and 39-152 records are collected for each univariate time series with sampling 200HZ. The dimension of this data set is 200 (MTS) × 2 (electrodes) × ~150 (time points).

4.2   Experiment Process and Evaluation

The dataset can be processed as described in the methodology section. For the purpose of performance evaluation, 80% data are selected randomly as training data and the rest 20% as testing data. Two evaluation measures are used F-measure [14] and Classification Accuracy (CA).

The traditional F-measure or balanced F-score (F1 score) is the harmonic mean of precision and recall [14] that can be defined as below. Let C_p, $p \in \{1, \ldots, k\}$, be a class after classification using MTSC and C_q, $q \in \{1, \ldots, k\}$, be the class that is previous known, (k is the number of classes), so the F-measure is defined for C_p and C_q as equal (4) shows [14].

$$F(C_p, C_q) = \frac{2 \times Recall\ (Cp,Cq) \times Precision(Cp,Cq)}{Recall(Cp,Cq) + Precision(Cp,Cq)},$$

$$Recall(C_p, C_q) = \frac{freq(C_p,C_q)}{freq(C_p)}, Precision(C_p, C_q) = \frac{freq(C_p,C_q)}{freq(C_q)} \qquad (4)$$

$freq(C_p, C_q)$ represents the number of MTS with the cluster label C_p in the discovered cluster, C_q, $freq(C_p)$ is the number of records with class label C_p and $freq(C_q)$, is the number of records in the predicted class label C_q. Given this definition, F-measure therefore takes on values in the interval [0, 1]. The large its value is, the better the classification quality it reflects. In addition to the F-measure, due to the classes are pre-known, classification accuracy (CA) can be used evaluate how accurate MTSC is and the definition of CA is shown in equation (5)

$$CA = \frac{Total\ Number\ of\ mvts\ in\ the\ correct\ classes}{Total\ number\ of\ MTS} \qquad (5)$$

4.3   Experimental Result for the Proposed Algorithm

In synthetic data set, we only insert rules for one time intervals. So, we consider the intra-temporal patterns within one variable and inter-temporal patterns among different variables only between the previous time point and next one time point (τ=1). We use the proposed feature extraction to process the MTS data firstly and then use SVM or ANN to classify them. Table 2 summarizes the classification result for synthetic dataset using different classifier. The result table shows the value of Mean Acc. (the average of classification accuracy), Highest Acc. (the highest classification accuracy) and F-measure (the average value of F-measure). When MTSC is applied for ANN classifier, the highest accuracy is 100% and the average of classification accuracy is 98.6% with F-measure is 0.99 for synthetic data set and the average of accuracy is more than 75% with F-measure is 0.72 for SVM classifier. Hence, in this experiment, ANN can get the higher value for both classification accuracy and F-measure than SVM.

Table 2. The result of synthetic dataset using the proposed algorithm

Evaluation	Mean Acc. (Highest Acc.)	F-measure
SVM	76.67% (87.5%)	0.72
ANN	**98.6% (100%)**	**0.99**

In addition, for real-world data sets, considering the dependency or relationship within one variable or between different variables may not only in one time interval, we set τ=1 to 5 for intra-/inter-temporal patterns. Table 3 and Table 4 show the result for the two real data sets using MTSC with SVM or ANN classification algorithm for different time intervals.

In summary, ANN can get higher classification accuracy and F-measure than SVM for the most of classification result. In EMG dataset, when τ=5, the proposed algorithm can achieve highest average classification accuracy of 90.6% with the average F-measure being 0.89 for SVM classifier, and accuracy of 91.78% with the average F-measure being 0.92 for ANN classifier. Hence, the intra- and inter- relationship between variables are the most significant for classification in 5 time intervals. In ECG data set, when τ=4, the proposed algorithm can achieve a slightly higher average classification accuracy of 77.37% with the average F-measure being 0.72 for SVM classifier, and accuracy of 75.89% with the average F-measure being 0.7 for ANN classifier. Hence, when setting time interval equal to 4, the relationship within the same variable and between different variables can distinguish different samples best.

Table 3. The result of EMG dataset using the proposed algorithm with different time intervals

	Evaluation	τ=1	τ=2	τ=3	τ=4	τ=5
SVM	Mean Acc.	73.92%	74.94%	74.97%	81.94%	**90.60%**
	(Highest Acc.)	(92.31%)	(76.92%)	(84.62%)	(100%)	**(100%)**
	F-measure	0.79	0.74	0.69	0.89	**0.89**
ANN	Mean Acc.	89.80%	88.41%	85.01%	83.00%	**91.78%**
	(Highest Acc.)	(100%)	(100%)	(87.50%)	(92.31%)	**(93.75%)**
	F-measure	0.85	0.88	0.81	0.83	**0.92**

Table 4. Theresult of ECG dataset using the proposed algorithm with different time intervals

	Evaluation	τ=1	τ=2	τ=3	τ=4	τ=5
SVM	Mean Acc.	75.45%	71.32%	70.30%	**77.37%**	73.75%
	(Highest Acc.)	(80.56%)	(73.17%)	(73.17%)	**(81.40%)**	(80.65%)
	F-measure	0.73	0.71	0.66	**0.72**	0.72
ANN	Mean Acc.	78.21%	68.12%	69.08%	**75.89%**	70.09%
	(Highest Acc.)	(88.89%)	(70.30%)	(73.17%)	**(80.00%)**	(72.97%)
	F-measure	0.783	0.60	0.57	**0.7**	0.70

4.4   Comparison Experimental Result

For performance benchmarking, we compare the proposed MTSC algorithm with 1) classification using SVM or ANN classification without feature extraction method (No FE), 2) SVM or ANN classification with PCA for MTS data. The Principle Component Analysis (PCA) method can reduce high dimensions of MTS data and transform original data into a feature vector. However, PCA can only applied into MTS with equal length. So we cut unequal length time series into minimum length. And then we use the same classifiers, SVM with RBF kernel function and MPL ANN to classify feature vectors. Table 5 and Table 6 summarize the comparison result in classification accuracy and F-measure.

Table 5. Comparison of the average of classification accuracy between different algorithms

Data Sets	SVM			ANN		
	No FE	PCA	MTSC	No FE	PCA	MTSC
Synthetic	50.46%	67.2%	**76.67%**	68%	76.45%	**98.6%**
EMG	85.71%	86.73%	**90.6%**	74.73%	78.64%	**91.78%**
ECG	75%	75.05%	**76.52%**	75%	74.96%	**77.37%**

Table 6. Comparison of the average of F-measure between different algorithms

Data Sets	SVM			ANN		
	No FE	PCA	MTSC	No FE	PCA	MTSC
Synthetic	0.46	0.60	**0.72**	0.69	0.77	**0.99**
EMG	0.71	0.85	**0.89**	0.70	0.78	**0.92**
ECG	0.707	0.67	**0.75**	0.727	0.66	**0.7**

For synthetic dataset, when no feature extraction method is applied, the classification accuracy is 50.46% and 68% with F-measure of 0.46 and 0.69 for SVM classifier and ANN classifier respectively. PCA+SVM can only achieve an accuracy of 67.2% with average F-measure of 0.41, and PCA+ANN can achieve a higher accuracy of 76.45% with F-measure of 0.67. When comparing with the proposed algorithm, the performance of traditional algorithm is worse. Similarly, for two real-world datasets, MTSC can get higher both classification accuracy and F-measure than two other traditional methods. We can conclude from the result table, the value of the average of classification accuracy using MTSC is higher than the result using PCA with SVM or ANN for all data sets.

Besides the classification performance comparison, the complexity analysis is another significant target. Suppose that we have m MTS with n variables and t time points for each univariate time series ($n<<t$). If there is no feature extraction method is used, the classifier has to process for mnt dimensions data. When some feature extraction method, such as PCA, is used, the run-time complexity of the PCA is $mO(t^2)$, and the run-time of complexity of our proposed algorithm could be $mO(t*n^2)$. Generally, the MTS contains very high dimensions of time points with less variables, so in this case, the advantage of MTSC just need to count once for the value of each time points.

5   Conclusion

This paper has presented a classification strategy that combining the proposed feature extraction method and classifier for classifying MTS data. Unlike many existing methods, it is able to handle multivariate time series that may consist of either continuous or discrete data or both. As the proposed feature extraction method can perform its tasks without requiring any special assumption about data models, it is generic and application-independent. Given that MTSC performs classification for MTS data by discovering patterns within each time series independently of the others, it can also handle time series of different length. For performance evaluation, FEMTS was tested with both artificial and real data. The results show that it can be a promising algorithm for multivariate time series classification. The future work could be investigated into the possibility of improving the current work in three aspects: 1) after discovering intra- and inter- patterns in MTS, the dimensions of each MTS need to be reduced using attribute selection method to make algorithm speed up; 2) for improving the classification accuracy, fuzzy data/classes can be investigated so that MTS which belongs to overlapping classes can be discovered; 3) the more verification to verify how the algorithm can be more generally applicable.

References

1. Spiegel, S., Gaebler, J., Lommatzsch, A.: Pattern recognition and classification for multivariate time series. In: 5th International Workshop on Knowledge Discovery from Sensor Data, pp. 34–42. ACM (2010)
2. Xing, Z., Pei, J., Keogh, E.: A Brief Survey On Sequence Classification. ACM SIGKDD Explorations Newsletter, 40–48 (2010)
3. Ho, P.G.P.: Multivariate Time Series Support Vector Machine or Multispectral Remote Sensing Image Classification. In: Geoscience and Remote Sensing, ch.16 (2009)
4. Yang, K., Shahabi, C.: A PCA-Based similarity measure for multivariate time series. In: 2nd ACM International Workshop on Multimedia Databases, pp. 65–74. ACM Press (2004)
5. Weng, X.: Classification of multivariate time series using supervised neighborhood preserving embedding. In: 25th Control and Decision Conference (CCDC), Guiyang, pp. 957–961 (2013)
6. Weng, X., Shen, J.: Classification of Multivariate Time Series Using Locality preserving projection. Knowledge-based System. 21(7), 581–587 (2008)
7. Li, C., Khan, L., Prabhakaran, B.: Feature selection for classification of variable length multi-attribute motions. In: Petrushin, V.A., Khan, L. (eds.) Multimedia Data Mining and Knowledge Discovery, ch. 7. Springer (2007)
8. Cai, D., He, X., Han, J.: Orthogonal Laplacianfaces for Face Recognition. IEEE Transactions on Image Processing. 15(11), 3608–3614 (2006)
9. Zhang, H., Ho, T., Huang, W.: Blind feature extraction for time-series classification using haar wavelet Transform. In: Wang, J., Liao, X., Yi, Z. (eds.) ISNN 2005. LNCS, vol. 3497, pp. 605–610. Springer, Heidelberg (2005)
10. Rodrgueza, J.J., Alonsob, C.J., Maestro, J.A.: Support Vector Machines Of Interval-Based Features For Time Series Classification. Knowledge-Based Systems. 18, 171–178 (2005)

11. Hayashi, A., Mizuhara, Y., Suematsu, N.: Embedding time series data for classification. In: Perner, P., Imiya, A. (eds.) MLDM 2005. LNCS (LNAI), vol. 3587, pp. 356–365. Springer, Heidelberg (2005)
12. Murph P.M., Aha, E.W.: UCI Repository of Machine Learning Databases.(1999). http://archive.ics.uci.edu/ml/datasets/EMG+Physical+Action+Data+Set
13. Olszewski, R.T.: ECG Database. http://www.cs.cmu.edu/~bobski/
14. Larsen, B., Aone, C.: Fast and effective text mining using linear-time document clustering. In: 5th ACM SIGKDD International Conference on Knowledge Discovery and Data Mining, pp. 16–22. ACM Press (1999)
15. Ding, S., Zhu, H., Jia, W., Su, C.: A Survey On Feature Extraction for Pattern Recognition. Artificial Intelligence Review **37**(3), 169–180 (2012)
16. Zhu, D., Wu, C., Qin, W.: Multivariate statistic analysis and software SAS. Southeast University Press, Nanjing (1999)
17. Wong, A.K.C., Wang, C.C.: DECA–A Discrete-Valued Ensemble Clustering Algorithm. IEEE Trans on Pattern Analysis and Machine Intelligence **PAMI-1**(4), 342–349 (1979)
18. Cortes, C.: VapnikV.: Support-Vector Networks. Machine Learning **20**(3), 273–297 (1995)
19. Haykin, S.: Neural Networks: A Comprehensive Foundation. Prentice Hall, Upper Saddle River (1998)
20. Han, J., Kamber M.: Data Mining: Concepts and Techniques, 2nd edn. Elsevier Inc. (2007)
21. Cohen, S., Intrator, N.: A study of ensemble of hybrid networks with strong regularization. In: 4th International Conference on Multiple Classifier Systems, pp. 227–235 (2003)
22. Liao, T.W.: Clustering of Time Series Data - A Survey. Pattern Recognition **38**, 1857–1874 (2005)
23. Ma, P.C.H., Chan, K.C.C., Chiu, D.K.Y.: Clustering and re-clustering for pattern discovery in gene expression data. Journal of Bioinformatics and Computational Biology **3**(2), 281–201 (2005)
24. Dimitrova, E.S., McGee, J.J., Laubenbacher, R.C.: Discretization of Time Course Data. Journal of Computational Biology. **17**(6), 853–868 (2010)

Scalable Outlying-Inlying Aspects Discovery via Feature Ranking

Nguyen Xuan Vinh[1(✉)], Jeffrey Chan[1], James Bailey[1], Christopher Leckie[1], Kotagiri Ramamohanarao[1], and Jian Pei[2]

[1] The University of Melbourne, Melbourne, Australia
{vinh.nguyen,jeffrey.chan,baileyj,caleckie,kotagiri}@unimelb.edu.au
[2] Simon Fraser University, Burnaby, Canada
jpei@cs.sfu.ca

Abstract. In outlying aspects mining, given a query object, we aim to answer the question as to what features make the query most outlying. The most recent works tackle this problem using two different strategies. (i) Feature selection approaches select the features that best distinguish the two classes: the query point vs. the rest of the data. (ii) Score-and-search approaches define an outlyingness score, then search for subspaces in which the query point exhibits the best score. In this paper, we first present an insightful theoretical result connecting the two types of approaches. Second, we present **OARank** – a hybrid framework that leverages the efficiency of feature selection based approaches and the effectiveness and versatility of score-and-search based methods. Our proposed approach is orders of magnitudes faster than previously proposed score-and-search based approaches while being slightly more effective, making it suitable for mining large data sets.

Keywords: Outlying aspects mining · Feature selection · Feature ranking · Quadratic programming

1 Introduction

In this paper, we are interested in the novel and practical problem of investigating, for a particular query object, the aspects that make it most distinguished compared to the rest of the data. In [5], this problem was coined *outlying aspect mining*, although it has also been known as *outlying subspaces detection* [15], *outlier explanation* [10], *outlier interpretation* [2], and *object explanation* [12]. Outlying aspects mining has many practical applications. For example, a home buyer would be highly interested in finding out the features that make a particular suburb of interest stand out from the rest of the city. A recruitment panel may be interested in finding out what are the most distinguishing merits of a particular candidate compared to others. An insurance specialist may want to find out what are the most suspicious aspects of a certain insurance claim. A natural complementary task to outlying aspects mining is *inlying aspects mining*, i.e., what features make the query most usual.

© Springer International Publishing Switzerland 2015
T. Cao et al. (Eds.): PAKDD 2015, Part II, LNAI 9078, pp. 422–434, 2015.
DOI: 10.1007/978-3-319-18032-8_33

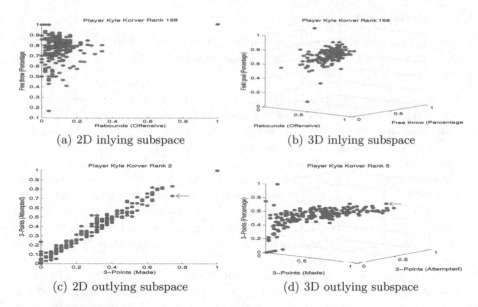

(a) 2D inlying subspace

(b) 3D inlying subspace

(c) 2D outlying subspace

(d) 3D outlying subspace

Fig. 1. OARank inlying/outlying subspaces for NBA player Kyle Korver (red circle)

A practical example of outlying aspects mining is given in Fig. 1, where we present the outlying-inlying aspects returned by our proposed approach–**OARank**–for player Kyle Korver in the NBA Guards dataset (data details given in Section 4.1). NBA sports commentators are interested in the features that make a player most unusual (and maybe also usual). If we take the kernel density rank as an outlyingness measure, then it can be observed that in the top 2D and 3D-inlying subspaces, Kyle has very low density ranking (198th and 168th over 220 players respectively). The attributes in which Kyle appears most usual are "Rebound (Offensive)", "Free throw (Percentage)" and "Field goal (Percentage)". On the other hand, in the top outlying subspaces, Kyle has very high density rank (2nd and 5th over 220 players respectively). Kyle is indeed very good at 3-points scoring: "3-points (Attempted)", "3-points (Made)", and "3-points (Percentage)".

Outlying aspects mining (or outlier explanation) has a close relationship with the traditional task of outlier detection, yet with subtle but critical differences. In this context, we only focus on the query object, which itself may or may not be an outlier. It is also not our main interest to verify whether the query object is an outlier or not. We simply are interested in subsets of features (i.e., subspaces) that make it most outlying. We are also not interested in other possible outliers in the data, if any. In contrast, outlier detection aims to identify all possible outlying objects in a given data set, often without explaining why such objects are considered as outliers. Outlier explanation is thus a complementary task to outlier detection, but could be used in principle to explain any object of interest. Thus, in this paper, we shall employ the term *outlying aspects mining*, which is more generic than outlier explanation.

1.1 Related work

The latest work on outlying aspects mining can be categorized into two main directions, which we refer to as feature selection approaches [4], and score-and-search approaches [5].

- In *feature selection approaches* [10,12], the problem of outlying aspects mining is first transformed into the classical problem of feature selection for classification. More specifically, the two classes are defined as the query point (positive class) and the rest of the data (negative class). In [4], to balance the classes, the positive class is over-sampled with samples drawn from a Gaussian distribution centred at the query point, while the negative class is under-sampled, keeping k full-space neighbors of the query point and some other data points from the rest of the data. Similarly in [12], the positive class is over-sampled while keeping all other data points as the negative class. The feature subsets that result in the best classification accuracy are regarded as outlying features and selected for user inspection.

   A similar approach to feature selection is feature transformation [2], which identifies a linear transformation that best preserves the locality around the neighborhood of the query point while at the same time distinguishing the query from its neighbors. Features with high absolute weights in the linear transformation are deemed to contribute more to the outlyingness of the query.
- In *score-and-search based methods*, a measure of outlyingness degree is needed. The outlyingness degree of the query object will be compared across all possible subspaces, and the subspaces that score the best will be selected for further user inspection. In [5], the kernel density estimate was proposed as an outlyingness measure. It is well known, however, that the raw density measure tends to be smaller for subspaces of higher dimensionality, as a result of increasing sparseness. For this reason, the rank statistic was used to calibrate the raw kernel density to avoid dimensionality bias. Having defined an outlyingness measure, it is necessary to search through all possible subspaces and enumerate the ones with lowest density rank. Given a dataset of dimension d, the number of subspaces is $(2^d - 1)$. If the user specifies a parameter d_{max} as the maximum dimensionality, then the number of subspaces to search through is in the order of $O(d^{d_{max}})$.

1.2 Contribution

In this paper, we advance the state of the art in outlying aspects mining by making two important contributions. First, we show an insightful theoretical result connecting the two seemingly different approaches of *density-based score-and-search* and *feature selection* for outlying aspects mining. In particular, we show that by using a relevant measure of mutual information for feature selection, namely the quadratic mutual information, density minimization can be regarded as contributing to maximizing the mutual information criterion. Second, as exhaustive search for subspaces is expensive, our most important contribution in this paper is to propose an alternative scalable approach, named

OARank, in which the features are ranked based on their potential to make the query point having low density. The top-ranked features are then selected either for direct user inspection, or for a more comprehensive score-and-search with the best-scored subspaces then reported to the user. The feature ranking procedure takes only quadratic time in the number of dimensions and scales linearly w.r.t the number of data points, making it much more scalable and suitable for mining large and high dimensional datasets, where a direct enumeration strategy is generally infeasible.

2 Connection Between Density-Based Score-and-Search and Feature Selection Based Approaches

In the feature selection approach, the problem of explaining the query is first posed as a two-class classification problem, in which we aim to separate the query $\mathbf{q}$ (positive class c_1) from the rest of the data $\mathbf{O}$ (negative class c_0) of n objects $\{\mathbf{o}_1, \ldots, \mathbf{o}_n\}$, $\mathbf{o}_i \in \mathbb{R}^d$. Let $\mathbf{D} = \{D_1, \ldots, D_d\}$ be a set of d numeric features (attributes). In the feature selection approaches [4,12], to balance the class distribution, the positive class c_1 is augmented with synthetic samples from a Gaussian distribution centred at the query. The task is then to select the top features that distinguish the two classes. These features are taken as outlying features for the query.

We now show that there exists a particular feature selection paradigm which has a close connection to density based approaches. Let us form a two-class data set

$$\mathbf{X} = \{\underbrace{\mathbf{x}_1 \equiv \mathbf{o}_1, \ldots, \mathbf{x}_n \equiv \mathbf{o}_n}_{c_0}, \underbrace{\mathbf{x}_{n+1} \equiv \mathbf{q}, \ldots, \mathbf{x}_{2n} \equiv \mathbf{q}}_{c_1}\}$$

Note that here we have over-sampled the positive class simply by duplicating $\mathbf{q}$ n times, so that the classification problem is balanced. Mutual information based feature selection aims to select a subset of m features such that the information shared between the data and the class variable is maximized, i.e., $\max_{\mathbf{S} \subset \mathbf{D}, |\mathbf{S}| = m} I(\mathbf{X}_{\mathbf{S}}; C)$, where $\mathbf{X}_{\mathbf{S}}$ is the projection of the data onto the subspace $\mathbf{S}$ and C is the class variable. We will show that by using a particular measure of entropy coupled with the Gaussian kernel for density estimation, we arrive at a formulation reminiscent of density minimization. In particular, we shall make use of the general Havrda-Charvat's α-structural entropy [6], defined as:

$$H_\alpha(\mathbf{X}) = (2^{1-\alpha} - 1)^{-1} \left[\int f(\mathbf{x})^\alpha d\mathbf{x} - 1 \right], \ \alpha > 0, \alpha \neq 1 \tag{1}$$

Havrda-Charvat's entropy reduces to Shanon's entropy in the limit when $\alpha \to 1$, hence it can be viewed as a generalization of Shannon's entropy [7,9].

In order to make the connection, we shall make use of a particular version of Havrda-Charvat's entropy with $\alpha = 2$, also known as quadratic Havrda-Charvat's entropy $H_2(\mathbf{X}) = 1 - \int f(\mathbf{x})^2 d\mathbf{x}$ (with the normalizing constant discarded for simplicity).

Using the Gaussian kernel $G(\mathbf{x} - \mathbf{X}_i, \sigma^2) = (2\pi\sigma^2)^{-d/2} \exp(\frac{-\|\mathbf{x}-\mathbf{X}_i\|^2}{2\sigma^2})$, the probability density of $\mathbf{X}$ is estimated as $\hat{f}(\mathbf{x}) = \frac{1}{2n} \sum_{i=1}^{2n} G(\mathbf{x} - \mathbf{X}_i, \sigma^2)$. The quadratic entropy of $\mathbf{X}$ can be estimated as:

$$H_2(\mathbf{X}) = 1 - \frac{1}{(2n)^2} \int_{\mathbf{x}} \left(\sum_{i=1}^{2n} G(\mathbf{x} - \mathbf{X}_i, \sigma^2) \right)^2 d\mathbf{x} = 1 - \frac{1}{(2n)^2} \sum_{i=1}^{2n} \sum_{j=1}^{2n} G(\mathbf{X}_i - \mathbf{X}_j, 2\sigma^2)$$

wherein we have employed a nice property of the Gaussian kernel, which is that the convolution of two Gaussian remains a Gaussian [14]:

$$\int_{\mathbf{x}} G(\mathbf{x} - \mathbf{X}_i, \sigma^2) G(\mathbf{x} - \mathbf{X}_j, \sigma^2) \, d\mathbf{x} = G(\mathbf{X}_i - \mathbf{X}_j, 2\sigma^2) \tag{2}$$

The conditional quadratic Havrda-Charvat's entropy of $\mathbf{X}$ given a (discrete) variable C is defined as $H_2(\mathbf{X}|C) = \sum_{k=1}^{K} p(c_k) H_2(\mathbf{X}|C = c_k)$. We have:

$$H_2(\mathbf{X}|C = c_0) = 1 - \frac{1}{n^2} \sum_{i=1}^{n} \sum_{j=1}^{n} G(\mathbf{X}_i - \mathbf{X}_j, 2\sigma^2)$$

$$H_2(\mathbf{X}|C = c_1) = 1 - \frac{1}{n^2} \sum_{i=n+1}^{2n} \sum_{j=n+1}^{2n} G(\mathbf{X}_i - \mathbf{X}_j, 2\sigma^2)$$

then $H_2(\mathbf{X}|C) = \frac{1}{2} H_2(\mathbf{X}|C = c_0) + \frac{1}{2} H_2(\mathbf{X}|C = c_1)$. Finally, the quadratic mutual information between $\mathbf{X}$ and C is estimated as:

$$I_2(\mathbf{X}; C) = H_2(\mathbf{X}) - H_2(\mathbf{X}|C) = \frac{1}{(2n)^2} (CC + CS)$$

where

$$CC = \sum_{i,j=1}^{n} G(\mathbf{X}_i - \mathbf{X}_j, 2\sigma^2) + \sum_{i,j=n+1}^{2n} G(\mathbf{X}_i - \mathbf{X}_j, 2\sigma^2)$$

$$CS = -2 \sum_{i=1}^{n} \sum_{j=n+1}^{2n} G(\mathbf{X}_i - \mathbf{X}_j, 2\sigma^2)$$

An interesting interpretation for the quadratic mutual information is as follows: the quantity $G(\mathbf{X}_i - \mathbf{X}_j, 2\sigma^2)$ can be regarded as a measure of interaction between two data points, which can be called the *information potential* [3]. The quantity $\sum_{i,j=1}^{n} G(\mathbf{X}_i - \mathbf{X}_j, 2\sigma^2)$ is the total strength of intra-class interaction within the negative class and $\sum_{i,j=n+1}^{2n} G(\mathbf{X}_i - \mathbf{X}_j, 2\sigma^2)$ is the total strength of intra-class interaction within the positive class (within-class total information potential), thus CC is a measure of *class compactness*. On the other hand, $\sum_{i=1}^{n} \sum_{j=n+1}^{2n} G(\mathbf{X}_i - \mathbf{X}_j, 2\sigma^2)$ measures inter-class interaction (cross-class information potential), thus CS is a measure of *class separability*. For maximizing $I_2(\mathbf{X}, C)$, we aim to maximize intra-class interaction while minimizing inter-class interaction.

Theorem 1. *Density minimization is equivalent to maximization of class separability in quadratic mutual information based feature selection.*

Proof. Note that since the positive class contains only $\mathbf{q}$ (duplicated n times), we have

$$CS = -2n \sum_{i=1}^{n} G(\mathbf{X}_i - \mathbf{q}, 2\sigma^2) = -2n^2 \hat{f}(\mathbf{q}),$$

where $\hat{f}(\mathbf{q}) = \frac{1}{n} \sum_{i=1}^{n} G(\mathbf{X}_i - \mathbf{q}, 2\sigma^2)$ is nothing but the kernel density estimate of $\mathbf{q}$. Thus, it can be seen that minimizing the density of $\mathbf{q}$ is equivalent to minimizing inter-class interaction (cross-class information potential), or equivalently maximizing class separability. $\qquad\square$

This theoretical result shows that there is an intimate connection between density-based score-and-search and feature selection based approaches for outlying aspects mining. Minimizing the density of the query will contribute to maximizing class separation, and thus maximizing the mutual information criterion. This insightful theoretical connection also points out that the mutual-information based feature selection approach is more comprehensive, in that it also aims to maximize the class compactness. The relevance of class-compactness to outlying aspects mining is yet to be explored.

3   Outlying Aspects Mining via Feature Ranking

We now present the main contribution of this paper—**OARank**—a hybrid approach for outlying aspects mining that leverages the strengths of both the feature selection and the score-and-search paradigms. This is a two-stage approach. In the first stage, we rank the features according to their potential to make the query outlying. In the second (and optional) stage, score-and-search can be performed on a smaller subset of the top-ranked $m \ll d$ features.

3.1   Stage 1: OARank–Outlying Features Ranking

We aim to choose a subset of m features $\mathbf{S} \subset \mathbf{D}$ such that the following criterion is minimized:

$$\text{SS} : \min_{\substack{\mathbf{S} \subset \mathbf{D} \\ |\mathbf{S}|=m}} \left\{ C(m) \sum_{i=1}^{n} \sum_{\substack{t,j \in \mathbf{S} \\ t<j}} K(\mathbf{q}_j - \mathbf{o}_{ij}, h_j) K(\mathbf{q}_t - \mathbf{o}_{it}, h_t) \right\} \tag{3}$$

where $K(x - \mu, h) = (2\pi h^2)^{-1/2} \exp\{-(x - \mu)^2/2h^2\}$ is the one dimensional Gaussian kernel with bandwidth h and center μ, and $C(m) = \frac{2}{nm(m-1)2^{m-2}}$ is a normalization constant.

We justify this *subset selection* (SS) objective function as follows: *the objective function in SS can be seen as a kernel density estimate at the query point* $\mathbf{q}$. To see this, we first develop a novel kernel function for density estimation, which is the sum of 2-dimensional kernels. Herein, we employ the Gaussian product kernel recommended by Scott [11], defined as:

$$K(\mathbf{q} - \mathbf{o}_i, \mathbf{h}) = \frac{1}{(2\pi)^{d/2} \prod_{j=1}^{d} h_j} \prod_{j=1}^{d} \exp(\frac{-(\mathbf{q}_j - \mathbf{o}_{ij})^2}{2h_j^2}) \qquad (4)$$

where h_j's are the bandwidth parameters in each individual dimension. We note that, in the product kernel (4), a particular dimension (feature) can be 'de-emphasized', by assigning its corresponding 1D-kernel to be the 'uniform' kernel, or more precisely, a rectangular kernel with sufficiently large bandwidth σ_u, i.e.,

$$K_u(\mathbf{q}_j - \mathbf{o}_{ij}, \sigma_u) = \begin{cases} 1/(2\sigma_u) & \text{if } \frac{|\mathbf{q}_j - \mathbf{o}_{ij}|}{\sigma_u} \leq 1 \\ 0 & \text{otherwise} \end{cases} \qquad (5)$$

Note that σ_u can be chosen to be arbitrarily large, so that we can assume that any query point of interest $\mathbf{q}$ will lie within σ_u-distance from any kernel center $\mathbf{o}_i$ in any dimension. In our work, we normalize the data (including the query) so that $\mathbf{o}_{ij} \in [-1, 1]$ in any dimension, thus σ_u could simply be chosen as $\sigma_u = 1$. From the product kernel (4), if we de-emphasize all dimensions, but keeping only two 'active' dimensions $\{t, j\}$, then we obtain the following d-dimensional kernel:

$$K_{tj}(\mathbf{q} - \mathbf{o}_i, \mathbf{h}) = \frac{1}{(2\sigma_u)^{d-2}} K(\mathbf{q}_j - \mathbf{o}_{ij}, h_j) K(\mathbf{q}_t - \mathbf{o}_{it}, h_t) \qquad (6)$$

Averaging over all $d(d-1)/2$ pairs of dimensions yields the following kernel:

$$K(\mathbf{q} - \mathbf{o}_i, \mathbf{h}) = \frac{2}{d(d-1)(2\sigma_u)^{d-2}} \sum_{j<t}^{d} K_{tj}(\mathbf{q} - \mathbf{o}_i, \mathbf{h}) \qquad (7)$$

Theorem 2. *The kernel function $K(\mathbf{q} - \mathbf{o}_i, \mathbf{h})$ as defined in (7) is a proper probability density function.*

Proof. It is straightforward to show that $K(\mathbf{q} - \mathbf{o}_i, \mathbf{h}) \geq 0$ and $\int_{\mathbb{R}^d} K(\mathbf{q} - \mathbf{o}_i, \mathbf{h}) = 1$ □

Employing this new kernel to estimate the density of the query point $\mathbf{q}$ in a subspace $\mathbf{S} \subset \mathbf{D}$, we obtain exactly the objective function in $\mathcal{SS}$, which when minimized will minimize the density at the query $\mathbf{q}$, i.e., making $\mathbf{q}$ most outlying.

3.2   Solving the Outlying-Inlying Aspects Ranking Problem

The subset selection problem $\mathcal{SS}$ can be equivalently formulated as a quadratic integer programming problem as follows:

$$\min_{\mathbf{w}} \{\mathbf{w}^T \mathbf{Q}' \mathbf{w}\} \text{ s.t. } w_i \in \{0, 1\}, \sum w_i = m \qquad (8)$$

where $\mathbf{Q}'_{tj} = \sum_{i=1}^{n} \frac{1}{h_j} K_j(\frac{q_j - o_{ij}}{h_j}) \times \frac{1}{h_t} K_t(\frac{q_t - o_{it}}{h_t})$, $t \neq j$ and $\mathbf{Q}'_{tt} = 0$. Equivalently, we can rewrite it in a maximization form as

$$QJP : \max_{\mathbf{w}} \left\{ \mathbf{w}^T \mathbf{Q} \mathbf{w} \right\} \text{ s.t. } w_i \in \{0,1\}, \sum w_i = m \tag{9}$$

where $\mathbf{Q}_{tj} = \Phi - \mathbf{Q}'_{tj}$ and $\Phi = \max_{i,j} \mathbf{Q}'_{ij}$. While (8) and (9) are equivalent, the Hessian matrix $\mathbf{Q}$ is entry-wise non-negative, a useful property that we will exploit shortly. The parameter m specifies the size of the outlying subspace we wish to find. It is noted that SS and QJP are not monotonic with respect to m, i.e., with two different m values, the resulting outlying subspaces are not necessarily subsets of one another.

As QJP is well known to be NP-hard [1], we relax the problem to the real domain, as follows. Note that with $w_i \in \{0,1\}, \sum w_i = m$, we also have $\|\mathbf{w}\|_2 = \sqrt{m}$. We shall now drop the integral 0-1 constraint, which in fact causes NP-hardness, while keeping the norm constraint:

$$\max_{\mathbf{w}} \left\{ \mathbf{w}^T \mathbf{Q} \mathbf{w} \right\} \text{ s.t. } \|\mathbf{w}\|_2 = \sqrt{m}, \ w_i \geq 0 \tag{10}$$

The additional non-negativity constraints $w_i \geq 0$ ensure that the relaxed solution can be reasonably interpreted as feature 'potential' in making $\mathbf{q}$ outlying. Also note that we can replace $\|\mathbf{w}\|_2 = \sqrt{m}$ with $\|\mathbf{w}\|_2 = 1$ without changing the optimal relative weight ordering (all the weights w_i are scaled by the same multiplicative constant $1/\sqrt{m}$). Thus, we arrive at

$$QP : \max_{\mathbf{w}} \left\{ \mathbf{w}^T \mathbf{Q} \mathbf{w} \right\} \text{ s.t. } \|\mathbf{w}\|_2 = 1, \ w_i \geq 0 \tag{11}$$

Observe that since $\mathbf{Q}_{ij} \geq 0$, the solution to this problem is simple and straightforward: it is the dominant eigenvector associated with the dominant eigenvalue of the Hessian matrix $\mathbf{Q}$ [13]. Note that with this relaxation scheme, the parameter m has been eliminated, thus OARank will produce a single ranking. The outcome of this quadratic program can be considered as feature potentials: features that have higher potentials contribute more to the outlyingness of the query. Features can be ranked according to their potentials. The top-ranked m features will be chosen for the next score-and-search stage.

Note that an interesting novel by-product of this ranking process is the *inlying aspects*, i.e., features with the lowest potentials. These inlying aspects are features (subspaces) in which *the query point appears to be most usual*.

3.3   Stage 2: OARank+Search

Having obtained the feature ranking in stage 1, there are two ways to proceed: (i) One can take the top k-ranked features as the single most outlying subspace of size k. A more flexible way is to (ii) perform a comprehensive score-and-search on the set of top-ranked $m \ll d$ features, and report a list of top-scored subspaces for user inspection. The search on the filtered feature set is however much cheaper than a search in the full feature space.

3.4   Complexity Analysis

Ranking stage: The cost of building the Hessian matrix $\mathbf{Q}$ is $O(d^2 n)$. The cost for finding the dominant eigenvector of $\mathbf{Q}$ is $O(d^2)$. Thus overall, the complexity of the ranking process is $O(d^2 n)$.

Score-and-Search stage: If we employ the density rank measure, the cost for scoring (i.e., computing the density rank for the query) in a single subspace is $O(n^2)$ time. The number of subspaces to score is $O(2^m - 1)$ for exhaustive search, or $O(m^{d_{max}})$ if a maximum subspace size is imposed. Note that for practical applications, we would prefer the subspaces to be of small dimensionality for improved interpretability, thus it is practical to set, for example, $d_{max} = 5$. The overall complexity of this stage is $O(n^2 m^{d_{max}})$

Overall, the complexity of both stages is $O(d^2 n + n^2 m^{d_{max}})$ for the proposed two-stage approach. For comparison, a direct density rank based score-and-search approach on the full space costs $O(n^2 d^{d_{max}})$, which is infeasible when d is moderately large.

Techniques for further improving scalability: While the ranking phase of OARank is generally very fast, the search phase can be slow, even on the reduced feature set. To further speed up the search phase, one can further prune the search space. In particular, the search space can be explored in a stage-wise manner, expanding the feature sets gradually. In exhaustive search, every feature set of size k will be expanded to size $k+1$ by adding 1 more feature. We can improve the efficiency of this process, sacrificing comprehensiveness, by only choosing a small subset of most promising subspaces (i.e., highly-scored) of size k to be expanded to size $k+1$.

4   Experimental Evaluation

In this section, we experimentally evaluate the proposed approaches, **OARank** and **OARank+Search**. We compare our approaches with the density rank based approach in [5] and Local Outlier with Graph Projection (LOGP) [2]. LOGP is the closest method in spirit to OARank, in that it also learns a set of weights: features with higher weights contribute more to the outlyingness of the query. These weights are from a linear transformation that aims to separate the query from its full-space neighborhood. LOGP was proposed as a method for both detecting and explaining outliers. We implemented all methods in Matlab/C++ except LOGP for which the Matlab code was kindly provided by the authors. The parameters for all methods were set as recommended in the original articles [2,5]. The bandwidth parameter for OARank was set according to [5], i.e., $h = 1.06 \min\{\sigma, \frac{R}{1.34}\} n^{-\frac{1}{5}}$ with σ being the standard sample deviation, and R being the difference between the first and third quartiles of the data distribution. In order to improve the scalability of score-and-search based methods, we apply a stage-wise strategy as discussed in Section 3.4 where only at most 100 top-scored subspaces are chosen for expansion at each dimension $k < d_{max} = 5$. All experiments were performed on an i7 quad-core PC with 16Gb of main memory. The source code for our methods will be made available via our website.

4.1   The NBA Data Sets

We first test the methods on the NBA data available at http://sports.yahoo.com/nba/stats. This data set was previously analyzed in [5], where the authors collected 20 attributes for all NBA guards, forwards and centers in the 2012-2013 season, resulting in 3 data sets. We compare the quality of the ranking returned by OARank and LOGP. More specifically, for each player, we find the top 1, 2 and 3 inlying and outlying features, and then compute the density rank for the player in his outlying-inlying subspaces.

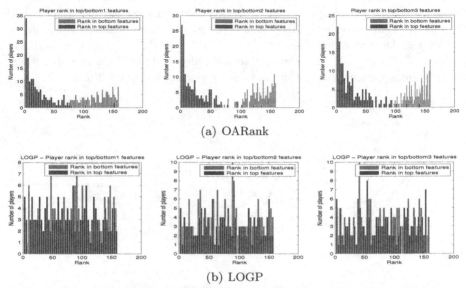

(a) OARank

(b) LOGP

Fig. 2. Outlying Feature Ranking: OARank vs. LOGP on NBA Forwards data set (best viewed in color)

The results of this analysis on the NBA Forwards data set are presented in Fig. 2. It can be clearly seen that OARank is able to differentiate between inlying and outlying aspects. More precisely, in the outlying subspaces (of the top-ranked 1/2/3 features), all players tend to have higher density rank than their ranks in the inlying subspaces (of the bottom-ranked 1/2/3 features). On the same data set, LOGP ranking does not seem to differentiate well between outlying and inlying features. In particular, the rank distribution appears to be uniform in both inlying and outlying subspaces. Thus, in this experiment, qualitatively we can see that OARank is more effective at identifying inlying-outlying aspects. The same conclusion applies for the NBA Guards and Centers data sets, for which we do not provide detailed results due to space restrictions. We have seen the detailed analysis for a specific player, Kyle Korver, in Figure 1. The feature weights and ranking returned by OARank for Kyle Korver can be inspected in Fig. 3(e).

4.2   Mining Non-Trivial Outlying Subspaces

For a quantitative analysis, we employ a collection of data sets proposed by Keller et al. [8] for benchmarking subspace outlier detection algorithms. This collection contains data sets of 10, 20, 30, 40, 50, 75 and 100 dimensions, each consisting of 1000 data points and 19 to 136 outliers. These outliers are challenging to detect, as they are only observed in subspaces of 2 to 5 dimensions but not in any lower dimensional projection. We note again that our task here is not outlier detection, but to explain why the annotated outliers are designated as such. For this data set, since the ground-truth (i.e., the outlying subspace for each outlier) is available as part of Keller et al.'s data, we can objectively evaluate the performance of all approaches. Let the true outlying subspace be T and the retrieved subspace be P. To evaluate the effectiveness of the algorithms, we employ the Jaccard index $Jaccard(T, P) \triangleq |T \cap P|/|T \cup P|$, and the precision, $precision \triangleq |T \cap P|/|P|$. The average Jaccard index and precision over all outliers for different approaches on all datasets are reported in Figure 3(a,b).

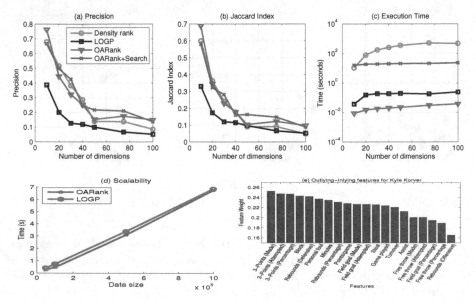

Fig. 3. (a)-(c): Performance on identifying non-trivial outlying high-dimensional subspaces; (d) Scalability; (e): OARank feature weights for Kyle Korver

We can observe that OARank and the density based score-and-search approach both outperform LOGP. OARank (without search) obtains relatively good results, slightly better than density rank score-and-search at higher dimensions. OARank (with search) did not seem to improve the results significantly on this data set. In terms of execution time (Fig. 3c), OARank is the fastest, being orders of magnitude faster than density rank score-and-search. It can also be observed that the OARank+Search approach admits a near-flat time complexity profile with regards to the number of dimensions. This agrees well with the theoretical complexity of $O(d^2 n)$ for ranking and $O(n^2 m^{d_{max}})$ for search. On these data sets, the ranking time was negligible compared to search time, while the search complexity of $O(n^2 m^{d_{max}})$ is independent of dimensionality.

4.3 Scalability

We tested the method on several large datasets. We pick the largest of Keller's data sets of 1000 points in 100 dimensions, and introduce more synthetic examples by drawing points from a Gaussian distribution centred at each data points, resulting in several large data sets of size ranging from 50,000 to 1 million data points. The run time for OARank and LOGP is presented in Figure 3(d). It is noted that for these large datasets, the search phase using the density score is computationally prohibitive, due to quadratic complexity in data size n. Both LOGP and OARank deal well with large datasets, with linear time complexity in the number of data points. This observation matches well with OARank's theoretical complexity of $O(d^2 n)$ and demonstrates that OARank is capable of handling large data sets on commodity PCs.

We shall note that another prominent feature of OARank is that it is suitable for applications on streaming data: as data come in, entries in the Hessian matrix $\mathbf{Q}$ can be updated gradually. Feature weights can also be updated on-the-fly in real time, given that there exist very efficient algorithms for computing the dominant eigenvector of symmetric matrices.

5 Conclusion

In this paper, we have made two important contributions to the outlying aspects mining problem. First, we have made an insightful connection between the density based score-and-search and the mutual information based feature selection approach for outlying aspects mining. This insight can inspire the development of further hybrid approaches, which leverage the strengths of both paradigms. Second, we proposed OARank, an efficient and affective approach for outlying aspects mining, which is inspired by the feature ranking problem in classification. We show that OARank is suitable for mining very large data sets.

Acknowledgments. This work is supported by the Australian Research Council via grant numbers FT110100112 and DP140101969.

References

1. Chaovalitwongse, A.W., et al.: Quadratic integer programming: complexity and equivalent forms quadratic integer programming: Complexity and equivalent forms. In: Floudas, C.A., Pardalos, P.M. (eds.) Encyclopedia of Optimization, pp. 3153–3159 (2009)
2. Dang, X.H., et al.: Discriminative features for identifying and interpreting outliers. In: ICDE 2014, pp. 88–99, March 2014

3. Dang, X.-H., Bailey, J.: A hierarchical information theoretic technique for the discovery of non linear alternative clusterings. In: 16th ACM SIGKDD, pp. 573–582. ACM, New York (2010)

4. Dang, X.H., Micenková, B., Assent, I., Ng, R.T.: Local outlier detection with interpretation. In: Blockeel, H., Kersting, K., Nijssen, S., Železný, F. (eds.) ECML PKDD 2013, Part III. LNCS, vol. 8190, pp. 304–320. Springer, Heidelberg (2013)

5. Duan, L., Tang, G., Pei, J., Bailey, J., et al.: Mining outlying aspects on numeric data. In: Data Mining and Knowledge Discovery (2014) (in press)

6. Havrda, J., Charvat, F.: Quantification method of classification processes. concept of structural α-entropy. Kybernetika 3, 30–35 (1967)

7. Jawaharlal, K.: Entropy Measures. Maximum Entropy Principle and Emerging Applications. Springer-Verlag New York Inc., Secaucus (2003)

8. Keller, F., Muller, E., Bohm, K.: HiCS: high contrast sub-spaces for density-based outlier ranking. In: ICDE 2012, pp. 1037–1048 (2012)

9. Mathai, A.M., Haubold, H.J.: On generalized entropy measures and pathways. Physica A: Statistical Mechanics and its Applications 385(2), 493–500 (2007)

10. Micenkova, B., Ng, R.T., Assent, I., Dang, X.-H.: Explaining outliers by subspace separability. In: ICDM (2013)

11. Scott, D.W.: Multivariate Density Estimation: Theory, Practice, and Visualization. Wiley (1992)

12. Vinh, N.X., Chan, J., Bailey, J.: Reconsidering mutual information based feature selection: A statistical significance view. In: AAAI 2014 (2014)

13. Vinh, N.X., Chan, J., Romano, S., Bailey, J.: Effective global approaches for mutual information based feature selection. In: KDD 2014 (2014)

14. Vinh, N.X., Epps, J.: Mincentropy: A novel information theoretic approach for the generation of alternative clusterings. In: 2010 IEEE 10th International Conference on Data Mining (ICDM), pp. 521–530 (2010)

15. Zhang, J., Lou, M., et al.: Hos-miner: a system for detecting outlyting subspaces of high-dimensional data. In: VLDB 2004, pp. 1265–1268 (2004)

A DC Programming Approach for Sparse Optimal Scoring

Hoai An Le Thi$^{(\boxtimes)}$ and Duy Nhat Phan

Laboratory of Theoretical and Applied Computer Science EA 3097,
University of Lorraine, Ile de Saulcy, 57045 Metz, France
{hoai-an.le-thi,duy-nhat.phan}@univ-lorraine.fr

Abstract. We consider the supervised classification problem in the high-dimensional setting. High-dimensionality makes the application of most classification difficult. We present a novel approach to the sparse linear discriminant analysis (LDA) based on its optimal scoring interpretation and the zero-norm. The difficulty in treating the zero-norm is overcome by using an appropriate continuous approximation such that the resulting problem can be formulated as a DC (Difference of Convex functions) program to which DCA (DC Algorithms) is investigated. The computational results on both simulated data and real microarray cancer data show the efficiency of the proposed algorithm in feature selection as well as classification.

Keywords: Classification · Feature selection · Optimal scoring · DC programming · DCA

1 Introduction

We focus on the multiclass classification problem. Let X be an $n \times p$ data matrix with observations x_i $(i = 1, ..., n)$ on the rows and features on the columns. Denote by n_i the number of observations in the cluster C_i. We assume that the features have been standardized to have mean 0 and variance 1.

Discriminant analysis (DA) is widely used in classification problems. The traditional way of DA was first introduced by Fisher in [3], known as the linear discriminant analysis (LDA). The basic idea of LDA is to find a linear transformation $W = [w_1, ..., w_{Q-1}]$ that best discriminates between classes, the classification is then performed in the transformed space based on some distance metrics such as Euclidean distance.

There are three different approaches to tackle LDA: the normal model, the optimal scoring problem, and the Fisher discriminant problem (see e.g [7,8]). In this paper we are interested in the optimal scoring problem which can be described as follows. Let $Y \in \mathbb{R}^{n \times Q}$ with $Y_{ik} = 1$ if $x_i \in C_k$ and 0 otherwise. To find the linear transformation $W = [w_1, ..., w_K]$, $K \leq Q - 1$, the optimal scoring criterion successively solves the problem

© Springer International Publishing Switzerland 2015
T. Cao et al. (Eds.): PAKDD 2015, Part II, LNAI 9078, pp. 435–446, 2015.
DOI: 10.1007/978-3-319-18032-8_34

$$\min_{w_k,\theta_k} \quad \{\|Y\theta_k - Xw_k\|_2^2\}$$

$$\text{subject to} \quad \frac{1}{n}\theta_k^T Y^T Y \theta_k = 1; \quad \theta_k^T Y^T Y \theta_l = 0, l = 1, ..., k - 1, \tag{1}$$

where θ_k is a Q-vector of *scores* and w_k is a p-vector of *discriminant vectors*.

LDA often performs quite well in simple, low-dimensional setting and it is known to fail when the number of features p is larger than the number of observations n. In many applications such as information retrieval, face recognition and microarray analysis, we often meet problems having a small number of observations but a very large number of features. In such cases, one challenge of the classical LDA is the difficulty in interpreting the classification rule. To overcome this, the most suitable approach is feature selection. A sparse classifier leads to easier model interpretation and may reduce overfitting of the training data.

In the literature, there exists a number of works to extend LDA to the high-dimensional setting in such a way that the resulting classifier involves a sparse linear combination of the features (see e.g. [14],[23],[2],[5],[21],[22],[6]). These approaches use the ℓ_1-norm to deal with sparsity. More precisely, the ℓ_1-regularization is added to the objective function of the normal model and/or the optimal scoring problem, and/or the Fisher discriminant problem.

In this work, we propose an algorithm for solving the sparse optimal scoring (SOS) problem. A natural way to deal with feature selection in machine learning is using ℓ_0-norm. Using $\ell_2 - \ell_0$ regularization for the optimal scoring problem (1) leads us to consider the following optimization problem

$$\min_{w_k,\theta_k} \quad \left\{ \frac{1}{2n}\|Y\theta_k - Xw_k\|_2^2 + \lambda\left[\frac{1}{2}(1-\gamma)\|w_k\|_2^2 + \gamma\|w_k\|_0\right]\right\}$$

$$\text{subject to} \quad \frac{1}{n}\theta_k^T Y^T Y \theta_k = 1; \quad \theta_k^T Y^T Y \theta_l = 0, l = 1, .., k - 1. \tag{2}$$

Here $\gamma \in [0,1]$ and $\lambda \geq 0$ are tuning parameters, and $\|w_k\|_0$ denotes the ℓ_0-norm of w_k, i.e. the number of non-zero components of vector w_k.

The problem (2) is nonconvex, discontinuous, and NP-hard. Optimization methods involving the ℓ_0-norm can be divided into three categories according to the way to treat the ℓ_0-norm: convex approximation, nonconvex approximation, and nonconvex exact reformulation. We refer to [13] for an excellent review on exact/approximation approaches to deal with the ℓ_0-norm. The best known and widely used convex approximation of ℓ_0-norm is ℓ_1-norm, called Lasso [20]. For the problem (2), Clemmensen et al. [2] replaced the ℓ_0-norm by the ℓ_1-norm and applied an alternating scheme for solving the resulting problem. DC (Difference of Convex functions) approximation approaches for ℓ_0-norm have been studied extensively on both theoretical and practical aspects in [13]. Theoretical results from [13] showed that the piecewise linear function [17] (Capped-ℓ_1) is the best approximation for ℓ_0-norm. This has been justified via the numerical results for the problem of feature selection in SVM. The success of Capped-ℓ_1 motivated us to apply it for the SOS problem.

Our method is based on DC programming and DCA (DC Algorithms) introduced by Pham Dinh Tao in their preliminary form in 1985. They have been extensively developed since 1994 by Le Thi Hoai An and Pham Dinh Tao and become now classic and increasingly popular (see e.g. [11],[12],[18],[19] and the list of references in http://lita.sciences.univ-metz.fr/~lethi/DCA.html). Our motivation is based on the fact that DCA is a fast and scalable approach which has been successfully applied to many large-scale (smooth or non-smooth) non-convex programs in various domains of applied sciences, in particular in data analysis and data mining, for which it provided quite often a global solution and proved to be more robust and efficient than standard methods (see e.g. [11],[12],[18],[19] and references therein).

The paper is organized as follows. In Section 2, we present DC programming and DCA for general DC programs, and show how to apply an alternating scheme based on DCA to solve the problem (2). The numerical experiments are reported in Section 3. Finally, Section 4 concludes the paper.

2   Our Algorithms

First let us describe the notations used in this paper. All vectors will be column vectors unless transposed to a row vector by a superscript T. For vectors $x, y \in \mathbb{R}^n$ and $1 \leq p < \infty$, the inner product and ℓ_p-norm are $\langle x, y \rangle = \sum_{i=1}^n x_i y_i$, $\|x\|_p = (\sum_{i=1}^n |x_i|^p)^{\frac{1}{p}}$ respectively.

For $\alpha > 0$, let η_α be the function defined by $\eta_\alpha(x) = \min\{1, \alpha|x|\}, x \in \mathbb{R}$. The piecewise linear approximation of the ℓ_0-norm [17] is given by

$$\|w_k\|_0 \approx \sum_{i=1}^p \eta_\alpha(w_{ki}). \tag{3}$$

Using the approximation (3), we can reformulate the problem (2) in the form

$$\min_{w_k, \theta_k} \left\{ \frac{1}{2n} \|Y\theta_k - Xw_k\|_2^2 + \lambda \left[\frac{1}{2}(1 - \gamma)\|w_k\|_2^2 + \gamma \sum_{i=1}^p \eta_\alpha(w_{ki}) \right] \right\} \tag{4}$$

$$\text{subject to} \quad \frac{1}{n}\theta_k^T Y^T Y \theta_k = 1; \quad \theta_k^T Y^T Y \theta_l = 0, l = 1, ..., k - 1.$$

We use an alternating scheme for finding a local optimum of (4). The algorithm consists of holding θ_k fixed and optimizing with respect to w_k based on DCA, and then holding w_k fixed and optimizing with respect to θ_k.

Let $D = \frac{1}{n} Y^T Y$ and Q_{k-1} be the $Q \times (k - 1)$ matrix containing the previous $k - 1$ solutions $\theta_1, ..., \theta_{k-1}$. For a fixed w_k, the problem (4) can then be formulated as

$$\min_{\theta_k \in \mathbb{R}^Q} \left\{ \|Y\theta_k - Xw_k\|_2^2 : \theta_k^T D\theta_k = 1; \theta_k^T D\theta_l = 0, l = 1, ..., k - 1 \right\}. \tag{5}$$

Let $s = (I - Q_{k-1}Q_{k-1}^T D)D^{-1}Y^T Xw_k$. Then $\widehat{\theta}_k = s/\sqrt{s^T Ds}$ solves (5) [2].

For a fixed θ_k, we have to solve the following problem

$$\min_{w_k}\left\{F(w_k) = \frac{1}{2n}\|Y\theta_k - Xw_k\|_2^2 + \lambda\left[\frac{1}{2}(1-\gamma)\|w_k\|_2^2 + \gamma\sum_{i=1}^{p}\eta_\alpha(w_{ki})\right]\right\}.$$

(6)

The problem (6) is nonconvex nondifferentiable and we will investigate DC programming and DCA for solving it. First, let us give a brief description of DC programming and DCA.

2.1   DC Programming and DCA

A general DC program is that of the form:

$$\alpha = \inf\{F(x) := G(x) - H(x)\,|\,x \in \mathbb{R}^n\}\quad(P_{dc}),$$

where G, H are lower semi-continuous proper convex functions on $\mathbb{R}^n$. Such a function F is called a DC function, and $G - H$ a DC decomposition of F while G and H are the DC components of F. Note that, the closed convex constraint $x \in C$ can be incorporated in the objective function of (P_{dc}) by using the indicator function on C denoted by χ_C which is defined by $\chi_C(x) = 0$ if $x \in C$, and $+\infty$ otherwise.

For a convex function θ, the subdifferential of θ at $x_0 \in \text{dom}\theta := \{x \in \mathbb{R}^n : \theta(x) < +\infty\}$, denoted by $\partial\theta(x_0)$, is defined by

$$\partial\theta(x_0) := \{y \in \mathbb{R}^n : \theta(x) \geq \theta(x_0) + \langle x - x_0, y\rangle, \forall x \in \mathbb{R}^n\},$$

and the conjugate θ^* of θ is

$$\theta^*(y) := \sup\{\langle x, y\rangle - \theta(x) : x \in \mathbb{R}^n\}, \quad y \in \mathbb{R}^n.$$

Then, the following program is called the dual program of (P_{dc}):

$$\alpha_D = \inf\{H^*(y) - G^*(y)\,|\,y \in \mathbb{R}^n\}\quad(D_{dc}).$$

One can prove (see, e.g. [18]) that $\alpha = \alpha_D$ and that there is a perfect symmetry between primal and dual DC programs: the dual to (D_{dc}) is exactly (P_{dc}). The necessary local optimality condition for the primal DC program (P_{dc}) is

$$\partial H(x^*) \subset \partial G(x^*).$$

(7)

The condition (7) is also sufficient for many important classes of DC programs, for example, for DC polyhedral programs, or when function F is locally convex at x^* ([11]).

A point x^* is called a *critical point* of $G - H$, or a generalized Karush-Kuhn-Tucker point (KKT) of (P_{dc})) if

$$\partial H(x^*) \cap \partial G(x^*) \neq \emptyset.$$

(8)

Based on local optimality conditions and duality in DC programming, the DCA consists in constructing two sequences $\{x^l\}$ and $\{y^l\}$ (candidates to be solutions of (P_{dc}) and its dual problem respectively). Each iteration l of DCA approximates the concave part $-H$ by its affine majorization (that corresponds to taking $y^l \in \partial H(x^l)$) and minimizes the resulting convex function (that is equivalent to determining $x^{l+1} \in \partial G^*(y^l)$).

Generic DCA scheme
Initialization: Let $x^0 \in \mathbb{R}^n$ be an initial guess, $l \leftarrow 0$.
Repeat
- Calculate $y^l \in \partial H(x^l)$
- Calculate $x^{l+1} \in \arg \min\{G(x) - \langle x, y^l \rangle : x \in \mathbb{R}^n\}$ (P_l)
- $l \leftarrow l + 1$
Until convergence of $\{x^l\}$.

Convergences properties of DCA and its theoretical basic can be found in [11],[18]. It is worth mentioning that

- DCA is a descent method (*without linesearch*): the sequences $\{G(x^l) - H(x^l)\}$ and $\{H^*(y^l) - G^*(y^l)\}$ are decreasing.
- If $G(x^{l+1}) - H(x^{l+1}) = G(x^l) - H(x^l)$, then x^l is a critical point of $G - H$ and y^l is a critical point of $H^* - G^*$. In such a case, DCA terminates at l-th iteration.
- If the optimal value α of problem (P_{dc}) is finite and the infinite sequences $\{x^l\}$ and $\{y^l\}$ are bounded then every limit point x (resp. y) of the sequences $\{x^l\}$ (resp. $\{x^l\}$) is a critical point of $G - H$ (resp. $H^* - G^*$).
- DCA has a *linear convergence* for general DC programs, and has a finite convergence for polyhedral DC programs.

A deeper insight into DCA has been described in [11]. For instant it is crucial to note the main feature of DCA: DCA is constructed from DC components and their conjugates but not the DC function f itself which has infinitely many DC decompositions, and there are as many DCA as there are DC decompositions. Such decompositions play a crucial role in determining the speed of convergence, stability, robustness, and globality of sought solutions. Therefore, it is important to study various equivalent DC forms of a DC problem. This flexibility of DC programming and DCA is of particular interest from both a theoretical and an algorithmic point of view.

For a complete study of DC programming and DCA the reader is referred to [11],[18],[19] and the references therein.

In the last decade, a variety of works in Machine Learning based on DCA have been developed. The efficiency and the scalability of DCA have been proved in a lot of works (see e.g. [9],[10],[16],[15] and the list of references in http://lita.sciences.univ-metz.fr/~lethi/DCA.html). These successes of DCA motivated us to investigate it for solving the problem (6).

2.2   DCA for Solving the Problem (6)

The approximation η_α can be presented as a DC function [9]:

$$\eta_\alpha(x) = g_\alpha(x) - h_\alpha(x), \tag{9}$$

where $g_\alpha(x) = \alpha|x|$ and $h_\alpha(x) = -1 + \max\{1, \alpha|x|\}$. Then, the problem (6) can be rewritten as follows

$$\min\{F(w_k) = G(w_k) - H(w_k)\}, \tag{10}$$

where

$$G(w_k) := \frac{1}{2n}\|Y\theta_k - Xw_k\|_2^2 + \lambda\left[\frac{1}{2}(1-\gamma)\|w_k\|_2^2 + \gamma\sum_{i=1}^{p} g_\alpha(w_{ki})\right], \tag{11}$$

and

$$H(w_k) := \lambda\gamma\sum_{i=1}^{p} h_\alpha(w_{ki}) \tag{12}$$

are clearly convex functions. According to the generic DCA scheme, at each iteration l, we have to compute a subgradient v^l of H at w_k^l and then solve the convex program of the form (P_l), namely

$$\min\{G(w_k) - \langle v^l, w_k\rangle\}. \tag{13}$$

H is differentiable and $v^l = \nabla H(w_k^l)$ is calculated as follows:

$$v_j^l = \begin{cases} \lambda\gamma\alpha.\mathrm{sgn}(w_{kj}) & \text{if } \alpha|w_{kj}^l| \geq 1 \\ 0 & \text{otherwise} \end{cases} j = 1, ..., p, \tag{14}$$

where $\mathrm{sgn}(w_{kj})$ is the sign of w_{kj}.

DCA for solving (10)-(12) can described as follows

Algorithm 1.. (DCA applied to (10)-(12))

Initialization: Let τ be a tolerance sufficient small, set $l = 0$ and choose $w_k^0 \in \mathbb{R}^p$.
repeat

1. Compute v^l by $v_j^l = \begin{cases} \lambda\gamma\alpha.\mathrm{sgn}(w_{kj}) & \text{if } \alpha|w_{kj}^l| \geq 1 \\ 0 & \text{otherwise} \end{cases} j = 1, ..., p.$

2. Solve the following convex problem to obtain w_k^{l+1}

$$\min_{w_k}\left\{\frac{1}{2n}\|Y\theta_k - Xw_k\|_2^2 + \lambda\left[\frac{1}{2}(1-\gamma)\|w_k\|_2^2 + \gamma\alpha\|w_{ki}\|_1\right] - \langle v^l, w_k\rangle\right\}. \tag{15}$$

3. $l \leftarrow l + 1$.
until $|F(w_k^{l+1}) - F(w_k^l)| \leq \tau(|F(w_k^l)| + 1)$.

Remark 1. For solving the convex problem (15), we use the coordinate descent method [4].

2.3   Description of the Main Algorithm

Finally, our main algorithm for finding K sparse discriminant vectors based on DCA can be described as follows.

Algorithm 2.

Let $D = \frac{1}{n}Y^T Y$.

for $k = 1$ to K, compute k-th discriminant vector w_k follows:

    Initialization:

        - Let $\theta_k = \tilde{\theta}\sqrt{\tilde{\theta}^T D \tilde{\theta}}$, where $\tilde{\theta}$ is a random Q-vector.

    repeat

        - For fixed θ_k, compute w_k by Algorithm 1.

        - For fixed w_k, compute $\tilde{\theta}_k = (I - Q_{k-1}Q_{k-1}^T D)D^{-1}Y^T X w_k$ and

        $\theta_k = \tilde{\theta}_k / \sqrt{\tilde{\theta}_k^T D \tilde{\theta}_k}$.

    until convergence.

end for

---

3   Numerical Experiments

We use the SOS for supervised classification problems in high dimension. The SOS transforms the set of labelled data points in the original space into a labelled set in a lower-dimensional space and selects relevant features. The classification is then performed in the transformed space based on Euclidean distance, i.e. the predicted class for a test observation x is

$$\arg \min_k ||x^T W - \mu_k^T W||_2^2, \tag{16}$$

where the linear transformation $W = [w_1, ..., w_K]$ is computed by Algorithm 2 and μ_k is the mean vector of the k-th class.

3.1   Comparative Algorithms

We will compare our proposed Algorithm 2 (AS_DCA) with the methods proposed in [2] (SDA) and [23] (PLDA). SDA applied an ℓ_1 penalty in the optimal scoring problem, namely

$$\min_{w_k,\theta_k} \left\{ \frac{1}{n}||Y\theta_k - Xw_k||_2^2 + \gamma w_k^T \Omega w_k + \lambda ||w_k||_1 \right\}$$

$$\text{subject to } \frac{1}{n}\theta_k^T Y^T Y \theta_k = 1; \quad \theta_k^T Y^T Y \theta_l = 0, l = 1, ..., k-1, \tag{17}$$

where Ω is a positive definite matrix, γ and λ are nonnegative tuning parameters. In [2], the author used an alternating scheme for finding a local optimum of this problem.

PLDA penalized the objective function of the Fisher's discriminant problem with an ℓ_1 penalty on the discriminant vectors. The R package **penalizedLDA** is available from CRAN (http://cran.r-project.org/).

3.2   Datasets

We evaluate the performance of the AS_DCA approach on three synthetic datasets and a collection of real world datasets. The three synthetic datasets are summarized in Table 1 and generated as follows:

The first setup: we generate a three classes classification problem. Each class is assumed to have a multivariate normal distribution $N(\mu_k, \Sigma)$, $k = 1, 2, 3$ with dimension $p = 500$. All elements on the main diagonal of covariance matrix Σ are equal to 1 and all other elements are equal to 0.6. The first 35 components of μ_1 are 0.7, $\mu_{2j} = 0.7$ if $36 \leq j \leq 70$ and $\mu_{3j} = 0.7$ if $71 \leq j \leq 105$ and 0 otherwise. For each class, we generate 100 training samples, 100 tuning samples and 500 test samples.

The second simulation setup: there are two classes of multivariate normal distributions $N(\mu_1, \Sigma)$ and $N(\mu_2, \Sigma)$, each of dimension $p = 500$. The components of μ_1 are assumed to be 0 and for μ_2, $\mu_{2j} = 0.6$ if $j \leq 200$ and 0 otherwise. The covariance matrix Σ is the block diagonal matrix with five blocks of dimension 100×100 whose element (j, j') is $0.6^{|j-j'|}$. For each class, 100 training samples, 100 tuning samples and 10000 test samples are generated.

In the last setup, we generate a three-class classification problem as follows: $i \in C_k$ then $X_{ij} \sim N((k-1)/2, 1)$ if $j \leq 100$, $k = 1, 2, 3$ and $X_{ij} \sim N(0, 1)$ otherwise, where $N(\mu, \sigma^2)$ denotes the Gaussian distribution with mean μ and variance σ^2. A total of 300 training samples, 300 tuning samples and 1500 test samples are generated with equal probabilities for each class.

Table 1. Synthetic datasets used in experiments

Dataset	#Features	#Train	#Test	#Classes
Simulation 1 (S1)	500	300	1500	3
Simulation 2 (S2)	500	200	20000	2
Simulation 3 (S3)	500	300	1500	3

The real world datasets consist of six real microarray gene expression datasets (SRBCT[1], ALL/AML[2], Leukemia microarray[3], MLL-Leukemia[4], Ovarian Cancer[5] and Nakayama[6]). All the datasets are preprocessed by normalizing each dimension of the data to zero mean and unit variance. The detailed information of these datasets is summarized in Table 2.

---

[1] http://research.nhgri.nih.gov/microarray/Supplement/
[2] http://www-genome.wi.mit.edu
[3] http://datam.i2r.a-star.edu.sg/datasets/krbd/
[4] http://research.dfci.harvard.edu/korsmeyer/Supp_pub/Supp_Armstrong_Main.html
[5] http://datam.i2r.a-star.edu.sg/datasets/krbd/
[6] http://www.ncbi.nlm.nih.gov/geo/query/acc.cgi?acc=GSE6481

Table 2. Microarray datasets used in experiments

Dataset	#Features	#Samples	#Classes
SRBCT (SRB)	2308	83	4
ALL/AML (ALL)	7129	72	4
Leukemia (LEU)	12558	248	6
MLL-Leukemia (MLL)	12582	72	3
Ovarian Cancer (OVA)	15154	253	2
Nakayama (NAK)	22283	86	5

3.3   Experimental Setups

AS_DCA and SDA were implemented in the Visual Studio 2012, and performed on a PC Intel i7 CPU3770, 3.40 GHz of 8GB RAM.

The value γ in (2) is set to 0.2 and the approximation parameter (3) is chosen as $\alpha = 5$ which gives a reasonable approximation of ℓ_0-norm as suggested in [1]. The stop tolerance of DCA is $\tau = 10^{-5}$ while the starting point of DCA is zero. We select relevant features as follows: feature i is deleted if $|w_{ki}| < 10^{-6}$ for all $k = 1, ..., K$. The values of parameters λ and K are chosen through a 5-fold cross-validation procedure on tuning or training set from the sets $\Lambda = \{0.01, 0.02, 0.04, 0.06, 0.1, 0.4, 2\}$ and $\{1, ..., Q - 1\}$, respectively.

For the experiment on synthetic data, we generate training, tuning, and test sets in the same manner as described in Sect. 3.2. The test sets are used to measure the accuracy of various classifiers trained on the training sets. We perform 10 trials for each experimental setting.

For the experiments on the real datasets, we use the cross-validation scheme to validate the performance of various classifiers. Each real dataset is split into a training set containing 2/3 of the samples and a test set containing 1/3 of the samples. This process is repeated 10 times, each with a random choice of training set and test set.

Table 3. Numerical results on synthetic data. The best results are in bold fonts.

		Selected features			Accuracy of classifiers		
		AS_DCA	SDA	PLDA	AS_DCA	SDA	PLDA
S1	(#)	**107.30 ± 1.55**	132.50 ± 3.65	170.2 ± 19.93	100	100	82.94 ± 9.77
	(%)	**21.46 ± 0.31**	26.50 ± 0.73	34.04 ± 3.98			
S2	(#)	**89.40 ± 4.35**	181.30 ± 6.70	159.5 ± 9.34	94.82 ± 0.58	93.72 ± 0.94	**96.62 ± 0.45**
	(%)	**17.88 ± 0.87**	36.26 ± 1.34	31.9 ± 1.86			
S3	(#)	**115.60 ± 6.25**	159.20 ± 7.55	293.7 ± 7.53	**96.97 ± 0.64**	96.73 ± 0.60	96.58 ± 0.34
	(%)	**23.12 ± 1.25**	31.84 ± 1.51	58.74 ± 1.5			
Average		**20.82%**	31.53 %	41.56 %	**97.26 %**	96.82%	92.04%

3.4   Numerical Results

The experimental results on synthetic data and real microarray data are given in Table 3-5. We are interested in the efficiency (the sparsity and the accuracy of classifiers) of the three approaches as well as the rapidity of the AS_DCA and SDA approaches. The numbers of discriminant vectors used K are reported in Table 5. The discriminant vectors can be used to visualize the datasets, for example, as in Figure 1.

Table 4. Numerical results on the real data. The best results are in bold fonts.

		Selected features			Accuracy of classifiers		
		AS_DCA	SDA	PLDA	AS_DCA	SDA	PLDA
SRB	(#)	**70.6±6**	246±9.69	1324.9 ± 147.15	**99.64±1.12**	97.46±3.88	97.44 ± 3.66
	(%)	**3.05±0.26**	10.66±0.42	57.4 ± 6.37			
ALL	(#)	**159±19.77**	240.5±32.5	4300.7 ± 699.06	**92.6±4.25**	91.75±5.21	86.81 ± 7.57
	(%)	**2.13±0.27**	3.37±0.4	60.32 ± 9.8			
LEU	(#)	**37.9±5.65**	94.7±3.89	4421.9 ± 29.04	95.52±2.49	95.39±2.17	**96.86 ± 1.33**
	(%)	**0.3±0.05**	0.75±0.03	35.21 ± 0.23			
MLL	(#)	**132.6±15**	216.9±8.81	6996.3 ± 172.56	**97.92±2.19**	97.08±2.81	87.42± 5.94
	(%)	**1.05±0,12**	1.72 ± 0.07	55.6 ± 1.37			
OVA	(#)	**27.7±2.35**	125.4±15.5	4160.8 ± 80.6	**100**	**100**	89.57 ± 4.27
	(%)	**0.18±0.01**	0.83±0.1	27.45 ± 0.53			
NAK	(#)	**520.1±24.1**	756.7±23.8	15242.4 ± 1587.61	83.7±4.53	83.35±4.17	**87.1 ± 7.13**
	(%)	**2.33±0.1**	3.39±0.1	68.4 ± 7.12			

Table 5. The number of discriminant vectors used K (second row) and comparative results of AS_DCA and SDA approaches in terms of the average of CPU time in second. The best results are in bold fonts.

	Datasets	S1	S2	S3	SRB	ALL	LEU	MLL	OVA	NAK
K	AS_DCA/SDA/PLDA	2	1	1	3	3	5	2	1	4
CPU	AS_DCA	**0.81**	**0.03**	**0.17**	**3.93**	**100.83**	**78.18**	**173.76**	**24.54**	**406.53**
	SDA	1.74	0.07	0.49	23.32	253.29	384.69	387.13	110.46	1726.02

We observe from computational results that:

Sparsity: In all the datasets, the AS_DCA approach gives much better results in terms of feature selection than the SDA and PLDA approaches. On average of 3 synthetic datasets, the AS_DCA approach selects 20.82% of features while the SDA and PLDA approaches select 31.53% and 41.56% of features, respectively. We also see that the AS_DCA approach is better than the SDA and PLDA approaches on all the real datasets. The AS_DCA approach selects from 0.18% to 3.05% of features while the SDA (resp. PLDA) approach selects from 0.75% to 10.66% (resp. 27.45% to 68.4%) of features.

Accuracy of classifiers: The AS_DCA approach not only provide a good performance in terms of feature selection, but also gives a high accuracy of classifiers on almost all the datasets. These results are higher than (or quite closed to) that of the SDA approach. The AS_DCA approach is better than the PLDA approach on the 6/9 datasets. The PLDA approach is slightly better than the AS_DCA approach on the three datasets (S2, LEU and NAK). This can be explained by the fact the PLDA approach selects 116 (resp. 29) times more features than the AS_DCA approach on the LEU (resp. NAK) dataset.

Training time: The training time given in Table 5 shows that the AS_DCA approach is faster than the SDA approach on both synthetic datasets and real datasets.

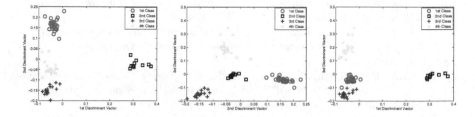

Fig. 1. The SRBCT dataset was projected onto the first three sparse discriminant vectors. The samples in each class are shown by using a distinct symbol.

4   Conclusions

We have proposed an efficient approach for solving the Sparse Optimal Scoring (SOS) problem. Using an appropriate approximation of the ℓ_0 regularization term, we reformulated the SOS problem as a continuous nonconvex optimization problem and then investigated an alternating scheme based on DCA for it. The computational results show the robustness, the effectiveness of the AS_DCA approach and it superiority with respect to the ℓ_1-norm approaches.

In the future we will study feature selection and dimension reduction for other models such as the partial least squares, or the principal component analysis, or the Fisher's discriminant problem.

References

1. Bradley, P.S., Mangasarian, O.L.: Feature selection via mathematical programming. In: Proceeding of International Conference on Machine Learning, ICML 1998 (2008)
2. Clemmensen, L., Hastie, T., Witten, D., Ersbøll, B.: Sparse discriminant analysis. Technometrics **53**(4), 406–413 (2011)
3. Fisher, R.A.: The use of multiple measurements in taxonomic problems. Annal of Eugenics **7**, 179–188 (1936)

4. Friedman, J., Hastie, T., Hoefling, H., Tibshirani, R.: Pathwise coordinate optimization. The Anals of Applied Statistics **1**, 302–332 (2007)
5. Grosenick, L., Greer, S., Knutson, B.: Interpretable classifers for fmri improve prediction of purchases. IEEE Transactions on Neural Systems and Rehabilitation Engineering **16**(6), 539–547 (2008)
6. Guo, Y., Hastie, T., Tibshirani, R.: Regularized linear discriminant analysis and its application in microarrays. Biostatistics **8**(1), 86–100 (2007)
7. Hastie, T., Buja, A., Tibshirani, R.: Penalized discriminant analysis. The Annals of Statistics **23**(1), 73–102 (1995)
8. Hastie, T., Tibshirani, R., Friedman, J.: The Elements of Statistical Learning. Springer, New York (2009)
9. Le Thi, H.A., Le Hoai, M., Nguyen, N.V., Pham Dinh, T.: A DC programming approach for feature selection in support vector machines learning. Journal of Advances in Data Analysis and Classification **2**(3), 259–278 (2008)
10. Le Thi, H.A., Le Hoai, M., Pham Dinh, T.: Optimization based DC programming and DCA for hierarchical clustering. European Journal of Operational Research **183**, 1067–1085 (2007)
11. Le Thi, H.A., Pham Dinh, T.: The DC (difference of convex functions) programming and DCA revisited with DC models of real world nonconvex optimization problems. Annals of Operations Research **133**, 23–46 (2005)
12. Le Thi, H.A., Pham Dinh, T., Huynh, V.N.: Exact penalty and error bounds in DC programming. Journal of Global Optimization **52**(3), 509–535 (2012)
13. Le Thi, H.A., Pham Dinh, T., Le Hoai, M., Vo Xuan, T.: DC approximation approaches for sparse optimization. To appear in European Journal of Operational Research (2014)
14. Leng, C.: Sparse optimal scoring for multiclass cancer diagnosis and biomarker detection using microarray data. Computational Biology and Chemistry **32**, 417–425 (2008)
15. Liu, Y., Shen, X.: Multicategory ψ-learning. Journal of the American Statistical Association **101**, 500–509 (2006)
16. Liu, Y., Shen, X., Doss, H.: Multicategory ψ-learning and support vector machine: Computational tools. Journal of Computational and Graphical Statistics **14**, 219–236 (2005)
17. Peleg, D., Meir, R.: A bilinear formulation for vector sparsity optimization. Signal Processing **88**(2), 375–389 (2008)
18. DPham Dinh, T., Le Thi, H.A.: Convex analysis approach to D.C. programming: Theory, algorithms and applications. Acta Mathematica Vietnamica **22**(1), 289–355 (1997)
19. Pham Dinh, T., Le Thi, H.A.: A DC optimization algorithm for solving the trust-region subproblem. SIAM. Journal of Optimization **8**(2), 476–505 (1998)
20. Tibshirani, R.: Regression shrinkage and selection via the lasso. J. Roy. Stat. Soc. **58**, 267–288 (1996)
21. Tibshirani, R., Hastie, T., Narasimhan, B., Chu, G.: Diagnosis of multiple cancer types by shrunken centroids of gene expression. Proc. Natl. Acad. Sci. **99**, 6567–6572 (2002)
22. Tibshirani, R., Hastie, T., Narasimhan, B., Chu, G.: Class prediction by nearest shrunken centroids, with applications to DNA microarrays. Statistical Science **18**(1), 104–117 (2003)
23. Witten, D., Tibshirani, R.: Penalized classification using Fisher's linear discriminant. Journal Royal Statistical Society B **73**, 753–772 (2011)

Graph Based Relational Features
for Collective Classification

Immanuel Bayer$^{(\boxtimes)}$, Uwe Nagel, and Steffen Rendle

University of Konstanz, 78457 Konstanz, Germany
{immanuel.bayer,uwe.nagel,steffen.rendle}@uni-konstanz.de

Abstract. Statistical Relational Learning (SRL) methods have shown
that classification accuracy can be improved by integrating relations
between samples. Techniques such as *iterative classification* or *relaxation
labeling* achieve this by propagating information between related samples during the inference process. When only a few samples are labeled
and connections between samples are sparse, *collective inference* methods
have shown large improvements over *standard feature-based* ML methods. However, in contrast to feature based ML, collective inference methods require complex inference procedures and often depend on the strong
assumption of label consistency among related samples. In this paper, we
introduce new *relational features* for standard ML methods by extracting information from *direct* and *indirect relations*. We show empirically
on three standard benchmark datasets that our relational features yield
results comparable to collective inference methods. Finally we show that
our proposal outperforms these methods when additional information is
available.

1   Introduction

Statistical relational learning (SRL) methods are used when samples are connected by one or more relation. These relations are helpful in tasks like scientific article classification where patterns ssuch as "connected samples have
similar labels" are very predictive. Feature based ML methods in contrast often
assume that samples are independently, identically distributed (iid). This approach is well established, allows for efficient parameter estimation and simplified prediction but ignores the relational information available in SRL settings.
Recently, a number of methods have been proposed [11,13,18,31] that significantly improves over *classical* methods by using joint inference. Exact joint inference have high runtime complexity which often requires approximate solutions
[30]. These approximated joint inference techniques introduce new difficulties
such as the need for specialized implementations that are expensive to run and
difficult to tune [30].

   In this paper we propose to transfer the relational information into classical
features. This allows a straightforward combination of relational and classical
(attribute) information. It also renders traditional, feature based ML methods

© Springer International Publishing Switzerland 2015
T. Cao et al. (Eds.): PAKDD 2015, Part II, LNAI 9078, pp. 447–458, 2015.
DOI: 10.1007/978-3-319-18032-8_35

competitive in settings where relational information is available and allows to leverage the large body of classical ML methods and their scalable algorithms.

We use three standard *collective classification* (CC) benchmark datasets to show that classical ML with relational features are strong competitors for state of the art SRL methods on this task. Note that on these datasets, collective classification are considered the best performing methods in the current literature [7,14]. In particular, we make the following contributions:

- We discuss how joint inference could be avoided by extending the sample description with relational features (Section 2.1).
- We extend relational features to indirect relations (Section 2.2). This is new and crucial to achive high accuracy when only few samples are labeled (Section 4.3).
- We introduce a new cluster based relational feature (Section 2.2) that provides strong results and is cheap to compute.
- We show that our approach improves state of the art collective classification even in network only settings (Section 4.2).

2 Problem Setting

We start by giving the necessary definitions with the traditional setting of samples $\mathcal{D} = \{(x_i, y_i)\}_{i=1}^{N}$ where y_i is the class label and x_i is a feature vector describing sample i. We assume that for the first u samples ($i \in \{1, \ldots, u\}$), the class label y_i is known and for the samples with index $i > u$, the class label is unknown. Relations among samples are represented by weighted, symmetric adjacency matrices $R_k \in \mathbb{R}^{n \times n}$ and the complete relational information is denoted as $\mathcal{R} = \{R_k\}_{k=1}^{K}$. While in the general case, each of the k relations could be complete, i.e. provide some similarity between each pair of samples, we explicitly consider the case of sparse, unweighted relations where only a minority of node pairs are connected by an edge. We start with a statistical argument to motivate the representation of relational information in a way that is compatible with the iid assumption.

2.1 Preserving IID

Many machine learning algorithms are based on the maximum likelihood principal to learn the optimal value θ^* for model parameters given a dataset $\mathcal{D}$ (c.f. [17])

$$\theta^* := \arg\max_{\theta} p(\theta|\mathcal{D}) = \arg\max_{\theta} p(\mathcal{D}|\theta)$$

where $l(\theta) := p(\mathcal{D}|\theta)$ is called the likelihood. A very common assumption in many ML approaches is that the samples in the dataset D are independent and identically distributed (iid). This assumption simplifies the likelihood to

$$p(\mathcal{D}|\theta) \overset{iid}{\propto} \prod_{i=1}^{N} p(y_i|x_i, \theta).$$

One of the central arguments of relational learning is that examples are not iid. In particular for any pair of examples (x_i, y_i) and (x_j, y_j) conditional independence does not hold

$$p((x_i, y_i), (x_j, y_j)|\theta) \not\propto p(y_i|x_i, \theta)\, p(y_j|x_j, \theta).$$

Note that in this formulation the relational information $\mathcal{R}$ is completely neglected. However if $\mathcal{R}$ is used, it can render the probabilities independent

$$
\begin{aligned}
p((x_i, y_i), (x_j, y_j), \mathcal{R}|\theta) \\
\propto p(y_i|x_i, \mathcal{R}, \theta)\, p(y_j|x_j, \mathcal{R}, \theta).
\end{aligned}
\tag{1}
$$

This formulation is close to standard non-relational ML with iid of samples.

Standard ML algorithms assume that all information about a sample can be encoded in a (usually real-valued) feature vector x. Let us assume that the influence of $\mathcal{R}$ on sample i can be described through a finite number of real valued variables x_i^r. We call x_i^r *relational features*. To simplify notation, we can define $\tilde{x}_i$ as the extended feature vector of an example i that combines both non-relational features x_i and relational features x_i^r. In total, this allows to rewrite Equation 1

$$p(\mathcal{D}|\theta) \stackrel{iid}{\propto} \prod_{i=1}^{N} p(y_i|\tilde{x}_i, \theta).$$

Note that due to the relational information in $\tilde{x}_i$, the iid assumption can be preserved. In the remainder of this section, we discuss several ways to generate relational features.

2.2   Graph Based Relational Features

Samples can be linked through multiple relations each of which can be described as a graph. Representing each relation by an independent set of features allows us to integrate an arbitrary number of relations per problem into a standard feature matrix. All of the proposed features have in common that the encoded relational information does not only consist of direct relations but in addition captures indirect relations which we found to be the key to their performance.

Neighbor Ids. Encoding the direct neighbors of a sample i in relation R_k can be achieved by treating each sample as a categorical variable which is true when samples are connected and false otherwise [1, 24, 25]. As this information is very local and yields limited information about i's position in R_k, we extend it by additionally including indirect neighbors at various distances to i. Distance refers to the number of edges $d_k(i, j)$ on a shortest path connecting i and j in R_k. In particular, for a (small) set of distances $d \geq 1$ we describe each relation R_k by the distance-d neighborhood matrices $D_k^d \in \{0, 1\}^{n \times n}$ with $\left(D_k^d\right)_{i,j} = 1$, if $d_k(i, j) = d$ and 0 otherwise. We illustrate the derivation of this feature in Figure 1.

Fig. 1. Feature matrix for the node neighborhoods of a relation. The edges of the original relation are depicted as gray lines, while connections in distance two and three are shown as blue/short dashed and red/long dashed curves.

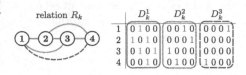

Aggregated Neighbor Attributes. Relational position can also be described by individual features of direct and indirect neighbors [20,21]. As before we extend the idea by calculating individual features at various distances. For a categorical attribute with categories $1,\ldots,c$, we define an $n \times c$ matrix L with $L_{i,j} = 1$ if x_i is labeled as j and $L_{i,j} = 0$ otherwise. Then the count matrix $C_k^d =: D_k^d L$ can be derived as the projection of the corresponding neighborhood matrix to the label matrix L such that $(C_k^d)_{i,j}$ yields the number of category j nodes in distance d of sample i. We denote this feature by *neighbor class counts* (NCC) and provide an illustration in Figure 2. An additional row normalization yields a probability matrix for the class labels in distance d, which we will denote by *neighbor class probabilities* (NCP).

Fig. 2. Aggregation for attribute counts on distances one, two and three in the relation R_k. Labels are given in text and as color (1-blue, 2-red). Sample 1 has a single node of each label as direct neighbors which is reflected in the first two columns(C_k^1) of its row, while the next two columns (C_k^2) encode the single 2-labeled node in distance 2. Note that unlabeled nodes (white) are ignored in the features.

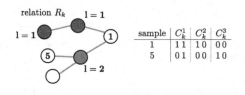

Random Walk Similarity. While the features described above are based strictly on shortest paths, random walks with restart (rwr) incorporate a different notion of connectivity. They have been proposed as a similarity measure in the context of auto-captioning of images [22].

The similarity between two nodes is measured as the probability of a random walk connecting them, i.e. the probability of the random walk process visiting one node when started from the other. To control for locality this includes a restart probability: in each step the walk will jump back to the starting point with probability r. This can be modeled as

$$p_i = (1 - r)W p_i + r e_i \qquad (2)$$

where the column vector p_i of matrix P describes the steady-state probability distribution over all samples for walks starting at i. W is the matrix encoding transition probabilities, i.e. a L_1 row normalized version of R_k, e_i is the vector of zeros and unit value at position i and the parameter r is the restart probability. P can be determined as the solution of a linear system or approximated efficiently [32], leaving r as free parameter. We derive $\hat{P}$ from P by column wise L_2 normalization and use analogous to the neighbor features row i of $\hat{P}$ as features for sample i.

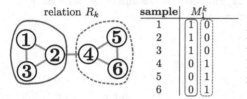

relation R_k	sample	M_1^k
	1	1 0
	2	1 0
	3	1 0
	4	0 1
	5	0 1
	6	0 1

Fig. 3. An example for encoding a clustering of a relation as a feature vector. Memberships of samples in clusters are represented as binary features.

Clustering Memberships. Clustering methods can be used to identify groups of similar samples. Clustering features encode this information by representing this group membership. Given a clustering of the graph representing relation R_k into c clusters, we obtain an $n \times c$ feature matrix M_k^c with $(M_k^c)_{i,m} = 1$ if sample i belongs to cluster m and zero otherwise. Since a single clustering yields limited information about the dense groups in the graph, we create features for various clusterings, i.e. different c. We limit c to $c = 2^j$ subject to $2 \leq c \leq n$ which limits the number of clusterings to $\lfloor \log_2(n) \rfloor$ while also providing a wide range of cluster sizes. This results in $O(n)$ features per relation that are very sparse with only $O(\log(n))$ non-zero features for per sample. The clusters can be calculated with negligible runtime using the METIS clustering framework[1] [8]. Note that the dense subgroups identified in the clusterings can be directly related to the homophily assumption often exploited in relational learning.

3   Related Work

We build on two main categories of related work. The first, in Section 3.1 uses features derived from the network structure to improve iid based inference. The second, discussed in Section 3.2 is work that views collective classification as a joint inference problem, simultaneously inferring the class label on every instance. The challenges specific to problems with few labeled data points have received special attention [4, 5, 14, 29] and helped us to understand the importance of indirect relations.

3.1   Relational Features

Relational features can be combined with collective inference or directly used with standard ML methods as we argue in Section 2.1. Models such as Relational

---

[1] We used the implementation available at http://glaros.dtc.umn.edu/gkhome/views/ metis with default parameters.

Probabilistic Trees [20], Relational Bayes Classifier [21] and the Link Based Classifier (LBC, [11]) concentrate primarily on the aggregation of attributes of connected samples. Others use rows from the (weighted) adjacency matrix as basis for feature construction [1,24,25]. We were especially inspired from the suggestion to extend the neighborhood of samples with few neighbors with distance-two neighborhoods [26] or ghost edges [5]. In contrast to previous work we keep the information from various neighborhood distances separated and introduce the concept of multiple indirect relations.

3.2   Collective Inference

Full relational models such as Markov Logic Networks (MLN, [28]) or the Probabilistic Relational Model [31] can be used for CC [3]. We refer to Sen et al. [30] for a comprehensive overview of collective inference based CC algorithms. Their strength in high label autocorrelation settings and the problem of error propagation has been examined [7,33] and improved inference schemes have been proposed [15]. Recently, stacking of non relational model has been introduced [9] as a fast approximation of Relational Dependency Networks [19].

4   Evaluation

In our experiments we investigate the following three questions:

1. Are classical ML methods with relational features competitive to MLN and Collective Classification approaches.
2. What are the main ingredients that make relational features effective.
3. Does the combination of relational and attribute information improve results?

4.1   Experimental Setup

Datasets. We use three standard benchmark SRL datasets. The *Cora* and *CiteSeer* scientific paper collections have been used in different versions, we chose the versions[2] presented in [30] and the *IMDb* dataset[3]. Both, Cora and CiteSeer include text features in form of bag of words (bow) representations. We give some statistics of these datasets in Table 1.

Benchmark Models. As baseline models we use the well established relational learning methods wvRN [12,13], nLB [11] and MLN [3]. We chose *relaxation labeling* [2] as the collective inference method for wvRN and nLB as it has been shown to outperform Gibbs sampling and iterative classification on our datasets [14]. For MLN we used the rules `HasWord(p,+w)=>Topic(p,+t)` and `Topic(p,t)^Link(p,p')=>Topic(p',t)` together with discriminative weight learning and MC-SAT inference as recommended for Cora and Citeseer in a previous study [3].

---

[2] http://linqs.cs.umd.edu/projects/projects/lbc/index.html
[3] http://netkit-srl.sourceforge.net/data.html

Table 1. Summary statistics of the datasets

	Nodes	Links	Classes	|Dictionary|	avg. Degree
Cora	2708	5278	7	1433	3.8981
CiteSeer	3312	4660	6	3703	2.8140
IMDb (all)	1377	46124	2	-	66.9920

Measures and Protocol. We follow [14] and remove samples that are not connected to any other sample (singletons) in all experiments. Each experiment is repeated 10 times with class-balanced splits into ratios of $0.1, 0.2, \ldots, 0.9$ for the train-test set (shown as percentage in the figures). The MLN experiments are only repeated 3 times due to their extremely high evaluation time. We calculate multi-class accuracies by micro averaging and plot them on the y-axis of each figure in this section. We used the netkit-srl framework [14] in version $(1.4.0)$[4] to evaluate the wvRN and nLB classifiers and the Alchemy 2.0 framework[5] to evaluate the MLN model. The graph clusterings were obtained using METIS [8] version 5.1.0. Relational features are learned with an L_2 penalized logistic regression model[6] included in scikit-learn [23] version 0.14.01. The penalty hyperparameter C is optimized via grid search over $\{0.001, 0.01, 0.1, 1, 10, 100, 1000\}$ on the training set.

4.2   Comparing Relational Features to SRL

This section examines whether feature based relational models (without collective inference) are able to compete with the prediction quality of specialized relational learning methods. Consequently, the benchmark is a task where only relational data is available. In the first experiment, we compare wvRN and nLB with two logistic regression models that use only our relational features. The first relational feature model (rwr) is based on a random walk. The second model uses both, neighborhood and aggregated (NCP) features with distances 1,2,3. We exclude distances higher than three, since almost every node can be reached over three edges from every other node and therefore further distances do not provide additional information.

Figure 4 illustrates two problems of SRL models: (i) nLB performs poorly[7] when labels are sparse and (ii) wvRN is sensitive to violations of its built in assumptions – i.e. if label consistency among neighbors is not met, as with the IMDb dataset.

The relational feature based models show a very consistent performance not much affected by the number of labeled samples. The results of the neighbor and NCP feature combination on IMDb illustrate the flexibility of relational features.

---

[4] http://netkit-srl.sourceforge.net/
[5] https://code.google.com/p/alchemy-2/
[6] We use a one-vs-all scheme for multiclass classification.
[7] This has been attributed to a lack of training data [14].

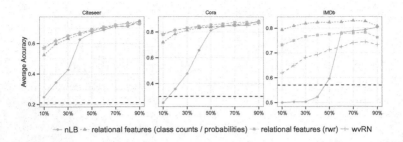

Fig. 4. Comparison between two SRL methods (nLB, wvRN) with relaxation labeling and two relational feature based models on network only data. The dashed black lines indicate base accuracy.

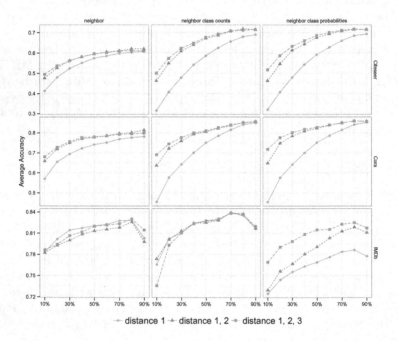

Fig. 5. Influence of the distance parameter for label dependent and independent relational features. Including information from indirect neighbors improves results especially if few samples are labeled.

4.3 Engineering Relational Features

We now examine the different relational features and the influence of their parameters. Two questions will be addressed: (i) *How important are indirect relations?* (ii) *Which of the proposed relational features lead to the best results?* All relational features that we consider can incorporate information about indirect neighbors. Each method has parameters that adjusts the locality of the

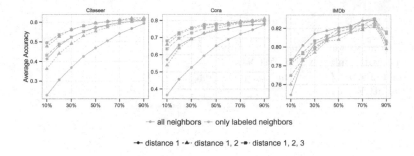

Fig. 6. Comparison of using only labeled vs all neighbors to construct relational neighbor features (NCC). Including unlabeled neighbors improves the results in all settings.

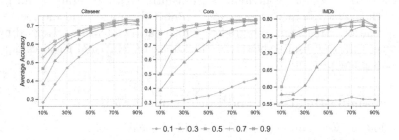

Fig. 7. Influence of various restart values on the prediction accuracy. While as a general trend larger values of r (c.f. Equation 2) lead to better results, the value $r = 0.9$ yield consistently good results.

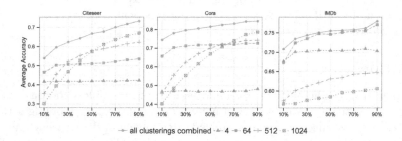

Fig. 8. Examination of various clusterings of the dependency network and their usage as relational feature. The optimal number of clusters strongly depends on the dataset and ratio of labeled samples. The combination of all clusterings, however, is consistently superior to individual clusterings.

resulting features. We first examine the effect of including indirect neighbors (Figure 5) and the importance of unlabeled neighbors (Figure 6) for relational neighbor features. The influence of the restart parameter r (c.f. Equation 2) for the rwr features can be seen in Figure 7 and an informative subset of results

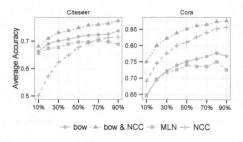

Fig. 9. Comparison of attribute only (bow), network only (NCC), MLN and our combination of relational features with local attributes

for various numbers of clusters is shown in Figure 8. The results suggest that the inclusion of indirect neighbors in the relational features is beneficial independently of whether they are used directly or for aggregation. Figure 6 shows that unlabeled neighbors contribute significantly to the overall performance. Together this answers our first question: unlabeled samples and indirect neighborhood relations are essential ingredients for relational features. Regarding the second question, the results show that the choice of relational features depends on the particular problem.

4.4 Combining Relational and Local Information

In the following, we examine the effect of adding local attributes. Figure 9 shows results with neighborhood count features (NCC) of distances 1,2,3. Interestingly, the bag of words model performs better than network only models on Citeseer but worse on Cora. Combining relational and local attributes on the other hand, improves results in both cases. The figure further shows that our features outperform MLN on both datasets. In summary, our experiments suggest that the combination of relational features and attributes is beneficial even with a simple model such as logistic regression.

4.5 Discussion

Our experiments indicate that relational feature based models compare well to specialized relational learners even in network only and sparse labeling settings. This has been verified on three standard SRL benchmark datasets and with three state of the art SRL methods for comparison. The inclusion of indirect neighbors has proven extremely important, especially in sparse label settings. We have further shown that the combination of relational features and local attributes is both straightforward and has the potential to improve considerably over both, feature only and network only models.

Note, that our relational features can lead to very high dimensional representations. Such feature spaces are, however, common in recommender systems, click-through rate prediction and websearch where regularized logistic regression has been shown to be very effective [6], [10]. In addition, we use a standard implementation of logistic regression and can consequently employ scalable versions that can be trained with billions of samples of high dimensions [16] [27].

We are further not committed to the logistic regression model as our features could be used as input for arbitrary vector space models.

5   Conclusion

We have shown that dependencies between samples can be exploited using relational feature engineering. Our method allows to combine relational information from various sources with attributes attached to individual samples. We tested this on standard SRL benchmark datasets, showing that even on network only data our features are competitive to specialized relational learning models. In addition, our features can outperform them when additional information is available. Note that in contrast to the SRL methods, our proposal achieves these results without collective inference. While we restricted our experiments to logistic regression as prediction model, the proposed features could be used as input to any other feature based learning algorithm such as SVM, neural networks or random forests. Extending the use of relational features to multi relational datasets would be straight forward and a interesting direction for further research.

Acknowledgments. This work was supported by the DFG under grants Re 3311/2-1 and Br 2158/6-1.

References

1. Bernstein, A., Clearwater, S., Provost, F.: The relational vector-space model and industry classification. In: IJCAI Workshop, vol. 266 (2003)
2. Chakrabarti, S., Dom, B., Indyk, P.: Enhanced hypertext categorization using hyperlinks. ACM SIGMOD Record **27**(2), 307–318 (1998)
3. Crane, R., McDowell, L.: Investigating markov logic networks for collective classification. In: ICAART (1), pp. 5–15 (2012)
4. Gallagher, B., Eliassi-Rad, T.: Leveraging Label-Independent Features for Classification in Sparsely Labeled Networks: An Empirical Study. In: Giles, L., Smith, M., Yen, J., Zhang, H. (eds.) SNAKDD 2008. LNCS, vol. 5498, pp. 1–19. Springer, Heidelberg (2010)
5. Gallagher, B., Tong, H., Eliassi-Rad, T., Faloutsos, C.: Using ghost edges for classification in sparsely labeled networks. In: KDD, pp. 256–264 (2008)
6. Graepel, T., Candela, J.Q., Borchert, T., Herbrich, R.: Web-scale bayesian clickthrough rate prediction for sponsored search advertising in microsoft's bing search engine. In: ICML, pp. 13–20 (2010)
7. Jensen, D., Neville, J., Gallagher, B.: Why collective inference improves relational classification. In: KDD, pp. 593–598 (2004)
8. Karypis, G., Kumar, V.: A fast and high quality multilevel scheme for partitioning irregular graphs. SIAM Journal on Scientific Computing **20**(1), 359–392 (1998)
9. Kou, Z., Cohen, W.W.: Stacked graphical models for efficient inference in markov random fields. In: SDM, pp. 533–538. SIAM (2007)
10. Liu, J., Chen, J., Ye, J.: Large-scale sparse logistic regression. In: KDD, pp. 547–556. ACM (2009)

11. Lu, Q., Getoor, L.: Link-based classification. In: ICML, vol. 3, pp. 496–503 (2003)
12. Macskassy, S.A.: Improving learning in networked data by combining explicit and mined links. AAAI **22**, 590–595 (2007)
13. Macskassy, S.A., Provost, F.: A simple relational classifier. In: KDD-Workshop, pp. 64–76 (2003)
14. Macskassy, S.A., Provost, F.: Classification in networked data: A toolkit and a univariate case study. JMLR **8**, 935–983 (2007)
15. McDowell, L.K., Gupta, K.M., Aha, D.W.: Cautious collective classification. JMLR **10**, 2777–2836 (2009)
16. Mukherjee, I., Canini, K., Frongillo, R., Singer, Y.: Parallel Boosting with Momentum. In: Blockeel, H., Kersting, K., Nijssen, S., Železný, F. (eds.) ECML PKDD 2013, Part III. LNCS, vol. 8190, pp. 17–32. Springer, Heidelberg (2013)
17. Murphy, K.P.: Machine learning: A probabilistic perspective. The MIT Press (2012)
18. Neville, J., Jensen, D.: Iterative classification in relational data. In: Proc. AAAI-2000 Workshop on Learning Statistical Models from Relational Data, pp. 13–20 (2000)
19. Neville, J., Jensen, D.: Collective classification with relational dependency networks. In: UAI, pp. 77–91 (2003)
20. Neville, J., Jensen, D., Friedland, L., Hay, M.: Learning relational probability trees. In: KDD, pp. 625–630 (2003)
21. Neville, J., Jensen, D., Gallagher, B.: Simple estimators for relational bayesian classifiers. In: ICDM, pp. 609–612 (2003)
22. Pan, J.Y., Yang, H.J., Faloutsos, C., Duygulu, P.: Automatic multimedia cross-modal correlation discovery. In: KDD, pp. 653–658. ACM (2004)
23. Pedregosa, F., Varoquaux, G., Gramfort, A., Michel, V., Thirion, B., Grisel, O., Blondel, M., Prettenhofer, P., Weiss, R., Dubourg, V., Vanderplas, J., Passos, A., Cournapeau, D., Brucher, M., Perrot, M., Duchesnay, E.: Scikit-learn: Machine learning in Python. JMLR **12**, 2825–2830 (2011)
24. Perlich, C., Provost, F.: Distribution-based aggregation for relational learning with identifier attributes. Machine Learning **62**(1), 65–105 (2006)
25. Perlich, C., Provost, F.: Aggregation-based feature invention and relational concept classes. In: KDD, pp. 167–176. ACM (2003)
26. Preisach, C., Schmidt-Thieme, L.: Relational ensemble classification. In: ICDM, pp. 499–509 (2006)
27. Rendle, S.: Scaling factorization machines to relational data. In: VLDB. vol. 6, pp. 337–348. VLDB Endowment (2013)
28. Richardson, M., Domingos, P.: Markov logic networks. Machine Learning **62**, 107–136 (2006)
29. Saar-Tsechansky, M., Provost, F.: Handling missing values when applying classification models. JMLR (2007)
30. Sen, P., Namata, G.M., Bilgic, M., Getoor, L., Gallagher, B., Eliassi-Rad, T.: Collective classification in network data. AI Magazine **29**(3), 93–106 (2008)
31. Taskar, B., Segal, E., Koller, D.: Probabilistic classification and clustering in relational data. IJCAI **17**, 870–878 (2001)
32. Tong, H., Faloutsos, C., Pan, J.Y.: Fast random walk with restart and its applications. In: ICDM, pp. 613–622 (2006)
33. Xiang, R., Neville, J.: Understanding propagation error and its effect on collective classification. In: ICDM, pp. 834–843. IEEE (2011)

A New Feature Sampling Method in Random Forests for Predicting High-Dimensional Data

Thanh-Tung Nguyen[1], He Zhao[2], Joshua Zhexue Huang[3],
Thuy Thi Nguyen[4], and Mark Junjie Li[3(✉)]

[1] Faculty of Computer Science and Engineering, Thuyloi University, Hanoi, Vietnam
`tungnt@tlu.edu.vn`
[2] Shenzhen Institutes of Advanced Technology, Chinese Academy of Sciences,
Shenzhen, People's Republic of China
`he.zhao@siat.ac.cn`
[3] College of Computer Science and Software Engineering, Shenzhen University,
Shenzhen, China
`{zx.huang,jj.li}@szu.edu.cn`
[4] Faculty of Information Technology, Vietnam National University of Agriculture,
Hanoi, Vietnam
`ntthuy@vnua.edu.vn`

Abstract. Random Forests (RF) models have been proven to perform well in both classification and regression. However, with the randomizing mechanism in both bagging samples and feature selection, the performance of RF can deteriorate when applied to high-dimensional data. In this paper, we propose a new approach for feature sampling for RF to deal with high-dimensional data. We first apply p-value to assess the feature importance on finding a cut-off between informative and less informative features. The set of informative features is then further partitioned into two groups, highly informative and informative features, using some statistical measures. When sampling the feature subspace for learning RFs, features from the three groups are taken into account. The new subspace sampling method maintains the diversity and the randomness of the forest and enables one to generate trees with a lower prediction error. In addition, quantile regression is employed to obtain predictions in the regression problem for a robustness towards outliers. The experimental results demonstrated that the proposed approach for learning random forests significantly reduced prediction errors and outperformed most existing random forests when dealing with high-dimensional data.

Keywords: Subspace feature selection · Regression · Classification · Random forests · Data mining · High-dimensional data

1 Introduction

High-dimensional data has become common in today's applications. State-of-the-art machine learning methods can work well for data sets of moderate size

T. Cao et al. (Eds.): PAKDD 2015, Part II, LNAI 9078, pp. 459–470, 2015.
DOI: 10.1007/978-3-319-18032-8_36

but they suffer when scaling for high-dimensional data. It is well-known that in a high-dimensional data set only a small portion of the predictor features are relevant to the response feature, the irrelevant features may even degrade the performance of the model. This requires methods for selecting good subsets of features for learning efficient prediction models.

Random forests (RF) [1] [2], an ensemble learning machine composed of decision trees for prediction, is defined as follow: Given a training data set $\mathcal{L} = \{(X_i, Y_i), X \in \mathbb{R}^M, Y \in \mathcal{Y}\}_{i=1}^N$, where X_i are features (also called predictor variables) and Y is the target (also called response feature), $\mathcal{Y} \in \mathbb{R}^1$ for a regression problem and $\mathcal{Y} \in \{1, 2, ..c\}$ for a classification problem ($c \geq 2$), N and M are the number of training samples and features, respectively. A standard version of RF independently and uniformly resamples observations from the training data $\mathcal{L}$ to draw a bootstrap data set $\mathcal{L}^*$ from which a decision tree T^* is grown. Repeating this process K times produces a series of bootstrap data sets $\mathcal{L}_k^*$ and corresponding decision trees T_k^* ($k = 1, 2, ..., K$), that form a RF.

Given an input $X = x$, the predicted value by the whole RF is obtained by aggregating the results given by individual trees. Let $\hat{f}_k(x)$ denote the prediction of unknown value y of input $x \in \mathbb{R}^M$ by kth tree, we have

$$\hat{f}(x) = \frac{1}{K} \sum_{k=1}^K \hat{f}_k(x) \quad \text{for regression problems, and} \tag{1}$$

$$\hat{f}(x) = argmax_{y \in \mathcal{Y}} \left\{ \sum_{k=1}^K \mathcal{I}[\hat{f}_k(x) = y] \right\} \text{for classification problems,} \tag{2}$$

where $\mathcal{I}(\cdot)$ and $\hat{f}(x)$ denote the indicator function and RF prediction, respectively.

RFs have shown to be a state-of-the-art tool in machine learning. RF model can be used for both feature selection and prediction, and it can perform well in both classification and regression problems. However, the performance of random forests suffers when applied to high-dimensional data, i.e., data with thousands to millions of features. The main cause is that in the process of growing a tree from the bagged sample data, the subspace of features randomly sampled from the thousands of features in the training data to split a node of the tree is often dominated by less important features. The tree grown from such randomly sampled subspace features will have low accuracy in prediction, hence affects the final prediction of the random forests.

In this paper, we propose a new approach for feature weighting subspace selection to improve the accuracy of prediction for RF, meanwhile maintaining the diversity and the randomness of the forest. Given a training data set $\mathcal{L}$, we first use a feature permutation technique [3] [4] to measure the importance of features and produce raw feature importance scores. Then we apply p-value assessment on finding the cut-off between informative and less informative features. For all informative features, the Spearman rank test is then used for regression problem and the χ^2 statistic is used for classification problem to find the subset of highly informative features. The separation forms three sub sets of features. When sampling the feature subspace for learning, features from these three groups of highly informative, informative and less-informative features are taken into account for splitting the data at a node. Since the subspace always contains highly informative features, it can guarantee a better split at a node, therefore assuring a qualified tree. This sampling method always provides enough highly informative features for the subspace feature at any levels of the decision tree. By using taking into account features from all three subsets, the diversity and the randomness of the forests in the Breiman's framework [1] are maintained.

The above feature subspace selection will be used for building trees in our new random forests algorithm, called ssRF, for dealing with both classification and regression problems. With the ssRF model, the quantile regression is employed to predict both point prediction and range prediction in regression problems. Our experimental results have shown that with the proposed feature sampling method, our random forests ssRF model outperformed existing random forests in reduction of prediction errors, even though a small feature subspace size of $\lfloor log_2(M) + 1 \rfloor$ is used, and especially they performed well in range prediction on high-dimensional data.

2   Feature Weighting Subspace Selection

2.1   Importance Measure of Features from a Random Forest

The feature importance measure obtained from the random forest is described as follows [5], [6]. At each node t in a decision tree, a split on feature X_j is determined by the decrease in node impurity $\Delta R(X_j, t)$. For a regression tree, the node impurity $R(t) = \sigma^2(t)p(t)$, where $p(t) = N(t)/N$ is the probability for the impurity reduction that an sample chosen at random from the underlying theoretical distribution falls into t, $N(t)$ is the total number of samples and $\sigma^2(t) = \sum_{x_i \in t}(Y_i - \bar{Y}_t)^2/N(t)$ is the sample variance of Y. Then the decrease of impurity in node t after splitting into t_L and t_R is

$$\begin{aligned} \Delta R(X_j, t) &= R(t) - [R(t_L) + R(t_R)] \\ &= \sigma^2(t)p(t) - [\sigma^2(t_L)p_L + \sigma^2(t_R)p_R], \end{aligned} \tag{3}$$

where p_L, p_R are the proportions of samples in t that go left and right, respectively.

For classification trees, the Gini index is used to reflect the node impurity $R(t)$. Suppose there are S categorical values in node t ($s \in S$). Let $\pi_t(s)$ be the proportion of the samples from the sth category in node t. The node impurity is defined as

$$R(t) = N(t) \sum_{s=1}^{S} \pi_t(s)[1 - \pi_t(s)].$$

The chosen split of feature X_j for each node t is the one that maximizes $\Delta R(X_j, t)$. Let $IS_k(X_j)$ denotes the importance score of feature X_j in a single decision tree T_k, we have

$$IS_k(X_j) = \sum_{t \in T_k} \Delta R(X_j, t).$$

Let IS_j be an importance score of feature X_j, IS_j is computed over all K trees in a random forest, defined as

$$IS_j = \sum_{k=1}^{K} IS_k(X_j)/K.$$

It is worth noting that a random forest uses *in-bag* samples (i.e. the set of the bagged samples used in building the trees) to produce importance scores IS_j. This is the main difference between this importance score and an *out-of-bag* measure, which requires so much computational time using OOB-permutation [7], [3]. We can normalize them into $[0, 1]$ using the min-max normalization as follows:

$$VI_j = \frac{IS_j - min(IS_j)}{max(IS_j) - min(IS_j)}. \tag{4}$$

Having the raw importance scores VI_j determined by Equation (4) we can evaluate the contributions of the features in predicting the response feature.

2.2 A New Feature Sampling Method for Subspace Selection

We first compute importance scores for all features according to Equation (4). Denote the feature set as $\mathcal{L}_X = \{X_j\}$, $j = 1, 2, ..., M$, we randomly permute all values in each feature to get a corresponding shadow feature set, denoted as $\mathcal{L}_A = \{A_j\}_1^M$. The shadow features do not have prediction power to the response feature. Following the feature permutation procedure recently presented in [3], we ran RF R times on the extended data set $\{\mathcal{L}_X \cup \mathcal{L}_A, Y\}$ to get importance scores $VI^r_{X_j}$ and $VI^r_{A_j}$, and the samples for comparison denoted as $V^* = max\{A_{rj}, r = 1, ..R\}$.

The unequal variance Welch's two-sample t-test [8] is then used to compare the importance score of each feature with the maximum importance scores of generated shadows. The non-parametric statistical test is required because the importance scores across the replicates are not normal distribution. Having computed the t statistic, we can compute the p-value for the features and perform hypothesis test on $\overline{VI}_{X_j} > \overline{V}^*$. This test confirms that if a feature is important, it consistently scores higher than the shadow over multiple permutations. Therefore, any feature whose importance score is smaller than the maximum importance score of noisy features, is considered less important, otherwise, it is considered important.

The p-value of a feature indicates the importance of the feature in prediction. The smaller the p-value of a feature, the more correlated the predictor feature to the response feature, and the more powerful the feature in prediction. Given a statistical significance level, we can identify informative features from low-informative ones. Given all p values of features, we set a significance level as a threshold λ, for instance $\lambda = 0.05$. Any feature whose p-value is greater than λ is added to the low-informative feature subset denoted as X_l, the direct relationship with the Y values is assessed otherwise.

The non-parametric Spearman ρ test is used to measure the strength of the relationship between X_j and $Y \in \mathbb{R}^1$ in regression problems. The value $|\rho| \in [0,1]$, where $|\rho| = 1$ means a perfect correlation, 0 means that there is no correlation. Spearman rank correlation coefficient performs well in cases when the conditional distribution is not normal, each pair (X_j, Y) is converted to ranks $(R(x_i), R(y_i)), (i = 1, .., N)$ and ρ is the absolute value, computed as follows:

$$\rho_j = \left| \frac{\sum_j (R(x_i) - \overline{X})(R(y_i) - \overline{Y})}{\sqrt{\sum_{i=1}^{N}(R(x_i) - \overline{X})^2 \sum_{i=1}^{N}(R(y_i) - \overline{Y})^2}} \right| \tag{5}$$

where $\overline{X}, \overline{Y}$ are the average values of important feature X_j and response feature Y, respectively. Given all ρ values in the remaining features $\{X \setminus X_l\}$, we take the mean of all ρ values as the threshold γ,

$$\gamma = \frac{1}{M_\lambda} \sum_{j=1}^{M_\lambda} \rho_j, \tag{6}$$

where M_λ is the number of numerical features in the important feature subset $\{X \setminus X_l\}$. Let X_h denote a subset of highly informative features, all features X_j are added to X_h whose ρ-value is greater than γ. The remaining features including categorical features are added to the informative feature subset, denoted as X_m.

For the classification problem, $\chi^2(X, Y)$ is used to test the association between the class label and each feature X_j. For the test of independence, a chi-squared probability of less than or equal to 0.05 is commonly interpreted for rejecting the hypothesis that the feature is independent of the response feature. All features X_j whose p-value is smaller than 0.05 from the results of χ^2-test are added into X_h, the remaining features are added to X_m otherwise.

Given X_h, X_m and X_l, at each node, we randomly select $mtry$ ($mtry > 1$) features from three separated groups. For a given subspace size, we can choose proportions between highly informative, informative and less-informative features depending on the size of the three groups. That is $mtry_{high} = \lceil mtry \times (M_{high}/M) \rceil$, $mtry_{mid} = \lceil mtry \times (M_{mid}/M) \rceil$ and $mtry_{low} = mtry - mtry_{high} - mtry_{mid}$, where M_{high} and M_{mid} are the number of features in X_h and X_m, respectively. These are merged to form the feature subspace for splitting nodes of trees.

3   The Proposed ssRF Algorithm

The new feature subspace sampling method is now used to grow decision trees for building RFs. In regression problem, we propose to use quantile regression to obtain both point and range prediction, this idea was introduced in [9]. Using the notations as in [1], let θ_k be the random parameter vector that determines the growth of the kth tree and $\Theta = \{\theta_k\}_1^K$ be the set of random parameter vectors for the forests generated from $\mathcal{L}$. In each regression tree T_k from $\mathcal{L}_k$, we compute a positive weight $w_i(x_i, \theta_k)$ for each case $x_i \in \mathcal{L}$. Let $l(x, \theta_k, t)$ be a leaf node t in T_k. The cases $x_i \in l(x, \theta_k, t)$ are assigned the same weight $w_i(x, \theta_k) = 1/N(t)$, where $N(t)$ is the number of cases in $l(x, \theta_k, t)$. In each classification tree,

$$w_i(x, \theta_k) = 1 \quad \text{if} \sum_{n=1}^{N(t)} \mathcal{I}(Y_n = Y_i) \geq \sum_{n=1}^{N(t)} \mathcal{I}(Y_n = Y_j) \forall Y_i \neq Y_j.$$

This means the prediction for a regression problem is simply the average and for the classification problem is the category received by a majority votes by all Y values in node t. In this way, all cases in $\mathcal{L}_k$ are assigned positive weights and the cases not in $\mathcal{L}_k$ are assigned zero weight.

For a single tree prediction, given $X = x$, the prediction value is

$$\hat{Y}^k = \sum_{i=1}^N w_i(x, \theta_k) Y_i = \sum_{x, X_i \in l(x, \theta_k, t)} w_i(x, \theta_k) Y_i. \tag{7}$$

The new random forests algorithm ssRF is summarized as follows.

1. Given $\mathcal{L}$, separate the highly informative features and the informative features from the less informative ones to obtain three feature subsets X_h, X_m and X_l as described in Section 2.2.
2. Sample the training set $\mathcal{L}$ with replacement to generate bagged samples $\mathcal{L}_k, k = 1, 2, .., K$.
3. For each $\mathcal{L}_k$, grow a regression tree T_k as follows:
   (a) At each node, select a subspace of $mtry$ ($mtry > 1$) features randomly and separately from X_l, X_m and X_h and use the subspace features as candidates for splitting the node.
   (b) Each tree is grown nondeterministically, without pruning until the minimum node size n_{min} is reached. At each leaf node, all $Y \in \mathbb{R}^1$ values of the samples in the leaf node are kept.
   (c) Compute the weights $w_i(x, \theta_k)$ of each X_i by individual tree T_k using out-of-bag samples.
4. Compute the weights $w_i(x)$ assigned by RF which is the average of weights by all trees:

$$w_i(x) = \frac{1}{K} \sum_{k=1}^K w_i(x, \theta_k) \tag{8}$$

5. Given an input $X = x$, use Equation (2) to predict the new sample for the classification problem. For the regression problem, we can find the leaf nodes $l_k(x, \theta_k)$ from all trees where X falls and the set of Y_i in these leaf nodes. Given all Y_i and the corresponding weights $w_i(x)$, the conditional distribution function of Y given X is estimated as $\hat{F}(y|X = x) = \sum_{i=1}^{N} w_i(x)\mathcal{I}(Y_i \leq y)$, where $\mathcal{I}(\cdot)$ is the indicator function that is equal to 1 if $Y_i \leq y$ and 0 otherwise. Given a probability α, the quantile $Q_\alpha(X)$ is estimated as $\hat{Q}_\alpha(X = x) = inf\{y : \hat{F}(y|X = x) \geq \alpha\}$. Given a probability τ, α_l and α_h for $\alpha_h - \alpha_l = \tau$, τ is the probability that prediction Y will fall in the range of $[Q_{\alpha_l}(X), Q_{\alpha_h}(X)]$, we have

$$[Q_{\alpha_l}(X), Q_{\alpha_h}(X)] = [inf\{y : \hat{F}(y|X = x) \geq \alpha_l\},$$
$$inf\{y : \hat{F}(y|X = x) \geq \alpha_h\}]$$
(9)

For the point regression, the median $\hat{Q}_{0.5}$ can be chosen in a range as the prediction of Y given input $X = x$.

4 Experiments and Evaluation

4.1 Data Sets

We conducted experiments to test our proposed system on high-dimensional data sets for both classification and regression problems. Table 1 lists the real data sets used to evaluate the performance of random forests models. The *Fbis* data set was compiled from the archive of the Foreign Broadcast Information Service and the *La1s, La2s* data sets were taken from the archive of the Los Angeles Times for TREC-5[1].

The *Rivers*[2] data set was used to predict the flow level of a river. It is based on a data set containing river discharge levels of $1,439$ Californian rivers for a period of $12,054$ days. This data set contains 48.6% missing values, all values were used to train the model. The level of the $1,440$-th river was predicted in our experiments, the target values were converted from $[0.062; 101,000]$ to $[0; 1]$. The *LOG1P* data set was used in [10]. The *Stock* data set was described in [11] to make a stock price prediction. This data set has about 8.35% missing values in the predictor features. The original Y value is between 880 and $82,710$, these target feature values were converted to $[0; 1]$ using linear scale. Regarding the characteristics of the data sets given in Table 1, the proportion of the sub-data sets for training was separately from the testing.

4.2 Experimental Setting

Evaluation Measure: We used Breiman's method of measurement as described in [1]. The accuracy of prediction of RF models was evaluated on test set.

---

[1] http://trec.nist.gov
[2] http://www.usgs.gov

Table 1. Description of high-dimensional data sets sorted by the number of features and grouped into two groups - for regression and classification problems, accordingly

Data set	#Train	#Test	#Features	#Classes
Stock	1,942	785	495	
Rivers	8,345	3,709	1,439	
LOG1P	16,087	3,308	4,272,227	
Fbis	1,711	752	2,000	17
La2s	1,855	845	12,432	5
La1s	1,963	887	13,195	5

In which, for the regression problem the *mean of square residuals* (*MSR*) measure was computed, for the classification problem the *test error* measure was used.

The latest RF [12], QRF [13], cRF (cForest) [14] and GRRF R-packages [15] in CRAN[3] were used in R environment to conduct these experiments. For the GRRF model, we used a value of 0.1 for the coefficient γ because GRRF(0.1) has shown competitive prediction performance in [16]. The novel SRF model [17] using the stratified sampling method was intended to solve the classification problem. The QRF and eQRF [18] models were developed for solving only regression problems. The ssRF model with the new subspace sampling method is a new implementation. In that implementation, we called the corresponding R/C++ functions in R environment.

From each training data set we built 10 random forest models and the average of MSRs and the test errors of the models were computed; each of the RF models had 200 and 500 trees, respectively. The number of the minimum node size n_{min} was 5 for regression and 1 for classification problems. The number of features-candidates was set with the default setting to $mtry = \lfloor log_2(M) + 1 \rfloor$. The parameters R, $mtry$ and λ for pre-computation of feature partition used in ssRF were 30, $\sqrt{M}$ and 0.05, respectively. In order to process the large-scale data set *LOG1P*, only 5% of the samples was used to train the eQRF and ssRF models for feature partition and subspace selection, since the computational time required for all the samples is too long.

To address the missing values in the data set, we separate all samples containing missing values and create an extra "missing" group for them. We then treat this "missing" class as a predictor feature of the response feature. If missing values occur in the response feature, those samples are routinely omitted. After separation, missing values are typically treated as if they were actually observed.

All experiments were conducted on the six 64-bit Linux machines, each one equipped with IntelR XeonR CPU E5620 2.40 GHz, 16 cores, 4 MB cache, and 32 GB main memory. The ssRF and eQRF models were implemented as multi-thread processes, while other models were run as single-thread processes.

---

[3] http://cran.r-project.org/

4.3   Results on Real Data Sets

The performance of RF models is evaluated when the number of trees and features are varied, those are two key parameters in the RF models. Figures 1(a), (b) show the regression errors of the random forest models varied with the number of K trees used with $mtry = \lfloor log_2(M) + 1 \rfloor$. Figures 1(c), (d) present the plots of curves when the number of random features $mtry$ in the subspace increases while the number of trees is fixed ($K = 200$), the vertical line in each plot indicates the size of a subspace of features $mtry = \lfloor log_2(M) + 1 \rfloor$, this subspace was suggested by Breiman [1] for the case when applying RF to low-dimensional data sets. Table 2 shows the test errors on the classification data sets against the number of trees and features. The RF, QRF and eQRF models were unable to build their models on the data sets *Stock* and *Rivers* containing missing values. The imputation function in *randomForest* R-package was used to recover missing values on the two data sets. The eQRF model was not considered in this experiment because its prediction accuracy is last in this ranking on imputed data sets. The cRF model was processed well on data set containing missing values, however this model crashed when applied to the large-size data sets. The results of RF models when applied to imputed data sets are denoted as *RF.i*, *QRF.i* in the plots, respectively.

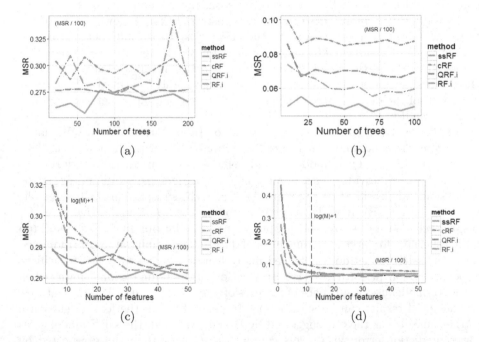

Fig. 1. The prediction performance of regression random forest models changes against the number of trees and features on real data sets. (a), (c) Stock data. (b), (d) Rivers data.

Table 2. The prediction test error of the RF models against the number of trees K and features $mtry$ on classification data sets. Numbers in bold are the best results.

Data set	Model	The number of trees					The number of features				
		K=50	100	150	200	300	mtry=10	20	30	40	50
Fbis	RF	.2307	.2241	.2254	.2261	.2279	.2434	.2351	.2156	.2303	.2187
	GRRF	.2394	.2407	.2287	.2314	.2340	.2527	.2101	.1955	.1862	.1981
	SRF	.1689	**.1649**	.1622	**.1569**	.1618	**.1569**	.1702	.1636	.1715	.1715
	ssRF	**.1676**	.1676	**.1543**	.1689	**.1569**	.1822	**.1556**	**.1503**	**.1503**	**.1522**
La2s	RF	.2303	.2363	.2256	.2315	.2280	.2536	.1611	.1586	.1432	.1402
	GRRF	.2476	.2121	.2180	.2156	.2192	.2820	.1860	.1540	.1505	.1386
	SRF	.1327	.1517	.1493	.1445	.1410	.1244	.1315	.1374	.1386	.1434
	ssRF	**.1078**	**.1066**	**.1102**	**.1185**	**.1090**	**.1149**	**.1102**	**.0995**	**.1002**	**.1014**
La1s	RF	.6708	.6697	.6731	.6742	.6488	.6776	.6032	.4543	.3337	.2052
	GRRF	.1928	.1759	.2063	.1849	.1966	.1905	.1691	.1612	.1577	.1409
	SRF	**.1308**	.1353	.1330	.1353	.1488	**.1330**	**.1375**	**.1387**	.1364	.1398
	ssRF	.1354	**.1321**	**.1322**	**.1321**	**.1264**	.1477	.1432	.1443	**.1319**	**.1387**

We can see that ssRF always provided good results and achieved lower prediction error in Figure 1 and Table 2 when varying K and $mtry$ on both kind of data sets. In some cases where the ssRF model did not obtain the best results compared with SRF on the data sets *Fbis* and *La1s*, the differences from the best results were minor. These results demonstrated that, at lower levels of the tree, the gain is reduced because of the effect of splits on different features at higher levels of the tree. The other random forests models increase prediction errors while the ssRF model always produces better results. This was because the selected subspace of features contains enough highly informative features at any levels of the decision tree. The effect of the new sampling method is clearly demonstrated in this result.

In Figures 1 (c), (d) and the right panel of Table 2, the RF and QRF models require larger number of features to achieve the lower prediction error. This means the RF and QRF models could achieve better prediction performance only if they are provided with a much larger feature subspace. For solving the regression and classification problem, the size of the subspace in the default settings of RF and QRF R-packages were set to $mtry = \lfloor M/3 \rfloor$ and $mtry = \lfloor \sqrt{M} \rfloor$, respectively. With this size, the computational time for building a RF is still too high, especially for large high-dimensional data. These empirical results indicated that, the ssRF model does not need many features in the subspace to achieve good prediction performance. For application on high-dimensional data, when the ssRF model uses a subspace of features of size $mtry = \lfloor log_2(M)+1 \rfloor$ features, the achieved results can be satisfactory. In general, when the feature subspace of the same size as the one suggested by Breiman is used, the ssRF model gives lower prediction error with a less computational time than those reported by Breiman. This achievement is considered to be one of the contributions in this work.

Figure 2 shows the point and 90% range prediction results of the large high-dimensional data set $LOG1P$ by the eQRF and ssRF models. The green and red points show the predictions inside and outside the predicted ranges, respectively. Figure 2 (a) shows the point and 90% range predictions of the eQRF model, we can see that the point prediction is more scattered than that of the ssRF model in the results. Significant improvement in the prediction results of the ssRF model can be observed in Figure 2 (b). We can see that, the predicted points are closer to the diagonal line which indicates that the predicted values were close to the true values in data, and there are less red points in the Figure 2 (b) which indicates that a large number of predictions were within the predicted ranges. These results clearly demonstrate the advantages of the ssRF model over very recently proposed eQRF model.

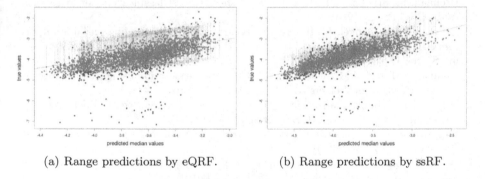

(a) Range predictions by eQRF.   (b) Range predictions by ssRF.

Fig. 2. Comparisons of range predictions by the regression eQRF and ssRF models on large high-dimensional data sets $LOG1P$

5   Conclusions

We have presented a new approach for feature subspace selection for efficient node splitting when building decision trees in random forests. Based on that, a new random forest algorithm, ssRF, has been developed for prediction high-dimensional data. The quantile regression is employed to obtain predictions in the regression problem, which makes the RF more robust towards outliers. With the new subspace feature selection, the small subspace size $mtry = \lfloor log_2(M)+1 \rfloor$ reported by Breiman can be used in our algorithm to get lower prediction error. With ssRF, the performance for both classification and regression problems (the point and range prediction) is preserved and improved. Experimental results have demonstrated the improvement of our ssRF in reduction of prediction errors in comparison with existing recent proposed random forests including eQRF, GRRF and SRF, and especially it performed well on large high-dimensional data.

Acknowledgments. This research is supported in part by NSFC under Grant NO. 61203294 and Natural Science Foundation of SZU(grant no. 201433). Joshua Huang was supported by The National Natural Science Foundation of China under Grant No. 61473194.

References

1. Breiman, L.: Random forests. Machine learning **45**(1), 5–32 (2001)
2. Breiman, L.: Manual on setting up, using, and understanding random forests v3. 1. (2002) (retrieved October 23, 2010)
3. Nguyen, T.T., Huang, J., Nguyen, T.: Two-level quantile regression forests for bias correction in range prediction. Machine Learning, 1–19 (2014)
4. Tuv, E., Borisov, A., Runger, G., Torkkola, K.: Feature selection with ensembles, artificial variables, and redundancy elimination. The Journal of Machine Learning Research **10**, 1341–1366 (2009)
5. Breiman, L., Friedman, J., Stone, C.J., Olshen, R.A.: Classification and regression trees. CRC Press (1984)
6. Louppe, G., Wehenkel, L., Sutera, A., Geurts, P.: Understanding variable importances in forests of randomized trees. In: Advances in Neural Information Processing Systems, pp. 431–439 (2013)
7. Genuer, R., Poggi, J.M., Tuleau-Malot, C.: Variable selection using random forests. Pattern Recognition Letters **31**(14), 2225–2236 (2010)
8. Welch, B.L.: The generalization ofstudent's' problem when several different population variances are involved. Biometrika, 28–35 (1947)
9. Meinshausen, N.: Quantile regression forests. The Journal of Machine Learning Research **7**, 983–999 (2006)
10. Ho, C.H., Lin, C.J.: Large-scale linear support vector regression. The Journal of Machine Learning Research **13**(1), 3323–3348 (2012)
11. Cai, Z., Jermaine, C., Vagena, Z., Logothetis, D., Perez, L.L.: The pairwise gaussian random field for high-dimensional data imputation. In: Data Mining (ICDM), pp. 61–70. IEEE (2013)
12. Liaw, A., Wiener, M.: Classification and regression by randomforest. R News **2**(3), 18–22 (2002)
13. Meinshausen, N.: quantregforest: quantile regression forests. R package version 0.2-3 (2012)
14. Hothorn, T., Hornik, K., Zeileis, A.: party: A laboratory for recursive part (y) itioning. r package version 0.9-9999 (2011). http://cran.r-project.org/package=party (date last accessed November 28, 2013)
15. Deng, H.: Guided random forest in the rrf package. arXiv preprint arXiv:1306.0237 (2013)
16. Deng, H., Runger, G.: Gene selection with guided regularized random forest. Pattern Recognition **46**(12), 3483–3489 (2013)
17. Ye, Y., Wu, Q., Zhexue Huang, J., Ng, M.K., Li, X.: Stratified sampling for feature subspace selection in random forests for high dimensional data. Pattern Recognition **46**(3), 769–787 (2013)
18. Tung, N.T., Huang, J.Z., Khan, I., Li, M.J., Williams, G.: Extensions to Quantile Regression Forests for Very High-Dimensional Data. In: Tseng, V.S., Ho, T.B., Zhou, Z.-H., Chen, A.L.P., Kao, H.-Y. (eds.) PAKDD 2014, Part II. LNCS, vol. 8444, pp. 247–258. Springer, Heidelberg (2014)

Mining Heterogeneous, High Dimensional, and Sequential Data

Seamlessly Integrating Effective Links with Attributes for Networked Data Classification

Yangyang Zhao[1]([⊠]), Zhengya Sun[1], Changsheng Xu[2], and Hongwei Hao[1]

[1] IDMTech, Institute of Automation, Chinese Academy of Sciences, Beijing, China
{yangyang.zhao,zhengya.sun,hongwei.hao}@ia.ac.cn
[2] NLPR, Institute of Automation, Chinese Academy of Sciences, Beijing, China
csxu@nlpr.ia.ac.cn

Abstract. Networked data is emerging with great amount in various fields like social networks, biological networks, research publication networks, etc. Networked data classification is therefore of critical importance in real world, and it is noticed that link information can help improve learning performance. However, classification of such networked data can be challenging since: 1) the original links (also referred as relations) in such networks, are always sparse, incomplete and noisy; 2) it is not easy to characterize, select and leverage effective link information from the networks, involving multiple types of links with distinct semantics; 3) it is difficult to seamlessly integrate link information with attribute information in a network. To address these limitations, in this paper we develop a novel Seamlessly-integrated Link-Attribute Collective Matrix Factorization (SLA-CMF) framework, which mines highly effective link information given arbitrary information network and leverages it with attribute information in a unified perspective. Algorithmwise, SLA-CMF first mines highly effective link information via link path weighting and link strength learning. Then it learns a low-dimension link-attribute joint representation via graph Laplacian CMF. Finally the joint representation is put into a traditional classifier such as SVM for classification. Extensive experiments on benchmark datasets demonstrate the effectiveness of our method.

Keywords: Networked data classification · Heterogeneous information fusion · Collective matrix factorization

1 Introduction

In recent years, with the advance of the World Wide Web and social networks such as Twitter, YouTube, Facebook and Flickr, more and more networked data are available on the web. Compared with traditional data, the networked data brings us a lot of extra meaningful link information besides their attribute (content) information. In the majority cases, such data contains more than one types of entities and links, and is always referred as Heterogeneous Information Network [5,11].

© Springer International Publishing Switzerland 2015
T. Cao et al. (Eds.): PAKDD 2015, Part II, LNAI 9078, pp. 473–484, 2015.
DOI: 10.1007/978-3-319-18032-8_37

The link information in these networks has been proved beneficial for classification [2,6,9,11] in many works. However, classification of such networked data can be challenging since: 1) the original links (also referred as relations) in such networks, are always sparse, incomplete and noisy; 2) it is not easy to characterize, select and leverage effective link information from the networks, involving multiple types of links with distinct semantics; 3) it is difficult to seamlessly integrate link information with attribute information in a network.

A great deal of recent works have shown their interests in networked data classification and try to utilize link information in a network to enhance the classification performance, unfortunately, none of them simultaneously address the three challenges well, as far as we know. Information fusion based methods

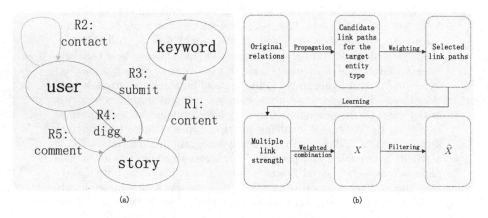

Fig. 1. (a)Entity types and original relations in Digg dataset;(b)The process of mining effective link information

[2] try to combine the link information and attribute information together and perform quite well because of their ability to exploit correlation between the two aspects of the network. These methods treat the link information as another kind of attribute which directly decompose the original link information and attribute information simultaneously. But their results are always limited to noises and the sparsity of the links. Graph-based techniques [6] treat the link information in the view of manifolds which assume that the linked nodes have similar labels. They are sensitive to the quality of the graph and inclined to fail when the network is intrinsically of low label consistency. Collective classification based methods [7–9,11] predict the unlabeled nodes with the help of related labeled nodes. What makes it powerful for networked data classification is its great ability to learn and make use of various kinds of dependency structures. But at the same time the performance might be largely degraded due to the lack of neighbours when given a sparse network. Another obvious limitation is that most of these methods use link information and attribute information non-synchronously so that the links and attributes cannot be integrated well.

To summarize, most of the existing research suffers from three limitations: 1) only utilizing the original link information but ignoring mining richer semantics conveyed by the link paths; or 2) intuitively selecting and indiscriminately utilizing different types of link paths; and/or 3) utilizing link information and attribute information in distinct perspectives.

In this study, we mine highly effective link information given arbitrary information network (as shown in Fig.1(b)) and leverage it with attribute information in a unified perspective. First, we learn the weights of different link paths under the guidance of a few labeled nodes sampled from the training set. Second, we learn link strength of the selected link paths, and get a weighted combination of all the selected link paths. Then in a unified perspective, we learn a low-dimension link-attribute joint representation via CMF for networked data classification. Finally the joint representation is put into a traditional classifier such as SVM for classification. Experiments on three real datasets demonstrate the effectiveness of our SLA-CMF method.

The primary contributions of SLA-CMF are as follows.

- We propose to integrate link path weighting, link strength learning and graph-regularized CMF to select, characterize and leverage effective link information, and it is finally seamlessly integrated with attribute information.
- We adopt a simple but effective strategy to learn the weights of different link paths, which can be used for link path selection. Then we propose a general link path based similarity computing method, through which the strength of either symmetric or asymmetric link path can be accurately characterized. These two processes not only mine richer semantics conveyed by link paths, but also alleviate noise and sparsity of link information. (Section 4.1, 4.2).
- We treat both the links and the attributes in a unified perspective. Precisely, we treat the attributes as a kind of link path when characterizing the link structure. Meanwhile, we treat the link information as a kind of attribute when directly decomposing it in an attribute perspective. This scheme ensures the seamless integration by mutual penetration. (Section 4.3).

2   Related Work

In this section, we review some of the research literatures related to networked data classification.

To combine text content and the explicit links for classification, information fusion based method [2] simultaneously decomposes the adjacency matrix of the original cite relation and the bag-of-word attribute matrix by collective matrix factorization and gets great enhancement on the results. However, it is under the assumption of only one type of links and therefore does not apply in heterogeneous information networks.

Unlike [2] directly learning from the link information, graph-based technique [6] interprets it in a manifolds view and utilizes it to build up a graph Laplacian regularization to constraint the encoding matrix of attribute information. This method is efficient and can outperform [2] in some cases. But it is still lack of mining link paths so that the result will inevitably be sensitive to the outliers in the original relation.

Recent researches [7,8] propose a tensor factorization method, which has the effect of collective classification and solve the classification problem by reconstructing the appropriate slice of the class relation. The three-way tensor puts all of the entities into the tensor and every frontal slice of the tensor is an observation matrix that describes one type of relation. So this model supports the representation of Heterogeneous Information Networks and linked data classification. Unfortunately, blindly utilizing different types of links without distinction and decomposing two many types of link information simultaneously leads to strong disturb from each other.

Collective classification based method [9] seeks to combine the explicit links with the the links mined from local attribute similarity to increase the information in the network, and it adopts a node-based assortativity coefficient to combine different edges with different weights, which greatly inspires us. However, the big difference is that, it adopts wvRN-RL as the classifier and just utilizes the link information and attribute information in a link perspective while we leverage both of them in both link perspective and attribute perspective.

Another collective classification method [11] is an two-stage method, which utilizes the attributes only for bootstrap, simply adopts the generation scheme of link path defined in [5] and iteratively infer the unlabeled entities based on the neighbours with respect to different link paths. It is limited because it pays no attention to link path weights and utilizes the links and attributes separately.

3   Problem Formulation

The networked data classification in this paper is in a supervised or semi-supervised setting. Given a network with known labels for the nodes in training set, we predict the labels of the nodes in testing set. In a certain network, the content matrix and the original link (relation) adjacency matrices of different types can be easily extracted. To leverage the link information and attribute information for an enhanced classification performance, the main idea behind this paper is selecting and characterizing effective link information, and then seamlessly integrating it with attribute information to get a link-attribute joint representation that is more appropriate for classification.

First, we need to specify the type of objects in the network we will classify, and it is called the **target entity type**. If the number of entity type or that of the relation type a certain network is more than one, the network can be called a **heterogeneous information network** [5], otherwise homogeneous information network. And we call the adjacency matrix of the original link path **original adjacency matrix**.In this paper, **Link path**, or referred as meta path [5], is a path that connects object types via a sequence of relations. It includes the original relations whose length is one. In this paper, in order to facilitate the weighted combination of different link paths, we integrate all the relations related to the target entity type, including **inter-type relations** (e.g., $R4 : User \xrightarrow{comment} Story$ in Fig.1) and **inner-type relations** (e.g., $R2 : User \xrightarrow{contact} User$ in Fig.1), to construct the link information of the **inner-type link paths** (e.g., $Story \xrightarrow{comment^{-1}} User \xrightarrow{comment} Story$ and $story \xrightarrow{comment^{-1}} user \xrightarrow{contact} user \xrightarrow{comment} story$, they can also be represented by multiplying adjacency matrices for each relation along the link path, i.e., $R_4^T \times R_4$ and $R_4^T \times R_2 \times R_4$). Note that, $comment^{-1}$ is the inverted relation of $comment$, which may means $commentedby$.

In the process of seamless link-attribute integration, $X_k(k = 1, \cdots, m)$ is the adjacency matrix of a certain selected link path with the size $n \times n$, where n is the number of objects of target entity type. and m is the number of the selected link paths. X is the weighted combination of selected link paths. D is the attribute information matrix with the size $n \times l$, where l is the number of attributes. Z is the latent relation space matrix, whose size is $r \times r$. V is the basis matrix in attribute information matrix factorization with the size $l \times r$. A is the encoding matrix in which every row present one object of target entity type, and the size of it is $n \times r$.

4   Details of SLA-CMF

In this section, we will detail the proposed $SLA - CMF$ model. We first introduce a link path weighting strategy by which we can characterize the importance of a certain link path and filter out less important ones whose weights are less than the threshold. Then we learn the link strength by the method proposed by us. Finally we get a weighted combination of the selected link paths, and utilize a Laplacian regularized CMF to integrate the effective link information with attribute information, and get a joint link-attribute representation.

4.1   Link Path Weighting and Selection

Different link paths always have different semantics so that have different degree of impact in label consistency. Unlike many existing work intuitively selects and indiscriminately utilizes different types of link paths, we employ a novel strategy to learn the weights of different link paths under the guidance of some sampled labeled nodes from the training set. The weight is evaluated by the correlation between a certain link path and the label consistency of the sampled nodes, which can also be seen as the conditional probability of the label consistency w.r.t. (with respect to) a certain link path. The strategy is very simple but effective, and the weighting function and the threshold function of the k-th link path are defined as follows.

$$Weight_k = \frac{\sum_{i=1}^{n_s} \sum_{j=1}^{n_s} Consistency(o_i, o_j) Strength(o_i \xrightarrow{LinkPath_k} o_j)}{\sum_{i=1}^{n_s} \sum_{j=1}^{n_s} Strength(o_i \xrightarrow{LinkPath_k} o_j)} \quad (1)$$

$$Threshold_k = \frac{\sum_{i=1}^{n_s} \sum_{j=1}^{n_s} Consistency(o_i, o_j) NotLink(o_i \xrightarrow{LinkPath_k} o_j)}{\sum_{i=1}^{n_s} \sum_{j=1}^{n_s} NotLink(o_i \xrightarrow{LinkPath_k} o_j)} \quad (2)$$

where n_s is the number of sampled nodes, $Strength(\ )$ is the link strength of a certain link between two nodes; $Consistency(\ )$ is a two-valued function that sets the value to be 1 if two nodes have the same label, 0 otherwise; and $NotLink(\ )$ is also is a two-valued function that sets the value to be 1 if two nodes are not linked by a certain link path, 0 otherwise. Note that, if $Weight_k$ is less than the $Threshold_k$, this link path has negative impact in label consistency and should be filtered out because it means the k-th link path makes the nodes linked by it less possible to have the same label than the nodes not linked by it.

4.2   Link Strength Learning

As the selected meta-paths are all inner-typed, we propose a link path based similarity calculation method g-$PathSim$ (general Path Similarity) to characterize the strength of pairwise interactions among the objects via calculating the similarity of the two objects connected along a certain link path. The basic idea is that similar objects are not only strongly connected but also have few connection with others. Given an arbitrary link path $P : P_1 \times P_2 \times \cdots \times P_n$ of length $n(n>1)$, it can be decomposed into two shorter link paths $P_L : P_1 \times \cdots \times P_m$ and $P_R : P_{m+1} \times \cdots \times P_n$. The g-$PathSim$ is defined as follows,

$$g\text{-}PathSim(o_i \xrightarrow{P} o_j) = \frac{2 * (|o_i \xrightarrow{P} o_j| + |o_j \xrightarrow{P} o_i|)}{|O(o_i \mid P_L)| + |O(o_j \mid P_L)| + |I(o_i \mid P_R)| + |I(o_j \mid P_R)|} \quad (3)$$

where $|\ |$ is a counter function, $|o_i \xrightarrow{P} o_j|$ is the number of link path instances from object o_i to o_j along the link path P_{LR}, $|O(o_i \mid P_L)|$ is the weighted out-degree of object o_i along the link path P_L, and $|I(o_i \mid P_R)|$ is the weighted in-degree of object o_i along the link path P_R. Note that if o_i and o_j are the same object, g-$PathSim(o_i, o_j \mid P_{LR})$ is directly set to 1.

W.r.t. objects of the same entity type, the g-$PathSim$ of them can be calculated in matrix or vector manner as follows, where $\| * \|$ is the L2-norm function, L and R are the adjacency matrices corresponding to the left and the right link paths respectively.

$$g\text{-}PathSim(o_i \xrightarrow{P} o_j) = \frac{2 * (L_{i,*} R_{*,j} + L_{j,*} R_{*,i})}{\|L_{i,*}\|^2 + \|L_{j,*}\|^2 + \|R_{*,i}\|^2 + \|R_{*,j}\|^2} \quad (4)$$

W.r.t. a single relation, or referred as one-length link path, we add an imaginary entity type between the real object type and decompose the atomic relation into two relations as applied in [10]. The g-$PathSim$ of one-length link path is calculated as follows, where P is the original adjacency matrix of the one-length link path.

$$g\text{-}PathSim(o_i \xrightarrow{P} o_j) = \frac{2(P_{i,j} + P_{j,i})}{\|P_{i,*}\|^2 + \|P_{j,*}\|^2 + \|P_{i,*}\|^2 + \|P_{j,*}\|^2} \quad (5)$$

PathSim [5] is a special case of *g-PathSim* with symmetry link paths. Compared with *PathSim* and *HeteSim* [10], the advantages of *g-PathSim* are as follows: 1) our method refers to the two objects' information of both the left and right sides, so the similarity search result is more synthesized; 2) the result maintains symmetry in arbitrary link path, so we only need to search the similarities of the upper triangular matrix at half of the computational cost.

4.3   Seamless Link-Attribute Integration

To seamlessly integrate the selected link information and attribute information, we design a graph Laplacian regularized CMF method, whose structure and the setting of X enables it to realize this goal. The objective function is

$$
\begin{aligned}
f &= f_{link}(A, Z) + f_{attrubute}(A, V) + f_{graphLapl.}(A) \\
&= \frac{1}{2}\|X - AZA^T\|_F^2 + \frac{\alpha}{2}\|D - AV^T\|_F^2 + \frac{\beta}{2}\sum_{i=1}^{n}\sum_{j=1}^{n}\widehat{X}_{i,j}\|A_{i,*} - A_{j,*}\|^2 \\
&= \frac{1}{2}tr(XX^T - 2AZA^TX^T + AZA^TAZ^TA^T) \\
&\quad + \frac{\alpha}{2}tr(DD^T - DVA^T - AV^TD^T + AV^TVA^T) + \frac{\beta}{2}tr(A^TLA)
\end{aligned}
\tag{6}
$$

where $X = \sum_{k=1}^{m} Weight_k \cdot g\text{-}PathSim_k (k = 1, \cdots, m)$. Note that, to ensure the weighted adjacency matrix for graph Laplacian regularization highly reliable, we choose the top-K highest-weighted links for every node in X and construct a filtered adjacency matrix $\widehat{X}$ to construct L [1].

This model simultaneously decomposes link information in attribute perspective and utilizes it as graph Laplacian regularization in link perspective. Meanwhile we treat the attributes as a kind of link path (i.e., the attribute similarity) when constructing the combined graph X, so attribute information is also taking part in graph Laplacian regularization and is utilized in link perspective. Therefore the seamless integration of link information and attribute information is ensured by mutual penetration. (Our regularization means that, the more effective links and similar attributes two nodes have, the more closer the encoding vectors of them will be.)

W.r.t. optimization, we adopt an alternating projection method to learn the parameters A, Z, V. More specifically, each time we update one parameter and fix the others. This procedure will be repeated for several iterations until the termination condition is satisfied. One straightforward way to learn the parameters is to set the gradient of f w.r.t. A, Z, V to 0 and solve the corresponding linear system or nonlinear system. And the gradients of the objective function w.r.t. variable A, Z and V are as follows.

---

[1] $L = S - \widehat{X}$ is known as the Laplacian matrix with S being a diagonal matrix whose diagonal elements $S_{i,i} = \sum_j \widehat{X}_{i,j}$.

$$\frac{\partial f}{\partial A} = \underline{A}Z^T A^T AZ + \underline{A}ZA^T AZ^T - XAZ^T - X^T AZ$$
$$+ \alpha(\underline{A}V^T V - DV) + \beta(L^T + L)A, \tag{7}$$
$$\frac{\partial f}{\partial Z} = A^T AZA^T A - A^T XA, \quad \frac{\partial f}{\partial V} = VA^T A - D^T A$$

R, V can be updated directly by solving the linear system as follows.

$$Z \Leftarrow (A^T A)^{-1} A^T XA(A^T A)^{-1}, \quad V \Leftarrow D^T A(A^T A)^{-1} \tag{8}$$

As this equation of A can not be solved directly, an alternative approach is to approximate this nonlinear problem by solving only for the left A with underlines while holding the right A constant in the same way as [7,8]. The experiments show the viability of the update of A in this situation. A can be updated by

$$A \Leftarrow [XAZ^T + X^T AZ + \alpha DV - \beta(L^T + L)A][Z^T A^T AZ + ZA^T AZ^T + \alpha V^T V]^{-1} \tag{9}$$

Alternatively, the update of A can also be implemented through gradient methods, such as the conjugate gradient method and quasi-Newton methods or just the gradient descent method. In this paper we choose the gradient descent method to update the coding matrix A since the equation solution has a bit large errors especially in the early steps of updating, and the trick initializing A from the eigendecomposition of X is adopted.

5   Experiments

5.1   Datasets and Evaluation Scheme

The Cora1 [3] and Cora2 [4] datasets contain research papers from the computer science community. And we adopt the whole Cora2 and the subset EC, OS, NW, DB of Cora1. In Cora2,there is only one type of original links and the original adjacency matrix M_{PP} describes the relation $Paper \xrightarrow{cite} Paper$. We first characterize Cora2 by the link paths of $M_{PP}, M_{PP} \times M_{PP}^T, M_{PP}^T \times M_{PP}, M_{PP} \times M_{PP}$ and augment them by $M_{PAttri} \times M_{PAttri}^T$. And then we do link path weighting and selection, link strength learning via g-PathSim and finally get a weighted combination of the selected link paths. Although Cora2 is originally a sparse homogeneous information network, we can mine abundant and effective link information through learning. Meanwhile, the subsets of Cora1 have one more original adjacency matrix M_{PA} which is corresponding to the relation $Paper \xrightarrow{write^{-1}} Author$. It is a simple heterogeneous information network, and we augment the candidate link paths by $M_{PA} \times M_{PA}^T$.

The Digg [1] dataset we utilize in this paper consists of stories, users and their actions ($submit, digg, comment$) w.r.t. the stories, as well as the explicit $contact$ relation among these users, and the attribute of Digg stories is made up of keywords extracted from the story titles. In this paper we choose stories of five topics (i.e., pc games, space, pets/animals, linux/unix, political news) as the objects of target entity type, 200 from each, as well as the related users. As described in Fig.1, $R_k(k = 1, \cdots, 5)$ are original adjacency matrices of the five original relations. The candidate link paths of this network are quite various, including symmetric link path such as $R_1 \times R_1^T, R_3^T \times R_3, R_3^T \times R_2 \times R_2 \times R_3$, as well as many asymmetric ones such as $R_3^T \times R_2, R_3^T \times R_2 \times R_3, R_3^T \times R_2 \times R_2^T \times R_4$. We traverse the link paths with the length constraint 4 according to cross-validation. And it is processed just via the schema described in Section 4.

W.r.t. evaluation scheme, we take accuracy as evaluation criteria, we adopt 5-fold cross validation to evaluate our method, set the rank of latent factor A to be 50, and put A into Linear SVM for classification (all as the same as [2,6]) after seamless link-attribute integration through graph regularized CMF.

5.2   Baselines and Parameters Setting

The compared approaches include the state-of-art information fusion based method Link-content MF [2], Graph based method RRMF [6], Collective classification based method RESCAL [7,8], HCC [11], wvRN-RL [9]. Meanwhile, we also compare SLA-CMF with several variants of it. Each variant differs from SLA-CMF just in one aspect while consistent in others. Among them, SLA-CMF(naive-link) is a variant with only one link path without link strength learning (just as the same as Link-content MF and RRMF), SLA-CMF(naive-link+attri.-simil.) is the variant with attribute similarity directly added to the naiva link described above, SLA-CMF(PathSim) and SLA-CMF(HeteSim) are the variants replacing g-PathSim with PathSim and HeteSim correspondingly for link strength learning, and SLA-CMF(NAC) is the one replacing our link path weighting method with node-based assortativity coefficient [9]. Note that, there is no naive link between $Story$, it is selected from $R_3^T \times R_3, R_4^T \times R_4, R_5^T \times R_5$.

To ensure the weighted adjacency matrix for graph Laplacian regularization highly reliable, we choose the top-K highest-weighted links for every node in X, where the K is set to be 6 through cross validation, and the relative importance parameters (i.e., α and β) are set by searching the grid of $\{0.01, 0.03, 0.1, 0.3, 1, 3\}$. And the parameters in baselines are set to respect the original settings as much as possible. In all the methods that need labeled seeds, the ratio of the sampled labeled nodes are set to be 20%.

5.3   Performance and Result Analysis

What motivates $SLA\text{-}CMF$ most is the assumption that both mining effective link information and seamlessly integrating links with attributes will enhance networked data classification. This is the primary hypothesis we want to verify. Second, we want to test the validity of the link path weighting method and the link strength learning method proposed by us. Finally, we try to observe its sensitivity to the ratio of the sampled labeled nodes and the rank of the latent latent factor A.

We independently repeat the experiments for 10 times and report the best average result in Fig.2 and Fig.3. As the degree of label consistency varies in dif-

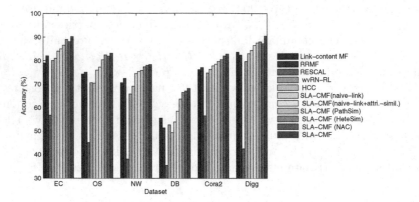

Fig. 2. Average classification accuracies of the compared methods

ferent networks, Link-content MF and RRMF have strong or weak performance respectively, which should be decided by their different mechanisms for utilizing link information. SLA-CMF(naive-link) outperforms Link-content MF and RRMF in all the datasets. As the link information and the attribute information of all the three methods are set in the same way, the result shows that, simultaneously decomposing the link information in attribute perspective and utilizing it as graph Laplacian regularization in link perspective, will lead to better result than separately processing it in either perspective. That is to say, the structure of graph Laplacian regularized CMF, which is designed to seamlessly integrating links with attributes, will indeed improve the robustness to label consistency and enhance networked data classification. Meanwhile, SLA-CMF(naive-link+attri.-simil.) performs a little better than SLA-CMF(naive-link), which indicates that utilizing attributes in link perspective can increase the information in the network and improve the performance. When it comes to other baselines, their performance is roughly consistent with the analysis of their strengths and weaknesses which can be referred in Section 2. RESCAL and HCC fail due to their ignorance of weighting different links, while HCC is also suffers from utilizing attributes and links separately. wvRN-RL is limited by only utilizing all the information in link perspective.

We can easily see that SLA-CMF, SLA-CMF(PathSim), SLA-CMF(HeteSim) and SLA-CMF(NAC) all outperform the former two SLA-CMFs, which confirms the validity of link path weighting and link strength learning. And the advantage of HeteSim and g-PtahSim to PathSim is obvious, because of their ability to computing the strength of asymmetric links and g-PathSim characterizes the link strength best. Moreover, the comparison between the performance of SLA-CMF and SLA-CMF(NAC) proves that our link path weighting method with the threshold check is more suitable for link path selection. In a word, the stable advantages shown in the experiments confirms that SLA-CMF indeed solve the 3 challenges well.

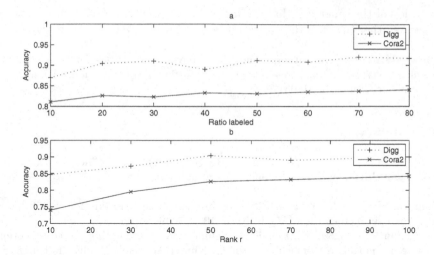

Fig. 3. (a) Average classification accuracies w.r.t. the ratio of sampled labeled nodes; (b) Average cassification accuracies w.r.t. the rank of the latent relation space

We can learn from Fig.3 that, the performance of SLA-CMF slightly improves with the increasing of the ratio of the sampled labeled nodes, but it changes very little, which may be due to the stability of our link path weighting method, or because we use a variety of information and they are related and redundancy. The performance of SLA-CMF w.r.t. the rank r of the latent factor A changes greatly when k is small, and then tend to remain unchanged. This feature indicates that SLA-CMF is suitable for networked data dimensionality reduction.

6   Conclusion and Discussion

In this paper, we propose a novel SLA-CMF framework for networked data classification, which mines highly effective link information given arbitrary information network by integrating link path weighting with link strength learning, and leverages it with attribute information in a unified perspective. First, we learn the weights of different link paths under the guidance of a few labeled nodes sampled from the training set, and utilizing these weights we select effective link paths. Second, we learn link strength of the selected link paths, and get a weighted combination of all the selected link paths. Finally, in a unified perspective, we learn a low-dimension link-attribute joint representation via CMF for networked data classification. Through these our method is enabled to solve the 3 challenges well and the experiments demonstrate its superiority for networked data classification compared with state-of-the-art approaches.

In our study, to facilitate the weighted combination of different link paths, we integrate the relations related to the target entity type to construct the link information of the inner-type link path. This schema works well, however, there may be the other schemas that work better, and it is worth deeper researching.

Acknowledgments. The authors thank the anonymous reviewers for their valuable comments. This research work was funded by the National Natural Science Foundation of China under Grant No. 61303179.

References

1. Lin, Y., Sun, J., Castro, P., Konuru, R., Sundaram, H., Kelliher, A.: Metafac: community discovery via relational hypergraph factorization. KDD **15**, 527–536 (2009)
2. Zhu, S., Yu, K., Chi, Y., Gong, Y.: Combining content and link for classification using matrix factorization. ACM SIGIR **30**, 487–494 (2007)
3. McCallum, A., Nigam, K., Rennie, J., Seymore, K.: Automating the construction of internet portals with machine learning. Kluwer Academic Publishers Hingham. Inf. Retr. **3**(2), 127–163 (2000)
4. Craven, M., DiPasquo, D., Freitag, D., McCallum, A., Mitchell, T.M., Nigam, K., Slattery, S.: Learning to extract symbolic knowledge from the world wide web. In: AAAI/IAAI, pp. 509–516 (1998)
5. Sun, Y., Han, J., Yan, X., Yu, P., Wu, T.: PathSim : Meta path-based top-k similarity search in heterogeneous information networks. In: VLDB (2011)
6. Li, W., Yeung, D.Y.: Relation regularized matrix factorization. In: IJCAI, pp. 1126–1131 (2009)
7. M. Nickel, V. Tresp and H. P. Kriegel: A three-way model for collective learning on multi-relational data. In: ICML, pp. 809–816 (2011)
8. M. Nickel, V. Tresp and H. P. Kriegel: Factorizing yago: scalable machine learning for linked data. In: WWW, pp. 271–280 (2012)
9. Sofus, A.: Macskassy: Improving Learning in Networked Data by Combining Explicit and Mined Links. In: AAAI (2007)
10. Shi, C., Kong, X., Huang, Y., Yu, P.S., Wu, B.: HeteSim: A General Framework for Relevance Measure in Heterogeneous Networks. IEEE TKDE (2013). doi:10.1109/TKDE.2013.2297920
11. Kong, X., Yu, P.S., Ding, Y., Wild, D.J.: Meta path-based collective classification in heterogeneous information networks. In: CIKM, pp. 1567–1571 (2012)
12. J. Liu, C. Wang, J. Gao and J. Han: Multi-view clustering via joint nonnegative matrix factorization. In Proceedings of the 13th SIAM International Conference on Data Mining, 252–260 (2013)

Clustering on Multi-source Incomplete Data via Tensor Modeling and Factorization

Weixiang Shao[1], Lifang He[2]([✉]), and Philip S. Yu[1,3]

[1] Department of Computer Science, University of Illinois at Chicago,
Chicago, IL, USA
[2] Institute for Computer Vision, Shenzhen University, Shenzhen, China
lifanghescut@gmail.com
[3] Institute for Data Science, Tsinghua University, Beijing, China

Abstract. With advances in data collection technologies, multiple data sources are assuming increasing prominence in many applications. Clustering from multiple data sources has emerged as a topic of critical significance in the data mining and machine learning community. Different data sources provide different levels of necessarily detailed knowledge. Thus, combining multiple data sources is pivotal to facilitate the clustering process. However, in reality, the data usually exhibits heterogeneity and incompleteness. The key challenge is how to effectively integrate information from multiple heterogeneous sources in the presence of missing data. Conventional methods mainly focus on clustering heterogeneous data with full information in all sources or at least one source without missing values. In this paper, we propose a more general framework T-MIC (**T**ensor based **M**ulti-source **I**ncomplete data **C**lustering) to integrate multiple incomplete data sources. Specifically, we first use the kernel matrices to form an initial tensor across all the multiple sources. Then we formulate a joint tensor factorization process with the sparsity constraint and use it to iteratively push the initial tensor towards a quality-driven exploration of the latent factors by taking into account missing data uncertainty. Finally, these factors serve as features to clustering. Extensive experiments on both synthetic and real datasets demonstrate that our proposed approach can effectively boost clustering performance, even with large amounts of missing data.

1 Introduction

Due to the recent advances of data collection technologies, many application fields are facing various data with complex multiple sources (*i.e.*, multi-source data), where different data sources provide different levels of necessarily detailed knowledge. For example, in a user-oriented recommendation system, data sources can be user profile database, users' log data, users' credit score and users' social connections (as shown in Fig. 1). Different from traditional data with a single source, these multi-source data commonly have the following properties:

1. Each data source can have its own feature sets. For example, users's credit score (Fig. 1c) has numerical features; the user's profile has mixed features

© Springer International Publishing Switzerland 2015
T. Cao et al. (Eds.): PAKDD 2015, Part II, LNAI 9078, pp. 485–497, 2015.
DOI: 10.1007/978-3-319-18032-8_38

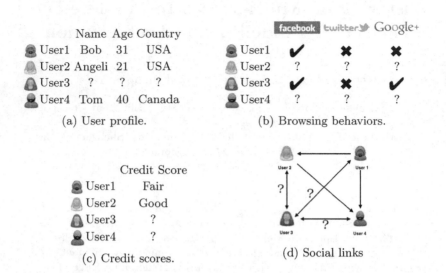

Fig. 1. Different sources for grouping the users in the recommendation systems

(numerical features and nominal features as shown in Fig. 1a) and the social relation (Fig. 1d) provides graph relational features. Such data is called heterogeneous data.

2. Each data source may be incomplete. For example, in Fig. 1a, User3 does not complete her profile. From Fig. 1b, we can see that User2 and User4 do not have browsing behavior history. In Fig. 1c, only User1 and User2 have credit scores. User3 does not have any social connection information in Fig. 1d.

3. Data can be with an arbitrary number of sources. In some applications, the number of available data sources may be small while in other applications, it may be quite large.

The above properties raise one fundamental challenge for data mining research: the complex structure and lack of consistent and complete representation of data. Multiple sources may provide complementary data, and multi-source data fusion can produce a better understanding of the observed situation by decreasing the uncertainty related to the individual sources. An effective model for multi-source data should be able to integrate or extract information from these multiple sources in order to perform analysis or management steps. Motivated by these challenges, multi-source mining, in particular multi-source clustering, have received considerable attention in the last decade.

In the literature, multi-source/multi-view clustering problem has been extensively studied [1,11,15,17]. Specifically, the majority of studies have focused on clustering heterogeneous data with an arbitrary number of sources. However, these studies either assumed that data have full information in all sources [11,15,17], or that there exists at least one source which is complete with no missing values [25]. Although the most advanced methods [14,21] do not require the completeness of each source, the application of these two methods is limited to two-source situation. Multi-source clustering is still challenging on multiple heterogeneous sources in the presence of missing data of each source.

A straightforward solution to this problem is that we can remove the data samples that suffer from missing information. However, this clearly contradicts with the target of clustering which aims at distributing all samples to their corresponding cluster. In this paper, we focus on clustering the multi-source incomplete data by handling the missing information with the help of tensor algebra, which is the extension of vector and matrix algebra. More specifically, we develop a tensor-based modeling and factorization framework, called T-MIC (**T**ensor based **M**ulti-source **I**ncomplete data **C**lustering). We first construct an initial tensor over the multi-source incomplete data based on kernel matrices. Then we formulate a joint tensor factorization process with the sparsity constraint and use it to iteratively push the initial tensor towards a quality-driven exploration of the latent factors by taking into account missing data uncertainty. This leads to obtain the common latent features across all the data sources, as well as a common latent feature space. We can then use any standard clustering algorithm for grouping the objects in this space. As compared with previous approaches, this paper has several advantages:

1. The proposed T-MIC framework can be used in situations even when all the sources are incomplete, in which the other methods are not applicable.
2. T-MIC uses kernel matrices to form the tensor across all multiple data sources. The use of kernel matrices makes it able to handle the heterogeneity of data, while the nature of tensor makes it easily extended to any arbitrary number of sources.
3. T-MIC produces a unified, non-redundant, consistent and complete set of data in the feature vector space, that can be fed into any standard clustering algorithm.

In order to evaluate the clustering results based on the proposed T-MIC, we conduct several experiments on synthetic data and real-world data. The proposed approach outperforms the comparison algorithms in all the scenarios in both normalized mutual information and average purity.

The rest of this paper is organized as follows: In the next section, we will describe the notation and background knowledge. In Section 3, we will describe the formulation of the problem and the proposed T-MIC framework. Experimental settings and result analysis are described in Section 4.

2   Notation and Background

Before proceeding, we introduce some related concepts and notation of tensors used throughout of the paper. For more details about tensor algebra, please refer to [7].

Tensors are higher-order arrays that generalize the notions of vectors (first-order tensors) and matrices (second-order tensors). The order of a tensor is the number of dimensions, also known as ways or modes. We denote tensors of order $N \geq 3$ by calligraphic letters ($\mathcal{X}$, $\mathcal{Y}$, $\mathcal{Z}$), matrices by boldface capital letters (**A**, **B**, **C**), vectors by boldface lowercase letters (**a**, **b**, **c**), scalars by lowercase letters (a, b, c). Columns of a matrix are denoted by boldface lower letters with a subscript, e.g., $\mathbf{a}_r$ is the rth column of matrix **A**. Entries of a matrix or a tensor are denoted by lowercase letters with subscripts, i.e., the $(i_1, i_2, \ldots, i_N)$ entry of an N-th order tensor $\mathcal{X}$ is denoted by $x_{i_1 i_2 \ldots i_N}$.

Given two equal-sized tensors $\mathcal{X}, \mathcal{Y} \in \mathbb{R}^{I_1 \times I_2 \times \cdots \times I_N}$, their Hadamard (elementwise) product is denoted by $\mathcal{X} * \mathcal{Y}$ and defined as

$$(\mathcal{X} * \mathcal{Y})_{i_1 i_2 \ldots i_N} = x_{i_1 i_2 \ldots i_N} y_{i_1 i_2 \ldots i_N}$$

for all values of the indices. Their inner product, denoted by $\langle \mathcal{X}, \mathcal{Y} \rangle$, is the sum of the products of their entries, i.e.,

$$\langle \mathcal{X}, \mathcal{Y} \rangle = \sum_{i_1=1}^{I_1} \sum_{i_2=1}^{I_2} \cdots \sum_{i_N=1}^{I_N} x_{i_1 i_2 \ldots i_N} y_{i_1 i_2 \ldots i_N.} \tag{1}$$

The Frobenius norm of a tensor $\mathcal{X} \in \mathbb{R}^{I_1 \times I_2 \times \cdots \times I_N}$ is defined as

$$\|\mathcal{X}\|_F = \sqrt{\langle \mathcal{X}, \mathcal{X} \rangle}.$$

Given a sequence of matrices $\mathbf{A}^{(n)} \in \mathbb{R}^{I_n \times R}$ for n=1, 2, ..., N, the notation $[\![\mathbf{A}^{(1)}, \mathbf{A}^{(2)}, \ldots, \mathbf{A}^{(N)}]\!]$ defines an N-th order tensor of size $I_1 \times I_2 \times \cdots \times I_N$ whose elements are given by

$$([\![\mathbf{A}^{(1)}, \mathbf{A}^{(2)}, \ldots, \mathbf{A}^{(N)}]\!])_{i_1 i_2 \ldots i_N} = \sum_{r=1}^{R} \prod_{n=1}^{N} a_{i_n r}^{(n)} \tag{2}$$

for $i_n \in \{1, 2, \ldots, I_n\}, n \in \{1, 2, \ldots, N\}$.

The outer product of two vectors $\mathbf{a} \in \mathbb{R}^{I_1}$ and $\mathbf{b} \in \mathbb{R}^{I_2}$, denoted by $\mathbf{a} \circ \mathbf{b}$, represents a matrix of size $I_1 \times I_2$ with the elements $(\mathbf{a} \circ \mathbf{b})_{ij} = a_i b_j$.

It is worth noting that mathematically a tensor is an element of the outer product of vector spaces, each of which has its own coordinate system. The outer product of vector spaces forms an elegant algebraic structure for the theory of tensors. Such structure endows the tensor with the inherent advantage in representing real-world data, which naturally results from the interactions of multiple factors or multiple sources.

The two most commonly used factorizations are the Tucker model and CAN-DECOMP/PARAFAC (CP) model, both of which can be regarded as higher-order generalizations of the matrix singular value decomposition (SVD). The CP model factorizes a tensor into a sum of rank-1 tensors, where each rank-1 tensor is the outer product of vector loadings in all modes, whereas in Tucker variants the factor interactions are modelled via a core tensor. This rank-1 component factorization of CP and its intrinsic axis property from parallel proportional profiles [4], along with uniqueness of solutions [9], gives it a very strong interpretive power. The Tucker model is more flexible, though, the complex interactions and non-uniqueness of solutions make its interpretation more difficult. Therefore, we adapt an underlying CP factorization for our model.

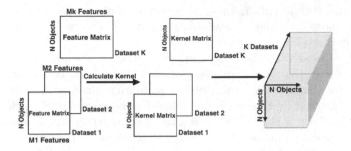

Fig. 2. The tensor construction

Formally, given an N-th order tensor $\mathcal{X} \in \mathbb{R}^{I_1 \times I_2 \times \cdots \times I_N}$, the CP factorization [7] is defined by factor matrices $\mathbf{A}^{(n)} \in \mathbb{R}^{I_n \times R}$ for $n = 1, 2, \ldots, N$, respectively, such that

$$\mathcal{X} = \sum_{r=1}^{R} \mathbf{a}_r^{(1)} \circ \mathbf{a}_r^{(2)} \circ \cdots \circ \mathbf{a}_r^{(N)} = [\![\mathbf{A}^{(1)}, \mathbf{A}^{(2)}, \ldots, \mathbf{A}^{(N)}]\!], \tag{3}$$

where R is the rank of the tensor $\mathcal{X}$, defined as the smallest number of rank-1 tensors in an exact CP factorization. The factor matrices $\mathbf{A}^{(1)}, \mathbf{A}^{(2)}, \ldots, \mathbf{A}^{(N)}$ can be viewed as the common latent feature matrices in different modes.

3 T-MIC framework

In this section, we discuss our approach T-MIC for multi-source incomplete clustering. The key challenge is how to effectively integrate information from multiple heterogeneous sources in the presence of missing data. Here we identify three broad goals: incorporating heterogeneous sources, extracting latent features, handling the incompleteness. In each of the following three subsections, we try to achieve one of the above goals respectively. Before we give more detailed discussions for each part, we first formulate the problem as below.

Given N distinct objects with K $(K > 2)$ related data sources, their object-feature matrices are denoted as $\mathbf{X}_1, \mathbf{X}_2, \ldots, \mathbf{X}_K$. We assume that none of the related data sources is complete. The goal is to derive a clustering solution based on all the incomplete sources.

3.1 Tensor Construction

Tensor has been used to model multiple data sources for many years. There are several ways to construct a tensor from multiple data sources. One natural way is stacking the object-feature matrices derived from multiple data sources in a tensor. This construction requires that all the object-feature matrices has the same dimension. However, in real-world problems, the data sources usually exhibit heterogeneous properties, $i.e.$, the dimension of the feature space for different source is different. Taking the recommendation system in Section 1 as an example, the dimensions of available sources are different. Just stacking the object-feature matrices cannot form a tensor.

In order to deal with the heterogeneity of the data sources, we construct a tensor in a way that is independent of data dimension. In [16], a tensor is constructed by stacking the object-object similarity matrices derived from all the sources. Inspired by this, we choose the homogeneous kernel as the similarity measure to form a shared latent tensor. Assume there are K data sources with different feature dimensions available. Each of the sources has the same N objects. Although the feature dimensions of each source are different, the kernel matrices are all with dimension of $N \times N$. A tensor can be constructed by stacking the K kernel matrices together as shown in Fig. 2. However, in our problem, all the sources are incomplete. Thus, we fill in the missing entries with average value or majority value. We denote the feature matrices with missing value filled as $\hat{\mathbf{X}}_1, \hat{\mathbf{X}}_2, ..., \hat{\mathbf{X}}_K$. The initial kernel matrices $\hat{\mathbf{S}}_1, \hat{\mathbf{S}}_2, ..., \hat{\mathbf{S}}_n \in \mathbb{R}^{N \times N}$ can be calculated and form an initial tensor $\mathcal{T} \in \mathbb{R}^{N \times N \times K}$ in the way as shown in Fig. 2.

3.2 Latent Features Extraction

Using the tensor construction approach described in Section 3.1, we can get an initial tensor $\mathcal{T}$. In order to get the common factor matrices across all the sources, we need to factorize the tensor. As mentioned above, we use the CP factorization model to achieve this goal. In order to impose the sparsity constraints on the latent factor matrices to get more concise latent factors, we add l_1 penalty on the latent factor matrices $\mathbf{A}$ and $\mathbf{B}$ to the objective function. The objective function for CP factorization with sparsity constraints can be stated as:

$$\min_{\mathbf{A},\mathbf{B},\mathbf{C}} \|\mathcal{T} - [\![\mathbf{A}, \mathbf{B}, \mathbf{C}]\!]\|_F^2 + \lambda_\alpha \|\mathbf{A}\|_1 + \lambda_\beta \|\mathbf{B}\|_1, \tag{4}$$

where $\mathbf{A} \in \mathbb{R}^{N \times R}$, $\mathbf{B} \in \mathbb{R}^{N \times R}$, $\mathbf{C} \in \mathbb{R}^{K \times R}$ are the latent factor matrices of $\mathcal{T}$, λ_α and λ_β are sparsity tradeoff parameters for latent factors $\mathbf{A}$ and $\mathbf{B}$ and $\|.\|_1$ is the l_1 norm. Here, $\mathbf{A}$ and $\mathbf{B}$ can be interpreted as the common latent factor across all the data sources. Note that the kernel matrix is symmetric, this will lead to the equality of $\mathbf{A}$ and $\mathbf{B}$. Thus, we can apply any standard clustering algorithm (*i.e.*, k-means) on $\mathbf{A}$ to get a clustering solution.

There exists quite a few algorithms to solve the optimization problems in Equation (4) [13,20]. We will not discuss the details of the optimization in this paper. In the following, We will illustrate how to use the optimization to refine the initial tensor $\mathcal{T}$ and learn the common latent factor simultaneously.

3.3 Quality-Driven Integration

By solving the optimization problem in Equation (4), we can get an "optimal" common factor $\mathbf{A}$ and the estimated tensor $\hat{\mathcal{T}}$:

$$\hat{\mathcal{T}} = [\![\mathbf{A}, \mathbf{B}, \mathbf{C}]\!]. \tag{5}$$

Algorithm 1.T-MIC for Clustering Multi-source Incomplete Data

Input: Incomplete sources $\mathbf{X}_1, ..., \mathbf{X}_K$, cluster number C, parameters $\lambda_\alpha, \lambda_\beta$ and α
Output: The clustering index
  1: Give initial values to the missing features in the K data sources.
  2: Calculate the kernel matrices $\mathbf{S}_1, \mathbf{S}_2, ..., \mathbf{S}_K$.
  3: Stack the kernel matrices to construct tensor $\mathcal{T}$.
  4: Calculate the weight tensor $\mathcal{W}$ according to Equation (6).
  5: **repeat**
  6:     Solve optimization problem in Equation (4).
  7:     Calculate the current estimated tensor by Equation (5).
  8:     Update the tensor by Equation (7).
  9:     Update the weight tensor by Equation (8).
10: **until** Convergence
11: Apply k-means to the common latent feature matrix $\mathbf{A}$ to get clustering index.

However, the "optimal" common factor $\mathbf{A}$ may not be exact because of the inaccuracy of the initial tensor $\mathcal{T}$. We use an iterative way to get a better result by taking into account the incompleteness of data sources. we define a weight tensor $\mathcal{W} \in \mathbb{R}^{N \times N \times K}$:

$$w_{ijk} = \begin{cases} 0 & \text{if } (\mathbf{X}_k)_{i,:} \text{ or } (\mathbf{X}_k)_{:,j} \text{ is a missing entry,} \\ 1 & \text{otherwise.} \end{cases} \tag{6}$$

Note that the weight tensor $\mathcal{W}$ represents the reliability of the entries in tensor $\mathcal{T}$. We want to retain the most reliable entries in tensor $\mathcal{T}$ and update those that are not reliable. Thus, we can use $\mathcal{W}$ to update the original tensor $\mathcal{T}$:

$$\mathcal{T} \leftarrow \mathcal{W} * \mathcal{T} + (\mathbf{1} - \mathcal{W}) * \hat{\mathcal{T}}, \tag{7}$$

where $*$ is Hadamard (elementwise) product and $\mathbf{1}$ is the tensor with all ones. After $\mathcal{T}$ is updated, $\mathbf{A}, \mathbf{B}$ and $\mathbf{C}$ can be updated using Equation (4). This procedure will repeat until convergence. The learned optimal common factor $\mathbf{A}$ can be used by any standard clustering algorithm (*i.e.*, k-means).

As the iteration continues, we can start to give the missing value more weight (the initial weight is 0 according to Equation (6)). We update the weight tensor at the end of each iteration as follows:

$$\mathcal{W} \leftarrow \mathbf{1} - (\alpha \times (\mathbf{1} - \mathcal{W})), \tag{8}$$

where α is a decay parameter. We set α to be 0.95 for all experiments. From Equation (8), we can see that W will converge to $\mathbf{1}$ as the number of iterations goes up. Thus, the whole framework will converge. The framework is shown in Algorithm 1.

4   Experimental Evaluation

In this section, we validate the effectiveness of the proposed T-MIC framework for multi-source incomplete clustering on both synthetic and real-world data.

4.1   Baselines and Metrics

We compare the proposed T-MIC framework with a number of baselines. In particular, we compare with:

- **Feature Concatenation (Concat)**: Concatenating the features of each view, and then running standard k-means clustering (after dimension reduction, if needed).
- **Canonical Correlation Analysis based Feature Extraction (CCA)**: Applying CCA for feature fusion from multiple views of the data [6], and then running standard k-means clustering using these extracted features.
- **Co-regularized Spectral Clustering (Co-regSC)**: Using co-regularized framework to spectral clustering [11]. We choose Gaussian kernel to build the affinity matrix for each view and set the parameter in this algorithm to be 0.01 as suggested.
- **Collective Kernel Learning (CoKL-KCCA)**: Collectively learn the kernel matrices for each source and applying kernel canonical correlation analysis [21]. This algorithm does not require the completeness of the data sources. However, CoKL-KCCA can only be applied to two sources. It also assumes two sources can cover all the instances, which may not be true in many real-world applications. This assumption makes it difficult to extend CoKL-KCCA to more than two sources.
- **Tensor based Multi-source Incomplete data Clustering (T-MIC)**: This is the proposed framework. In the experiments, we empirically set $\lambda_\alpha = \lambda_\beta = 1$. Considering the fact that there is no known closed-form solution to determine the rank R of a tensor a priori, and rank determination of a tensor is still an open problem [23], we set R equal to the number of clusters C. In addition, Gaussian kernel is used and the width parameter is set as the median pairwise distance between instances.

Note that Concat, CCA and Co-regSC only work for complete data sources. We first fill the incomplete sources with average values, then we apply these approaches.

The clustering result is evaluated by comparing the clustering index of each data point with the label provided by the data sources. Two metrics, the average purity (Purity) and the normalized mutual information (NMI) are used to measure the clustering performance. Since k-means is sensitive to initial seed selection, we run k-means 30 times on each parameter setting, and report the averaged NMI and purity. All the data we used in the experiments are complete, but we randomly delete some of the instances in each data source. The incomplete percentage for each source is equal in the experiments. To generate 30% incomplete data, we randomly delete 30% of the instances from every sources. We compare the clustering results for different incomplete percentage (from 5% to 50%).

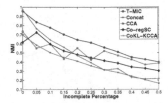

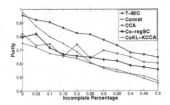

Fig. 3. NMIs for Synthetic

Fig. 4. Purities for Synthetic

4.2 Dataset Collection

One synthetic and two real-world data are used in the experiments.

- **Synthetic data (Synthetic):** The synthetic data consists of two sources for generated data in a two dimensional space. Each source contains samples randomly selected from three different distributions. In either source, two of the three randomly generated distributions are made to be very close, and therefore difficult to distinguish. In this experiment, we randomly select 500 samples from each of the three different distributions. Each source contains 1500 samples.
- **Internet Advertisements (ADs):** This data represents a set of possible advertisements on Internet pages [12]. 772 instances are randomly selected in a balanced manner from two classes. Two sources are considered in our experiments. The first source is the size of this advertisement (4 continuous features) and the second source is the url terms (457 binary features).
- **Handwritten Dutch Digit Recognition (Digit):** This data contains 2000 handwritten numerals ("0"-"9") extracted from a collection of Dutch utility maps [5]. The following feature spaces (sources) with different vector-based features are available for the numbers: (1) 76 Fourier coefficients of the character shapes, (2) 216 profile correlations, (3) 64 Karhunen-Love coefficients, (4) 240 pixel averages in 2×3 windows, (5) 47 Zernike moments, and (6) 6 morphological features. All these features are conventional vector-based features but in different feature spaces.

4.3 Experimental Results

Fig. 3 - Fig. 8 demonstrate the clustering performance of different methods on all the three data. In order to randomize the experiments, 30 test runs on every parameter settings were conducted and the average performance are reported. From the results, we can observe that as the incomplete percentage goes up, the clustering performance goes down. This is because when the incomplete percentage gets larger, the amount of useful information remained in the incomplete sources gets less. However, we can still observe that T-MIC outperforms all the other four comparison methods for all the settings.

On the synthetic data, T-MIC outperforms the best baseline method with a margin as large as 17% in NMI and 16.8 % in average purity. We can also observe

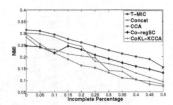

Fig. 5. NMIs for Ads

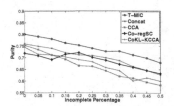

Fig. 6. Purities for Ads

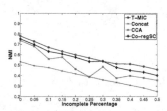

Fig. 7. NMIs for Digit

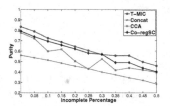

Fig. 8. Purities for Digit

from Fig. 3 and Fig. 4, that when the incomplete percentage is 0 (*i.e.*, both sources are complete), CoKL-KCCA is slightly better than T-MIC. However, as the incomplete percentage increases, the performance of CoKL-KCCA gets worse. It is also worth mentioning that when the incomplete percentage is large (30%-50%), CCA becomes the best baseline (closest to T-MIC).

On the Internet advertisements data, we also observe that T-MIC achieves the best performance in both NMI and average purity compared with the baselines. One interesting observation from Fig. 5 and Fig. 6 is that Concat performs better than other baselines in many situations. The reason behind this may be that the data itself does not contain too much useful information (with complete sources, we only get around 0.3 for NMI). For a less informative data, the efforts to find latent spectral feature (Co-regSC) or maximize the correlation (CCA and CoKL-CCA) may lead to worse performance.

For the handwritten digit data, we cannot apply CoKL-KCCA since it can only be applied to two-source incomplete data. From Fig. 7 and Fig. 8, we can observe that T-MIC still achieves the best performance in NMI and average purity compared with the baselines. It is also worth mentioning that Co-regSC remains the best baseline method for almost every settings, while for Synthetic data and Ads data, Co-regSC is not that good. The reason why Co-regSC is better than other baselines is that Co-regSC co-regularizes the latent feature matrices for all the sources towards a common consensus while the other two baselines don't. Thus, for a complex task (more sources and more clusters) Co-regSC can get better performance. However, T-MIC uses tensor to model the multi-source incomplete data, which can better capture the relationship between sources. So the iteratively tensor factorization with sparsity constraint can learn the common latent features more accurately.

5   Related Work

Our work is related to both multi-view learning and data fusion techniques. We briefly discuss both of them. Multi-view learning [3,8,18,22], is proposed to learn from instances which have multiple representations in different feature spaces. Specifically, Multi-view clustering [1,17] is most related to our work. For example, [1] developed and studied partitioning and agglomerative, hierarchical multi-view clustering algorithms for text data. [10,11] are among the first works proposed to solve the multi-view clustering problem via spectral projection. [17] proposed a novel approach to use mapping function to make the clusters from different pattern spaces comparable and hence an optimal cluster can be learned from the multiple patterns of multiple views. Linked Matrix Factorization [24] is proposed to explore clustering of a set of entities given multiple graphs. [15] used nonnegavite matrix factorization to integrate multi-view data. Recently, [25] proposed a kernel based approach which allows clustering algorithms to be applicable when there exists at least one complete view with no missing data. As far as we know, [14,21] are the only two works that do not require the completeness of any view.

Data fusion from multiple sources have been investigated by many researchers. The goal of such approaches is to integrate information from multiple data sources to create a single representation that are more complete and accurate than those derived from any individual data source, *i.e.*, the whole is greater than the sum of its parts. So far, many integration architectures have been proposed, a thorough survey of these techniques can be found in [2]. Here we only focus on tensor based data fusion architecture. The work most closed to our method are [16] and [19]. In [16], a clustering method based on Tucker decomposition was proposed to integrate multiple view data without the sparsity constraint. In [19], a multi-view clustering method based on CP factorization and sparsity constraint was proposed for multi-view graph data. However, all of the above mentioned methods can only deal with complete multi-source data.

There are some differences between our work and the previous approaches. First, to the best of our knowledge, we take the first step toward studying the problem of multi-source clustering, in which each data source suffers from missing information and data. Second, T-MIC has no constraints on the feature space of the datasets. T-MIC constructs a tensor by only using the kernel matrices, which is independent of the types/dimensions of the data sources. Third but not last, T-MIC does not have limitation for the number of available data sources. The high order/dimension nature of tensor makes it suitable for modeling multi-source problems. Clustering via T-MIC framework can be easily extended to K ($K > 2$) data sources.

6   Conclusions

In this paper, we study the problem of clustering on multi-source incomplete data. We proposed a tensor-based modeling and factorization framework,

T-MIC, to integrate multiple incomplete data sources. We first construct an initial tensor from the multi-source incomplete data using kernel matrices. Then a joint tensor factorization process with sparsity constraint is applied to iteratively factorize the tensor. A common latent feature space across all the sources can be learned simultaneously. The experiments on both synthetic and real-world data were conducted to evaluate the clustering performance. It can be clearly observed that the proposed approach outperforms the comparison algorithms in almost every scenario.

Acknowledgments. This work is supported in part by NSF through grants CNS-1115234, and OISE-1129076, Google Research Award, the Pinnacle Lab at Singapore Management University, the NSFC 61273295 and 61472089 and the Science and Technology Plan Project of Guangzhou City 2013Y2-00034.

References

1. Bickel, S., Scheffer, T.: Multi-view clustering. In: ICDM, pp. 19–26 (2004)
2. Bleiholder, J., Naumann, F.: Data fusion. ACM Comput. Surv. **41**(1), 1–41 (2009)
3. Blum, A., Mitchell, T.: Combining labeled and unlabeled data with co-training. In: COLT, New York, NY, USA, pp. 92–100 (1998)
4. Cattell, R.B.: Parallel proportional profiles and other principles for determining the choice of factors by rotation. Psychometrika **9**(4), 267–283 (1944)
5. Duin, R.P.: Handwritten-Numerals-Dataset
6. Kettenring, J.R.: Canonical Analysis of Several Sets of Variables. Biometrika **58**(3), 433–451 (1971)
7. Kolda, T.G., Bader, B.W.: Tensor Decompositions and Applications. SIAM REVIEW **51**, 455–500 (2009)
8. Kriegel, H.P., Kunath, P.,Pryakhin, A., Schubert, M.: MUSE: multi-represented similarity estimation. In: ICDE, pp. 1340–1342. IEEE Computer Society, Washington (2008)
9. Kruskal, J.B.: Three-way arrays: rank and uniqueness of trilinear decompositions, with application to arithmetic complexity and statistics. Linear Algebra and its Applications **18**(2), 95–138 (1977)
10. Kumar, A., Daume III, H.: A co-training approach for multi-view spectral clustering. In: ICML, New York, NY, USA, pp. 393–400, June 2011
11. Kumar, A., Rai, P., Daume III, H.: Co-regularized multi-view spectral clustering. In: NIPS, pp. 1413–1421 (2011)
12. Kushmerick, N.: Learning to Remove Internet Advertisements, pp. 175–181. ACM Press (1999)
13. De Lathauwer, L., De Moor, B., Vandewalle, J.: On the Best Rank-1 and Rank-(R1, R2, RN) Approximation of Higher-Order Tensors. SIAM J. Matrix Anal. Appl. **21**(4), 1324–1342 (2000)
14. Li, S., Jiang, Y., Zhou, Z.: Partial Multi-View Clustering (2014)
15. Liu, J., Wang, C., Gao, J., Han, J.: Multi-view clustering via joint nonnegative matrix factorization. In: SDM (2013)
16. Liu, X., Ji, S., Glanzel, W., De Moor, B.: Multiview Partitioning via Tensor Methods. IEEE Trans. Knowl. Data Eng. **25**(5), 1056–1069 (2013)
17. Long, B., Yu, P.S., Zhang, Z.M.: A general model for multiple view unsupervised learning. In: SDM, pp. 822–833 (2008)

18. Nigam, K., Ghani, R.: Analyzing the effectiveness and applicability of co-training. In: CIKM, pp. 86–93. ACM, New York (2000)
19. Papalexakis, E.E., Akoglu, L., Ience, D.: Do more views of a graph help? community detection and clustering in multi-graphs. In: FUSION, pp. 899–905. IEEE (2013)
20. Papalexakis, E.E., Sidiropoulos, N.D.: Co-clustering as multilinear decomposition with sparse latent factors. In: ICASSP, pp. 2064–2067. IEEE (2011)
21. Shao, W., Shi, X., Yu, P.S.: Clustering on multiple incomplete datasets via collective kernel learning. In: ICDM, pp. 1181–1186 (2013)
22. Shi, X., Paiement, J., Grangier, D., Yu, P.S.: Learning from heterogeneous sources via gradient boosting consensus. In: SDM (2012)
23. Silva, V., Lim, L.-H.: Tensor rank and the ill-posedness of the best low-rank approximation problem. SIAM J. Matrix Anal. Appl. **30**(3), 1084–1127 (2008)
24. Tang, W., Lu, Z., Dhillon, I.S.: Clustering with multiple graphs. In: ICDM, Miami, Florida, USA, pp. 1016–1021, December 2009
25. Trivedi, A., Rai, P., Daumé III, H., DuVall, S.L.: Multiview clustering with incomplete views. In: NIPS Workshop, Whistler, Canada (2010)

Locally Optimized Hashing for Nearest Neighbor Search

Seiya Tokui[1]([⊠]), Issei Sato[2], and Hiroshi Nakagawa[2]

[1] Preferred Networks, Inc., Tokyo, Japan
tokui@preferred.jp
[2] The University of Tokyo, Tokyo, Japan
sato@r.dl.itc.u-tokyo.ac.jp, nakagawa@dl.itc.u-tokyo.ac.jp

Abstract. Fast nearest neighbor search (NNS) is becoming important to utilize massive data. Recent work shows that hash learning is effective for NNS in terms of computational time and space. Existing hash learning methods try to convert neighboring samples to similar binary codes, and their hash functions are globally optimized on the data manifold. However, such hash functions often have low resolution of binary codes; each bucket, a set of samples with same binary code, may contain a large number of samples in these methods, which makes it infeasible to obtain the nearest neighbors of given query with high precision. As a result, existing methods require long binary codes for precise NNS. In this paper, we propose *Locally Optimized Hashing* to overcome this drawback, which explicitly partitions each bucket by solving optimization problem based on that of Spectral Hashing with stronger constraints. Our method outperforms existing methods in image and document datasets in terms of quality of both the hash table and query, especially when the code length is short.

Keywords: Similarity search · Nearest neighbor search · Hashing

1 Introduction

Nearest neighbor search (NNS) is a fundamental task with various applications in machine learning, data mining and information retrieval. Exact search is done by linear scan through all data for each query, which costs prohibitive time and space when the dataset is large and cannot fit into the main mamory. Consequently, we should approximately find the nearest neighbors (NNs) of the query in short time with less memory.

Binary hashing is an approximate method for NNS which meets the time and space requirements. It consists of r hash functions which translates each datum as an r-bit code preserving the local structure of the data. We call a set of data with same binary code a *bucket*. The NNs of a novel query are approximately found by searching buckets corresponding to binary codes similar to that of the query. Hamming distance, for instance, is used as similarity measure of binary codes, because its computation is far more efficient than similarity calculation

© Springer International Publishing Switzerland 2015
T. Cao et al. (Eds.): PAKDD 2015, Part II, LNAI 9078, pp. 498–509, 2015.
DOI: 10.1007/978-3-319-18032-8_39

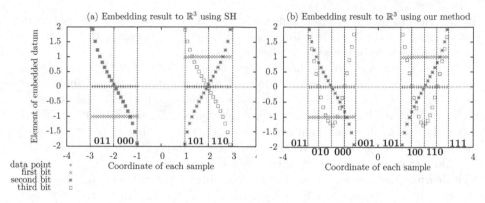

Fig. 1. Embedding results of one dimensional artificial data (blue points) into $\mathbb{R}^3$ using (a) Spectral Hashing and (b) our method. Positive and negative elements are binarized to 1 and -1, respectively. Each dashed box represents a bucket. In (a), data are only divided into four buckets, while three-bit binary codes could represent eight buckets. On the other hand, data are divided into eight small buckets using our method in (b).

of the original data pair. If each binary code is more compact compared to the original datum, hashing enables NNS of a large amount of data on memory.

Hashing methods are categorized into data independent ones and data dependent ones. Data independent methods such as Locality Sensitive Hashing [2] and b-Bit Minwise Hashing [5] simulate the objective similarity measure by randomized algorithms. Their approximation quality is theoretically guaranteed, though they practically require very long binary codes. On the other hand, hash functions optimized to given data achieve high search precision with shorter binary codes. This approach is called *learning to hash*.

Spectral Hashing (SH) [11] is a hash learning method to generate binary codes globally optimized over the data manifold. In SH, the data are first embedded into $\mathbb{R}^r$ by dimensionality reduction, and then binarized to r-bit codes using the sign of each element. The embedding is learned to be smooth, i.e. it converts neighboring samples to similar r-vectors. Most hash learning methods other than SH are also designed to convert neighboring samples to similar binary codes.

SH has a problem that its hash functions may produce a *large* bucket, which contains a large number of samples. To verify this statement, we show a toy example of one dimensional artificial data illustrated in Fig. 1(a). x axis is coordinate of data, and y axis is each element of data embedded into $\mathbb{R}^3$. A positive element is binarized to 1 and negative one to -1. Here the third bit does not divide any buckets, which causes each bucket to be large.

Once the dataset is converted to a set of buckets, we cannot distinguish samples in a same bucket. Therefore, large buckets degrade the precision of NNS. For instance, in the above toy example, the third bit does not improve search precision within top-ranked samples. Since three-bit codes potentially have ability to divide data into eight buckets, this result suggests SH may be inefficient in terms of search precision. Existing hash learning methods [11][12][6] are designed to make hash functions smooth on the data manifold, while they are not constrained to divide neighbors of each example. Therefore, these methods may potentially have the same inefficiency as SH.

In this paper, we propose *Locally Optimized Hashing* (LOH), formulated as an optimization problem based on that of SH with stricter constraints which force each bit to divide buckets. For instance, the toy data described above are embedded by LOH as Fig. 1(b), where it generates eight small buckets. LOH is solved bit-by-bit in linear time and space of data size using a sparse similarity matrix [12] or Anchor Graph as same as Anchor Graph Hashing [6].

The rest of the paper is organized as follows. In Sect. 2, we review existing hash learning methods and explain the formulation of SH and its extensions. In Sect. 3, we formulate LOH and propose its algorithm. In Sect. 4, we show experimental results. In Sect. 5, we make conclusion.

2   Related Work

Hashing methods proposed in early stage of study such as Locality Sensitive Hashing [2] (LSH) are randomized algorithms that simulate some fixed similarity measure. b-Bit Minwise Hashing [5] also approximates fixed similarity measure, Jaccard coefficients, by randomized way. Shift Invariant Kernel Hashing [8] and Kernelized LSH [3] are data dependent randomized methods to simulate kernel on the given data. These randomized methods are asymptotically guaranteed to be precise, which though require very long binary codes to be usable.

Binary codes optimized to data can be compact and accurate. Therefore, hash learning methods have recently been attracting much attention. Hash functions of such methods convert neighboring samples in the dataset to similar binary codes. For instance, in [9], hash functions are constructed one-by-one using pairwise penalty determined by the previous ones, while [7] proposes a method to learn hash functions using pairwise "near" and "far" labels.

Spectral Hashing [11] (SH) is one of the hash learning methods. Here we describe its formulation and extensions related to our methods. Let $A \in \mathbb{R}^{n \times n}$ be a similarity matrix of n data points $x^{(1)}, \ldots, x^{(n)}$. One converts the data to $y^{(1)}, \ldots, y^{(n)} \in \mathbb{R}^r$ and then thresholds their positive and negative elements to 1 and -1, respectively. We denote entire embedding of n samples by $Y = (y_1, \ldots, y_r) \in \mathbb{R}^{n \times r}$, where $y_j \in \mathbb{R}^n$ is the j-th elements through all samples. The binary codes of SH are obtained by solving the problem

$$\min_{Y} \frac{1}{2} \sum_{ij} \left\| y^{(i)} - y^{(j)} \right\|^2 A_{ij} = \mathrm{tr}(Y^\top L Y) \quad \text{s.t. } Y^\top \mathbf{1} = 0, \ Y^\top Y = nI, \quad (1)$$

where $\mathbf{1} = (1, \ldots, 1)^\top \in \mathbb{R}^n$ and $L = D - A$ ($D = \mathrm{diag}(A\mathbf{1})$) is the graph Laplacian of a graph whose weight matrix is A. The objective function is designed to penalize a pair of neighboring samples whose corresponding codes are far apart, which makes the embedding smooth. *Balance constraints* $Y^\top \mathbf{1} = 0$ maximize the information of each bit and off-diagonal elements of $Y^\top Y = nI$ called *orthogonality constraints* maximize the "mutual" information of each pair of bits. The constraints $Y^\top Y = nI$ also contain normalization constraints $\|y_j\|^2 = n$, without which the optimum degenerates to 0. The global optimum of (1) is $Y = \sqrt{n}(u_1, \ldots, u_r)$, where $u_0 = \mathbf{1}, u_1, \ldots, u_{n-1}$ are normalized eigenvectors of L whose corresponding eigenvalues are $0 = \sigma_0 \le \sigma_1 \le \cdots \le \sigma_{n-1}$.

One can obtain binary codes of given data by (1), while it cannot be directly used to obtain corresponding hash functions. [11] solves (1) in the limit of $n \to \infty$ under the assumption that the data are uniformly distributed on the hyperrectangle whose axes are aligned to the principal components of the data, which is unrealistic in most cases. Alternatively, [12] proposes a two-step framework called Self-Taught Hashing (STH) to obtain hash functions, where r Support Vector Machines are learned to obtain r hash functions using each bit of n codes as a supervisor. We use this approach in our method.

Similarity matrix A, in SH, is the gram matrix with heat kernel $k(x^{(i)}, x^{(j)}) = \exp(-\|x^{(i)} - x^{(j)}\|^2 / t)$, constructing which is though intractable on dataset with very large n. [12] uses a sparse symmetric matrix using the undirected k-NN graph. Eigenvectors corresponding to the smallest eigenvalues of a large sparse matrix can be efficiently computed by Lanczos method with implicit restart [10]. [6] uses another efficient similarity matrix constructed as follows. First, one placed anchors $a_1, \ldots, a_m$ on data space by k-means algorithm. Then, sample-anchor similarity matrix $Z \in \mathbb{R}^{n \times m}$ is computed as

$$Z_{ij} \propto \begin{cases} \exp(-\|x^{(i)} - a_j\|^2 / t) & \text{if } a_j \text{ is one of the } s \text{ nearest anchors of } x^{(i)}, \\ 0 & \text{otherwise,} \end{cases}$$

where each row of Z is normalized to be a stochastic vector. s is chosen to be much smaller than m. The similarity matrix is formulated as $A = Z\Lambda^{-1}Z$, where $\Lambda = \mathrm{diag}(Z^\top \mathbf{1})$ is the column-wise normalizer of Z. The rank of A is at most m, which enables us to efficiently compute its eigenvectors.

As we saw in the previous section, the optimum of (1) may contain a small number of large buckets, which are suboptimal in terms of search precision. We will introduce stronger constraints to overcome this drawback in the next section.

3   Locally Optimized Hashing

We define novel constraints to explicitly divide the buckets in this section. They are transformed into symmetric form which induces a unified optimization problem. We also introduce an algorithm to solve our problem.

3.1   Buckets Partitioning Constraints

We start by analyzing a simple toy example of eight samples indexed by $1, \ldots, 8$ with three-bit codes. Suppose the first two bits are already fixed as

$$\begin{pmatrix} y_1^\top \\ y_2^\top \end{pmatrix} = \begin{pmatrix} 1 & 1 & 1 & 1 & -1 & -1 & -1 & -1 \\ 1 & 1 & -1 & -1 & 1 & 1 & -1 & -1 \end{pmatrix}.$$

Eight samples are currently divided into four buckets, $\{1, 2\}, \{3, 4\}, \{5, 6\}$ and $\{7, 8\}$. Note that balance constraints and orthogonality constraints of (1) are not enough to divide these buckets; e.g. the third bit $(1, 1, -1, -1, -1, -1, 1, 1)^\top$ satisfies them, though it does not divide any buckets. On the other hand, $(1, -1, 1, -1, 1, -1, 1, -1)^\top$ does divide all the buckets. Therefore, we should constrain y_3 to explicitly divide each bucket. It means y_3 should be chosen to be orthogonal to the normalized indicator vectors of buckets, which we call *bucket vectors* of $\{y_1, y_2\}$. We define *bucket matrix* of $\{y_1, y_2\}$ as a matrix which has all bucket vectors as its columns. In the current example, the bucket matrix is written as

$$\frac{1}{\sqrt{2}} \begin{pmatrix} 1 & 1 & 0 & 0 & 0 & 0 & 0 & 0 \\ 0 & 0 & 1 & 1 & 0 & 0 & 0 & 0 \\ 0 & 0 & 0 & 0 & 1 & 1 & 0 & 0 \\ 0 & 0 & 0 & 0 & 0 & 0 & 1 & 1 \end{pmatrix}^\top .$$

The buckets partitioning constraints can be expressed in cases of general code length r and data size n. Suppose $y_1, \ldots, y_{r-1}$ are fixed, and let B be a matrix whose columns are the normalized indicator vectors of current buckets. We call B a *bucket matrix* of $y_1, \ldots, y_{r-1}$, and its columns *bucket vectors* of them. Then the constraints $B^\top y_r = 0$ force the r-th bit to partition current buckets.

3.2 Combinatorial Orthogonality Constraints

Our method is based on the buckets partitioning constraints, while it is not obvious whether a corresponding unified optimization problem exists. We transform them into symmetric form of $y_1, \ldots, y_r$. It shows that our strict constraints can be seen as an extension of the balance and orghotonality constraints (1).

Here we start by above toy example again. The whole constraints are $\mathbf{1}^\top y_1 = \mathbf{1}^\top y_2 = y_1^\top y_2 = B^\top y_3 = 0$. Note that the subspace spanned by $\{\mathbf{1}, y_1, y_2, y_1 \odot y_2\}$ equals to the image of B, where $\odot$ is the elementwise product operator. Therefore, the constraints can be rewritten as

$$\mathbf{1}^\top y_1 = \mathbf{1}^\top y_2 = \mathbf{1}^\top y_3 = y_1^\top y_2 = y_1^\top y_3 = y_2^\top y_3 = (y_1 \odot y_2)^\top y_3 = 0. \quad (2)$$

Since inner product $a^\top b$ is also written as $\mathbf{1}^\top (a \odot b)$, (2) is farther rewritten as

$$\forall J \in \mathcal{P}(3) \setminus \{\emptyset\}, \ \mathbf{1}^\top \bigodot_{j \in J} y_j = 0,$$

where $\mathcal{P}(S)$ is the power set of a set S and $[r] = \{1, \ldots, r\}$ for integer r. The constraints are now expressed in symmetric form of y_1, y_2 and y_3, which means that y_1 and y_2 also divide buckets consisting of other two bits.

This symmetrization is easily generalized to arbitrary code length as follows.

Theorem 1. *Suppose $r \geq 2$, $\mathbf{1}^\top y_1 = 0$ and $y_1, \ldots, y_r \in \{1, -1\}^n$. Then y_j are orthogonal to bucket vectors of $\{y_1, \ldots, y_{j-1}\}$ for all $j \in \{2, \ldots, r\}$ if and* *only if* $y_1, \ldots, y_r$ *satisfies following conditions* (**Combinatorial Orthogonality Constraints, COC**).

$$\forall J \in \mathcal{P}([r]) \setminus \{\emptyset\}, \; \mathbf{1}^\top \bigodot_{j \in J} y_j = 0. \tag{3}$$

Proof. We prove it by induction. The case of $r = 2$ is trivial. Suppose $r > 2$. Buckets partitioning constraints of $y_2, \ldots, y_{r-1}$ and COC of $y_1, \ldots, y_{r-1}$ are equivalent by induction hypothesis, therefore we only need to show the equivalence of buckets partitioning constraints of y_r and following subset of COC.

$$\mathbf{1}^\top y_r = 0 \text{ and } \forall J \in \mathcal{P}([r-1]) \setminus \{\emptyset\}, \; y_r^\top \bigodot_{j \in J} y_j = 0. \tag{4}$$

$\mathbf{1}^\top y_r = 0$ is common to each other. Since elementwise products of any subset of $\{y_1, \ldots, y_{r-1}\}$ can be written as linear combinations of bucket vectors of $\{y_1, \ldots, y_{r-1}\}$, y_r satisfying buckets partitioning constraints also satisfies (4).

We show the converse. Suppose S is an arbitrary bucket of $\{y_1, \ldots, y_{r-1}\}$ and denote the corresponding bucket vector by $\mathbf{1}_S$. Then, $\mathbf{1}_S$ is written as a linear combination of elementwise products of nonempty subsets of $\{\mathbf{1}, y_1, \ldots, y_{r-1}\}$. We can see it by observing that the bucket S is a subset of the bucket $\hat{S}$ of $\{y_1, \ldots, y_{r-2}\}$ and $\mathbf{1}_S$ is proportional to either of $\mathbf{1}_{\hat{S}} \odot (y_{r-1} \pm \mathbf{1})$. Therefore, $\mathbf{1}_S^\top y_r = 0$ is derived from (4). $\qquad\square$

Theorem 1 shows that the buckets partitioning constraints through all bits are equivalent to COC. Note that the balance constraints $\mathbf{1}^\top y_j = 0$ and the orthogonality constraints $y_j^\top y_k = 0$ are special cases of COC with $|J| = 1, 2$, respectively. COC thus completely contains these constraints used in SH.

The problem to obtain locally optimized binary codes is then formulated as

$$\min_Y \text{tr}(Y^\top L Y) \tag{5}$$

$$\text{s.t. } \forall j \in [r], \; \|y_j\|^2 = n \text{ and } \forall J \in \mathcal{P}([r]) \setminus \{\emptyset\}, \; \mathbf{1}^\top \bigodot_{j \in J} y_j = 0.$$

Here L is either the graph Laplacian $D - A$ or its normalized version $I - D^{-1/2} A D^{-1/2}$, where the similarity matrix A is either sparse one used in STH or low rank one used in AGH. Note that when we use the similarity matrix of AGH, these two Laplacians point to same one. We call this method *Locally Optimized Hashing* (LOH). Above derivation of (5) supposes that Y is a binarized matrix. Since (5) is difficult to solve as a discrete problem, we consider its real value relaxation and binarize the solution as same as SH.

3.3   Sequential Buckets Partitioning Algorithm

The optimization problem of SH (1) is solved by eigenvalue decomposition, while that of LOH (5) cannot be solved at one time using this technique. Since COC indicates highly nonconvex manifold in the space of Y where at most r variables interact via multiplication, it is probably hard to directly optimize all variables. Therefore, we consider to optimize $y_1, \ldots, y_r$ one by one in this order by solving the following problem for each $y = y_k, k \in \{1, \ldots, r\}$.

$$\min_{y} \; y^\top L y \quad \text{s.t.} \; \|y\|^2 = n, \; B^\top y = 0, \tag{6}$$

where B is the bucket matrix of $\{y_1, \ldots, y_{k-1}\}$. In the case $k = 1$, we simply ignore the buckets partitioning constraints, i.e. y_1 is same as the first bit of SH.

The problem (6) is transformed into an eigenproblem as follows. Let $\beta > 0$ be the largest eigenvalue of L and let $\hat{A} = \beta I - L$. Note that $\beta = 1$ and $\hat{A} = A$ when we use the normalized graph Laplacian. $\hat{A}$ is positive semidefinite by definition. Since $y^\top L y = \beta \|y\|^2 - y^\top \hat{A} y = \beta n - y^\top \hat{A} y$, minimizing $y^\top L y$ is equivalent to maximizing $y^\top \hat{A} y$. One can transform it using the constraints $B^\top y = 0$ as

$$y^\top \hat{A} y = y^\top (I - BB^\top)\hat{A}(I - BB^\top)y.$$

Note that $I - BB^\top$ is the projection matrix onto $\text{Ker } B^\top = (\text{Im } B)^\perp$, since $B^\top B = I$ by the definition of buckets matrix. The problem (6) is then equivalent to the following problem.

$$\max_{y} \; y^\top (I - BB^\top)\hat{A}(I - BB^\top)y \quad \text{s.t.} \; \|y\|^2 = n. \tag{7}$$

Here the buckets partitioning constraints are suppressed, since y that violates them cannot be optimal due to the projection matrix $I - BB^\top$ in the objective function. The global optimum of (7) is the first eigenvector of $(I - BB^\top)\hat{A}(I - BB^\top)$. It is efficiently computed by Lanczos method. Alternatively, when one uses the low rank similarity matrix $A = Z\Lambda^{-1}Z^\top$ as same as AGH, the optimal solution y is also calculated using the top eigenvector v of

$$\Lambda^{-1/2}Z^\top(I - BB^\top)(I - BB^\top)Z\Lambda^{-1/2} = \Lambda^{-1/2}Z^\top(I - BB^\top)Z\Lambda^{-1/2}$$

as $y = \frac{\sqrt{n}}{\sigma}(I - BB^\top)Z\Lambda^{-1/2}v$, where σ is the eigenvalue corresponding to v. Here we used the fact that $I - BB^\top$ is idempotent, i.e. $(I - BB^\top)^2 = I - BB^\top$, since it is a projection matrix as we noted before.

3.4 Slow Partitioning for Singletons Problem

After solving (7) for some bits, buckets may become consisting of only one sample. We call such buckets *singletons*. The buckets partitioning constraints degenerate subsequent bits of singletons to zero; if a singleton $\{i\}$ appears right after k-bit code is generated, then the corresponding constraint of the next bit is $y_{k+1}^{(i)} = 0$. Once there appear too many singletons, the search precision cannot improve with longer codes anymore. We therefore introduce a method which weakly divides each bucket to fully utilize a longer code aiming at high precision.

Conventional orthogonality constraints are too weak to generate small buckets, as discussed in Sect. 1. On the other hand, the buckets partitioning constraints are too aggressive to achieve high precision with longer codes, since they do not consider available code length. In each step of LOH, the embedding of samples in each bucket is improved only by one bit. In order to utilize longer codes, one can use two or more bits to optimize the embedding of samples inside the buckets. This idea induces us to an intermediate method of SH and strict LOH which we call *slow partitioning*. As we stated before, the top eigenvector of $(I - BB^\top)\hat{A}(I - BB^\top)$ is used in LOH with strict constraints. Instead, we use its top q eigenvectors for each step, where q adjusts the redundancy of codes. Such vectors for each step are the optimum of the following problem.

$$\min_{Y_k} \ \mathrm{tr}(Y_k^\top L Y_k) \quad \text{s.t.} \ Y_k^\top Y_k = nI, \ B^\top Y_k = O, \tag{8}$$

where $Y_k = (y_{(k-1)q+1}, \ldots, y_{kq-1})$ is a matrix whose columns are codes corresponding to the k-th step. We use the orthogonality constraints in (8), which makes the q bits different from each other. This method can be seen as a greedy algorithm for the optimization problem with a subset of COC which is still stronger than the convensional constraints of (1). Slow partitioning produces more redundant and globally better codes than strict partitioning, while the codes are still locally optimized compared to those of the existing methods.

3.5 Computational Cost

LOH requires computation of graph Laplacian L with its top eigenvalue β and r times computation of eigenvectors of $(I - BB^\top)\hat{A}(I - BB^\top)$. Multiplication by $(I - BB^\top)$ runs in $O(n)$ time just by adjusting mean of each bucket to zero. On the other hand, multiplication by $\hat{A}$ costs $O(n^2)$ time in general. In practice, we can circumvent this inefficiency by using k-NN graph [12] or Anchor Graph [6], where the cost is reduced to linear time. Hence, LOH runs in linear time of rn using Lanczos method. Note that in case of using k-NN graph, its construction costs $O(n^2)$ time, which is dominant in the whole procedure.

4 Experiments

We conducted experiments on image and document datasets. In this section, we state the objective of experiments, describe the detail of datasets and experimental settings, and show the results.

4.1 Objective of Experiments

We design LOH by focusing on dividing buckets to be smaller than those of existing methods. Therefore, we show that our method produces a larger number of smaller buckets. Furthermore, since hashing is aiming at efficient NNS with high precision, we also show that our method achieves higher precision with very

Table 1. Summary of datasets

Name	Number of samples	Dimension	Contents
MNIST	69,000	784	Handwritten digits
RCV1	518,571	47,236	Newswire documents

short codes. We use image and document datasets as dense and sparse examples, respectively. We compare LOH with STH (Self-Taught Hashing), AGH (Anchor Graph Hashing) and its extension called 2-AGH [6], all of which are improved extensions to SH.

4.2   Experimental Settings

We used two similarity graphs in our method, undirected k-NN graph and Anchor Graph. We call these variants LOH and ALOH (Anchor LOH), respectively. In our method and STH, we used Support Vector Machines with linear kernel as hash functions as same as [12]. We used LIBLINEAR [1] as the implementation of linear kernel SVM. The raw samples $x^{(1)}, \ldots, x^{(n)}$ were used as training vectors for SVM in LOH and STH, while we used Z in ALOH. In other words, we used sample-anchor similarities as a feature vector in ALOH.

The experiment is conducted as follows for each method and dataset. We first learned hash functions, which then converted training data to binary codes. Note that these binary codes may be different from binarized Y in LOH, ALOH and STH, since they use two-step framework of STH [12]. We finally converted test data to binary codes as same as training data, and evaluated the results.

We use two evaluation measures: mean number of samples in each bucket and search precision. We labeled the nearest 100 samples as positive for each query, and use the mean ratio of positive ones in top ten samples retrieved by hamming distance of binary codes as the search precision.

4.3   Datasets and Detailed Settings

We conducted experiments on two datasets, **MNIST** and **RCV1**, which are summarized at Table 1. Details of datasets and related settings are follows.

MNIST[1] is a dataset consisting of 70,000 handwritten digit images. Each image is a 784 dimensional real vector corresponding to 28 x 28 gray scale pixels, each of which is normalized into the interval $[0, 1]$. They are randomly divided into 69,000 training samples and 1,000 queries. We use the heat kernel of Euclidean distance with scaling parameter $t = 10$ as similarity measure in all methods. In LOH and STH, we use the weighted k-NN graph of $k = 25$ as a similarity matrix. In ALOH and AGHs, we use $m = 500$ anchors chosen by 10 iterations of k-means, and $s = 5$ as the sparsity parameter of Anchor Graph.

RCV1 is Reuters newswire dataset [4]. Each document is represented by a 47,236 dimensional bag-of-words vector of tf-idf scores. 518,571 samples are

---

[1] http://yann.lecun.com/exdb/mnist/

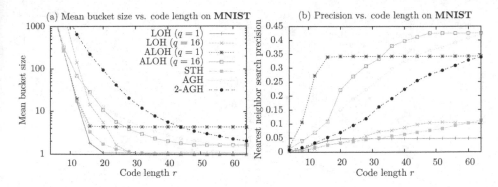

Fig. 2. Results on **MNIST**. (a) Mean size of buckets. (b) NNS precision at ten samples.

used as training data and 15,564 samples as queries. We use the heat kernel of Hellinger distance with scaling parameter $t = 1$ as similarity measure in all methods. Hellinger distance, defined by

$$\mathcal{D}_{\mathrm{H}}(x, y) = \sqrt{\sum_{i=1}^{d}(\sqrt{x_i/\|x\|} - \sqrt{y_i/\|y\|})^2} \in [0, 2],$$

measures divergence between two distributions $x/\|x\|$ and $y/\|y\|$. In LOH and STH, we use the weighted k-NN graph of $k = 25$ as a similarity matrix. In ALOH and AGHs, we use $m = 800$ anchors chosen by 10 iterations of k-means, and $s = 10$ as the sparsity parameter of Anchor Graph.

4.4 Experimental Results

We first review the results on **MNIST**. Mean bucket size is shown at Fig. 2(a). q is the redundancy parameter of slow partitioning described in Sect. 3.4. LOH, ALOH of $q = 1$ and STH achieve small buckets in short code lengths. Note that small buckets are meaningful only when the precision is high, which we review later. ALOH of $q = 1$ stops the improvement at $r = 16$, since all buckets become singletons at $r = 16$ within the algorithm. Since y_{16} is too complex as a supervisor and not linearly separatable, SVM cannot learn it completely and the resulting mean bucket size is not 1. On the other hand, ALOH of $q = 16$ slowly improves the bucket size, which is same or better than that of AGHs.

The precision of NNS is shown at Fig. 2(b). ALOH of $q = 1$ produces short codes of high precision. In particular, its precision at $r = 16$ is almost same as that of AGH at $r = 44$. On the other hand, ALOH of $q = 16$ slowly improves the precision, which overtakes that of $q = 1$ at the code length $r = 36$. Since the samples of **MNIST** are dense vectors, LOH and STH fails to correctly learn hash functions by linear SVM of raw features, which causes very low precision.

We next review the results on **RCV1**. Mean bucket size is shown at Fig. 3(a). This result is similar to that on **MNIST**. Here LOH of $q = 16$ produces small buckets with short codes.

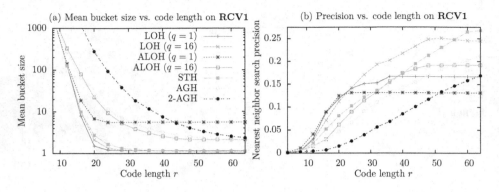

Fig. 3. Results on **RCV1**. (a) Mean bucket size. (b) NNS precision at ten samples.

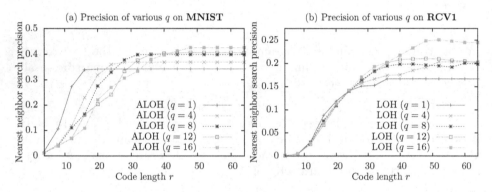

Fig. 4. NNS precision of various redundancy parameter q. (a) The results of ALOH on **MNIST**. (b) The results of LOH on **RCV1**.

The NNS precision is shown at Fig. 3(b). LOH and ALOH of $q = 1$ achieve relatively high precision at small r, while LOH of $q = 16$ achieves high precision at medium r. Since **RCV1** is a dataset of high dimensinoal sparse vectors, linear SVMs succeed to learn hash functions, which results in high precision of LOHs.

Finally, we show the NNS precision of various redundancy parameter q illustrated at Fig. 4. (a) is the results on **MNIST**, where only those of ALOH are shown, since they are superior to LOH on the dataset. On the other hand, we only show at (b) the results of LOH on **RCV1**, since they are superior to ALOH on the dataset. We can see from (a) that smaller q makes the precision higher at short codes, while larger q makes it higher at relatively long codes. In (b), the precision with smaller q is a bit better at small r than that of larger q. On the other hand, when the code length gets long, larger q achieves higher precision.

5   Conclusion

We showed that existing hash learning methods have a problem that they produce a small number of large buckets which degrade the precision of NNS.

Buckets partitioning constraints enabled one to generate a large number of small buckets, which are transformed into symmetric form named combinatorial orthogonality constraints (COC). The optimization problem with COC is solved bit-by-bit in greedy way by solving the simple eigenproblems. We described that each of resulting buckets may contain only one sample, which causes the subsequent bits to be useless. We overcame this singletons problem by introducing slow partitioning, which we saw as an intermediate method between COC and conventional balance and orthogonality constraints. Experimental results showed that our Locally Optimized Hashing method and its variant using Anchor Graph achieve superior precision with short binary codes compared to existing methods on both image and document datasets. We would like to find more detailed strategy of solving the trade-off between dividing buckets and optimizing the embedding in the future.

References

1. Fan, R.-E., Chang, K.-W., Hsieh, C.-J., Wang, X.-R., Lin, C.-J.: Liblinear: A library for large linear classification. Journal of Machine Learning Research **9**, 1871–1874 (2008)
2. Indyk, P., Motwani, R.: Approximate nearest neighbors: towards removing the curse of dimensionality. In: Proceedings of the 30th Annual ACM Symposium on Theory of Computing, pp. 604–613 (1998)
3. Kulis, B., Grauman, K.: Kernelized locality-sensitive hashing for scalable image search. In: IEEE 12th International Conference on Computer Vision, pp. 2130–2137 (2009)
4. Lewis, D.D., Yang, Y., Rose, T.G., Li, F.: Rcv1: A new benchmark collection for text categorization research. Journal of Machine Learning Research **5**, 361–397 (2004)
5. Li, P., König, A.C.: b-Bit minwise hashing. In: Proceedings of the 19th International Conference on World Wide Web, pp. 671–680 (2010)
6. Liu, W., Wang, J., Kumar, S., Chang, S.-F.: Hashing with graphs. In: Proceedings of the 28th International Conference on Machine Learning, pp. 1–8 (2011)
7. Norouzi, M., Fleet, D.: Minimal loss hashing for compact binary codes. In: Proceedings of the 28th International Conference on Machine Learning, pp. 353–360 (2011)
8. Raginsky, M., Lazebnik, S.: Locality-sensitive binary codes from shift-invariant kernels. In: Advances in Neural Information Processing Systems, vol. 22, pp. 1509–1517 (2009)
9. Wang, J., Kumar, S., Chang, S.-F.: Sequential projection learning for hashing with compact codes. In: Proceedings of the 27th International Conference on Machine Learning, pp. 1127–1134 (2010)
10. Watkins, D.S.: Fundamentals of Matrix Computations. Pure and Applied Mathematics: A Wiley Series of Texts, Monographs and Tracts. Wiley, third edition edition (2010)
11. Weiss, Y., Torralba, A., Fergus, R.: Spectral hashing. In: Advances in Neural Information Processing Systems, vol. 21, pp. 1753–1760 (2009)
12. Zhang, D., Wang, J., Cai, D., Lu, J.: Self-taught hashing for fast similarity search. In: Proceedings of the 33rd International ACM SIGIR Conference on Research and Development in Information Retrieval, pp. 18–25 (2010)

Do-Rank: DCG Optimization for Learning-to-Rank in Tag-Based Item Recommendation Systems

Noor Ifada[1,2(✉)] and Richi Nayak[1]

[1] School of Electrical Engineering and Computer Science,
Queensland University of Technology, Brisbane, Australia
{noor.ifada,r.nayak}@qut.edu.au
[2] Informatics Department, University of Trunojoyo Madura, Bangkalan, Indonesia
noor.ifada@if.trunojoyo.ac.id

Abstract. Discounted Cumulative Gain (DCG) is a well-known ranking evalua-tion measure for models built with multiple relevance graded data. By handling tagging data used in recommendation systems as an ordinal relevance set of $\{negative, null, positive\}$, we propose to build a DCG based recommenda-tion model. We present an efficient and novel learning-to-rank method by opti-mizing DCG for a recommendation model using the tagging data interpretation scheme. Evaluating the proposed method on real-world datasets, we demon-strate that the method is scalable and outperforms the benchmarking methods by generating a quality top-N item recommendation list.

Keywords: Tagging data · Tag-based item recommendation · Discounted cu-mulative gain · Top-N recommendation

1 Introduction

In a tag-based recommendation system, users annotate items of their interest with freely-defined tags, hence the ternary relation of $\langle user, item, tag \rangle$ is naturally formed. This system must interpret the observed and non-observed entries in tagging data efficiently in order to generate quality recommendations [1]. The observed, or positive, entries reveal the user interest by indicating that the user has annotated an item using certain tags. The non-observed entries can reveal two types of information: (1) negative entries that indicate users are not interested with the items; or (2) null values that indicate users might be interested in them in the future and they need to be predicted [2]. Accordingly, tagging data can be labelled using the ordinal relevance set of $\{negative, null, positive\}$ for a tuple of $\langle user, item, tag \rangle$.

The task of a tag-based recommendation system is to generate the list of items, which may be of interest to a user, by learning the users past tagging behavior. The list of recommended items is ordered in descending order based on the predicted pre-ference score. Users usually show more interest to the fewer items at the top of the list than the ones further down the list [3]. The order of items in the recommendation list is crucial and, therefore, the recommendation task can be considered as a ranking problem in which the item preference score needs to be determined and sorted for

© Springer International Publishing Switzerland 2015
T. Cao et al. (Eds.): PAKDD 2015, Part II, LNAI 9078, pp. 510–521, 2015.
DOI: 10.1007/978-3-319-18032-8_40

generating the list. A "learning-to-rank" approach optimizes the recommendation model with respect to the evaluation measure such that it can generate a quality top-N recommendation list [3-5].

The measures widely used to evaluate the performance of a ranking model are Discounted Cumulative Gain (DCG), Mean Average Precision (MAP), and Mean Reciprocal Rank (MRR) [6]. Compared to other two measures, DCG is more widely used especially for models that include multiple relevance graded data [4]. Therefore, for the tagging data that is labelled using ordinal relevance set, DCG becomes the most suitable evaluation measure for measuring the ranking quality [4]. DCG emphasizes on getting the correct order of higher ranked items than that of lower ranked items. Since the users only refer to the top-N items in the list, the higher positions have more influence on the score than the lower positions. DCG consists of two functions: (1) a discount function which makes items lower down in the ranked list contribute less to the DCG score; and (2) a gain function which gives the significance of the known relevance grade [4, 7]. These two functions make DCG significantly useful for solving the recommendation task which generates a quality top-N recommendation list.

In this paper, we propose an efficient and novel learning-to-rank method, called as *Do-Rank*, by optimizing DCG for a recommendation model as the ranking measure. This method generates an optimal list of recommended items from the DCG perspective for all users. However, directly optimizing DCG across all users in the recommendation model is computationally expensive. To deal with this, we propose a fast learning algorithm with an efficient tagging data interpretation scheme [1] for the learning-to-rank model. This interpretation scheme has shown good performance when implemented on a *pair-wise ranking model* [1, 2, 8], i.e. the objective function of the model is formed based on *the pairs of items* on each user-tag set. In this paper, we design this scheme on a *list-wise ranking model*. This means that we must form the objective function only based on the list of *all items* on each user-tag set. Experimental results on real-world datasets show that *Do-Rank* is scalable and outperforms state-of-the-art recommendation methods in recommendation quality. It ascertains that optimizing DCG while building the learning-to-rank recommendation model improves the recommendation performance.

Recent works have proposed the optimized recommendation models based on MAP [9] and MRR [10], dealing with binary relevance data. Since the tag-based recommendation systems use the ordinal relevance data, these models are not suitable. Weimer et.al [11] have proposed a recommendation method by optimizing the Normalized DCG (NDCG) using rating data as *explicit* feedback. Our problem is quite different and difficult in comparison to this work as the tag-based recommendation systems use tags as *implicit* feedback. A recommendation system with explicit rating data builds its model by collecting the ratings which represent the preference level of each $\langle user, item \rangle$ binary relation. The list of recommendations is then generated by ranking the predicted preference scores inferred from the unobserved $\langle user, item \rangle$ relations [7, 11]. In contrast, a tag-based recommendation system builds its model by using the user tagging history as data entries. The key challenges it faces are modeling the $\langle user, item, tag \rangle$ ternary input data, inferring the latent relationships, and predicting each entry with a score which indicates its relevance degree [1, 2]. The recommendation list needs to be generated by ranking the predicted preference scores of list of items under all tags which may be interest to a user.

Recently researchers have used tensor models to represent and analyze the latent relationships inherent in a three-dimensional tagging data model [12, 13]. A tensor model can be used as a predictor function which maps the $\langle user, item, tag \rangle$ latent relationship inherent in tagging data to a predicted preference score, and enables optimization of the evaluation measure [14] such as DCG.

The proposed item recommendation method *Do-Rank* comes closest to the tag recommendation method PITF [8] that implements tensor approach and applies the ordinal relevance set labelling to build a learning-to-rank model. *Do-Rank* has four significant differences compared to PITF. Firstly, *Do-Rank* employs a *list-wise ranking model*, whereas PITF employs a *pair-wise ranking model*. PITF focuses on getting the ranking order within each item pair correctly, instead of getting the correct order in the entire items recommendation list as in *Do-Rank*. Secondly, the objective function of PITF is an AUC-based optimization. In contrast to DCG as utilized in *Do-Rank* that assigns higher penalty to lower ranked items, AUC assigns equal penalty to the mistakes done at the top or bottom positions in the recommendation list [9]. Thirdly, PITF infers the negative values of non-observed tagging data from all items that have not been annotated by the user using a tag, whereas, *Do-Rank* implements an efficient scheme which interprets the negative values from items that have never been annotated by the user using any other tags. Lastly, PITF is a tag recommendation method and *Do-Rank* is an item recommendation method. The tag and item recommendations are two distinct tasks. Tag recommendations are generated with two specified dimensions, i.e. user and item, while the item recommendations are made with specified users only and, therefore, the item predicted preference scores must be calculated for the whole available tags before being sorted as a list of top-N recommendation.

To the best of our knowledge, this is the first work on tag-based item recommendation system that directly optimizes the ranking evaluation measure. Our contributions can be summarized as follows: (1) We propose a novel tag-based item recommendation method that directly optimizes a (smoothed) DCG for building the learning-to-rank model, (2) *Do-Rank* is the first tensor-based item recommendation ranking method with ordinal relevance set tagging data, and (3) We propose a fast learning-to-rank algorithm that implements an efficient tagging data interpretation scheme.

The remainder of this paper is organized as follows. Section 2 details the *Do-Rank* learning method. Section 3 presents the experimental results based on real-world datasets. Section 4 concludes the paper.

2   *Do-Rank*: DCG Optimization for Learning to Rank

Let $U = \{u_1, u_2, u_3, ..., u_Q\}$ be the set of Q users, $I = \{i_1, i_2, i_3, ..., i_R\}$ be the set of R items, and $T = \{t_1, t_2, t_3, ..., t_S\}$ be the set of S tags. The observed tagging data can be denoted as $A \subseteq U \times I \times T$, where a vector of $(u, i, t) \in A$ represents the observed tagging activity of user u using tag t to annotate item i.

2.1   Tensor Based Recommendation Prediction Model

The user tagging data can be naturally modelled as a three-dimensional tensor of $\mathcal{Y} \in \mathbb{R}^{Q \times R \times S}$. Figure 1 illustrates a tensor model representing a toy example of the observed tagging data, $\mathcal{Y} \in \mathbb{R}^{3 \times 4 \times 5}$ where $U = \{u_1, u_2, u_3\}$, $I = \{i_1, i_2, i_3, i_4\}$, and $T = \{t_1, t_2, t_3, t_4, t_5\}$. Each slice of the tensor represents a user matrix which contains the user tag usage for an item. The latent relationship between users, items, and tags can be inferred after decomposition. We use CP [15] as the predictor function in our model since Tucker, the other well-known decomposition technique, is more expensive in both memory and time [15]. As illustrated in Figure 2, CP factorizes a third-order tensor $\mathcal{Y} \in \mathbb{R}^{Q \times R \times S}$ into three factor matrices $\hat{U} \in \mathbb{R}^{Q \times F}$, $\hat{I} \in \mathbb{R}^{R \times F}$, $\hat{T} \in \mathbb{R}^{S \times F}$, and a diagonal core tensor $\mathcal{C} \in \mathbb{R}^{F \times F \times F}$, where F is the column size of the corresponding reduced factor matrices. The predicted preference score is calculated as:

$$\hat{y}_{u,i,t} := \sum_{f=1}^{F} \hat{u}_{u,f} \cdot \hat{\iota}_{i,f} \cdot \hat{t}_{t,f} = [\![\hat{U}_u, \hat{I}_i, \hat{T}_t]\!] \tag{1}$$

A predicted score reflects the preference level of a user in choosing an item for a tag. Research has shown that simply ranking the scored entries of tensor model does not produce quality recommendation [13]. Assuming that users show interest to a few top recommended items only [3], a recommendation model can be optimized with respect to the evaluation measure during the learning procedure [4] in order to generate quality recommendation by ranking the tensor entries most effectively.

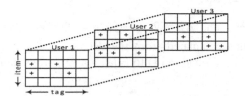

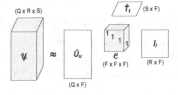

Fig. 1. A Tensor model showing a sample tagging data with observed (i.e. positive) entries

Fig. 2. The CP factorization model for third-order tensor

The task now becomes to recommend an optimal (from the DCG perspective) item list to users using the latent factor matrices after decomposing input tensor model. In DCG, the correct order of higher ranked items is more important than that of the lower ranked items and, therefore, the higher positions have more influence on the score. The DCG score consists of discount and gain functions formulated as denominator and numerator in Equation (2) [4, 7]. The discount function makes items lower down in the ranked list contribute less to DCG score while the gain function gives weight to the items based on their relevance label. The DCG score for a user u across all items i under tag t can be defined as:

$$DCG_{u,t} := \sum_{i \in I} \frac{2^{y_{u,i,t}} - 1}{\log_2(1 + r_{u,i,t})} \tag{2}$$

where $y_{u,i,t}$ is the relevance label and, is assigned as one of element in the ordinal relevance set of $\{negative, null, positive\}$ (or $\{-1, 0, 1\}$) from the initial tensor model. The $r_{u,i,t}$ is the ranking position of item i for user u with tag t and, is approximated using $\hat{y}_{u,i,t}$ (which is calculated from the factor matrices using Equation (1)). The DCG score of all users over all items under all tags can be defined as:

$$DCG := \frac{1}{QS} \sum_{u \in U} \sum_{t \in T} \sum_{i \in I} \frac{2^{y_{u,i,t}} - 1}{log_2(1 + r_{u,i,t})} \tag{3}$$

The item recommended list is generated for user u by ranking all the items in descending order of the computed preference scores over all items under all tags.

2.2   Smoothed DCG and Optimization

It can be seen from Equation (3) that DCG is dependent on the ranking positions. The rankings change in a non-smooth way with respect to predicted relevance scores calculated based on the model parameters (i.e. factor matrices). The non-smooth function of DCG makes the application of standard optimization methods difficult [4, 16] which require smoothness in the objective function such as the gradient based approaches [4]. In this paper, we solve this problem by approximating the ranking position $r_{u,i,t}$ by a smoothed function with respect to the model parameters.

Inspired by the Information Retrieval approach of "learning-to-rank" [4, 16], we approximate $r_{u,i,t}$ by the following smoothing function:

$$r_{u,i,t} \approx 1 + \sum_{j \neq i} \sigma(\Delta \hat{y}) \tag{4}$$

where $\sigma(x)$ is the logistic function $\frac{1}{1 + e^{-x}}$, and $\Delta \hat{y} = \hat{y}_{u,i,t} - \hat{y}_{u,j,t}$ is the predicted relevance scores difference for two items calculated from the decomposed tensor model (i.e. using Equation (1)). Substituting Equation (4) to Equation (3), we obtain the smoothed approximation of DCG:

$$sDCG := \frac{1}{QS} \sum_{u \in U} \sum_{t \in T} \sum_{i \in I} \frac{2^{y_{u,i,t}} - 1}{1 + log_2(\sum_{j \neq i} \sigma(\Delta \hat{y}))} \tag{5}$$

The resulted objective function can now be formulated as:

$$L(\Theta) := \sum_{u \in U} \sum_{t \in T} \sum_{i \in I} \frac{2^{y_{u,i,t}} - 1}{1 + log_2(\sum_{j \neq i} \sigma(\Delta \hat{y}))} - \lambda_\Theta \|\Theta\|_F^2 \tag{6}$$

where λ_Θ is the regularization coefficient corresponding to σ_Θ as model parameters that controls overfitting. Note that the constant coefficient $(\frac{1}{QS})$ in $sDCG$ can be neglected since it has no influence on the optimization. We can now perform gradient descent to optimize the objective function in Equation (6). Given a case (u, i, t) with respect to the model parameters $\{\hat{U}_u, \hat{I}_i, \hat{T}_t\}$, the gradient of $sDCG$ can be computed as:

$$\frac{\partial L}{\partial \theta} = \sum_{u \in U} \sum_{t \in T} \sum_{i \in I} \frac{-(2^{y_{u,i,t}} - 1) \left[\frac{1}{((\sum_{j \neq i} \sigma(\Delta \hat{y})) \ln 2} \left(\frac{\partial}{\partial \theta} (\sum_{j \neq i} \sigma(\Delta \hat{y})) \right) \right]}{\left[1 + log_2 \left((\sum_{j \neq i} \sigma(\Delta \hat{y})) \right) \right]^2} - \lambda_\theta \theta \tag{7}$$

$$\frac{\partial L}{\partial \theta} = \sum_{u \in U} \sum_{t \in T} \sum_{i \in I} \frac{-(2^{y_{u,i,t}}-1)\left[\frac{1}{\left(\left(\sum_{j \neq i}\sigma(\Delta\hat{y})\right)\ln 2\right)}\left(\left(\sum_{j \neq i}\left(-\sigma(\Delta\hat{y})+(\sigma(\Delta\hat{y}))^2\right)\right)\frac{\partial}{\partial\theta}\Delta\hat{y}\right)\right]}{\left[1+\log_2\left(\sum_{j \neq i}\sigma(\Delta\hat{y})\right)\right]^2} - \lambda_\theta\theta \quad (8)$$

By substituting $\sigma(\Delta\hat{y})$ with δ we get:

$$\frac{\partial L}{\partial \theta} = \sum_{u \in U} \sum_{t \in T} \sum_{i \in I} \frac{-(2^{y_{u,i,t}}-1)\left[\frac{1}{\left(\ln 2 \sum_{j \neq i}\delta\right)}\left(\sum_{j \neq i}(-\delta+\delta^2)\frac{\partial}{\partial\theta}\Delta\hat{y}\right)\right]}{\left[1+\log_2\left(\sum_{j \neq i}\delta\right)\right]^2} - \lambda_\theta\theta \quad (9)$$

Based on Equation (9), we can see that we only have to compute $\frac{\partial}{\partial\theta}\Delta\hat{y}$ with respect to the model parameters to implement the $sDCG$ optimization. However, we can also see that directly optimizing $sDCG$ across all users is an expensive task as we need to compute the predicted relevance score difference $\Delta\hat{y}$ for all items pair-wise across all tags. Therefore, in the next subsection, we propose a fast learning algorithm that implements the efficient tagging data interpretation scheme so that the computation of $\Delta\hat{y}$ of all users is only required on observed or positive items across all tags.

2.3   Fast Learning

As per Equation (5), in order to optimize $sDCG$, we need to calculate the predicted relevance score difference $\Delta\hat{y}$ between each item with all other items in the system. We propose to solve this expensive process by employing a fast learning algorithm. The basic idea of the fast learning algorithm is to optimize $sDCG$, on each (u,t) set, by calculating only the predicted relevance score difference $\Delta\hat{y}$ between items which have observed to show user interest and items which are not of user interest, i.e. positive and negative items, respectively. The key challenge here is to efficiently infer the positive and negative items from the tagging data, on each (u,t) set. Our previous work [1] has demonstrated that the *User-Tag Set* (*UTS*) scheme efficiently interprets the observed and non-observed tagging data and labels each entry as one of element in the ordinal relevance set of $\{negative, null, positive\}$ when implemented on a pair-wise ranking model. In this paper, we propose to approximate $\Delta\hat{y}$ in Equation (5) using the *UTS* scheme that is implemented on a *list-wise ranking model*.

User-Tag Set (*UTS*) **Scheme.** The *UTS* scheme, based on $(u,t) \in A$, interprets the positive and negative entries amongst items. The items of positive entries are derived from the observed data, while the items of negative entries are interpreted from items that have not been annotated by user u using any other tags [1]. As illustrated in Figure 3(a) and (b), the positive entries show that u_1 has used t_1 to reveal his interest for items $\{i_2, i_3\}$, t_3 for $\{i_2\}$, and t_4 for $\{i_3\}$. This means that, on (u_1, t_1) set, the positive and negative entries are $\{i_2, i_3\}$ and $\{i_1, i_4\}$, respectively, given $I = \{i_1, i_2, i_3, i_4\}$. Using the scheme, the input set consists of a list of tag assignment A labelled with the corresponding relevance score $y_{u,i,t}$ using the following rules:

$$y_{u,i,t} := \begin{cases} 1 & if\ (u,i,t) \in A \\ -1 & if\ (u,i,t) \notin A\ and\ i \in I\backslash\{i|(u,i,*) \in A\} \\ 0 & otherwise \end{cases}$$

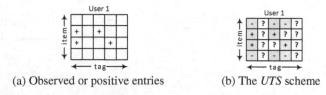

(a) Observed or positive entries   (b) The *UTS* scheme

Fig. 3. Toy example for u_1

1: **Algorithm: *Do-Rank* Learning**
2: **Input :** Training set $D_{train} \subseteq U \times I \times T$, learning rate α, factor matrix column
            size F, regularization λ, maximal iteration $iterMax$
3: **Output:** Learned factor matrices $\hat{U}, \hat{I}, \hat{T}$
4:   $Q = |U|, R = |I|, S = |T|, \mathcal{Y} \in \mathbb{R}^{Q \times R \times S}$
5:   $ZP = \{i | y_{u,i,t} = 1\}, ZN = \{i | y_{u,i,t} = -1\}$
6:   Initialize $\hat{U}^{(0)} \in \mathbb{R}^{Q \times F}, \hat{I}^{(0)} \in \mathbb{R}^{R \times F}, \hat{T}^{(0)} \in \mathbb{R}^{S \times F}, h = 0$
7:   $g_0 = DCG$ based $\mathcal{Y}$ and $\hat{U}^{(0)}, \hat{I}^{(0)}, \hat{T}^{(0)}$
8: **repeat**
9:    **for** $u \in U$ **do**
10:         $\hat{U}_u \leftarrow \hat{U}_u + \alpha \frac{\partial L}{\partial \hat{U}_u}$ based on Equation (11)
11:   **for** $t \in T$ **do**
12:         $\hat{T}_t \leftarrow \hat{T}_t + \alpha \frac{\partial L}{\partial \hat{T}_t}$ based on Equation (11)
13:   **for** $u \in U$ **do**
14:      **for** $t \in T$ **do**
15:         **for** $i \in ZP$ **do**
16:            **for** $j \in ZN$ **do**
17:                  $\hat{I}_i \leftarrow \hat{I}_i + \alpha \frac{\partial L}{\partial \hat{I}_i}$ based on Equation (11)
18:                  $\hat{I}_j \leftarrow \hat{I}_j + \alpha \frac{\partial L}{\partial \hat{I}_j}$ based on Equation (11)
19:    $++h$
20:    $g = DCG$ based $\mathcal{Y}$ and $\hat{U}^{(h)}, \hat{I}^{(h)}, \hat{T}^{(h)}$
21:    **if** $g - g_0 \leq 0$
22:         break
23: **until** $h \geq iterMax$

Fig. 4. Do-Rank Learning Algorithm

The fast learning algorithm based on the *UTS* scheme infers the user's positive and negative items on each $(u, t) \in A$ and defines them as: (1) $ZP = \{i | y_{u,i,t} = 1\}$, positive items derived from the observed data, and (2) $ZN = \{i | y_{u,i,t} = -1\}$, negative items derived from the items that have not been tagged by u using any other tags. The resultant objective function can be formulated by:

$$L(\Theta) := \sum_{u \in U} \sum_{t \in T} \sum_{i \in ZP} \frac{2^{y_{u,i,t}} - 1}{1 + \log_2(\sum_{j \in ZN} \sigma(\Delta \hat{y}))} - \lambda_\Theta \|\Theta\|_F^2 \qquad (10)$$

where $\Delta \hat{y} = \hat{y}_{u,i,t} - \hat{y}_{u,j,t}$. The gradient of $sDCG$ given a case (u, i, j, t) with respect to the model parameter $= \{\hat{U}_u, \hat{I}_i, \hat{I}_j, \hat{T}_t\}$ is given by Equation (11).

$$\frac{\partial L}{\partial \theta} = \sum_{u \in U} \sum_{t \in T} \sum_{i \in ZP} \frac{-(2^{y_{u,i,t}}-1)\left[\frac{1}{(\ln 2 \sum_{j \in ZN} \delta)}\left(\sum_{j \in ZN}(-\delta+\delta^2)\frac{\partial}{\partial \theta}\Delta \hat{y}\right)\right]}{[1+\log_2(\sum_{j \in ZN} \delta)]^2} - \lambda_\theta \theta \qquad (11)$$

where $\delta = \sigma(\Delta \hat{y})$. To apply the $sDCG$ optimization, we only have to compute the gradient of $\frac{\partial}{\partial \theta}\Delta \hat{y}$. The gradients for the model based on its parameters are: $\frac{\partial \Delta \hat{y}}{\partial \hat{U}_u} = (\hat{I}_i \odot \hat{T}_t - \hat{I}_j \odot \hat{T}_t)$, $\frac{\partial \Delta \hat{y}}{\partial \hat{I}_i} = (\hat{U}_u \odot \hat{T}_t)$, $\frac{\partial \Delta \hat{y}}{\partial \hat{I}_j} = -(\hat{U}_u \odot \hat{T}_t)$, $\frac{\partial \Delta \hat{y}}{\partial \hat{T}_t} = (\hat{U}_u \odot \hat{I}_i - \hat{U}_u \odot \hat{I}_j)$, where $\odot$ denotes element-wise product. From Equation (11), we can see that for optimizing $sDCG$ across all users and under all tags, we only need to compute $\Delta \hat{y}$ for each ZP that is less computationally expensive than computing $\Delta \hat{y}$ for each R, since $|ZP| \ll R$. The *Do-Rank* learning algorithm is outlined in Figure 4.

2.4   Complexity Analysis and Convergence

We analyze the complexity of learning process for a single iteration. The initial *Do-Rank* complexity is $O(F|A|(Q + S + QSR^2))$. When the fast learning approach is implemented, the *Do-Rank* complexity (illustrated in Figure 4), becomes $O(F|A|(Q + S + QS\tilde{p}\tilde{n}))$, where $\tilde{p}$ and $\tilde{n}$ denote the average number of ZP and ZN per (u, t) set. Since $\tilde{p}, \tilde{n} \ll R$, the *Do-Rank* complexity now becomes $O(F|A|(Q + S + QSR))$.

The objective function of *Do-Rank* is optimizing $sDCG$ as given in Equation (5) that is the smoothed approximation of DCG as given in Equation (3). We use the iterated DCG scores during the optimization process as the termination criterion [9], instead of using the conventional criteria such as the number of iterations [10] and the convergence rate [8]. Optimization process is terminated when DCG scores start declining.

3   Empirical Analysis

We used two real-world tagging datasets, detailed in Table 1, to implement the proposed tag-based item recommendation method *Do-Rank*. Adapting standard practice of eliminating noise and decreasing data sparsity [12], datasets are refined by using p-core technique, i.e. selecting users, items, and tags that have occurred in at least p number posts. Post is the set of distinct user-item sets in the observed tagging data.

Table 1. The details of datasets

Dataset	#Users (Q)	#Items (R)	#Tags (S)	#Observed Tagging data	p-core
LastFM	867	1,715	1,423	99,211	10
MovieLens	571	1,684	1,559	25,103	5

We implemented the 5-fold cross-validation. For each fold, we randomly divide into a training set D_{train} (80%) and a test set D_{test} (20%) based on the number of posts data. D_{train} and D_{test} do not overlap in posts, i.e., there exist no triplets for a user-item combination in the training set if a triplet $(u, i, *)$ is present in the test set. The recommendation task is to predict and rank the Top-N items for the users present in D_{test} according to DCG. The evaluation metrics [4] used to measure the recommendation performance are (1) NDCG, Normalized DCG score, presented at various top-N positions, and (2) MAP, Mean Average Precision (AP).

Following the initialization approach proposed for a ranking model [7], we randomly initialized the factor matrices of our model. In order to empirically tune the parameters in *Do-Rank*, we randomly selected 25% of all the observed data available in D_{train}. The best performances were achieved using the following values: learning rate parameter $\alpha = 0.01$ and regularization parameter $\lambda = 1e^{-05}$. We used $F = 128$ as the size of latent factor matrix.

3.1   Scalability of *Do-Rank*: Impact of Fast Learning Approach

We first investigated the impact of implementing the fast learning approach on optimizing $sDCG$ as defined in Equation (10), by measuring the running time for learning the model on a single iteration at different scales, i.e. 10% to 100% of D_{train}. Figures 5(a) and 5(b) show that the learning time of "fast learning" approach, on the LastFM and Movielens datasets respectively, is linear to the size of data, i.e. determined by the size of items R. The "regular" approach, i.e. optimizing $sDCG$ without implementing fast learning approach requires more learning time as the computational complexity is determined by R^2, as previously described in Section 2.4. These results confirm that the *User-Tag Set (UTS)* scheme employed in this algorithm efficiently interprets the tagging data and, the pair-wise difference between the positive and negative item entries is sufficient for determining effective ranking instead of calculating difference between all pair of items in the dataset.

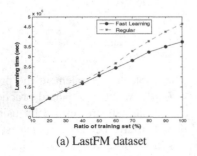

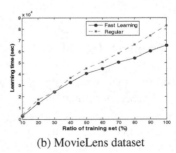

(a) LastFM dataset	(b) MovieLens dataset

Fig. 5. Impact of fast learning approach

3.2   Effectiveness of Convergence Criterion

Figures 6(a) and 6(b) show the evolution of DCG@10 across iterations on the training and test sets respectively. DCG increases through early iterations on both sets, before the performance is declined. It ascertains that *Do-Rank* is able to effectively optimize DCG. We can notice that the DCG measure drops after a few iterations (less than 15) which indicates that using a measure score as termination criterion is a useful approach in order to avoid the model to overfit [9].

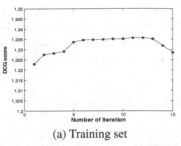
(a) Training set

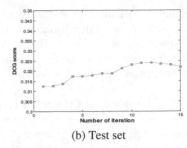
(b) Test set

Fig. 6. Effectiveness of convergence criterion in the learning procedure

3.3   Performance Comparison

The performance of *Do-Rank* is compared with the following methods.

- **PITF [8].** We have adapted PITF [8] to generate item recommendations based on the user-tag sets. PITF is the pair-wise tensor factorization model for generating tag recommendations that implements the ordinal relevance set labelling to build the *pair-wise ranking model* on each user-item set. The tuned parameters are $\alpha = 0.01$, $\lambda = 5e^{-05}$, and $F = 128$.
- **CP-TRPR [13].** A probabilistic ranking tensor-based method that ranks the predicted preference scores, calculated from the decomposed models, by utilizing the users past collaborative tagging data. The tuned parameters for this method are $tolerance = 1e^{-04}$, $\lambda = 0$, $voc\_size = 20$, and $F = 128$.
- **CTS [17].** The state-of-the-art matrix-based method that ranks the recommendations by using the users past tagging activities in forming users' likelihood. The tuned parameters are $neighborhood\_size = 50$ and $model\_size = 20$.

The recommendation performance comparisons of the proposed *Do-Rank* and the benchmarking methods on LastFM and MovieLens datasets are listed in Table 2 and Table 3, respectively. We can observe that *Do-Rank* outperforms the benchmarking methods in terms of NDCG (at any top-N positions) on both datasets. It can be noted that the higher the top-N position, the less the NDCG score is. In terms of MAP, *Do-Rank* is still able to outperform other methods on the LastFM dataset.

Compared to PITF, an AUC-based optimization approach which gives equal penalty to the mistakes at the top and bottom list of recommendations [9], *Do-Rank* enhances the top-N recommendation performance by optimizing the top-biased measure DCG. This confirms that optimizing top-N recommendation evaluation measure for building the learning model will improve the recommendation performance. Additionally, PITF is *pair-wise ranking model* that aims to get the ranking order within each pair correctly, while a DCG measure objective is to get the correct order of all lists, i.e. list-wise measure.

Both *Do-Rank* and CP-TRPR utilize the users past tagging history to correctly rank the order of items that might interest users. However, CP-TRPR interprets the tagging data as a binary relevance set to build the model. In other word, the method simply considers the observed entries as "1" and overfits the negative and null values inferred from the non-observed entries as "0". This inappropriate tagging data interpretation impacts the recommendation performance [2]. Moreover, the model was designed to solve the classification problem by minimizing the Mean Square Error (MSE) [13]. Finally, *Do-Rank* outperformance towards CTS is again proving that three-dimensional characteristic of tagging data must be captured so that the many-to-many relationships that exist among the dimensions can be kept rather than projecting the three-dimension into two-dimensions [12].

Table 2. NDCG at top-N position and MAP on the LastFM dataset

Methods	Score (%)							
	NDCG@1	NDCG@2	NDCG@3	NDCG@4	NDCG@5	NDCG@10	NDCG@20	MAP
PITF	9.06	9.01	8.17	7.90	7.56	6.45	5.45	5.96
CP-TRPR	7.12	7.67	7.08	6.67	6.56	5.68	4.79	5.20
CTS	6.60	5.65	5.11	4.90	4.87	4.17	3.50	3.87
Do-Rank	**9.57**	**9.37**	**8.66**	**8.38**	**8.15**	**7.05**	**5.98**	**6.50**

Table 3. NDCG at top-N position and MAP on the MovieLens dataset

Methods	Score (%)							
	NDCG@1	NDCG@2	NDCG@3	NDCG@4	NDCG@5	NDCG@10	NDCG@20	MAP
PITF	3.02	2.77	2.63	2.70	2.60	2.42	2.26	2.70
CP-TRPR	5.18	4.85	4.27	4.02	3.72	3.17	2.60	**3.54**
CTS	4.10	3.44	3.14	3.19	3.25	2.74	2.43	3.31
Do-Rank	**5.40**	**4.86**	**4.29**	**4.04**	**3.97**	**3.26**	**2.68**	3.21

4 Conclusion and Future Work

We have presented *Do-Rank*, a top-N tag-based item recommendation method that directly optimizes the (smoothed) DCG in building the learning model for generating an ordered list of items that might interest the user. Entries in the tensor-based model are generated from the ordinal relevance set tagging data. We presented a fast learning approach that implements the *User-Tag Set (UTS)* tagging data interpretation scheme, and enables efficient execution of *Do-Rank*. The experimental results on real datasets show that *Do-Rank* is scalable and outperforms all benchmarking methods on the NDCG measure. This ascertains that optimizing DCG for building the learning model improves the recommendation performance. For the future work, we are planning to investigate the implementation of DCG optimization on other ranking models and the potential of optimizing other measures for tag-based item recommendation as different measures possibly yield different recommendation performances.

References

1. Ifada, N., Nayak, R.: An efficient tagging data interpretation and representation scheme for item recommendation. In: The 12th Australasian Data Mining Conference. ACS, Brisbane, Australia. (2014)
2. Rendle, S., Balby Marinho, L., Nanopoulos, A., Schmidt-Thieme, L.: Learning optimal ranking with tensor factorization for tag recommendation. In: The 15th ACM SIGKDD International Conference on Knowledge Discovery and Data Mining, pp. 727–736. ACM, Paris, France (2009)

3. Cremonesi, P., Koren, Y., Turrin, R.: performance of recommender algorithms on top-N recommendation tasks. In: The 4th ACM Conference on Recommender Systems, pp. 39–46. ACM Barcelona, Spain (2010)
4. Chapelle, O., Wu, M.: Gradient Descent Optimization of Smoothed Information Retrieval Metrics. Information Retrieval 13(3), 216–235 (2010)
5. Xu, J., Li, H.: AdaRank: a boosting algorithm for information retrieval. In: The 30th Annual International ACM SIGIR Conference on Research and Development in Information Retrieval, pp. 391–398. ACM, Amsterdam, The Netherlands (2007)
6. Liu, T.-Y.: Learning to Rank for Information Retrieval. Foundations and Trends in Information Retrieval 3(3), 225–331 (2009)
7. Balakrishnan, S., Chopra, S.: Collaborative ranking. In: The 5th ACM International Conference on Web Search and Data Mining, pp. 143–152. ACM, Seattle, Washington, USA (2012)
8. Rendle, S., Schmidt-Thieme, L.: Pairwise interaction tensor factorization for personalized tag recommendation. In: The 3rd ACM International Conference on Web Search and Data Mining, pp. 81–90. ACM, New York, USA (2010)
9. Shi, Y., Karatzoglou, A., Baltrunas, L., Larson, M., Hanjalic, A., Oliver, N.: TFMAP: optimizing MAP for top-N context-aware recommendation. In: The 35th International ACM SIGIR Conference on Research and Development in Information Retrieval, pp. 155–164. ACM, Portland, Oregon, USA (2012)
10. Shi, Y., Karatzoglou, A., Baltrunas, L., Larson, M., Oliver, N., Hanjalic, A.: CLiMF: learning to maximize reciprocal rank with collaborative less-is-more filtering. In: The 6th ACM Conference on Recommender Systems, pp. 139–146. ACM, Dublin, Ireland (2012)
11. Weimer, M., Karatzoglou, A., Le, Q.V., Smola, A.: Maximum Margin Matrix Factorization for Collaborative Ranking. Advances in Neural Information Processing Systems. (2007)
12. Symeonidis, P., Nanopoulos, A., Manolopoulos, Y.: A Unified Framework for Providing Recommendations in Social Tagging Systems Based on Ternary Semantic Analysis. IEEE Transactions on Knowledge and Data Engineering 22(2), 179–192 (2010)
13. Ifada, N., Nayak, R.: A two-stage item recommendation method using probabilistic ranking with reconstructed tensor model. In: Dimitrova, V., Kuflik, T., Chin, D., Ricci, F., Dolog, P., Houben, G.-J. (eds.) UMAP 2014. LNCS, vol. 8538, pp. 98–110. Springer, Heidelberg (2014)
14. Karatzoglou, A., Amatriain, X., Baltrunas, L., Oliver, N.: Multiverse recommendation: n-dimensional tensor factorization for context-aware collaborative filtering. In: The 4th ACM Conference on Recommender Systems, pp. 79–86. ACM, Barcelona, Spain (2010)
15. Kolda, T., Bader, B.: Tensor Decompositions and Applications. SIAM Review 51(3), 455–500 (2009)
16. Wu, M., Chang, Y., Zheng, Z., Zha, H.: Smoothing DCG for learning to rank: a novel approach using smoothed hinge functions. In: The 18th ACM Conference on Information and Knowledge Management, pp. 1923–1926. ACM, Hong Kong, China (2009)
17. Kim, H.-N., Ji, A.-T., Ha, I., Jo, G.-S.: Collaborative Filtering based on Collaborative Tagging for Enhancing the Quality of Recommendation. Electronic Commerce Research and Applications 9(1), 73–83 (2010)

Efficient Discovery of Recurrent Routine Behaviours in Smart Meter Time Series by Growing Subsequences

Jin Wang[1,2]($\boxtimes$), Rachel Cardell-Oliver[1,2], and Wei Liu[1,2]

[1] CRC for Water Sensitive Cities, PO Box 8000, Clayton, VIC 3800, Australia
[2] School of Computer Science and Software Engineering, The University of Western Australia, 35 Stirling Highway, Crawley, WA 6009, Australia
{jin.wang,rachel.cardell-oliver,wei.liu}@uwa.edu.au

Abstract. Data mining techniques have been developed to automatically learn consumption behaviours of households from smart meter data. In this paper, recurrent routine behaviours are introduced to characterize regular consumption activities in smart meter time series. A novel algorithm is proposed to efficiently discover recurrent routine behaviours in smart meter time series by growing subsequences. We evaluate the proposed algorithm on synthetic data and demonstrate the recurrent routine behaviours extracted on a real-world dataset from the city of Kalgoorlie-Boulder in Western Australia.

Keywords: Smart metering · Routine behaviour · Motif detection · Temporal pattern

1 Introduction

Smart meters are being deployed by utility companies to monitor water and energy use in real-time. Hourly consumption data is recorded and wirelessly reported to a central server by smart meters. This real-time consumption data is useful for better understanding water and energy consumption behaviours. However, one challenge is to efficiently analyse these time series data to characterize consumption behaviours. Manually inspecting and analysing is time-consuming and labour intensive. Data mining techniques are needed to automatically extract information and knowledge about consumption behaviours from smart meter time series [1].

In this paper, we introduce recurrent routine behaviours to characterize regular water and energy use activities, and propose a novel algorithm to efficiently discover recurrent routines in smart meter time series. A recurrent routine is a group of similar subsequences that occurs frequently within its parent sequence. Figure 1 shows an example routine (marked in the red dotted-line boxes) in a hourly observation smart water meter data, which records 350 hours water consumption for a household in the area of Kalgoorlie-Boulder in Western Australia.

© Springer International Publishing Switzerland 2015
T. Cao et al. (Eds.): PAKDD 2015, Part II, LNAI 9078, pp. 522–533, 2015.
DOI: 10.1007/978-3-319-18032-8_41

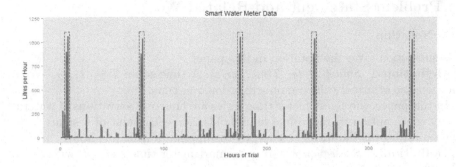

Fig. 1. An example routine in a smart water meter time series

This recurrent routine corresponds to a 2-hour water use activity, occurring 5 times during the 350 hours. This routine uses about 880 litres water in the first hour and approximately 1000 litres water in the second hour.

The problem of discovering routines is to find all repeated subsequences with various lengths in a smart meter time series, which is similar to motif discovery in previous work [2–6]. The work in [2][3] defined motifs as previous unknown patterns that frequently occur in a time series. Efficient motif detection algorithms were proposed by encoding subsequences as symbolic representation. Exact motif discovery was computationally intractable until the work in [4][5] proposed an efficient way to early terminate searching. Later, Mueen [6] proposed a novel algorithm to efficiently enumerate motifs of all possible lengths. However, instead of finding the most frequently occurring patterns, the work in [4–6] defined the motif as the most similar subsequence pair in a time series.

Detecting routines in smart meter time series is more challenge in three aspects. First, we are not only interested in segment shapes but also raw values of segments, since the segment shapes reflect water and energy use trend while the raw values provide exact water and energy consumption. Second, the subsequence length is much shorter than that in previous work, since routines normally last for only a few hours. Finally, since we do not have prior knowledge about the length of routines, we need to detect all repeated subsequences with variable length.

In this paper, we first present a brute-force algorithm to detect routines in smart meter time series. Then, we propose a novel subsequence growing algorithm to efficiently discover all routines with various lengths in smart meter time series. The contribution of the paper is three-fold. First, we formally formulate the problem of discovering routines in smart meter time series. Second, a novel algorithm is proposed to efficiently detect routines by growing subsequences. Third, we demonstrate the application of the proposed algorithm on a real-world dataset and evaluate the proposed algorithm using a synthetic dataset.

2   Problem Statement and Related Work

2.1   Notation

This subsection gives the notation in this paper.

Definition 1. *Smart Meter Time Series*: A time series $T = (t_1, t_2, \cdots, t_n)$ is a sequence of n real valued numbers ordered in time.

In this paper, the smart meter time series are hourly observations of water or energy consumption by individual households, i.e., each point of a smart meter time series records the water or energy consumption during one hour.

Definition 2. *Subsequence*: Given a smart meter time series T with length n, a subsequence S of T is a subset of m consecutive observations from T, i.e., $S_p^m = (t_p, \cdots, t_{p+m-1})$, where $1 \leq p \leq n - m + 1$, and $m < n$.

In the context of smart meter time series, a subsequence with length m represents a consumption activity over m consecutive hours.

Definition 3. *Magnitude*: Given a subsequence S_p^m with length of m, the magnitude of S_p^m is the maximum of all the elements in the subsequence i.e., $Mag(S_p^m) = max(t_k)$, where t_k is the kth element of S_p^m.

Definition 4. *Match*: Given two subsequences, S_i^m and S_j^m, with the same length of m from a smart meter time series T, if the distance between the two subsequences is less than a threshold R, i.e., $Dist(S_i^m, S_j^m) < R$, then the two subsequences are *matched*. If two subsequences are matched, they are also called *similar* with each other.

Definition 5. *Trivial Match*: Given a subsequence S_i^m and its matched subsequence S_j^m ($i < j$), if for all k between i and j, $Dist(S_i^m, S_k^m) \leq R$, then S_j^m is a *trivial match* of S_i^m.

Trivial match rarely occurs in smart meter time series because smart meter time series are always spiky.

The distance function $Dist(.,.)$ is critical to the definition of *match*. We define the distance measurement based on Euclidean distance.

Definition 6. *Distance*: Given two subsequences, S_i^m and S_j^m, with the same length of m, the distance between S_i^m and S_j^m is defined based on the Euclidean distance between S_i^m and S_j^m, i.e., $Dist(S_i^m, S_j^m) = \sqrt{\sum_{k=0}^{m-1}(t_{i+k} - t_{j+k})^2/m}$.

We incorporate the subsequence length m into the distance metric to fairly compare distances of subsequence pairs with different lengths, since longer subsequence pair has larger standard Euclidean distance than their corresponding shorter subsequence pair.

Definition 7. *Motif*: Given a smart meter time series T, a subsequence length m, and a distance threshold R, the most significant motif is the subsequence with length m that has most number of matched occurrences under the distance threshold, i.e., $\forall i, j : Dist(S_i^m, S_j^m) < R$. The Kth significant motif in T is the motif that has the Kth most number of matched occurrences.

Definition 8. *Recurrent Routine*: Given a frequency threshold C and a magnitude threshold G, a recurrent routine in a smart meter time series is a motif that has at least C matched occurrences in the time series, each of which has at least G magnitude.

The parameters magnitude threshold G and frequency threshold C guarantee that water or energy used by the routines is significant enough to be interesting and these behaviours occur regularly. These parameters are determined in terms of domain application requirement.

2.2 Why the Problem is Difficult

One big challenge of our problem is that it is difficult to represent smart meter segments (behaviours) using low dimensional representations (e.g., symbolic conversion) without losing exact consumption information. Representing segments into a low dimension in previous motif discovery work [2][3] requires to discretize and normalize the segments, which only keeps the consumption trend but removes exact consumption information. Furthermore, encoding a short smart meter segment that only consists of several data points into a low dimensional representation (such as symbolic string [2][3]) does not save computation cost, as the dimension of the segment is already very low. Therefore, the previous *approximate* motif discovery algorithms are not well suited to the problem of routine discovery in smart meter time series.

Our problem is similar to the problem of exactly finding motifs in time series, which is thought to be intractable until the work in [4] shows a representational trick to efficiently estimate lower bound of subsequence distance. Unlike approximation motif discovery algorithms that convert subsequences into a low dimension, exact algorithms [4–6] calculate distance based on raw values of subsequences, which is able to keep exact consumption information in smart meter time series. However, instead of finding the most frequent patterns, the existing exact algorithms define the motif as the most similar subsequence pair in a time series, i.e., the exact algorithms only find the most similar pair in a time series [4–6]. Although it is argued that the exact algorithms can be extended to find the most frequent subsequence patterns by finding other occurrences of the most similar pair within a distance threshold, the extension is not easy and efficient.

3 Brute-Force Algorithm

3.1 Discovering Routines with Fixed Length

The algorithm to discover routines with fixed length (DRFL) is illustrated in algorithm 1. The output of algorithm 1 is a set of m-length routines B^m, each of which comprises a cluster centre (motif) and its corresponding subsequence instances. Functions Cent(B_i) andInst(B_i) in the following algorithms return the cluster centre of B_i and the list of corresponding subsequence instances, respectively. First, a sub-window with length m is slid along the time series to extract a group of candidate subsequences (line 1-2). Then, these candidate subsequences are input into a sequential cluster algorithm for grouping (line 3), which is given in algorithm 2. Since the sub-window is slid point by point, multiple clusters may actually correspond to the same routine. If two clusters

---

Algorithm 1. Discovering routine with fixed length (DRFL)

Input   : A n length time series
Parameters: routine length m, distance threshold R; frequency threshold C
              and magnitude threshold G
Output   : m-length routines B^m

1 **for** $i = 1$ **to** $n - m - 1$ **do**
2 $\quad$ extract subsequence S_i^m;

3 $\texttt{SubGroup}(S^m, R, C, G) \rightarrow$ output B^m;
4 **for** $i = 1$ **to** $|B^m| - 1$ **do**
5 $\quad$ **for** $j = i$ **to** $|B^m|$ **do**
6 $\quad\quad$ $\texttt{OverlapTest}(\texttt{Inst}(B_i^m), \texttt{Inst}(B_j^m), \epsilon) \rightarrow$ output K_i, K_j;

7 **for** $i = 1$ **to** $|B^m|$ **do**
8 $\quad$ **if** $K_i == FALSE$ **then** remove B_i^m;

---

correspond to the same routine, their instance occurrences will overlap with each other. Therefore, an overlap testing algorithm is introduced to determine if two clusters correspond to the same routine (line 4-6), the detail of this step is given in algorithm 3.

Algorithm 2 gives the detail of sequential clustering of a set of subsequences. First, we initialize the first cluster centre (potential routine) $\texttt{Cent}(B_1)$ as the first subsequence S_1. When a new subsequence S_i arrives, if the magnitude of S_i is not larger than the magnitude threshold G, this subsequence will be discarded. Otherwise, if it is larger than the magnitude threshold G, the distance between the new subsequence with each of cluster centres $\texttt{Cent}(B_j)$ is calculated (line 4-5). The cluster centre $\hat{j}$ that has minimum distance to the new subsequence is selected (line 6). If the minimum distance is smaller than the given distance threshold R and S_i is not a trivial match with any of the instances of $B_{\hat{j}}$ (function $\texttt{NTM}()$), the new subsequence S_i is grouped into the $\hat{j}$ cluster (line 7-9). Otherwise, a new cluster centre is created by inserting the new subsequence S_i (line 11-12). Finally, the cluster centres that have small number of instances are removed (line 13-14), since only the behaviours that occur multiple (regular) times are routines.

Algorithm 3 gives the detail of overlap testing for two lists of subsequences (i.e., instances from two clusters). Suppose a p-length subsequence S_i^p with start point at i, a q-length subsequence S_j^q with start point at j in a time series, and $p < q$, if the inequality equation, $((i + p) < j) \vee ((j + q) < i)$, holds, then the two subsequences are not overlapped with each other.

Since subsequences are extracted by sliding a sub-window along the time series, the subsequences in a cluster are temporally ordered. Suppose that we are trying to insert a subsequence in the first list into the second list according to its temporal order (line 3-4, I_i^m is the temporal start point of subsequence S_i^m), the inserted subsequence in the first list can only be overlapped with its adjacent subsequences in the second list. Therefore, we test if the inserted subsequence

---

Algorithm 2. Subsequence grouping (SubGroup)

    Input   : A list of subsequences S
    Parameters: distance threshold R; frequency threshold C and magnitude
               threshold G
    Output   : routines B

1 **Initialization:** $\text{Cent}(B_1) = S_1$; $|B| = 1$;
2 **for** $i = 1$ **to** $|S|$ **do**
3   **if** $\text{Mag}(S_i) > G$ **then**
4   **for** $j = 1$ **to** $|B|$ **do**
5   calculate $\text{Dist}(S_i, \text{Cent}(B_j))$;
6         $\hat{j} = \arg\min_j \text{Dist}(S_i, \text{Cent}(B_j))$;
7   **if** $\forall S' \in \text{Inst}(B_{\hat{j}}): \text{NTM}(S_i, S')$ **then**
8   append S_i into $B_{\hat{j}}$: $S_i \rightarrow \text{Inst}(B_{\hat{j}})$;
9   update centre $\text{Cent}(B_{\hat{j}})$;
10   **else**
11   create a new center $\text{Cent}(B_{|B|+1})$;
12   append S_i into $B_{|B|+1}$, i.e., $S_i \rightarrow \text{Inst}(B_{|B|+1})$;

13 **for** $k = 1$ **to** $|B|$ **do**
14   **if** $|\text{Inst}(B_k)| < C$ **then** remove B_k;

---

is overlapped with its adjacent subsequences. If it is overlapped, the number of overlapped instance will increase by 1 (line 5). This process is repeated for all the subsequences in the first list. Finally, if the number of overlapped subsequences is larger than the frequency threshold, then the two lists of subsequences are determined duplicated (line 7), and only the subsequence list that is longer or has larger magnitude will be kept (line 8-12).

3.2 Discovering Routines with Various Lengths

Extending the routine discovery algorithm with fixed length to find all routines with variable length between L_{min} and L_{max} is straightforward. We can repeat the DRFL algorithm by incrementally increasing the subsequence length from L_{min} to L_{max} to enumerate all the possible routines.

4 Subsequence Growing Algorithm

4.1 The Intuition Behind Subsequence Growing Algorithm

The motivation of the proposed algorithm is to make use of shorter motif information to simplify the length enumeration. Figure 2 illustrates the intuition behind the proposed algorithm. The three 4-point length subsequences in the time series are matched with each other. Our target is to find these three similar subsequences and their corresponding motif (cluster) without the knowledge

Algorithm 3. Efficient overlap testing (OverlapTest)

Input	: Two lists of subsequences S^m and S^n
Parameters:	Overlap threshold ϵ
Output	: keep tag K^m and K^n

1   **Initialization**: $K^m = TRUE$; $K^n = TRUE$; $N = 0$; $i = 1$; $j = 2$;
2   **while** $i \leq |S^m| \wedge j \leq |S^n|$ **do**
3   **while** $I_i^m > I_j^n \wedge j \leq |S^n|$ **do**
4         $\lfloor \; j = j + 1$;
5   **if** $IsOverlap(S_i^m, S_{j-1}^n) \vee IsOverlap(S_i^m, S_j^n)$ **then** $N = N + 1$;
6       $i = i + 1$
7   **if** $N > \epsilon * \min(|S^m|, |S^n|)$ **then**
8   **if** $m > n$ **then** $K^m = TRUE$; $K^n = FALSE$;
9   **else if** $m < n$ **then** $K^m = FALSE$; $K^n = TRUE$;
10   **else**
11   **if** $\text{Mag}(S^m) > \text{Mag}(S^n)$ **then** $K^m = TRUE$; $K^n = FALSE$;
12   **else** $K^m = FALSE$; $K^n = TRUE$;

of subsequence length. Three shorter segments that only contains 2 points can be extracted from each of the longer subsequences. For the shorter segments extracted from the longer subsequences with the same location offset (such as the red segments), they are also similar to each other as well, i.e., they belong to a motif with length of 2. Three motifs with length of 2 can be detected by grouping the shorter segments. The key insight of our algorithm is that shorter motifs and their subsequence instances contain the location information of all potential longer motifs and their subsequences, which are interesting routines. The observation is that if two longer subsequences are similar with each other, then segments extracted from the two subsequences, respectively, must also be similar.

This observation leads to incrementally grow short segments to longer subsequences and only group the candidate longer subsequences into clusters (motifs). This subsequence growing method avoids repeatedly sliding different size sub-windows along a time series to extract candidate subsequences with different lengths, which has a high computational cost. In another words, instead of sliding a sub-window to extract longer subsequences, the segment growing method grows candidate longer subsequences from shorter motifs, which can be detected with a minimum length parameter.

4.2 Formal Statement of Subsequence Growing Algorithm

The proposed subsequence growing algorithm consists of three steps: detecting routines with minimum length; clustering candidate subsequences grown from shorter subsequences; and overlap testing to remove duplicate routines. Algorithm 4 describes the detail of discovering routines by growing subsequences.

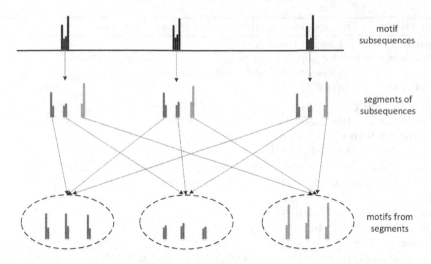

Fig. 2. The intuition behind the subsequence growing algorithm

The algorithm discovers all routines with lengths between L_{min} and L_{max} from a smart meter time series.

First, DRFL is performed with the shortest subsequence length L_{min} (line 1), which extracts all the routines that have the minimum length L_{min}. Then, we incrementally increase the subsequence length from $L_{min} + 1$ to L_{max} (line 2). For a certain longer subsequence length i, there are $i - L_{min} + 1$ possible ways to grow an i-length subsequence from a L_{min}-length subsequence. Therefore, for each of the $i - L_{min} + 1$ possible growing ways (line 3), we grow the instances (L_{min}-length subsequences) of a routine $B_k^{L_{min}}$ to extract candidate longer subsequences $S^{i,j,k}$ (Line 4-5). Each set of the longer subsequences $S^{i,j,k}$ are then input into the subsequence grouping algorithm (algorithm 2) so that similar longer subsequences are grouped into a cluster. The subsequence grouping algorithm guarantees that the discovered longer routines have a large number of instances and correspond to high consumption. Once the routines are discovered for each subsequence length and growing way, all of the discovered routines are concatenated with the routines with shortest length to form a final routine list B.

Since subsequence instances of multiple routines in B may be grown from subsequence instances of the same shortest routine, there maybe duplicated routines in B that actually correspond to the same routine. In order to remove the duplicated routines from B, the overlap test algorithm (algorithm 3) is performed to identify duplicate routines that should be removed. Similar to the overlap test step in algorithm 1, each routine in B is compared with the other routines to determine if they are duplicated or if it will be kept. Finally, only one copy of duplicated routines that is most interesting in terms of longer length and higher magnitude will be kept and the remaining duplicated copies will be removed (line 8-12).

Algorithm 4. Discovering routines by growing subsequences

Input   : A n length time series
Parameters: routine length range $[L_{min}, L_{max}]$, distance threshold R;
                frequency threshold C and magnitude threshold G
Output   : routines B

1 DRFL(L_{min}, R, C, G);
2 **for** $i = L_{min} + 1$ **to** L_{max} **do**
3   **for** $j = 0$ **to** $(i - L_{min})$ **do**
4   **for** $k = 1$ **to** $|B^{L_{min}}|$ **do**
5   grow instances of $B_k^{L_{min}}$ to i-length subsequences: $S^{i,j,k}$;
6   SubGroup$(S^{i,j,k}, R, C, G) \rightarrow$ output $B^{i,j,k}$;

7 concatenate all $B^{i,j,k}$ with $B^{L_{min}}$;
8 **for** $i = 1$ **to** $|B| - 1$ **do**
9   **for** $j = i$ **to** $|B|$ **do**
10   OverlapTest$(\text{Inst}(B_i), \text{Inst}(B_j), \epsilon) \rightarrow$ output K_i, K_j;

11 **for** $i = 1$ **to** $|B|$ **do**
12   **if** $K_i ==FALSE$ **then** remove B_i;

---

5   Experiment

5.1   Case Study on Real Dataset

Real Dataset. The dataset compromises one year of hourly smart water meter readings from 500 households in the city of Kalgoorlie-Boulder, 600 km inland from the capital Perth in Western Australia.

Discovered Routine Behaviours. Figure 3 shows five routine behaviours discovered by the proposed algorithm for one household in the Kalgoorlie data set. These routines occur over 367 days with an average of 24 recurrences per routine. The black, red, and blue routines occur only on Wednesdays and Saturdays at the same hour of the day: either 4am or 8pm, with a few occurrences at 5pm. These routines are most likely associated with an automated garden watering system. The green and cyan routines have less regular timing, occurring on any day of the week except Sunday and at different times of day between 7am and 11pm. The total water use of these routines is 184 kL which is 34% of all water use for this household. These results demonstrate that identifying routines provides useful information for water providers and users. They capture different types of regularity, different length patterns, and identify patterns with potential for significant water savings.

5.2   Evaluation on Synthetic Dataset

Synthetic Dataset. A synthetic dataset with ground truth is generated to fairly evaluate the performance of the proposed algorithm. Four groups of primi-

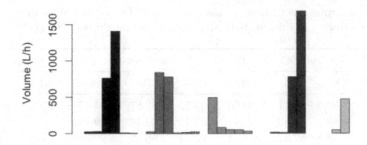

Fig. 3. Five recurrent routines for a Kalgoorlie-Boulder household

tive subsequences with random small variations (within 30) are randomly planted onto a long base sequence with different number of instances. The four groups of primitive subsequences are as below.

Frequent and interesting: Primitive subsequences in this group have large magnitude G and they occur from 2.5 to 5 times of C (frequency threshold). The primitive subsequences in this group are $\langle 824, 726 \rangle$, $\langle 580, 691, 472, 575 \rangle$, $\langle 1242, 1985, 1058 \rangle$, $\langle 886, 2422, 2380, 2465, 1718 \rangle$ and $\langle 1272, 1607, 1182, 794 \rangle$.

Frequent but uninteresting: The magnitude of primitive subsequences in this group are not large enough but their instance number is from 1 to 2 times of the frequency threshold C. The primitive subsequences in this group are $\langle 123, 305 \rangle$, $\langle 233, 246, 289 \rangle$ and $\langle 324, 56, 152, 203 \rangle$.

Not frequent but interesting: Primitive subsequences in this group have large magnitude but their instance number is less than the frequency threshold C. The primitive subsequences in this group are $\langle 612, 212 \rangle$, $\langle 165, 854, 328 \rangle$ and $\langle 1422, 132, 68, 534 \rangle$.

Combined: Primitive subsequences combined from interesting and uninteresting primitive subsequences. Their instance number is below the frequency threshold C. The primitive subsequences in this group are $\langle 233, 246, 289, 1242, 1985, 1058 \rangle$ and $\langle 824, 726, 123, 305 \rangle$.

Only the subsequences in the group of *frequent and interesting* are the routines that need to be detected. The ground truth of the synthetic dataset is the location of the planted primitive subsequences in the group of *frequent and interesting*.

Correctness on Synthetic Dataset. Precision, recall and F-measure [7] are used to quantitatively evaluate the performance of the proposed algorithm. Let P be a set of planted subsequence instances (patterns) and D be the set of detected motif instances, the precision, recall and F-measure are defined as:
$$precision = \frac{|P \cap D|}{|D|}; \ recall = \frac{|P \cap D|}{|P|} \text{ and } F\text{-}measure = 2 \cdot \frac{precision \cdot recall}{precision + recall}.$$
100 synthetic time series with length of 10000 are generated to evaluate the performance of the subsequence growing algorithm. Table 1 gives the precision, recall, F-measure and running time of the subsequence growing algorithm and the brute-force algorithm with different distance thresholds on the synthetic dataset.

Table 1. Correctness and running time (seconds) of brute-force algorithm (BF) and subsequence growing algorithm (SG) with various distance thresholds (DT)

DT	Precision (%)		Recall (%)		F-Measure (%)		Time	
	BF	SG	BF	SG	BF	SG	BF	SG
20	27.82	30.11	40.60	39.51	32.87	34.07	27.4	8.1
50	98.88	98.88	99.41	99.41	99.14	99.14	16.2	6.3
60	97.48	97.48	99.74	99.74	98.59	98.59	15.3	5.9
100	94.45	97.01	99.00	98.47	96.65	97.71	14.8	5.7
150	91.23	95.68	99.87	100	95.31	97.80	17.1	6.5
200	90.46	94.66	98.49	99.86	96.65	97.71	14.2	5.9

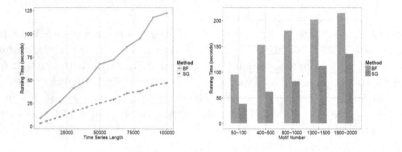

(a) Different sequence lengths   (b) Different number of instances

Fig. 4. Scalability experiment on synthetic dataset

Since the small variations added to the primitive subsequences are within 30, the optimal distance threshold should be approximate 60. We intentionally varied the distance threshold to test the sensitivity of the algorithm to the distance threshold.

It can be seen that both the subsequence growing algorithm and brute-force algorithm achieve very high accuracy (F-measure over 95%) when the distance threshold is between 50 and 200. The accuracies obtained by the subsequence growing algorithm and the brute-force algorithm are comparable. However, the subsequence growing algorithm takes significantly lower computation time than the brute-force algorithm. This is mainly because the subsequence growing algorithm avoids repeatedly sliding a sub-window with variable size to extract candidate subsequences.

Scalability Experiment. A scalability experiment is conducted to compare the computational cost of the subsequence growing algorithm and brute-force algorithm. All the experiments are implemented in R on a computer with an Intel I7-4700 2.4 GHz processor with 8 GB RAM. First, we produce 11 sets of synthetic datasets with different lengths from 5000 to 100,000. We ran the algorithms for 10 times. The average execution time on a time series with respect to time series length is shown in Figure 4a. Second, we fix the length of base sequences to 80,000 and planted different number of subsequence primitives onto the base sequences. Five sets of time series are generated with different number of subsequence instances from 50 to 100 (2.5 to 5 times of frequency threshold C), from 400 to 500, from 800 to 1000, from 1300 to 1500, and from 1800 to 2000, respectively. The average execution time of the two algorithms on a time series with respect to the number of subsequence instances are shown in Figure 4b. From the figure, it can be seen that the average execution time of the subsequence growing algorithm is much less than that of the brute-force algorithm, which demonstrates that the subsequence growing scales better than the brute-force algorithm with respect to the length of time series and number of instances.

6   Conclusion

In this paper recurrent routine behaviours were introduced to characterize regular consumption behaviours from smart meter time series. we proposed a novel algorithm to efficiently discover recurrent routine behaviours in smart meter time series by growing subsequences. The proposed algorithm incrementally grows longer subsequences from shorter subsequences to avoid enumerating all possible subsequence lengths. We demonstrated the discovered routine behaviours in a real-world dataset, and evaluated the performance of the proposed algorithm on a synthetic dataset.

References

1. Cardell-Oliver, R.: Water use signature patterns for analyzing household consumption using medium resolution meter data. Water Resources Research **49**(12), 8589–8599 (2013)
2. Lin, J., Keogh, E., Lonardi, S., Patel, P.: Finding motifs in time series. In: Proceedings of the 2nd Workshop on Temporal Data Mining, at the 8th ACM SIGKDD, Alberta, Canada, pp. 53–68 (2002)
3. Patel, P., Keogh, E., Lin, J., Lonardi, S.: Mining motifs in massive time series databases. In: 2002 IEEE ICDM, pp. 370–377 (2002)
4. Mueen, A., Keogh E.J., Zhu, Q., Cash, S., Westover, B.: Exact discovery of time series motifs. In: SIAM International Conference on Data Mining. American Statistical Association (ASA) (2009)
5. Mueen, A., Keogh, E.: Online discovery and maintenance of time series motifs. In: 16th ACM SIGKDD, New York, USA, pp. 1089–1098 (2010)
6. Mueen, A.: Enumeration of time series motifs of all lengths. In: 2013 IEEE 13th International Conference on Data Mining (ICDM), pp. 547–556, December 2013
7. Powers, D.M.W.: Evaluation: from precision, recall and f-measure to roc, informedness, markedness and correlation. International Journal of Machine Learning Technology **2**(1), 37–63 (2011)

Convolutional Nonlinear Neighbourhood Components Analysis for Time Series Classification

Yi Zheng[1,2], Qi Liu[1], Enhong Chen[1(✉)], J. Leon Zhao[2],
Liang He[1], and Guangyi Lv[1]

[1] School of Computer Science and Technology,
University of Science and Technology of China, Hefei, China
{xiaoe,hsh105,gylv}@mail.ustc.edu.cn,
{qiliuql,cheneh}@ustc.edu.cn
[2] Department of Information Systems, City University of Hong Kong,
Hong Kong, China
jlzhao@cityu.edu.hk

Abstract. During last decade, tremendous efforts have been devoted to the research of time series classification. Indeed, many previous works suggested that the simple nearest-neighbor classification is effective and difficult to beat. However, we usually need to determine the distance metric (e.g., Euclidean distance and Dynamic Time Warping) for different domains, and current evidence shows that there is no distance metric that is best for all time series data. Thus, the choice of distance metric has to be done empirically, which is time expensive and not always effective. To automatically determine the distance metric, in this paper, we investigate the distance metric learning and propose a novel Convolutional Nonlinear Neighbourhood Components Analysis model for time series classification. Specifically, our model performs supervised learning to project original time series into a transformed space. When classifying, nearest neighbor classifier is then performed in this transformed space. Finally, comprehensive experimental results demonstrate that our model can improve the classification accuracy to some extent, which indicates that it can learn a good distance metric.

1 Introduction

Among time series data mining tasks, the classification has attracted amount of interest during last decade. Actually, many studies on time series classification methods have been proposed and it is suggested that Nearest Neighbor classifier (especially, 1-NN) is difficult to beat [1,3]. Since the performance of 1-NN algorithm depends critically on the distance metric given for specific tasks, the subsequent question then becomes how to determine the distance metric for so many applications.

A number of different distance metrics have been proposed. Among them, two of the most widely used are Euclidean distance and Dynamic Time Warping

© Springer International Publishing Switzerland 2015
T. Cao et al. (Eds.): PAKDD 2015, Part II, LNAI 9078, pp. 534–546, 2015.
DOI: 10.1007/978-3-319-18032-8_42

(DTW) [1,3,22]. Euclidean distance is simple and efficient, and it could achieve a good performance for certain applications. In contrast, DTW introduces the alignment of two sequences and allows the points of different time stamps to match, which leads to even better performance than Euclidean distance for some scenarios. However, one of the deficiencies of DTW is that it needs more time cost when calculating the distance. Also, even though 1-NN with DTW can achieve best performance in many domains, for some other applications, it performs no better than other distance metrics. In summary, current evidence shows that there is no distance metric that is best for all time series data [3]. Typically, the choice of distance metric has to be determined empirically, which is time expensive and not always effective. Hence, we believe that it's a challenge to choose a suitable distance metric for the specific data set automatically.

Inspired by the learning perspective, we investigate to use distance metric learning to obtain better distance metric and further to improve the classification performance for time series data. Indeed, many distance metric learning methods have been proposed. For instance, [4] provided a linear transformation model named Neighbourhood Components Analysis (NCA) to optimize the performance of k-NN in the learnt low-dimensional space. As [19] noted, the linear transformation has a limitation that "it cannot model higher-order correlations between the original data dimensions". Hence, [19] proposed a nonlinear distance metric learning model named Nonlinear NCA (NNCA). The discovered low-dimensional representations could work better than previous linear NCA. Unfortunately, both Linear NCA (LNCA) and NNCA models cannot capture the intrinsic property of the time series data, i.e., time shift.

To capture the time shift property, in this paper, we consider the merit of Convolutional Neural Network (CNN), e.g., invariance of spatial-temporal, and propose a novel distance metric learning method for time series. Specifically, we follow NNCA model [19] and propose a novel Convolutional Nonlinear Neighbourhood Components Analysis (CNNCA) model, which could not only learn a nonlinear transformation from the data but also naturally capture the time shift of sequences. Based on the learnt distance metric, 1-NN classifier would be used to perform the classification. Moreover, we conduct comprehensive experiments on the data sets from UCR Time Series repository [7]. By comparing to conventional Euclidean distance, DTW and window constraint DTW, the experimental results reveal the classification performance is improved for many data sets, especially for the data sets that have sufficient training samples for each class. On the other hand, we also evaluate the efficiency of each method. It reveals that CNNCA is more efficient for larger data set and long time series. We summarize the contributions of this paper in these parts:

- Though there are several studies that have explored the distance metric learning for time series data [12,15], to the best of our knowledge, we are the first to consider the time shift property when learning distance metric for the time series classification task.
- Along this line, we propose a novel distance metric learning method CNNCA for time series data, which can obtain combined feature representation by

concatenating CNN and Multiple Perceptron (MLP), and then learn a distance metric based on the scheme of stochastic neighbour assignments.
- We conduct comprehensive experiments on amount of public data sets, then compare the performance of CNNCA with other distance metrics, including not only three conventional distance metrics, but also two learnt by LNCA and NNCA. The results prove that CNNCA can improve classification accuracy to some extent, especially for the relatively large scale data sets.

The rest of this paper is organized as follows. Section 2 shows the related studies. Definitions of time series and relevant distance metric learning methods are given in section 3. In section 4, the CNNCA is introduced and comprehensive experiments are presented in section 5. Finally, we conclude the paper and give the future work in section 6.

2   Related Work

We group the related studies into two categories. In the first category, researchers focus on improving the performance of time series classification by choosing distance metrics combined with 1-NN classifier. As [1,16,22] claimed, the Nearest Neighbour (NN) classification algorithm (especially 1-NN) has been empirically proven as the current state-of-the-art [1,16,22]. Then the challenge of 1-NN is how to determine the distance metric for specific data sets. Extensive experiments have been conducted by [3] on amount of time series data sets and many distance metrics have been evaluated, i.e., Manhattan distance, Euclidean distance, L_∞-norm, DISSIM, DTW, LCSS, EDR, Swale, ERP, TQuEST, SpADe [3]. According to the experimental results, they concluded that there is no clear evidence that there exists one similarity measure that is superior to others for most of data sets. Hence, for specific data set, it is challenging to determine a suitable distance metric for better performance.

In the second category, researchers concentrate on the distance metric learning (or manifold learning). Essentially, the aim of distance metric learning is to learn either a linear or nonlinear transformation based on the original data for further tasks (e.g., classification, clustering or visualization) [4,12,15,19]. For instance, [4] proposed a method by optimizing the expected leave-one-out error of a stochastic nearest neighbor classifier in the projection space, which can learn a linear distance metric to be used for data visualization and fast classification. [19] said that the linear transformation cannot capture the higher-order correlations between original data dimensions and proposed a nonlinear NCA model, which stacks multiple neural networks to learn the nonlinear transformation for handwritten digit recognition task. To the best of our knowledge, there are only several existing studies using distance metric learning on time series classification. For instance, [15] considered to learn a variation of Mahalanobis distance and performed the time series classification with 1-NN algorithm. They concluded that such a kind of distance is inferior to DTW in accuracy but it is more efficient. Recently, [12] proposed two novel models to learn a task-specific similarity measure for time series data, however, the transformation is still linear.

In general, existing distance metric learning methods either linear or nonlinear cannot capture the time shift property well, thus the performance of time series classification are suffered. Motivated from the nonlinear distance metric learning and utilizing the merit of CNN, we will propose a convolutional nonlinear NCA model to learn a better distance metric for time series, and further improve the performance of classification.

3   Preliminaries

In this section, we provide preliminaries for our work. Specifically, we first give the definitions of time series and subsequence. Then, two related distance metric learning models are explained.

3.1   Definitions of Time Series and Subsequence

Definition 1 *A time series (denoted as T) is a sequence of data points, measured typically at successive points in time spaced at uniform time intervals. A time series can be denoted as $T = t_1, t_2, ..., t_n$, and n is the length of T.*

Following the previous works [5], we first extract some subsequences from the long time series instead of classifying time series with the whole sequence. Then, we proceed the classification with these subsequences, since the pattern or shape in the subsequences of time series could be a key feature to distinguish different classes of time series. The subsequence is defined as follows.

Definition 2 *Subsequence is a series of consecutive points which are extracted from a long time series T and could be denoted as $s = t_i, t_{i+1}, ..., t_{i+k-1}$, where k is the length of subsequence, and we have $1 \leq i \leq n$, $1 \leq k \leq n$ and $i + k - 1 \leq n$.*

Three conventional distance metrics are most widely used: Euclidean distance, DTW and window constraint DTW. Due to space limitations, we skip the details of these distance metrics (which could be found in [3,22]).

3.2   Distance Metric Learning

Many distance metric learning methods have been proposed during last decade [4,19]. In this paper, we concentrate on two preliminary methods on Neighborhood Components Analysis (NCA), i.e., linear and nonlinear NCAs.

Linear Neighbourhood Components Analysis (LNCA). Based on stochastic neighbour assignments in the transformed space, [4] introduced a differentiable cost function for learning neighbour components analysis. Specifically, for each point x_i, it selects another point x_j as its neighbour with the probability p_{ij}, and furthermore, x_i would be classified as the label of point x_j with the same probability. In the softmax scheme, the definition of p_{ij} with Euclidean distance is shown in Equation 1.

$$p_{ij} = \frac{exp(-\|Ax_i - Ax_j\|^2)}{\sum_{k \neq i} exp(-\|Ax_i - Ax_k\|^2)}, \quad p_{ii} = 0, \tag{1}$$

where A is the matrix that needs to be learnt for transforming the input data linearly. Based on such a stochastic neighbour assignments scheme, the probability that point x_i would be classified correctly is computed as follows.

$$p_i = \sum_{j \in C_i} p_{ij}, \tag{2}$$

where C_i represents the set of points that have same class label as point x_i, and c_i denotes the class label of point x_i then we define this set as $C_i = \{j | c_i = c_j\}$. The objective function of LNCA is shown in Equation 3, which is also the expected number of points that is correctly classified.

$$\mathcal{L} = \sum_i \sum_{j \in C_i} p_{ij} = \sum_i p_i. \tag{3}$$

To maximize the objective function, the common method is to use a gradient based optimizer according to the derivative of $\mathcal{L}$. When we denote that $x_{ij} = x_i - x_j$, then the derivative of $\mathcal{L}$ with respect to A is derived in Equation 4.

$$\frac{\partial \mathcal{L}}{\partial A} = -2A \sum_i \left(p_i \sum_k p_{ik} x_{ik} x_{ik}^\top - \sum_{j \in C_i} p_{ij} x_{ij} x_{ij}^\top \right). \tag{4}$$

Nonlinear Neighborhood Components Analysis (NNCA). The limitation of linear transformation is that it cannot capture the higher-order correlations between original data dimensions [19]. Based on LNCA and by introducing a multilayer neural network, [19] proposed a Nonlinear Neighborhood Components Analysis (NNCA) model.

In contrast to Equation 1, for NNCA model, the probability that point x_i selects one of its neighbours x_j and inherits the class label of x_j is defined in Equation 5.

$$p_{ij} = \frac{exp(-\|f(x_i) - f(x_j)\|^2)}{\sum_{k \neq i} exp(-\|f(x_i) - f(x_k)\|^2)}, \quad p_{ii} = 0, \tag{5}$$

where $f(\cdot)$ is the nonlinear transformation learnt by a multilayer neural network, which is different from the linear transformation of LNCA in Equation 1 (i.e., Ax_i). Besides that, the subsequent process of NNCA model is similar to LCNA as shown in Equation 2 and 3, which includes the probability that point x_i belongs to a certain class z and the objective function. The optimization of the objective function is performed with gradient ascent method. Denote $x_{ij} = x_i - x_j$, the derivative of $\mathcal{L}$ with respect to $f(x_i)$ is derived as:

$$\frac{\partial \mathcal{L}}{\partial f(x_i)} = 2 \left(\sum_{l:c^l=c^i} p_{li} d_{li} - \sum_{l \neq i} \left(\sum_{q:c^l=c^q} p_{lq} \right) p_{li} d_{li} \right)$$

$$-2 \left(\sum_{j:c^i=c^j} p_{ij} d_{ij} - \sum_{j:c^i=c^j} p_{ij} \left(\sum_{z \neq i} p_{iz} d_{iz} \right) \right). \tag{6}$$

Through computing gradient and iterating to update the parameters, then we could obtain the nonlinear transformation when the model convergent.

Even though NNCA can learn a nonlinear transformation of the input space, it does not consider the time shift and still cannot capture the intrinsic property of time series. Therefore, for time series classification, both of LNCA and NNCA cannot achieve good performance. We will verify this in the experiments.

4 Convolutional Nonlinear NCA (CNNCA)

In this section, we show the novel distance metric learning model CNNCA, including the architecture and the learning procedure. Meanwhile, we explain how to perform classification with CNNCA at the end of this section.

4.1 Architecture

We follow the scheme of NCA model and extend the nonlinear NCA model for subsequent classification. Specifically, we propose a novel Convolutional Nonlinear Neighborhood Components Analysis (CNNCA) model to learn a better distance metric for time series. By consideration of time shift property of time series, the motivation of introducing CNN into distance metric learning is that convolutional and pooling operations can preserve the spatial and temporal locality, i.e.,

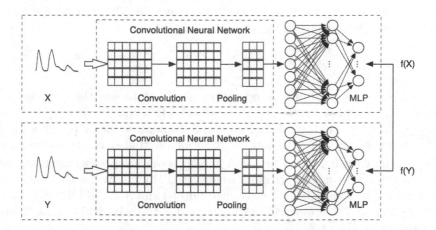

Fig. 1. Architecture of Convolutional Nonlinear Neighborhood Components Analysis. X and Y represent two time series that have identical class label. $f(X)$ and $f(Y)$ denote the nonlinear transformation.

CNN has the advantage of time shift invariance to some extent [9], which may improve the performance of subsequent classification. Furthermore, MLP can combine the feature representations learnt by CNN and perform nonlinear transformation for better classification. Hence, CNNCA extends LNCA by combining CNN and MLP, in other words, the distance d_{ij} between two projected points with respect to x_i and x_j, is calculated in this form: $d_{ij} = \|f(x_i) - f(x_j)\|^2$, where $f(\cdot)$ defines a nonlinear transformation through convolutional neural networks and multilayer perceptron. We illustrate the architecture of CNNCA model in Fig. 1. The probability that point i belongs to class z depends on the relative proximity for all other points that belongs to class z, which is the same as NNCA that was shown in Equation 2. Moreover, similarly, the distribution of distance p_{ij} is formalized and was shown in Equation 5. The objective function of CNNCA is identical to that of linear and nonlinear NCA in Equation 3. Our aim is to maximize this function, from another perspective, $\mathcal{L}$ is the expected number of correctly classified points for the training data.

4.2 Optimization

Based on conventional backpropagation algorithm, to update the parameters iteratively, feedforward computation and backpropagation need to be performed alternatively until the model converges.

Feedforward Pass. The feedforward pass aims to perform the nonlinear transformation from the input time series to the final low-dimensional space. Concretely, we use CNN to learn the features and then feed the output feature maps into a MLP, the purpose of which is to combine of the learnt features and obtain a good distance metric at the final layer. For the traditional CNN, it could consist of multiple stages and each stage contains three cascaded layers [8,9,11,21,24]. We briefly recall the process of these three layers, i.e., filter (convolutional), activation and pooling layers.

$$z_j^l = \sum_i x_i^{l-1} * k_{ij}^l + b_j^l, \quad x_j^l = \phi(z_j^l), \quad x_j^{l+1} = pool(x_j^l),$$

where $*$ denotes the convolutional operation, $pool(\cdot)$ represents the function used in pooling layer, and $\phi(\cdot)$ represents the activation function. Besides, x_i^{l-1} and z_j^l denote the input and output of filter layer and the superscript l represents which layer they involve. z_j^l and x_j^l denote the input and output of activation layer, x_j^l and x_j^{l+1} denote the input and output of pooling layer. For pooling layer, *average* and *max* pooling strategies are most widely used [13,20]. While the activation function could be considered as $sigmoid(\cdot)$, $tanh(\cdot)$ and ReLU [14,23]. We adopt *max* pooling and ReLU function in this paper due to their good generality and fast convergence [13,14,20,23].

After CNN, we also use a 2-layers fully-connected MLP to combine the learnt features, since the feedforward pass of MLP is standard and the space consumption is limited. More details of MLP can be referred to [10].

Backpropagation Pass. In this paper, we utilize the backpropagation algorithm to train the CNNCA model. Specifically, once the loss function $\mathcal{L}$ is acquired, then based on the chain-rule of derivatives, the error can be propagated back from layer to layer reversely. Here, the derivative of $\mathcal{L}$ with respect to $f(x_i)$ is the same as that of NNCA model, which is already shown in Equation 6. Then the error could be propagated back to the conventional MLP based on $\frac{\partial \mathcal{L}}{\partial f(x_i)}$ layer-wise. After that, the backpropagation of conventional CNN is performed layer by layer reversely [2,24].

4.3   Classification with Distance Metric Learning

We adopt an objective and widely used evaluation method in this work [6], which uses 1-NN classifier on labeled training data to evaluate the classification accuracy of the distance metric used. Each time series has been labeled with correct class in both of training and test sets. 1-NN classifier tries to find the nearest neighbour of input and predict its class label as that of nearest neighbour. For distance metric learning framework, once we have learnt the transformations, according to specific models (LNCA, NNCA, CNNCA), we first transform the test data and training data. Then, 1-NN classifier would be applied on the transformed training and test data for further classification. In this way, the better the distance metric the lower the classification error should be observed.

5   Experiments

In this section, we conduct experiments on a bunch of public time series data sets, and we demonstrate: 1) the classification accuracies/errors with respect to different distance metrics i.e., CNNCA and other existing distance metrics; 2) the comparison of classification performance on the largest 9 data sets with more training samples; 3) the efficiency analysis and discussion.

5.1   Experimental Setup

We conduct comprehensive experiments on 39 diverse time series data sets, provided by UCR Time Series repository [7], which is shown in the first column of Table 1. As claimed by [3], these 39 diverse data sets could make up approximately more than 90% of all publicly available, labeled time series data sets. Besides, the preprocessing was also applied, e.g., standard normalization was formed for each data set and the maximum scale of each time series is 1.0.

Previous studies observed that Euclidean distance (ED), Dynamic Time Warping (DTW) and window constraint DTW (denoted as DTW(r), r represents the percentage of time series length) are competitive distance metrics for time series classification [3,17,18,22]. Following them, we consider five distance metrics as baseline methods of our CNNCA, and they include ED, DTW, DTW(r), and two of the related distance metric learning models LNCA and NNCA. All the six distance metrics combine 1-NN to perform classification.

5.2 Experimental Results

Overall effectiveness. The experimental results are shown in Table 1. Six rightmost columns of this table exhibit the classification error with respect to these different distance metrics. Bold number accompanied with star symbol of each row indicates the best result for the corresponding data set. For all 39 data sets, our CNNCA model achieves the best results on 13 out of them, which is more than that of ED (3), DTW (9), LNCA (2), NNCA (6) and equals to that of DTW(r). It reveals that our CNNCA model is competitive not only to conventional ED, DTW and DTW(r) but also to LNCA and NNCA models. Especially, it is superior to ED, LNCA and NNCA for most of the data sets.

To illustrate the performance of these different distance metrics more intuitively compared to Table 1, we also provide some scatter plots in Fig. 2 to depict the pair-wise comparisons between CNNCA and the baseline distance metrics. For each of the scatter plots, the vertical (y) and horizontal (x) axes represent CNNCA and the compared distance metrics, which are denoted as "C" and "O"

Table 1. Classification Error of Different Distance Metrics on 1-NN Classifier

	classes	Size	Ratio	DTW	DTW(r)	ED	LNCA	NNCA	CNNCA
wafer	2	1,000	500	0.020	0.005	0.005	0.007	0.005	**0.004**(*)
StarLight	3	1,000	333.3	0.093	0.095	0.151	0.155	0.091	**0.090**(*)
Two-Patterns	4	1,000	250	**0**(*)	0.002	0.090	0.359	0.085	0.048
Chlorine	3	467	155.6	0.352	0.350	0.350	0.471	0.436	**0.250**(*)
yoga	2	300	150	0.164	0.155	0.170	0.227	0.232	**0.151**(*)
ECG200	2	100	50	0.230	0.120	0.120	0.130	0.100	**0.070**(*)
synthetic-control	6	300	50	0.007	0.017	0.120	0.033	0.047	**0.003**(*)
Thorax1	42	1,800	42.8	0.171	0.185	0.209	0.297	0.295	**0.131**(*)
Thorax2	42	1,800	42.8	0.120	0.129	0.135	0.166	0.264	**0.101**(*)
FaceAll	14	560	40	**0.192**(*)	**0.192**(*)	0.286	0.410	0.301	0.231
MedicalImages	10	381	38.1	0.263	**0.253**(*)	0.316	0.379	0.321	0.321
ItalyPowerDemand	2	67	33.5	0.050	0.045	0.045	0.038	**0.031**(*)	0.044
OSULeaf	6	200	33.3	0.409	**0.384**(*)	0.483	0.579	0.533	0.463
SwedishLeaf	15	500	33.3	0.210	**0.157**(*)	0.213	0.320	0.166	0.168
Haptics	5	155	31	0.623	0.588	0.630	0.653	**0.558**(*)	0.617
Lighting2	2	60	30	**0.131**(*)	**0.131**(*)	0.246	0.197	0.279	0.213
FISH	7	175	25	0.167	**0.160**(*)	0.217	0.520	0.229	0.166
Gun-Point	2	50	25	0.093	0.087	0.087	0.047	0.100	**0.033**(*)
Trace	4	100	25	**0**(*)	0.010	0.240	0.350	0.280	0.130
FacesUCR	14	200	14.2	0.095	**0.088**(*)	0.231	0.363	0.241	0.191
InlineSkate	7	100	14.2	0.616	**0.613**(*)	0.658	0.769	0.707	0.660
Coffee	2	28	14	0.179	0.179	0.250	**0**(*)	**0**(*)	0.036
SonyAIBORobotSurfaceII	2	27	13.5	0.169	**0.141**(*)	**0.141**(*)	0.163	0.172	0.142
ECGFiveDays	2	23	11.5	0.232	0.203	0.203	0.273	**0.046**(*)	0.056
TwoLeadECG	2	23	11.5	0.096	0.132	0.253	0.368	**0.068**(*)	0.223
WordsSynonyms	25	267	10.6	0.351	**0.252**(*)	0.382	0.633	0.541	0.401
Adiac	37	390	10.5	0.396	0.391	0.389	0.575	0.783	**0.340**(*)
CBF	3	30	10	**0.003**(*)	0.004	0.148	0.141	0.050	0.141
CinC-ECG-torso	4	40	10	0.349	**0.070**(*)	0.103	0.469	0.251	0.164
Lighting7	7	70	10	**0.274**(*)	0.288	0.425	0.521	0.425	0.301
MoteStrain	2	20	10	0.165	0.134	**0.121**(*)	0.141	0.137	0.160
SonyAIBORobotSurface	2	20	10	0.275	0.305	0.305	**0.186**(*)	0.236	0.195
50words	50	450	9	0.310	**0.242**(*)	0.369	0.629	0.409	0.338
OliveOil	4	30	7.5	**0.133**(*)	0.167	**0.133**(*)	0.300	0.167	**0.133**(*)
MALLAT	8	55	6.8	0.066	0.086	0.086	0.107	**0.061**(*)	0.157
Beef	5	30	6	0.500	0.467	0.467	0.333	0.367	**0.267**(*)
FaceFour	4	24	6	0.170	**0.114**(*)	0.216	0.261	0.227	**0.114**(*)
Symbols	6	25	4.1	**0.050**(*)	0.062	0.100	0.228	0.139	0.122
DiatomSizeReduction	4	16	4	**0.033**(*)	0.065	0.065	0.176	0.046	0.082

(e.g., ED), respectively. The classification error ratio of two distance metrics under comparison for certain data set is a point that locates at certain coordinates (x, y). Considering that we use classification error but not accuracy to compare the performance, if the classification error ratio (i.e., the point (x, y)) locates above the diagonal line (red line in 2), then it indicates that "O" is more accurate than "C", i.e., $x < y$. Moreover, the further point (x, y) is away from the diagonal line, the greater the margin of classification accuracy being improved. Otherwise, when "C" is more accurate than "O", and point (x, y) would locate below the diagonal line, i.e., $x > y$. All the points that locate at diagonal line indicate that they achieve identical classification error on these data sets, i.e., $x = y$. Besides, more points on one side of the diagonal line indicates that one distance metric is more superior to the other.

Through comparing the classification accuracy between CNNCA and conventional ED, DTW, DTW(r), the results in Fig. 2 reveal that CNNCA is superior to ED on most of the data sets, which demonstrates that such a learnt distance metric can improve the classification accuracy to some extent. However, on total 39 data sets, there is no evidence that either DTW (or DTW(r)) is better or worse than CNNCA, even though window constraints DTW is a little better than DTW. Moreover, by comparing the classification accuracy between CNNCA and LNCA, NNCA, the results show that CNNCA outperforms both of LNCA and NNCA on most of the data sets, which provides the evidence that CNNCA is more effective than previous NCAs just as our expectation.

Effectiveness on Large Data Sets. We provide both the number of classes and the size of training set in Table 1 (in the second and the third columns) and furthermore the average number of training samples per class is calculated as shown in the forth column. As is well known, if the training samples of each class are too few then neural networks cannot capture the features well and may obtain a poor performance. Motivated by this, we filter out the data sets that has few training samples per class, i.e., eliminating the ratio that is no larger than 40 as shown in Table 1. Finally, we obtain 9 data sets (the top 9 rows of Table 1). Likewise, we provide Fig. 3 to depict the comparisons between CNNCA and the baseline distance metrics on these 9 data sets. From both the top rows of Table 1 and the results in Fig. 3, we could observe that our CNNCA model is

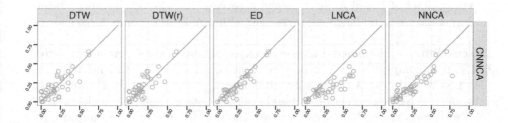

Fig. 2. Comparison of classification accuracy between CNNCA and other baselines

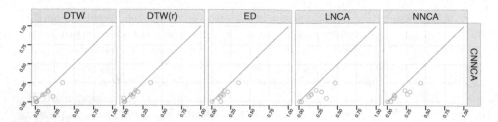

Fig. 3. Comparison of classification accuracy between CNNCA and other baselines. The average number of training samples per class is more than 40.

superior to all other methods on 8 out of these 9 data sets, which demonstrates that if the training samples are sufficient then our CNNCA model could achieve good performance and outperform the baseline methods.

Efficiency Analysis. Supposed that the size of training set is $\mathcal{N}$, and given two time series of length $\mathcal{D}$, then the time complexity of ED, DTW and DTW(r) are $O(\mathcal{N}\mathcal{D})$, $O(\mathcal{N}\mathcal{D}^2)$ and $O(\mathcal{N}\mathcal{D}^2 r)$, respectively, when we apply dynamic programming to compute DTW. Usually, r is no larger than 10% for most applications. Before analyzing the time complexity for LNCA, NNCA and CNNCA, we should note that we only focus on analyzing the online classification of them and skip the offline training process due to the limited space, and it is necessary to define some notations. One hidden layer NNCA is considered for convenience and the number of its hidden neurons sets to $\tilde{n}_h$. For CNNCA, the number of kernels in filer layer and hidden neurons in MLP are denoted as n_k and n_h, respectively. Moreover, the size of kernel and the pooling factor are usually set to 5 and 2. We use d to represent the dimensions of transformed space. To classify each test case, the time complexity of LNCA, NNCA and CNNCA are $O(\mathcal{D}d + \mathcal{N}d)$, $O(\mathcal{D}\tilde{n}_h + \tilde{n}_h d + \mathcal{N}d)$ and $O(5n_k\mathcal{D} + \frac{1}{2}(\mathcal{D} - 5 + 1)n_kn_h + n_hd + \mathcal{N}d)$, respectively, where $O(\mathcal{N}d)$ is the time cost of 1-NN on the transformed data and the remainder is the transformational cost. After reduction, the time cost of CNNCA is $O(n_k\mathcal{D} + \mathcal{D}n_kn_h + n_hd + \mathcal{N}d)$. When $d \ll \mathcal{D}$ and $\mathcal{N}$ is large enough, it is efficient for LNCA, NNCA and CNNCA compared to conventional ED, DTW and DTW(r) if we fix $\tilde{n}_h$, n_h, n_k to constants. We also provide the real time cost of classification on the top 9 data sets in Fig. 4, the data sets in which are ordered by the product of $\mathcal{D}$ and $\mathcal{N}$ increasingly. It reveals that CNNCA is more efficient for larger data set and long time series, i.e., either $\mathcal{N}$ or $\mathcal{D}$ becomes large enough.

Discussion. In summary, the overall experimental results demonstrate that CNNCA is competitive to not only conventional ED, DTW and DTW(r) but also LNCA and NNCA. Especially, after filtering out the relatively small data sets, our CNNCA is superior to all the other distance metrics, which verifies the motivation that CNNCA can capture the intrinsic features and improve the classification performance if the training samples per class are sufficient. By comparison with both of LNCA and NNCA, we also demonstrate that CNNCA

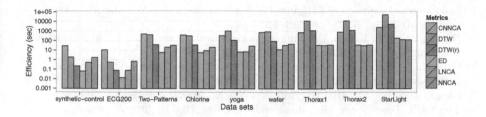

Fig. 4. Time cost of each distance metric combined with 1-NN classifier

is more effective than them to some extent, which is benefited from the capability of capturing time shift property. On the other hand, CNNCA is more efficient than DTW and DTW(r) when the data set grows large enough and time series is long, e.g., for three large data sets, Thorax1, Thorax2 and StarLight in Fig. 4.

6   Conclusions and Future Work

In this paper, we proposed a novel CNNCA model for time series classification. Specifically, we extended the NCA model with CNN and MLP to learn distance metric and then combined 1-NN to classify time series. The benefit of introducing CNN into NCA is to get good feature representations for further classification, and MLP is used to combine these learnt features and obtain nonlinear transformation for better distance metric. For evaluation, we conducted experiments on a bunch of public time series data sets, and observed encouraging results. In particular, CNNCA is superior to current state-of-the-art methods when the training samples are sufficient. We hope this work could lead to many future studies. Actually, we plan to investigate better methods based on CNNCA and further improve the performance (e.g., efficiency) of time series classification.

Acknowledgments. This research was partially supported by grants from the National Science Foundation for Distinguished Young Scholars of China (Grant No. 61325010), the Natural Science Foundation of China (Grant No. 61403358), the Fundamental Research Funds for the Central Universities of China (Grant No. WK0110000042) and the Anhui Provincial Natural Science Foundation (Grant No. 1408085QF110). Qi Liu gratefully acknowledges the support of the Youth Innovation Promotion Association, CAS.

References

1. Batista, G., Wang, X., Keogh, E.J.: A complexity-invariant distance measure for time series. In: SIAM Conf. Data Min. (2011)
2. Bouvrie, J.: Notes on convolutional neural networks (2006)
3. Ding, H., Trajcevski, G., Scheuermann, P., Wang, X., Keogh, E.: Querying and mining of time series data: experimental comparison of representations and distance measures. Proc. VLDB Endow. **1**(2), 1542–1552 (2008)

4. Goldberger, J., Roweis, S., Hinton, G., Salakhutdinov, R.: Neighbourhood components analysis. In: NIPS, pp. 513–520 (2005)
5. Hu, B., Chen, Y., Keogh, E.: Time series classification under more realistic assumptions. In: SIAM Int. Conf. Data Min., p. 578 (2013)
6. Keogh, E., Kasetty, S.: On the Need for Time Series Data Mining Benchmarks: A Survey and Empirical Demonstration. DMKD 7(4), 349–371 (2003)
7. Keogh, E., Zhu, Q., Hu, B., Hao, Y., Xi, X., Wei, L., Ratanamahatana, C.A.: The UCR Time Series Classification/Clustering (2011). http://www.cs.ucr.edu/~eamonn/time_series_data/
8. Krizhevsky, A., Sutskever, I., Hinton, G.: Imagenet classification with deep convolutional neural networks. In: NIPS, vol. 25, pp. 1106–1114 (2012)
9. LeCun, Y., Bengio, Y.: Convolutional networks for images, speech, and time series. Handb. Brain Theory Neural Networks 3361 (1995)
10. LeCun, Y., Bottou, L., Orr, G.B., Müller, K.-R.: Efficient backprop. In: Orr, G.B., Müller, K.-R. (eds.) NIPS-WS 1996. LNCS, vol. 1524, pp. 9–50. Springer, Heidelberg (1998)
11. LeCun, Y., Kavukcuoglu, K., Farabet, C.: Convolutional networks and applications in vision. In: IEEE ISCS, pp. 253–256, May 2010
12. Lu, Y., Zhao, W.X., Yan, H., Li, X.: A metric learning based approach to evaluate task-specific time series similarity. In: Wang, J., Xiong, H., Ishikawa, Y., Xu, J., Zhou, J. (eds.) WAIM 2013. LNCS, vol. 7923, pp. 314–325. Springer, Heidelberg (2013)
13. Nagi, J., Ducatelle, F., et al.: Max-pooling convolutional neural networks for vision-based hand gesture recognition. In: IEEE ICSIPA, pp. 342–347 (2011)
14. Nair, V., Hinton, G.E.: Rectified linear units improve restricted boltzmann machines. In: ICML, vol. (3), pp. 807–814. Omnipress Madison, WI (2010)
15. Prekopcsák, Z., Lemire, D.: Time series classification by class-specific Mahalanobis distance measures. ADAC 6(3), 185–200 (2012)
16. Rakthanmanon, T., Campana, B., Mueen, A., Batista, G., Westover, B., Zhu, Q., Zakaria, J., Keogh, E.: Searching and mining trillions of time series subsequences under dynamic time warping. In: ACM SIGKDD, p. 262 (2012)
17. Ratanamahatana, C., Keogh, E.: Making time-series classification more accurate using learned constraints. In: SIAM Int. Conf. Data Min., p. 11 (2004)
18. Ratanamahatana, C.A.: Three Myths about Dynamic Time Warping Data Mining. In: SIAM Int. Conf. Data Min., pp. 506–510 (2005)
19. Salakhutdinov, R., Hinton, G.E.: Learning a nonlinear embedding by preserving class neighbourhood structure. In: ICAIS, pp. 412–419 (2007)
20. Scherer, D., Müller, A., Behnke, S.: Evaluation of pooling operations in convolutional architectures for object recognition. In: Diamantaras, K., Duch, W., Iliadis, L.S. (eds.) ICANN 2010, Part III. LNCS, vol. 6354, pp. 92–101. Springer, Heidelberg (2010)
21. Sutskever, I., Martens, J., Dahl, G., Hinton, G.: On the importance of initialization and momentum in deep learning. In: ICML, p. 28 (2013)
22. Xi, X., Keogh, E.J., Shelton, C.R., Wei, L., Ratanamahatana, C.A.: Fast time series classification using numerosity reduction. In: ICML, pp. 1033–1040 (2006)
23. Zeiler, M.D., Ranzato, M., Monga, R., et al.: On rectified linear units for speech processing. In: IEEE ICASSP, pp. 3517–3521 (2013)
24. Zheng, Y., Liu, Q., Chen, E., Ge, Y., Zhao, J.L.: Time series classification using multi-channels deep convolutional neural networks. In: Li, F., Li, G., Hwang, S., Yao, B., Zhang, Z. (eds.) WAIM 2014. LNCS, vol. 8485, pp. 298–310. Springer, Heidelberg (2014)

Entity Resolution and Topic Modelling

Clustering-Based Scalable Indexing for Multi-party Privacy-Preserving Record Linkage

Thilina Ranbaduge[⊠], Dinusha Vatsalan, and Peter Christen

Research School of Computer Science, College of Engineering and Computer Science,
The Australian National University, Canberra, ACT 0200, Australia
{thilina.ranbaduge,dinusha.vatsalan,peter.christen}@anu.edu.au

Abstract. The identification of common sets of records in multiple databases has become an increasingly important subject in many application areas, including banking, health, and national security. Often privacy concerns and regulations prevent the owners of the databases from sharing any sensitive details of their records with each other, and with any other party. The linkage of records in multiple databases while preserving privacy and confidentiality is an emerging research discipline known as privacy-preserving record linkage (PPRL). We propose a novel two-step indexing (blocking) approach for PPRL between multiple (more than two) parties. First, we generate small mini-blocks using a multi-bit Bloom filter splitting method and second we merge these mini-blocks based on their similarity using a novel hierarchical canopy clustering technique. An empirical study conducted with large datasets of up-to one million records shows that our approach is scalable with the size of the datasets and the number of parties, while providing better privacy than previous multi-party indexing approaches.

Keywords: Hierarchical canopy clustering · Bloom filters · Scalability

1 Introduction

Many real-world applications require data from multiple databases to be integrated and combined to improve data analysis and mining. Record linkage (also known as entity resolution or data matching) is the process of identifying matching records that refer to the same entity from multiple databases [3].

The lack of unique entity identifiers in databases requires the use of quasi-identifiers (QIDs) [16], such as first name, last name, address details, etc. to link records across databases. However, due to privacy and confidentiality concerns organizations generally do not want to share any sensitive information regarding

Clustering-based Scalable Indexing for Multi-party Privacy-preserving Record Linkage - This research is funded by the Australian Research Council under Discovery Project DP130101801.

T. Cao et al. (Eds.): PAKDD 2015, Part II, LNAI 9078, pp. 549–561, 2015.
DOI: 10.1007/978-3-319-18032-8_43

their entities with other data sources. Finding records in multiple databases that relate to the same entity or having approximately the same values for a set of QIDs without revealing any private or sensitive information is the research area known as 'privacy-preserving record linkage' (PPRL) [8,16].

The naive pair-wise comparison of multiple data sources is of exponential complexity in terms of the number of parties. This makes record linkage applications not scalable to large databases and increasing number of participating parties. Applying indexing is a possible solution aimed at improving scalability [4]. Indexing reduces the large number of potential comparisons by removing as many record sets as possible that correspond to non-matches, such that expensive similarity comparisons are only required on a smaller number of candidate record sets. Indexing for PPRL needs to be conducted such that no sensitive information that can be used to infer individual records and their attribute values is revealed to any party involved in the process, or to an external adversary [18].

We propose a novel indexing mechanism for multi-party PPRL that can provide better scalability, blocking quality, and privacy compared to previous approaches. Our approach efficiently creates blocks across multiple parties without revealing any private information. Specific contributions of our paper are (1) a two-step blocking protocol which consists of (2) a multi-bit splitting approach for Bloom filters and (3) a hierarchical canopy clustering algorithm; and (4) an empirical evaluation using large datasets, and a comparison with other multi-party approaches in terms of efficiency, effectiveness and privacy.

2   Related Work

Can we do efficient and effective indexing for record linkage? This is a problem that has been considered for several decades. According to a survey by Christen [4], a variety of indexing approaches have been developed. Some of these have been adapted for PPRL [16], including standard blocking [1], mapping based indexing [8], clustering [8,10,19], and locality sensitive hashing [7]. However, performing scalable record linkage that provides high linkage quality while preserving privacy is an open research question that needs further investigation.

Bloom filters are commonly used for encoding of records in PPRL due to their capability of computing similarities. Lai et al. [11] proposed a multi-party approach that uses Bloom filters to securely transfer data between multiple parties for private set intersection. This approach was recently adapted by Vatsalan et al. [17] for multi-party PPRL, however their approach does not address blocking.

Schnell [15] introduced a new blocking method for record linkage, based on the concept of a multi-bit tree data structure [9] to hold a set of Bloom filters. This approach was further extended by Ranbaduge et al. [13] as a blocking mechanism for multi-party PPRL. Their experimental results showed that the proposed approach is scalable with the dataset size and the number of parties, and provides better linkage quality and privacy than a phonetic based indexing approach. However, the blocks generated using this approach might miss some true matches due to the recursive splitting of Bloom filter sets.

Clustering is the process of grouping records such that records within a cluster are similar to each other, while dissimilar records are in different clusters. Several clustering approaches have been adapted for private blocking [8,10,19], however neither of these techniques considers blocking of more than two databases.

Canopy clustering [6,12] is a technique for clustering large high dimensional datasets. It can generate candidate record sets by efficiently calculating similarities between blocking key values. Records are inserted into one or more overlapping clusters based on their similarity to the cluster centroid. Each cluster then forms a block from which candidate sets are generated. However, the use of canopy clustering for indexing in PPRL has so far not been studied.

3   Clustering Based Indexing for Multi-party PPRL

We now introduce the building blocks required for our clustering based PPRL indexing approach, and then study the indexing protocol in detail.

3.1   Building Blocks

Bloom Filters: are bit vectors proposed by Bloom [2]. In a Bloom filter initially all the bits are set to 0. To map an element of a set into the domain between 0 and $m - 1$ of a Bloom filter of length m, k independent hash functions $h_1, \ldots, h_k$ are used. Furthermore, to store n elements of the set $S = \{s_1, s_2, \ldots, s_n\}$ into a Bloom filter, each element $s_i \in S$ is encoded using the k hash functions and all bits having index positions $h_j(s_i)$ for $1 \leq j \leq k$ are set to 1.

Q-grams: are character sub-strings of length q in a string [3]. For example, the string "PETER" can be padded with character '_' and the resulting q-gram set (for $q = 2$) is $\{\_P, PE, ET, TE, ER, R\_\}$. In our approach we encode q-gram sets of the quasi-identifiers (QIDs) into Bloom filters. First, the selected QIDs of a given record are converted into a q-gram set. Then each q-gram set is encoded into a Bloom filter by using k hash functions.

Secure Summation Protocol: is a method used in secure multi-party computation [5], and has been used in several PPRL approaches [13,17]. The basic idea is to compute a summation of private inputs of P parties without revealing individual values to any other parties, and at the end of the computation no party knows anything except its own input and the final sum [5].

3.2   Basic Multi-bit Indexing Protocol

We now describe our indexing approach for multi-party PPRL. The construction of the index of an individual party consists of two main phases:

1. *Multi-bit Bloom filter splitting*: This phase can be further extended into three steps, which are:

Algorithm 1. Multi-bit Bloom filter splitting by party P_i, $1 \leq i \leq P$

Input:
- **D_i:**   Dataset belonging to party P_i - **A:**   Set of selected attributes
- s_{min}: Minimum *mini-block* split size   - d_{max}: Maximum split degree
- s_{max}: Maximum *mini-block* split size   - bs_t: Bit selection threshold

Output:
- **C:**   Set of Bloom filter mini-blocks

```
 1: B = generateBloomfilters(Dᵢ, A)
 2: Q = [B]                                    // Initialization of queue
 3: while Q ≠ ∅ do:
 4:     b = Q.pop()                            // Get the current block
 5:     R = generateRatios(b)                  // Generate local bit ratios
 6:     R_g = secureSummation(R)               // Get ratios globally
 7:     BC_l = getCombinations(R_g,d_max,bs_t) // Get bit combinations
 8:     BC_g = secureSummation(BC_l)           // Get combinations globally
 9:     mb_j = splitData(BC_g,b)               // Create mini-blocks
10:     if all (|mb_j| ≥ s_min) then:          // If mini-blocks large enough
11:         if any (|mb_j| > s_max) then:      // If mini-blocks too large
12:             Q.push(mb_j)                   // Add to queue for further splitting
13:     else:                                  // Block sizes are acceptable
14:         C.add(mb_j)                        // Add mini-blocks to final set
15: return C
```

   (a) *Generate Bloom filters for the records in the dataset.*
   (b) *Perform secure summation to find the best splitting bit position.*
   (c) *Split the set of Bloom filters into mini-blocks.*
2. *Merge mini-blocks using clustering.*

Generating large blocks of different sizes makes the comparison step more problematic and requires more computational time. With our approach, a user has control over the block sizes by merging blocks until the size of blocks reaches an acceptable lower limit suitable for comparison. The aim of our protocol is to divide (split) the set of records in the datasets into *mini-blocks* (phase 1) and merge these *mini-blocks* based on their similarity (phase 2). Merging of *mini-blocks* by clustering reduces the overall running time requirement compared to using a clustering technique for blocking the datasets [10,19]. These merged blocks can then be compared using private comparison and classification techniques to determine the matching record sets in different databases [17]. Each party follows these phases to construct the blocks from their own dataset.

3.3   Multi-bit Bloom Filter Splitting

Before performing the clustering algorithm, the set of records needs to be encoded into Bloom filters, and split into sets of smaller blocks (which we call *mini-blocks*). All P parties need to agree upon a bit array length m (length of the Bloom filter); the length (in characters) of grams q, the k hash functions, and a set of QID attributes that are used to block the records. The parameters s_{min} and s_{max} specify the lower and upper bound of the acceptable size of a *mini-block*, respectively. The overall splitting approach for each party P_i, where $1 \leq i \leq P$, is outlined in Algorithm 1.

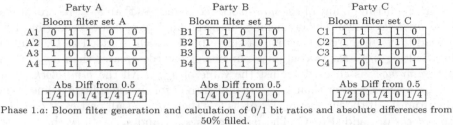

Phase 1.a: Bloom filter generation and calculation of 0/1 bit ratios and absolute differences from 50% filled.

Random vector (R) = | 10 | 5 | 12 | 13 | 6 |

| 10.25 | 0 | 12.25 | 13.25 | 6.25 | → | 10.5 | 0 | 12.5 | 13.25 | 6.25 | → | 11 | 0 | 12.75 | 13.25 | 6.5 |

Secure sums: | 1 | 0 | 0.75 | 0.25 | 0.5 |
Final absolute differences from 50% filled: | 0.33 | 0 | 0.25 | **0.083** | **0.167** |

Selected bit positions: {**2, 4**}

Phase 1.b: Secure summation of absolute differences and selecting best bit positions for splitting ($d_{max} = 2$, and $bs_t = 0.2$)

Fig. 1. Selecting $d_{max} = 2$ best bit positions for multi-bit Bloom filter splitting

In line 1, each party iterates over its dataset to encode each record into a Bloom filter. Once all the parties have generated their sets of Bloom filters they are added into a queue Q as a single block. At each iteration the first block of Bloom filters b that is available at the front of Q is processed. In line 5, each party calculates a vector of length m that contains the ratios between the number of 0's and 1's for each bit position in the Bloom filters, using $f_{ij} = abs(0.5 - \frac{o_{ij}}{l})$, where f_{ij} is the ratio value of bit position j of party P_i, o_{ij} is the number of 1's in position j, and l is the number of Bloom filters processed in a given block.

Once all parties have computed their own ratio vector, we use an extended secure summation protocol to compute common bit positions suitable for splitting (line 6). The globally summed ratio vector R_g is used to find the d_{max} best splitting bit positions.

The bit positions with a sum less than the bit selection threshold bs_t are then selected into the set I_j of splitting bit positions as $I_j = \{j \mid (\frac{P}{2} - \sum_{i=1}^{P} f_{ij}) < bs_t\}$. The d_{max} bit positions in I_j with the lowest ratio values (which we call *match-bits*) are selected as the best splitting bit positions. Fig. 1 illustrates an example of selecting best splitting bits.

Based on the selected bit positions, each party generates all possible bit combinations BC_l (line 7). An example of bit positions $\{j_x, j_y, j_z\}$ would generate the set of combinations $\{\{j_x, j_y, j_z\}, \{j_x, j_y\}, \{j_y, j_z\}, \{j_x, j_z\}, \{j_x\}, \{j_y\}, \{j_z\}\}$. The value of d_{max} needs to be kept small as this generation grows exponentially with d_{max}. For each combination, the Bloom filters in the current block are processed to analyze the sizes of resulting *mini-blocks* with different bit patterns. For a bit combination $\{j_x, j_y\}$, for example, the set of bit patterns is $\{00, 01, 10, 11\}$. Once the current block is processed for all possible bit combinations, we find the common best bit combination BC_g in line 8 in Algorithm 1. The current block is split into *mini-blocks* according to this globally accepted bit combination.

After splitting, if any of the *mini-blocks* contains less than s_{min} records, then these *mini-blocks* will not be included into the final *mini-block* set $\mathbf{C}$ (line 10). Instead they are merged back into a single block which is added to $\mathbf{C}$. If all of the resulting *mini-blocks* contain a number of records greater than s_{min}, then each *mini-block* is checked against the value of s_{max}. All the *mini-blocks* that contain records greater than s_{max} are added to Q for future splitting (line 12 in Algorithm 1). If the number of records is less than s_{max} at all parties, then these *mini-blocks* are added to $\mathbf{C}$. The parameters s_{min} and s_{max} allow the user to control the number of iterations that occur in the multi-bit splitting algorithm.

4   Merge Mini-Blocks by Clustering

Our clustering of *mini-blocks* is based on a canopy technique [6,12]. The generated *mini-blocks* are merged into larger clusters by inserting records into one or more overlapping clusters based on their similarity to their nearest cluster centroid. This allows us to perform computationally efficient indexing. We use a normalized Hamming distance based similarity calculation for computing the similarity of clusters: $sim_H(\boldsymbol{x}, \boldsymbol{y}) = 1 - \frac{|i|}{m} \mid \boldsymbol{x}_i \neq \boldsymbol{y}_i$ and $1 \leq i \leq m$, where $sim_H(\boldsymbol{x}, \boldsymbol{y})$ is the normalized Hamming distance similarity between the two bit vectors $\boldsymbol{x}$ and $\boldsymbol{y}$, $sim_H(\boldsymbol{x}, \boldsymbol{y}) = 1$ if and only if $\boldsymbol{x} = \boldsymbol{y}$.

Each *mini-block* generated by the multi-bit splitting algorithm is initially considered as a separate cluster, which we refer to as a *mini-cluster*. A Bloom filter a_c is selected as the centroid for each *mini-cluster* which has the highest similarity to all other Bloom filters in the cluster. For this selection process we use a maximum average Hamming distance based similarity calculation which is shown in (1).

$$a_c = \underset{a_i}{\operatorname{argmax}} \left\{ \frac{\sum sim_H(a_i, a_j)}{|\mathbf{C}|} \mid a_i, a_j \in \mathbf{C}, \ 1 \leq i, j \leq |\mathbf{C}|, and \ i \neq j \right\} \quad (1)$$

We use threshold-based canopy clustering for merging similar *mini-clusters*. Before starting the clustering algorithm, all parties need to agree upon the tight similarity threshold s_t, loose similarity threshold s_l, and the maximum merge size ms_{max} which controls the size of merged clusters (blocks) and indirectly controls the number of iterations of the merging process. For merging of *mini-clusters* we suggest two canopy based clustering algorithms, which are:

- *Standard canopy clustering (SCC)*: Mini-clusters are merged until the resulting cluster size increases to ms_{max}. This algorithm merges a set of similar *mini-clusters* greedily in a given iteration.
- *Hierarchical canopy clustering (HCC)*: The merging of *mini-clusters* is based on an agglomerative clustering approach until ms_{max} is met. In a given iteration, the two most similar *mini-clusters* are merged.

Algorithm 2. Merge mini-clusters using standard canopy clustering	**Algorithm 3.** Merge mini-clusters using hierarchical canopy clustering
Input: - s_t: Tight similarity threshold - s_l: Loose similarity threshold - ms_{max}: Maximum merge size - C: Set of mini-clusters ($[c_1, \ldots, c_l]$) - $sim_H(\cdot, \cdot)$: Similarity comparison function	**Input:** - s_t: Tight similarity threshold - s_l: Loose similarity threshold - ms_{max}: Maximum merge size - C: Set of mini-clusters ($[c_1, \ldots, c_l]$) - $sim_H(\cdot, \cdot)$: Similarity comparison function
Output: - O: Set of merged clusters	**Output:** - O: Set of merged clusters

1: $\mathbf{O} = \emptyset$	1: $\mathbf{O} = \emptyset$
2: **while** $\mathbf{C} \neq \emptyset$ **do:**	2: **while** $\mathbf{C} \neq \emptyset$ **do:**
3: $c_x = \mathbf{C}.\text{pop}()$	3: $c_x = \mathbf{C}.\text{pop}()$
4: $a_x = \text{getCentroid}(c_x)$	4: $a_x = \text{getCentroid}(c_x)$
5: $c_{xy} = c_x$	5: **for** $c_y \in \mathbf{C}$ **do:**
6: **while** $|c_{xy}| \leq ms_{max}$ **do:**	6: $a_y = \text{getCentroid}(c_y)$
7: $c_y = \mathbf{C}.\text{next}()$	7: $s = sim_H(a_x, a_y)$
8: $a_y = \text{getCentroid}(c_y)$	8: **if** $s \leq s_t$ **then:**
9: $s = sim_H(a_x, a_y)$	9: $c_{xy} = c_x + c_y$
10: **if** $s \leq s_t$ **then:**	10: del c_x, c_y
11: $c_{xy} = c_{xy} + c_y$	11: **if** $s \leq s_l$ **then:**
12: del c_y	12: $c_{xy} = c_x + c_y$
13: **if** $s \leq s_l$ **then:**	13: del c_x
14: $c_{xy} = c_{xy} + c_y$	14: **if** $|c_{xy}| \geq ms_{max}$ **then:**
15: del c_x	15: $\mathbf{O}.\text{add}(c_{xy})$
16: $\mathbf{O}.\text{add}(c_{xy})$	16: **else:**
	17: $\mathbf{C}.\text{add}(c_{xy})$

4.1 Merge Mini-Clusters with Standard Canopy Clustering

The suggested SCC approach for merging of *mini-clusters* is shown in Algorithm 2. In line 2, each party iterates over the set of *mini-clusters* C. At each iteration the *mini-cluster* at the front of C is processed as the initial cluster (line 3). As discussed above, the centroid is computed for the initial cluster (line 4).

The initial cluster is compared and merged with other *mini-clusters* in the set until the size of the cluster reaches ms_{max} as shown in lines 5 to 15. At the merging step, the computed similarity value s (line 9) is checked against s_t and s_l. As per lines 10 to 12, if $s \leq s_t$, then the *mini-clusters* will be merged. The merged *mini-clusters* are removed from the set C if they are within s_t (line 12).

If $s \leq s_l$, then the merge is performed between the *mini-clusters* but only the initial cluster is deleted from the set C (line 15). Once the size of the resulting merged cluster c_{xy} reaches ms_{max}, the cluster is added to the set of merged clusters O (line 16). Therefore at each iteration a set of *mini-clusters* which are similar to the initial cluster are merged until the overall cluster size reaches ms_{max}. Each cluster generated by this approach will become a block to be used in the comparison step in the PPRL pipeline.

4.2 Merge Mini-Clusters with Hierarchical Canopy Clustering

In the SCC algorithm described in Sect. 4.1, depending on the sizes of the *mini-clusters* that are merged, the final cluster size can grow beyond ms_{max} which will result in more record set comparisons. We propose a novel threshold-based hierarchical canopy clustering (HCC) approach which guarantees that clusters are only merged up-to the size limit ms_{max} as shown in Algorithm 3.

To merge clusters, each party iterates over its set of *mini-clusters* $\mathbf{C}$ (line 2). At each iteration one *mini-cluster* is selected and the centroid is computed as discussed in Sect. 4 (lines 3 and 4). A similarity value s is computed between the initial cluster and other *mini-clusters* in $\mathbf{C}$ (line 7). The computed value s is checked against s_t and s_l for merging (lines 8 to 13). Similar to Algorithm 2, *mini-clusters* are merged if the value s satisfies the threshold values.

After each iteration, the size of the resulting merged cluster is checked against ms_{max}. If the size of the merged cluster is less than ms_{max}, the cluster is added back into $\mathbf{C}$ as a new *mini-cluster* c' (line 17). This enables c' to be merged further with other close *mini-clusters*. Therefore, at each iteration the two most similar *mini-clusters* are merged into one. Once the size of a merged cluster reaches ms_{max}, it is added to the final set of merged clusters $\mathbf{O}$.

5   Analysis of the Protocol

We now analyze our protocol in terms of complexity, privacy, and quality.

Complexity: By assuming there are N records in a dataset with each having an average of n q-grams, we analyze the computational and communication complexities in terms of a single party.

In the first phase of our protocol all records are encoded using k hash functions. The Bloom filter generation for a single party is of $O(k \cdot n \cdot N)$ complexity. In the multi-bit splitting step, the parameters s_{min}, s_{max} and d_{max} are used to control the size of *mini-blocks* generated. Suitable values for the parameters s_{min} and s_{max} need to be set as the size of the *mini-blocks* decides the number of iterations that occur in the splitting phase. At each iteration, a set of Bloom filters is split into $2^{d_{max}}$ *mini-blocks*. The number of iterations in the splitting phase can be calculated as $log_2(N/s_{max})/d_{max}$. Therefore, the splitting of N Bloom filters into a set of *mini-blocks* is of $O(N \cdot log_2(N/s_{max})/d_{max})$.

In the second phase of our protocol, merging *mini-clusters* requires the processing of $|\mathbf{C}|$ merged clusters. The computation of the centroid for all *mini-clusters* is of $O(s_{max}^2 \cdot |\mathbf{C}|)$ complexity. Merging *mini-clusters* using the SCC approach requires a total computation of $O(ms_{max}/s_{max} \cdot |\mathbf{C}|)$, where at each iteration ms_{max}/s_{max} clusters are merged. At each iteration in the HCC approach the two most similar *mini-clusters* are merged which requires a total of $O(|\mathbf{C}|^2)$ computations.

The parties only need to communicate with each other to perform the secure summation protocol in the phases of multi-bit splitting and the merging of *mini-clusters*, with each message of length m and $|\mathbf{C}|$, respectively. By assuming each party directly connects to all other parties, the P parties require P messages to be sent in each iteration. Therefore the entire protocol has a communication complexity of $O(m \cdot P \cdot log_2(N/s_{max})/d_{max} + m \cdot |C|)$ for P parties.

Quality: We analyze the quality of our protocol in terms of effectiveness and efficiency [19]. The SCC approach merges *mini-clusters* greedily which can generate clusters with sizes larger than ms_{max}. This results in the SCC approach to have higher effectiveness and lower efficiency compared to the HCC approach. Both the SCC and HCC approaches retrieve more similar records compared to a previous approach [13], as the similarity between *mini-clusters* is used to merge the clusters up-to ms_{max}.

For $|C|$ merged clusters, with each containing ms_{max} records, the number of candidate record sets generated for each party is $ms_{max} \cdot |C|$. The parameter ms_{max} limits the size of the clusters generated by one of the clustering algorithms, which indirectly determines the number of merged clusters generated by the protocol. In the worst case scenario a merged cluster can be of size $2(ms_{max} - 1)$ if the two *mini-clusters* merged are each of size $ms_{max} - 1$. A suitable value for ms_{max} therefore needs to be set by considering factors such as the dataset size and the number of parties, such that both high effectiveness and high efficiency are achieved while guaranteeing sufficient privacy as well.

Privacy: We assume that each party follows the honest-but-curious adversary model [18]. All parties participate in a secure summation protocol for exchanging of ratio values of Bloom filters with other parties. During these summations, each party computes the sum of its ratio values but neither of the parties is capable of deducing anything about the other parties' private inputs [5].

Our protocol performs a generalization strategy on clusters that makes re-identification not possible. The parameter ms_{max} is used to guarantee that every resulting cluster contains at least ms_{max} records. This ensures all clusters that are generated have the same minimum number of records, which guarantees k-anonymous mappings ($k = ms_{max}$) privacy [8,19]. The merging of *mini-blocks* makes the protocol more secure and harder for dictionary and frequency attacks [18]. A higher value for ms_{max} provides stronger privacy guarantees but requires more computations as more candidate record sets will be generated.

6 Experimental Evaluation

We evaluated our protocol using the North Carolina Voter Registration (NCVR) database[1]. We based our experiments on the datasets used in and provided by [13], which contain from $5,000$ to $1,000,000$ records for 3, 5, 7 and 10 parties. In each of these sub-sets, 50% of records were matches. Some of these datasets included corrupted records which allowed us to evaluate how our approach works with 'dirty' data. The corruption levels were set to 20% and 40%.

We implemented a prototype of our protocol using Python (version 2.7.3). All experiments were run on a server with 64-bit Intel Xeon (2.4 GHz) CPUs, with 128 GBytes of main memory and Ubuntu 14.04. The programs and test datasets are available from the authors. We used four attributes commonly used for record linkage as QIDs: Given name, Surname, Suburb (town) name, and Postcode. We

---

[1] Available from: ftp://alt.ncsbe.gov/data/

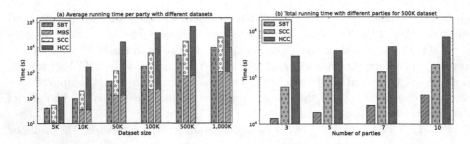

Fig. 2. (a) Average blocking runtime per party for different dataset sizes and (b) total blocking runtime for the $500K$ dataset with different number of parties

set the Bloom filter parameters as $m = 1000$ bits, $k = 30$ hash functions, and $q = 2$ by following earlier Bloom filter work in PPRL [13,14]. For comparative evaluation we used the single-bit tree (SBT) multi-party PPRL blocking approach by Ranbaduge et al. [13] as to our knowledge there are no other blocking approaches for multi-party PPRL available. We also used a phonetic based blocking approach (PHO) as a baseline [8,17] to comparatively evaluate the level of privacy. We named our multi-bit splitting, standard canopy clustering, and hierarchical canopy clustering as MBS, SCC, and HCC, respectively.

In the PHO approach we used Soundex [4] as encoding function for all QIDs except Postcode where the first three digits of the value were used as the blocking key. Based on a set of parameter evaluation experiments we set the MBS parameters $d_{max} = 3$, $bs_t = 0.1$, $s_{max} = 50$ and $s_{min} = s_{max}/2$. We set the SCC and HCC parameters of s_t, s_l, and ms_{max} to 0.9, 0.8, and 500, respectively, as these values gave us the minimum overlap between clusters.

We measured the average total runtime for the protocol to evaluate the complexity of blocking. The reduction ratio (RR), which is the fraction of record sets that are removed by a blocking technique, and pairs completeness (PC), which is the fraction of true matching record sets that are included in the candidate record sets generated by a blocking technique, were used to evaluate the blocking quality. These are standard measures to assess indexing in record linkage [3].

Fig. 2 illustrates the scalability of our approach in terms of the average time required with different dataset sizes and number of parties. As expected the MBS-SCC approach requires less runtime than the MBS-HCC approach but both show a linear scalability with the size of the datasets and number of parties.

Fig. 3(a) illustrates that RR remains close to 1 for all dataset sizes and for different number of parties. This shows our approach significantly reduces the total number of candidate record sets that need to be compared. Fig. 3(b) to (d) illustrate the PC of our approach with different dataset sizes and corruption levels, indicating that our approach can provide significantly better blocking quality than the earlier proposed SBT approach [13].

To evaluate the privacy of our approach we use the measure *probability of suspicion* (P_s) [18], which is defined for a value in an encoded dataset as $1/n_g$,

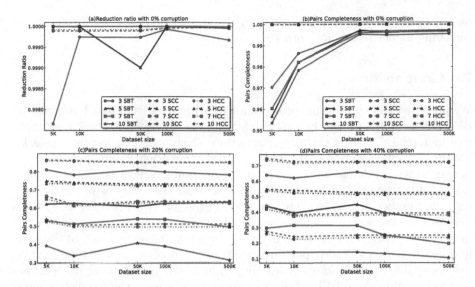

Fig. 3. (a) Reduction ratio with 0% corruption and (b) to (d) pairs completeness with 0%, 20%, and 40% corruption for different dataset sizes. Note the different y-axis scales.

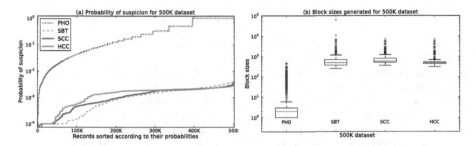

Fig. 4. (a) Probability of suspicion (P_s) and (b) block sizes generated by the different approaches using the $500K$ dataset

where n_g is the number of values in a global dataset (**G**) that match with the corresponding value in the encoded dataset **D**. As shown in Fig. 4(a), the MBS-SCC and MBS-HCC approaches both provide significantly better privacy compared to the PHO approach which has a maximum P_s of 1 (under the worst case assumption of the global dataset **G** being equal to the linkage dataset i.e. $\mathbf{G} \equiv \mathbf{D}$). By increasing the parameter ms_{max} stronger privacy can be guaranteed in our approach. Fig. 4(b) shows that the PHO approach [8,17] creates a large number of blocks of size 1 which makes this approach not suitable for PPRL. According to Fig. 4(b), the MBS-HCC approach provides clusters within the acceptable size limit of ms_{max} which results in better block structures compared to the SBT and MBS-SCC approaches. This illustrates that our novel MBS-HCC technique provides better privacy than the other compared approaches while achieving higher results for both RR and PC.

7   Conclusion

We proposed a novel blocking protocol for multi-party PPRL based on multi-bit Bloom filter splitting and canopy clustering. We suggested a novel agglomerative hierarchical canopy clustering algorithm which generates canopies (blocks) within a specific size range. We demonstrated the efficiency and effectiveness of our approach on datasets containing up-to one million records. The evaluation results indicated that our approach is scalable with both the size of the datasets and the number of parties. Our approach outperforms a previous multi-party private blocking and a phonetic based indexing approach in terms of blocking quality and privacy. A limitation in our approach is the assumption of the semi-honest adversary model. We plan to extend our protocol to adversary models that are applicable for malicious parties and evaluate the privacy over other attack methods applicable to PPRL [18]. We will also investigate the parallelization of our approach to improve its performance.

References

1. Al-Lawati, A., Lee, D., McDaniel, P.: Blocking-aware private record linkage. In: ACM IQIS, pp. 59–68, Baltimore (2005)
2. Bloom, B.: Space/time trade-offs in hash coding with allowable errors. Communications of the ACM **13**(7), 422–426 (1970)
3. Christen, P.: Data Matching - Concepts and Techniques for Record Linkage, Entity Resolution, and Duplicate Detection. Springer (2012)
4. Christen, P.: A survey of indexing techniques for scalable record linkage and deduplication. IEEE TKDE **24**(9), 1537–1555 (2012)
5. Clifton, C., Kantarcioglu, M., Vaidya, J., Lin, X., Zhu, M.: Tools for privacy preserving distributed data mining. SIGKDD Explorations **4**(2), 28–34 (2002)
6. Cohen, W.W., Richman, J.: Learning to match and cluster large high-dimensional data sets for data integration. In: ACM SIGKDD, pp. 475–480, Edmonton (2002)
7. Durham, E.: A framework for accurate, efficient private record linkage. Ph.D. thesis, Faculty of the Graduate School of Vanderbilt University, Nashville, TN (2012)
8. Karakasidis, A., Verykios, V.: Secure blocking+secure matching = secure record linkage. Journal of Computing Science and Engineering **5**, 223–235 (2011)
9. Kristensen, T.G., Nielsen, J., Pedersen, C.N.: A tree-based method for the rapid screening of chemical fingerprints. Algo. for Molecular Biology **5**(1), 9 (2010)
10. Kuzu, M., Kantarcioglu, M., Inan, A., Bertino, E., Durham, E., Malin, B.: Efficient privacy-aware record integration. In: ACM EDBT, Genoa, Italy (2013)
11. Lai, P., Yiu, S., Chow, K., Chong, C., Hui, L.: An efficient Bloom filter based solution for multiparty private matching. In: SAM, Las Vegas (2006)
12. McCallum, A., Nigam, K., Ungar, L.H.: Efficient clustering of high-dimensional data sets with application to reference matching. In: ACM SIGKDD, Boston (2000)
13. Ranbaduge, T., Vatsalan, D., Christen, P.: Tree based scalable indexing for multi-party privacy-preserving record linkage. In: AusDM, CRPIT 158, Brisbane (2014)

14. Schnell, R., Bachteler, T., Reiher, J.: Privacy-preserving record linkage using Bloom filters. BMC Medical Informatics and Decision Making **9**(1) (2009)
15. Schnell, R.: Privacy-preserving record linkage and privacy-preserving blocking for large files with cryptographic keys using multibit trees. In: JSM, Montreal (2013)
16. Vatsalan, D., Christen, P., Verykios, V.: A taxonomy of privacy-preserving record linkage techniques. JIS **38**(6), 946–969 (2013)
17. Vatsalan, D., Christen, P.: Scalable privacy-preserving record linkage for multiple databases. In: ACM CIKM, pp. 1795–1798, Shanghai (2014)
18. Vatsalan, D., Christen, P., O'Keefe, C.M., Verykios, V.: An evaluation framework for privacy-preserving record linkage. JPC **6**(1), 35–75 (2014)
19. Vatsalan, D., Christen, P., Verykios, V.: Efficient two-party private blocking based on sorted nearest neighborhood clustering. In: ACM CIKM, San Francisco (2013)

Efficient Interactive Training Selection
for Large-Scale Entity Resolution

Qing Wang$^{(\boxtimes)}$, Dinusha Vatsalan, and Peter Christen

Research School of Computer Science, The Australian National University,
Canberra, ACT 0200, Australia
{qing.wang,dinusha.vatsalan,peter.christen}@anu.edu.au

Abstract. Entity resolution (ER) has wide-spread applications in many areas, including e-commerce, health-care, the social sciences, and crime and fraud detection. A crucial step in ER is the accurate classification of pairs of records into matches (assumed to refer to the same entity) and non-matches (assumed to refer to different entities). In most practical ER applications it is difficult and costly to obtain training data of high quality and enough size, which impedes the learning of an ER classifier. We tackle this problem using an interactive learning algorithm that exploits the cluster structure in similarity vectors calculated from compared record pairs. We select informative training examples to assess the purity of clusters, and recursively split clusters until clusters pure enough for training are found. We consider two aspects of active learning that are significant in practical applications: a limited budget for the number of manual classifications that can be done, and a noisy oracle where manual labeling might be incorrect. Experiments using several real data sets show that manual labeling efforts can be significantly reduced for training an ER classifier without compromising matching quality.

Keywords: Data matching · Record linkage · Deduplication · Active learning · Noisy oracle · Hierarchical clustering · Interactive labeling

1 Introduction

Entity resolution (ER), also known as data matching, record linkage, or duplicate detection, is the process of identifying and matching records that correspond to the same entities from one or more databases [6]. As the databases to be matched generally do not include entity identifiers, ER has to be based on the available attributes, for example, personal names, addresses and dates of birth. In the past decades, ER has attracted much interest from various application domains, the most prominent being health, census statistics, e-commerce, national security and digital libraries. For recent surveys see [6,14].

The core steps in ER in their most basic form consist of the pair-wise comparison of records using functions that calculate numerical similarities between

This research was partially funded by the Australian Research Council (ARC), Veda, and Funnelback Pty. Ltd., under Linkage Project LP100200079.

© Springer International Publishing Switzerland 2015
T. Cao et al. (Eds.): PAKDD 2015, Part II, LNAI 9078, pp. 562–573, 2015.
DOI: 10.1007/978-3-319-18032-8_44

attribute values, followed by either an unsupervised or supervised classification of pairs of records into matches and non-matches [6,14]. The comparison of attribute values used in ER is commonly based on approximate string comparison functions that return a normalized similarity between 0 (totally different values) and 1 (exact matching values). For each compared record pair, a *weight vector* is calculated with the similarities over the different attributes of that pair [6].

Various supervised and unsupervised learning techniques [3,5,7,9,13] have been proposed for ER in past years. While supervised techniques generally result in much better matching quality, these techniques require training data in the form of labeled examples of true matching pairs of records that refer to the same entity, and true non-matching pairs of records that refer to different entities. While in certain, mostly academic, situations such training data are available, in most practical applications of ER actual truth data are difficult to obtain. In many cases training data have to be manually generated, a task that is known to be difficult both in terms of cost and quality [6]. The traditional way of selecting training data is to use random sampling. However, from a robust statistical point of view, random sampling needs to select a significantly large number of examples for guaranteeing the quality of training data, which was also verified in our experiments discussed in Sect. 5. Another difficulty of using random sampling is caused by the imbalance of the ER problem, as the vast majority of record pairs will correspond to non-matches [6]. Two challenges thus stand out in particular when training data are to be manually generated over large-scale data sets: (1) How can we ensure "good" examples are selected for training? (2) How can we minimize the user's burden of labeling examples?

Active learning is a promising approach for selecting training data [1,19,22]. The central idea is to reduce the labeling efforts through *actively* choosing *informative* or *representative* examples [16]. In doing so, instead of choosing a large quantity of examples to label as is required for fully supervised learning, active learning only selects examples based on the hints from previously labeled examples, which can often yield a training set that is small but still sufficient for supporting accurate classification.

Although successful, existing active learning methods for ER have limitations in achieving efficient training for large-scale data sets. Most of these methods are grounded on a monotonicity assumption – a record pair with higher similarity is more likely to represent the same entity than a pair with lower similarity. This assumption is valid in some real-world applications but does not generally hold, as we will illustrate in Sect. 3. Thus, two difficult issues arise in selecting training data: (1) How do we know whether the monotonicity assumption holds on a data set since training data are not available? (2) How can we effectively select training data when the monotonicity assumption does not hold?

In this paper we develop a generic active learning method for efficiently selecting ER training data over large data sets. Unlike other works, we do not rely on the monotonicity assumption. Instead, our method exploits the cluster structure in data through active learning, which can circumvent the first issue above, meanwhile solving the second issue. The basic idea of our method is illustrated in

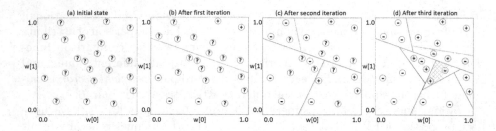

Fig. 1. Example of the training selection process with 2-dimensional weight vectors. Weight vectors that have been labeled as matches are shown with a $\oplus$, non-matches with a $\ominus$, and unlabeled ones with a circled question mark. The shaded areas in (d) represent fuzzy clusters, while the other areas in (d) represent pure clusters, which are to be used as training data for an ER classifier.

Fig. 1, where (a) shows weight vectors that are generated from pair-wise record comparisons, and the labels of these weight vectors are unknown. Then, (b) to (d) show how the weight vectors are interactively selected and manually classified, and how the set of weight vectors is recursively split into smaller clusters until each cluster is classified as being *pure* or *fuzzy* (to be formally defined in Sect. 4) based on the label *purity* of its informative weight vectors. During this process, the training set is interactively constructed by gathering the weight vectors from pure clusters.

We make the following contributions in this paper. (1) We develop an interactive training method which can be applied to ER training tasks without prior knowledge of the match and non-match distributions of the underlying data sets. (2) Our training method incorporates a budget-limited noisy human oracle, which ensures: (i) the overall labeling efforts can be controlled at an acceptable level and as specified by the user, and (ii) the accuracy of labeling provided by human experts can be simulated. This is in contrast to existing active learning methods for ER which often assume a perfect and unlimited labeling process [20]. (3) We experimentally evaluate our method on four real-world data sets from different application domains.

In the following section we discuss related work. In Sect. 3 we present the problem and building blocks of our approach, which we describe in detail in Sect. 4. We experimentally evaluate our approach in Sect. 5, and conclude the paper in Sect. 6 with an outlook to future work.

2   Related Work

Active learning has previously been studied in many problem domains, such as text classification and speech recognition [20]. In the area of ER, active learning has been explored for learning ER classifiers, which classify pairs of records as matches or non-matches through actively selecting a reduced number of examples for labeling [1,2,10,19,22]. In the following we provide a brief overview of work that relates to our study in this paper.

Early work on active learning strategies for finding informative or representative examples typically used disagreement between multiple classifiers. For example, a committee of classifiers was used to identify the most representative examples for labeling [19,22], i.e., labeling is iteratively required for pairs of records where the classifiers return contradictory labels for the same example. Sampling based active learning and its bias have been discussed in [11].

Later work has concentrated on the learning quality guarantee, that typically has a linear combination of the two measures precision and recall as the learning objective. For example, in [1,2], given a minimum precision specified by the user, a learned classifier aims to have a precision greater than the minimum precision and a recall close to the best possible. Compared with these active learning techniques, our algorithm has several interesting properties: (1) providing an integrated view on labeling budget control and quality guarantee, (2) using interactive purity-based classification to reduce examples for labeling, and (3) not relying on the monotonicity assumption for improving quality.

To improve the efficiency of active learning, two techniques have commonly been used. One is to incorporate blocking or indexing [6] into the learning process with parameters that are tuned manually [1,2] or semi-automatically [10]. Blocking in ER is the process of dividing data sets into smaller blocks according to some criteria so that only records within the same block are compared with each other. In principle, existing blocking techniques can be easily incorporated into active learning algorithms as a pre-processing step before learning. The second technique is to optimize active learning algorithms under certain distribution assumptions, such as the monotonicity [1] and low noise [2] assumptions. Our training method is completely independent of any assumption concerning the data set or any blocking technique used, which makes our method more generally applicable.

A number of studies have attempted to control labeling noise using certain strategies. Repeated labeling strategies were investigated in [21], including round-robin repeated labeling and selective repeated labeling based on the uncertainty of labels. In [12], a combined strategy was proposed, which selects examples that are more likely to be correctly labeled yet still provide high quality information, and examples that are most likely to have been incorrectly labeled. In [23] the most reliable oracle was selected among multiple noisy oracles for labeling. In this paper, we explore active learning in the presence of a noisy human oracle, which allows us to simulate the challenging manual clerical labeling process in real-world ER applications.

3   Problem Statement and Building Blocks

We study the problem of reducing the labeling costs for selecting training data, while keeping the quality of ER classification at a high level. In contrast to the works of [1,2], we do not rely on the monotonicity assumption since it does

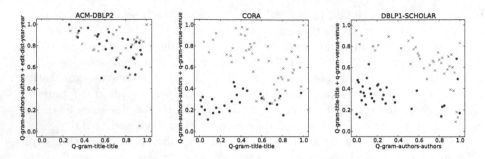

Fig. 2. Examples where the monotonicity assumption of similarities does not hold: non-matches with the highest similarity (denoted by light green crosses) and matches with the lowest similarity (denoted by dark blue dots)

not generally hold for ER. This is evident from the plots in Fig. 2, which show the non-matches with the highest similarity and matches with the lowest similarity from three of the real-world data sets we used in our experimental evaluation in Sect. 5. To address this problem, we propose an active learning approach that, given a set of weight vectors and a classifier, recursively splits the weight vectors into clusters, and classifies these as being matches or non-matches if the purity of informative weight vectors in a cluster is higher than a specified threshold. In the following we present the building blocks of our proposed approach.

Let $\mathbf{R}$ be a set of records from one or more data sets, each $r \in \mathbf{R}$ having a set of attributes. We use $r.A$ to refer to the value of an attribute A in a record r. Given two records $r_1, r_2 \in \mathbf{R}$ and an attribute A of r_1 and r_2, a *similarity weight* of A between r_1 and r_2 is a value in $[0, 1]$, denoted as $f(r_1.A, r_2.A)$, where f is a similarity function [6] that quantifies the similarity between $r_1.A$ and $r_2.A$. Taking the edit distance similarity function f_{ed} for example [6], $f_{ed}(r_1.fname, r_2.fname) = 1.0 - \frac{3}{6} = 0.5$, where $r_1.fname = Rob$ and $r_2.fname = Robert$.

For a set $\mathbf{A} = \{A_1, \ldots, A_n\}$ of attributes selected for performing ER tasks, each compared pair (r_1, r_2) of records that has the attributes $\mathbf{A}$ results in a *weight vector* $\langle a_1, \ldots, a_n \rangle \in [0, 1]^n$ over $\mathbf{A}$, where a_i is the similarity weight of A_i between r_1 and r_2 ($i = 1, \ldots, n$). For example, the pair (r_1, r_2) of records over the attributes $\{fname, sname, age\}$ with $r_1.fname = Rob$, $r_1.sname = Smith$, $r_1.age = 30$, $r_2.fname = Robert$, $r_2.sname = Smith$ and $r_2.age = 31$ may correspond to a weight vector $\langle 0.5, 1, 0.5 \rangle$. A *weight vector set* $\mathbf{W}$ over $\mathbf{A}$ consists of all the weight vectors over $\mathbf{A}$ to which the pairs of records in $\mathbf{R}$ correspond. A *cluster* $\mathbf{W}_i \subseteq \mathbf{W}$ is a subset of weight vectors in $\mathbf{W}$. A *partition* of $\mathbf{W}$ is a set $\{\mathbf{W}_1, \ldots, \mathbf{W}_m\}$ of pairwise disjoint clusters whose union contains all the weight vectors in $\mathbf{W}$, i.e. $\mathbf{W}_i \cap \mathbf{W}_j = \emptyset$ for $1 \leq i \neq j \leq m$, and $\bigcup_{1 \leq i \leq m} \mathbf{W}_i = \mathbf{W}$.

We consider a noisy human oracle that simulates a non-perfect manual clerical labeling process. The main reason behind such noisy human oracles is due to the fact that human experts often have different levels of expertise for labeling matches and non-matches [6]. Thus, depending on which human expert is asked for labeling an example, the labeling accuracy varies. A human oracle takes a set of record pairs and their corresponding weight vectors as input, and based on manual inspection of the attribute values of these records assigns each weight vector with a label. Let $\mathbf{W}_i$ be a weight vector set. Then a *human oracle* over $\mathbf{W}_i$ is a function $\zeta : \mathbf{W}_i \mapsto \{M, N\}$, where M and N are the two labels indicating *match* and *non-match* of a weight vector, respectively. Moreover, each human oracle ζ is associated with a pair $\langle bud(\zeta), acc(\zeta) \rangle$, where $bud(\zeta) > 0$ is a budget limit (b_{tot}) indicating the maximal number of weight vectors that can be labeled by ζ, and $acc(\zeta) \in [0, 1]$ is indicating the accuracy of labels provided by ζ. If $acc(\zeta) = 1$ then the oracle is *perfect*.

We view an *ER classifier* as a black-box that classifies record pairs into matches and non-matches through their corresponding weight vectors [6]. More specifically, an *ER classifier* takes as input a weight vector set $\mathbf{W}_i$ and a subset of labeled (with M and N) weight vectors $\mathbf{W}_i^T \subseteq \mathbf{W}_i$ as the training set, and generates a partition of $\mathbf{W}_i$ into $\mathbf{W}_i^M$ of matches and $\mathbf{W}_i^N$ of non-matches, with $\mathbf{W}_i^M \cap \mathbf{W}_i^N = \emptyset$. A variety of classifiers have previously been used for ER, such as decision trees [22], SVMs [7,19] and k-nearest neighbor [4], any of which can be used in our approach.

4 Recursive Interactive Training Algorithm

In this section we discuss the details of our approach. A high-level description of our interactive training approach is provided in Algorithm 1. Let $\mathbf{W}$ be a weight vector set, and $\mathbf{T}^M$ and $\mathbf{T}^N$ be the subsets of $\mathbf{W}$ that are selected into the match and non-match training sets, respectively, with $\mathbf{T}^M \cap \mathbf{T}^N = \emptyset$.

After initialization, the algorithm starts with $\mathbf{W}$ being inserted into an empty queue $\mathbf{Q}$ of clusters to be processed (line 2). The main iteration (line 4) loops as long as the queue is not empty (i.e. there are clusters to process) and the total oracle budget b_{tot} has not been fully used ($b \leq b_{tot}$). In each iteration, the first cluster $\mathbf{W}_i$ in the queue is being processed. In the first loop (with $b = 0$ indicating no manual labeling has been done), the INIT_SELECT() function is used to select a first set of weight vectors $\mathbf{S}_i \subseteq \mathbf{W}_i$ to be manually classified by the oracle, while in sub-sequent iterations the MAIN_SELECT() function is used. Different approaches for these selection functions will be described in Sect. 4.1. In general, a selection function selects k informative weight vectors $\mathbf{S}_i$ from a cluster $\mathbf{W}_i$ (lines 7 or 9).

The weight vectors in $\mathbf{S}_i$ are then manually classified by the human oracle (line 11) into a match set $\mathbf{T}_i^M$ and non-match set $\mathbf{T}_i^N$, which are added to the final training sets $\mathbf{T}^M$ and $\mathbf{T}^N$, respectively (line 12). The used budget is also increased (line 10) by the number of manually classified weight vectors $|\mathbf{S}_i|$. Then, the purity p_i of the cluster is calculated (line 13), as will be described further below. All weight vectors in the cluster are added into one of the training sets (lines 14 to 17) if the cluster is pure enough ($p_i \geq p_{min}$); otherwise, the cluster requires further splitting if it is larger than a minimum cluster size c_{min}, and

---
Algorithm 1. *Recursive interactive training algorithm*

Input:
- A weight vector set: $\mathbf{W}$
- Budget limit: b_{tot}
- Minimum purity threshold: p_{min}
- Initial selection function: INIT_SELECT$(\cdot)$
- Main selection function: MAIN_SELECT$(\cdot)$
- Human oracle for labeling: ORACLE$(\cdot)$
- Number of weight vectors to select for labeling: k
- Minimum size of a cluster: c_{min}
- Classifier function used for splitting clusters: CLASSIFIER

Output:
- Match and non-match training set $\mathbf{T}^M$ and $\mathbf{T}^N$

1:   $\mathbf{T}^M = \emptyset, \mathbf{T}^N = \emptyset$ // Initialize training sets as empty
2:   $\mathbf{Q} = [\mathbf{W}]$ // Initialize queue of clusters
3:   $b = 0$ // Initialize number of manually labeled examples
4:   **while** $\mathbf{Q} \neq \emptyset$ and $b \leq b_{tot}$ **do:**
5:      $\mathbf{W}_i = \mathbf{Q}.pop()$ // Get first cluster from queue
6:      **if** $b = 0$ **then:**
7:         $\mathbf{S}_i = $ INIT_SELECT$(\mathbf{W}_i, k)$ // Initial selection of weight vectors
8:      **else:**
9:         $\mathbf{S}_i = $ MAIN_SELECT$(\mathbf{W}_i, k)$ // Select informative weight vectors
10:     $b = b + |\mathbf{S}_i|$ // Update number of manual labeling done so far
11:     $\mathbf{T}_i^M, \mathbf{T}_i^N, p_i = $ ORACLE$(\mathbf{S}_i)$ // Manually classify selected weight vectors
12:     $\mathbf{T}^M = \mathbf{T}^M \cup \mathbf{T}_i^M; \mathbf{T}^N = \mathbf{T}^N \cup \mathbf{T}_i^N; \mathbf{W}_i = \mathbf{W}_i \setminus (\mathbf{T}_i^M \cup \mathbf{T}_i^N)$
13:     **if** $p_i \geq p_{min}$ **then:**
14:        **if** $|\mathbf{T}_i^M| > |\mathbf{T}_i^N|$ **then:**
15:           $\mathbf{T}^M = \mathbf{T}^M \cup \mathbf{W}_i$ // Add whole cluster to match training set
16:        **else:**
17:           $\mathbf{T}^N = \mathbf{T}^N \cup \mathbf{W}_i$ // Add whole to non-match training set
18:     **else if** $|\mathbf{W}_i| > c_{min}$ and $b \leq b_{tot}$ **then:**   // Low purity, split cluster further
19:        **if** $\mathbf{T}_i^M \neq \emptyset$ and $\mathbf{T}_i^N \neq \emptyset$ **then:**
20:           CLASSIFIER$.train(\mathbf{T}_i^M, \mathbf{T}_i^N)$ // Train classifier
21:           $\mathbf{W}_i^M, \mathbf{W}_i^N = $ CLASSIFIER$.classify(\mathbf{W}_i)$ // Classify current cluster
22:           $\mathbf{Q}.append(\mathbf{W}_i^M); \mathbf{Q}.append(\mathbf{W}_i^N)$ // Append new clusters to queue
23: **return** $\mathbf{T}^M$ and $\mathbf{T}^N$

---

the total oracle budget b_{tot} has not been fully used, and if $\mathbf{T}_i^M$ and $\mathbf{T}_i^N$ are not empty (lines 18 to 22). If $\mathbf{T}_i^M$ and $\mathbf{T}_i^N$ are both not empty, they will be used to train a classifier for the current cluster $\mathbf{W}_i$ (line 20). The splitting of $\mathbf{W}_i$ (line 21) leads two smaller clusters $\mathbf{W}_i^M$ and $\mathbf{W}_i^N$ of matches and non-matches, respectively, which are then added to the queue (line 22). In principle, the two smaller clusters $\mathbf{W}_i^M$ and $\mathbf{W}_i^N$ should have a higher purity compared to $\mathbf{W}_i$. Clusters that are small ($|\mathbf{W}_i| \leq c_{min}$) and not pure are not considered for inclusion into the final training sets.

The algorithm thus generates a partition of $\mathbf{W}$ such that the weight vectors in each pure cluster are selected into the training sets, i.e., the weight vectors from a match cluster into $\mathbf{T}^M$ and the ones from a non-match cluster into $\mathbf{T}^N$, while the weight vectors in fuzzy clusters (those too small for further splitting and not pure enough) are discarded.

The *purity* p_i of a cluster $\mathbf{W}_i$ is calculated based on the classification done by the human oracle (line 11) using the manually classified weight vector set $\mathbf{S}_i$ as the proportion of classified weight vectors that have the majority label:

$$p_i = purity(\mathbf{W}_i) = max \left(\frac{|\mathbf{T}_i^M|}{|\mathbf{T}_i^M \cup \mathbf{T}_i^N|}, \frac{|\mathbf{T}_i^N|}{|\mathbf{T}_i^M \cup \mathbf{T}_i^N|} \right), \tag{1}$$

where $|\mathbf{T}_i^M \cup \mathbf{T}_i^N| = |\mathbf{S}_i|$. For a given purity threshold $p_{min} \in [0.5, 1]$, a cluster $\mathbf{W}_i$ is labeled as *pure* if $purity(\mathbf{W}_i) > p_{min}$; otherwise $\mathbf{W}_i$ is labeled as *fuzzy*.

4.1   Weight Vector Selection Methods

The informativeness of selected weight vectors crucially influences the quality of the final generated training sets $\mathbf{T}^M$ and $\mathbf{T}^N$. Therefore, the selection methods in Algorithm 1 need to be carefully chosen. Here we propose three methods for the INIT_SELECT function and four methods for the MAIN_SELECT functions.

Let $\mathbf{W}_i$ be a set of weight vectors over the predefined set $\mathbf{A}$ of attributes. For the initial selection (line 7 in Algorithm 1) using INIT_SELECT we consider: (1) *Far:* Farthest-first weight vectors with random initialization [15], based on the farthest first clustering algorithm which selects the k weight vectors from $\mathbf{W}_i$ that are farthest apart from each other. The idea of this approach is to start with a selection of weight vectors with the highest possible variety. (2) *01:* Weight vectors that are closest to the two corners $[1]^{|\mathbf{A}|}$ and $[0]^{|\mathbf{A}|}$. These are most likely to represent matches and non-matches, respectively, as they correspond to weight vectors closest to exact matching and totally different record pairs [7]. (3) *Corner:* Weight vectors that are closest to all corners $\{[a_1, \ldots, a_{|\mathbf{A}|}] | a_i \in \{0, 1\}$ for $i = 1, \ldots, |\mathbf{A}|\}$, where there are $2^{|\mathbf{A}|}$ corners in total. This approach combines the ideas of both **Far** and **01**, selecting weight vectors with the highest possible variety in terms of all the attributes in $\mathbf{A}$.

Analogously, for follow-up selection of weight vectors from a cluster $\mathbf{W}_i$ during the main iteration of Algorithm 1 (line 9) using MAIN_SELECT we consider: (1) *Ran:* A random selection of k weight vectors. We use this as a baseline in our experiments to evaluate the effectiveness of the other selection methods. (2) *Far:* Farthest-first weight vectors selection within a cluster, as done in the *Far* initial selection method. This will give us weight vectors at the outer boundary of a cluster. (3) *Far-Med:* Here we select the $k-1$ farthest apart weight vectors from $\mathbf{W}_i$, and additionally we add the medoid weight vector closest to the center of the cluster. The idea is to not just manually classify pairs at the boundary of a cluster, but also one weight vector in its center to get a better picture of the distribution of matches and non-matches in the cluster.

In the following section we evaluate these different methods in combination with different parameter settings on several real-world data sets.

5   Experimental Evaluation

We conducted experiments on four data sets: ACM-DBLP [17], CORA[1], DBLP-Google Scholar (DBLP-GS) [17], and the North Carolina Voter Registration (NCVR) database[2]. The characteristics of these data sets are summarized in Table 1. As can be seen, all data sets exhibit a high to very high class imbalance

---

[1] Available from: http://secondstring.sourceforge.net
[2] Available from: ftp://alt.ncsbe.gov/data/

Table 1. Characteristics of data sets used in experiments

Data set name(s)	Number of records	Number of unique weight vectors	Class imbalance	Time for pair-wise comparisons
NCVR	224,073 / 224,061	3,495,580	1 : 27	441.6 sec
CORA	1,295	286,141	1 : 16	47.0 sec
DBLP-GS	2,616 / 64,263	8,124,258	1 : 3273	963.1 sec
ACM-DBLP	2,616 / 2,294	687,910	1 : 1785	95.3 sec

between true matches and true non-matches. We used the *Febrl* open source record linkage system for the pair-wise linkage step, together with a variety of blocking/indexing and string comparison functions [8]. The output of this step are sets of weight vectors of the compared record pairs, and their known true labels (match or non-match).

The following parameter variations were used in our experiments: minimum purity threshold $p_{min} = [0.95, 0.9, 0.85, 0.8, 0.75]$, oracle accuracy $acc(\zeta) = [1.0, 0.95, 0.9, 0.85, 0.8, 0.75]$, total budget $b_{tot} = [100, 200, 500, 1,000, 2,000, 5,000, 10,000]$, number of weight vectors selected $k = [9, 19, 49, 69, 99]$, and the different initial selection (*Far*, *01* and *Corner*) and selection (*Ran*, *Far* and *Far-Med*) methods discussed in the previous section. The classifiers used for splitting weight vectors were decision trees (DTree) with entropy and information gain [18]. Default values for the parameters were set to minimum purity $p_{min} = 0.95$, oracle accuracy $acc(\zeta) = 1.0$, number of weight vectors selected $k = 49$ for the CORA data set and $k = 69$ for the other data sets, total budget $b_{tot} = 1,000$ for the CORA data set and $b_{tot} = 5,000$ for the other data sets, minimum cluster size $c_{min} = 50$, initial selection method *01* and selection method *Far*, as these settings resulted in the best quality based on a set of pre-experiments.

We evaluated the effectiveness of our approach using the F-measure [6], and the efficiency using the time required for the classification. The baseline approaches we used to compare with our approach (which we refer as DTree-AL) were: (1) fully supervised decision tree (DTree-S), (2) fully supervised support vector machines with linear and polynomial kernels (SVM-S), (3) unsupervised automatic k-nearest neighbor clustering (kNN-US) [7], (4) unsupervised k-means clustering (kMeans-US), and (5) unsupervised farthest first clustering (Far-US) [8]. Our proposed active learning approach and the baseline approaches are implemented in Python 2.7.3, and we ran all experiments on a server with 6-core 64-bit Intel Xeon 2.4 GHz CPUs, 128 GBytes of memory and running Ubuntu 14.04. The programs and test data sets are available from the authors.

We first evaluated how different values for the six main parameters of our approach (i.e., p_{min}, $acc(\zeta)$, b_{tot}, k, INIT_SELECT methods, and MAIN_SELECT methods) affect the quality of the classification results. Fig. 3 (a) shows the F-measure of our approach for different minimum purity thresholds (p_{min}). F-measure increases with an increasing p_{min} since a higher purity of cluster requirement results in more accurately classified clusters. As expected the F-measure also increases when the accuracy of the oracle ($acc(\zeta)$) increases (Fig. 3 (b)).

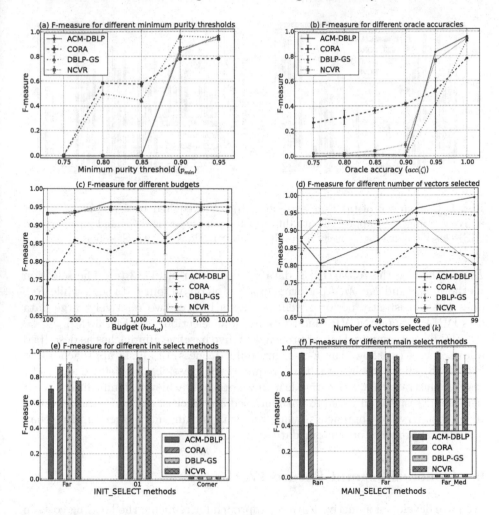

Fig. 3. F-measure against (a) minimum purity threshold, (b) oracle accuracy, (c) budget, (d) number of weight vectors selected, (e) different initial selection methods and (f) different main selection methods averaged over the results of all classifiers

F-measure increases with larger budgets (b_{tot}) and larger number of weight vectors selected (k) as can be seen from Fig. 3 (c) and (d), respectively. Larger budgets allow more vectors to be manually labeled, and a larger number of weight vectors selected from each cluster can represent the clusters more effectively, resulting in increased F-measure. However, as can be seen when $k = 99$, a smaller number of clusters can be manually assessed with larger k, potentially leading to lower F-measure. An interesting result is that a high F-measure (of ≥ 0.8) is achieved on all data sets even with a small budget size of $b_{tot} = 200$.

Among the three initial selection methods, *01* comparatively performs well, though all three methods achieve high F-measure on all four data sets except the

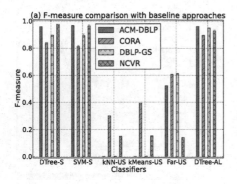
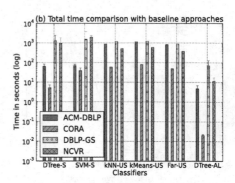

Fig. 4. Comparison of (a) F-measure and (b) total required time (log scale) of our active learning (AL) approach with different baseline supervised (S) and unsupervised (US) classifiers, averaged over the results of all variations of each classifier

Far method on the ACM-DBLP data set, as shown in Fig. 3 (e). The selection methods *Far* and *Far-Med* perform equally well on all four data sets, while *Ran* does not consistently perform well, particularly over two relatively large data sets DBLP-GS and NCVR, due to its random selection. (see Fig. 3 (f)).

Finally, we compared our approach with five baseline approaches as described above. Fig. 4 (a) shows the F-measure (effectiveness) of all six approaches and Fig. 4 (b) shows their total time required for classification (efficiency). The results illustrate that our active learning approach achieves significantly higher F-measure results compared to unsupervised approaches, and comparable results to fully supervised approaches, while requiring significantly lower runtime than all other approaches on all four data sets.

6   Conclusions and Future Work

We have developed an active learning approach for reducing the labeling costs in ER while achieving high linkage quality results. Experiments conducted on four real data sets validate the efficiency and effectiveness of our approach compared to both existing fully supervised and unsupervised ER classifiers.

As future work we plan to study the following two issues. First, how does the ordering of clusters (line 5 in Algorithm 1) in the queue affect the training quality? Since only a limited labeling budget is available, the number of weight vectors a human oracle can manually label is restricted. Once the labeling budget is run out, the training selection process terminates. Thus, the cluster selected for manual labeling at each iteration should be the one that can provide an optimal improvement in the quality, coverage and representativeness of the training data set. Second, how can our approach be improved if the accuracy of a human oracle is known? Knowing this accuracy may significantly affect the purity calculation of clusters. It is thus plausible to enhance the performance of our approach by taking the accuracy of a human oracle into account.

References

1. Arasu, A., Götz, M., Kaushik, R.: On active learning of record matching packages. In: ACM SIGMOD, Indianapolis, pp. 783–794 (2010)
2. Bellare, K., Iyengar, S., Parameswaran, A.G., Rastogi, V.: Active sampling for entity matching. In: ACM SIGKDD, Beijing, pp. 1131–1139 (2012)
3. Bilenko, M., Mooney, R.J.: Adaptive duplicate detection using learnable string similarity measures. In: ACM SIGKDD, Washington DC, pp. 39–48 (2003)
4. Chaudhuri, S., Ganti, V., Motwani, R.: Robust identification of fuzzy duplicates. In: IEEE ICDE, Tokyo, pp. 865–876 (2005)
5. Chen, Z., Kalashnikov, D.V., Mehrotra, S.: Exploiting context analysis for combining multiple entity resolution systems. In: ACM SIGMOD, Providence, pp. 207–218 (2009)
6. Christen, P.: Data Matching. Data-Centric Systems and Applications. Springer (2012)
7. Christen, P.: Automatic training example selection for scalable unsupervised record linkage. In: Washio, T., Suzuki, E., Ting, K.M., Inokuchi, A. (eds.) PAKDD 2008. LNCS (LNAI), vol. 5012, pp. 511–518. Springer, Heidelberg (2008)
8. Christen, P.: Development and user experiences of an open source data cleaning, deduplication and record linkage system. SIGKDD Explorations 11(1) (2009)
9. Cochinwala, M., Kurien, V., Lalk, G., Shasha, D.: Efficient data reconciliation. Information Sciences 137(1), 1–15 (2001)
10. Dal Bianco, G., Galante, R., Heuser, C.A., Gonçalves, M.A.: Tuning large scale deduplication with reduced effort. In: SSDBM, Baltimore, p. 18 (2013)
11. Dasgupta, S., Hsu, D.: Hierarchical sampling for active learning. In: IEEE ICML, Helsinki, pp. 208–215 (2008)
12. Du, J., Ling, C.X.: Active learning with human-like noisy oracle. In: IEEE ICDM, Sydney, pp. 797–802 (2010)
13. Elfeky, M.G., Verykios, V.S., Elmagarmid, A.K.: TAILOR: a record linkage toolbox. In: IEEE ICDE, San Jose, pp. 17–28 (2002)
14. Elmagarmid, A., Ipeirotis, P., Verykios, V.: Duplicate record detection: A survey. IEEE TKDE 19(1), 1–16 (2007)
15. Hochbaum, D.S., Shmoys, D.B.: A best possible heuristic for the k-center problem. Mathematics of Operations Research 10(2), 180–184 (1985)
16. Huang, S.J., Jin, R., Zhou, Z.H.: Active learning by querying informative and representative examples. In: NIPS, Vancouver, pp. 892–900 (2010)
17. Köpcke, H., Thor, A., Rahm, E.: Evaluation of entity resolution approaches on real-world match problems. VLDB Endowment 3(1–2), 484–493 (2010)
18. Pedregosa, F., Varoquaux, G., Gramfort, A., et al.: Scikit-learn: Machine learning in Python. The Journal of Machine Learning Research 12, 2825–2830 (2011)
19. Sarawagi, S., Bhamidipaty, A.: Interactive deduplication using active learning. In: ACM SIGKDD, Edmonton, pp. 269–278 (2002)
20. Settles, B.: Active learning literature survey, vol. 52, pp. 55–66. University of Wisconsin, Madison (2010)
21. Sheng, V.S., Provost, F., Ipeirotis, P.G.: Get another label? improving data quality and data mining using multiple, noisy labelers. In: ACM SIGKDD, Las Vegas, pp. 614–622 (2008)
22. Tejada, S., Knoblock, C.A., Minton, S.: Learning domain-independent string transformation weights for high accuracy object identification. In: ACM SIGKDD, Edmonton, pp. 350–359 (2002)
23. Wu, W., Liu, Y., Guo, M., Wang, C., Liu, X.: A probabilistic model of active learning with multiple noisy oracles. Neurocomputing 118, 253–262 (2013)

Unsupervised Blocking Key Selection
for Real-Time Entity Resolution

Banda Ramadan$^{(\boxtimes)}$ and Peter Christen

Research School of Computer Science, College of Engineering and Computer Science,
The Australian National University, Canberra, ACT 0200, Australia
{banda.ramadan,peter.christen}@anu.edu.au

Abstract. Real-time entity resolution (ER) is the process of matching
query records in sub-second time with records in a database that represent the same real-world entity. Indexing is a major step in the ER
process, aimed at reducing the search space by bringing similar records
closer to each other using a blocking key criterion. Selecting these keys
is crucial for the effectiveness and efficiency of the real-time ER process. Traditional indexing techniques require domain knowledge for optimal key selection. However, to make the ER process less dependent on
human domain knowledge, automatic selection of optimal blocking keys
is required. In this paper we propose an unsupervised learning technique
that automatically selects optimal blocking keys for building indexes that
can be used in real-time ER. We specifically learn multiple keys to be
used with multi-pass sorted neighbourhood, one of the most efficient and
widely used indexing techniques for ER. We evaluate the proposed approach using three real-world data sets, and compare it with an existing
automatic blocking key selection technique. The results show that our
approach learns optimal blocking/sorting keys that are suitable for real-
time ER. The learnt keys significantly increase the efficiency of query
matching while maintaining the quality of matching results.

Keywords: Record linkage · Unsupervised learning · Automatic blocking · Key selection · Sorted neighbourhood indexing

1 Introduction

Massive amounts of data are being collected by most business and government
organisations. Given that many of these organisations rely on information in their
day-to-day operations, the quality of the collected data has a direct impact on
the quality of the produced outcomes [4]. Data validation and cleaning are often
employed to improve data quality [4]. One important practice in data cleaning is
entity resolution (ER), which is the task of identifying records that refer to the same
real-world entity.

This research was funded by the Australian Research Council (ARC), Veda, and
Funnelback Pty. Ltd., under Linkage Project LP100200079.

© Springer International Publishing Switzerland 2015
T. Cao et al. (Eds.): PAKDD 2015, Part II, LNAI 9078, pp. 574–585, 2015.
DOI: 10.1007/978-3-319-18032-8_45

The ER process encompasses several steps [4]: *data preprocessing*, which cleans and standardizes the data to be used; *indexing* or (*blocking*), which reduces the search space; *record comparison*, which compares candidate records in detail using a set of similarity matching functions [8]; *classification*, where pairs or groups of candidate records are classified into matches (records that are assumed to refer to the same entity) and non-matches (records that are assumed to refer to different entities); and finally, *evaluation*, where the ER process is evaluated with regard to matching accuracy and completeness [4]. Since many services in both the private and public sectors are moving online, organisations increasingly require real-time ER (with sub-second response times) on query records that need to be matched with existing entity databases [7,15].

Indexing is a vital step in the ER process especially for large databases as it reduces the number of candidate records to be compared in detail to find matching records. This can be achieved by two main approaches. The first is to partition a database to be matched into several blocks according to a *blocking key* criterion, where only records that are inserted into the same block are compared with each other [9]. The second approach is to sort the records in a database according to a *sorting key* criterion that brings similar records close to each other, so that only records that are close to each other will be compared [11].

A good indexing technique should group similar records into one block or close to each other in the index [5]. This depends mainly on the blocking/sorting key used to partition/sort the records in a database. An optimal key needs to find all the true matching records, while keeping to a minimum the number of true non-matching records. However, an optimal key for one domain will likely not work for another domain [5].

Moreover, an optimal key for batch ER might not be suitable for real-time ER, because for real-time ER we need to have small block sizes to achieve fast query matching. Selecting an optimal key needs expert knowledge of the nature of the data and the requirements of the domain. To the best of our knowledge, no existing learning technique for indexing considers real-time ER. Therefore, there is a need for novel techniques that learn optimal keys for different real-time ER domains without the need for manual intervention.

Contribution: In this paper, we propose a general learning technique that automatically selects optimal keys for building indexes to be used in real-time ER in order to find matches in a database effectively and efficiently. Our approach can be used with different indexing techniques. We demonstrate how this automatic key selection can be used with an existing sorted neighbourhood-based real-time indexing technique [19]. We learn more than one key to be used with multi-pass sorting or blocking techniques. We evaluate the proposed technique on three real-world databases and compare it with an existing technique [12].

2   Related Work

The earliest proposed indexing approach is standard blocking [9], which inserts records into blocks according to a *blocking key* criterion. This criterion is usually based on one or more attribute values. Only records within the same block are compared with each other. This approach has the disadvantage of assigning records into the wrong block in case of errors in the attributes used as blocking keys (i.e. dirty data). To prevent this from occurring, iterative blocking [24] can be applied where multiple blocking keys are used and each record can be inserted into more than one block.

The sorted neighbourhood method (SNM) [11] arranges all records in the database(s) to be matched into a sorted array using a *sorting key*. Then a fixed-size widow is used to scan over the sorted records comparing only records within the window at any step. The main drawback of this method is its sensitivity to errors and variations at the beginning of the attribute values that are used as sorting keys, which can significantly affect the quality of the matching results [5]. This drawback is handled by performing a multi-pass approach [11] where different sorting keys are used in each pass to improve the matching quality of the approach. Various other indexing techniques that are based on either one of the above main approaches have been proposed [1,13,15,17,20,21]. However, for all of these techniques blocking or sorting keys need to be defined manually by an expert who has domain and application knowledge.

Various automatic techniques were proposed that allow learning optimal blocking/sorting keys based on supervised learning which requires the use of gold standard data for training. Bilenko et al. [2] proposed an approach that deals with the learning process as an approximation problem that is based on the red-blue set cover problem. Michelson et al. [18] proposed a related approach for learning which attributes are more suitable as blocking keys, and which similarity measures should be used for comparing these attributes.

Another supervised approach was recently proposed by Vogel and Naumann [23]. The authors use unigrams of attribute values (i.e. a combination of single characters from different attributes) as blocking keys. Both accuracy and efficiency of the generated blocks are used to learn the set of optimal blocking keys. They also improved their approach by taking the length of attribute values into consideration when generating the unigrams to be used as keys. All of the above automatic approaches require labeled training data. However, such labeled data is not always available and is usually expensive to generate.

To overcome this problem, several un-supervised automatic blocking key selection techniques have been developed [3,10,12,16]. Ma et al. [16] has proposed an approach that is based on type semantics, where the authors consider the type of entities when learning the blocking keys for data from the web. Another unsupervised learning approach was proposed by Kejriwal and Miranker [12] where the authors automatically generate a weakly labeled data set. This labeled data is then used as a training set to learn the optimal blocking keys using the Fisher discrimination criterion [12]. Giang [10] on the other hand proposed a technique that learns the blocking keys in context of the classifier function that is used in the classification step of the ER process. The classifier is used to generate labeled data. The authors then use the Probably Approximately Correct (PAC) approach to learn the blocking keys.

All mentioned unsupervised approaches focus mainly on the quality of the generated blocks and do not consider the block sizes when selecting blocking keys. However, a blocking key that can be used with real-time ER must also ensure that the sizes of the generated blocks are small enough to be able to resolve queries in real-time. In this paper, we propose an automatic blocking key selection technique that considers the coverage of a key, the maximum size of the generated blocks, as well as the distribution of the size of the generated blocks. Our aim is to learn blocking keys that are suitable for real-time ER.

3   Preliminaries and Overview

We use the following notation to present our approach. We assume a database $R = \{r_1, r_2, ..., r_{|R|}\}$, where each $r_i \in R$ contains several attributes $A = (a_1, a_2, ..., a_{|A|})$. We denote the attribute value a_j in r_i with $r_i.a_j$, where $1 \leq i \leq |R|$ and $1 \leq j \leq |A|$.

A *blocking key* (BK), denoted as $k_{j,l} = \langle a_j, f_l \rangle$, is a pair consisting of an attribute $a_j \in A$ and a blocking function $f_l \in F$, with F being the set of candidate blocking functions. Examples of such functions include exact value (isExact), same first character (sameFirst1), or same last three characters (sameLast3). The blocking function f_l is applied on attribute a_j, and the resulting value for a record r_i is called a blocking key value (BKV) and denoted as $k_{j,l}(r_i) = f_l(r_i, a_j)$. We denote the set of all candidate BKs with K. We assume the functions in F are manually selected by domain and ER experts, but for future work we aim to investigate techniques to automatically identify suitable blocking functions based on the content of a database. Our optimal key selection approach will identify the best BKs for real-time ER based on three criteria, as described in Sect. 4.

A *block* $b \in B_{j,l}$ is a set of records R_b where all $r_i \in R_b$ have the same BKV: $R_b = \{r_i \in R : k_{j,l}(r_i) = f_l(r_i, a_j)\}$. $B_{j,l}$ is the set of all blocks generated by a BK $k_{j,l}$ on all records in R.

Current approaches for learning blocking/sorting keys [3,10,12,16] do not learn keys that are suitable for real-time ER. In real-time ER, the selected BK(s) should generate block sizes within a controllable range to make sure that the number of detailed comparisons needed to match a query record (denoted as q) is within an allocated time. Also, keys that generate blocks of similar sizes are more suitable for real-time ER than keys that generate blocks of different sizes, as the time required to resolve different query records will be the same [6].

In our work, we aim to learn a list of optimal blocking/sorting keys, $O \subset K$, to be used with multi-pass indexing techniques [19] to perform ER and deliver high quality matching results in real-time. Following [12], our approach does not require existing training data sets to learn these optimal keys.

The overall framework of the proposed approach contains the following steps, as illustrated in Fig. 1. In step (1) we generate positive and negative training data sets (R_P and R_N) to be used in the learning process [12], as detailed below. In step (2) the set of candidate blocking keys K is generated. The proposed learning

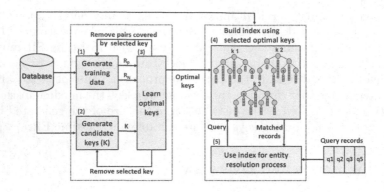

Fig. 1. Framework of the proposed approach

algorithm, described in Sect. 4, is employed in step (3) using the generated training data sets R_P and R_N to select a set of optimal blocking keys $O \subset K$. The selected optimal keys O are used in step (4) to index (block) all records from the database R. Any real-time indexing technique can be used for this step [19]. Finally, in step (5), the built index is used for matching query records q with records within the index in real-time.

3.1   Generating Training Data Sets

As in most practical applications of ER no training data sets (gold standard data) are available, such data can be generated using classification functions as in [10,12]. In this step we use Kejriwal's [12] approach to generate weakly labeled training data sets using a TF-IDF weighting scheme to calculate the similarity between record pairs $(r_x, r_y) \in R$ as follows.

A lower and upper thresholds $0 < l < u < 1$ are used to generate the training data sets. Record pairs (r_x, r_y) that have a TF-IDF similarity value $sim(r_x, r_y)$ below l are labeled as negative matches, and all pairs that have a TF-IDF value above u are labeled as positive matches. We generate a positive training set $R_P \subset R$ where the similarity between record pairs is greater than or equal to the upper threshold u: $R_P = \{r_x, r_y \in R : sim(r_x, r_y) \geq u\}$; and a negative training set $R_N \subset R$ where the similarity between record pairs is less than or equal to the lower threshold l: $R_N = \{r_x, r_y \in R : sim(r_x, r_y) \leq l\}$ with $R_P \cap R_N = \emptyset$.

Both R_P and R_N are then used to generate a set of *blocking key vectors*, V_P and V_N respectively, by applying all keys $k_{j,l} \in K$ on the record pairs in R_P and R_N [12]. Each record pair is converted into a vector of Boolean values (i.e 0 or 1 bits), with one value for each candidate key $k_{j,l} \in K$. If a record pair (r_x, r_y) in R_P or R_N for a certain candidate key $k_{j,l}$ results in having the same key value, i.e. $k_{j,l}(r_x) = k_{j,l}(r_y)$, then the corresponding element in the pair's vector is set to 1 and the pair is said to be *covered* by this key. Otherwise, the corresponding vector element is set to 0, and the pair is said to be *uncovered* by that key.

The generated set of key vectors V_P and V_N are then used in our key selection algorithm to learn the optimal keys, as described in the following section. Alternatively, if a truth training set is available, step (1) of our framework is not required. The rest of the steps of our framework are described in more detail in the following sections.

4 Optimal Key Selection

The indexing step of real-time ER should bring similar records close to each other while maintaining small block sizes to be able to match query records in real-time. The BKs used in the indexing step have an impact on the quality and efficiency of query matching. To make sure that the keys we select are suitable for real-time ER we use three criteria:

- **Key coverage:** The coverage C of a key $k_{j,l}$ that is applied on record pairs (r_x, r_y) in database R is defined as the number of record pairs that evaluate to the same key value: $C_{k_{j,l}} = |\{r_x, r_y \in R : k_{j,l}(r_x) = k_{j,l}(r_y)\}|$. A key with a high coverage value leads to grouping a high number of true positive matches into the blocks generated using that key while having a minimal number of negative matches in the blocks. We use the blocking key vectors V_P and V_N (described in Sect. 3.1) to measure the coverage of a BK by calculating its Fisher score [12] as follows:

$$C_k = \frac{|V_P|(\mu_{p,k} - \mu_k)^2 + |V_N|(\mu_{n,k} - \mu_k)^2}{|V_P|\sigma_{p,k}^2 + |V_N|\sigma_{n,k}^2} \tag{1}$$

  where $\mu_{p,k}$ and $\mu_{n,k}$ are the mean of all bits generated from evaluating the key $k \in K$ on all pairs in V_P and V_N respectively, $\sigma_{p,k}^2$ and $\sigma_{n,k}^2$ are the variance of corresponding bits of k in V_P and V_N respectively, and μ_k is the mean of the corresponding bits of key k in $V_P \cup V_N$. Note that a key will have a high Fisher score if it has high coverage in V_P and low coverage in V_N. The aim of using this measure is to select keys that produce high quality blocks (with mostly true matches, and only few negative matches grouped within the generated blocks).

- **Block Size:** The size of a block b is the number of records that are inserted into that block and it is denoted as $S_b = |R_b|$. In this criterion we use two measures: the maximum block size denoted as $S_{b_{(max)}} = max\{S_b : b \in B\}$, and the average block size denoted as $S_{b_{(ave)}} = ave\{S_b : b \in B\}$. The aim of using the size criterion is to control the number of candidate records that are required to be compared with a query record within a desired time range.

- **Distribution of blocks:** For the set of blocks B generated from applying a key k on all records in database R, the distribution of k is measured by calculating the variance V_k of the sizes of all blocks in B (which reflects how far the generated block sizes are spread) using:

$$V_k = \frac{\sum_{b=1}^{|B|}(S_b - \mu_S)^2}{|B|} \tag{2}$$

where S_b is the size of a block $b \in B$ and μ_S is the mean of all block sizes in B. A variance value that is equal to 0 means that all generated blocks have exactly the same size. For real-time ER it is better to generate blocks of similar sizes where the time required to match a query record is similar for different query records. Therefore, a BK is more suitable for real-time ER if its variance of the sizes of the generated blocks is close to 0.

Generating Candidate Keys: The candidate key list is the list of all keys which we select our optimal keys from. The candidate BKs can differ based on the domain, the used indexing technique, and the databases to be matched. Because we are evaluating our key selection algorithm using a sorted neighbourhood indexing technique (as will be discussed in Sect. 5), we generate a list of candidate BKs K that capture the beginning of attribute values (i.e. isExact, sameFirst1, sameFirst2, sameFirst3, sameFirst4, sameFirst5, and concatenatedIs-Exact). To generate a set of K blocking keys we apply all blocking functions in F on all $r_i.a_j \in R$. The generated set of blocking keys K is given to the proposed learning algorithm along with V_P and V_N to select optimal keys as follows.

Learning Optimal Keys: Our learning algorithm (see Algorithm 1) automatically selects the list of optimal keys O based on the three criteria discussed earlier (key coverage, generated block sizes, and distribution of block sizes) to ensure that the selected keys can be used with real-time ER to provide matching results efficiently. We start the algorithm by initialising the valid key list (K_v) to be empty. This list is then filled with keys that cover less than n_m pairs in V_N, where n_m is the maximum allowed number of covered vectors from negative key vectors (lines 1-4). Then, for each key in the valid key list K_v, if the key has a maximum block size $S_{b_{max}}$ that is greater than the maximum allowed block size s_m, it is removed from K_v (lines 5-8). In lines 9-13, for all keys k left in K_v, we calculate an overall score SC_k to determine which keys should be added to the optimal key list O based on the following equation:

$$SC_k = \alpha \cdot (1 - C_k) + \beta \cdot S_{b_{(ave)_k}} + (1 - \alpha - \beta) \cdot V_k \tag{3}$$

where C_k is the coverage of k (as calculated in Equation 1), $S_{b_{(ave)_k}}$ and V_k are the average block size and the variance between the block sizes, respectively. We assume that the blocks are generated by applying the blocking key k on all records in R. The aim is to select a set of blocking keys that have high coverage, low average block size, and low variance between block sizes (note that keys with large maximum block size $S_{b_{(max)}}$ were removed earlier from K_v in line 8). The parameters α and β are used to control the weights of the three criteria based on the domain and application area. Each weight parameter is a value between 0 and 1 where the total of all weights is equal to 1. Regardless of the weight parameters used, the lower the overall score for a key is, the more this key is suited for real-time ER.

Algorithm 1. *LearnOptimalBK(V_P, V_N, K, n_m, s_m, L)*

Input:
- Positive key vectors: V_P
- Negative key vectors: V_N
- Candidate blocking keys: K
- Maximum allowed covered vectors from negative key vectors: n_m
- Maximum allowed block size: s_m
- Number of blocking keys to be selected: L

Output:
- List of optimal blocking keys: O

```
0:  while l ≤ L do
1:      Kᵥ := [ ], O := [ ], scores := [ ]        // Initialise valid keys, optimal keys,
                                                   // and scores to be empty
2:      for k ∈ K do
3:          if k covers pairs in V_N that are < n_m then
4:              Kᵥ.add(k)                          // Add k to the valid key list Kᵥ
5:      for k ∈ Kᵥ do
6:          S_b(max) := GetMaxBlockSize(k)         // Get the maximum block size for k
7:          if S_b(max) > s_m then                 // Remove keys with large block size
8:              Kᵥ.remove(k)
9:      for k ∈ Kᵥ do
10:         Vₖ := GetVariance(k)                   // Get the variance for k
11:         S_b(ave) := GetAveBlockSize(k)         // Get the average block size for k
12:         SCₖ := CalcScore(V_P, V_N, S_b(ave), Vₖ) // Calculate overall score for k
13:         scores.add(SCₖ)                        // Add this key's score to the list of scores
14:     Sort scores ascending
15:     o := scores[0]                             // This optimal key has the lowest score value
16:     O.add(o)                                   // Add this optimal key to optimal key list
17:     Vc := getCoveredPairs(k, V_P)              // Get pairs from V_P that are covered by k
18:     V_P.remove(Vc)                             // Remove all pairs covered by k from V_P
19:     K.remove(o)                                // Remove the selected optimal key from K
20:     l := l + 1
21: Return O                                       // Return the optimal key list
```

After calculating the overall score SC_k for all keys in K_v, these scores are sorted in an ascending order, since a lower overall score is better (line 14). The first key in the overall score list is then added to the optimal key list O (lines 15-16). In lines $17 - 18$, all positive vectors that are covered by the selected optimal key are removed from V_P, and the optimal key is also removed from K in line 19. This process continues until the required number of optimal keys L is reached or until there are no positive vectors left in V_P. The selected optimal keys O are evaluated by performing the ER process on database R using the keys selected in the indexing step as described next.

5   Experimental Evaluation

In our experiments, we use three data sets (see Table 1). The OZ data set contains personal information that is generated by randomly selecting records from an Australian telephone directory (a clean data set). Duplicates are added to this data set by randomly modifying attribute values based on typing, scanning and OCR errors, or phonetic variations [22]. Both the Cora[1] and DBLP/ACM [14] data sets contain bibliographic information and are commonly used in ER research.

---

[1] Available from: http://secondstring.sourceforge.net

Table 1. Data sets used in our experiments

Data set	Type	Number of records	Number of entities
OZ	Real-world (modified)	34,588	30,292
Cora	Real-world	1,295	112
DBLP2/ACM	Real-world	2,616/2,294	2,686

We use the blocking key selection approach proposed in [12] as a baseline. The authors in [12] propose a Fisher Disjunctive algorithm (FDJ) that uses the Fisher discrimination criterion to select optimal blocking keys. Unlike our approach (that considers key coverage, block sizes, and blocks distribution), this approach only considers key coverage when selecting optimal blocking keys. We compare our approach with the baseline approach using recall (the fraction of relevant instances that are retrieved) to measure the quality of the compared approaches, and query time (the time required to resolve a single query record) to measure efficiency. In addition, we generate various statistics about the number of candidate records required by both approaches to resolve a query record.

For generating the training data sets (described in Sect. 3.1) we used a lower threshold $l = 0.1$ and an upper threshold $u = 0.7$ to weakly label record pairs into positive and negative pairs. The generated training data sets are then used in our learning algorithm as described in Sect. 4. To learn the optimal blocking keys we used $n_{max} = 100$ for the maximum allowed number of covered vectors from V_N, we used a maximum block size of $sm = 100$, and for the weight parameters we used $\alpha = 0.2$ and $\beta = 0.4$. Weights and thresholds used are selected based on an experimental investigation of using different values. We aim to investigate learning these values to produce blocks with high quality and small size in our future work.

To conduct the evaluation we use the keys selected by our approach and the keys selected by the FDJ approach to build indexes that can be used to resolve query records in real-time. A real-time forest-based dynamic sorted neighbourhood index (F-DySNI) is used for this purpose [19]. The index consists of multiple tree data structures where each tree is built using a different sorting key. The F-DySNI has two phases: a build phase where index trees are built using records from an existing entity database, and a query phase where the built index can be queried by retrieving candidate records for a query record from all index trees, and the index is updated by inserting the query record.

When a query record arrives it is inserted into all trees in the index. Then, in each tree, a window of size w is used to generate a set of candidate records from the tree node that contains the query record and the neighbouring tree nodes that fall within the window w. The query is then compared (using an approximating string similarity function [4]) with the generated candidate records. Candidate records with similarities above a specific threshold are considered to be matches. We use 50% of the records in each data set to build the indexes, and the remaining records are used as query records. For generating the candidate records we use a window of size $w = 2$ (the same window size used in [19]).

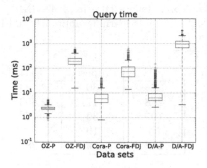 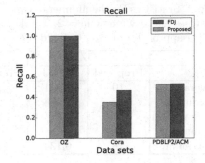

Fig. 2. Time and recall measures for the different data sets generated using the keys selected by our approach and the Fisher Disjunctive (FDJ) approach proposed in [12]. D/A in the left plot refers to DBLP/ACM and P refers to the proposed approach.

6   Results and Discussions

The aim of the experiments is to investigate if the optimal keys selected by our learning approach are suitable for real-time ER. Results in Fig. 2 illustrate the query time required to resolve a single query in milliseconds (ms) and recall values for the three data sets. The results show that the blocking keys selected by our approach improve the efficiency of query matching significantly. The selected keys using the proposed approach achieved an average query time of 2, 8, and 9 ms for the OZ, Cora, and DBLP-ACM respectively, while the selected keys using the baseline achieved an average query time of 206, 175, and 938 ms for OZ, Cora and DBKP-ACM respectively. This significant improvement in query time is achieved while maintaining recall for the OZ and DBLP-ACM but with a 5% decrease in recall value for Cora. Note that in our experiments the weight ($\alpha = 0.2$) that we use for the quality (i.e. key coverage) is half of the weight ($\beta = 0.4, \gamma = 0.4$) that we use for the block size and the distribution.

The results in Table 2 show various statistical measures for the number of candidate records generated using the proposed and the FDJ approaches using the F-DySNI with a window of size $w = 2$. It is clear from the table that the proposed approach has decreased the number of candidate records greatly which is the reason behind the significant decrease in query times.

Fig. 3 illustrates the distribution of the block sizes generated using the keys selected by the proposed and the FDJ approaches on the OZ data set (the other two data sets were too small to clearly show how the blocks are distributed). The results show that the keys selected by our approach lead to having block sizes that do not exceed the maximum allowed block size. It also shows that the generated blocks using our approach have similar sizes. In contrast, the keys selected by the FDJ approach lead to blocks of various sizes. These results are also supported by Table 2 where the standard deviation values (which measure the amount of variation from the average) of the number of candidate records required using our selected keys are small (compared to the values of the baseline) for the three data sets. This means that the block sizes tend to be close to

Table 2. Statistics for the number of candidate records generated using F-DySNI with three trees and a window $w = 2$. The keys selected by our learning approach and by the FDJ approach are used to build the trees in the index.

	OZ		Cora		DBLP-ACM	
	Proposed	FDJ	Proposed	FDJ	Proposed	FDJ
Average	10	2,041	28	274	30	2,643
Median	10	1,936	26	268	17	2,853
St.deviation	2	800	11	107	30	1,170
Minimum	6	162	10	54	10	13
Maximum	18	6,134	70	583	235	4,389

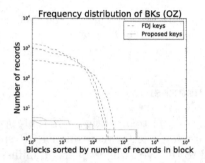

Fig. 3. Frequency distribution of the sizes of the blocks generated using the blocking keys selected by the proposed approach and the FDJ approach on the OZ data set

the average block size. We can conclude that the blocking keys selected by our proposed approach are suitable for use with real-time ER.

7   Conclusion and Future Work

We proposed an unsupervised blocking key selection algorithm that automatically selects optimal blocking keys for building indexes that can be used with real-time ER. We specifically learnt multiple keys to be used with multi-pass sorted neighbourhood indexing. We evaluated our approach using three real-world data sets and compared it with an existing automatic blocking key selection technique. The results show that our approach can learn keys that are suitable for real-time ER. The keys selected by our approach reduced query times significantly while maintaining matching quality. For future work we aim to investigate how we can automatically identify candidate blocking functions based on the content of the database. We also aim to investigate learning the weights that are used in our key selection algorithm to produce blocks with high quality and small size. Additionally, we plan to compare our proposed approach with other existing blocking key selection approaches.

References

1. Aizawa, A., Oyama, K.: A fast linkage detection scheme for multi-source information integration. In: WIRI, Tokyo (2005)

2. Bilenko, M., Kamath, B., Mooney, R.J.: Adaptive blocking: learning to scale up record linkage. In: IEEE ICDM, Hong Kong (2006)
3. Cao, Y., Chen, Z., Zhu, J., Yue, P., Lin, C.Y., Yu, Y.: Leveraging unlabeled data to scale blocking for record linkage. In: IJCAI, Barcelona (2011)
4. Christen, P.: Data Matching. Springer (2012)
5. Christen, P.: A survey of indexing techniques for scalable record linkage and deduplication. IEEE Transactions on Knowledge and Data Engineering 24(9) (2012)
6. Das Sarma, A., Jain, A., Machanavajjhala, A., Bohannon, P.: An automatic blocking mechanism for large-scale de-duplication tasks. In: ACM CIKM, Hawaii (2012)
7. Dong, X.L., Srivastava, D.: Big data integration. In: IEEE ICDE, Brisbane (2013)
8. Elmagarmid, A.K., Ipeirotis, P.G., Verykios, V.S.: Duplicate record detection: A survey. IEEE Transactions on Knowledge and Data Engineering 19(1) (2007)
9. Fellegi, I., Sunter, A.: A theory for record linkage. Journal of the American Statistical Association 64(328) (1969)
10. Giang, P.H.: A machine learning approach to create blocking criteria for record linkage. Health Care Management Science (2014)
11. Hernandez, M.A., Stolfo, S.J.: The merge/purge problem for large databases. In: ACM SIGMOD, San Jose (1995)
12. Kejriwal, M., Miranker, D.P.: An unsupervised algorithm for learning blocking schemes. In: IEEE ICDM, Dallas (2013)
13. Kim, H., Lee, D.: HARRA: fast iterative hashed record linkage for large-scale data collections. In: ICDT, Lausanne, Switzerland (2010)
14. Köpcke, H., Thor, A., Rahm, E.: Evaluation of entity resolution approaches on real-world match problems. VLDB Endowment 3(1–2) (2010)
15. Liang, H., Wang, Y., Christen, P., Gayler, R.: Noise-tolerant approximate blocking for dynamic real-time entity resolution. In: Tseng, V.S., Ho, T.B., Zhou, Z.-H., Chen, A.L.P., Kao, H.-Y. (eds.) PAKDD 2014, Part II. LNCS (LNAI), vol. 8444, pp. 449–460. Springer, Heidelberg (2014)
16. Ma, Y., Tran, T.: Typimatch: type-specific unsupervised learning of keys and key values for heterogeneous web data integration. In: ACM WSDM, Rome (2013)
17. McCallum, A., Nigam, K., Ungar, L.: Efficient clustering of high-dimensional data sets with application to reference matching. In: ACM SIGKDD, Boston (2000)
18. Michelson, M., Knoblock, C.A.: Learning blocking schemes for record linkage. In: AAAI, Boston (2006)
19. Ramadan, B., Christen, P.: Forest-based dynamic sorted neighborhood indexing for real-time entity resolution. In: ACM CIKM, Shanghai (2014)
20. Ramadan, B., Christen, P., Liang, H.: Dynamic sorted neighborhood indexing for real-time entity resolution. In: Wang, H., Sharaf, M.A. (eds.) ADC 2014. LNCS, vol. 8506, pp. 1–12. Springer, Heidelberg (2014)
21. Ramadan, B., Christen, P., Liang, H., Gayler, R.W., Hawking, D.: Dynamic similarity-aware inverted indexing for real-time entity resolution. In: Li, J., Cao, L., Wang, C., Tan, K.C., Liu, B., Pei, J., Tseng, V.S. (eds.) PAKDD 2013 Workshops. LNCS (LNAI), vol. 7867, pp. 47–58. Springer, Heidelberg (2013)
22. Tran, K.N., Vatsalan, D., Christen, P.: Geco: an online personal data generator and corruptor. In: ACM CIKM, New York (2013)
23. Vogel, T., Naumann, F.: Automatic blocking key selection for duplicate detection based on unigram combinations. In: VLDB Workshops, Istanbul (2012)
24. Whang, S.E., Menestrina, D., Koutrika, G., Theobald, M., Garcia-Molina, H.: Entity resolution with iterative blocking. In: ACM SIGMOD, Providence (2009)

Incorporating Probabilistic Knowledge into Topic Models

Liang Yao, Yin Zhang$^{(\boxtimes)}$, Baogang Wei, Hongze Qian, and Yibing Wang

College of Computer Science and Technology,
Zhejiang University, Hangzhou 310027, People's Republic of China
{yaoliang,yinzh,wbg,azureqianhz,bingtel}@zju.edu.cn

Abstract. Probabilistic Topic Models could be used to extract low-dimension aspects from document collections. However, such models without any human knowledge often produce aspects that are not interpretable. In recent years, a number of knowledge-based models have been proposed, which allow the user to input prior knowledge of the domain to produce more coherent and meaningful topics. In this paper, we incorporate human knowledge in the form of probabilistic knowledge base into topic models. By combining latent Dirichlet allocation, a widely used topic model with Probase, a large-scale probabilistic knowledge base, we improve the semantic coherence significantly. Our evaluation results will demonstrate the effectiveness of our method.

Keywords: Topic model · Knowledge base · Probase

1 Introduction

The explosion of online text content, such as news, blogs, Twitter messages and product reviews has given rise to the challenge to understand the very dynamic sea of text. To address the challenge, we need to extract the concepts from the sea of text, for the reason that "Concepts are the glue that holds our mental world together" [19] and "Without concepts, there would be no mental world in the first place" [6].

Most text mining tasks, especially aspects extraction tasks, use statistical topic models such as PLSA and Latent Dirichlet Allocation (LDA) [5,17]. However, these unsupervised models without any human knowledge often result in topics that are not interpretable, in other words, could not generate semantically coherent concepts [8,18].

In this paper, we propose a new probabilistic method, called Probase-LDA, which combines topic model and a probabilistic knowledge base, in particular LDA model and Probase. The proposed method explicitly models text content with large-scale knowledge, could extract more coherent topics with interpretable expression. The main contribution of this work is proposing a simple but effective method for combining a probabilistic topic model and a large-scale probabilistic knowledge base to improve the coherence of topic models.

© Springer International Publishing Switzerland 2015
T. Cao et al. (Eds.): PAKDD 2015, Part II, LNAI 9078, pp. 586–597, 2015.
DOI: 10.1007/978-3-319-18032-8_46

We begin this paper by reviewing some related works, including studies which devote to improving the interpretability of topic models mainly by incorporating domain knowledge into topic models, studies which focus on acquiring knowledge for machines, and studies which measure the coherence of topic models. In the remainder of the paper, we first describe our framework, then do experiments on real world data sets and analyze experimental results. Evaluations show the effectiveness of our method. Finally, we conclude our work.

2   Related Work

To overcome the shortcoming of semantic coherence in topic model, especially in LDA, some previous studies incorporate domain knowledge into the LDA model. Andrzejewski and Zhu proposed topic-in-set knowledge [1] which restricts topic assignment of terms to a subset of topics. They improved the topic-in-set knowledge in [3] by incorporating general knowledge specified by first-order logic. Similarly, in [9], Concept model was proposed by using ontologies like Open Directory Project (ODP) or The Cambridge International Dictionary of English (CIDE). The DF-LDA (Dirichlet Forest LDA) model in [2] can incorporate knowledge in the form of must-links and cannot links input by human beings. A must-link states that two words should belong to the same topic, while a cannot-link means two words should not be in the same topic.

Recently, Chen et al. [14] presented LDA with Multi-Domain Knowledge (MDK-LDA) which is capable of using prior knowledge from multiple domains. In [13], a more advanced topic model, called MC-LDA (LDA with must-link set and cannot-link set), was proposed as an extension of MDK-LDA. MC-LDA is a model that uses must-link and cannot-link knowledge like DF-LDA, it assumes that all knowledge is correct. GK-LDA (General Knowledge based LDA) is another model that uses the ratio of word probabilities under each topic to reduce the effect of wrong knowledge [12].

More recently, AKL (Automated Knowledge LDA) and LTM (Lifelong Topic model) were proposed [10,11], which learn knowledge automatically from multiple domains to improve topics in each domain.

Although these knowledge-based topic models use knowledge in many ways, they either suffer from the lack of sufficient knowledge, or use knowledge in a black or white form, but knowledge often associates with probability [24]. Inspired by this, we incorporate large-scale probabilistic knowledge into LDA model.

There are two major efforts on acquiring knowledge for machines. One is constructing fact-oriented knowledge bases, the other is building term-based knowledge bases. The fact-oriented knowledge bases focus on collecting black or white facts as more as possible from experts, communities, or collaboratively built semi-structured content like Wikipedia. YAGO [22] and Freebase [7] are two typical fact-oriented knowledge bases. The term-based knowledge bases focus on extracting concepts, instances, and their relations from web pages by using natural language processing and information extraction techniques. The most typical work is KnowItAll [16], TextRunner [4] and Probase [25], which proposes some kind of scores to serve taxonomy inference.

To measure the coherence of topic models, Mimno et al. [18] presented an automatic coherence measure of topic models using word co-occurrence in the training corpus, which automates the human judging method in [8]. At the same time, they put forward an unsupervised method which improves the coherence score by considering the word co-occurrence in the corpus. Newman et al. [20] showed that an automated evaluation metric based on word co-occurrence statistics gathered from Wikipedia could predict human evaluations of topic quality. Chuang et al. [15] measured the correspondence between a set of latent topics and a set of reference concepts when applying topic models for domain-specific tasks, which provides another way to assess the coherence.

3   The Proposed Probase-LDA Model

3.1   Probase: A Large Scale IsA Taxonomy

Probase[1] is a probabilistic knowledge base consisting of more than 2.7 million concepts automatically extracted using syntactic patterns (such as the Hearst patterns) from 1.68 billion Web pages [25]. For example, from the sentence "... presidents such as Obama ...", it extracts a piece of evidence for the claim that "Obama" is an instance of the concept president. A unique advantage of Probase is that the concept-instance relationships are probabilistic such that for each concept c and each instance e, Probase specifies the probability of each instance belonging to that concept, $P(e|c)$. It also specifies the probabilities of the reverse direction, $P(c|e)$. This probabilistic nature of Probase allows us to compute and combine the probabilities of concepts and entities in given text.

3.2   Exploiting Probase Knowledge

In Probase, one of the most basic scores is the typicality score. This score widely exists in human minds, given a concept "country", more people may think of "USA" or "China" instead of "Nepal"; given an item "apple", people will treat concept "fruit" or "company" more important than "music track". It looks like human beings assign a typicality score for each instance in a concept or each concept an entity belongs to, and ranked them automatically when they think of the concept or the entity. Typicality score is very useful for machines. It can make machines do reasoning like human beings.

Formally, typicality score can be driven from co-occurrences of concept and instance pairs as follows:

$$P(e|c) = \frac{n(c,e)}{\sum_{e_i \in c} n(c, e_i)} \tag{1}$$

$$P(c|e) = \frac{n(c,e)}{\sum_{e \in c_i} n(c_i, e)} \tag{2}$$

---

[1] availiable at http://probase.msra.cn/dataset.aspx

where $n(c, e)$ is the co-occurrence of concept c and instance e in Hearst patterns' sentences from the whole Web documents.

Based on typicality score, Wang et al. [24] define the representativeness score as:

$$Rep(e, c) = P(c|e)P(e|c) \tag{3}$$

Their experimental results show that the representativeness score performs best for entities conceptualization.

In this work, we leverage the scores of Probase as prior knowledge to improve topic modeling. First, we map each sentence in the training corpus to its top concepts by using a Naive Bayes approach as [21]:

$$P(c_k|E) = \frac{P(E|c_k)P(c_k)}{P(E)} \propto P(c_k) \prod_{i=1}^{M} P(e_i|c_k) \tag{4}$$

where $E = \{e_i, i \in 1, \dots, M\}$ is the set of observed instances in a sentence, $P(c_k) \propto \sum_{e_i \in c_k} n(c_k, e_i)$ is the observed frequency of c_k. Laplace smoothing is used to filter out noise and introduce concept diversities when calculating the likelihood $P(e_i|c_k)$ in Equation (1).

After mapping, we remove stop concepts and vague concepts (vagueness > 0.7) defined in Probase, and get a set of concepts of the corpus. The number of concepts is usually larger than the number of topic K, therefore, we employ a k-Medoids clustering algorithm to cluster these concepts (we tried k-Means and k-Medoids, and found k-Medoids performs slightly better).

The basic idea is that if two concepts share many entities, they are similar to each other. We use the entity distribution to represent each concept and calculate the semantic distance between two concepts c_1 and c_2 as

$$d_{sem}(c_1, c_2) = 1 - cosine(\Delta_{c_1}, \Delta_{c_2}) \tag{5}$$

where Δ_{c_i} represents the vector of entity distribution of concept c_i as defined as

$$\Delta_c = \langle (e_1, s_1), \dots, (e_{|c|}, s_{|c|}) \rangle \tag{6}$$

In each (e, s) of Δ_c, e is the entity and s is $Rep(e, c)$ of e and c in Probase.

In this work, we treat a concept cluster as a topic in LDA, so we set the number of clusters to be K when performing k-Medoids clustering.

3.3 Asymmetric Dirichlet Priors

After concept clustering, we use a simple method to generate asymmetric Dirichlet priors. In LDA, the asymmetric Dirichlet prior α_{dk} can be interpreted as a prior observation count for the number of times a topic k is sampled in a document d before having observed any actual words from that document. Similarly, β_{kw} is a prior observation count on the number of times a vocabulary word w is sampled from a topic k before any actual observations. Based on these assumptions, we compute asymmetric Dirichlet priors as following sub-sections.

Topic-Word Priors. For each word w (entity) in vocabulary under each topic (concept cluster) t, the corresponding β_{tw} in asymmetric topic-word prior matrix is defined as

$$\beta_{tw} = \sum_{c_i \in C_w \cap t} Rep(w, c_i) \tag{7}$$

where C_w is the set of top concepts w belongs to, c_i is a concept in C_w, $c_i \in t$ means concept c_i is in concept cluster t.

Then for each topic t, $\boldsymbol{\beta}_t = (\beta m_{w_1}, \ldots, \beta m_{w_V})$, where β is the symmetric Dirichlet prior, m_{w_i} is the normalized value under topic t given by

$$m_{w_i} = \frac{\beta_{tw_i} - min}{max - min} \tag{8}$$

In Equation (8) max and min represent the maximum and minimum values of $\{\beta_{tw} | w = 1, \ldots, V\}$, respectively.

Document-Topic Priors. The asymmetric Dirichlet prior vector for each document d is computed as follows:

$$\boldsymbol{\alpha}_d = tfidf_{w_{d1}} \boldsymbol{\beta}_{.w_{d1}} + \ldots + tfidf_{w_{dn}} \boldsymbol{\beta}_{.w_{dn}} \tag{9}$$

where w_{di} is the i-th word in document d, $tfidf_{w_{di}}$ is the TFIDF value of w_{di}, $\boldsymbol{\beta}_{.w_{di}}$ is the corresponding column of w_{di} in topic-word prior matrix.

We also normalize the document-topic vector as $\boldsymbol{\alpha}_d = (\alpha m_1, \ldots, \alpha m_K)$, where $m_t = \frac{\alpha_{dt} - min}{max - min}$, α is the symmetric Dirichlet prior, max and min represent the maximum and minimum values of $\{\alpha_{dt} | t = 1, \ldots, K\}$, respectively.

3.4 Inference

Finally, we simply use asymmetric Dirichlet priors to infer the posterior probability of the topic assignment at each position: $P(z_{dn} = k | \mathbf{w}, \mathbf{z}_{-dn}, \boldsymbol{\alpha}, \boldsymbol{\beta}) \propto$

$$(n_{dk} + \alpha_{dk}) \times \frac{n_{kw_{dn}} + \beta_{kw_{dn}}}{\sum_{w=1}^{V}(n_{kw} + \beta_{kw})} \tag{10}$$

where w_{dn} is the n-th word in document d, z_{dn} is the topic assignment of w_{dn}, $n_{kw_{dn}}$ is the number of times w_{dn} is assigned to topic k, and n_{dk} is the number of times topic k is assigned to a word in d.

Although the equation above looks exactly the same as that of LDA, the important distinction is we encode probabilistic knowledge into Dirichlet priors.

4 Experiments

This section evaluates the proposed Probase-LDA model and compares it with four state-of-the-art baseline models:

1. **LDA** [5]: The classic unsupervised topic model.
2. **DF-LDA** [2]: A topic model that can use the user-provided knowledge.
3. **GK-LDA** [12]: A knowledge-based topic model that uses the ratio of word probabilities under each topic to reduce the effect of wrong knowledge.
4. **LTM** [10]: A lifelong learning topic model that learns the must-link type of knowledge automatically. It outperforms AKL [11].

4.1   Dataset and Settings

Dataset. We use a large dataset from [10][2]. This dataset contains 50 review collections from 50 product domains crawled from Amazon.com. Each domain has 1,000 reviews. The dataset has been pre-processed by the authors, each review has been divided into sentences, and each sentence is treated as a document.

Settings. For comparison, we use the same parameter settings as [10]. All models are trained using 2,000 iterations with an initial burn-in of 200 iterations, the sampling lag is 20. The symmetric priors of all models are set as $\alpha = 1$, $\beta = 0.1$, the number of topics $K = 15$ (when comparing priors for LDA, we change the number of topics and symmetric priors). The other parameters for baselines were set as suggested in their original papers. When running LTM, we use test setting 1 in the original paper (mining prior knowledge from topics of all domains including the test domain). Since DF-LDA and GK-LDA cannot mine any prior knowledge, we feed them the knowledge produced by LTM at each iteration. For Probase-LDA, we use the top three concepts of each sentence to form concept set when performing Bayesian inference.

4.2   Topic Coherence

We evaluate topics generated by each model based on Topic Coherence [18]. Traditionally, topic models have been evaluated using perplexity. However, perplexity on the heldout test set does not reflect the semantic coherence of topics and may be contrary to human judgments [8]. Instead, the metric Topic Coherence has been shown in [18] to correlate well with human judgments. As our goal is to discover meaningful or coherent topics, Topic Coherence is more suitable for our evaluation. A higher Topic Coherence value indicates a higher quality of topics.

Table 1 shows the average Topic Coherence of each model using knowledge learned at different learning iterations. Each value is the average over all 50 domains. Note that the results of Probase-LDA are produced by 7 independent runs, and are basically the same. Since LDA cannot use any prior knowledge, its results remain the same. From Table 1, we can see that Probase-LDA performs the best and has the highest Topic Coherence values in general, which shows that Probase-LDA finds higher quality topics than the baselines. Both LTM

---

[2] The dataset and the GK-LDA and LTM code are available at http://www.cs.uic.edu/~zchen/

Table 1. Average Topic Coherence values of each model at different learning iterations

Model \ Iteration	0	1	2	3	4	5	6
Probase-LDA	-1245.55	-1245.95	-1244.48	-1246.96	-1243.67	-1248.01	-1247.65
LTM	-1465.32	-1442.19	-1428.71	-1421.38	-1417.24	-1418.58	-1418.91
GK-LDA	-1465.32	-1455.61	-1456.85	-1458.52	-1459.24	-1458.87	-1460.90
LDA	-1465.32	-1465.32	-1465.32	-1465.32	-1465.32	-1465.32	-1465.32
DF-LDA	-1465.32	-1487.36	-1492.27	-1493.66	-1492.46	-1490.50	-1491.26

and GK-LDA perform better than LDA but worse than Probase-LDA, showing their ability of dealing with wrong knowledge to some extent. DF-LDA does not perform well. Without an automated way to deal with each piece of (correct or incorrect) knowledge specifically for each individual domain, its performance is actually worse than LDA.

In summary, we can say that the proposed Probase-LDA model can generate better quality topics than all baseline models. Improvements of Probase-LDA are all significant ($p < 10^{-6}$) based on 2-tailed paired t-test.

4.3   Human Evaluation

As our objective is to discover more coherent topics, here we evaluate the topics based on human judgment. Following [10], we recruited two human judges who are fluent in English and familiar with Amazon products and reviews to label the generated topics. Since we have a lot of domains (50), we selected 10 domains for labeling. The selection was based on the knowledge of the products of the two human judges. Without enough knowledge, the labeling will not be reliable. We labeled the topics generated by Probase-LDA, LTM and LDA, the topics generated by LTM are at learning iteration 5. For labeling, we followed the instructions in [18].

Topic Labeling. We first asked the judges to label each topic as coherent or incoherent. Each topic was presented as a list of 10 most probable words. In general, a topic was labeled as coherent if it had more than half (5) of its words coherently related to each other representing a semantic concept together; otherwise incoherent.

Word Labeling. The topics labeled as coherent by both judges were used for word labeling. Each topical word was labeled as correct if it was coherently related to the concept represented by the topic (identified in the topic labeling step); otherwise incorrect.

The Cohens Kappa agreement scores for topic labeling and word labeling are 0.879 and 0.827 respectively.

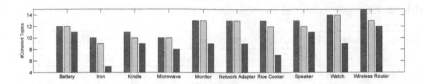

Fig. 1. Number of coherent topics discovered by each model. The bars from left to right in each group are for Probase-LDA, LTM and LDA

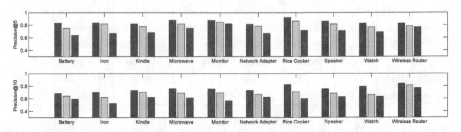

Fig. 2. Topical words *Precision*@5 and *Precision*@10 of coherent topics of each model

Since topics are rankings of words based on their probabilities, without knowing the exact number of correct topical words, a natural way to evaluate these rankings is to use *Precision*@n (or $p@n$) which was also used in [10,11].

Figure 1 shows that Probase-LDA discovers more coherent topics than LTM and LDA. On average, Probase-LDA discovers 0.6 more coherent topics than LTM and 3.3 more coherent topics than LDA over the 10 domains.

Figure 2 gives the average *Precision*@5 ($p@5$) and *Precision*@10 ($p@10$) of topical words of only coherent topics (incoherent topics are not considered) for each model in each domain. It is clear that Probase-LDA achieves the highest $p@5$ and $p@10$ values for all 10 domains. LTM is also better than LDA in general, but clearly worse than Probase-LDA. This is consistent with the Topic Coherence results in Table 1. On average, for $p@5$ and $p@10$, Probase-LDA improves LTM by 5.8% and 9.1%, and LDA by 19.7% and 21.8% respectively. Significance testing using 2-tailed paired t-tests shows that the improvements of Probase-LDA are significant over LTM ($p < 0.00003$) and LDA ($p < 0.00001$) on $p@5$ and $p@10$.

4.4   Example Topics

This section shows some example topics produced by Probase-LDA, LTM, and LDA in several domains to give an intuitive feeling of improvements made by Probase-LDA. Each topic is shown with its top 10 terms. Errors are italicized and marked in red. From Table 2 ("Monitor (Color)" means topic *Color* in domain "Monitor"), we can see that Probase-LDA discovers many more correct and meaningful topical terms at the top than the baselines. Note that for Probase-LDA's topics that were not discovered by the baseline models, we tried to find the best possible matches from the topics of the baseline models. From the table, we can clearly see that Probase-LDA discovers more coherent topics than

Table 2. Example Topics of Probase-LDA (Pro-LDA for short), LTM and LDA

Monitor (Color)			Watch (Material)			Wireless Router (Access)		
Pro-LDA	LTM	LDA	Pro-LDA	LTM	LDA	Pro-LDA	LTM	LDA
black	screen	screen	plastic	*band*	*band*	address	network	network
white	white	*side*	glass	*strap*	plastic	password	*home*	*setting*
screen	black	black	steel	*comfortable*	*pin*	ip	device	guest
race	*touch*	*line*	stainless	metal	*link*	ssid	computer	password
blue	color	white	*strap*	leather	steel	firewall	connection	security
green	*side*	*top*	gold	*link*	stainless	network	guest	*feature*
kind	line	*space*	leather	*pin*	metal	encryption	ssid	*home*
gray	*bezel*	*bottom*	metal	steel	*bracelet*	device	*separate*	key
sensitive	dark	*inch*	*black*	clasp	case	access	internet	ssid
contrast	glare	*vertical*	rubber	stainless	*clasp*	wep	*ability*	*default*

LTM and LDA. Apart from Table 2, many topics are significantly improved by Probase-LDA, including some commonly shared aspects such as *Brand* and *User experience*. In summary, we can say that Probase-LDA produces better results.

4.5 Comparing Priors for LDA

We follow [23] to investigate the effects of the different priors. We compared the four combinations of symmetric and asymmetric Dirichlets: symmetric priors over both θ and ϕ (LDA), a symmetric prior over θ and an asymmetric prior over ϕ (denoted SA), an asymmetric prior over θ and a symmetric prior over ϕ (denoted AS), and asymmetric priors over both θ and ϕ (Probase-LDA).

The four models (LDA, SA, AS, Probase-LDA), GK-LDA and LTM are performed with different number of topics. Each model was run with $T \in \{10, 15, 20, 25, 30\}$ for 2000 Gibbs sampling iterations with an initial burn-in of 200 iterations, others are all the same as in experimental settings.

Figure 3 shows the average Topic Coherence of each model given different number of topics. Each value is the average over all 50 domains. Results of LTM and GK-LDA are at learning iteration 3 (have stabilized). From Figure 3, we note the following:

1. AS achieves higher Topic Coherence scores than SA and LDA, this is consistent with the discovery in [23], which shows the advantages of asymmetric Dirichlet prior over the document-topic distributions.
2. SA is worse than all others at $K = 15$ and 20, but with the increase of the number of topics K, SA outperforms LDA, and GK-LDA and LTM. This shows that when the number of concept clusters increases from 10 to 30, the effectiveness of concept clustering becomes better.
3. Given different number of topics, Probase-LDA consistently achieves higher Topic Coherence scores than the baseline models. Among them, LTM performs best with the highest Topic Coherence score, but not as good as

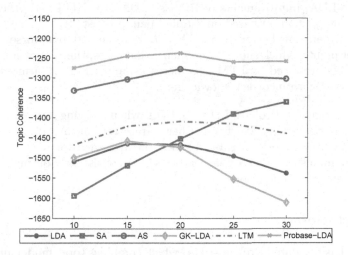

Fig. 3. Average Topic Coherence score of each model across different number of topics

Table 3. Average Topic Coherence of Probase-LDA and LDA using different symmetric priors

β / α		0.1	0.2	0.3	0.5	1	10	100
0.1	Probase-LDA	-1272.49	-1268.62	-1274.85	-1274.72	-1281.01	-1278.54	-1283.49
	LDA	-1498.32	-1495.05	-1496.70	-1499.72	-1542.37	-1663.63	-1649.34
0.5	Probase-LDA	-1251.59	-1253.14	-1250.49	-1253.12	-1253.14	-1260.39	-1262.83
	LDA	-1462.35	-1471.19	-1487.48	-1523.56	-1588.65	-1649.89	-1649.39
1	Probase-LDA	-1245.55	-1238.28	-1237.19	-1239.24	-1242.34	-1246.07	-1245.10
	LDA	-1465.32	-1493.53	-1528.42	-1575.33	-1629.84	-1649.57	-1649.50
2	Probase-LDA	-1226.87	-1221.86	-1225.08	-1225.19	-1225.09	-1235.76	-1197.16
	LDA	-1514.22	-1565.65	-1599.71	-1631.18	-1647.11	-1649.57	-1649.49
3	Probase-LDA	-1217.05	-1215.08	-1214.73	-1215.71	-1222.83	-1224.09	-1219.10
	LDA	-1565.91	-1608.97	-1628.32	-1645.99	-1648.41	-1649.18	-1649.53
4	Probase-LDA	-1214.74	-1213.94	-1211.70	-1210.24	-1216.35	-1219.74	-1213.87
	LDA	-1603.49	-1630.41	-1641.34	-1646.69	-1648.75	-1649.63	-1649.15
5	Probase-LDA	-1213.01	-1206.37	-1206.88	-1212.33	-1211.44	-1215.75	-1202.88
	LDA	-1627.80	-1641.52	-1645.90	-1648.86	-1649.35	-1649.38	-1649.67
10	Probase-LDA	-1200.68	-1201.93	-1201.69	-1204.87	-1205.35	-1206.69	-1195.46
	LDA	-1662.10	-1649.22	-1646.36	-1648.13	-1648.20	-1649.03	-1649.29
100	Probase-LDA	-1193.75	-1194.39	-1196.62	-1192.75	-1191.34	-1180.86	-1177.34
	LDA	-1665.58	-1650.44	-1648.82	-1649.50	-1649.48	-1649.26	-1649.16
1000	Probase-LDA	-1186.97	-1184.87	-1179.07	-1178.59	-1172.82	-1171.17	-1171.85
	LDA	-1664.73	-1649.06	-1648.76	-1649.42	-1649.06	-1649.34	-1649.25

Probase-LDA, improvements of Probase-LDA over LTM and others are all significant ($p < 10^{-5}$) based on 2-tailed paired t-test.

4. GK-LDA performs better than LDA at $K = 15$ and 20, but worse than LDA at other points, which shows that when K increases, the proportion of wrong knowledge produced by LTM also increases, and LTM has a more effective way to deal with incorrect knowledge.

We also compare Probase-LDA to LDA when varying the concentration parameter on the priors. We fixed other parameters as in *Settings* part and change the symmetric priors. Table 3 shows the average Topic Coherence scores over all 50 domains, we can see that Probase-LDA consistently outperforms LDA with different symmetric priors.

5 Conclusion

This paper has presented Probase-LDA, which combines topic model and a probabilistic knowledge base, in particular LDA model and Probase. The proposed method explicitly models text content with large-scale knowledge, could extract more coherent topics with interpretable expression. Experimental results on real world datasets show the effectiveness of the proposed method.

Acknowledgments. This work is supported by China Knowledge Centre for Engineering Sciences and Technology (No. CKCEST-2014-1-2), Specialized Research Fund for the Doctoral Program of Higher Education (SRFDP) (20130101110136), and China Academic Digital Associative Library (CADAL).

References

1. Andrzejewski, D., Zhu, X.: Latent dirichlet allocation with topic-in-set knowledge. In: Proceedings of the NAACL HLT 2009 Workshop on Semi-Supervised Learning for Natural Language Processing, pp. 43–48. Association for Computational Linguistics (2009)
2. Andrzejewski, D., Zhu, X., Craven, M.: Incorporating domain knowledge into topic modeling via dirichlet forest priors. In: ICML, pp. 25–32. ACM (2009)
3. Andrzejewski, D., Zhu, X., Craven, M., Recht, B.: A framework for incorporating general domain knowledge into latent dirichlet allocation using first-order logic. In: IJCAI (2011)
4. Banko, M., Cafarella, M.J., Soderland, S., Broadhead, M., Etzioni, O.: Open information extraction for the web. In: IJCAI, vol. 7, pp. 2670–2676 (2007)
5. Blei, D.M., Ng, A.Y., Jordan, M.I.: Latent dirichlet allocation. JMLR **3**, 993–1022 (2003)
6. Bloom, P.: Glue for the mental world (2003)
7. Bollacker, K., Evans, C., Paritosh, P., Sturge, T., Taylor, J.: Freebase: a collaboratively created graph database for structuring human knowledge. In: SIGMOD. ACM (2008)
8. Chang, J., Gerrish, S., Wang, C., Boyd-graber, J.L., Blei, D.M.: Reading tea leaves: how humans interpret topic models. In: NIPS, pp. 288–296 (2009)

9. Chemudugunta, C., Holloway, A., Smyth, P., Steyvers, M.: Modeling documents by combining semantic concepts with unsupervised statistical learning. In: Sheth, A.P., Staab, S., Dean, M., Paolucci, M., Maynard, D., Finin, T., Thirunarayan, K. (eds.) ISWC 2008. LNCS, vol. 5318, pp. 229–244. Springer, Heidelberg (2008)
10. Chen, Z., Liu, B.: Topic modeling using topics from many domains, lifelong learning and big data. In: ICML, pp. 703–711 (2014)
11. Chen, Z., Mukherjee, A., Liu, B.: Aspect extraction with automated prior knowledge learning. In: ACL, pp. 347–358 (2014)
12. Chen, Z., Mukherjee, A., Liu, B., Hsu, M., Castellanos, M., Ghosh, R.: Discovering coherent topics using general knowledge. In: CIKM, pp. 209–218. ACM (2013)
13. Chen, Z., Mukherjee, A., Liu, B., Hsu, M., Castellanos, M., Ghosh, R.: Exploiting domain knowledge in aspect extraction. In: EMNLP, pp. 1655–1667 (2013)
14. Chen, Z., Mukherjee, A., Liu, B., Hsu, M., Castellanos, M., Ghosh, R.: Leveraging multi-domain prior knowledge in topic models. In: IJCAI. AAAI Press (2013)
15. Chuang, J., Gupta, S., Manning, C., Heer, J.: Topic model diagnostics: assessing domain relevance via topical alignment. In: ICML, pp. 612–620 (2013)
16. Etzioni, O., Cafarella, M., Downey, D., Kok, S., Popescu, A.M., Shaked, T., Soderland, S., Weld, D.S., Yates, A.: Web-scale information extraction in knowitall:(preliminary results). In: WWW, pp. 100–110. ACM (2004)
17. Hofmann, T.: Probabilistic latent semantic indexing. In: SIGIR, pp. 50–57. ACM (1999)
18. Mimno, D., Wallach, H.M., Talley, E., Leenders, M., McCallum, A.: Optimizing semantic coherence in topic models. In: EMNLP, pp. 262–272 (2011)
19. Murphy, G.L.: The big book of concepts. MIT press (2002)
20. Newman, D., Lau, J.H., Grieser, K., Baldwin, T.: Automatic evaluation of topic coherence. In: NAACL-HLT, pp. 100–108. Association for Computational Linguistics (2010)
21. Song, Y., Wang, H., Wang, Z., Li, H., Chen, W.: Short text conceptualization using a probabilistic knowledgebase. In: IJCAI, pp. 2330–2336. AAAI Press (2011)
22. Suchanek, F.M., Kasneci, G., Weikum, G.: Yago: a core of semantic knowledge. In: WWW, pp. 697–706. ACM (2007)
23. Wallach, H.M., Minmo, D., McCallum, A.: Rethinking lda: why priors matter. In: NIPS (2009)
24. Wang, Z., Wang, H., Xiao, Y., Wen, J.R.: How to make a semantic network probabilistic. In: TechReport. MSR-TR-2014-59 (2014)
25. Wu, W., Li, H., Wang, H., Zhu, K.Q.: Probase: a probabilistic taxonomy for text understanding. In: SIGMOD, pp. 481–492. ACM (2012)

Learning Focused Hierarchical Topic Models with Semi-Supervision in Microblogs

Anton Slutsky[✉], Xiaohua Hu, and Yuan An

College of Computing and Informatics, Drexel University, Philadelphia, USA
as3463@drexel.edu

Abstract. Topic modeling approaches, such as Latent Dirichlet Allocation (LDA) and Hierarchical LDA (hLDA) have been used extensively to discover topics in various corpora. Unfortunately, these approaches do not perform well when applied to collections of social media posts. Further, these approaches do not allow users to focus topic discovery around subjectively interesting concepts. We propose the new Semi-Supervised Microblog-hLDA (SS-Micro-hLDA) model to discover topic hierarchies in short, noisy microblog documents in a way that allows users to focus topic discovery around interesting areas. We test SS-Micro-hLDA using a large, public collection of Twitter messages and Reddit social blogging site and show that our model outperforms hLDA, Constrained-hLDA, Recursive-rCRP and TSSB in terms of Pointwise Mutual Information (PMI) Score. Further, we test our model in terms of information entropy of held-out data and show that the new approach produces highly focused topic hierarchies.

1 Introduction

Modern applications of text mining often deal with large collections of documents that cover diverse sets of topics. Various topic modeling techniques have been developed in recent decades to discover these topics automatically and present visualizations that capture the spectrum of themes in a corpus. In a real-world setting, however, analysts are often interested in grasping the nature of the discourse around a particular concept or entity rather than understanding the corpus as a whole.

Such a task may be difficult to perform when dealing with social media texts. Social microblog systems are populated with millions of noisy, content-poor documents discuss large variety of subjects and concepts. The short, unedited nature of social media texts complicates applications of common topic modeling approaches [1] [2] [3][4][5][6] and makes extraction of interesting patterns especially difficult.

In this paper, we propose the new Semi-Supervised Microblog-hLDA (SS-Micro-hLDA) model that learns topics (defined as probability distributions over words) from microblog data in a way that allows for sets of interesting keywords (referred to as *supervisory word sets* from here on) to influence the topic learning process. To make the job of interpreting the learned topics easier, we require our

© Springer International Publishing Switzerland 2015
T. Cao et al. (Eds.): PAKDD 2015, Part II, LNAI 9078, pp. 598–609, 2015.
DOI: 10.1007/978-3-319-18032-8_47

approach to organize topics as hierarchies. This is motivated by well-known works in cognitive research that suggest that hierarchies may be instrumental in enhancing human sense making [7,8].

We test the new approach using the standard Tweets2011 data set made public by the TREC project and show that our model produces more interpretable and coherent topic models when measured in terms of PMI-Score against TSSB, Recursive-CRP, Constrained-hLDA and hLDA. Further, we test our approach and related approaches using information entropy and show that our model learns topic hierarchies that are more subject-focused than those produced by TSSB, Recursive-CRP, Constrained-hLDA and hLDA.

The paper is organized as follows. Section 2 discusses the current state of research in the area of topic modeling in general and microblog topic modeling in particular. Section 3 offers an analysis of topic modeling challenges in social stream data and describes the new Semi-Supervised Microblog-hLDA model designed to overcome these challenges. In Section 4, we discuss data sets and experiments that were used to evaluate how well our new topic modeling approach performed as compared with other approaches. Section 5 concludes the paper and outlines future work.

2   Related Works

Discovering hidden relationships between words may be accomplished using a number of different techniques. Matrix factorization approaches such as Latent Semantic Indexing (LSI) [9] and Non-Negative Matrix Factorization (NMF) [10] have been used to infer latent relationships between terms. While matrix factorization may be employed for topic discovery, approaches based on Latent Dirichlet Allocation (LDA) [11] have become very popular in recent years. This popularity has often been attributed to the flexibility and modularity of LDA, which easily lends itself to extensions and generalizations that accommodate many types of relationships in data [12].

LDA is a generative probabilistic model that makes the "Bag-of-Words" assumption and represents documents as probability distributions over K topics. These topics are, in turn, viewed as probability distributions over W words.

While LDA has enjoyed much popularity serving as basis for numerous extensions and generalizations, one of its major limitations is that users must select the number of topics K before the approach can be used. This requirement makes the approach quite rigid, as it cannot accommodate influx of new data [13]. To make topic modeling more flexible, LDA machinery was modified in [13] to use the Chinese Restaurant Process (CRP) [14]. CRP relaxes the fixed K constraint of LDA by assuming an infinite number of topics and postulating that words are generated from topics chosen according to the following distribution:

$$p(existing\ topic\ i|previous\ words) = \frac{m_i}{\lambda + m - 1}$$
$$p(new\ topic|previous\ words) = \frac{\lambda}{\lambda + m - 1} \tag{1}$$

where m_i is the number of words assigned to topic i, λ is a parameter and m is the total number of words seen so far. The formulation in Equation 1 removes the need to know K *apriori* as it assigns a non-zero probability to choosing a new topic. This allows the number of discovered topics to grow as new data arrives.

To improve interpretability of discovered topics, the work by Blei et al. on Hierarchical LDA (hLDA)[13] attempted to learn organized topic hierarchies. The hLDA generative probabilistic model assumes that words in a document are generated from an infinitely branched tree of height L according to a document-specific mixture model. In hLDA, each node of the tree is associated with a single topic.

To learn topic and tree structure from data, sampling is often used by first choosing an L-level path c_d for each document d according to Equation 2.

$$p(c_d|w, c_{-d}, z) \propto p(w_d|c, w_{-d}, z)p(c_d|c_{-d}) \tag{2}$$

where w_{-d} and c_{-d} are words and paths of documents other than d; w and c are respectively words and paths of all documents; z is the topic assignments. Once the path is found, topic assignments for words are approximated by sampling [1].

Including partial supervision in topic modeling has been an active area of research in recent years. Approaches such as the one proposed in [15] and extended in [16] work by defining sets of possible topic assignments for each word type (referred to as *topic-in-sets*) and modifying the Gibbs sampler to constrain possible topic choices for words according to these *topic-in-sets*. These approaches are flexible in that they allow certain "seed words" to help focus topic discovery for words and documents from underrepresented or noisy topics. Unfortunately, specifying *topic-in-sets* early in the topic discovery process is only meaningful if the number of topics is known ahead of time. This limits the usefulness of these approaches as they cannot be applied in a non-parametric settings, such as the nested CRP or the Hierarchical LDA.

Semi-Supervised hLDA (SSHLDA) proposed in [17] introduces partial supervision into the Hierarchical LDA learning process by restricting the initial structure of the topic tree to known hierarchies of labels and then allowing the nested CRP process to discover new branches in the tree with stochastic sampling, as in hLDA.

Constrained-hLDA is particularly relevant to this work because, as in this paper, it focuses on hierarchical topic modeling in microblogs. Specifically, Constrained-hLDA experimented with Chinese microblogs and showed significant improvement in terms of held-out log-likelihood. As noted by the authors, much of the improvements were realized by an additional heuristic aimed specifically at microblog data, which restricted word-level assignments during sampling. Their relied on the document frequency function, which returned the number of documents containing a word in a corpus, as well as upper and lower inclusion boundary thresholds and part-of-speech indicators.

The novel model discussed in the next section improves upon *topic-in-set*-based approaches, such as the one proposed in [15], by allowing for partial supervision and guidance to be applied to hierarchical topic learning in a way that is non-parametric with respect to the number of topics. The new model further improves on recently proposed hierarchical semi-supervised approaches in that it incorporates supervision in a way that does not require an existing label hierarchy (as in SSHLDA) nor does it necessitate the initial supervisory hierarchy to be learned by other means (such as FP-Tree in Constrained-hLDA).

3   Semi-Supervised Microblog-hLDA Model

We motivate our model by imagining that topics are not atomic constructs, but are rather comprised of levels of topic specificity. That is, we consider that microblog posts pertain to a single conceptual theme (such as the presidential election or the World Cup), and that each theme contains a number of stages or levels of specificity. For example, when discussing the World Cup, one microblog message may express excitement about the fact of the World Cup's existence, while another post may speak about an outcome of a particular match or the role of a specific player. In both cases, messages may be set to belong to a "World Cup" theme, but the former message is surely more general than the latter one.

The above intuition may be captured by making a simple modification to hLDA to sample levels according to a uniform distribution, rather than a multinomial one.

3.1   Generative Process for Semi-Supervision

We, then, take our approach a step further and attempt to discover a way to allow semi-supervised focus to be introduced into topic modeling. That is, we imagine a user interested in a particular subject area supplies a topic modeling algorithm with few keywords or phrases about the subject. The user, then, expects the algorithm to highlight her keywords and phrases by restricting them to a single position in the resulting topic tree. Further, the user may expect the approach to discover topics *around* the given subject area (siblings, parents, etc.) providing the user with further insights into her area of interest.

The novel approach, which we term Semi-Supervised Microblog hLDA, is outlined in Figure 1. It assumes that for each social media collection, there exists a parallel corpus of short phrases, which has a bearing on how microblog posts are generation. We treat the parallel corpus as a collection of word phrases and refer to these phrases as *supervisory word-sets*. We, then, imagine that these supervisory word sets are themselves generated with a random generative process.

Figure 1 depicts the resulting algorithm. There, $W_{sup} = \{\mathbf{w}_1, ..., \mathbf{w}_S\}$ is a collection of supervisory word-sets, such that $\mathbf{w}_s \in W_{sup} = \{w | w \in V\}$, and $S = |W_{sup}|$ is the number of supervisory word-sets. The process starts by generating S supervisory word-sets in step 3. In step 3(c)c, supervisory words are aggregated into the set $\mathbf{W}_{sup}$, which is used in later steps to ensure that supervisory words may only be generated on paths associated with supervisory word sets. In step 3d, leaf nodes of paths chosen for each of the supervision word sets are aggregated into a set L_{sup}. The resulting collection of S paths is used in later steps to ensure that words in supervisory sets may only emerge from paths associated with those supervisory sets.

1. Let $L_{sup} = \emptyset$ be a collection of paths
2. Let $\mathbf{W}_{sup} = \emptyset$ be a collection of words
3. For supervisory word-set $\mathbf{w}_s \in W_{sup}$
   a. Let c_1 be the root
   b. For each level $l \in \{2, \ldots, L\}$:
      a) Draw a child node form c_{l-1} using Equation 1. Set c_l to be that node
   c. For each word $n_{sup} \in \{1, \ldots, |\mathbf{w}_s|\}$:
      a) Draw $z \in \{1, \ldots, L\}$ from $Uniform(L)$
      b) Draw w from the topic associated with c_z
      c) Set $\mathbf{W}_{sup} = \mathbf{W}_{sup} \cup \{w\}$
   d. Set $L_{sup} = L_{sup} \cup \{c_l\}$
4. For each document
   a. Draw $x \in \{1, ..., L\}$ from $Mult(\sigma)$
   b. Draw $s \in \{1, ..., |L_{sup}|\}$ from $Mult(\omega)$
   c. Select s^{th} node c_s from L_{sup}
   d. Let c_1 be the root
   e. For each level $l \in \{2, \ldots, x\}$:
      a) Select the lth node c_l from the path to node c_s
   f. For each level $l \in \{x+1, \ldots, L\}$:
      a) Draw a child node form c_{l-1} using Equation 1. Set c_l to be that node
   g. For each word $n \in \{1, \ldots, N\}$:
      a) Draw $z \in \{1, \ldots, L\}$ from $Uniform(L)$
      b) Let β_{c_z} be the word-topic proportions vector associated with node c_z
      c) If($z < x$):
         1. Construct set $V' = (V \backslash \mathbf{W}_{sup}) \cup \mathbf{w}_s$
      d) If($z >= x$):
         1. Construct set $V' = V \backslash \mathbf{W}_{sup}$
      e) Construct a $|V| \times |V|$ diagonal matrix M^s such that for each $i = j$, $M_{ij}^s = cell(i, V')$ (see Equation 3)
      f) Let $\beta'_{c_z} = M^s \times (\beta_{c_z})$
      g) Draw $w \in V$ from $Mult(\beta'_{c_z})$

Fig. 1. Semi-Supervised Microblog-hLDA generative process

Having generated the supervisory sets, the process begins to produce document content in step 4. First, the process draws a number $x \in \{1, ..., L\}$ from a multinomial distribution parameterized by an L-sized vector σ (step 4a). Then, an index s into the set L_{sup} is drawn from a distribution parameterized by an $|L_{sup}|$-dimensional vector w. Then, the process chooses a path for each document by deterministically selecting the first x nodes from the sth path in L_{sup} (step 4e) and then allowing the CRP to randomly choose nodes from $(s+1)^{th}$ level to the leaf level L (step 4f). It is important to note that the realization $x = 1$ in step 4a amounts to no supervision, since all paths share the root node.

The resulting path is used to generate words in the document. For each word, the process draws $z \in \{1, \ldots, L\}$ from a uniform distribution. Once the level assignment is known, set V' is constructed in step 4(g)c and initially contains all words in vocabulary V except for all supervisory terms of set $\mathbf{W}_{sup}$. If the chosen node assignment is on the path associated with some supervisory word-set, the supervisory words of that set are added to V'. Then, the $|V|$-dimensional word proportions vector associated with the chosen topic multiplies a diagonal matrix, which contains zeros in elements of the diagonal that correspond to indices of words not found in vector V'. The multiplication in step 4(g)f has the effect of allowing supervisory words to be generated only from a single hierarchy path. The resulting unnormalized parameter vector is used to randomly select words in a way identical to hLDA.

$$cell(i, V) = \begin{cases} 1 & \text{if } w_i \in V \\ 0 & \text{otherwise} \end{cases} \tag{3}$$

3.2 Inference

Posterior inference tries to visualize hidden process structures by repeatedly adjusting its mental vision of them to better fit the actual observations. In our application, observations consist of two collections – 1) a corpus of microblog messages and, 2) a number of user-specified supervisory word sets. Given these observations, we are interested in learning the shape of the hidden topic hierarchy and the word proportions for the nodes in the hierarchy.

With that, the posterior inference for our model is conducted as follows. First, a random tree scaffolding is constructed by many hierarchical random walks. Then, for each microblog message, an L-level path through the scaffolding is selected. This is followed by a path node selection and subsequent counters updated for both path and the node. These counters are used in later stages to approximate hierarchy makeup and parameters.

The L-level path is selected according to Equation 2. Once the path is chosen, the algorithm knows that all words in the message were generated by the particular path, but still needs to determine which member of the path was responsible for which words. This is a challenge as our model postulates a uniform distribution over levels for each document, which means that the posterior inference algorithm cannot learn level assignments from data, as in hLDA. To have a reasonable chance of intelligently approximating the hidden structure, the inference procedure may consider the following argument.

The uniform distribution postulate of our approach implies that each document draws equally many words from each of the levels, but gives no guidance as to how to determine which level of the hierarchy generated which of the words. If the hidden word distributions at each level were known, the inference algorithm could simply choose a node with the highest probability of a given word. However, since these distributions are unknown, the inference procedure may consider the following dichotomy regarding word proportions in nodes of the hidden tree – 1) all distributions are of the same (or similar) shape and, 2) all distributions are **not** of the same (or similar) shape.

The notion that all the distributions are the same or similar contradicts with everyday common sense – obviously, language texts, such as social media posts, discuss a variety of subjects. Therefore, we must conclude that words are distributed unequally among paths and, consequently, path nodes. With that, words that are favored by 'popular' nodes (nodes that appear on many paths) must appear more frequently than words from unpopular nodes. Again arguing from the observations, because empirical laws (e.g.: the Zipf's Law [18]) suggest that words are distributed according to the inverse power-law, there must be few 'popular' nodes and many 'unpopular' ones. Then, in graph-theoretic terms, since, by definition, there are fewer higher-level nodes than lower-level ones, the higher-level nodes (those closer to the root) must be the 'popular' ones and the lower-level (towards leafs) nodes must be relatively 'unpopular'.

With that, the level assignment task is straight forward. Given a word, the algorithm may simply consult a frequency table and determine its corpus-level rank. Then, if the word ranks first, the word must be associated with the root node, whereas if its ranked last, it gets assigned to the leaf node of the given document path.

Naturally, the above raises the question of what to do if the rank is somewhere between first and last. We tackle this challenge by partitioning the corpus frequency table into L *buckets* (one for each hierarchy level) in such a way as to place few highly ranked words into the top-level bucket and many very infrequent terms into leave level one. This intuition is quantified by assigning words to levels during sampling according to the following equation:

$$Level_w = \lfloor log_{\sqrt[L]{N+1}}(rank(w)) \rfloor + 1 \qquad (4)$$

where N is the number of distinct words and $rank(w)$ is the rank of word w in the corpus frequency table. The equation captures our intuition by exponentially increasing bucket sizes towards the leaf level.

For an illustrative example, when considering a 3-level hierarchy ($L = 3$) and a 1000 term vocabulary ($|V| = 1000$), Equation 4 will associate the 10 most frequent terms with the root level, next 90 with the intermediate level, and 900 least frequent ones with the leaf nodes. Then, for the same hierarchy, if the word *"the"* were the most frequent word in a corpus containing 1000 unique terms, its rank would necessarily be 1 and $Level_{\text{"the"}} = \lfloor log_{\sqrt[3]{1001}}(1) \rfloor + 1 = 1$, which is the root level. If, however, the word *"unique"* were the only word to appear just once in the corpus, it would be ranked 1000 and its level would be computed as $Level_{\text{"unique"}} = \lfloor log_{\sqrt[3]{1001}}(1000) \rfloor + 1 = 3$, which is the leaf.

During inference, we approximate each word's position with the value of $Level_w$ for each observed word w by deterministically selecting level assignments with the help of Equation 4.

3.3   Inference with Semi-Supervision

We outline the supervised inference procedure by recalling that, in addition to a document corpus, observations in the SS-Micro-hLDA also contain collections of supervisory words. Since SS-Micro-hLDA uses the same generative approach for both the supervisory and the document corpora, same sampling procedure may apply. The restriction that supervisory words may originate from only a single path (step 4(g)f in Figure 1) implies that documents containing supervisory words must have been generated from paths that share a prefix with paths to leafs associated with supervisory word sets.

To introduce supervision into the sampling process, we start by randomly and without replacement selecting a node from a set of hierarchy leafs for each supervisory set $\mathbf{w}_s \in W_{sup}$. This results in a collection of tuples $\mathbf{S}_{sup} = \{< \mathbf{w}_1, c_1 >, ..., < \mathbf{w}_S, c_S >\}$ such that each c_i is a leaf node and $|\mathbf{S}_{sup}| = |W_{sup}|$. Then, for each ith tuple $S_i \in \mathbf{S}_{sup}$, words in its word set $\mathbf{w}_i$ are assigned to nodes on the path to c_i according to word ranks as specified by Equation 4.

Then, for each observed document, topic hierarchy path is selected by first checking whether any words in the document are found in any supervisory set and constraining the path selection to go through the corresponding node. Once, the path is known, word assignments are sampled according to Equation 4.

4   Evaluation

The proposed model was tested with two datasets – the Tweets2011 Twitter Collection made available through the TREC project [19] and a collection of user comments on a popular Reddit news and social networking site, which we manually collected by monitoring the site's programmatic API end-points. The Twitter data set consisted of 16 million Twitter messages sampled in early months of 2011. The Reddit collection was comprised of 51, 563 user comments to articles posted in Reddit subsections (known as *subreddits*) labeled */gaming*, */politics* and */sports*.

It is common knowledge that many social media messages are tagged with special topical annotations known as *hashtags*. While users often misplace or misspell hashtags or abuse the hashtag notation (i.e.: some messages may contain more hashtags than actual text), with no standard corpus available, our approach was tested on collections of carefully select tagged messages.

To construct a test corpus, we parsed the English language messages in the Tweets2011 collection and assembled corpus-level hashtag counts. We then selected those hashtags that appeared in at least 1000 messages in the corpus. The resulting 34 hashtags were used to construct the corpus by retaining only those messages that contained the frequent tags. The data set was further restricted to those Twitter messages that contained only a single hashtag. This was done to control noise with the intuition that messages with just a single hashtag are more likely to be focused on a particular subject.

4.1   PMI-Score Evaluation

To compare performance of our approach to others, we expressed our interest in the Egyptian revolution and the major American Football sporting event by constructing two supervisory sets – { *'protests'*, *'egypt'*} and { *'super'*, *'bowl'*, *'packers'*, *'steelers'*}. We then trained topic models using semi-supervised and unsupervised variants of our approach[1] as well as the Constrained-hLDA and hLDA (to serve

---

[1] Unsupervised variant of SS-Micro-hLDA is achieved trivially by providing an empty collection of supervisory word sets.

as a baseline) and compared resulting models in terms of the PMI-Score [20]. The PMI-Score measure was chosen in favor of other metrics, such as such as perplexity or log-likelihood, as this measure has been reported by numerous researchers ([21],[20],[22]) to correlate well with human interpretation of topic models.

PMI-Score is motivate by the observation that human evaluation of topic models is often conducted by considering the top n representative words for each topic. The PMI-Score aims to provide quantitative approximation of human evaluation by considering the Pointwise Mutual Information for the top n words as quantified by Equation 5.

$$PMI - Score(\mathbf{w}) = median\{PMI(w_i, w_j), ij \in \{1, ..., n\}\} \qquad (5)$$

where $\mathbf{w}$ is the topic, w_i and w_j are i^{th} and j^{th} ranked words in topic $\mathbf{w}$, n is the number of 'top words' selected (for example, n=10 top words), $PMI(w_i, w_j) = \frac{p(w_i, w_j)}{p(w_i)p(w_j)}$.[20]

Evaluation results using ten-fold cross-validation are outlined in Figure 2. The figure reports PMI-Scores for hierarchies of different heights and shows that SS-Micro-hLDA outperforms other approaches for deeper hierarchies. All models appeared to perform similarly in terms of the PMI-Score for shallower hierarchies (number of levels less than 5). This is expected as shallow hierarchies do not allow for deep specialization in topic structures.

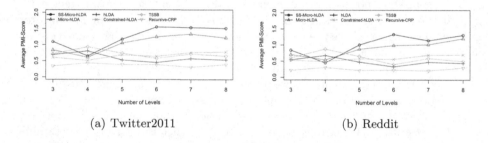

(a) Twitter2011   (b) Reddit

Fig. 2. Average PMI-Score evaluation results

4.2   Information Entropy Evaluation

While the PMI-Score evaluation presented above tested topic models in terms of their interpretability, the metric did not measure how well sections of hierarchies focused on particular topical areas. That is, in hierarchical topic learning, it is expected that siblings are somehow conceptually related to one another. For example, topics on "dogs" and "cats" may be expected to appear under the general topic on "mammals", while "apples" and "oranges" should occur under the general topic heading on "fruits". If a hierarchical topic modeling approach were to place the "dogs" topic under the "fruits" heading, a human analyst would likely find such a placement in error even if the top words of the topic were coherent and interpretable.

To evaluate how closely the model places related documents, we estimated probabilities of each node as proportional to the number of times a node appeared on any document's path. We then computed Shannon's information entropy [23] given as $H(C) = -\sum_d p(c_d|testdata)log(p(c_d|testdata))$ where $C = \{c_1, ..., c_{N_{test}}\}$ is a random variable taking on values of all possible paths. The information entropy quantity may be interpreted by considering that, in a *focused* hierarchy, test documents on the same topic would likely be concentrated in a particular area of the hierarchy, their placement being more predictable and implying lower entropy. On the other hand, classification using an *unfocused* hierarchical model would place documents more evenly across the entire hierarchy, resulting in higher entropy. Therefore, we would expect the information entropy of a focused hierarchy to be lower than that of an unfocused one.

Results of the information entropy evaluation are presented in Figure 3. We only present results for the Twitter #*egypt* and Reddit #*sports* test samples because of space considerations. In Figure 3, information entropy for the test data using SS-Micro-HLDA model is lower than that of other models for deeper hierarchies, suggesting a more focused topic tree. This is particularly encouraging as deeper hierarchies provide a way for analysts to focus on a particular area among a potentially large number of topics.

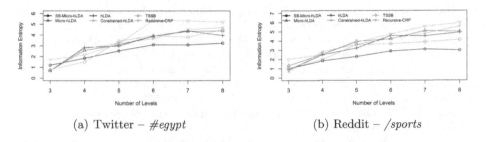

(a) Twitter – #*egypt*   (b) Reddit – /*sports*

Fig. 3. Entropy results for Twitter #egypt and #superbowl and Reddit /sports and /politics data

5   Conclusions and Future Work

In this paper, we developed an algorithm to infer hierarchical topic models around specific concepts that may be of interest to analysts. We evaluated our new algorithm using a large, publicly available collection of microblog messages and showed that the proposed method outperformed other approaches in terms of the PMI-Score. As PMI-Score has been shown to relate favorably to topic interpretability by humans, this evaluation suggests that our new approach produces highly meaningful topic models.

While we were able to show that our new approach preforms better than existing state-of-the-art topic modeling on a static data set, our approach is not designed for continuous operation on stream data. In our future work, we will focus on developing an approach to handle streaming social media messages with the goal of tracking and monitoring social discourse over time.

References

1. Mehrotra, R., Sanner, S., Buntine, W., Xie, L.: Improving lda topic models for microblogs via tweet pooling and automatic labeling. In: Proceedings of the 36th International ACM SIGIR Conference on Research and Development in Information Retrieval, pp. 889–892. ACM (2013)
2. Zhao, W.X., Jiang, J., Weng, J., He, J., Lim, E.-P., Yan, H., Li, X.: Comparing twitter and traditional media using topic models. In: Clough, P., Foley, C., Gurrin, C., Jones, G.J.F., Kraaij, W., Lee, H., Mudoch, V. (eds.) ECIR 2011. LNCS, vol. 6611, pp. 338–349. Springer, Heidelberg (2011)
3. Bellaachia, A., Al-Dhelaan, M.: Ne-rank: a novel graph-based keyphrase extraction in twitter. In: Proceedings of the 2012 IEEE/WIC/ACM International Joint Conferences on Web Intelligence and Intelligent Agent Technology, vol. 1, pp. 372–379. IEEE Computer Society (2012)
4. Jin, O., Liu, N.N., Zhao, K., Yu, Y., Yang, Q.: Transferring topical knowledge from auxiliary long texts for short text clustering. In: Proceedings of the 20th ACM International Conference on Information and Knowledge Management, pp. 775–784. ACM (2011)
5. Hu, Y., John, A., Seligmann, D.D., Wang, F.: What were the tweets about? topical associations between public events and twitter feeds. In: ICWSM (2012)
6. Zhao, W.X., Jiang, J., He, J., Song, Y., Achananuparp, P., Lim, E.-P., Li, X.: Topical keyphrase extraction from twitter. In: Proceedings of the 49th Annual Meeting of the Association for Computational Linguistics: Human Language Technologies, vol. 1, pp. 379–388. Association for Computational Linguistics (2011)
7. Conrad, C.: Cognitive economy in semantic memory (1972)
8. Collins, A.M., Loftus, E.F.: A spreading-activation theory of semantic processing. Psychological Review **82**(6), 407 (1975)
9. Deerwester, S.: Improving information retrieval with latent semantic indexing (1988)
10. Dhillon, I.S., Sra, S.: Generalized nonnegative matrix approximations with bregman divergences. In: NIPS, vol. 18 (2005)
11. Blei, D.M., Ng, A.Y., Jordan, M.I.: Latent dirichlet allocation. The Journal of Machine Learning Research **3**, 993–1022 (2003)
12. Hong, L., Davison, B.D.: Empirical study of topic modeling in twitter. In: Proceedings of the First Workshop on Social Media Analytics, pp. 80–88. ACM (2010)
13. Blei, D.M., Griffiths, T.L., Jordan, M.I., Tenenbaum, J.B.: Hierarchical topic models and the nested chinese restaurant process. In: NIPS, vol. 16 (2003)
14. Aldous, D.: Exchangeability and related topics. In: École d'Été de Probabilités de Saint-Flour XIII1983, pp. 1–198 (1985)
15. Andrzejewski, D., Zhu, X.: Latent dirichlet allocation with topic-in-set knowledge. In: Proceedings of the NAACL HLT 2009 Workshop on Semi-Supervised Learning for Natural Language Processing, pp. 43–48. Association for Computational Linguistics (2009)

16. Bodrunova, S., Koltsov, S., Koltsova, O., Nikolenko, S., Shimorina, A.: Interval semi-supervised LDA: classifying needles in a haystack. In: Castro, F., Gelbukh, A., González, M. (eds.) MICAI 2013, Part I. LNCS, vol. 8265, pp. 265–274. Springer, Heidelberg (2013)
17. Mao, X.L., Ming, Z.Y., Chua, T.S., Li, S., Yan, H., Li, X.: Sshlda: a semi-supervised hierarchical topic model. In: Proceedings of the 2012 Joint Conference on Empirical Methods in Natural Language Processing and Computational Natural Language Learning, pp. 800–809. Association for Computational Linguistics (2012)
18. Zipf, G.K.: Selected studies of the principle of relative frequency in language. Harvard University Press, Cambridge (1932)
19. Tweets2011 corpus, Online, TREC (2011). https://sites.google.com/site/microblogtrack/2011-guidelines
20. Newman, D., Karimi, S., Cavedon, L., Kay, J., Thomas, P., Trotman, A.: External evaluation of topic models. In: Australasian Document Computing Symposium (ADCS), pp. 1–8. School of Information Technologies, University of Sydney (2009)
21. Newman, D., Lau, J.H., Grieser, K., Baldwin, T.: Automatic evaluation of topic coherence. In: Human Language Technologies: The 2010 Annual Conference of the North American Chapter of the Association for Computational Linguistics, pp. 100–108. Association for Computational Linguistics (2010)
22. Newman, D., Noh, Y., Talley, E., Karimi, S., Baldwin, T.: Evaluating topic models for digital libraries. In: Proceedings of the 10th Annual Joint Conference on Digital libraries, pp. 215–224. ACM (2010)
23. Shannon, C.: A mathematical theory of communication. The Bell System Technical Journal **27**(3), 379–423 (1948)

Predicting Future Links Between Disjoint Research Areas Using Heterogeneous Bibliographic Information Network

Yakub Sebastian[1]([✉]), Eu-Gene Siew[2], and Sylvester Olubolu Orimaye[1]

[1] School of Information Technology, Monash University Malaysia,
Bandar Sunway, Selangor, Malaysia
{yakub.sebastian,sylvester.orimaye}@monash.edu
[2] School of Business, Monash University Malaysia,
Bandar Sunway, Selangor, Malaysia
siew.eu-gene@monash.edu

Abstract. Literature-based discovery aims to discover hidden connections between previously disconnected research areas. Heterogeneous bibliographic information network (HBIN) provides a latent, semi-structured, bibliographic information model to signal the potential connections between scientific papers. This paper introduces a novel literature-based discovery method that builds meta path features from HBIN network to predict co-citation links between previously disconnected literatures. We evaluated the performance of our method in predicting future co-citation links between fish oil and Raynaud's syndrome papers. Our experimental results showed that HBIN meta path features could predict future co-citation links between these papers with high accuracy (0.851 F-Measure; 0.845 precision; 0.857 recall), outperforming the existing document similarity algorithms such as LDA, TF-IDF, and Bibliographic Coupling.

Keywords: Literature-based discovery · Co-citation link prediction · Heterogeneous information network · Features · Classification

1 Introduction

Literature-based discovery (LBD) is one of the grand challenges in data mining [1]. Its goal is to discover interesting, novel connections between previously disconnected research areas [2,3]. Associations between literatures often take the form of co-citation links between them [4]. Figure 1 shows the evolution of paper clusters[1] *before* and *after* the publication of Swanson's seminal hypothesis on previously unknown relationships between dietary fish oil and Raynaud's syndrome [5]. It can be observed that new co-citation links emerged between fish oil (FO) and Raynaud's syndrome (RS) papers following the announcement of Swanson's discovery in 1986. Consequently, the LBD problem can be formulated

---

[1] Sci²Tool (https://sci2.cns.iu.edu/user/index.php)

© Springer International Publishing Switzerland 2015
T. Cao et al. (Eds.): PAKDD 2015, Part II, LNAI 9078, pp. 610–621, 2015.
DOI: 10.1007/978-3-319-18032-8_48

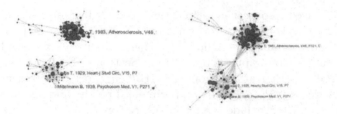

Fig. 1. Left: FO cluster (*top*) and RS cluster (*bottom*) were disjoint prior to 1986. **Right**: New co-citation links emerged between FO and RS clusters after 1986.

as predicting future co-citation links between papers from previously disconnected literature clusters, and the challenge for data mining is to find which features can be used to reliably predict the future formation of these links.

Most of early LBD algorithms used lexical statistics to postulate hidden associations between disjoint literatures based on the number of co-occurring terms between them [6,7]. Unfortunately, a major limitation of these methods is that merely relying on lexical analysis tends to produce high recall but low precision results [8]. Many false associations were generated because term co-occurrence can only send a weak signal about meaningful associations between different papers [3]. The more sophisticated LBD methods incorporate knowledge-based resources [9,10] and natural language processing (NLP) techniques [11,12] to increase the LBD accuracy. For example, Hristovski et al. [11] extracted logical assertions from text as subject-predicate-object triplets and assembled them to make inferences about hidden relationships between papers. Unfortunately, it has been argued that most logical associations in text may not exist in such a simple linguistic template [2], and such approaches are likely to suffer from low recall.

In this paper, we propose a new LBD method that aims to predict future co-citation links between papers in disconnected research areas by exploiting latent, semi-structured data found in heterogeneous bibliographic information network (HBIN) [13]. We first constructed a HBIN network from which we built a number of meta path features. We evaluated the performance of these meta path features using five machine learning algorithms for predicting three types of co-citation links: *inter-cluster link*, *within-cluster link*, and *no link*. The evaluation involved reproducing Swanson's classic dietary fish oil and Raynaud's syndrome (DFO-RS) hypothesis [5]. We found that meta path features outperformed several well-known document similarity features.

Our contributions are as follows. In contrast to previous LBD methods, our method combines lexical and citation information. HBIN meta paths allow lexical associations (e.g. common terms between papers) to propagate through the citation structures between papers (e.g. via common authors or shared references). We argue that this approach has increased both the precision and recall of our LBD method. Results show that we could accurately predict future co-citation links between fish oil and Raynaud's syndrome papers with 0.851 F-Measure. More importantly, the performance of our method is achieved without relying on knowledge-based resources nor sophisticated NLP techniques. To the best of our knowledge, this is the first work that uses HBIN to solve LBD problems.

This paper is organized as follows. Section 2 highlights some related work followed by detailed descriptions of HBIN in Section 3. Section 4 explains how we learned meta path features from an HBIN network to predict future co-citation links. Section 5 presents our experimental results and Section 6 presents some discussions. We conclude the paper in Section 7.

2   Related Work

A few authors have used HBIN to enhance the accuracy of citation recommendation algorithms [14–16]. Yu et al. [14] and Ren et al. [16] constructed meta path features to predict direct citation link between papers in DBLP and PubMed datasets. Given a query paper, they learned the probabilities that candidate papers will be cited by the query paper based on the similarity of their meta paths. These methods first reduce their search space by grouping candidate papers into 'buckets' of papers that belong to similar research areas or interests. This grouping allows these algorithms to focus their search on candidate papers that have *similar* research areas to the query paper. Liu et al. [15] combined meta path features with additional features that capture citation count and citation topic motivations in order to predict direct citation links between papers. These features, however, require that the abstracts or full-text of papers be available. Our method does not assume that full-text papers are available and only requires basic bibliographic meta data, e.g. title, author, publisher, and citations.

Although related, current citation recommendation systems such as the above are fundamentally different from the LBD work presented in this paper. These algorithms assume that a paper will only cite papers from the same or closely related research areas [14]. In contrast, LBD is concerned with finding hidden relationships between papers from seemingly unrelated research areas. In this respect, our LBD problem is harder than than citation recommendation problem. Another important difference is that the existing works [14–16] aim at predicting direct citation links, whereas our work aims at predicting co-citation links. In addition, we also study more types of meta path features (in total 87 distinct features) compared to only 9 in [15], 14 in [14] and 15 in [16]. We describe our features in more detail in Section 4.

3   Heterogeneous Bibliographic Information Network

A heterogeneous information network is a directed graph $G = (V, E)$ composed of multiple-typed entities and links [13]. HBIN is a type of heterogeneous information network that allows one to model rich interactions between various types of bibliographic entities. As shown in Figure 2, HBIN consists of four types of bibliographic entities A with many possible connections among them: paper (P), author (AU), venue (journal or conference) (V), and term (T), $A = \{P, AU, V, T\}$. Unlike previous work [14–16], we further categorize the entity type P into core paper (P_{core}), citing paper (P_{cite}), and reference paper (P_{ref}) sub-types, $P = \{P_{core}, P_{cite}, P_{ref}\}$. *Core papers* are papers that belong to either one of the disconnected literatures being studied. For example, in the case of Swanson's DFO-RS hypothesis, papers that belong to either the fish oil or Raynaud's syndrome literature are considered as core papers. *Citing papers* are papers that cite core papers while *reference papers* are papers cited by core papers. Each core paper is associated with zero or more citing papers and reference papers. Using these entity sub-types, we hope to capture more specific relational semantics between instances of entity P.

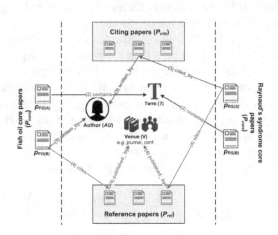

Fig. 2. Heterogeneous bibliographic information network (HBIN)

We identify four types of edges $\mathcal{R}$ to represent relationships between entities in HBIN, $\mathcal{R} = \{P—AU, P \longleftrightarrow P, P—V, P—T\}$. We will describe these edge types in sections that follow. A *meta path MP* is a path over HBIN, $MP = A_1 \xrightarrow{R_1} A_2 \xrightarrow{R_2} \ldots \xrightarrow{R_l} A_{l+1}$; $A \in \mathcal{A}$; $R_i \in \mathcal{R}$ [13]. We use meta paths to model latent, semi-structured relationships among bibliographic entities in HBIN which could signal previously unknown connections between core papers that have never physically cited each other or have never been co-cited before (isolated). For example, the red meta path "(3)" and green meta path "(4)" in Figure 2 suggest some connections between core papers $p_{FO(B)}$ and $p_{RS(A)}$. Likewise $p_{FO(B)}$ is written by the same author of a paper that cites $p_{RS(A)}$, (path "(3)"). Both papers also cite papers published in a similar venue (path "(4)"), also signaling their potential associations.

4 Learning Predictive Meta Path Features from HBIN

Given a pair of disconnected core papers p_i and p_j, we define 16 different types of n-degree meta-paths between the core papers, where n is the number of edges separating p_i from p_j. These sixteen meta paths consist of 4 *two*-degree meta paths, 6 *three*-degree meta paths, and 6 *four*-degree meta paths. Refer again to Figure 2, meta path "(2)" is a two-degree meta path connecting $p_{FO(A)}$ (i.e. p_i) and $p_{RS(B)}$ (i.e. p_j) via a common term T. The same meta path type can be established via AU, V, or P_{ref}. Meta path "(3)" is a three-degree meta path where $p_{FO(B)}$ shares a common author AU with paper that cites $p_{RS(A)}$. Likewise the same meta path can be established through the sharing of T and V via P_{cite} and P_{ref}. Finally, path "(4)" is a four-degree meta path through which $p_{FO(B)}$ and $p_{RS(A)}$ are connected by their cited references which are published at the same venue V. The same path may exist via P_{cite} that share T and AU.

We measure the association strength of a meta path using a meta path score. This score is determined by the weights of a path's component edges. We propose five edge weighting schemes which measures the *local importance* and the *global importance* of a meta path edge. For example, the local importance of edge P—AU is the importance of author AU with respect to paper P. AU is more important to P if he or she is the sole author of P than if he or she is just one of many authors of P. The global importance of AU is its importance with respect to the research areas being studied. It is measured by the frequency of his or her co-authorships with other authors in those research areas. If the author has a high number of co-authorships in a particular research area, we assume he or she is an important contributor to that area.

We compute the weight of an edge between paper p_i and author a_j, $w(p_i$—$a_j)$:

$$w(p_i{-}a_j) = \frac{1}{count(p_i, AU_{p_i})} \cdot \frac{1}{N} \left(\sum_{k=1}^{N} \frac{freq(a_j, AU_{P_k})}{freq(AU_{P_k})} \right); 0 \le w \le 1 \qquad (1)$$

where $count(p_i, AU_{p_i})$ is the total authors in paper p_i, $freq(a_j, AU_{P_k})$ is the total non-unique co-authorship pairs between author a_j and other authors in cluster k, and $freq(AU_{P_k})$ is the total non-unique co-authorship pairs of all authors in cluster k. This weighting score measures the importance of p_i—a_j association with respect to paper p_i (local importance) as well as author a_j's global importance in N number of disconnected research areas being studied. Next, we compute the weight of edge $P \longleftrightarrow P$ as follows:

$$w(p_i \longrightarrow p_j) = \frac{1}{\overrightarrow{count}(p_i, P_{ref}^i)} \cdot \frac{1}{N} \left(\sum_{k=1}^{N} \frac{\overrightarrow{count}(P_k, p_j)}{count(P_k)} \right); 0 \le w \le 1, or \qquad (2)$$

$$w(p_i \longleftarrow p_j) = \frac{1}{\overleftarrow{count}(p_i, P_{cit}^i)} \cdot \frac{1}{N} \left(\sum_{k=1}^{N} \frac{\overleftarrow{count}(P_k, p_j)}{count(P_k)} \right); 0 \le w \le 1 \qquad (3)$$

where $\overrightarrow{count}(p_i, P_{ref}^i)$ and $\overleftarrow{count}(p_i, P_{cit}^i)$ are the total references cited by p_i and the total papers that cite p_i, respectively. The less references in p_i, the more important p_j is to p_i. This is because, given limited opportunities to cite references in their papers, most authors would only include important references. The $\overrightarrow{count}(P_k, p_j)$ and $\overleftarrow{count}(P_k, p_j)$ are the total papers in cluster k that cite p_j and the total papers in cluster k cited by p_j, respectively, whereas $count(P_k)$ is the total number of papers in cluster k. The more frequently p_j is cited in cluster

k or the more frequently p_j cites papers in the cluster k, the more important it is to the cluster. Next, the weight of edge P—T is:

$$w(p_i\text{—}t_j) = \frac{freq(p_i, t_j)}{freq(T_{p_i})} \cdot \frac{1}{N}\left(\sum_{k=1}^{N} \frac{freq(P_k, t_j)}{freq(T_{P_k})}\right); 0 \le w \le 1 \qquad (4)$$

where $\frac{freq(p_i, t_j)}{freq(T_{p_i})}$ is the frequency of occurrence of t_j in p_i divided by the total number of terms in p_i, and $\frac{freq(P_k, t_j)}{freq(T_{P_k})}$ is the appearance frequency of t_j in all papers of cluster k divided by the total all terms in the cluster. The more frequently term t_j appears in a paper or a cluster, the more important the term is. Lastly, we define the weight of edge P—V as:

$$w(p_i\text{—}v_j) = \frac{1}{N}\left(\sum_{k=1}^{N} \frac{count(P_k, v_j)}{count(P_k)}\right); 0 \le w \le 1 \qquad (5)$$

where $count(P_k, v_j)$ is the total papers in cluster k published by venue v_j.

4.1   Constructing HBIN Matrices

Having defined the edge weighting schemes above, we constructed HBIN network using a set of matrices in a manner similar to Ren et al. [16]. Given n number of papers in P, $P = \{P_{core}, P_{cite}, P_{ref}\}$, we first built adjacency matrix $M \in \mathbb{R}^{n \times n}$ that stores citation relationships between instances of P, where $M_{ij} = 1$ if p_i cites p_j, and $M_{ij} = 0$ if no citation exists between them. Based on M, we subsequently built two weighted adjacency matrices:

$$W^{(P_{ref})} \in \mathbb{R}^{m \times n}; \text{where } m = |P_{core}|, n = |P|, \text{and } W_{ij}^{(P_{ref})} = w(p_i \longrightarrow p_j) \qquad (6)$$

$$W^{(P_{cit})} \in \mathbb{R}^{m \times n}; \text{where } m = |P_{core}|, n = |P|, \text{and } W_{ij}^{(P_{cit})} = w(p_i \longleftarrow p_j) \qquad (7)$$

We also built bi-adjacency matrices that represent the relationships between instances of P and instances of AU, V, and T. Bi-adjacency matrix $B^{(AU)} \in \mathbb{R}^{n \times |AU|}$ represents bi-partite relationships between instances of P and AU, where $B_{ij}^{(AU)} = 1$ if p_i is written by a_j, and $B_{ij}^{(AU)} = 0$ if otherwise. From $B^{(AU)}$, we built a weighted bi-adjacency matrix:

$$W^{(AU)} \in \mathbb{R}^{n \times |AU|}; \text{where } W_{ij}^{(AU)} = w(p_i\text{—}a_j) \qquad (8)$$

Next, we built bi-adjacency matrix $B^{(V)} \in \mathbb{R}^{n \times |V|}$, where $B_{ij}^{(V)} = 1$ if p_i is published in venue v_j, and $B_{ij}^{(V)} = 0$ if not. Based on $B^{(V)}$, we built a weighted bi-adjacency matrix:

$$W^{(V)} \in \mathbb{R}^{n \times |V|}; \text{where } W_{ij}^{(V)} = w(p_i\text{—}v_j) \qquad (9)$$

Finally, given a set of unique terms $T = \{t_1, \ldots, t_{|T|}\}$ in P, we built a non-binary bi-adjacency matrix $B^{(T)} \in \mathbb{R}^{n \times |T|}$ where the value of $B_{ij}^{(T)}$ is the frequency of term t_j appearing in the title and abstract of paper p_i. We removed English stop-words[2] and applied *Porter Stemmer*[3] to discard morphological and inflexional endings from English terms. From $B^{(T)}$ we built the following matrix:

$$W^{(T)} \in \mathbb{R}^{n \times |T|}; \text{where } W_{ij}^{(T)} = w(p_i - t_j) \tag{10}$$

4.2 Meta Path Features

We used 16 adjacency matrices $MTM^{(mp)} \in MP^{mm}$ to store all meta paths between m core papers, where $MTM_{ij}^{(mp)}$ is a set of zero or more meta path instances between core papers p_i and p_j. Given a set of all meta paths between p_i and p_j, we compute five different meta path scores: (1) *Path count*, a simple count of meta paths; (2) *Sum of weights*, the sum of edge weights of all meta paths; (3) *Average of weights*, the average edge weights of all meta paths; (4) *Minimum weight*, the smallest edge weight of all meta paths; and (5) *Maximum weight*, the largest edge weight of all meta paths.

From these 16 types of meta path and 5 types of meta path scores, we derived 80 distinct meta path features. We also added 7 meta path features consisting of 3 features based on the total count of n-degree meta paths, 3 features based on total count of meta paths according to their meta data type, and 1 feature which is the total count of all meta paths between p_i and p_j. Table 1 presents a compacted list of all meta path features we used. For example, feature 7 is read as *2_term_sum*, which is the sum of all edge weights of all two-degree meta path instances of type $P-T-P$ between p_i and p_j. Feature 26 is read as *3_term_ref_count* which is the number of distinct three-degree meta path instances of type $P-T-P_{ref} \longleftarrow P$. Feature 74 is read as *4_term_cite_min*, which is the minimum weight of all edges of all four-degree meta path instances of type $P \longleftarrow P_{cite} - T - P_{cite} \longrightarrow P$.

Table 1. List of the eighty-seven meta path features being studied

Meta path	Cnt	Sum	Avg	Min	Max	Meta path	Cnt	Sum	Avg	Min	Max	Meta path	Feat.
			Feat.						Feat.				
2_author_	1	2	3	4	5	3_term_cite_	41	42	43	44	45	2_tot	81
2_term_	6	7	8	9	10	3_venue_cite_	46	47	48	49	50	3_tot	82
2_venue_	11	12	13	14	15	4_author_ref_	51	52	53	54	55	4_tot	83
2_ref_	16	17	18	19	20	4_term_ref_	56	57	58	59	60	author_tot	84
3_author_ref_	21	22	23	24	25	4_venue_ref_	61	62	63	64	65	term_tot	85
3_term_ref_	26	27	28	29	30	4_author_cite_	66	67	68	69	70	venue_tot	86
3_venue_ref_	31	32	33	34	35	4_term_cite_	71	72	73	74	75	all_tot	87
3_author_cite_	36	37	38	39	40	4_venue_cite _	76	77	78	79	80		

---

[2] http://weka.sourceforge.net/doc.dev/weka/core/Stopwords.html
[3] http://tartarus.org/martin/PorterStemmer/java.txt

5   Experiment and Results

5.1   Benchmark

We formulated LBD as a multi-class classification problem and evaluated the performance of HBIN meta path features in predicting co-citation links between disconnected research areas. Our goal is to replicate Swanson's DFO-RS hypothesis [5]. We emphasize that many previous LBD methods have used this hypothesis as an evaluation benchmark [3,6,7,10,12]. Each instance in the learning set is a vector of features and a class label. There are three classes: inter-cluster co-citation link (+1), within-cluster co-citation link (-1), and no link (0). The performance of our model was evaluated in two ways. First, we evaluated the ability of HBIN meta path features to accurately classify instances of all three classes. Secondly, and more importantly, we evaluated their ability to predict future inter-cluster co-citation links between FO and RS papers (i.e. class +1). The actual co-citation links formed between FO and RS papers following Swanson's discovery was used as the evaluation ground truth.

We compared the performance of HBIN meta path features against three existing document similarity measures: *LDA topic model* [18], *term frequency-inverse document frequency* (TF-IDF) [19] and *bibliographic coupling* (BC) [20]. Similarity algorithms underpinned many existing LBD methods [2] because it is commonly assumed that highly similar documents or terms tend to form associations. We used Mallet's implementation of LDA[4] to discover topics in title and abstract text of core papers (10, 30, 50, and 100 topics) and computed the cosine similarity between their topic probability distribution to give us 8 LDA-based features. We also computed the cosine similarity between TF-IDF vectors of terms in core papers' titles and abstracts, obtaining 2 TF-IDF features. Finally, using Sci^{2}Tool[5], we calculated the bibliographic coupling strength between core papers through simple coupling count and cosine similarity, giving us 2 features.

5.2   Experimental Setup

First, we retrieved bibliographic records of fish oil and Raynaud's syndrome papers from Thomson Reuter's *Web of Science* (WoS)[6]. Each record includes a paper's title, author(s), publication venue, cited references, and citing articles. To be consistent with [5], we used the same search keywords originally used by Swanson and restricted the query results to year 1900 to 1985 prior to his discovery (Figure 1). We filtered the query results to obtain only records that had abstracts. We obtained 485 records (352 FO and 133 RS core papers).

From these core papers, we extracted 117,370 unique core paper pairs to become instances in our learning set. Using HBIN matrices constructed in Section 4.1, we indexed all distinct meta paths between all unique core paper pairs. To assign a class label to each instance, we retrieved another set of FO and RS records from

---

[4] http://mallet.cs.umass.edu/
[5] https://sci2.cns.iu.edu/user/index.php
[6] http://thomsonreuters.com/thomson-reuters-web-of-science/

WoS but this time included papers published in 1986 after Swanson's discovery. We labeled an instance as class +1 if it consists of a pair of FO paper and RS papers that had no co-citation link before 1986 but which became co-cited in 1986. An instance is labeled as -1 if it consists of two papers in the same cluster that were co-cited in 1986. We labeled an instance as 0 if there is no co-citation link at all. We obtained 210 instances of +1, 677 instances of -1, and 116,483 instances of 0. To obtain a balanced learning set, we randomly selected 210 instances from class -1, 210 instances from class 0, and kept all +1 instances. In total we had 630 instances and 99 features as the final learning set.

We evaluated the performance of our model in predicting instances of the three classes using five classifiers implemented in *Weka* [17] (default parameter settings): the SMO-variant of *Support Vector Machine* (SVM), *Neural Networks* (NN), *Nave Bayes* (NB), *Bayesian Network* (BN), and *C4.5 Decision Tree* (C4.5). We applied a 10-time repeated holdout validation at 70:30, 60:40, and 50:50 percent split as follows:

1. Randomize 630 instances in our learning set and select the first 70% of instances as the training set and the remaining 30% as the test set. Repeat this step ten times.
2. Repeat Step 1 for 60:40 and 50:50 split to obtain 30 training-test sets.
3. Apply each classifier on each training-test set and measure their performance in terms of Accuracy % and Weighted Average F-Measure ($\beta = 1$).
4. Calculate each classifier's average performance in all 30 training-test sets.

In the following phase of our experiment, we also evaluated the performance of meta path features against LDA, TF-IDF and BC features using the three best performing classifiers previously learned. In particular, we analyzed their performance in predicting future co-citation links (+1) between FO and RS papers, which is the main goal of literature-based discovery.

5.3   Results

Using *InfoGain* feature selection, we removed 5 meta path features that had zero contribution to the classification performance, leaving us with 94 features. *BN*, *NN*, and *SVM* emerged as top performers. Due to page limitation, in this paper we only present results based on these three classifiers. Using 94 features, BN performed the best with the average accuracy ranging between 86.27% and 87.14% in classifying instances of the three classes at 70:30, 60:40, and 50:50 ten-time repeated holdout settings. It also achieved the highest mean weighted average F-Measure (in the range of 0.863 and 0.872).

Subsequently we found that using HBIN meta path features alone could achieve up to 86.67% accuracy and 0.87 weighted average F-Measure in classifying instances of the three classes, outperforming LDA, TF-IDF, and Bibliographic Coupling (Table 2). We also analyzed the performance contributions of different meta path feature categories. The table also shows that 4-degree meta path features outperformed other lower degree meta path features in terms of

Table 2. Performance comparison between meta path features and the competing document similarity features on a 10-fold cross validation

Features	Accuracy %			Wgt. Avg. F1			F1 (+1)
	SVM	*NN*	*BN*	*SVM*	*NN*	*BN*	*BN*
LDA	49.048	51.587	54.762	0.462	0.515	0.533	0.608
TF-IDF	48.730	52.381	55.556	0.423	0.500	0.543	0.457
BC	54.127	57.619	57.619	0.557	0.586	0.480	0.612
Meta Path	**83.333**	**85.238**	**86.667**	**0.834**	**0.852**	**0.867**	**0.851**
Meta-2	67.143	67.460	68.254	0.660	0.651	0.669	0.683
Meta-3	64.286	70.318	75.397	0.605	0.700	0.756	0.723
Meta-4	**78.889**	**84.444**	**83.968**	**0.791**	**0.844**	**0.839**	**0.798**
Meta-Author	66.032	69.524	71.111	0.663	0.688	0.692	0.481
Meta-Term	**76.191**	**78.254**	**81.905**	**0.762**	**0.781**	**0.818**	**0.812**
Meta-Venue	66.032	73.333	73.333	0.662	0.734	0.735	0.683
Meta-Paper	55.556	56.667	57.302	0.469	0.578	0.478	0.611
Meta all path count	47.778	71.429	72.222	0.425	0.714	0.727	0.657

Accuracy % and F-Measure for predicting instances of all classes. The results also show that, regardless of the number of path degrees, meta paths formed by the term-sharing between papers performed better than other types of meta paths. Table 3 shows that among all types of 4-degree meta path features, meta paths via term-sharing (*4_term_ref_max* and *4_term_cite_max*) led to the highest F-Measure (0.704). Similarly, in terms of accuracy, even though the 4-degree author-sharing features achieved the highest accuracy (71.40%), these were followed closely by the same term-sharing features (70.64%).

Table 3. Performance comparison of 4-degree meta path features (10-fold cross val.)

Meta path scores	Accuracy %			Wgt. Avg. F1		
	Author	*Term*	*Venue*	*Author*	*Term*	*Venue*
Path count	68.06	68.15	64.48	0.646	0.681	0.641
Sum of weights	**71.40**	70.06	62.49	0.693	0.701	0.617
Avg of weights	61.75	53.27	68.72	0.544	0.421	0.688
Minimum weight	64.43	55.60	66.78	0.561	0.520	0.670
Maximum weight	61.36	70.64	69.21	0.540	**0.704**	0.691

More importantly, we also found that our meta path features performed very well in predicting future inter-cluster co-citation links (+1) with 0.851 F-Measure (*precision*: 0.845, *recall*: 0.857). This performance is much better than BC (0.612 F-Measure), LDA (0.608 F-Measure), and TF-IDF (0.457 F-Measure). Figure 3 shows that HBIN meta path features gained the largest area under curve (AUC) compared to its competing algorithms. Lastly, consistent with our findings earlier, 4-degree meta path features and term-sharing meta path features also performed the best (F-Measure 0.798 and 0.812, respectively) in predicting instances of class +1 compared to other meta paths (last column of Table 2).

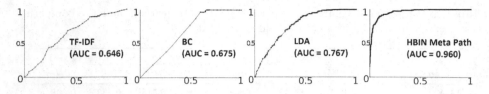

Fig. 3. ROC of TF-IDF, BC, LDA, and HBIN for predicting +1 class using BN

6   Discussions

Our results suggest that the rich relational features of HBIN meta paths could be used to discover latent connections between papers in disconnected research literatures better than the existing document similarity measures that rely only on homogeneous information (LDA, TF-IDF, BC). The higher performance of our 4-degree meta path features indicates that many hidden connections between core papers run through their '*peripheral*' papers, i.e. P_{ref} and P_{cite}, implying that it is useful for future LBD methods to incorporate citation information into their models. In addition, the high precision and recall achieved by HBIN meta path features also suggest that our method can mitigate the performance trade-off normally suffered by the exclusive use of lexical and citation analysis in LBD. Lexical analysis is known to suffer from *low precision-high recall*, whereas citation analysis usually suffers from *high precision-low recall* [8]. This improvement may attributable to the combinations of lexical and citation information in HBIN meta path features. For example, the best performing meta path features in predicting class +1, *4_term_ref_max* and *4_term_cite_max*, are the measures of lexical similarity (i.e. term-sharing) between core papers that propagate through the papers' citation structures (P_{ref} and P_{cite}). Further studies can be conducted to verify this intuition.

7   Conclusion and Future Work

In this paper, we formulate literature-based discovery as a inter-cluster co-citation link prediction problem. Unlike the previous LBD methods, our method exploited the rich, semi-structured information in heterogeneous bibliographic information network. The performance of our HBIN meta path features was compared to well-established document similarity algorithms. Experimental results showed that meta path features outperformed the competing algorithms in predicting future co-citation links between papers from fish oil and Raynaud's syndrome literature. Future work includes extending our method to other LBD cases, e.g. the discovery of novel alternatives to water purification [3].

References

1. Piatetsky-Shapiro, G., Dheraba, C., Getoor, L., Grossman, R., Feldman, R., Zaki, M.: What are the grand challenges for data mining?: KDD-2006 panel report. SIGKDD Explor. Newslett. **8**(2), 70–77 (2006)

2. Smalheiser, N.R.: Literature-based discovery: Beyond the ABCs. J. Am. Soc. Inform. Sci. Tech. **63**(2), 218–224 (2012)
3. Kostoff, R.N., Block, J.A., Solka, J.L., Briggs, M.B., Rushenberg, R.L., Stump, J.A., Johnson, D., Lyons, T.J., Wyatt, J.R.: Literature-related discovery. Annu. Rev. Inform. Sci. Tech. **43**(1), 1–71 (2009)
4. Small, H.: Co-citation in the scientific literature: A new measure of the relationship between two documents. J. Am. Soc. Inform. Sci. **24**(4), 265–269 (1973)
5. Swanson, D.R.: Fish oil, Raynaud's syndrome, and undiscovered public knowledge. Persp. Bio. Med. **30**(1), 7–18 (1986)
6. Swanson, D.R., Smalheiser, N.R.: An interactive system for finding complementary literatures: a stimulus to scientific discovery. Artif. Intell. **91**(2), 183–203 (1997)
7. Yetisgen-Yildiz, M., Pratt, W.: Using statistical and knowledge-based approaches for litera-ture-based discovery. J. Biomed. Inform. **39**(6), 600–611 (2006)
8. Bassecoulard, E., Zitt, M.: Patents and publications. In: Moed, H.F., Glanzel, W., Schmoch, U. (eds.) Handbook of Quantitative Science and Technology Research, Chap. 30, pp. 665–694. Springer (2005)
9. Wei, C.-P., Chen, K.-A., Chen, L.-C.: Mining biomedical literature and ontologies for drug repositioning discovery. In: Tseng, V.S., Ho, T.B., Zhou, Z.-H., Chen, A.L.P., Kao, H.-Y. (eds.) PAKDD 2014, Part II. LNCS (LNAI), vol. 8444, pp. 373–384. Springer, Heidelberg (2014)
10. Cheng, L., Lin, H., Zhou, F., Yang, Z., Wang, J.: Enhancing the accuracy of knowl-edge dis-covery: a supervised learning method. BMC Bioinform. **15**, S9 (2014)
11. Hristovski, D., Friedman, C., Rindflesch, T., Peterlin, B.: Literature-based knowl-edge discovery using natural language processing. In: Bruza, P., Weeber, M. (eds.) Literature-Based Discovery, pp. 133–152. Springer (2008)
12. Cameron, D., Bodenreider, O., Yalamanchili, H., Danh, T., Vallabhaneni, S., Thirunarayan, K., Sheth, A.P., Rindflesch, T.C.: A graph-based recovery and decomposition of Swanson's hypothesis using semantic predications. J. Biomed. Inform. **46**(2), 238–251 (2013)
13. Sun, Y., Han, J.: Mining heterogeneous information networks: principles and methodologies. Morgan & Claypool (2012)
14. Yu, X., Gu, Q., Zhou, M., Han, J.: Citation prediction in heterogeneous bibli-ographic net-works. In: 2012 SIAM Conference on Data Mining, Anaheim, pp. 1119–1130 (2012)
15. Liu, X., Yu, Y., Guo, C., Sun, Y., Gao, L.: Full-text based context-rich heteroge-neous network mining approach for citation recommendation. In: 2014 ACM IEEE Joint Conference on Digital Libraries, London, pp 361–370 (2014)
16. Ren, X., Liu, J., Yu, X., Khandelwal, U., Gu, Q., Wang, L., Han, J.: ClusCite: effective cita-tion recommendation by information network-based clustering. In: 20th ACM International Conference on Knowledge Discovery and Data Mining, New York, pp. 821–830 (2014)
17. Hall, M., Frank, E., Holmes, G., Pfahringer, B., Reutemann, P., Witten, I.H.: The WEKA data mining software: an update. SIGKDD Explor. Newslett. **11**(1), 10–18 (2009)
18. Blei, D.M., Ng, A.Y., Jordan, M.I.: Latent dirichlet allocation. J. Mach. Learn. Res. **3**, 993–1022 (2003)
19. Salton, G., McGill, M.J.: Introduction to modern information retrieval. McGraw-Hill (1983)
20. Kessler, M.M.: Bibliographic coupling between scientific papers. Am. Doc. **14**(1), 10–25 (1963)

Itemset and High Performance Data Mining

CPT+: Decreasing the Time/Space Complexity of the Compact Prediction Tree

Ted Gueniche[1], Philippe Fournier-Viger[1]([⊠]), Rajeev Raman[2],
and Vincent S. Tseng[3]

[1] Department of computer science, University of Moncton, Moncton, Canada
{etg8697,philippe.fournier-viger}@umoncton.ca
[2] Department of Computer Science, University of Leicester, Leicester, UK
r.raman@leicester.ac.uk
[3] Department of computer science and inf. eng.,
National Cheng Kung University, Tainan City, Taiwan
tsengsm@mail.ncku.edu.tw

Abstract. Predicting next items of sequences of symbols has many applications in a wide range of domains. Several sequence prediction models have been proposed such as DG, All-k-order markov and PPM. Recently, a model named Compact Prediction Tree (CPT) has been proposed. It relies on a tree structure and a more complex prediction algorithm to offer considerably more accurate predictions than many state-of-the-art prediction models. However, an important limitation of CPT is its high time and space complexity. In this article, we address this issue by proposing three novel strategies to reduce CPT's size and prediction time, and increase its accuracy. Experimental results on seven real life datasets show that the resulting model (CPT+) is up to 98 times more compact and 4.5 times faster than CPT, and has the best overall accuracy when compared to six state-of-the-art models from the literature: All-K-order Markov, CPT, DG, Lz78, PPM and TDAG.

Keywords: Sequence prediction · Next item prediction · Accuracy · Compression

1 Introduction

Sequence prediction is an important task with many applications [1,11]. Let be an alphabet of items (symbols) $Z = \{e_1, e_2, ..., e_m\}$. A sequence s is an ordered list of items $s = \langle i_1, i_2, ...i_n \rangle$, where $i_k \in Z$ $(1 \leq k \leq n)$. A prediction model is trained with a set of training sequences. Once trained, the model is used to perform sequence predictions. A prediction consists in predicting the next items of a sequence. This task has numerous applications such as web page prefetching, consumer product recommendation, weather forecasting and stock market prediction [1,3,8,11].

Many sequence predictions models have been proposed. One of the most popular is Prediction by Partial Matching (PPM)[2]. It is based on the Markov

© Springer International Publishing Switzerland 2015
T. Cao et al. (Eds.): PAKDD 2015, Part II, LNAI 9078, pp. 625–636, 2015.
DOI: 10.1007/978-3-319-18032-8_49

property and has inspired a multitude of other models such as Dependancy Graph (DG)[8], All-K-Order-Markov (AKOM)[10], Transition Directed Acyclic Graph (TDAG)[7], Probabilistic Suffix Tree (PST)[1] and Context Tree Weighting (CTW)[1]. Although, much work has been done to reduce the temporal and spatial complexity of these models (e.g. [1,3]), few work attempted to increase their accuracy. Besides, several compression algorithms have been adapted for sequence predictions such as LZ78 [12] and Active Lezi [4]. Moreover, machine learning algorithms such as neural networks and sequential rule mining have been applied to perform sequence prediction [6,11].

However, these models suffer from some important limitations [5]. First, most of them assume the Markovian hypothesis that each event solely depends on the previous events. If this hypothesis does not hold, prediction accuracy using these models can severely decrease [3,5]. Second, all these models are built using only part of the information contained in training sequences. Thus, these models do not use all the information contained in training sequences to perform predictions, and this can severely reduce their accuracy. For instance, Markov models typically considers only the last k items of training sequences to perform a prediction, where k is the order of the model. One may think that a solution to this problem is to increase the order of Markov models. However, increasing the order of Markov models often induces a very high state complexity, thus making them impractical for many real-life applications [3].

CPT[5] is a recently proposed prediction model which compress training sequences without information loss by exploiting similarities between subsequences. It has been reported as more accurate than state-of-the-art models PPM, DG, AKOM on various real datasets. However, a drawback of CPT is that it has an important spatial complexity and a higher prediction time than these models. Therefore, an important research problem is to propose strategies to reduce the size and prediction time of CPT. Reducing the spatial complexity is a very challenging task. An effective compression strategy should provide a huge spatial gain while providing a minimum overhead in terms of training time and prediction time. Furthermore, it should also preserve CPT's lossless property to avoid a decrease in accuracy. Reducing prediction time complexity is also very challenging. An effective strategy to reduce prediction time should access as little information as possible for making predictions to increase speed but at the same time it should carefully select this information to avoid decreasing accuracy.

In this paper, we address these challenges by proposing three strategies named FSC (Frequent Subsequence Compression), SBC (Simple Branches Compression) and PNR (Prediction with improved Noise Reduction). The two first strategies are compression strategies that reduce CPT size by up to two orders of magnitude while not affecting accuracy. The third strategy reduces the prediction time by up to 4.5 times and increases accuracy by up to 5%. This paper is organized as follows. Section 2 introduces CPT. Sections 3 and 4 respectively describes the two compression strategies (FSC and SBC) and the prediction time reduction strategy (FNR). Section 5 presents an experimental evaluation of each strategy on seven real datasets against five state-of-the-art prediction models. Finally, Section 6 draws conclusion.

2   Compact Prediction Tree

The Compact Prediction Tree (CPT) is a recently proposed prediction model [5]. Its main distinctive characteristics w.r.t to other prediction models are that (1) CPT stores a compressed representation of training sequences with no loss or a small loss and (2) CPT measures the similarity of a sequence to the training sequences to perform a prediction. The similarity measure is noise tolerant and thus allows CPT to predict the next items of subsequences that have not been previously seen in training sequences, whereas other proposed models such as PPM and All-K-order-markov cannot perform prediction in such case.

The training process of CPT takes as input a set of training sequences and generates three distinct structures: (1) a Prediction Tree (PT), (2) a Lookup Table (LT) and (3) an Inverted Index. During training, sequences are considered one by one to incrementally build these three structures. For instance, Fig. 1 illustrates the creation of the three structures by the successive insertions of sequences $s_1 = \langle A, B, C \rangle$, $s_2 = \langle A, B \rangle$, $s_3 = \langle A, B, D, C \rangle$, $s_4 = \langle B, C \rangle$ and $s_5 = \langle E, A, B, A \rangle$, where the alphabet $Z = \{A, B, C, D, E\}$ is used. The *Prediction*

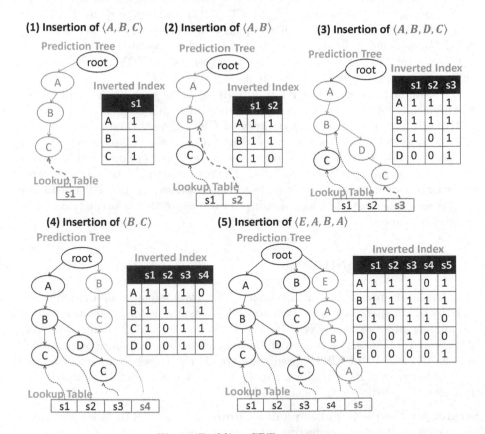

Fig. 1. Building CPT structures

Tree is a type of *prefix tree* (aka *trie*). It contains all training sequences. Each tree node represents an item and each training sequence is represented by a path starting from the tree root and ending by an inner node or a leaf. Just like a prefix tree, the prediction tree is a compact representation of the training sequences. Sequences sharing a common prefix share a common path in the tree. The *Lookup Table* is an associative array which allows to locate any training sequences in the prediction tree with a constant access time. Finally the *Inverted Index* is a set of bit vectors that indicates for each item i from the alphabet Z, the set of sequences containing i.

CPT's prediction process relies on the three aforementioned data structures. For a sequence $s = \langle i_1, i_2, ... i_n \rangle$ of n elements, the suffix of s of size y with $1 \leq y \leq n$ is defined as $P_y(s) = \langle i_{n-y+1}, i_{n-y+2} ... i_n \rangle$. Predicting the next items of s is performed by identifying the sequences similar to $P_y(s)$, that is the sequences containing all items in $P_y(s)$ in any order. The suffix length is a parameter similar to the model's order of All-k-order Markov and DG. Identifying the optimal value is done empirically by starting with a length of 1. CPT uses the *consequent* of each sequence similar to s to perform the prediction. Let $u = \langle j_1, j_2, ... j_m \rangle$ be a sequence similar to s. The *consequent* of u w.r.t to s is the longest subsequence $\langle j_v, j_{v+1}, ... j_m \rangle$ of u such that $\bigcup_{k=1}^{v-1} \{j_k\} \subseteq P_y(s)$ and $1 \leq v \leq m$. Each item found in the consequent of a similar sequence of s is stored in a data structure called *Count Table* (CT). The count table stores the support (frequency) of each of these items, which is an estimation of $P(e|P_y(s))$. CPT returns the most supported item(s) in the CT as its prediction(s).

The similarity measure in CPT is initially strict for each prediction task but is dynamically loosened to ignore noise. Identifying similar sequences, and more particularly the noise avoidance strategy of CPT, is very time consuming and account for most of CPT's prediction time [5]. For a given sequence, if CPT cannot find enough similar sequences to generate a prediction, it will implicitly assume the sequence contains some noise. The prediction process is then repeated but with one or more items omitted from the given sequence. CPT's definition of noise is implicit and has for sole purpose to ensure that a prediction can be made every time.

3   Compression Strategies

CPT has been presented as one of the most accurate sequence prediction model [5] but its high spatial complexity makes CPT unsuitable for applications where the number of sequences is very large. CPT's size is smaller than All-k-Order Markov and TDAG but a few orders of magnitude larger than popular models such as DG and PPM. CPT's prediction tree is the largest data structure and account for most of its spatial complexity. In this section, we focus on strategies to reduce the prediction tree's size.

Strategy 1 Frequent Subsequence Compression (FSC). In a set of training sequences, frequently occurring subsequences of items can be found. For some datasets, these subsequences can be highly frequent. The FSC strategy identifies these frequent subsequences and replace each of them with a single item. Let be a sequence $s = \langle i_1, i_2, ..., i_n \rangle$. A sequence $c = \langle i_{m+1}, i_{m+2}, ..., i_{m+k} \rangle$ is a subsequence of s, denoted as $c \sqsubseteq s$, iff $1 \leq m \leq m + k \leq n$. For a set of training sequences S, a subsequence d is considered a *frequent subsequence* iff $|\{t|t \in S \land d \sqsubseteq t\}| \geq minsup$ for a minimum support threshold $minsup$ defined per dataset.

Frequent subsequences compression is done during the training phase and is performed in three steps: (1) identification of frequent subsequences in the training sequences, (2) generation of a new item in the alphabet Z of items for each frequent subsequence, and (3) replacement of each frequent subsequence by the corresponding new item when inserting training sequences in the prediction tree. Identifying frequent subsequence in a set of sequences is a known problem in data mining for which numerous approaches have been proposed. In FSC, we use the well known PrefixSpan [9] algorithm. PrefixSpan is one of the most efficient sequential pattern mining algorithm. It has been adapted by incorporating additional constraints to fit the problem of sequence prediction. Subsequences have to be contiguous, larger than a minimum length $minSize$ and shorter than a maximum length $maxSize$. Both $minSize$ and $maxSize$ are parameters of this compression strategy that are defined per application.

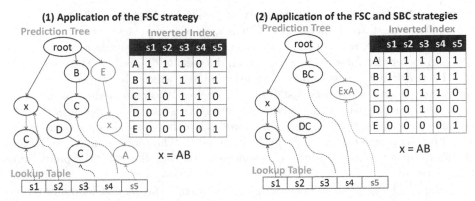

Fig. 2. Application of the FSC and SBC compression strategies

A new data structure, *Subsequence Dictionary* (DCF), is introduced to store the frequent subsequences. This dictionary associates each frequent subsequence with its corresponding item. The DCF offers a fast way to translate each subsequence into its respective item and vice-versa, $O(1)$. When inserting training sequences into the prediction tree, the DCF is used to replace known frequent subsequences with single items. For example, figure 2 illustrates the resulting prediction tree after applying FSC to the tree shown in Fig. 1. The frequent subsequence $\langle A, B \rangle$ has been replaced by a new symbol x, thus reducing the number of nodes in the prediction tree. The FSC compression strategy influences the shape of the prediction tree by reducing its height and number of nodes. With respect to the prediction process, FSC only influences execution time. The additional cost is the on-the-fly decompression of the prediction tree, which is fast and non intrusive because of the DCF structure.

Strategy 2: Simple Branches Compression (SBC). Simple Branches Compression is an intuitive compression strategy that reduces the size of the prediction tree. A *simple branch* is a branch leading to a single leaf. Thus, each node of a simple branch has between 0 and 1 children. The SBC strategy consists of replacing each simple branch by a single node representing the whole branch. For instance, part (2) of Fig. 2 illustrates the prediction tree obtained by applying the DCF and SBC strategies for the running example. The SBC strategy has respectively replaced the simple branches D, C, B, C and E, x, A by single nodes DC, BC and ExA. Identifying and replacing simple branches is done by traversing the prediction tree from the leafs using the inverted index. Only the nodes with a single child are visited. Since the Inverted Index and Lookup Table are not affected by this strategy, the only change that needs to be done to the prediction process is to dynamically uncompress nodes representing simple branches when needed.

4   Time Reduction Strategy

Strategy 3: Prediction with Improved Noise Reduction (PNR). As previously explained, to predict the next item i_{n+1} of a sequence $s = \langle i_1, i_2, ..., i_n \rangle$, CPT uses the suffix of size y of s denoted as $P_y(s)$ (the last y items of s), where y is a parameter that need to be set for each dataset. CPT predicts the next item of s by traversing the sequences that are similar to its suffix $P_y(s)$. Searching for similar sequences is very fast $(O(y))$. However, the noise reduction mechanism used for prediction (described in Section 2) is not. The reason is that it considers not only $P_y(s)$ to perform a prediction, but also all subsequences of $P_y(s)$ having a size $t > k$, where k is a parameter. The more y and k are large, the more subsequences need to be considered, and thus the more the prediction time increases.

For a prediction task, items in a training sequence may be considered as noise if their sole presence negatively impact a prediction's outcome. The PNR strategy is based on the hypothesis that noise in training sequences consists of items having a low frequency, where an item's frequency is defined as the number of training sequences containing the item. For this reason, PNR removes only items having a low frequency during the prediction process. Because the definition of noise used in CPT+ is more restrictive than that of CPT, a smaller number of subsequences are considered. This reduction has a positive and measurable impact on the execution time, as it will be demonstrated in the experimental evaluation (see Section 5).

The PNR strategy (Algorithm 1) takes as parameter the prefix $P_y(s)$ of a sequence to be predicted s, CPT's structures and the noise ratio TB and a minimum number of updates, MBR, to be performed on the count table (CT) to perform a prediction. The noise ratio TB is defined as the percentage of items in a sequence that should be considered as noise. For example, a noise ratio of 0 indicates that sequences do not contain noise, while a ratio of 0.2 means that 20% of items in a sequence are considered as noise. PNR is a recursive procedure. To perform a prediction, we require that PNR consider a minimum number of subsequences derived from $P_y(s)$. PNR first removes noise from each subsequence. Then, the CT is updated using these subsequences. When the mimimum number of updates is reached, a prediction is performed as in CPT using the CT. The PNR strategy is a generalization of the noise reduction strategy used by CPT. Depending on how the parameters are set, PNR can reproduce the behavior of CPT's noise reduction strategy. The three main contributions brought by PNR are to require a minimum number of updates on the CT to perform a prediction, and to define noise based on the frequency of items, and to define noise proportionally to a sequence length. Finding the appropriate values for both TB and MBR can be achieved empirically.

Algorithm 1. The prediction algorithm using PNR

---

input : $P_y(s)$: a sequence suffix, CPT: CPT's structures, TB: a noise ratio,
        MBR: minimum number of CT updates
output: x: the predicted item(s)

queue.add($P_y(s)$);
while *updateCount* $<$ *MBR* $\wedge$ *queue.notEmpty()* **do**
    suffix = queue.next();
    noisyItems = selectLeastFrequentItems(TB);
    foreach *noisyItem* $\in$ *noisyItems* **do**
        suffixWithoutNoise = removeItemFromSuffix(suffix, noisyItem);
        if *suffixWithoutNoise.length* $>$ 1 **then**
           | queue.add(suffixWithoutNoise);
        end
        updateCountTable(CPT.CT, suffixWithoutNoise);
        updateCount++;
    end
    return performPrediction(CPT.CT);
end

---

5   Experimental Evaluation

We have performed several experiments to compare the performance of CPT+ against five state-of-the-art sequence prediction models: All-K-order Markov, DG, Lz78, PPM and TDAG. We picked these models to match the models used in the original paper describing CPT [5] and added both Lz78 and TDAG. To implement CPT+, we have obtained and modified the original source code of CPT [5]. To allow reproducing the experiments, the source code of the prediction models and datasets are provided at http://goo.gl/JwbVEM. All models are implemented using Java 8. Experiments have been performed on a computer with a dual core 4th generation Intel i5 with 8 GB RAM and a SSD drive connected with SATA 600. For all prediction models, we have empirically attempted to set their parameters to optimal values. PPM and LZ78 do not have parameters. DG and AKOM have respectively a window of four and an order of five. To avoid consuming an excessive amount of memory, TDAG has a maximum depth of 7. CPT has two parameters, *splitLength* and *maxLevel* and CPT+ has six parameters; three for the FSC strategy, two for the PNR strategy and *splitLength* from CPT. The values of these parameters have also been empirically found and are provided in the project source code. Experiment specific parameters are the minimum and maximum length of sequences used, the number of items to be considered to perform a prediction (the suffix length) and the number of items used to verify the prediction (called the consequent length). Let be a sequence $s = \langle i_1, i_2, ...i_n \rangle$ having a suffix $S(s)$ and a consequent $C(s)$. Each model takes the suffix as input and outputs a predicted item p. A prediction is deemed successful if p is the first item of $C(s)$. Datasets having various charac-

Table 1. Datasets

Name	Sequence count	Distinct item count	Average length	Type of data
BMS	15,806	495	6.01	webpages
KOSARAK	638,811	39,998	11.64	webpages
FIFA	573,060	13,749	45.32	webpages
MSNBC	250,697	17	3.28	webpages
SIGN	730	267	93.00	language
BIBLE Word	42,436	76	18.93	sentences
BIBLE Char	32,502	75	128.35	characters

teristics have been used (see Table 1) such as short/long sequences, sparse/dense sequences, small/large alphabets and various types of data. The BMS, Kosarak, MSNB and FIFA datasets consist of sequences of webpages visited by users on a website. In this scenario, prediction models are applied to predict the next webpage that a user will visit. The SIGN dataset is a set of sentences in sign language transcribed from videos. Bible Word and Bible Char are two datasets originating from the Bible. The former is the set of sentences divided into words. The latter is the set of sentences divided into characters. In both datasets, a sentence represents a sequence.

To evaluate prediction models, a prediction can be either a *success* if the prediction is accurate, a *failure* if the prediction is inaccurate or a *no match* if the prediction model is unable to perform a prediction. Four performance measures are used in experiments: *Coverage* is the ratio of sequences without prediction against the total number of test sequences. *Accuracy* is the number of successful predictions against the total number of test sequences. *Training time* is the time taken to build a model using the training sequences *Testing time* is the time taken to make a prediction for all test sequences using a model.

Experiment 1: Optimizations Comparison. We have first evaluated strategies that aims to reduce the space complexity of CPT (cf. Section 3) by measuring the compression rate and the amount of time spent for training. Other measures such as prediction time, coverage and accuracy are not influenced by the compression. For a prediction tree A having s nodes before compression and $s2$ nodes after compression, the compression rate tc_a of A is defined as $tc = 1 - (s2/s)$, a real number in the $[0,1[$ interval. A larger value means a higher compression. The two strategies are first evaluated separately (denoted as FSC and SBC) and then together (denoted as CPT+). Compression provides a spatial gain but increases execution time. Figure 3 illustrates this relationship for each compression strategy.

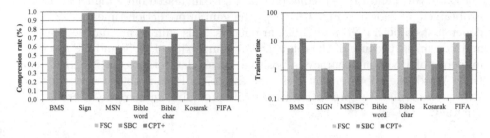

Fig. 3. Compression rate and training time of the compression strategies

It can also be observed in Fig. 3 (left) that the compression rate varies depending on the dataset from 58.90% to 98.65%. FSC provides an average compression rate of 48.55% with a small standard deviation (6.7%) and SBC offers an average compression rate of 77.87% with a much larger standard deviation (15.9%). For each tested dataset, SBS has a compression rate very similar to CPT+ with the exception of MSNBC and Bible Char. It accounts for most of CPT+ compression rate and thus making FSC relevant only in applications requiring a smaller model. MSNBC is the least affected by the compression strategies. The reason is that MSNBC has a very small alphabet and thus that the tree is naturally compressed because its branches are highly overlapping. In fact, MSNBC has only 17 distinct items, although the length of its sequences is similar to the other datasets. The dataset where the SBC and SFC strategies provide the highest compression rate is SIGN. Even though SIGN contains a small number of sequences, each sequence is very long (an average of 93 items). It causes the branches of SIGN's prediction tree to rarely overlap, and a large amount of its nodes only have a single child. SIGN is a very good candidate for applying the SBC strategy. Using only SBC, a compression rate of 98.60 % is attained for SIGN.

Figure 3 also illustrates the training time for the FSC and SBC strategies. It is measured as a multiplicative factor of CPT's training time. For example, a factor x for SBC means that the training time using SBC was x times longer than that of CPT without SBC. A small factor, close to one, means the additional training time was small. It is interesting to observe how the training time when both strategies are combined is less than the sum of their respective training time. Althought SBC and FSC are independently applied to CPT, SBC reduces the execution time of FSC because less branches have to be compressed. Overall, SBC has a small impact on the training time while providing most of the compression gain, this makes SBS the most profitable strategy. While SFC has a higher impact on the training time, SFC enhances the compression rate for each dataset.

We also evaluated the performance improvement in terms of execution time and accuracy obtained by applying the PNR strategy. Fig. 4 (left) compares the prediction time of CPT+ (with PNR) with that of CPT. The execution time is reduced for most datasets and is up to 4.5 times smaller for SIGN and MSNBC. For Bible Word and FIFA, prediction times have increased but a higher accuracy is obtained, as shown in Fig. 4 (right). The gain in prediction time for CPT+ is dependant on both PNR and CPT parameters. This gain is thus dataset specific and non linear because of the difference in complexity of CPT and CPT+. The influence of PNR on accuracy is positive for all datasets except MSNBC. For Bible Word, the improvement is as high as 5.47%. PNR is thus a very effective strategy to reduce the prediction time while providing an increase in prediction accuracy.

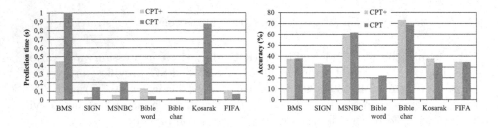

Fig. 4. Execution time and accuracy gains provided by the PNR strategy

Experiment 2: Scalability. In this experiment we compared the spatial complexity of CPT+ (with both compression strategies) against CPT, All-K-order Markov, DG, Lz78, PPM and TDAG. Only the FIFA and Kosarak datasets were used in this experiment because of their large number of sequences. In Fig. 5, the spatial size of each model is evaluated against a quadratically growing set of training sequences - up to 128,000 sequences. Both PPM and DG have a sub linear growth which makes them suitable for large datasets. CPT+'s growth is only an order of magnitude larger than PPM and DG and a few orders less than CPT, TDAG and LZ78. The compression rate of CPT+ tends to slightly diminish as the number of sequences grows. This is due to more branches overlapping in the prediction tree, a phenomenon that can generally be observed in tries.

An interesting observation is that for both datasets, when the number of training sequences is smaller than the number of items, CPT+ has a smaller footprint than the other prediction models. For the FIFA dataset, when 1000 sequences are used, CPT+'s node count is 901 compared to the 1847 unique items in the alphabet. Results are similar for Kosarak. Models such as PPM and DG can't achieve such a small footprint in these use cases because they have a least one node per unique item.

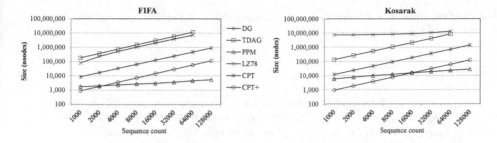

Fig. 5. Scalability of the prediction models

Experiment 3: Comparison with Other Prediction Models. In experiment 1, we have compared the prediction accuracy of CPT+ and CPT to assess the improvement obtained by applying the PNR strategy. In this experiment, we compare the accuracy of CPT+ with five state-of-the-art prediction models commonly used in the literature, All-K-order Markov, DG, Lz78, PPM et TDAG, on the same datasets. Each prediction model has been trained and tested using k-fold cross validation with $k = 14$ to obtain a low variance for each run. Table 2 shows the prediction accuracy obtained by each model. Results indicates that CPT+ offers a generally higher accuracy than the compared models from the literature while also being more consistent across the various datasets.

Table 2. Predictions models and their accuracy

Datasets	CPT+	CPT	AKOM	DG	LZ78	PPM	TDAG
BMS	**38.25**	37.90	31.26	36.46	33.46	31.06	6.95
SIGN	**33.01**	32.33	8.63	3.01	4.79	4.25	0.00
MSNBC	61.50	**61.64**	47.88	55.68	43.64	38.06	31.14
Bible word	27.52	22.05	**38.68**	24.92	27.39	27.06	23.33
Bible char	**73.52**	69.14	7.96	0.00	3.02	0.10	9.90
Kosarak	**37.64**	33.82	20.52	30.82	20.50	23.86	1.06
FIFA	**35.94**	34.56	25.88	24.78	24.64	22.84	7.14

6   Conclusion

In this paper we presented CPT+, a variation of CPT that includes two novel compression strategies (FSC and SBC) to reduce its size and a strategy to improve prediction time and accuracy (PNR). Experimental results on seven real datasets shows that CPT+ is up to 98 times smaller than CPT, performs predictions up to 4.5 times faster, and is up to 5% more accurate. A comparison with six state-of-the art sequence prediction models (CPT, All-K-Order Markov, DG, Lz78, PPM and TDAG) shows that CPT+ is on overall the most accurate model. To allow reproducing the experiments, the source code of the prediction models and datasets are provided at http://goo.gl/JwbVEM.

Acknowledgments. This project was supported by a NSERC grant from the Government of Canada.

References

1. Begleiter, R., El-yaniv, R., Yona, G.: On prediction using variable order markov models. Journal of Artificial Intelligence Research **22**, 385–421 (2004)
2. Cleary, J., Witten, I.: Data compression using adaptive coding and partial string matching. IEEE Trans. on Inform. Theory **24**(4), 413–421 (1984)
3. Deshpande, M., Karypis, G.: Selective markov models for predicting web page accesses. ACM Transactions on Internet Technology **4**(2), 163–184 (2004). https://www.developers.google.com/prediction, Accessed: 2014-02-15
4. Gopalratnam, K., Cook, D.J.: Online sequential prediction via incremental parsing: The active lezi algorithm. IEEE Intelligent Systems **22**(1), 52–58 (2007)
5. Gueniche, T., Fournier-Viger, P., Tseng, V.S.: Compact prediction tree: a lossless model for accurate sequence prediction. In: Motoda, H., Wu, Z., Cao, L., Zaiane, O., Yao, M., Wang, W. (eds.) ADMA 2013, Part II. LNCS, vol. 8347, pp. 177–188. Springer, Heidelberg (2013)
6. Fournier-Viger, P., Gueniche, T., Tseng, V.S.: Using partially-ordered sequential rules to generate more accurate sequence prediction. In: Zhou, S., Zhang, S., Karypis, G. (eds.) ADMA 2012. LNCS, vol. 7713, pp. 431–442. Springer, Heidelberg (2012)
7. Laird, P., Saul, R.: Discrete sequence prediction and its applications. Machine Learning **15**(1), 43–68 (1994)
8. Padmanabhan, V.N., Mogul, J.C.: Using Prefetching to Improve World Wide Web Latency. Computer Communications **16**, 358–368 (1998)
9. Pei, J., Han, J., Mortazavi-Asl, B., Wang, J., Pinto, H., Chen, Q., Dayal, U., Hsu, M.: Mining sequential patterns by pattern-growth: the PrefixSpan approach. IEEE Trans. Known. Data Engin. **16**(11), 1424–1440 (2004)
10. Pitkow, J., Pirolli, P.: Mining longest repeating subsequence to predict world wide web surng. In: Proc. 2nd USENIX Symposium on Internet Technologies and Systems, Boulder, CO, pp. 13–25 (1999)
11. Sun, R., Giles, C.L.: Sequence Learning: From Recognition and Prediction to Sequential Decision Making. IEEE Intelligent Systems **16**(4), 67–70 (2001)
12. Ziv, J., Lempel, A.: Compression of individual sequences via variable-rate coding. IEEE Transactions on Information Theory **24**(5), 530–536 (1978)

Mining Association Rules in Graphs Based on Frequent Cohesive Itemsets

Tayena Hendrickx[✉], Boris Cule, Pieter Meysman, Stefan Naulaerts,
Kris Laukens, and Bart Goethals

University of Antwerp, Antwerp, Belgium
{Tayena.Hendrickx,Boris.Cule,Pieter.Meysman,Stefan.Naulaerts,
Kris.Laukens,Bart.Goethals}@uantwerp.be

Abstract. Searching for patterns in graphs is an active field of data mining. In this context, most work has gone into discovering subgraph patterns, where the task is to find strictly defined frequently re-occurring structures, i.e., node labels always interconnected in the same way. Recently, efforts have been made to relax these strict demands, and to simply look for node labels that frequently occur near each other. In this setting, we propose to mine association rules between such node labels, thus discovering additional information about correlations and interactions between node labels. We present an algorithm that discovers rules that allow us to claim that if a set of labels is encountered in a graph, there is a high probability that some other set of labels can be found nearby. Experiments confirm that our algorithm efficiently finds valuable rules that existing methods fail to discover.

1 Introduction

Discovering interesting patterns in graphs is a popular data mining task, with wide applications in social network analysis, bioinformatics, etc. In traditional approaches, the dataset consists of a (very large) single graph and the task is to find frequent subgraphs, i.e., reoccuring stuctures consisting of labelled nodes frequently interconnected in exactly the same way. However, the concept of frequent subgraphs is not flexible enough to capture all patterns. Another problem with the subgraph mining approaches is that they are also typically computationally complex since they are forced to deal with the graph isomorphism problem. For small graphs, isomorphism checking is still manageable, but for large graphs such checks become computationally very expensive.

In this work, we therefore base our study on the Frequent Cohesive Itemset (*FCI*) approach proposed by Hendrickx et al. [6]. A frequent cohesive itemset is defined as a set of node labels that are all, as individual items, frequent, and are, on average, tightly connected to each other, but not necessarily always connected in exactly the same way. In this paper, we introduce the concept of association rules into this setting. More formally, the aim is to generate rules of the form *if X occurs, Y occurs nearby*, where $X \cap Y = \emptyset$ and $X \cup Y$ is a frequent cohesive itemset. Such rules provide the end user with much more information about

© Springer International Publishing Switzerland 2015
T. Cao et al. (Eds.): PAKDD 2015, Part II, LNAI 9078, pp. 637–648, 2015.
DOI: 10.1007/978-3-319-18032-8_50

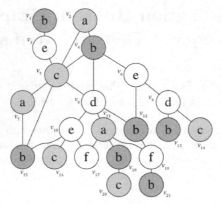

Fig. 1. A toy graph dataset

correlations and interactions between node labels than itemsets alone. Consider, for example, the graph given in Figure 1, and suppose ab is a frequent cohesive itemset (for simplicity, we denote an itemset $\{a, b\}$ as ab in our examples). Note that this pattern will not necessarily be discovered by subgraph mining, as nodes labelled a and b are not always connected by an edge. Based on this itemset, we can generate rule *if a node labelled 'a' occurs, a node labelled 'b' occurs nearby*, which is, as we can see, true for every node labelled a. However, we can also generate rule *if a node labelled 'b' occurs, a node labelled 'a' occurs nearby*, which does not necessarily hold, since some nodes labelled b are relatively far from the nearest node labelled a. If we only mine frequent cohesive itemsets, this type of information would remain undiscovered.

Given an association rule $X \Rightarrow Y$, we define its confidence as the average distance from an occurrence of X to the nearest occurrence of Y. Inspired by a similar approach to mining association rules in sequential data by Cule and Goethals [4], we develop an efficient algorithm to mine association rules based on frequent cohesive itemsets. The main two features of the algorithm are that it allows us to mine itemsets and association rules in parallel (rather than first finding all itemsets, and then all association rules, as is the case in most traditional approaches), and that we only need to compute the confidence of a simple class of rules, which can then be used to quickly evaluate all other rules. Experiments show that our algorithm finds interesting rules within acceptable runtime.

2   Related Work

Mining knowledge from graph structured data is an active research topic in data mining. A survey of the early graph-based data mining methods is given by Washio and Motada [16]. The traditional way of finding patterns in graphs is limited to searching for frequent subgraphs, i.e., reoccuring patterns within which nodes with certain labels are frequently interconnected in exactly the same way. The first attempts to find frequent subgraphs in a single graph were made by Cook and Holder [3]. Meanwhile, Motoda and Indurkhya [18] developed an approach for a setting where the dataset consists of many (typically smaller) graphs, rather than a single large one. Further attempts at subgraph mining have been made by Yan and Han [17], Nijssen and Kok [14], and Dehaspe and Toivonen [5]. Although algorithms for discovering frequent subgraphs find patterns that are useful in many applications, they suffer from two main drawbacks — costly isomorphism testing and an enormous number of generated candidates since both edges and nodes must be added to the pattern.

To tackle these problems, Hendrickx et al. [6] recently proposed the FCI algorithm for mining frequent cohesive itemsets in graphs. An itemset is defined as a set of node labels which often occur in the graph and are, on average, tightly connected to each other. By searching for itemsets rather than subgraphs, the costly isomorphism tests are avoided, and the number of candidate patterns is reduced. Khan et al. [12] proposed another itemset mining approach for graphs, where node labels are, according to given probabilities, propagated to neighbouring nodes. Labels are thus aggregated, and can be mined as itemsets in the resulting graph.

The concept of association rule mining was first introduced by Agrawal et al. [2]. An association rule, denoted $X \Rightarrow Y$, expresses an observation that an occurrence of pattern X implies an occurrence of pattern Y. In the setting of transaction databases, the strength of a rule was measured in terms of its confidence which determines how frequently itemset Y appears in transactions that contain X. Since then, association rules have been applied in a variety of settings, with various definitions of the confidence measure. Cule and Goethals [4] successfully applied association rule mining on sequential data, where they defined the confidence of a rule in terms of how far on average from an occurrence of itemset X one has to look in order to find itemset Y.

In this paper, however, we apply the concept of association rules to graph data. In this setting, Inokuchi et al. [9] proposed an Apriori-based algorithm, called AGM, for mining frequenty appearing induced subgraphs in a given graph dataset and association rules among such subgraphs. In contrast with our work, they mine rules of the form $G_b \Rightarrow G_h$ where G_b and G_h are subgraphs, rather than itemsets. In another work, Inokuchi et al. [10] proposed an approach to mine frequent graph structures and the association rules among them embedded in massive graph structured transaction data. Here, the graph data is transformed in such a way that traditional association rule mining techniques can be applied, but the resulting rules still contain subgraph patterns.

3   Problem Setting

In this section, we define the concept of association rules in graph data based on frequent cohesive itemsets. For simplicity of presentation, we assume that the graph consists of labelled nodes and unlabelled egdes, and we focus on connected graphs with at most one label per node. However, we can also handle input graphs that contain nodes with multiple labels, by transforming each such node into multiple nodes each carrying one label and connecting all of them to a central dummy node.

As our work is based on the Frequent Cohesive Itemset mining approach of Hendrickx et al. [6], we start with a reproduction of some of the necessary definitions and notations. We define a graph G as a set of nodes $V(G)$ and a set of edges $E(G)$. Each node $v \in V(G)$ carries a label $l(v) \in S$, where S is the set of all labels. For a label $i \in S$, we denote the set of nodes in the graph carrying this label as $L(i) = \{v \in V(G)|l(v) = i\}$. We define the frequency of a label $i \in S$ as the probability of encountering that label in G, or $freq(i) = \frac{|L(i)|}{|V(G)|}$. From now on, we will refer to labels as items, and sets of labels as itemsets.

For an itemset X, the set of all nodes labelled by an item in X is denoted by $N(X) = \{v \in V(G)|l(v) \in X\}$. In order to compute the cohesion of an itemset X we must look, for each occurence of an item in X, for the nearest occurence of all other items in X. For a node v, we define the sum of all these smallest distances as $W(v, X) = \sum_{x \in X} min_{w \in N(\{x\})} d(v, w)$, where $d(v, w)$ is the length of the shortest path from node v to node w. The average of such sums for all occurences of items making up itemset X is expressed as

$$\overline{W}(X) = \frac{\sum_{v \in N(X)} W(v, X)}{|N(X)|}.$$

The cohesion of an itemset X, where $|X| \geq 2$, is defined as the ratio of the itemset size and the average sum of the smallest distances defined above:

$$C(X) = \frac{|X|-1}{\overline{W}(X)}.$$

Note that the itemset size is reduced by one, since each considered node already carries one item in X. If $|X| < 2$, we define $C(X)$ to be equal to 1.

Cohesion measures how close to each other the items making up itemset X are on average. If the items are always directly interconnected by an edge, the sum of these distances for each occurence of an item in X will be equal to $|X|-1$, as will the average of such sums, and the cohesion of X will be equal to 1.

Finally, an itemset X is considered *frequent cohesive* if, given user defined thresholds for frequency (*min_freq*) and cohesion (*min_coh*), it holds that $\forall x \in X : freq(x) \geq min\_freq$ and $C(X) \geq min\_coh$. Two optional parameters, *minsize* and *maxsize*, can be used to limit the size of the discovered itemsets.

In this setting, our goal is to generate rules of the form *if X occurs, Y occurs nearby*, where $X \cap Y = \emptyset$ and $X \cup Y$ is a frequent cohesive itemset. In the rest of this paper, we will denote a rule in the traditional way as $X \Rightarrow Y$, with X the body of the rule and Y the head of the rule. It is clear that the closer the items of Y occur to the items of X in the graph, the higher the confidence value of the rule should be. More formally, in order to compute the confidence of the rule, we must compute the average sum of minimal distances for $X \cup Y$, but only from the point of view of items making up itemset X, i.e.,

$$\overline{W}(X, Y) = \frac{\sum_{v \in N(X)} W(v, X \cup Y)}{|N(X)|}.$$

The confidence of a rule can then be defined as

$$c(X \Rightarrow Y) = \frac{|X \cup Y| - 1}{\overline{W}(X,Y)}. \tag{1}$$

Given a confidence threshold min_conf, we consider rule $X \Rightarrow Y$ confident if $c(X \Rightarrow Y) \geq min\_conf$.

Let us now apply the definitions introduced above on the graph depicted in Figure 1. Assume that min_freq, min_coh and min_conf are set to $0.1, 0.5$ and 0.6, respectively. Looking at itemset ab we first note that $freq(a) = \frac{3}{21} \approx 0.14$ and $freq(b) = \frac{7}{21} \approx 0.33$, which shows us that both a and b are frequent. Now let us have a look at the cohesion of itemset ab. According to our defintions, $|N(ab)| = 10$. To compute the cohesion, we now search the neighbourhood of each node in $N(ab)$ which gives us $W(v_2, ab) = W(v_4, ab) = W(v_7, ab) = W(v_{11}, ab) = W(v_{15}, ab) = W(v_{18}, ab) = 1$, $W(v_{12}, ab) = W(v_{21}, ab) = 2$, $W(v_1, ab) = 3$ and $W(v_{13}, ab) = 4$. Subsequently, computing the average of the above minimal distances, we get $\overline{W}(ab) = \frac{17}{10} = 1.7$. We can now compute the cohesion of ab as $C(ab) = \frac{2-1}{1.7} \approx 0.58$, and conclude that itemset ab is sufficiently cohesive.

If we look at the occurences of a and b in the graph, we see that every node labelled a has a node labelled b nearby, but not vice versa. Therefore, the confidence of rule $a \Rightarrow b$ should be higher than that of rule $b \Rightarrow a$. According to our definitions, we see that for each node v labelled a, $W(v, ab) = 1$. Therefore, $\overline{W}(a, b) = \frac{3}{3} = 1$, and $c(a \Rightarrow b) = 1$ which means that rule $a \Rightarrow b$ has, as expected, a confidence of 100%. Meanwhile, the sum of distances from each b to the nearest a is $\sum_{v \in N(\{b\})} W(v, ab) = 14$. It follows that $\overline{W}(b, a) = \frac{14}{7} = 2$ and $c(b \Rightarrow a) = 0.5$. We see that $a \Rightarrow b$ is a confident association rule, while $b \Rightarrow a$ is not, and we conclude that while an occurence of a b does not imply finding an a nearby, when we find an a, however, we can expect to find a b nearby.

4   Algorithm

In this section we present a detailed description of our algorithm for discovering confident association rules based on frequent cohesive itemsets. Unlike the traditional approaches, which need all frequent itemsets to be found before the generation process of the association rules can begin, we will generate the rules in parallel with the frequent cohesive itemsets.

We begin by giving a quick overview of the FCI algorithm [6], which will serve as a basis for our algorithm. The FCI algorithm, given in Algorithm 1, discovers frequent cohesive itemsets. Candidate itemsets are generated by applying depth-first search, using recursive enumeration. During this process, a candidate consists of two elements — items that make up the candidate, and all frequent items which still have to be enumerated, denoted as X and Y, respectively. The algorithm uses the UBC pruning function, which computes an upper bound on the cohesion of all candidates yet to be generated in a given branch of the search tree, and decides whether to proceed deeper into the search tree, or to prune the complete branch. When a frequent cohesive itemset is found in line 3, rather

---

Algorithm 1. FCI($\langle X, Y \rangle$) finds frequent cohesive itemsets

---

1: **if** $UBC(\langle X, Y \rangle) \geq min\_coh$ **then**
2:     **if** $Y = \emptyset$ **then**
3:         **if** $X \neq \emptyset$ and $|X| \geq minsize$ **then** GENERATE_RULES(X)
4:     **else**
5:         Choose a in Y
6:         **if** $|X \cup \{a\}| \leq maxsize$ **then** FCI($\langle X \cup \{a\}, Y \setminus \{a\} \rangle$)
7:         **if** $|X \cup (Y \setminus \{a\})| \geq minsize$ **then** FCI($\langle X, Y \setminus \{a\} \rangle$)
8:     **end if**
9: **end if**

---

than outputting it, we proceed to mine association rules that can be generated from this itemset, as discussed below.

The most computationally costly step of the FCI algorithm is the computation of $\sum_{v \in N(X)} W(v, X)$, needed to evaluate the cohesion of itemset X. Computing this at each node in the search tree would be unfeasible, but it has been shown that such a sum for an itemset X can be expressed as a sum of separate sums of such distances for each pair of items individually [6]. FCI stores these sums between individual items in a matrix which is generated in the beginning of the algorithm. Since we are only interested in itemsets consisting of frequent items, the matrix will only contain the minimal distances between each pair of frequent items. Therefore, the matrix that will be generated will only contain $|F| \times |F|$ sums of smallest distances, where F are the frequent items.

On top of being used in the generation of the frequent cohesive itemsets, this matrix can help us easily compute the confidence of association rules $X \Rightarrow Y$. More formally, for each $x \in X$, we can find the sum of smallest distances to each $y \in Y$ seperately. This would allow us to efficiently compute $\overline{W}(X, Y)$, and thus $c(X \Rightarrow Y)$. However, we can optimise this process further, taking advantage of the fact that it is sufficient to limit our computations to rules of the form $x \Rightarrow X \setminus \{x\}$ (i.e., rules where the body consists of a single item), with $x \in X$, which can be generated from an itemset X. To compute the confidence of all other rules, we first note that

$$\sum_{v \in N(X)} W(v, X \cup Y) = \sum_{x \in X} \left(\sum_{v \in N(\{x\})} W(v, X \cup Y) \right).$$

The last part of the above expression can be rewritten as

$$\sum_{v \in N(\{x\})} W(v, X \cup Y) = \overline{W}(x, X \cup Y \setminus \{x\}) |N(\{x\})|,$$

which allows us to reformulate the formula for the average sum of minimal distances as

$$\overline{W}(X, Y) = \frac{\sum_{x \in X} \left(\overline{W}(x, X \cup Y \setminus \{x\}) |N(\{x\})| \right)}{|N(X)|}.$$

This, in turn, implies that

$$c(X \Rightarrow Y) = \frac{(|X \cup Y| - 1)|N(X)|}{\sum_{x \in X} \left(\overline{W}(x, X \cup Y \setminus \{x\})|N(\{x\})| \right)}.$$

For the case where X contains just one element, i.e., $X = \{x\}$, this gives us

$$c(x \Rightarrow Y \cup X \setminus \{x\}) = \frac{|X \cup Y| - 1}{\overline{W}(x, Y \cup X \setminus \{x\})},$$

and it follows that

$$c(X \Rightarrow Y) = \frac{|N(X)|}{\sum_{x \in X} \frac{|N(\{x\})|}{c(x \Rightarrow X \cup Y \setminus \{x\})}}. \tag{2}$$

As a result, once we have computed the confidence of all the rules of the form $x \Rightarrow X \setminus \{x\}$, with $x \in X$, we can evaluate all other rules $Y \Rightarrow X \setminus Y$, with $Y \subset X$, without looking at either the dataset or the distance matrix.

Let us now return to our example graph shown in Figure 1 and compute the confidence of rule $ab \Rightarrow c$, where abc is a frequent cohesive itemset. First we will compute the confidence of rules $a \Rightarrow bc$ and $b \Rightarrow ac$. We find that $\sum_{v \in N(\{a\})} W(v, abc) = 7$ and $\sum_{v \in N(\{b\})} W(v, abc) = 27$. Consequently, $c(a \Rightarrow bc) = \frac{3-1}{\frac{7}{3}} \approx 0.85$ and $c(b \Rightarrow ac) = \frac{3-1}{\frac{27}{7}} \approx 0.51$. Following Equation 2, we can compute $c(ab \Rightarrow c)$ as $\frac{10}{\frac{3}{0.85} + \frac{7}{0.51}} \approx 0.58$. If we use Equation 1 defined in section 3 for computing the confidence of the same rule, we get $c(ab \Rightarrow c) = \frac{3-1}{\frac{34}{10}} \approx 0.58$, which shows us that Equations 1 and 2 are equivalent. For reasons explained above, we will use Equation 2 in our algorithm.

The algorithm for generating the confident association rules is given in Algorithm 2. Having found a frequent cohesive itemset X, we first generate rules of the form $x \Rightarrow X \setminus \{x\}$, store their confidence values in memory and send all confident rules to the output (lines 1 - 4). We then generate all other rules, compute their confidence based on the stored confidences computed during the first stage, and output all confident rules (lines 5 - 7).

Algorithm 2. GENERATE_RULES(X) generates rules based on itemset X

1: **for all** $x \in X$ **do**
2:     **compute and store** $c(x \Rightarrow X \setminus \{x\})$
3:     **if** $c(x \Rightarrow X \setminus \{x\}) \geq min\_conf$ **then output** $x \Rightarrow X \setminus \{x\}$
4: **end for**
5: **for all** $Y \subset X$ with $|Y| \geq 2$ **do**
6:     **if** $c(Y \Rightarrow X \setminus Y) \geq min\_conf$ **then output** $Y \Rightarrow X \setminus Y$
7: **end for**

---

5 Experiments

In this section, we present the experimental results of our association rule mining algorithm. For our experiments, we used two different biological graph datasets. The first dataset concerns the human interactome. In this graph, each node is a protein and each edge is a direct interaction between two proteins. This network was obtained by fusing the BioGRID [15] database with the IntAct [11] and MINT [13] networks. Next, the protein nodes were labelled with the annotations contained in InterPro [7], which concerns common domains, protein families and active sites that are present in the corresponding protein. Note that each protein can contain multiple domains and can thus have more than one annotation label. The second dataset concerns the transcriptional regulatory network of Saccharomyces cerevisiae, commonly known as baker's yeast, which serves as an important model organism in life sciences. The yeast regulatory network was obtained from the YEASTRACT [1] database. Within this graph, each node is a yeast gene, and each edge corresponds to a regulatory interaction between a transcription factor gene and a target gene. Each node was labelled with gene ontology annotation information with regards to the biological process, molecular function and cellular location as obtained from UniProtKB [8]. As a result, here, too, each node can have multiple labels.

Since both datasets consist of nodes that can carry multiple labels, we first had to transform the given graphs. More formally, we expanded the given graph structure by replacing each node with a so-called dummy node carrying a unique label, and then connecting this node to a set of new nodes, each carrying one of the original labels. Note that due to the nature of this reconstruction, any two labels in the resulting graph will be at a distance of at least 2 from each other, as there will always be a dummy node in between. As a result, the maximal cohesion for any itemset in such a graph will be 0.5, and the same holds for the maximal confidence for any possible association rule. For the human interactome graph dataset, the transformed graph consisted of 64 090 nodes and 141 828 edges. The yeast regulatory network was transformed into a graph consisting of 21 314 nodes and 62 991 edges.

For all the experiments discussed below, we chose a frequency threshold that was low enough to guarantee a large number of frequent items, since we were only interested in the performance of our association rule miner. For this reason, we do not include the time needed to set up the distance matrix in our analysis, since this only needs to be done once, after which the matrix can be reused for various cohesion and confidence thresholds. The frequency threshold, *minsize* and *maxsize* were set to 0.002, 2 and 7, respectively, for the human interactome network, and to 0.001, 2 and 5, respectively, for the yeast network. In our first set of experiments, we also fixed the cohesion threshold in order to evaluate the effect of varying the confidence threshold. For the human interactome network we used $min\_coh = 0.2$, and for the yeast network $min\_coh = 0.3$. Figure 2 shows, for both datasets, the number of discovered rules and the runtimes needed to generate these rules for varying min_conf values. As expected, the number of discovered rules grows as we lower the confidence threshold, but the runtimes

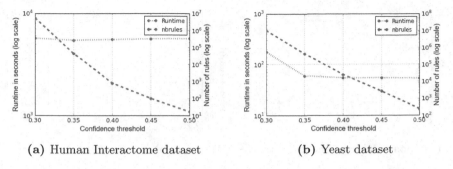

(a) Human Interactome dataset

(b) Yeast dataset

Fig. 2. Runtime and output size for varying confidence thresholds

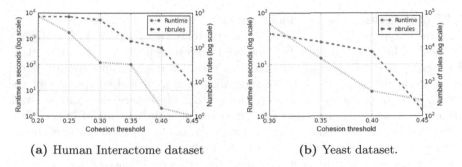

(a) Human Interactome dataset

(b) Yeast dataset.

Fig. 3. Runtime and output size for varying cohesion thresholds

stay stable, since most time is spent mining itemsets and generating candidate rules, regardless of how many rules are actually found to be confident.

In the second set of experiments, we only vary the cohesion threshold, while we set the confidence threshold to 0.4 for both datasets. The results are shown in Figure 3 and confirm that varying the cohesion threshold does have an effect on runtimes. Naturally, the number of rules decreases as the cohesion threshold increases, since the number of frequent cohesive itemsets on which we base the rules also decreases.

Note that the cohesion of an itemset can be interpreted as a weighted average of the confidences of all association rules that can be generated from that itemset. In our third set of experiments, we varied both the cohesion and confidence thresholds, by setting $min\_conf = min\_coh$, thus finding those rules that had an above average confidence within the set of rules originating from the same itemset. The results are shown in Figure 4 and are similar to those given in Figure 3. Naturally, the number of rules now decreases faster, as the confidence threshold is raised together with the cohesion threshold. Finally, we note that in all experiments we managed to find a large number of confident association rules quickly and efficiently.

Let us now have a closer look at some of the discovered rules. Having shown the output to domain experts, it appeared that, for the human protein-protein interaction network, the majority of the discovered rules concern protein kinases.

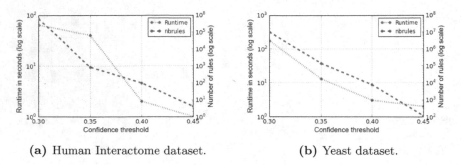

(a) Human Interactome dataset.   (b) Yeast dataset.

Fig. 4. Runtime and output size for varying cohesion and confidence thresholds, where $min\_conf = min\_coh$

99.4% of the rules included an annotation with the term 'kinase'. Kinase proteins are enzymes that can modify other proteins by phosphorylating specific residues on the target protein. Phosphorylation results in the addition of a phosphate group to the target protein. This is a common mechanism for many signal transduction pathways in all living cells, and thus a critical component of the protein-protein interaction network subjected to analysis here. For example, with min_coh and min_conf both set to 0.3 one example rule, existing of items which occur often in data but which form a non-trivial set, is *Tyrosine-protein kinase, catalytic domain* ⇒ *Protein kinase, ATP binding site, Serine-threonine/tyrosine-protein kinase catalytic domain, SH2 domain* with a confidence of 0.42. Proteins that contain a catalytic domain for tyrosine-protein kinase activity will phosphorylate a tyrosine residue on their target. The phosphorylation reaction typically requires ATP, and thus kinases will often co-occur with ATP-binding sites within the network. The annotation of 'Serine-threonine/tyrosine-protein kinase catalytic domain' is a higher level one that contains all instances of 'Tyrosine-protein kinase, catalytic domain' and thus is redundant with this annotation. The presence of the 'SH2 domain' in the head of the association rule is much more interesting. While the SH2 domain is not directly related to protein kinase activity itself, it allows a protein to dock phosphorylated tyrosine residues.

In the yeast dataset, with min_coh and min_conf set to 0.3, we found rule *heterocycle catabolic process* ⇒ *cellular nitrogen compound catabolic process*, with a confidence of 0.5. Genes annotated with 'heterocycle catabolic process' code for proteins that catalyse reactions for breaking down heterocyclic compounds. Heterocyclic molecules contain ring structures that consist of at least two types of atoms. The high confidence of this rule suggests that all genes involved in the degradation of heterocycles are also associated with 'cellular nitrogen compound catabolic process' in our network. This is to be expected as almost all heterocycles in living organisms consist of carbon and nitrogen atoms. Thus any gene involved in the breakdown of heterocycles will also be associated in the breakdown of compounds that include carbon and nitrogen. Another logical rule is *sulfur compound biosynthetic process* ⇒ *cellular nitrogen compound biosynthetic*, with a confidence of 0.48. This association rule suggests that there is a direct link between genes involved in the synthesis of compounds containing sulfur and genes involved in the synthesis of nitrogen compounds. This can be explained by the fact that the primary carriers of sulfur in yeast are the amino acids methionine and cysteine, both of which also contain nitrogen. We conclude that our algorithm discovers both expected and interesting rules, but produces no spurious output.

6   Conclusions

In this work, we presented a novel method for mining association rules among node labels in a graph dataset. We relax the structural constraint used in subgraph mining, and discover rules that consist of sets of labels on both sides. A discovered rule $X \Rightarrow Y$ tells us that if we encounter all the labels in X in a tightly connected form somewhere in the input graph, there is a high probability that all the labels in Y will be encountered nearby. We evaluate the rules by looking at how far from itemset X, on average, do we need to look in order to find itemset Y. This approach provides more insight into the data than merely mining itemsets or subgraphs. We developed an algorithm for mining association rules, and experimentally confirmed its efficiency and usefulness. In future work, we intend to look at the possibility of extending this work to a setting where the dataset consists of multiple graphs, rather than a single graph.

Acknowledgments. This work was supported by the Fund for Scientific Research–Flanders (FWO- Vlaanderen) projects "Evolving graph patterns" and "Data mining for privacy in social networks".

References

1. Abdulrehman, D., Monteiro, P.T., Teixeira, M.C., Mira, N.P., Lourenço, A.B., dos Santos, S.C., Cabrito, T.R., Francisco, A.P., Madeira, S.C., Aires, R.S., Oliveira, A.L., Sá-Correia, I., Freitas, A.T.: YEASTRACT: providing a programmatic access to curated transcriptional regulatory associations in Saccharomyces cerevisiae through a web services interface. Nucleic Acids Research **39** (2011)
2. Agrawal, R., Imieliński, T., Swami, A.: Mining association rules between sets of items in large databases. In: Proc. of the ACM SIGMOD Int. Conf. on Managemant of Data, pp. 207–216 (1993)
3. Cook, D.J., Holder, L.B.: Substructure discovery using minimum description length and background knowledge. Journal of Artificial Intelligence Research **1**, 231–255 (1994)
4. Cule, B., Goethals, B.: Mining association rules in long sequences. In: Zaki, M.J., Yu, J.X., Ravindran, B., Pudi, V. (eds.) PAKDD 2010, Part I. LNCS, vol. 6118, pp. 300–309. Springer, Heidelberg (2010)
5. Dehaspe, L., Toivonen, H.: Discovery of frequent datalog patterns. Data Mining and Knowledge Discovery **3**, 7–36 (1999)
6. Hendrickx, T., Cule, B., Goethals, B.: Mining cohesive itemsets in graphs. In: Džeroski, S., Panov, P., Kocev, D., Todorovski, L. (eds.) DS 2014. LNCS, vol. 8777, pp. 111–122. Springer, Heidelberg (2014)

7. Hunter, S., Jones, P., Mitchell, A., Apweiler, R., Attwood, T.K., Bateman, A., Bernard, T., Binns, D., Bork, P., Burge, S., de Castro, E., Coggill, P., Corbett, M., Das, U., Daugherty, L., Duquenne, L., Finn, R.D., Fraser, M., Gough, J., Haft, D., Hulo, N., Kahn, D., Kelly, E., Letunic, I., Lonsdale, D., Lopez, R., Madera, M., Maslen, J., McAnulla, C., McDowall, J., McMenamin, C., Mi, H., Mutowo-Muellenet, P., Mulder, N., Natale, D., Orengo, C., Pesseat, S., Punta, M., Quinn, A.F., Rivoire, C., Sangrador-Vegas, A., Selengut, J.D., Sigrist, C.J.A., Scheremetjew, M., Tate, J., Thimmajanarthanan, M., Thomas, P.D., Wu, C.H., Yeats, C., Yong, S.Y.: InterPro in 2011: new developments in the family and domain prediction database. Nucleic Acids Research **40**(D1), D306–D312 (2012)

8. Huntley, R.P., Sawford, T., Mutowo-Meullenet, P., Shypitsyna, A., Bonilla, C., Martin, M.J., O'Donovan, C.: The goa database: gene ontology annotation updates for 2015. Nucleic Acids Research p. gku1113 (2014)

9. Inokuchi, A., Washio, T., Motoda, H.: Complete mining of frequent patterns from graphs: Mining graph data. Machine Learning **50**(3), 321–354 (2003)

10. Inokuchi, A., Washio, T., Motoda, H., Kumasawa, K., Arai, N.: Basket analysis for graph structured data. In: Zhong, N., Zhou, L. (eds.) PAKDD 1999. LNCS (LNAI), vol. 1574, pp. 420–432. Springer, Heidelberg (1999)

11. Kerrien, S., Aranda, B., Breuza, L., Bridge, A., Broackes-Carter, F., Chen, C., Duesbury, M., Dumousseau, M., Feuermann, M., Hinz, U., Jandrasits, C., Jimenez, R.C., Khadake, J., Mahadevan, U., Masson, P., Pedruzzi, I., Pfeiffenberger, E., Porras, P., Raghunath, A., Roechert, B., Orchard, S., Hermjakob, H.: The IntAct molecular interaction database in 2012. Nucleic Acids Research **40**, D841–D846 (2012)

12. Khan, A., Yan, X., Wu, K.L.: Towards proximity pattern mining in large graphs. In: Proc. of the 2010 ACM SIGMOD Int. Conf. on Management of Data, pp. 867–878 (2010)

13. Licata, L., Briganti, L., Peluso, D., Perfetto, L., Iannuccelli, M., Galeota, E., Sacco, F., Palma, A., Nardozza, A.P., Santonico, E., Castagnoli, L., Cesareni, G.: MINT, the molecular interaction database: 2012 update. Nucleic Acids Research **40**(D1), D857–D861 (2012)

14. Nijssen, S., Kok, J.: The gaston tool for frequent subgraph mining. Electronic Notes in Theoretical Computer Science **127**, 77–87 (2005)

15. Stark, C., Breitkreutz, B.J., Chatr-aryamontri, A., Boucher, L., Oughtred, R., Livstone, M.S., Nixon, J., Auken, K.V., Wang, X., Shi, X., Reguly, T., Rust, J.M., Winter, A., Dolinski, K., Tyers, M.: The BioGRID interaction database: 2011 update. Nucleic Acids Research p. gkq1116, November 2010

16. Washio, T., Motoda, H.: State of the art of graph-based data mining. ACM SIGKDD Explorations Newsletter **5**, 59–68 (2003)

17. Yan, X., Han, J.: gspan: Graph-based substructure pattern mining. In: Proc. of the 2002 IEEE Int. Conf. on Data Mining, pp. 721–724 (2002)

18. Yoshida, K., Motoda, H., Indurkhya, N.: Graph-based induction as a unified learning framework. Journal of Applied Intelligence **4**, 297–316 (1994)

Mining High Utility Itemsets in Big Data

Ying Chun Lin[1](✉), Cheng-Wei Wu[2], and Vincent S. Tseng[2]

[1] Department of Computer Science and Information Engineering,
National Cheng Kung University, Tainan, Taiwan, Republic of China
`yclin@idb.csie.ncku.edu.tw`
[2] Department of Computer Science, National Chiao Tung University,
Tainan, Taiwan, Republic of China
`silvemoonfox@hotmail.com, vtseng@cs.nctu.edu.tw`

Abstract. In recent years, extensive studies have been conducted on *high utility itemsets (HUI)* mining with wide applications. However, most of them assume that data are stored in centralized databases with a single machine performing the mining tasks. Consequently, existing algorithms cannot be applied to the big data environments, where data are often distributed and too large to be dealt with by a single machine. To address this issue, we propose a new framework for *mining high utility itemsets in big data*. A novel algorithm named *PHUI-Growth (Parallel mining High Utility Itemsets by pattern-Growth)* is proposed for parallel mining HUIs on Hadoop platform, which inherits several nice properties of Hadoop, including easy deployment, fault recovery, low communication overheads and high scalability. Moreover, it adopts the MapReduce architecture to partition the whole mining tasks into smaller independent subtasks and uses Hadoop distributed file system to manage distributed data so that it allows to parallel discover HUIs from distributed data across multiple commodity computers in a reliable, fault tolerance manner. Experimental results on both synthetic and real datasets show that *PHUI-Growth* has high performance on large-scale datasets and outperforms state-of-the-art non-parallel type of HUI mining algorithms.

Keywords: High utility itemset mining · Big data analytics · Hadoop platform

1 Introduction

The rapid growth of data generated and stored has led us to the new era of Big Data [3, 4, 14, 18, 19]. Nowadays, we are surrounded by different types of big data, such as enterprise data, sensor data, machine-generated data and social data. Extracting valuable information and insightful knowledge from big data has become an urgent need in many disciplines. In view of this, *big data analytics* [3, 4, 14, 18, 19] has emerged as a novel topic in recent years. This technology is particularly important to enterprises and business organizations because it can help them to increase revenues, retain customers and make more intelligent decisions. Due to its high impact in many areas, more and more systems and analytical tools have been developed for big data analytics, such as *Apache Mahout* [14], *MOA* [3], *SAMOA* [19] and *Vowpal Wabbit* [20]. However, to the best of our knowledge, no existing studies have incorporated the concept of *utility mining* [2, 6, 7, 8, 11, 12, 13] into big data analytics.

© Springer International Publishing Switzerland 2015
T. Cao et al. (Eds.): PAKDD 2015, Part II, LNAI 9078, pp. 649–661, 2015.
DOI: 10.1007/978-3-319-18032-8_51

Utility mining is an important research topic in data mining. The main objective of utility mining is to extract valuable and useful information from data by considering profit, quantity, cost or other user preferences. *High utility itemset (HUI) mining* is one of the most important tasks in utility mining, which can be used to discover sets of items carrying high utilities (e.g., high profits). This technology has been applied to many applications such as *market analysis, web mining, mobile computing* and even *bioinformatics*. Due to its wide range of applications, many studies [2, 6, 7, 8, 11, 12, 13] have been proposed for mining HUIs in databases. However, most of them assume that data are stored in centralized databases with a single machine performing the mining tasks. However, in big data environments, data may be originated from different sources and highly distributed. A large volume of data also makes it difficult to be moved to a centralized database. Thus, existing algorithms are not suitable for the applications of big data.□

Although mining HUIs from big data is very desirable for many applications, it is a challenging task due to the following problems posed: First, due to a large amount of transactions and varied items in big data, it would face the large search space and the combination explosion problem. This leads the mining task to suffer from very expensive computational costs in practical. Second, pruning the search space in HUI mining is more difficult than that in frequent pattern mining because the *downward closure property* [1] does not hold for the utility of itemsets. Therefore, many search space pruning techniques developed for frequent pattern mining cannot be directly transferred to the scenario of HUI mining. Third, a large amount of data cannot be efficiently processed by a single machine. A well-designed algorithm incorporated with parallel programming architecture is needed. However, implementing a parallel algorithm involves several problematic issues, such as search space decomposition, avoidance of duplicating works, minimization of synchronization and communication overheads, fault tolerance and scalability problems.

In this paper, we address all of the above challenges by proposing a new framework for *mining high utility itemsets in big data*. To our knowledge, this topic has not yet been explored. The contributions of this work are summarized as follows:

- First, we propose a novel algorithm named *PHUI-Growth* (*Parallel mining High Utility Itemsets by pattern-Growth*) for parallel mining HUIs in big data. It is implemented on a Hadoop platform [14] and thus it inherits several nice properties from Hadoop, such as easy deployment in high level language, fault tolerance, low communication overheads and high scalability on commodity hardware.
- Second, PHUI-Growth adopts the MapReduce architecture to partition the whole mining task into smaller independent subtasks and uses *HDFS (Hadoop Distributed File System)* to process distributed data. Thus, it can parallel mine HUIs from distributed databases across multiple commodity computers in a reliable manner.
- Third, PHUI-Growth adopts a novel strategy called *DLU-MR (Discarding local unpromising items in MapReduce framework)* to effectively prune the search space and unnecessary intermediate itemsets produced during the mining process, which further enhances the performance of PHUI-Growth.

- Experimental results on both synthetic and real datasets show that PHUI-Growth outperforms the state-of-the-art algorithms developed for mining HUIs on a single machine and that it has good scalability on large-scale datasets.

The remaining of this paper is organized as follows. Section 2 and Section 3 respectively introduce the basic concepts of HUI mining and related works. Section 4 presents the proposed methods. Experimental results and conclusion are presented in Section 5 and Section 6, respectively.

2   Basic Concept and Definitions

In this section, we introduce the related definitions about HUI mining. Let $I^* = \{I_1, I_2, ..., I_N\}$ be a finite set of distinct *items*. A *transactional database* $D = \{T_1, T_2, ..., T_M\}$ is a set of transactions, where each transaction $T_c \in D$ ($1 \leq c \leq M$) is a set of items has a unique transaction identifier c, called its *TID*. Each item $I_j \in I^*$ ($1 \leq j \leq N$) in D has a global positive real number $p(I_j, D)$, called its *external utility* (e.g., unit profit). The external utilities of items are stored in a *utility table*. Every item I_j in a transaction T_c has a local positive real number $q(I_j, T_c)$, called its *internal utility* (e.g., quantity). An *itemset X* is a set of items $\{I_1, I_2,..., I_k\}$, where $k=|X|$ is called *the length of X*. An itemset of length k is called k-itemset.

Definition 1 (Utility of an itemset in a transaction). *The utility of an itemset X in a transaction $T_c \in D$ is* denoted as $u(X, T_c)$ and defined as $\sum_{I_j \in X \wedge I_j \in T_c} p(I_j, D) \times q(I_j, T_c)$.

Definition 2 (Utility of an itemset in a database). *The utility of an itemset X in a database D is the summation of the utilities of X in all the transactions containing X,* which is denoted as $u(X)$ and defined as $\sum_{T_c \in D \wedge X \subseteq T_c} u(X, T_c)$.

Definition 3 (Transaction utility of a transaction). *The transaction utility (abbreviated as TU) of a transaction $T_c \in D$ is the summation of the utility of each item in T_c,* which is denoted as $tu(T_c)$ and defined as $\sum_{T_c \in D} u(T_c, T_c)$.

Definition 4 (Total utility of a database). *The total utility of a database D is the* summation of the transaction utility of each transaction in the database, which is denoted as λ and defined as $\sum_{T_c \in D} tu(T_c)$.

Table 1. Transactional database

TID	Transaction
T_1	A(4), B(2), C(8), D(2)
T_2	A(4), B(2), C(8)
T_3	C(4), D(2), E(2), F(2)
T_4	E(2), F(2), G(1)

Table 2. Utility table

Item	A	B	C	D	E	F	G
External Utility	2	3	1	3	4	4	8

Definition 5 (Relative utility of an itemset in a database). *The relative utility of an itemset X in a database D is denoted as ru(X) and defined as the ratio of u(X) to λ.*

Definition 6 (High utility itemset). *Let θ (0 < θ ≤ λ) be a user-specified minimum utility threshold. An itemset X is called a high utility itemset* (abbreviated as *HUI*) *iff* $u(X) \geq \theta$. An equivalent definition is that X is a HUI iff $ru(X) \geq \theta/\lambda$, where θ/λ is called *relative minimum utility threshold*. Otherwise, X is called *low utility itemset.*

Notice that the well-known *downward closure property* [1] does not held for the utility of itemsets. For example, {A} is low utility, but its superset {AC} is high utility. As a consequence, the search space of HUI mining cannot be effectively pruned as it is done in traditional frequent itemset mining. To effectively prune the search space, the concept of *transaction-weighted utilization model* (abbreviated as *TWU model*) [6] was proposed, which is based on the following definitions.

Definition 7 (TWU of an itemset). The *transaction-weighted utilization* (*abbreviated as TWU*) *of an itemset X* is the summation of transaction utility of each transaction containing X, which is denoted as *TWU(X)* and defined as $\sum_{T_c \in D \wedge X \subseteq T_c} tu(T_c)$.

Definition 8 (High TWU itemset). An itemset X is called *high TWU* itemset iff $TWU(X) \geq \theta$. Otherwise, X is called *low TWU itemset.*

Definition 9 (TWU downward closure property). *The TWU downward closure property* states that any superset of a low TWU itemset is low utility. By this property, the *downward closure property* can be applied to the TWU of itemsets for effectively prune the search space. The detailed proof of this property can be found in [6].

3   Related Work

In this section, we review some studies that are related to parallel programming, HUI mining, and parallel HUI mining.

3.1   Parallel Programming

Parallel programming has become a necessity for handling big data. The parallel algorithms can be generally categorized into two types: *shared-memory algorithm* and *shared-nothing algorithm* (also called *distributed algorithm*) [14]. The main feature of shared-memory algorithms is that it allows all processing units to concurrently access a shared memory. In general, it is easier to adapt algorithms to the shared-memory architecture. However, the resulting algorithms are usually not scalable enough and still suffer from the bottleneck of huge memory requirement. On the other hand, the main feature of shared-nothing algorithms is that it allows different processors that have their own memories to communicate with each other by passing messages. Although it is not easy to adapt algorithms to the shared-nothing architecture, a well-designed distributed algorithm usually has better scalability.

The *message passing interface* (*MPI*) is one of the most well-known framework based on shared-nothing architecture, but it works efficiently only on low-level programming languages (e.g., *C* and *Fortran*). Nowadays, high-level programming languages (e.g., *Java*) have become more and more important and popular in many domains. *Apache Foundation* has developed a Jave-based open-source library named *Hadoop* [14] for parallel processing big data. It consists of two key services: *HDFS* (*Hadoop Distributed File System*) and *MapReduce software*. HDFS is a reliable distributed file system designed to efficiently access and store large-scale data. On the other hand, MapReduce software is designed to process vast amounts of data in parallel. The combination of HDFS and MapReduce software allows to parallel process large-scale datasets across multiple clusters of commodity hardware in a reliable, fault-tolerant manner. A MapReduce program consists of two stages: *map stage* and *reduce stage*. In map stage, each Mapper processes a distinct chunk of data and produces several key-value pairs. In reduce stage, these key-value pairs are aggregated and transformed. The transformed key-value pairs are fed to Reducers. Reducers further process these transformed pairs and then output the final or intermediate results.

3.2   High Utility Itemset Mining

Extensive studies have been proposed for efficiently mining HUIs in centralized databases. These algorithms can be generally categorized into two types: *two-phase* and *one-phase* algorithms. The main characteristic of two-phase algorithms is that they consist of two phases. In the first phase, they generate a set of candidates (e.g., high TWU itemsets) for HUIs. In the second phase, they calculate the utility of each candidate found in the first phase to identify HUIs. For example, *Two-Phase* [6], *IHUP* [2], *IIDS* [9] and *UP-Growth*$^+$ [12] are typical two-phase algorithms. On the contrary, the main feature of one-phase algorithms [7, 8] is that they discover HUI using only one phase. For example, *HUI-Miner* [7] is one of the state-of-the-art one-phase algorithms. Although these algorithms are very efficient for mining HUIs from a centralized database, they have not been parallelized for handling big data.

3.3   Parallel High Utility Itemset Mining

In utility mining, only few preliminary studies [11, 13] have been proposed for parallel mining HUIs in distributed databases. Vo et al. proposed the *DTWU-Mining* algorithm [13] for parallel mining HUIs from vertical partitioned distribute databases. Subramanian et al. proposed the *FUM-D* algorithm [11] to extract HUIs from distributed horizontal databases. FUM-D enumerates local HUIs in each local database and then uses local HUIs to infer all global HUIs in the global database. Although these two approaches are parallel HUI mining algorithms, they are not implemented in Hadoop platform and do not integrated with the MapReduce framework. Therefore, they do not support fault tolerance. However, fault tolerance is a very important issue in parallel mining algorithms because the probability that none of computers of cluster crashes is very small when handling big data. Therefore, DTWU-Mining and FUM-D are not reliable and practical enough for handling big data.

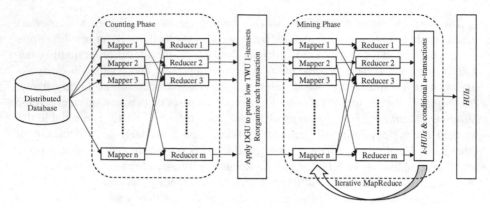

Fig. 1. The system architecture of the PHUI-Growth algorithm

4   The Proposed Method: PHUI-Growth

In this section, we propose a novel algorithm named *PHUI-Growth* (*Parallel mining High Utility Itemsets by pattern-Growth*) for efficiently parallel mining HUIs based on Hadoop MapReduce architecture. It has three input parameters: (1) a distributed database *DD*, (2) a utility table and (3) a user-specified minimum utility threshold *min_util*. After the whole mining process, the algorithm outputs the complete set of HUIs in   *DD*. PHUI-Growth consists of three main phases: (1) *counting phase*, (2) *transformation phase* and (3) *mining phase*. Fig. 1 shows the system architecture of PHUI-Growth.

4.1   Counting Phase

The input database *DD* can be viewed as a set of transactions that are stored in several computers. In counting phase, the algorithm takes one MapReduce pass to parallel counts TWU of items in *DD*. The whole process in this phase can be divided into map stage and reduce stage.

Map Stage. In map stage phase, each Mapper is fed with a transaction $T_c = \{I_1, I_2,..., I_L\}$ in *DD*. For each item I_j in T_c $(1 \leq j \leq L)$, the Mapper outputs a key-value pair $<I_j, TU(T_c)>$, called *Item-TU pair*.

Reduce Stage. In reduce stage, Item-TU pairs outputted by Mappers are fed to Reducers.   The Item-TU pairs having the same key are collected into the same Reducer. Let $R = \{<I, v_1>, <I, v_2>,..., <I, v_n>\}$ be the set of all the Item-TU pairs collected by a Reducer. The Reducer calculates TWU of *I* by summing up each value of a pair and outputs a key-value pair $<I, TWU(I)>$, called *Item-TWU pair*.

Fig. 1 shows the process of map and reduce stages in counting phase. In Fig.2(a), the input of the Mapper 1 is $T_1 = \{A, B, C, D\}$ of Table 1. After the process, the Mapper outputs four Item-TU pairs $<\{A\}, 28>$, $<\{B\}, 28>$, $<\{C\}, 28>$ and $<\{D\}, 28>$. All the Item-TU pairs having the same key {A} are collected into the Reducer 1. After the process, Reducer 1 outputs an Item-TWU pair $<\{A\}, 50>$.

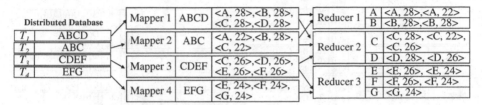

Fig. 2. Map and reduce stages in counting phase

4.2   Database Transformation Phase

In database transformation phase, the algorithm removes all low TWU 1-itemsets from DD (Definition 7, 8 and 9) and sorts remaining items in a TWU ascending order. A transaction after the above process is called *reorganized transaction*. Then, the algorithm transforms each reorganized transaction $T' = \{I_1, I_2,..., I_L\}$ in DD into a special structure called *u-transaction*. A *u-transaction* is of the form $<I_1(u_1), I_2(u_2),..., I_L(u_L)>$, where u_j is called *the utility of I_j in T'*. Initially, u_j is set to $p(I_j, DD) \times q(I_j, T')$. The *transaction utility of T'* is denoted as $TU(T')$ and defined as the summation of utilities of all the items in T'.

4.3   Mining Phase

In mining phase, the algorithm parallel discovers HUIs through several iterations. Initially, a variable k is set to 0. Then, the algorithm starts to generate HUIs having a length greater than k. In the k-th iteration, all the HUIs of length k are discovered by performing a MapReduce pass. The process of this phase can be divided into two cases: (1) $k = 1$ and (2) $k \geq 2$. We first explain the former case and then describe the latter case.

4.3.1   Map Stage in the First Iteration

When $k = 1$, each Mapper is fed with a *u-transaction* $T' = <I_1(u_1), I_2(u_2),..., I_L(u_L)>$. For each item I_j in T' ($1 \leq j \leq L$), the Mapper generates a special structure called *conditional u-transaction*. A conditional *u*-transaction has three fields: *Prefix, PrefixUtility* and *UTrans*. When $k = 1$, I_j and u_j are respectively stored in *Prefix* and *PrefixUtility* fields. For *UTrans* field, it stores the set of items appearing after I_j in T' according to TWU ascending order, that is, $<I_{j+1}(u_{j+1}), I_{j+2}(u_{j+2}),..., I_L(u_L)>$. The *prefix utility of T'* is denoted as $TU(T')$ and defined as the value in *PrefixUtility* field.

For example, in Fig. 3, the Mapper 1 is fed with the *u*-transaction {A(8), B(6), D(6), C(8)}. After the process, it generates four conditional *u*-transactions <{A}, 8, {B(6), D(6), C(8)}>, <{B}, 6, {D(6), C(8)}>, <{D}, 6, {C(8)}>, <{C}, 8, ϕ >. The PU of the first conditional *u*-transaction is 8.

Mapper 1	A(8) B(6) D(6) C(8)	<{A}, 8, {B(6)D(6)C(8)}> <{B}, 6, {D(6)C(8)}> <{D}, 6, {C(8)}> <{C}, 8, {φ}>
Mapper 2	A(8) B(6) C(8)	<{A}, 8, {B(6)C(8)}> <{B}, 6, {C(8)}> <{C}, 8, {φ}>
Mapper 3	E(8) F(8) D(6) C(4)	<{E}, 8, {F(8)D(6)C(4)}> <{F}, 8, {D(6)C(4)}> <{D}, 6, {C(4)}> <{C}, 4, {φ}>
Mapper 4	E(8) F(8)	<{E}, 8, {F(8)}> <{F}, 8, {φ}>

Reducer 1	A	<{A}, 8, {B(6)D(6)C(8)}>, <{A}, 8, {B(6)C(8)}>
	B	<{B}, 6, {D(6)C(8)}>, <{B}, 6, {C(8)}>
Reducer 2	C	<{C}, 8, {φ}>, <{C}, 8, {φ}>, <{C}, 4, {φ}>
	D	<{D}, 6, {C(8)}>, <{D}, 6, {C(4)}>
Reducer 3	E	<{E}, 8, {F(6)D(6)C(4)}> <{E}, 8, {F(8)}>
	F	<{F}, 8, {D(6)C(4)}>, <{F}, 8, {φ}>

Fig. 3. Map and reduce stages in mining phase when $k = 1$

Mapper 1	<{A},8,{B(6)D(6)C(8)}>	<{AB}, 14, {D(6)C(8)}> <{AD}, 14, {C(8)}> <{AC}, 16, {φ}>
	<{A},8,{B(6)C(8)}>	<{AB}, 14, {C(8)}> <{AC}, 16, {φ}>
Mapper 2	<{B}, 6, {D(6)C(8)}>	<{BD}, 12, {C(8)}> <{BC}, 14, {φ}>
	<{B}, 6, {C(8)}>	<{BC}, 14, {φ}>
Mapper 3	<{D}, 6, {C(8)}>	<{DC}, 14, {φ}>
	<{D}, 6, {C(4)}>	<{DC}, 10, {φ}>
Mapper 4	<{E}, 8, {F(6)D(6)C(4)}>	<{EF}, 14, {D(6)C(4)}> <{ED}, 14, {C(4)}> <{EC}, 12, {φ}>
	<{E}, 8, {F(8)}>	<{EF}, 16, {φ}>
Mapper 5	<{F}, 8, {D(6)C(4)}>	<{FD}, 14, {C(4)}> <{FC}, 12, {φ}>

Reducer 1	AB	<{AB}, 14, {D(6)C(8)}>, <{AB}, 14, {C(8)}>
	AC	<{AC}, 16, {φ}>, <{AC}, 16, {φ}>
	AD	<{AD}, 14, {C(8)}>
Reducer 2	BD	<{BD}, 12, {C(8)}>
	BC	<{BC}, 14, {φ}>, <{BC}, 14, {φ}>
	DC	<{DC}, 14, {φ}>, <{DC}, 10, {φ}>
Reducer 3	EF	<{EF}, 14, {D(6)C(4)}>, <{EF}, 16, {φ}>
	ED	<{ED}, 14, {C(4)}>, <{ED}, 14, {C(4)}>
	EC	<{EC}, 12, {φ}>
Reducer 4	FD	<{FD}, 14, {C(4)}>
	FC	<{FC}, 12, {φ}>

Fig. 4. Map and reduce stages in mining phase when $k = 2$

4.3.2   Reduce Stage in Mining Phase

In reduce stage, each Reducer is fed with a conditional u-transaction. The conditional u-transactions having the same prefix are collected into the same Reducer. Let R' be the set of conditional u-transactions collected by a Reducer and X be the prefix of these conditional u-transactions. Then, the Reducer calculates the utility of X by summing up PUs of all the u-transactions in R'. After the process, if the utility of X is no less than the min_util threshold, the Reducer outputs X and its utility because X is a HUI (Definition 6). Then, the algorithm applies a proposed strategy called *DLU-MR* (*Discarding local unpromising items in MapReduce framework*) to further reduce the search space. The main idea of this strategy is based on the following definitions.

Definition 10 (Local TWU of an item in k-th iteration). Let R' be the set of conditional u-transactions collected by a Reducer in k-th iteration, *the local TWU of an item Y* is the summation of transaction utility of each u-transaction containing X, which is denoted as $LTWU(X, k)$ and defined as $\sum_{T \in R \wedge Y \in T} [TU(T) + PU(T)]$.

Definition 11 (Local unpromising items in k-th iteration). Let R' be the set of conditional u-transactions collected by a Reducer in k-th iteration, an item is called *local unpromising items* in k-th iteration iff $LTWU(X, k) < \theta$.

Definition 12 (Local TWU downward closure property). *The local TWU downward closure property* (or simply called *LTWU-DC Property*) states that $\forall X$ iff $LTWU(X, k) < \theta$, all the L-itemsets containing X are low utility ($L \geq k$).

The DLU-MR strategy is performed by scanning R' twice. In the first scan of R', the Reducer calculates local TWU of items in u-transactions of R'. In the second scan of R', the Reducer removes local unpromising items in each u-transaction T' of R' and discards their utilities from $TU(T)$. When the Reducer has finished its work, each trimmed conditional u-transaction is outputted as the input of $(k+1)$-th iteration.

4.3.3   Map-Reduce Stages in the k-th Iteration $(k \geq 2)$

In the k-th iteration $(k \geq 2)$, the algorithm takes a MapReduce pass to find all the HUIs of length k. In map stage, each Mapper is fed with a conditional u-*transaction* $CT =$ $<X, u_x, T'>$ produced from $(k-1)$-th iteration, where X is the prefix of CT, u_x is the *PU* of CT and $T' = <I_1(u_1), I_2(u_2),\ldots, I_L(u_L)>$ is a u-transaction related to X. For each item I_j in T' $(1 \leq j \leq L)$, the Mapper outputs a conditional u-transaction $<\{X\cup I_j\}, (u_x+u_j),$ $<I_{j+1}(u_{j+1}), I_{j+2}(u_{j+2}),\ldots, I_L(u_L)>>$. When all the Mappers have finished their work, those conditional u-transactions having the same prefix are fed to the same Reducer. When $k \geq 2$, the process of the reduce stage is the same as that in subsection 4.3.2. Fig. 3 and Fig. 4 respectively show the running examples when $k =1$ and $k = 2$.

5   Experimental Results

In this section, we compare the performance of PHUI-Growth with HUI-Miner [7], a state-of-the-art non-parallel type of HUI mining algorithms. To evaluate the effectiveness of the DLU-MR strategy, we prepared two versions of PHUI-Growth, respectively called *PHUI-Growth(Baseline)* and *PHUI-Growth(DLU-MR)*. We also evaluate the number of intermediate itemsets produced by the algorithms. For HUI-Miner, intermediate itemsets refers to the itemsets having an estimated utility no less than the *min_util* threshold. Thus, the number of intermediate itemsets produced by HUI-Miner can be regarded as that of utility-lists constructed by HUI-Miner during the mining process. For the proposed algorithms, the number of intermediate itemsets is the number of conditional u-transactions produced by Reducers during the mining process. In this section, the two kinds of intermediate itemsets are called *candidates*. All experiments were conducted on a five-node Hadoop Cluster. Each node is equipped with Intel® Celeron® CPU G1610 @ 2.60GHz CPU and 4 GB main memory. All the algorithms are implemented in Java. Both synthetic and real datasets were used in the experiments. Chainstore [10] is a real-life dataset acquired from [10], which already contain unit profits and purchase quantities. Retail dataset was obtained from FIMI Repository [15]. A synthetic dataset T10I4N10K|D|2,000K was generated from the IBM data generator [1]. The parameters of the dataset are described as follows: |D| is the total number of transactions, T is the average size of transactions; N is the number of distinct items; I is the average size of potential maximal itemsets. Internal and external utilities of items are generated as the settings of [6, 12]. In Retail and T10I4N10K|D|2,000K datasets, external utilities of items are generated between 1 and 1,000 by using a log-normal distribution and internal utilities of items are generated randomly between 1 and 5, as the settings of [6, 12]. To evaluate the performance of the algorithms on a larger dataset, we duplicate each transaction in Chainstore five times to form a dataset named Chainstore×5. Table 3 shows characteristics of the datasets.

Table 3. Characteristic of Datasets

Dataset	#Trans.	# Items	Average Trans. Length	Maximum Trans. Length
Retail	88,162	16,470	10	76
Chainstore	1,112,949	46,086	7	170
T10I4N10K|D|2,000K	2,000,000	10,000	10	33
Chainstore×5	5,564,745	46,086	7	170

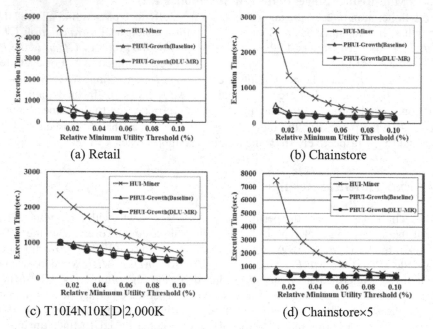

(a) Retail

(b) Chainstore

(c) T10I4N10K|D|2,000K

(d) Chainstore×5

Fig. 5. Execution time of the algorithms on different datasets

5.1 Performance Evaluation on Small Dataset

In this subsection, we evaluate the performance of the algorithms on Retail dataset under varied relative *min_util* thresholds. The execution time of the algorithms and the number of candidates are respectively shown in Fig. 5(a) and Fig 6(a). In Fig. 5(a), PHUI-Growth(Baseline) and Growth(DLU-MR) generally run slightly slower than HUI-Miner when relative *min_util* thresholds are higher than 0.02%. This is because that PHUI-Growth(Baseline) and Growth(DLU-MR) use five-node Hadoop Cluster to parallel process the mining tasks and they need to pass necessary data and messages across different machines via networks, which requires additional communication overheads. Therefore, they take more time than HUI-Miner for high thresholds. However, when the threshold decreases, HUI-Miner starts to suffer from long execution time. For example, for relative *min_util* = 0.01%, HUI-Miner takes 4,429 seconds, while PHUI-Growth(DLU-MR) only takes 566 seconds. When the threshold decreases, the number of HUIs dramatically increases and HUI-Miner need to produce a large amount of utility-lists for intermediate itemsets. However, the number of candidates produced by PHUI-Growth(DLU-MR) is up to two orders of magnitude smaller than that produced by HUI-Miner.

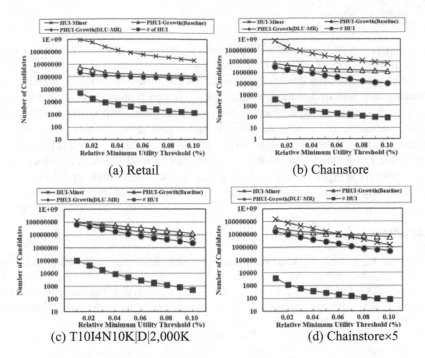

Fig. 6. Number of intermediate itemsets produced by the algorithms

5.2   Performance Evaluation on Large-Scale Datasets

In this subsection, we evaluate the performance of the algorithms on large datasets, including Chainstore, T10I4N10K|D|2000K and Chainstore×5. Execution time of the algorithms and the number of candidates are respectively shown in Fig.5 and Fig. 6. Results show that PHUI-Growth(Baseline) and Growth(DLU-MR) outperform HUI-Miner significantly. The reason why PHUI-Growth(Baseline) and Growth(DLU-MR) perform so well is that they effectively use nodes of a cluster to parallel process HUIs across multiple machines, while HUI-Miner is executed on non-parallel single machine. In Fig.5, PHUI-Growth(DLU-MR) generally runs much faster than PHUI-Growth(Baseline) on all the datasets. This is because that PHUI-Growth(DLU-MR) integrates the DLU-MR strategy for effectively prune the candidates and hence enhances its mining performance. Then, we compare the scalability of the algorithms on large datasets Chainstore×5 and T10I4N10K|D|2000K. As shown in Fig 5(c) and Fig. 5(d), PHUI-Growth(DLU-MR) has very good scalability on large datasets. On the contrary, the execution time of HUI-Miner increases dramatically on large datasets. In Fig 6(d), as the relative *min_util* is set to 0.01%, PHUI-Growth(DLU-MR) only takes about 592 seconds, while HUI-Miner takes more than 7,500 seconds.

6   Conclusion

In this paper, we propose a new framework for *mining high utility itemsets in big data*. A novel algorithm *PHUI-Growth* is proposed for efficiently parallel mining high utility itemsets from distributed data across multiple commodity computers. It is implemented on a shared-nothing Hadoop platform and thus inherits several merits of Hadoop, including easy deployment in high level language, supporting fault recovery and fault tolerance, low communication overheads and high scalability on commodity hardware. A novel strategy called *DLU-MR* is proposed to effectively prune the search space and greatly improve the performance of *PHUI-Growth*. Empirical evaluations of different types of real and synthetic datasets show that PHUI-Growth has good scalability on large datasets and outperforms the state-of-the-art algorithms.

References

1. Agrawal, R., Srikant, R.: A fast algorithms for mining association rules. In: Proceedings of VLDB, pp. 487–499 (1994)
2. Ahmed, C.F., Tanbeer, S.K., Jeong, B.-S., Lee, Y.-K.: Efficient Tree Structures for High-utility Pattern Mining in Incremental Databases. IEEE Trans. Knowl. Data Eng. **21**(12), 1708–1721 (2009)
3. Bifet, A., Holmes, G., Kirkby, R., Pfahringer, B.: MOA: Massive Online Analysis. JMLR (2010). http://moa.cms.waikato.ac.nz/
4. Fan, W., Bifet, A.: Mining Big Data: Current Status, and Forecast to the Future. SIGKDD Explorations **14**(2), 1–5 (2012)
5. Han, J., Pei, J., Yin., Y.: Mining frequent patterns without candidate generation. In: Proceedings of ACM SIGMOD, pp. 1–12 (2000)
6. Liu, Y., Liao, W., Choudhary, A.: Fast high-utility itemsets mining algorithm. In: Proceedings of UBDM (2005)
7. Liu, M., Qu, J.: Mining high utility itemsets without candidate generation. In: Proceedings of CIKM, pp. 55–64 (2012)
8. Liu, J., Wang, K., Fung, B.C.M.: Direct discovery of high utility itemsets without candidate generation. In: Proceedings of IEEE ICDM, pp. 984–989 (2012)
9. Li, Y.-C., Yeh, J.-S., Chang, C.-C.: Isolated Items Discarding Strategy for Discovering High-utility Itemsets. DKE **64**(1), 198–217 (2008)
10. Pisharath, J., Liu, Y., Ozisikyilmaz, B., Narayanan, R., Liao, W.K., Choudhary, A., Memik, G.: NU-MineBench version 2.0 dataset and technical report
11. Subramanian, K., Kandhasamy, P., Subramanian, S.: A Novel Approach to Extract High Utility Itemsets from Distributed Databases. Computing and Informatics **31**, 1597–1615 (2012)
12. Tseng, V.S., Shie, B.-E., Wu, C.-W., Yu, P.S.: Efficient Algorithms for Mining High Utility Itemsets from Transactional Databases. IEEE Trans. Knowl. Data Eng. **25**(8), 1772–1786 (2013)
13. Vo, B., Nguyen, H., Ho, T.B., Le, B.: Parallel method for mining high utility itemsets from vertically partitioned distributed databases. In: Velásquez, J.D., Rios, S.A., Howlett, R.J., Jain, L.C. (eds.) KES 2009, Part I. LNCS, vol. 5711, pp. 251–260. Springer, Heidelberg (2009)

14. Apache Hadoop. http://hadoop.apache.org
15. Frequent itemset mining implementations repository. http://fimi.cs.helsinki.fi/
16. Microsoft Corporation: Example database FoodMart of Microsoft SQL Server Analysis Server
17. SAMOA. http://samoa-project.net
18. Vowpal Wabbit. http://hunch.net/~vw/

Decomposition Based SAT Encodings
for Itemset Mining Problems

Said Jabbour, Lakhdar Sais[✉], and Yakoub Salhi

CRIL - CNRS, Université d'Artois, Rue Jean Souvraz, SP-18,
62307 Lens Cedex 3, France
{jabbour,sais,salhi}@cril.fr

Abstract. Recently, several constraint programming (CP)/propositional satisfiability (SAT) based encodings have been proposed to deal with various data mining problems including itemset and sequence mining problems. This research issue allows to model data mining problems in a declarative way, while exploiting efficient and generic solving techniques. In practice, for large datasets, they usually lead to constraints network/Boolean formulas of huge size. Space complexity is clearly identified as the main bottleneck behind the competitiveness of these new declarative and flexible models w.r.t. specialized data mining approaches. In this paper, we address this issue by considering SAT based encodings of itemset mining problems. By partitioning the transaction database, we propose a new encoding framework for SAT based itemset mining problems. Experimental results on several known datasets show significant improvements, up to several orders of magnitude.

Keywords: Declarative data mining · Itemset mining

1 Introduction

Recently, a constraint programming (CP) based data mining (DM) framework was proposed by Luc De Raedt et al. in [10] for itemset mining (CP4IM). This new framework offers a declarative and flexible representation model. New constraints often require new implementations in specialized approaches, while they can be easily integrated in such a CP framework. It allows data mining problems to benefit from several generic and efficient CP solving techniques. The authors show how some typical constraints (e.g. frequency, maximality, monotonicity) used in itemset mining can be formulated for use in CP [5]. This study leads to the first CP approach for itemset mining displaying nice declarative opportunities. Encouraged by these promising results, several contributions addressed other data mining problems using the two well-known AI models: constraint programming and propositional satisfiability. For example, in [7], the authors proposed a SAT based formulation of the problem of enumerating the Top-k frequent closed itemsets. In [2], the authors solve the frequent itemset mining problem by compiling the set of all itemset into a binary decision diagram (BDD)

© Springer International Publishing Switzerland 2015
T. Cao et al. (Eds.): PAKDD 2015, Part II, LNAI 9078, pp. 662–674, 2015.
DOI: 10.1007/978-3-319-18032-8_52

(augmented with counts). Frequent itemset are then extracted by querying the BDD. By considering the relationship between local constraint-based mining and constraint satisfaction problems, Khiari et al. [8] proposed a model for mining patterns combining several local constraints, i.e., patterns defined by n-ary constraints. Also, several constraint-based language for modeling and solving data mining problems have been designed. Let us mention, the constraint-based language, defined in [9], which enables the user to define queries in a declarative way addressing pattern sets and global patterns. All primitive constraints of the language are modeled and solved using the SAT framework. More recently, Guns at al. [3], introduced a general-purpose, declarative mining framework called MiningZinc. Compared to CP4IM framework [4], MiningZinc supports a wide variety of different solvers (including DM algorithms and general purpose solvers) and employe a significantly more high-level modeling language.

The above non exhaustive description of some recent works on constraint programming based data mining shows an important research activity in this new research trend. In most of these contributions, particularly for SAT based data mining framework, despite the nice declarative aspects, the authors pointed out that the main challenge concern their competitiveness against specialized data mining algorithms. As mentioned in [3], "this is non-trivial because data mining algorithms are highly optimized for specific tasks and large datasets, while generic constraint solvers may struggle in particular with the size of the problems". The work presented in this paper fit into this framework. Our goal is to provide a step in that direction by pushing forward the effectiveness of SAT-based data mining approaches.

In this paper, we consider the SAT-based encodings of itemset mining problems proposed in [7]. An enhancement of the encoding is provided. We propose an original partition based approach allowing us to partition the whole problem into sub-problems of reasonable size while maintaining solving incrementality. It takes as input a transaction database and a partition of the set of items, then it incrementally generates and solves a sequence of sub-problems while ensuring completeness. This partition based SAT encoding improves Jabbour et al. [7] SAT-based itemset mining approach by several orders of magnitude on several well-known datasets.

2   Preliminaries

Let Ω be a finite non empty set of symbols, called *items*. From now on, we assume that this set is fixed. We use the letters a, b, c, etc to range over the elements of Ω. An *itemset* I over Ω is defined as a subset of Ω, i.e., $I \subseteq \Omega$. We use 2^{Ω} to denote the set of itemsets over Ω and we use the capital letters I, J, K, etc to range over the elements of 2^{Ω}.

A *transaction* is an ordered pair (i, I) where i is a natural number, called *transaction identifier*, and I an itemset, i.e., $(i, I) \in \mathbb{N} \times 2^{\Omega}$. A *transaction database* $\mathcal{D}$ is defined as a finite non empty set of transactions $(\mathcal{D} \subseteq \mathbb{N} \times 2^{\Omega})$ where each transaction identifier refers to a unique itemset.

Let $\mathcal{D}$ be a transaction database and I an itemset. The *cover* of I in $\mathcal{D}$, denoted $\mathcal{C}(I, \mathcal{D})$, is the following set of transaction identifiers:

$$\{i \in \mathbb{N} \mid (i, J) \in \mathcal{D} \text{ and } I \subseteq J\}$$

Moreover, the *support* of I in $\mathcal{D}$, denoted $\mathcal{S}(I, \mathcal{D})$, corresponds to the cardinality of $\mathcal{C}(I, \mathcal{D})$, i.e., $\mathcal{S}(I, \mathcal{D}) = |\mathcal{C}(I, \mathcal{D})|$.

Table 1. A Transaction Database $\mathcal{D}$

Tid	Itemset
1	a, b, c, d
2	a, b, e, f
3	a, b, c
4	a, c, d, f
5	g
6	d
7	d, g

For instance, consider the transaction database $\mathcal{D}$ in Table 1. In this case, we have $\mathcal{C}(\{a, b\}, \mathcal{D}) = \{1, 2, 3\}$ and $\mathcal{S}(\{a, b\}, \mathcal{D}) = 3$ while $\mathcal{S}(\{f\}, \mathcal{D}) = 2$.

In this work, we are mainly interested in the problem of finding frequent itemsets (FIM problem). Given a transaction database $\mathcal{D}$ and a natural number n greater than 0, solving this problem consists in computing the following set of itemsets: $FIM(\mathcal{D}, n) = \{I \subseteq \Omega \mid \mathcal{S}(I, \mathcal{D}) \geqslant n\}$. In this context, we call n a *minimal support threshold*.

The number of frequent itemsets in a transaction database can be significant. Indeed, this problem is #P-hard [15]. The complexity class #P corresponds to the counting problems associated with a decision problem in NP. In order to partially face this problem, condensed representations have been proposed:

Definition 1 (Closed Frequent Itemsets). *Let $\mathcal{D}$ be a transaction database and n a minimal support threshold and I an itemset in $FIM(\mathcal{D}, n)$. The itemset I is closed iff, for all $J \supset I$, $\mathcal{S}(I, \mathcal{D}) > \mathcal{S}(J, \mathcal{D})$.*

Definition 2 (Maximal Frequent Itemsets). *Let $\mathcal{D}$ be a transaction database and n a minimal support threshold and I an itemset in $FIM(\mathcal{D}, n)$. The itemset I is maximal iff, for all $J \supset I$, $\mathcal{S}(J, \mathcal{D}) < n$.*

For instance, in the previous example, for $n = 2$, the sets of closed and maximal frequent itemsets are respectively $\{\{a, b\}, \{a, b, c\}, \{d\}, \{a, c, d\}, \{a, f\}, \{g\}\}$ and $\{\{a, b, c\}, \{a, c, d\}, \{a, f\}, \{g\}\}$. We use $CFIM(\mathcal{D}, n)$ (resp. $MFIM(\mathcal{D}, n)$) to denote the set of closed (resp. maximal) frequent itemsets.

3   Propositional Logic and SAT Problem

In this section, we define the syntax and the semantics of propositional logic. Let Prop be a countably set of propositional variables. We use the letters p,

q, r, etc to range over Prop. The set of *propositional formulæ*, denoted Form, is defined inductively started from Prop, the constant $\perp$ denoting false, the constant $\top$ denoting true, and using the logical connectives $\neg$, $\wedge$, $\vee$, $\rightarrow$. We use $\mathcal{P}(A)$ to denote the set of propositional variables appearing in the formula A. The equivalence connective $\leftrightarrow$ is defined by $A \leftrightarrow B \equiv (A \rightarrow B) \wedge (B \rightarrow A)$.

A *Boolean interpretation* $\mathcal{I}$ of a formula A is defined as a function from $\mathcal{P}(A)$ to $\{0, 1\}$ (0 corresponds to *false* and 1 to *true*). It is inductively extended to propositional formulæ as usual:

$$\mathcal{I}(\perp) = 0 \quad \mathcal{I}(\top) = 1 \quad \mathcal{I}(A \wedge B) = min(\mathcal{I}(A), \mathcal{I}(B))$$
$$\mathcal{I}(\neg A) = 1 - \mathcal{I}(A) \quad \mathcal{I}(A \vee B) = max(\mathcal{I}(A), \mathcal{I}(B))$$
$$\mathcal{I}(A \rightarrow B) = max(1 - \mathcal{I}(A), \mathcal{I}(B))$$

A *model* of a formula A is a Boolean interpretation $\mathcal{I}$ that satisfies A, i.e. $\mathcal{I}(A) = 1$. A formula A is satisfiable if there exists a model of A. A is *valid* or *a theorem*, if every Boolean interpretation is a model of A. We use $Mod(A)$ to denote the set of models of A.

Let us now define the *conjunctive normal form* (CNF) representation of propositional formulæ. A CNF formula is a conjunction ($\wedge$) of clauses, where a *clause* is a disjunction ($\vee$) of literals. A *literal* is a propositional variable (p) or a negated propositional variable ($\neg p$). The two literals p and $\neg p$ are called *complementary*. A CNF formula can also be seen as a set of clauses, and a clause as a set of literals. The size of the CNF formula A corresponds to the value $\sum_{c \in A} |c|$ where $|c|$ is the number of literals in the clause c. Let us mention that any propositional formula can be translated to a CNF formula equivalent w.r.t. satisfiability, using linear Tseitin's encoding [13]. The *SAT problem* consists in deciding if a given CNF formula admits a model or not.

4   SAT Encoding of Itemset Mining

In this section, we describe SAT encodings for itemset mining which are mainly based on the encodings proposed in [7]. In order to do this, we fix, without loss of generality, a transaction database $\mathcal{D} = \{(1, I_1), \ldots, (m, I_m)\}$ and a minimal support threshold n.

The SAT encoding of itemset mining that we consider is based on the use of propositional variables representing the items and the transaction identifiers in $\mathcal{D}$. More precisely, for each item a (resp. transaction identifier i), we associate a propositional variable, denoted p_a (resp. q_i). These propositional variables are used to capture all possible itemsets and their covers. Formally, given a model $\mathcal{I}$ of the considered encoding, the candidate itemset is $\{a \in \Omega \mid \mathcal{I}(p_a) = 1\}$ and its cover is $\{i \in \mathbb{N} \mid \mathcal{I}(q_i) = 1\}$.

The first propositional formula that we describe allows us to obtain the cover of the candidate itemset:

$$\bigwedge_{i=1}^{m} (\neg q_i \leftrightarrow \bigvee_{a \in \Omega \setminus I_i} p_a) \tag{1}$$

This formula expresses that q_i is true if and only if the candidate itemset is supported by the ith transaction. In other words, the candidate itemset is not supported by the ith transaction (q_i is false), when there exists an item a (p_a is true) that does not belong to the transaction ($a \in \Omega \setminus I_i$).

The following propositional formula allows us to consider the itemsets having a support greater than or equal to the minimal support threshold:

$$\sum_{i=1}^{m} q_i \geqslant n \tag{2}$$

This formula corresponds to 0/1 linear inequalities, usually called cardinality constraints. The first linear encoding of general 0/1 linear inequalities to CNF have been proposed by J. P. Warners in [14]. Several authors have addressed the issue of finding an efficient encoding of cardinality (e.g. [1,11,12]) as a CNF formula. Efficiency refers to both the compactness of the representation (size of the CNF formula) and to the ability to achieve the same level of constraint propagation (generalized arc consistency) on the CNF formula.

We use $\mathcal{E}_{FIM}(\mathcal{D}, n)$ to denote the encoding corresponding to the conjunction of the two formulæ (1) and (2).

Proposition 1 ([7]). *Let $\mathcal{D}$ be a transaction database and n a minimal support threshold. $\mathcal{I}$ is a model of $\mathcal{E}_{FIM}(\mathcal{D}, n)$ iff $I = \{a \in \Omega \mid \mathcal{I}(p_a) = 1\}$ is a frequent itemset where $\mathcal{C}(I, \mathcal{D}) = \{i \in \mathbb{N} \mid \mathcal{I}(q_i) = 1\}$.*

We now describe the propositional formula allowing to force the candidate itemset to be closed:

$$\bigwedge_{a \in \Omega} (\bigwedge_{i=1}^{m} q_i \rightarrow a \in I_i) \rightarrow p_a \tag{3}$$

This formula means that if we have $\mathcal{S}(I, \mathcal{D}) = \mathcal{S}(I \cup \{a\}, \mathcal{D})$ then $a \in I$ holds. This condition is necessary and sufficient to force the candidate itemset to be closed. Let us note that the expressions of the form $a \in I_i$ correspond to constants, i.e., $a \in I_i$ corresponds to $\top$ if the item a is in I_i, to $\bot$ otherwise.

Note that the formula (3) can be simply reformulated as a conjunction of clauses as follows:

$$\bigwedge_{a \in \Omega} ((\bigvee_{1 \leqslant i \leqslant m, a \notin I_i,} q_i) \vee p_a) \tag{4}$$

This reformulation is obtained using the equivalence $A \rightarrow B \equiv \neg A \vee B$.

For illustration purposes, and to show the generality of our proposed framework, we give below the encoding of the maximality constraint. Indeed, the following formula corresponds to a constraint forcing the candidate itemset to be maximal:

$$\bigwedge_{a \in \Omega} (\sum_{i=1}^{m} (q_i \wedge a \in I_i) \geqslant n) \rightarrow p_a \tag{5}$$

Since the expressions of the form $a \in I_i$ correspond to constants, the formulæ of the form $\sum_{i=1}^{m} (q_i \wedge a \in I_i) \geqslant n$ correspond to cardinality constraints. Indeed, for every propositional variable q_i, we have $q_i \wedge \bot = \bot$ and $q_i \wedge \top = q_i$. In the same way as in the case of (3), the formula (5) provide a necessary and sufficient condition to force the candidate itemset to be maximal, since we have, for every $a \in \Omega$, $\mathcal{S}(I \cup \{a\}, \mathcal{D}) \geqslant n$ implies $a \in I$. It is worth noticing that this formula is not provided in [7]. However, an equivalent constraint is provided in De Raedt et al's encoding proposed in [10].

We use $\mathcal{E}_{CFIM}(\mathcal{D}, n)$ (resp. $\mathcal{E}_{MFIM}(\mathcal{D}, n)$) to denote the encoding corresponding to the conjunction of the formulæ (1), (2) and (4) (resp. (1), (2) and (5)).

Proposition 2. *Let $\mathcal{D}$ be a transaction database and n a minimal support threshold. $\mathcal{I}$ is a model of $\mathcal{E}_{CFIM}(\mathcal{D}, n)$ iff $I = \{a \in \Omega \mid \mathcal{I}(p_a) = 1\}$ is a closed frequent itemset where $\mathcal{C}(I, \mathcal{D}) = \{i \in \mathbb{N} \mid \mathcal{I}(q_i) = 1\}$.*

Proposition 3. *Let $\mathcal{D}$ be a transaction database and n a minimal support threshold. $\mathcal{I}$ is a model of $\mathcal{E}_{MFIM}(\mathcal{D}, n)$ iff $I = \{a \in \Omega \mid \mathcal{I}(p_a) = 1\}$ is a maximal frequent itemset where $\mathcal{C}(I, \mathcal{D}) = \{i \in \mathbb{N} \mid \mathcal{I}(q_i) = 1\}$.*

5   A Partition Based Method

One of the significant problems of the declarative approaches in data mining is the large size of the encodings. Indeed, even if the sizes of the encodings is polynomial in the size of the input, this does not mean that these sizes are reasonable. In order to tackle this problem, we propose an approach which allows us to decompose the encodings described previously.

Let $\mathcal{D}$ be a transaction database and S an itemset. We use $\mathcal{D}_{|S}$ to denote the transaction database $\{(i, I) \in \mathcal{D} \mid I \cap S \neq \emptyset\}$. Let k be a natural number smaller than or equal to $|\Omega|$. A k-*partition* of Ω is a structure of the form $(\{S_1, \ldots, S_k\}, \prec)$ where $\{S_1, \ldots, S_k\}$ is a partition of Ω into k subsets and $\prec$ is an ordering on $\{S_1, \ldots, S_k\}$.

Let $\mathcal{D}$ be a transaction database, $\mathcal{P} = (W, \prec)$ a k-partition of Ω, $S \in W$ and n a minimal support threshold. We use $\Sigma_{DM}(\mathcal{D}, n, S, \mathcal{P})$ to denote the following propositional formula:

$$\mathcal{E}_{DM}(\mathcal{D}_{|S}, n) \wedge (\bigvee_{a \in S} p_a) \wedge (\bigwedge_{b \in \bigcup_{S' \prec S} S'} \neg p_b) \tag{6}$$

where $DM \in \{FIM, CFIM, MFIM\}$.

In formula 6, the subformula $(\bigvee_{a \in S} p_a)$ expresses that the itemsets must contains at least one item from S. The second subformula $(\bigwedge_{b \in \bigcup_{S' \prec S} S'} \neg p_b)$ avoid generating itemsets previously generated using the previous elements of the partition sequence.

Proposition 4. *Let $\mathcal{D}$ be a database, $\mathcal{P} = (W, \prec)$ a k-partition of Ω, $S \in W$ and n a minimal support threshold. $\mathcal{I}$ is a model of $\Sigma_{DM}(\mathcal{D}, n, S, \mathcal{P})$ iff the following properties are satisfied:*

1: **procedure** $SAT_{DM}^{\mathcal{P}}(\mathcal{D},n,\mathcal{P})$ ▷ $\mathcal{P} = (\{S_1,\ldots,S_k\},\prec)$
2:     $R \leftarrow \emptyset$
3:     **for** $i \leftarrow 1, k$ **do**   ▷ $S_1 \prec \cdots \prec S_k$
4:         $R \leftarrow R \cup enumModels(\Sigma_{DM}(\mathcal{D},n,S_i,\mathcal{P}))$
5:     **end for**
6:     **return** $itemsets(R)$
7: **end procedure**

Fig. 1. Algorithm $SAT_{DM}^{\mathcal{P}}$

(i) $I = \{a \in \Omega \mid \mathcal{I}(p_a) = 1\} \in DM(\mathcal{D}, n)$;
(ii) $\mathcal{C}(I, \mathcal{D}) = \{i \in \mathbb{N} \mid \mathcal{I}(q_i) = 1\}$;
(iii) $I \cap S \neq \emptyset$; and
(iv) $I \cap \bigcup_{S' \prec S} S' = \emptyset$.

Proof. Part $\Rightarrow$. Using Propositions 1, 2 and 3, we obtain the properties (i) and (ii). The properties (iii) and (iv) are directly obtained from the formulæ $\bigvee_{a \in S} p_a$ and $\bigwedge_{b \in \bigcup_{S' \prec S} S'} \neg p_b$ respectively.

Part $\Leftarrow$. Using the properties (i) and (ii), we obtain that $\mathcal{I}$ is a model of $\mathcal{E}_{DM}(\mathcal{D}_{|S}, n)$. This is a consequence of Propositions 1, 2 and 3. Since $I \cap S \neq \emptyset$ (resp. $I \cap \bigcup_{S' \prec S} S' = \emptyset$), we get that $\mathcal{I}$ is a model of $\bigvee_{a \in S} p_a$ (resp. $\bigwedge_{b \in \bigcup_{S' \prec S} S'} \neg p_b$).

Proposition 5. *Let $\mathcal{D}$ be a transaction database, $\mathcal{P} = (W, \prec)$ a k-partition of Ω, $S, S' \in W$ such that $S \neq S'$ and n a minimal support threshold. Then $Mod(\Sigma_{DM}(\mathcal{D}, n, S, \mathcal{P})) \cap Mod(\Sigma_{DM}(\mathcal{D}, n, S', \mathcal{P})) = \emptyset$ holds.*

Proof. Since $S \neq S'$, we have either $S \prec S'$ or $S' \prec S$. We here consider, without loss of generality, that $S \prec S'$. Since all models of $\Sigma_{DM}(\mathcal{D}, n, S', \mathcal{P})$ satisfy $\bigwedge_{a \in S} \neg p_a$ because of $S' \prec S$ and all models of $\Sigma_{DM}(\mathcal{D}, n, S, \mathcal{P})$ satisfy $\bigvee_{a \in S} p_a$, we deduce that $Mod(\Sigma_{DM}(\mathcal{D}, n, S, \mathcal{P})) \cap Mod(\Sigma_{DM}(\mathcal{D}, n, S', \mathcal{P})) = \emptyset$.

We provide in Figure 1 a simple algorithm using our partition based method for enumerating all (closed/maximal) frequent itemsets. This algorithm takes as input a transaction database $\mathcal{D}$, a minimal support threshold n and a k-partition $\mathcal{P} = (\{S_1, \ldots, S_k\}, \prec)$ such that $S_1 \prec \cdots \prec S_k$. The procedure $enumModels(A)$ enumerate all Boolean models of the propositional formula A. The procedure $itemsets(R)$ returns the itemsets corresponding to the Boolean models in the set R.

Theorem 1 (Soundness). *The algorithm $SAT_{DM}^{\mathcal{P}}$ is sound w.r.t. the data mining task DM.*

Proof. The soundness of $SAT_{DM}^{\mathcal{P}}$ is a direct consequence of Proposition 4.

Theorem 2 (Completeness). *The algorithm $SAT_{DM}^{\mathcal{P}}$ is complete w.r.t. the data mining task DM.*

Proof. The completeness of $SAT_{DM}^{\mathcal{P}}$ comes from the fact that $\Omega = \bigcup_{1 \leqslant i \leqslant k} S_i$ and Proposition 4.

It is worth noticing that, using Proposition 5, we know that there is no itemset computed in two different steps in the for-loop.

In the propositional formula (6), the main difficulty is in the sub-formula $\mathcal{E}_{DM}(\mathcal{D}_{|S}, n)$. Indeed, if the number of transactions of $\mathcal{D}_{|S}$ is close to that of $\mathcal{D}$, then the gain is minimal. For this reason, it is more interesting to consider orderings $\prec$ in k-partitions that are based on the number of transactions in the sense that $S \prec S'$ if the number of transactions of $\mathcal{D}_{|S}$ is smaller than that of $\mathcal{D}_{|S'}$. Let us note that the number of transactions of $\mathcal{D}_{|S}$ depends directly on the size of S: if S is included in S', then the number of transactions of $\mathcal{D}_{|S}$ is smaller than that of $\mathcal{D}_{|S'}$.

6 Experiments

In this section, we carried out an experimental evaluation of the performance of our partition based approach for enumerating frequent closed itemsets. We considered a variety of datasets taken from the FIMI[1] and CP4IM[2] repositories.

All the experiments were done on Intel Xeon quad-core machines with 32GB of RAM running at 2.66 Ghz. For each instance, we used a timeout of 2 hours of CPU time. As described in the previous section, the Algorithm 1 takes a transaction database and a k-partition of the set of items as inputs, and returns the set of patterns of interest (frequent, closed or maximal). In our experiments, we consider a partition $\mathcal{P}$ where each $S_i \in \mathcal{P}$ contains a single item and $S_i \prec S_{i+1}$ if S_i contains a less frequent item than S_{i+1}. Obviously, in this case k is equal to $|\Omega|$. This ordering allows us to first generate encodings of smaller size. Our goal is to show the feasibility of our proposed approach even when a basic partitioning scheme is considered. Finding a k-partition of the set of items that leads to better

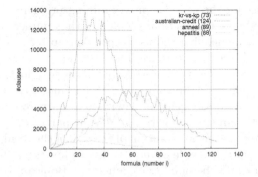

Fig. 2. $SAT_{CFIM}^{\mathcal{P}}$ Formulas: Evolution of the Number of Clauses

---

[1] FIMI: http://fimi.ua.ac.be/data/
[2] CP4IM: http://dtai.cs.kuleuven.be/CP4IM/datasets/

improvements is an interesting optimisation problem. This issue is out of the scope of this paper. We compare our partition based approach noted $SAT^{\mathcal{P}}_{CFIM}$ (*SAT-P-CFIM* in the figures) with SAT_{CFIM}, the encoding of frequent closed itemsets mining problem without partition (named *SAT-CFIM* in the figures).

We implemented the Algorithm 1 in C. For the enumeration of all models of the Boolean formula encoding the corresponding itemset mining problem, we use the extension of modern SAT solvers described in [6]. The implementation is based on an extension of MiniSAT 2.2 [3].

Note that, the time needed to generate the partition does not exceed 1 seconds on the majority of the instances, except for `splice-1` and `connect`, which takes 17 and 60 seconds respectively. First, we present in Table 2, the charac-

Table 2. Characteristics of the instances & Encoding size

instance	#trans	#items	#vars	#clauses	min	max
zoo-1	101	36	173	2196	4	387
Hepatitis	137	68	273	4934	193	934
Lymph	148	68	284	6355	28	724
audiology	216	148	508	17575	4	657
Heart-cleveland	296	95	486	15289	192	2534
Primary-tumor	336	31	398	5777	117	801
Vote	435	48	531	14454	225	2591
Soybean	650	50	730	22153	4	1847
Australian-credit	653	125	901	48573	108	5896
Anneal	812	93	990	39157	4	4149
Tic-tac-toe	958	27	1012	18259	485	3619
german-credit	1000	112	1220	73223	319	6957
Kr-vs-kp	3196	73	3342	121597	4	13879
Hypothyroid	3247	88	3419	143043	4	14410
chess	3196	75	3346	124797	4	14853
splice-1	3190	287	3764	727897	4	105540
mushroom	8124	119	8348	747635	23	34695
connect	67558	129	67815	5877720	297	291139

teristics of the dataset and the size of the CNF formula encoding the whole transaction database for the closed itemsets mining problem. For each instance, we mention the number of transactions (*#trans*), the number of items (*#items*), the size of the formula encoding the whole problem in terms of number of variables (*#vars*) and clauses (*#clauses*). The last two columns give the number of clauses of the smallest (*min*) respectively the largest formula (*max*) generated using our partition based encoding. On all instances, the number of clauses of the largest formula (*max*) generated with our partition based encoding is significantly smaller than those generated without partition (*#clauses*).

In Figure 2, we represent the evolution of the size of the resulting CNFs during the partition process for a sample of representative instances. For each instance, we mention its name and the number of generated Boolean formulas (in parenthesis) corresponding to the number of elements in the partition (k). As we can observe, our approach allows to generate small CNFs compared to the approach

---

[3] MiniSAT: http://minisat.se/

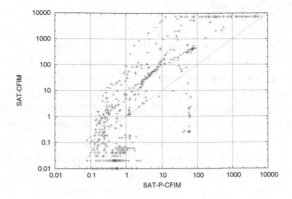

Fig. 3. $SAT^{\mathcal{P}}_{CFIM}$ vs SAT_{CFIM}

encoding the whole transaction database (see Table 2). For example, if we consider the Kr-vs-kp instance, the largest generated CNF formula does not exceed 14 000 clauses while it reaches 121 597 clauses without partitioning. Moreover, the size of the different formulas evolves as follows. During the first steps of the partitioning process, the size of the obtained CNFs is smaller, as the first elements of the partition contains less frequent items. For these formulas, the enumeration process takes relatively short time. The same observation can be made for the latest steps of the partitioning process (last elements of the partition). We recall that at iteration i, the transactions containing the items from S_j $(j < i)$ are removed. A peak in the size of the generated formulas can be observed in the middle of the partitioning process. Indeed, for the first generated formulas, the considered items appears in less transactions leading to an encoding of a smaller transaction database. For higher elements of the partition, even if they correspond to frequent items, some of their occurrences are avoided by the previous removed items. The peak in size corresponds to the steps of the partitioning process, where items occurs a reasonable number of times.

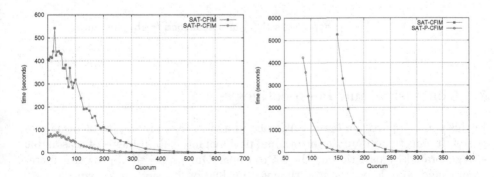

Fig. 4. $SAT^{\mathcal{P}}_{CFIM}$: anneal (left) and australian (right) instances

To evaluate the performances of our proposed approach, we compare the time needed for our partition based approach $SAT^{\mathcal{P}}_{CFIM}$ with SAT_{CFIM} (without partitioning) to enumerate all models corresponding to all closed frequent itemsets. The comparison is depicted by the the scatter plot of Figure 3. Each dots (x, y) represents an instance (see Table 2) with a fixed minimal support threshold n. The scatterplot represent all the instances described in Table 2 (18 instances), where for each instance, we tested 70 different values of n. The total number of instances is 1260. The x-axis (respectively y-axis) represents the CPU time (seconds) needed for the enumeration of all closed frequent itemsets using $SAT^{\mathcal{P}}_{CFIM}$ (respectively SAT_{CFIM}). Clearly, the partition based approach outperforms those without partition on the majority of instances. For several instances and values of n, our partition based approach allows to enumerate all models while the classical method without partitioning is not able to enumerate them under the time limit. For instance, for `connect` data, SAT_{CFIM} is able to enumerate all closed frequent itemsets for only $n = 50000$, while our partition based approach enumerates all models for 8 different minimal support threshold (for all $n \geqslant 5500$).

Figures 4 and 5, highlight the results obtained by $SAT^{\mathcal{P}}_{CFIM}$ and SAT_{FCIM} on `anneal`, `australian`, `hepatitis` and `mushroom` instances while varying the minimum support threshold (quorum). These figures highlights the great potential of our partitioning based approach.

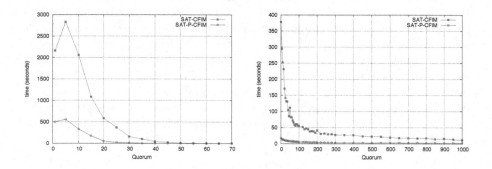

Fig. 5. $SAT^{\mathcal{P}}_{CFIM}$: `hepatitis` (left) and `mushroom` (right) instance

7   Conclusion and Perspectives

In this paper, a partition based SAT encoding of itemset mining problems is proposed. It aims to decompose the original problem into subproblems of reasonable size generated by partitioning the set of items. Our proposed approach is incremental and complete. The experimental evaluation on several known datasets shows significant improvements, up to several orders of magnitude. The results obtained in this paper, open several interesting paths for future works, including the design of parallel based approaches. Handling the different subproblems in parallel, will leads to substantial additional improvements. Finding the "best" partition of the set of items is another interesting issue.

Acknowledgments. This work is supported by the French ANR Agency under the TUPLES project. We thank the reviewers for their comments that helped us to improve the paper.

References

1. Asin, R., Nieuwenhuis, R., Oliveras, A., Rodriguez-Carbonell, E.: Cardinality networks: a theoretical and empirical study. Constraints **16**(2), 195–221 (2011)
2. Cambazard, H., Hadzic, T., O'Sullivan, B.: Knowledge compilation for itemset mining. In: ECAI 2010, pp. 1109–1110 (2010)
3. Guns, T., Dries, A., Tack, G., Nijssen, S., De Raedt, L.: Miningzinc: A modeling language for constraint-based mining. In: Proceedings of the Twenty-Third International Joint Conference on Artificial Intelligence, IJCAI 2013, pp. 1365–1372 (2013)
4. Guns, T., Nijssen, S., De Raedt, L.: Itemset mining: A constraint programming perspective. Artificial Intelligence **175**(12–13), 1951–1983 (2011)
5. Guns, T., Nijssen, S., Raedt, L.D.: Itemset mining: A constraint programming perspective. Artif. Intell. **175**(12–13), 1951–1983 (2011)
6. Jabbour, S., Lonlac, J., Sais, L., Salhi, Y.: Extending modern sat solvers for models enumeration. In: Proceedings of the 11th IEEE International Conference on Information Reuse and Integration (IEEE-IRI 2014), San Francisco, 13–15 September 2014 (2014) (to appear). http://arxiv.org/abs/1305.0574, CoRR 2013
7. Jabbour, S., Sais, L., Salhi, Y.: The top-k frequent closed itemset mining using top-k sat problem. In: European Conference on Machine Learning and Knowledge Discovery in Databases (ECML/PKDD 2013), pp. 403–418 (2013)
8. Khiari, M., Boizumault, P., Crémilleux, B.: Combining csp and constraint-based mining for pattern discovery. In: Taniar, D., Gervasi, O., Murgante, B., Pardede, E., Apduhan, B.O. (eds.) ICCSA 2010, Part II. LNCS, vol. 6017, pp. 432–447. Springer, Heidelberg (2010)
9. Metivier, J.P., Boizumault, P., Crémilleux, B., Khiari, M., Loudni, S.: A constraint-based Language for Declarative Pattern Discovery. In: 2011 IEEE 11th International Conference on Data Mining Workshops (ICDMW), Vancouver, Canada, pp. 1112–1119 (2011)
10. Raedt, L.D., Guns, T., Nijssen, S.: Constraint programming for itemset mining. In: ACM SIGKDD, pp. 204–212 (2008)
11. Marques-Silva, J., Lynce, I.: Towards robust cnf encodings of cardinality constraints. In: Bessière, C. (ed.) CP 2007. LNCS, vol. 4741, pp. 483–497. Springer, Heidelberg (2007)

12. Sinz, C.: Towards an optimal cnf encoding of boolean cardinality constraints. In: van Beek, P. (ed.) CP 2005. LNCS, vol. 3709, pp. 827–831. Springer, Heidelberg (2005)
13. Tseitin, G.: On the complexity of derivations in the propositional calculus. In: Structures in Constructives Mathematics and Mathematical Logic, Part II, pp. 115–125 (1968)
14. Warners, J.P.: A linear-time transformation of linear inequalities into conjunctive normal form. Information Processing Letters (1996)
15. Yang, G.: The complexity of mining maximal frequent itemsets and maximal frequent patterns. In: KDD 04: Proceedings of the tenth ACM SIGKDD International Conference on Knowledge Discovery and Data mining, pp. 344–353. ACM Press (2004)

A Comparative Study on Parallel LDA Algorithms in MapReduce Framework

Yang Gao[1], Zhenlong Sun[2], Yi Wang[2], Xiaosheng Liu[1],
Jianfeng Yan[1], and Jia Zeng[1]([✉])

[1] School of Computer Science and Technology, Soochow University,
Suzhou 215006, China
j.zeng@ieee.org
[2] Tencent, Peking 100080, China

Abstract. Although several parallel latent Dirichlet allocation (LDA) algorithms have been implemented to extract topic features from large-scale text data sets, very few studies compare their performance in real-world industrial applications. In this paper, we build a novel multi-channel MapReduce framework to compare fairly three representative parallel LDA algorithms such as parallel variational Bayes (PVB), parallel Gibbs sampling (PGS) and parallel belief propagation (PBP). Experimental results confirm that PGS yields the best application performance in search engine and online advertising system of Tencent, one of the biggest Internet companies in China, while PBP has the highest topic modeling accuracy. Moreover, PGS is more scalable in MapReduce framework than PVB and PBP because of its low memory usage and efficient sampling technique.

Keywords: Latent Dirichlet allocation · Parallel inference algorithms · MapReduce · Search engine · Online advertising system

1 Introduction

Latent Dirichlet allocation (LDA) [1], as a generative probabilistic topic model, represents each document by a mixture of latent topic distributions over the vocabulary words. To get the best topic labeling configuration over observed words, LDA maximizes the joint probability of latent topics and observed words by inference algorithms, which calculate the posterior distribution of topics conditioned on words. In the past decade, LDA inference algorithms can be broadly categorized into two types. The first is based on the idea of stochastic gradient descent (SGD) [2], which uses the topic posterior distribution of each word to update the LDA parameters sequentially. Typical SGD-based inference algorithms include Gibbs sampling (GS) [3–5], zero-order approximation of collapsed variational Bayes (CVB0) [6,7], and asynchronous belief propagation (BP) [8]. The second is based on the idea of coordinate descent (CD), which fixes the

© Springer International Publishing Switzerland 2015
T. Cao et al. (Eds.): PAKDD 2015, Part II, LNAI 9078, pp. 675–689, 2015.
DOI: 10.1007/978-3-319-18032-8_53

LDA parameters to infer the posterior of topics over all words, and then fixes the inferred posterior to estimate LDA parameters iteratively. Typical CD-based inference algorithms include *maximum a posterior* (MAP), variational Bayes (VB) [1,9], and synchronous BP [8].

In the big data era, scaling out batch or online LDA inference algorithms on a cluster of computers is a necessary step to web-scale industrial applications [10–16]. However, previous researches discuss little about the effectiveness of different parallel LDA algorithms in real-world industrial applications. Therefore, within the unified MapReduce framework, we compare three representative parallel LDA algorithms (PVB [1], PGS [3] and PBP [8]) in terms of complexity, accuracy and practical performance in search engine and online advertising system of Tencent, one of the biggest Internet companies in China. Through experimental results, we obtain the following observations:

– PGS is more scalable than PVB and PBP in the MapReduce framework because of its low memory usage in storing LDA parameters and fast speed by efficient sampling techniques. It achieves the best performance in two industrial applications such as search engine and online advertising system.
– PBP has the best predictive performance in terms of held-out log-likelihood. However, it consumes more memory space than PGS, and has difficulty in scaling out to the larger number of topics. Moreover, its performance in two real-world applications is worse than PGS, which indicates that the best topic modeling performance does not mean the best application performance.
– PVB has the slowest speed and the lowest held-out log-likelihood. Its application performance is also the worst among three algorithms.

MapReduce is a simple framework for dealing with large-scale data. Traditionally, the intercommunication of machines using MapReduce is based on the shared disk files like Hadoop distributed file (HDF) system[1], which is not suitable for iterative algorithm due to repeatedly load from and write to disk. However, the advantage of MapReduce is redundancy and fault tolerance mechanism. If the training time per iteration is long and errors occur during one iteration, the fault tolerance is useful to recover model parameters from the last successful iteration. Therefore, MapReduce has been widely used in industry and can be viewed as a fair framework for this comparative study. Prior works have implemented PGS [17] and PVB [15] within the MapReduce framework. In PGS, each mapper processes the subset of documents and the reducer collects the output of all mappers and synchronizes the model parameters. Mahout[2] implements a MapReduce version of collapsed variational Bayes (CVB) algorithm [6], which has a close relation to BP [8] and MAP inference. Mr.LDA [15] develops PVB based on the MapReduce framework. Although parallel LDA algorithms have been implemented within other framework [13,14,18], we focus on MapReduce for a fair comparison of three representative parallel LDA algorithms.

---

[1] http://hadoop.apache.org/
[2] http://mahout.apache.org/users/clustering/latent-dirichlet-allocation.html

Table 1. Define of Notations

$1 \leq d \leq D$	Document index
$1 \leq w \leq W$	Vocabulary word index
$1 \leq k \leq K$	Topic index
$1 \leq m \leq M$	Shard index on vocabulary dimension
$1 \leq n \leq N$	Shard index on document dimension
NNZ	Number of non-zero elements
$ntokens = \sum_{w,d} x_{w,d}$	Number of word tokens
$\mathbf{x}_{W \times D} = \{x_{w,d}\}$	Document-word matrix
$\mathbf{z}_{W \times D} = \{z_{w,d}^k\}$	Topic labels for words
$\boldsymbol{\theta}_{K \times D}$	Document-topic distribution
$\boldsymbol{\phi}_{K \times W}$	Topic-word distribution
$\boldsymbol{\mu}_{K \times NNZ}$	Responsibility matrix (messages)
α, β	Dirichlet hyperparameters

The rest of this paper is organized as follows. In Section 2, we discuss three representative batch LDA inference algorithms like VB [1], GS [3] and BP [8]. We show that they infer different posterior probability of LDA with different time and space complexities. Section 3 introduces our multi-channel MapReduce framework for a fair comparison among PVB, PGS and PBP. Section 4 describes their topic modeling results, and compares the performance of topic features learned by three inference algorithms in search engine and online advertising system of Tencent. We conclude this paper in Section 5.

2 Three Representative Batch Inference Algorithms for LDA

LDA allocates a set of thematic topic labels, $\mathbf{z} = \{z_{w,d}^k\}$, to explain non-zero elements in the document-word co-occurrence matrix $\mathbf{x}_{W \times D} = \{x_{w,d}\}$. Table 1 summarizes the important notations in this paper. The nonzero element $x_{w,d} \neq 0$ denotes the number of word counts at the index $\{w, d\}$. For each word token $x_{w,d,i} = \{0, 1\}, x_{w,d} = \sum_i x_{w,d,i}$, there is a topic label $z_{w,d,i}^k = \{0, 1\}$, $\sum_{k=1}^{K} z_{w,d,i}^k = 1, 1 \leq i \leq x_{w,d}$. The full joint probability of LDA is $p(\mathbf{x}, \mathbf{z}, \boldsymbol{\theta}, \boldsymbol{\phi} | \alpha, \beta)$ [6], where $\boldsymbol{\theta}_{K \times D}$ and $\boldsymbol{\phi}_{K \times W}$ are two non-negative matrices of multinomial parameters for document-topic and topic-word distributions, satisfying $\sum_k \theta_d(k) = 1$ and $\sum_w \phi_w(k) = 1$. Both multinomial matrices are generated by two Dirichlet distributions with hyperparameters α and β. For each non-zero element $x_{w,d} \neq 0$, there is a normalized vector $\sum_k \mu_{w,d}(k) = 1$ representing the **responsibility** that the topic k takes for word index $\{w, d\}$ (also called *message* vector in [8]). For simplicity, we consider the smoothed LDA with fixed symmetric hyperparameters [3].

The objective of inference algorithms is to infer posterior probability from the full joint probability [19]. We choose three representative LDA algorithms

Table 2. Time and Space Complexities of Batch Algorithms

	Posterior	Time	Space (Memory)
VB [1]	$p(\boldsymbol{\theta}, \mathbf{z}|\mathbf{x}, \boldsymbol{\phi}, \alpha, \beta)$	$2 \times K \times NNZ \times digamma$	$2 \times K \times (D + W)$
GS [5]	$p(\mathbf{z}|\mathbf{x}, \alpha, \beta)$	$\delta_1 \times K \times ntokens$	$\delta_2 \times K \times W + ntokens$
BP [8]	$p(\boldsymbol{\theta}, \boldsymbol{\phi}|\mathbf{x}, \alpha, \beta)$	$2 \times K \times NNZ$	$2 \times K \times (D + W)$

for comparison because they infer slightly different posterior probability of LDA. Also, other inference algorithms have close relation to these three representative algorithms. We shall see that these three representative algorithms all originate from the expectation-maximization (EM) framework for maximum-likelihood estimation.

The first inference algorithm is VB [1] that infers the posterior,

$$p(\boldsymbol{\theta}, \mathbf{z}|\mathbf{x}, \boldsymbol{\phi}, \alpha, \beta) = \frac{p(\mathbf{x}, \mathbf{z}, \boldsymbol{\theta}, \boldsymbol{\phi}|\alpha, \beta)}{p(\mathbf{x}, \boldsymbol{\phi}|\alpha, \beta)}. \tag{1}$$

However, computing this posterior is intractable because the denominator contains intractable integration, $\int_{\boldsymbol{\theta}, \mathbf{z}} p(\mathbf{x}, \mathbf{z}, \boldsymbol{\theta}, \boldsymbol{\phi}|\alpha, \beta)$. Therefore, VB infers an approximate variational posterior based on the variational EM algorithm:

– Variational E-step:

$$\mu_{w,d}(k) \propto \exp[\Psi(\hat{\theta}_d(k) + \alpha)] \times \frac{\exp[\Psi(\hat{\phi}_w(k) + \beta)]}{\exp[\Psi(\sum_w [\hat{\phi}_w(k) + \beta])]}, \tag{2}$$

$$\hat{\theta}_d(k) = \sum_w x_{w,d} \mu_{w,d}(k). \tag{3}$$

– Variational M-step:

$$\hat{\phi}_w(k) = \sum_d x_{w,d} \mu_{w,d}(k). \tag{4}$$

In variational E-step, we update $\mu_{w,d}(k)$ and $\hat{\theta}_d(k)$ until convergence, which makes the variational posterior approximate the true posterior $p(\boldsymbol{\theta}, \mathbf{z}|\mathbf{x}, \boldsymbol{\phi}, \alpha, \beta)$ by minimizing the Kullback-Leibler (KL) divergence between them. However, the variational posterior cannot touch the true posterior for inaccurate solutions [6]. In addition, the calculation of exponential digamma function $\exp[\Psi(\cdot)]$ is computationally complicated. As shown in Table 2, the time complexity of VB for one iteration is $\mathcal{O}(2 \times K \times NNZ \times digamma)$, where $digamma$ is the computing time for exponential digamma function. For each non-zero element, we need K iterations for variational E-step and K iterations for normalizing $\mu_{w,d}(k)$. The space complexity is $\mathcal{O}(2 \times K \times (D + W))$ for two multinomial parameters and temporary storage for variational M-step.

The second is the GS [3] algorithm that infers the posterior,

$$p(\mathbf{z}|\mathbf{x}, \alpha, \beta) = \frac{p(\mathbf{x}, \mathbf{z}|\alpha, \beta)}{p(\mathbf{x}|\alpha, \beta)} \propto p(\mathbf{x}, \mathbf{z}|\alpha, \beta). \tag{5}$$

Maximizing the joint probability $p(\mathbf{x}, \mathbf{z}|\alpha, \beta)$ is intractable, an approximate inference called Markov chain Monte Carlo (MCMC) EM is used as follows:

- MCMC E-step:

$$\mu_{w,d,i}(k) \propto [\hat{\theta}_d^{-z_{w,d,i}^{k,old}}(k) + \alpha] \times \frac{\hat{\phi}_w^{-z_{w,d,i}^{k,old}}(k) + \beta}{\sum_w [\hat{\phi}_w^{-z_{w,d,i}^{k,old}}(k) + \beta]}, \qquad (6)$$

$$\text{Random Sampling } z_{w,d,i}^{k,new} = 1 \text{ from } \mu_{w,d,i}(k). \qquad (7)$$

- MCMC M-step:

$$\hat{\theta}_d(k) = \hat{\theta}_d^{-z_{w,d,i}^{k,old}}(k) + z_{w,d,i}^{k,new}, \qquad (8)$$

$$\hat{\phi}_w(k) = \hat{\phi}_w^{-z_{w,d,i}^{k,old}}(k) + z_{w,d,i}^{k,new}. \qquad (9)$$

In the MCMC E-step, GS infers the posterior $\mu_{w,d,i}(k) = p(z_{w,d,i}^{k,new}|\mathbf{z}_{w,d,-i}^{k,old}, \mathbf{x}, \alpha, \beta)$ for each word token, and randomly samples a new topic label $z_{w,d,i}^{k,new} = 1$ from this posterior. Here, we use the notation $\hat{\phi}(k) = \sum_w \hat{\phi}_w(k)$ for the denominator in (6). The notation $-z_{w,d,i}^{k,old}$ means excluding the old topic label from the corresponding matrices $\{\hat{\theta}, \hat{\phi}\}$. In the MCMC M-step, GS updates immediately $\{\hat{\theta}, \hat{\phi}\}$ by the new topic label of each word token. In this sense, GS can be viewed as a kind of SGD algorithm that learns parameters by processing data point sequentially. Normalizing $\{\hat{\theta}, \hat{\phi}\}$ yields the multinomial parameters $\{\theta, \phi\}$. In Table 2, the time complexity of GS for one iteration is $\mathcal{O}(\delta_1 \times K \times ntokens)$, where $\delta_1 \ll 2$. The reason is that we require K iterations in MCMC E-step and less K iterations for normalizing $\mu_{w,d,i}(k)$. According to sparseness of $\mu_{w,d,i}(k)$, efficient sampling techniques [5] can make δ_1 even smaller. Practically, when K is larger than 1000, $\delta_1 \approx 0.05$. Generally, we do not need to store $\hat{\theta}_{K \times D}$ in memory because $\mathbf{z}$ can recover $\hat{\theta}_{K \times D}$. So, the space complexity is $\mathcal{O}(\delta_2 \times K \times W + ntokens)$ because $\hat{\phi}_{K \times W}$ can be compressed due to sparseness [5]. Practically, when K is larger than 1000, $\delta_2 \approx 0.8$. Note that all parameters in GS are stored in integer type, saving lots of memory space than double type used by both VB and BP.

The third is BP [8] that infers the posterior,

$$p(\boldsymbol{\theta}, \boldsymbol{\phi}|\mathbf{x}, \alpha, \beta) = \frac{p(\mathbf{x}, \boldsymbol{\theta}, \boldsymbol{\phi}|\alpha, \beta)}{p(\mathbf{x}|\alpha, \beta)} \propto p(\mathbf{x}, \boldsymbol{\theta}, \boldsymbol{\phi}|\alpha, \beta). \qquad (10)$$

we integrate out the labeling configuration $\mathbf{z}$ in full joint probability, and use the standard EM algorithm to optimize this objective (10):

- E-step:

$$\mu_{w,d}(k) \propto [\hat{\theta}_d(k) + \alpha - 1] \times \frac{\hat{\phi}_w(k) + \beta - 1}{\sum_w [\hat{\phi}(k) + \beta - 1]}, \qquad (11)$$

– M-step:

$$\hat{\theta}_d(k) = \sum_w x_{w,d} \mu_{w,d}(k), \tag{12}$$

$$\hat{\phi}_w(k) = \sum_d x_{w,d} \mu_{w,d}(k). \tag{13}$$

In the E-step, BP infers the responsibility $\mu_{w,d}(k)$ conditioned on parameters $\{\hat{\theta}, \hat{\phi}\}$. In the M-step, BP updates parameters $\{\hat{\theta}, \hat{\phi}\}$ based on the inferred responsibility $\mu_{w,d}(k)$. Here, we use the notation $\hat{\phi}(k) = \sum_w \hat{\phi}_w(k)$ for the denominator in (11). Normalizing $\{\hat{\theta}, \hat{\phi}\}$ yields the multinomial parameters $\{\theta, \phi\}$. Unlike VB, BP can touch the true posterior distribution $p(\boldsymbol{\theta}, \boldsymbol{\phi} | \mathbf{x}, \alpha, \beta)$ in the E-step for maximization. Besides BP [8,20], CVB0 [7] and MAP are all like EM algorithms. When compared with VB, the time complexity of BP for one iteration is $\mathcal{O}(2 \times K \times NNZ)$ without calculating exponential digamma functions. The space complexity of BP is the same as VB with $\mathcal{O}(2 \times K \times (D+W))$ because of storing $\{\hat{\theta}, \hat{\phi}\}$ as well as the temporary variables in the M-step.

3   Multi-channel MapReduce Framework

Currently, both Hadoop[3], Spark [21] are distributed computation systems. Hadoop stores and processes data in hard disks by Hadoop distributed file system (HDFS), while Spark processes data in memory. Spark integrates the MapReduce of Hadoop, primarily targeted at speeding up batch analysis jobs, iterative machine learning jobs, interactive query and graph processing. Due to the memory limitation of our cluster, in this paper we focus on MapReduce in Hadoop to process big data with fault toleration, though the training time will be longer than that with Spark. Note that these parallel LDA algorithms can be easily deployed in the MapReduce framework of Spark.

In real-world applications, the vocabulary size W is often larger than 10^5. Taking polysemy and synonyms under consideration, a rough estimate of the number of topics is close to the same magnitude of vocabulary words, i.e., $K \geq 10^4$. If we store the full topic-word matrix $\hat{\phi}_{K \times W}$ in memory, the entire space needs to be larger than 10GBytes, which is impractical with general parallel framework like PLDA [17] and Mr.LDA [15], where each machine in the cluster has only 2GBytes memory. Also, an industrial computer cluster often processes multiple tasks, and the available memory space for parallel LDA algorithms is significantly less than 2GB for each machine. Therefore, we propose a new multi-channel MapReduce framework to reduce the memory demand of each task.

3.1   Overview

To reduce the memory demand, we partition the matrix $\hat{\phi}_{K \times W}$ into M column shards $\hat{\phi}_w^{m,n}(k)$ with $1/M$ size of $\hat{\phi}_{K \times W}$. To reduce the memory demand for

---

[3] http://hadoop.apache.org/

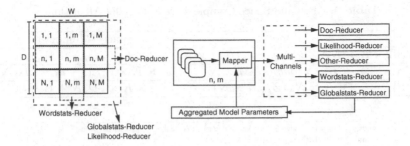

Fig. 1. Data partition and a new multi-channel MapReduce framework

$\hat{\boldsymbol{\theta}}_{K \times D}$, we partition it into N column shards $\hat{\theta}_d^{m,n}(k)$ with $1/N$ size of $\hat{\boldsymbol{\theta}}_{K \times D}$. In the meanwhile, we partition the input document-word matrix $D \times W$ into $N \times M$ small shards $\mathbf{x}_{w,d}^{m,n}$ with $1/(M \times N)$ size of $\mathbf{x}_{w,d}$ on both document and vocabulary dimensions, where the D and W denote the number of documents and the size of vocabulary, respectively. Fig. 1 shows the partition of input document-word matrix into small shards. As with [22], we use the random shuffling method in both document and vocabulary dimensions for data shard load balancing, i.e., making the number of words almost equal in each data shard. In MapReduce, each mapper uses one core of CPU in a machine, and takes one small data shard and two small parameter shards in memory for computation. Therefore, each mapper only consumes $1/(N \times M)$ data size and $1/M + 1/N$ parameter size when compared with the original batch algorithms on a single machine. We set the values of M and N for fitting the topic-word distribution and document-topic distribution in the memory. For example, when $M = 10$, we need only 1GB memory of each machine to hold a small shard of 10GB LDA $\hat{\phi}_{K \times W}$ parameters.

All mappers perform local E-step and M-step in parallel on $N \times M$ data blocks by corresponding parameter shards using equations (2), (6) and (11). After we obtain the updated local parameters $\{\hat{\theta}_d^{m,n}(k), \hat{\phi}_w^{m,n}(k)\}$, we use multiple channels of reducers to aggregate them into the updated global parameters called synchronization. Each channel contains multiple reducers for the same kind of tasks. For example, to get the document-topic distribution of one document $\hat{\theta}_d^{·,n}(k)$, we aggregate the distribution $\hat{\theta}_{w,d}^{m,n}(k)$ of M shards with the same document index d using N reducers illustrated as the Doc-Reducer channel in Fig. 1. To get the topic-word distribution of one topic $\hat{\phi}_w^{m,·}(k)$, we also aggregate the distribution $\hat{\phi}_w^{m,n}(k)$ of N shards with the same word index w using M reducers illustrated as the Wordstats-Reducer channel in Fig. 1. Similarly, we use the Globalstats reducer to aggregate $\hat{\phi}_w^{m,·}(k)$ into $\hat{\phi}(k) = \sum_{w,m} \hat{\phi}_w^{m,·}(k)$, which constitutes the denominator in the E-step (2), (6) and (11) of all algorithms. Finally, we use the Likelihood-Reducer to check the convergence of training process. The held-out log-likelihood is defined as follows [1,7,8]:

$$\mathcal{L} = \sum_{w,d} \left(x_{w,d}^{\text{held-out}} \log \left[\sum_k \theta_d(k) \phi_w(k) \right] \right), \tag{14}$$

Table 3. Time and Space Complexities of Parallel Algorithms

	Time	Space (Memory)
PVB	$2 \times K \times NNZ \times digamma/(M \times N)$	$2 \times K \times (D/N + W/M)$
PGS	$\delta_1 \times K \times ntokens/(M \times N)$	$\delta_2 \times K \times W/M + ntokens/N$
PBP	$2 \times K \times NNZ/(M \times N)$	$2 \times K \times (D/N + W/M)$

---

input : Key: shard index (m, n); Value: document content $x_{w,d}^{m,n}$; Parameters: K, α, β.
output : $\hat{\phi}_w^{m,n}(k), \hat{\theta}_d^{m,n}(k), \hat{\phi}^{m,n}(k), \mathcal{L}^{m,n}$.

1 **if** $t = 1$ **then**

2 | random initialization and normalization($\{\hat{\theta}_d^{m,n}(k), \hat{\phi}_w^{m,n}(k), \hat{\phi}(k)\}$);

3 **else**

4 | read($aggregated \{\hat{\theta}_d^{\cdot,n}(k), \hat{\phi}_w^{m,\cdot}(k), \hat{\phi}(k)\}$) from HDFS;

5 | Copy these aggregated parameters to each mapper;

6 **end**

7 **for** *each mapper in parallel* **do**

8 | E-step: Calculate $\mu_{w,d}^{m,n}(k)$ based on $\{\hat{\theta}_d^{m,n}(k), \hat{\phi}_w^{m,n}(k), \hat{\phi}(k)\}$;

9 | M-step: Update local parameters $\{\hat{\theta}_d^{m,n}(k), \hat{\phi}_w^{m,n}(k), \hat{\phi}^{m,n}(k)\}$ based on $\mu_{w,d}^{m,n}(k)$;

10 | Calculate local log-likelihood $\mathcal{L}^{m,n} = \sum_{w,d} \left(x_{w,d}^{m,n} \log \left[\sum_k \hat{\theta}_d^{m,n}(k) \hat{\phi}_w^{m,n}(k) \right] \right)$;

11 **end**

12 output($\{\hat{\theta}_d^{m,n}(k), \hat{\phi}_w^{m,n}(k), \hat{\phi}^{m,n}(k), \mathcal{L}^{m,n}\}$) to Multi-channel Reducers;

Fig. 2. The mapper algorithm

where $x_{w,d}^{\text{held-out}}$ denotes the held-out test data. If the data is changed to $x_{w,d}^{\text{training}}$, the training log-likelihood can be used to check the convergence condition. When the absolute difference ratio of training log-likelihood between two successive iterations is smaller than a threshold, i.e., 0.1%, we terminate the training process. In total, there are $M + N + 2$ reducers for aggregation as shown in the left panel of Fig. 1. The updated model parameters are copied back to each mapper for the next training iteration. Similar to Table 2, Table 3 shows the time and space complexities of PVB, PGS and PBP without considering the synchronization and I/O costs. We see that both time and space complexities have been significantly reduced. In practice, M and N can be set according to each machine's memory constraint.

3.2 Mapper and Reducer Algorithms

In the MapReduce framework, all data are organized in key-value pairs. Fig. 2 shows the mapper algorithm, which performs the local E-step and M-step based on local parameters. In the first iteration $t = 1$, each mapper randomly initialize the local parameters in line 2. Then, all mappers run E-step/M-step in parallel, and calculate local log-likelihood (lines $8 - 10$). In the successive iterations, all mappers read the aggregated parameters from HDFS (lines 4 and 5), and continues to do E-step and M-step in parallel. Finally, all mappers output the updated local parameters and log-likelihood to multi-channel reducers (line 12). Fig. 3 shows the multi-channel reducer algorithm, which aggregates the local parameters and log-likelihood by sum operations.

---

input : Key: shard index (m, n); Value: local parameters $\{\hat{\theta}_d^{m,n}(k), \hat{\phi}_w^{m,n}(k), \hat{\phi}^{m,n}(k)\}$.
output : Aggregated parameters $\{\hat{\theta}_d^{\cdot,n}(k), \hat{\phi}_w^{m,\cdot}(k), \hat{\phi}(k)\}$ and log-likelihood $\mathcal{L}$.
1 **for** *each reducer in parallel* **do**
2   M Doc-Reducers: Aggregate $\hat{\theta}_d^{\cdot,n}(k) = \sum_m \hat{\theta}_d^{m,n}(k)$;
3   N Wordstats-Reducers: Aggregate $\hat{\phi}_w^{m,\cdot}(k) = \sum_n \hat{\phi}_w^{m,n}(k)$;
4   Globalstats-Reducer: Aggregate $\hat{\phi}(k) = \sum_{m,n,w} \hat{\phi}_w^{m,n}(k)$;
5   Likelihood-Reducer: Calculate $\mathcal{L} = \sum_{m,n} \mathcal{L}^{m,n}$;
6 **end**
7 write(*aggregated parameters* $\{\hat{\theta}_d^{\cdot,n}(k), \hat{\phi}_w^{m,\cdot}(k), \hat{\phi}(k)\}$) to HDFS;

Fig. 3. The reducer algorithm

The major difference between the multi-channel MapReduce with previous solutions [15,17] lies in the parallel reducers in Fig. 3. In practice, mappers can speed up parallel LDA algorithms linearly while reducers introduce additional synchronization time when compared with the batch algorithms on a single machine. By splitting the document-word matrix into $M \times N$ blocks, we can not only save the memory consumption for model parameters, but also implement parallel reducers for a further speedup. In this way, the synchronization time can be reduced linearly with the number of reducers. Future work includes the multi-channel MapReduce in Spark, which can save the I/O cost of each mapper and reducer iteration in parallel LDA algorithms.

4 Experimental Results

We compare PVB, PGS and PBP within the unified multi-channel MapReduce framework in three tasks: topic modeling accuracy measured by held-out log-likelihood, search matching measured by mean average precision (MAP), and online advertising system by area under curve (AUC) of click-through-rate (CTR) prediction. We compose the training corpus of search queries received in recent months from Tencent (called SOSO data set). The SOSO corpus contains one billion search queries with 4.5 word tokens per query on average and takes 17.2GB storage space. The vocabulary size of SOSO is around 2.1×10^5. We use the subset of SOSO data set in our comparative study. The subset contains about 10 million queries and takes 216MB storage space. Each query is a document in the LDA model. We set $M = N = 10$ to partition the data into 100 data blocks, and set the same Dirichlet hyper-parameters $\alpha = 5/K$ and $\beta = 0.01$ in parallel LDA algorithms. We carried out experiments on a cluster of 257 machines with XFS of Tencent, where each machine has 16 cores.

4.1 Topic Modeling Performance

We evaluate the topic modeling performance by two measures: 1) Scalability: the ability to handle a growing number of topics; 2) Accuracy: the held-out log-likelihood achieved by increasing number of iterations (We randomly select 10^4 queries as held-out set for calculating the held-out log-likelihood (14)). Fig. 4A shows the training time per iteration with the increasing number of topics $K =$

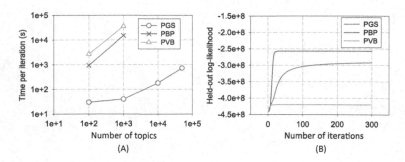

Fig. 4. (A)The runtime (second) as a function of number of topics; (B)The held-out log-likelihood as a function of number of training iterations

$\{10^2, 10^3, 10^4, 5 \times 10^4\}$. The training time per iteration of PVB/PBP is too long to be shown when $K = \{10^4, 5 \times 10^4\}$. Clearly, PGS is much more scalable to the larger number of topics than PVB and PBP (less per-iteration training time). There are three reasons why PGS can scale out to more number of topics. First, PGS uses integer type as well as sparse storage for model parameters [5], so that its memory requirement is insensitive to the number of topics as shown in Table 3. Using the same amount of memory, PGS can process the larger number of topics. Second, PGS uses the efficient sampling technique [5], whose time complexity is insensitive to the number of topics. Third, the reducers of PGS consume significantly less cost in synchronizing model parameters in the sparse format and integer type. Since the processed queries are very short (less than 5 words), it is memory-efficient to store the labeling configuration **z** rather than the document-topic distribution $\hat{\theta}_d(k)$. As a comparison, PVB and PBP have to sore parameters in floating type and full/dense format, leading to more synchronization time, memory consumption and I/O costs. PBP is faster than PVB due to the lack of complicated digamma function in Table 3. The results are consistent with our complexity analysis in Table 3.

Fig. 4B shows the held-out log-likelihood as a function of the number of training iterations (fixing $K = 2000$). We see that PBP reaches the highest value using around 20 iterations, PGS converges at the middle level value in around 100 iterations, while PVB converges at the lowest level value in around 10 iterations. Taking per-iteration training time into account, it takes PGS 1.13 hours, PBP 16 hours and PVB 33 hours for convergence. The reasons why PBP reaches the highest held-out log-likelihood are as follows. First, PBP's objective is to infer the posterior $p(\boldsymbol{\theta}, \boldsymbol{\phi}|\mathbf{x}, \alpha, \beta)$, which determines the log-likelihood (14). Second, PBP's E-step can touch the lower-bound of the true posterior without approximation. Through Fig. 4, we may make a conclusion that PVB is neither scalable nor accurate in topic modeling when compared with PGS and PBP. Although PGS is more scalable than PBP, its topic modeling accuracy measured by held-out log-likelihood is lower than that of PBP.

Fig. 5 to compare the top 5 words in 5 topics learned by PGS, PBP and PVB. We select the top 5 words by ranking $\hat{\phi}_w(k)$ (in descending order) in terms of

Methods	Soso-query with $K = 2000$				
PGS	工作(0.0735683) work	建设(0.031067) development	方案(0.0176651) project	报告(0.0151843) report	基层(0.0136017) substratum
	魔兽世界(0.0666045) World of Warcraft	wow(0.045114) wow	天赋(0.0206433) talent	魔兽(0.0192115) warcraft	装备(0.012199) equipment
	公司(0.140368) company	有限(0.118591) corporation	科技(0.0643346) technology	email(0.0505498) email	北京(0.0217445) Peking
	公司(0.144686) company	有限(0.125053) corporation	email(0.0674655) email	科技(0.0526834) technology	电子(0.0200016) electronic
	考试(0.0953906) examination	期末(0.0933077) end of semester	学期(0.0603608) semester	年级(0.0569946) grade	答案(0.0529762) answer
PBP	有限(0.292777) corporation	公司(0.276463) company	email(0.0685029) email	科技(0.0639485) technology	深圳市(0.0217818) Shenzhen
	公司(0.255899) company	有限(0.237999) corporation	email(0.133271) email	上海(0.0514269) Shanghai	贸易(0.0377472) trading
	答案(0.119748) answer	试题(0.0819238) question	考试(0.0587892) examination	试卷(0.046424) examination paper	数学(0.0419715) math
	歌(0.30235) song	首(0.204176) a	唱(0.120012) sing	歌词(0.0972891) lyric	歌曲(0.0322284) tune
	年级(0.249817) grade	小学(0.103819) primary school	下册(0.0949679) next volume	语文(0.090856) language	数学(0.0655755) math
PVB	qq(0.00390474) qq	钱(0.00374481) money	公司(0.00364183) company	吃(0.00343261) eat	手机(0.00331416) telephone
	qq(0.00533542) qq	手机(0.00475849) telephone	钱(0.00470225) money	公司(0.0041574) company	吃(0.00387107) eat
	吃(0.00508503) eat	公司(0.00501708) company	qq(0.00460769) qq	手机(0.00401276) telephone	图片(0.00346598) picture
	公司(0.0087744) company	吃(0.00625295) eat	有限(0.00549546) corporation	钱(0.00463175) money	怀孕(0.00419475) pregnant
	公司(0.00724935) company	有限(0.00380177) corporation	钱(0.00374403) money	吃(0.00344524) eat	dnf(0.00304602) dnf

Fig. 5. Topic top words: top 5 words of 5 topics inferred by PGS, PVB, PBP

Table 4. Comparison of topic features learned by PGS, PVB, PBP in information retrieval ($K = 2000$).

Model	baseline	PVB	PBP	PGS
MAP	0.1603	0.0005	0.1668	0.2122

w in each topic k. For a better illustration, we translate these Chinese words into English, while "qq" represents an instant message software from Tencent. We see that the topic topic words generated by PBP have highest probability and they are much more correlative (subjective judgement). The top topic words produced by PGS are partly the same as those by PBP, but the word probability is smaller and less concentrated than those by PBP. PVB generates almost the same words in these 5 topics and these words have almost the same probabilities, which are small and dispersive. These duplicated topics are unfavorable in many applications such as search engine and online advertising, because given one query it is hard to decide which topic this query belongs to.

4.2 Industrial Applications

We compare the topic features produced by PVB, PGS and PBP. The topic feature is the likelihood $p(w|d)$ of a vocabulary word w given a query d [23]:

$$p(w|d) = \sum_{k=1}^{K} \phi_w(k)\theta_d(k). \tag{15}$$

The W-length vector $p(w|d)$ is compatible with the standard word vector space model. Ranking $p(w|d)$ in descending order yields top likely topic features of the query. In the following experiments, we fix $K = 2000$ to obtain the topic features. The difference between word vectors and topic features is that word vectors are sparse while topic features are compact in latent semantic space. In real-world applications, we prefer compact features because they can be used by different classifiers.

4.3   Search Engine

Search engines use the well-known vector space model in information retrieval and compute cosine similarity between queries and documents in their vector representations. For information retrieval, our test-bed is a real search engine in Tencent, www.soso.com, which ranks the fourth largest in China market. The test set comes from human-labeled data used for relevance evaluation of the search engine. This set consists of 922 common queries and 58, 853 query-document pairs with human labeled relevance rate. Each query-document pair was rated by three human editors and the average rate was taken. We use cosine similarity to measure the similarity between one query and one document both represented by word vectors. Top similar documents are retrieved to be relevant to the given query. We use the similarity in word vector space to do information retrieval as the baseline in Table 4. We compute MAP using TREC evaluation tool [24]. The higher MAP means the better retrieval performance. We use topic features (15) generated by PVB, PGS and PBP to replace the word vector in the retrieval system.

Table 4 shows the MAP values of all algorithms in information retrieval. The MAP values of PBP and PGS are higher than that of baseline (PBP 4% improvement and PGS 32.4% improvement), while the MAP of PVB is significantly lower. This result implies that the topic features learned by PVB cannot provide differentiable semantics in information retrieval. Indeed, these topic features significantly reduce the relevance between queries and retrieved documents when compared with the baseline. From the top words shown in Fig. 5, we can see that PVB find many duplicated topics that are semantically ambiguous. Although PBP can recover topic features with the highest likelihood in Fig. 4B, it has a smaller MAP value than PGS in Table 4. Analyzing the topic features by PBP and PGS, we find that PBP cannot differentiate similar queries in latent topic space. For example, two queries such as "fashion clothes" and "new clothes" will have the similar topic feature vectors, where the top likely 40 topic features are almost the same. On the contrary, PGS, due to its random sampling process, can better understand the subtle semantic difference between these two queries, where the top 5 likely topic features of two queries are different but useful. The experimental results confirm that PGS is more suitable to learn the latent semantic information for short documents like queries. One major reason is that PGS infers the posterior $p(\mathbf{z}|\mathbf{x}, \alpha, \beta)$ to find the best labeling configuration $\mathbf{z}^*$ over words, which provides more explainable topic features for each

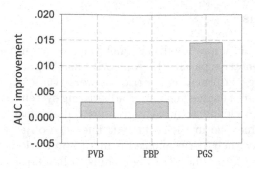

Fig. 6. The AUC improvement of CTR prediction using topic features by PGS, PVB and PBP

word. As a comparison, PBP integrates out **z** during inference and cannot find the best topic labeling configuration.

4.4 Online Advertising System

Online advertising has been a fundamental financial support of the many free Internet services. Most contemporary online advertising systems follow the Generalized Second Price auction model, which requires that the system is able to predict the CTR of an ad. One of the key questions is the availability of suitable input features that allow accurate CTR prediction for a given ad impression. These features can be grouped into three categories: ad features including bid phrases, ad title, landing page, and a hierarchy of advertiser account, campaign, ad group and ad creative. User features include recent search queries, and user behavior data. Context features include display location, geographic location, content of page under browsing, and time. Most of these features are text data in word vector space. We learn an L_1-regularized log-linear model as the baseline, which uses a set of text and other features such as ad title, content of page under browsing, content of landing page, ad group id, demographic information of users, categories of ad group and categories of the page under browsing. Besides the traditional ad features mentioned above, we add their topic features (15) by PVB, PGS and PBP as input to L_1-regularized log-linear model.

In Fig. 6, we compare the topic features with the baseline model, provided by the Task 2 in KDD Cup 2012, in performance measure Area Under ROC Curve (AUC). The higher AUC value the better prediction performance. The AUC value of the baseline model is 0.74393. Fig. 6 shows the relative improvement of AUC over the baseline model. PGS gains the highest AUC improvement (around 1.5% improvement). The AUC improvements of PVB and PBP are slight when compared with the baseline model. Analyzing the weights of the added topic features, we find that the weights of topic features by PVB and PBP are almost zeros, implying that these topic features do not influence the prediction result. These results are consistent with those in search engine.

5   Conclusions

In this study, we build a novel multi-channel MapReduce framework to compare fairly three representative parallel LDA algorithms. Although PBP achieves the best topic modeling performance, it performs worse than PGS in two industrial applications like search engine and online advertising system. PVB does not work well in both topic modeling accuracy and industrial applications. We think that the major reason may lie in their different inference objectives. In our future work, we will study their performance on more applications, and improve the parallel architecture for online LDA algorithms to handle streaming data.

Acknowledgments. This work was supported by NSFC(Grant No. 61373092, 61272449, 61202029, 61003154 and 61033013), National Science Foundation of the Jiangsu Higher Education Institutions of China (Grant No. 12KJA520004), Jiangsu Key Technology R&D Program (Grant No. BE2014005) and Guangdong Province key Laboratory Open R&D Program of Popular High Performance Computers (Grant No. SZU-GDPHPCL-2012-09).

References

1. Blei, D.M., Ng, A.Y., Jordan, M.I.: Latent Dirichlet allocation. Journal of Machine Learning Research **3**, 993–1022 (2003)
2. Robbins, H., Monro, S.: A stochastic approximation method. The Annals of Mathematical Statistics **22**(3), 400–407 (1951)
3. Griffiths, T.L., Steyvers, M.: Finding scientific topics. Proc. Natl. Acad. Sci. **101**, 5228–5235 (2004)
4. Porteous, I., Newman, D., Ihler, A.T., Asuncion, A.U., Smyth, P., Welling, M.: Fast collapsed gibbs sampling for latent Dirichlet allocation. In: KDD, pp. 569–577 (2008)
5. Yao, L., Mimno, D.M., McCallum, A.: Efficient methods for topic model inference on streaming document collections. In: KDD, pp. 937–946 (2009)
6. Teh, Y.W., Newman, D., Welling, M.: A collapsed variational bayesian inference algorithm for latent Dirichlet allocation. In: NIPS, pp. 1353–1360 (2006)
7. Asuncion, A.U., Welling, M., Smyth, P., Teh, Y.W.: On smoothing and inference for topic models. In: UAI, pp. 27–34 (2009)
8. Zeng, J., Cheung, W.K., Liu, J.: Learning topic models by belief propagation. IEEE Trans. Pattern Anal. Mach. Intell. **35**(5), 1121–1134 (2013)
9. Hoffman, M.D., Blei, D.M., Bach, F.R.: Online learning for latent Dirichlet allocation. In: NIPS, pp. 856–864 (2010)
10. Yan, F., Xu, N., Qi, Y.: Parallel inference for latent Dirichlet allocation on graphics processing units. In: NIPS, pp. 2134–2142 (2009)
11. Asuncion, A.U., Smyth, P., Welling, M.: Asynchronous distributed learning of topic models. In: NIPS, pp. 81–88 (2008)
12. Wang, Y., Bai, H., Stanton, M., Chen, W.-Y., Chang, E.Y.: PLDA: parallel latent dirichlet allocation for large-scale applications. In: Goldberg, A.V., Zhou, Y. (eds.) AAIM 2009. LNCS, vol. 5564, pp. 301–314. Springer, Heidelberg (2009)
13. Smola, A.J., Narayanamurthy, S.M.: An architecture for parallel topic models. PVLDB **3**(1), 703–710 (2010)

14. Liu, Z., Zhang, Y., Chang, E.Y., Sun, M.: PLDA+: Parallel latent Dirichlet allocation with data placement and pipeline processing. ACM TIST **2**(3), 26 (2011)
15. Zhai, K., Boyd-Graber, J.L., Asadi, N., Alkhouja, M.L.: Mr. LDA: a flexible large scale topic modeling package using variational inference in MapReduce. In: WWW, pp. 879–888 (2012)
16. Ahmed, A., Aly, M., Gonzalez, J., Narayanamurthy, S.M., Smola, A.J.: Scalable inference in latent variable models. In: WSDM, pp. 123–132 (2012)
17. Wang, C., Blei, D.M., Li, F.F.: Simultaneous image classification and annotation. In: CVPR, pp. 1903–1910 (2009)
18. Yan, J., Zeng, J., Liu, Z.Q., Gao, Y.: Towards big topic modeling. arXiv:1311.4150 (2013)
19. Blei, D.M.: Introduction to probabilistic topic models. Communications of the ACM, 77–84 (2012)
20. Zeng, J.: A topic modeling toolbox using belief propagation. J. Mach. Learn. Res. **13**, 2233–2236 (2012)
21. Zaharia, M., Chowdhury, M., Franklin, M.J., Shenker, S., Stoica, I.: Spark: cluster computing with working sets. In: HotCloud (2010)
22. Zhuang, Y., Chin, W.S., Juan, Y.C., Lin, C.J.: A fast parallel sgd for matrix factorization in shared memory systems. In: RecSys, pp. 249–256 (2013)
23. Wang, Y., Zhao, X., Sun, Z., Yan, H., Wang, L., Jin, Z., Wang, L., Gao, Y., Law, C., Zeng, J.: Towards topic modeling for big data. ACM Transactions on Intelligent Systems and Technology (2014)
24. Voorhees, E.M., Harman, D.K. (eds.): TREC: Experiment and evaluation in information retrieval. MIT Press, Cambridge (2005)

Distributed Newton Methods for Regularized Logistic Regression

Yong Zhuang, Wei-Sheng Chin, Yu-Chin Juan, and Chih-Jen Lin[✉]

Department of Computer Science, National Taiwan University, Taipei, Taiwan
{r01922139,d01944006,r01922136,cjlin}@csie.ntu.edu.tw

Abstract. Regularized logistic regression is a very useful classification method, but for large-scale data, its distributed training has not been investigated much. In this work, we propose a distributed Newton method for training logistic regression. Many interesting techniques are discussed for reducing the communication cost and speeding up the computation. Experiments show that the proposed method is competitive with or even faster than state-of-the-art approaches such as Alternating Direction Method of Multipliers (ADMM) and Vowpal Wabbit (VW). We have released an MPI-based implementation for public use.

1 Introduction

In recent years, distributed data classification becomes popular. However, it is known that a distributed training algorithm may involve expensive communication cost between machines. The aim of this work is to construct a scalable distributed training algorithm for large-scale logistic regression.

Logistic regression is a binary classifier that has achieved a great success in many fields. Given a data set with l instances $(y_i, \boldsymbol{x}_i)$, $i = 1, \ldots, l$, where $y_i \in \{-1, 1\}$ is the label and $\boldsymbol{x}_i$ is an n-dimensional feature vector, we consider regularized logistic regression by solving the following optimization problem to obtain the model $\boldsymbol{w}$.

$$\min_{\boldsymbol{w}} \quad \tfrac{1}{2}\|\boldsymbol{w}\|^2 + C\sum_{i=1}^{l} \log \sigma(y_i \boldsymbol{w}^T \boldsymbol{x}_i), \tag{1}$$

where $\sigma(y_i \boldsymbol{w}^T \boldsymbol{x}_i) = 1 + \exp(-y_i \boldsymbol{w}^T \boldsymbol{x}_i)$ and C is the regularization parameter.

In this paper, we design a distributed Newton method for logistic regression. Many algorithmic and implementation issues are addressed in order to reduce the communication cost and speed up the computation. For example, we investigate different ways to conduct parallel matrix-vector products and discuss data formats for storing feature values. Our resulting implementation is experimentally shown to be competitive with or faster than ADMM and VW, which were considered state-of-the-art for distributed machine learning. Recently, Lin et al. [12] implement a variant of our approach on Spark[1] through private communication, but they mainly focus on the efficient use of Spark.

---

[1] https://spark.apache.org/

© Springer International Publishing Switzerland 2015
T. Cao et al. (Eds.): PAKDD 2015, Part II, LNAI 9078, pp. 690–703, 2015.
DOI: 10.1007/978-3-319-18032-8_54

This paper is organized as follows. Section 2 discusses existing approaches for distributed data classification. In Section 3, we present our implementation of a distributed Newton method. Experiments are in Section 4 while we conclude in Section 5.

2   Existing Methods for Distributed Classification

Many distributed algorithms have been proposed for linear classification. ADMM has recently emerged as a popular method for distributed convex optimization. Although ADMM is a known optimization method for decades, only recently some (e.g., [3,19]) show that it is particularly suitable for distributed machine learning. Zinkevich et al. [20] proposed a way to parallelize stochastic gradient methods. Besides, parallel coordinate descent methods have been considered for linear classification (e.g., [2,4,14]). Agarwal et al. [1] recently report that the package VW [9] is scalable and efficient in distributed environments. VW applies a stochastic gradient method in the beginning, then switches to parallel LBFGS [13] for a faster final convergence.

In the rest of this section, we describe ADMM and VW because they are considered state-of-the-art and therefore are involved in our experiments.

2.1   ADMM for Logistic Regression

Zhang et al. [19] apply ADMM on linear support vector machine with squared hinge loss. Here we modify it for logistic regression. Assume data indices $\{1, \ldots, l\}$ are partitioned to J sets $M_1, \ldots, M_J$ to indicate data on J machines. We can rewrite the problem (1) to the following equivalent form.

$$\min_{\boldsymbol{w}_1, \ldots, \boldsymbol{w}_J, \boldsymbol{z}} \quad \frac{1}{2}\|\boldsymbol{z}\|^2 + C\sum_{j=1}^{J}\sum_{i \in M_j} \log \sigma(y_i \boldsymbol{w}_j^T \boldsymbol{x}_i) \tag{2}$$
$$\text{subject to} \quad \boldsymbol{z} = \boldsymbol{w}_j, \quad j = 1, \ldots, J.$$

All optimal $\boldsymbol{z}, \boldsymbol{w}_1, \ldots, \boldsymbol{w}_J$ are the same as the solution of the original problem (1). ADMM repeatedly performs (3)-(5) to update primal variables $\boldsymbol{w}$ and $\boldsymbol{z}$, and Lagrangian dual variable $\boldsymbol{\mu}$ using the following rules.

$$\boldsymbol{w}_j^{k+1} = \arg\min_{\boldsymbol{w}_j} \frac{\rho}{2}\|\boldsymbol{w}_j - \boldsymbol{z}^k - \boldsymbol{\mu}_j^k\|^2 + C\sum_{i \in M_j} \log \sigma(y_i \boldsymbol{w}_j^T \boldsymbol{x}_i), \tag{3}$$

$$\boldsymbol{z}^{k+1} = \arg\min_{\boldsymbol{z}} \frac{1}{2}\|\boldsymbol{z}\|^2 + \frac{\rho}{2}\sum_{j=1}^{J}\|\boldsymbol{z} - \boldsymbol{w}_j^{k+1}\|^2 + \rho\sum_{j=1}^{J}(\boldsymbol{\mu}_j^k)^T(\boldsymbol{z} - \boldsymbol{w}_j^{k+1}), \tag{4}$$

$$\boldsymbol{\mu}_j^{k+1} = \boldsymbol{\mu}_j^{k+1} + \boldsymbol{z}^{k+1} - \boldsymbol{w}_j^{k+1}, \tag{5}$$

where $\rho > 0$ is a chosen penalty parameter. Depending on local data, the local model $\boldsymbol{w}_j$ on the jth machine can be independently updated using (3). To calculate the closed-form solution of $\boldsymbol{z}$ in (4), each machine must collect local models

$\boldsymbol{w}_j, \forall j$, so an $\mathcal{O}(n)$ amount of local data from each machine is communicated across the network. The iterative procedure ensures that under some assumptions, as $k \to \infty$, $\{\boldsymbol{z}^k\}$ approaches an optimum of (1).

Recall that the communication in (4) involves $\mathcal{O}(n)$ data per machine. Obviously the cost is high for a data set with a huge number of instances (i.e., $n \gg l$). Fortunately, Boyd et al. [3] mention that splitting the data set across its features can transform the scale of the communicated data to $\mathcal{O}(l)$. For example, if we have J machines, the data matrix X is partitioned to $X_1, \ldots, X_J$. Note that

$$X = [\boldsymbol{x}_1, \ldots, \boldsymbol{x}_l]^T = [X_1, \ldots, X_J].$$

However, the optimization process becomes different from (2)-(5) [3].

2.2 VW for Logistic Regression

VW [1] is a machine learning package supporting distributed training. Firstly, by using only local data at each machine, it applies stochastic gradient method with adaptive learning rate [5]. Then, to get a faster convergence, VW weightedly averages the model as the initial solution for the subsequent quasi Newton method [13] on the whole data. The stage of applying stochastic gradient updates goes through all local data once for approximately solving the following sub-problem.

$$\sum_{i \in M_j} (\tfrac{1}{2} \|\boldsymbol{w}\|^2 + C \log \sigma(y_i \boldsymbol{w}^T \boldsymbol{x}_i)).$$

In the second stage, the objective function of (1) is considered and the quasi Newton method applied is LBFGS, which uses m vectors to approximate inverse Hessian (m is a user-specific parameter). To support both numerical and string feature indices in the input data, VW uses feature hashing to have a fast feature lookup. That is, it applies a hash function on the feature index to generate a new index for that feature value.

We discuss the communication cost of VW, which supports only the instance-wise data split. For the stage of running a stochastic gradient method, there is no communication until VW weightedly averages local $\boldsymbol{w}_j, \forall j$, where $\mathcal{O}(n)$ data on each machine must be aggregated. For LBFGS, it collects $\mathcal{O}(n)$ results on each machine to calculate the function value and the gradient. Therefore, the communication cost per LBFGS iteration is similar to that of each ADMM iteration under the instance-wise data split.

3 Distributed Newton Methods

In this section, we describe our proposed implementation of a distributed Newton method.

3.1 Newton Methods

We denote the objective function of (1) as $f(\boldsymbol{w})$. At each iteration, a Newton method updates the current model $\boldsymbol{w}$ by

$$\boldsymbol{w} \leftarrow \boldsymbol{w} + \boldsymbol{s}, \tag{6}$$

where $\boldsymbol{s}$, the Newton direction, is obtained by minimizing

$$\min_{\boldsymbol{s}} \quad q(\boldsymbol{w}), \quad q(\boldsymbol{w}) = \nabla f(\boldsymbol{w})^T \boldsymbol{s} + \tfrac{1}{2}\boldsymbol{s}^T \nabla^2 f(\boldsymbol{w})\boldsymbol{s}. \tag{7}$$

Because the Hessian matrix $\nabla^2 f(\boldsymbol{w})$ is positive definite, we can solve the following linear system instead.

$$\nabla^2 f(\boldsymbol{w})\boldsymbol{s} = -\nabla f(\boldsymbol{w}). \tag{8}$$

For data with a huge number of features, $\nabla^2 f(\boldsymbol{w})$ becomes too large to be stored. Hessian-free methods have been developed to solve (8) without explicitly forming $\nabla^2 f(\boldsymbol{w})$. For example, Keerthi and DeCoste [8] and Lin et al. [11] apply conjugate gradient (CG) methods to solve (8). CG is an iterative procedure that requires a Hessian-vector product $\nabla^2 f(\boldsymbol{w})\boldsymbol{v}$ at each step. For logistic regression, we note that

$$\nabla^2 f(\boldsymbol{w})\boldsymbol{v} = \boldsymbol{v} + CX^T(D(X\boldsymbol{v})), \tag{9}$$

where D is a diagonal matrix with

$$D_{ii} = \frac{\sigma(y_i \boldsymbol{w}^T \boldsymbol{x}_i) - 1}{\sigma(y_i \boldsymbol{w}^T \boldsymbol{x}_i)^2}.$$

From (9), we can clearly see that a sequence of matrix-vector products is sufficient to finish the Hessian-vector product. Because $\nabla^2 f(\boldsymbol{w})$ is not explicitly formed, the memory difficulty is alleviated.

The update (6) does not guarantee the convergence to an optimum, so we apply a trust region method [11]. A constraint $\|\boldsymbol{s}\|_2 \leq \Delta$ is added to (7), where Δ is called the radius of the trust region. At each iteration, we check the ratio ρ in (10) to ensure the reduction of the function value.

$$\rho = \frac{f(\boldsymbol{w} + \boldsymbol{s}) - f(\boldsymbol{s})}{q(\boldsymbol{w})}. \tag{10}$$

If ρ is not large enough, then $\boldsymbol{s}$ is rejected and $\boldsymbol{w}$ is kept. Otherwise, $\boldsymbol{w}$ is updated by (6). Then, the radius Δ is adjusted based on ρ [11].

Although Hessian-vector products are the main computational bottleneck of the above Hessian-free approach, we note that in Newton methods other operations such as function and gradient evaluations are also important. For example, the function value in (1) requires the calculation of $X\boldsymbol{w}$. Based on this discussion, we can conclude that a scalable distributed Newton method relies on the effective parallelization of multiplying the data matrix with a vector. We will discuss more details in the rest of this section.

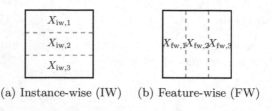

(a) Instance-wise (IW)   (b) Feature-wise (FW)

Fig. 1. Two methods to distributedly store training data

3.2   Instance-Wise and Feature-Wise Data Splits

Following the discussion in Section 2.1, we may split training data instance-wisely or feature-wisely to different machines. An illustration is in Figure 1, in which $X_{\mathrm{iw},j}$ or $X_{\mathrm{fw},j}$ represents the jth segment of data stored in the jth machine. We will show that the two splits lead to different communication cost. Subsequently, we use "IW/iw" and "FW/fw" to denote instance-wise and feature-wise splits.

To discuss the distributed operations easily, we use vector notation to rewrite (1) as

$$f(\boldsymbol{w}) = \frac{1}{2}\|\boldsymbol{w}\|^2 + C\log(\sigma(YX\boldsymbol{w})) \cdot \boldsymbol{e}, \qquad (11)$$

where X and Y are defined in Section 2.1, $\log(\cdot)$ and $\sigma(\cdot)$ are component-wisely applied to a vector, and "$\cdot$" stands for a dot product. In (11), $\boldsymbol{e} \in \mathbb{R}^{n \times 1}$ is a vector of all ones. Then the gradient can be represented as

$$\nabla f(\boldsymbol{w}) = \boldsymbol{w} + C(YX)^T(\sigma(YX\boldsymbol{w})^{-1} - \boldsymbol{e}). \qquad (12)$$

Next we discuss details of distributed operations under the two different data splits.

Instance-wise Split: Based on (9)-(12), the distributed form of function, gradient values, and Hessian-vector products can be written as

$$f(\boldsymbol{w}) = \tfrac{1}{2}\|\boldsymbol{w}\|^2 + C\bigoplus_{j=1}^{J}\log(\sigma(Y_j X_{\mathrm{iw},j}\boldsymbol{w})) \cdot \boldsymbol{e}_j, \qquad (13)$$

$$\nabla f(\boldsymbol{w}) = \boldsymbol{w} + C\bigoplus_{j=1}^{J}(Y_j X_{\mathrm{iw},j})^T(\sigma(Y_j X_{\mathrm{iw},j}\boldsymbol{w})^{-1} - \boldsymbol{e}_j), \qquad (14)$$

and

$$\nabla^2 f(\boldsymbol{w})\boldsymbol{v} = \boldsymbol{v} + C\bigoplus_{j=1}^{J}X_{\mathrm{iw},j}^T D_j X_{\mathrm{iw},j}\boldsymbol{v}, \qquad (15)$$

where Y_j, $\boldsymbol{e}_j$, and D_j are respectively the sub-matrix, the sub-vector, the sub-matrix of Y, $\boldsymbol{e}$, and D corresponding to instances in the jth machine. We use $\bigoplus$ to denote an *all-reduce* operation [15] that collects results from all machines and redistributes the sum to them. For example, $\bigoplus_{j=1}^{J}\log(\sigma(Y_j X_{\mathrm{iw},j}\boldsymbol{w}))$ means that each machine calculates its own $\log(\sigma(Y_j X_{\mathrm{iw},j}\boldsymbol{w}))$, and then an *all-reduce* summation is performed.

Feature-wise Split: We notice that

$$Xw = \bigoplus_{j=1}^{J} X_{\mathrm{fw},j} w_j, \tag{16}$$

where w_j is a sub-vector of w corresponding to features stored in the jth machine. Therefore, in contrast to IW, each machine maintains only a sub-vector of w. The situation is similar to other vectors such as s. However, to calculate the function value for checking the sufficient decrease (10), $\|w_j\|^2, \forall j$ must be summed and then distributed to all machines. Therefore, the function value is calculated by

$$\tfrac{1}{2} \bigoplus_{j=1}^{J} \|w_j\|^2 + C \log(\sigma(Y \bigoplus_{j=1}^{J} X_{\mathrm{fw},j} w_j)) \cdot e. \tag{17}$$

Similarly, each machine must calculate part of the gradient and the Hessian-vector product:

$$\nabla f(w)_{\mathrm{fw},p} = w_p + C(Y X_{\mathrm{fw},p})^T (\sigma(Y \bigoplus_{j=1}^{J} X_{\mathrm{fw},j} w_j)^{-1} - e) \tag{18}$$

and

$$(\nabla^2 f(w) v)_{\mathrm{fw},p} = v_p + C X_{\mathrm{fw},p}^T D \bigoplus_{j=1}^{J} X_{\mathrm{fw},j} v_j,$$

where, like w, only a sub-vector v_p of v is needed at the pth machine.

We notice that some *all-reduce* operations are needed for inner products in Newton methods. For example, to evaluate the value in (7) we must obtain

$$\nabla f(w)^T s = \bigoplus_{j=1}^{J} \nabla f(w)_{\mathrm{fw},j}^T s_j, \tag{19}$$

where $\nabla f(w)_{\mathrm{fw},j}$ and s_j are respectively the sub-vectors of $\nabla f(w)$ and s stored at the jth machine. The communication cost is not a concern because (19) sums J values rather than vectors in (16). Another difference from the IW approach is that the label vector y is stored in every machine because of the diagonal matrix Y in (17) and (18). It is worth mentioning that to save computational and communication cost, we can cache $\bigoplus_{j=1}^{J} X_{\mathrm{fw},j} w_j$ obtained in (17) for (18).

Analysis: To compare the communication cost between IW and FW, in Table 1 we show the number of operations at each machine and the amount of data sent to all others. From Table 1, to minimize the communication cost, IW and FW should be used for $l \gg n$ and $n \gg l$, respectively. We will confirm this property through experiments in Section 4.

3.3   Other Implementation Techniques

Load Balancing: The parallel matrix-vector product requires that all machines finish their local tasks first. To reduce the synchronization cost, we should let machines have a similar computational load. Now the computational cost is related to the number of non-zero elements, so we split data in a way such that each machine contains data of a similar number of non-zero values.

Table 1. A comparison between IW and FW on the amount of data distributed from one machine to all others. J is the number of machines and nnz is the number of non-zero elements in the data matrix. For FW, the calculation of $\nabla f(\boldsymbol{w})$ does not involve communication because we mentioned that $\bigoplus_{j=1}^{J} X_{\text{fw},j}\boldsymbol{w}_j$ is available while calculating $f(\boldsymbol{w})$.

	# of operations		# data sent to other machines	
	IW	FW	IW	FW
$f(\boldsymbol{w})$	$\mathcal{O}(\text{nnz}/J)$	$\mathcal{O}(\text{nnz}/J)$	$\mathcal{O}(1)$	$\mathcal{O}(l)$
$\nabla f(\boldsymbol{w})$	$\mathcal{O}(\text{nnz}/J)$	$\mathcal{O}(\text{nnz}/J)$	$\mathcal{O}(n)$	0
$\nabla^2 f(\boldsymbol{w})\boldsymbol{v}$	$\mathcal{O}(\text{nnz}/J)$	$\mathcal{O}(\text{nnz}/J)$	$\mathcal{O}(n)$	$\mathcal{O}(l)$
inner product	$\mathcal{O}(n)$	$\mathcal{O}(n/J)$	0	$\mathcal{O}(1)$

Data Format: Lin et al. [11] discuss two approaches to store the sparse data matrix X in an implementation of Newton methods: compressed sparse row (CSR) and compressed sparse column (CSC). They conclude that because of the possibility of storing a whole row or column into a higher-level memory (e.g., cache), CSR and CSC are more suitable for $l \gg n$ and $n \gg l$, respectively. Because we split data so that each machine has a similar number of non-zero elements, the number of columns/rows of the sub-matrix may vary. In our implementation, we dynamically decide the sparse format based on if the sub-matrix's number of rows is greater than columns.

Speeding Up Hessian-vector Product: Instead of sequentially calculating $X\boldsymbol{v}$, $D(X\boldsymbol{v})$, and $X^T(D(X\boldsymbol{v}))$ in (9), we can re-write $X^T DX\boldsymbol{v}$ as

$$\sum_{i=1}^{l} D_{ii}(\boldsymbol{x}_i(\boldsymbol{x}_i^T\boldsymbol{v})). \tag{20}$$

Then the data matrix X is accessed only once rather than twice. However, this technique can only be applied for instance-wisely split data in the CSR format. If the data is split by features, then calculating each $\boldsymbol{x}_i^T\boldsymbol{v}$ needs an *all-reduce* operation. The l *all-reduce* summations in (20) cause too high communication cost in practice.

4   Experiments

In this section, we begin with describing a specific Newton method used for our implementation. Then we evaluate techniques proposed in Section 3, followed by a detailed comparison between ADMM, VW, and the distributed Newton method on objective function values and test accuracy.

4.1   Truncated Newton Method

In Section 3.1, using CG to find the direction $\boldsymbol{s}$ may struggle with too many iterations. To alleviate this problem, we consider an approach by Steihaug [16] that employs CG to approximately minimize (7) under the constraint of $\boldsymbol{s}$ being in the trust region. With the Newton direction $\boldsymbol{s}$ obtained approximately, the

Table 2. Left: The statistics of each data set. **Right:** Communication and total running time (in seconds) of the distributed Newton method using IW and FW strategies.

Data set	l	n	#nonzeros	C	Communication IW	FW	Total IW	FW
yahoo-japan	140,963	832,026	18,738,315	0.5	166.22	13.78	169.73	15.15
yahoo-korea	368,444	3,052,939	125,190,807	2	1,185.32	189.72	1,215.76	207.53
url	2,396,130	3,231,961	277,058,644	2	9,727.89	6,995.23	10,194.23	7,286.63
epsilon	500,000	2,000	1,000,000,000	2	2.16	184.03	11.37	195.26
webspam	350,000	16,609,143	1,304,697,446	32	8,909.51	159.83	9,179.42	199.36

technique is called a truncated Newton method. Lin et al. [11] have applied this method for logistic regression on a single machine. Here we follow their detailed settings for our distributed implementation.

4.2 Experimental Settings

1. **Data Sets:** We use five data sets listed in Table 2 for experiments, and randomly split them into 80%/20% as training set and test set respectively (an exception is epsilon because it has official training/test sets). Except yahoo-japan and yahoo-korea, all others are publicly available.[2]
2. **Platform:** We use 32 machines in a local cluster. Each machine has an 8-core processor with computing power equivalent to 8×1.0GHz 2007 Xeon, and 16GB memory. The network bandwidth is approximately 1Gb/s.
3. **Parameters:** For C in (1), we tried different values in $\{2^{-5}, 2^{-3}, \ldots, 2^5\}$, and use the one that leads to the best performance on the test set.

4.3 IW versus FW

This subsection presents a comparison between two different data splits. We run the Newton method until the following stopping condition is satisfied:

$$\|\nabla f(\boldsymbol{w})\| \leq \epsilon \frac{\min(\text{pos}, \text{neg})}{l} \|\nabla f(\boldsymbol{0})\|, \tag{21}$$

where $\epsilon = 10^{-6}$ and pos/neg are the number of positive and negative data.

In Table 2, we present both communication and total training time. Results are consistent with our analysis in Table 1. For example, IW is better than FW for epsilon, which has $l \gg n$. On the contrary, FW is superior for webspam because $l \ll n$. This experiment reflects that a suitable data split strategy can significantly reduce the communication cost.

By the difference between total and communication time in Table 2, we have the time for computation and synchronization. It is only similar but not the same for IW and FW because of the variance of the cluster's run-time behavior and the algorithmic differences.

---

[2] http://www.csie.ntu.edu.tw/~cjlin/libsvmtools/datasets

4.4 Comparison Between State-of-the-art Methods on Function Values

We include the following methods for the comparison.

- **the distributed trust-region Newton method (TRON)**: It is implemented in C++ with OpenMPI [7].
- **ADMM**: We implement ADMM using C++ with OpenMPI [7]. Following Zhang et al. [19], the sub-problem (3) is approximately solved by a coordinate descent method [17] with a fixed number of iterations, where the number of iterations is decided by a validation procedure. The parameter ρ in (3) follows the setting in [19].
- **VW (version 7.6)**: The package uses sockets for communications and synchronization. Because VW uses the hashing trick for the features indices, hash collisions may cause not only different training/test sets but also worse accuracy. To reduce the possibility of hash collisions, we set the size of the hashing table larger than the original data sets.[3] Although the data set may be slightly different from the original one, we will see that the conclusion of our experiments is not affected. Besides, we observe that the default setting of running a stochastic gradient method and then LBFGS is very slow on webspam, so for this problem, we apply only LBFGS.[4] Except C and the size of features hashing, we use the default parameters in VW.

In Figure 2, we compare VW, ADMM, and the distributed Newton method by checking the relation between the training time and the relative distance from the current function value to the optimum:

$$\frac{|f(\boldsymbol{w}) - f(\boldsymbol{w}^*)|}{|f(\boldsymbol{w}^*)|}.$$

The reference optimal $\boldsymbol{w}^*$ is obtained approximately by running the Newton method with a very small stopping tolerance.[5] We present results of ADMM and distributed Newton using both instance-wise and feature-wise splits. For the sparse format in Section 3.3, the distributed Newton method dynamically chooses CSC or CSR, while ADMM uses CSC because of the requirement of the coordinate descent method for the sub-problem (3).

Results in Figure 2 indicate that with suitable data splits, the distributed Newton method converges faster than ADMM and VW. Regarding IW versus FW, when $l \gg n$, an instance-wise split is more suitable for the distributed

---

[3] We select 2^{20}, 2^{22}, 2^{22}, 2^{12}, and 2^{25} as the size of the hashing table for yahoo-japan, yahoo-korea, url, epsilon, and webspam respectively.

[4] We observe that the vector $\boldsymbol{w}$ obtained after running stochastic gradient methods is a poor initial point for LBFGS to have slow convergence. It is not entirely clear what happened, so further investigation is needed.

[5] The optimal solution of VW with hashing tricks may be different from the other two methods, so we obtain its $\boldsymbol{w}^*$ separately. Because of using the relative distance, we can still compare the convergence speed of different methods.

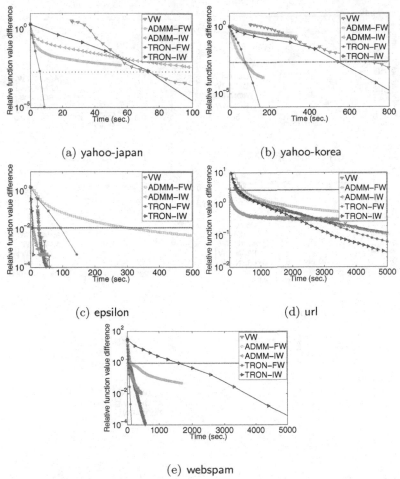

(a) yahoo-japan

(b) yahoo-korea

(c) epsilon

(d) url

(e) webspam

Fig. 2. A comparison on the relative difference to the optimal objective function value. The dotted line indicates the stopping conditions (21) with $\epsilon = 0.01$ has been achieved.

Newton method, while a feature-wise split is better for $n \gg l$. This observation is consistent with Table 2. However, the same result does not hold for ADMM. One possible explanation is that regardless of data splits, the optimization processes of the distributed Newton method are exactly the same. In contrast, ADMM's optimization processes are different under IW and FW strategies [3].

In Figure 2, a horizontal line indicates that the stopping condition of (21) with $\epsilon = 0.01$ has been satisfied. This condition, used as the default condition in the Newton-method implementation of the software LIBLINEAR [6], shows that a model having similar prediction capability to the optimal solution has been obtained. In Figure 2, ADMM can quickly reach the horizontal line for some problems, but is slow for others.

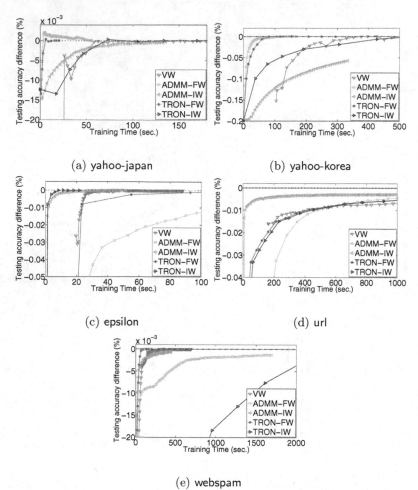

(a) yahoo-japan   (b) yahoo-korea

(c) epsilon   (d) url

(e) webspam

Fig. 3. A test-accuracy comparison among ADMM, VW and the distributed Newton method with different data split strategies. The dotted horizontal line indicates 0 difference to the final accuracy.

4.5   Comparison Between State-of-the-art Methods on Test Accuracy

We present in Figure 3 the relation between the training time and

$$\frac{\text{accuracy} - \text{best\_accuracy}}{\text{best\_accuracy}},$$

where best_accuracy is the best final accuracy obtained by all methods. In the early stage ADMM is better than the other two, while the distributed Newton method gets the final stable performance more quickly on all problems except url.

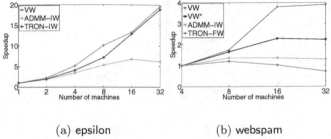

(a) epsilon   (b) webspam

Fig. 4. The speedup of different training methods.

4.6   Speedup

Following Agarwal et al. [1], we compare the speedup of ADMM, VW, and the distributed Newton method for obtaining a fix test accuracy by varying the number of machines. Results arc in Figure 4.

We consider the two largest sets epsilon and webspam for experiments. They have $l \gg n$ and $n \gg l$, respectively. For ADMM and the distributed Newton method, we use the data split strategy leading to the better convergence in Figure 2. Figure 4 shows that for epsilon ($l \gg n$), VW and the distributed Newton method yield a better speedup than ADMM as the number of machines increases, while for webspam ($n \gg l$), the distributed Newton method is better than the other two.

For the problem webspam, the speedup of VW is much worse than epsilon and problems in [1], so we conduct some investigation. For epsilon and data in [1], they have $l \gg n$. Thus we suspect that because $n \gg l$ for webspam and VW considers an instance-wise split, VW suffers from high communication cost. To confirm this point, we conduct an additional experiment in Figure 4(b). A special property of the problem webspam is that it actually has only 680,715 non-zero feature columns, although its feature indices are up to 16 million. We generate a new set by removing zero columns and rerun VW.[6] The result, indicated as VW* in Figure 4(b), clearly shows that the speedup is significantly improved.

Next, we investigate more about the unsatisfactory speedup of ADMM. Because of parallelizing only the matrix-vector products, the numbers of iterations in VW and the distributed Newton method are independent of the number of machines. In contrast, ADMM's number of iterations may significantly vary, because the reformulation in (2) is related to the number of machines. In Figure 5, we present the relation between the number of ADMM iterations and the relative difference to the optimum. It can be seen that as the number of machines increases, the higher number of iterations comes with more computational and communication costs. Thus the speedup of ADMM in Figure 4 is not satisfactory.

---

[6] The hash size is reduced correspondingly from 2^{25} to 2^{20}.

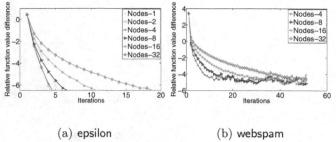

(a) epsilon   (b) webspam

Fig. 5. The relation between the number of ADMM iterations and the decrease of the function value. We show results of using different numbers of machines.

5   Conclusion

To the best of our knowledge, this work is the first comprehensive study on the distributed Newton method for regularized logistic regression. We carefully address important issues for distributed computation including communication and memory locality. An advantage of the proposed distributed Newton method and VW over ADMM is that the optimization processes are independent of the distributed configuration. That is, the number of iterations remains the same regardless of the number of machines. Our experiment shows that in a practical distributed environment, the distributed Newton method is faster and more scalable than ADMM and VW that are considered state-of-the-art for real-world problems.

However, because of requiring differentiability, Newton methods are more restrictive than ADMM. For example, if we consider L1-regularization by replacing $\|w\|^2/2$ in (1) with $\|w\|_1$, the optimization problem becomes non-differentiable, so Newton methods cannot be directly applied.[7]

Our experimental code is available at (removed for the blind review requirements).

References

1. Agarwal, A., Chapelle, O., Dudik, M., Langford, J.: A reliable effective terascale linear learning system. JMLR **15**, 1111–1133 (2014)
2. Bian, Y., Li, X., Cao, M., Liu, Y.: Bundle CDN: A highly parallelized approach for large-scale l1-regularized logistic regression. In ECML/PKDD
3. Boyd, S., Parikh, N., Chu, E., Peleato, B., Eckstein, J.: Distributed optimization and statistical learning via the alternating direction method of multipliers. Found. and Trend. in ML **3**(1), 1–122 (2011)
4. Bradley, J.K., Kyrola, A., Bickson, D., Guestrin, C.: Parallel coordinate descent for l1-regularized loss minimization. In: ICML
5. Duchi, J., Hazan, E., Singer, Y.: Adaptive subgradient methods for online learning and stochastic optimization. JMLR **12**, 2121–2159 (2011)

---

[7] An extension for L1-regularized problems is still possible (e.g., [10,18]), though the algorithm becomes more complicated.

6. Fan, R.-E., Chang, K.-W., Hsieh, C.-J., Wang, X.-R., Lin, C.-J.: LIBLINEAR: A library for large linear classification. JMLR **9**, 1871–1874 (2008)
7. Gabriel, E., Fagg, G.E., Bosilca, G., Angskun, T., Dongarra, J.J., Squyres, J.M., Sahay, V., Kambadur, P., Barrett, B., Lumsdaine, A., Castain, R.H., Daniel, D.J., Graham, R.L., Woodall, T.S.: Open MPI: Goals, concept, and design of a next generation MPI implementation. In: European PVM/MPI Users' Group Meeting, pp. 97–104 (2004)
8. Keerthi, S.S., DeCoste, D.: A modified finite Newton method for fast solution of large scale linear SVMs. JMLR **6**, 341–361 (2005)
9. Langford, J., Li, L., Strehl, A.: Vowpal Wabbit (2007). https://github.com/JohnLangford/vowpal_wabbit/wiki
10. Lin, C.-J., Moré, J.J.: Newton's method for large-scale bound constrained problems. SIAM J. Optim. **9**, 1100–1127 (1999)
11. Lin, C.-J., Weng, R.C., Keerthi, S.S.: Trust region Newton method for large-scale logistic regression. JMLR **9**, 627–650 (2008)
12. Lin, C.-Y., Tsai, C.-H., Lee, C.-P., Lin, C.-J.: Large-scale logistic regression and linear support vector machines using spark. In: IEEE BigData (2014)
13. Liu, D.C., Nocedal, J.: On the limited memory BFGS method for large scale optimization. Math. Program. **45**(1), 503–528 (1989)
14. Richtárik, P., Takáč, M.: Parallel coordinate descent methods for big data optimization. Math. Program (2012) (Under revision)
15. Snir, M., Otto, S.: MPI-The Complete Reference: The MPI Core. MIT Press, Cambridge (1998)
16. Steihaug, T.: The conjugate gradient method and trust regions in large scale optimization. SIAM J. on Num. Ana. **20**, 626–637 (1983)
17. Yu, H.-F., Huang, F.-L., Lin, C.-J.: Dual coordinate descent methods for logistic regression and maximum entropy models. MLJ **85**, 41–75 (2011)
18. Yuan, G.-X., Chang, K.-W., Hsieh, C.-J., Lin, C.-J.: A comparison of optimization methods and software for large-scale l1-regularized linear classification. JMLR **11**, 3183–3234 (2010)
19. Zhang, C., Lee, H., Shin, K.G.: Efficient distributed linear classification algorithms via the alternating direction method of multipliers. In: AISTATS (2012)
20. Zinkevich, M., Weimer, M., Smola, A., Li, L.: Parallelized stochastic gradient descent. In NIPS (2010)

Recommendation

Coupled Matrix Factorization Within Non-IID Context

Fangfang Li$^{(\boxtimes)}$, Guandong Xu, and Longbing Cao

Advanced Analytics Institute, University of Technology, Sydney, Australia
Fangfang.Li@student.uts.edu.au,
{Guandong.Xu,Longbing.Cao}@uts.edu.au

Abstract. Recommender systems research has experienced different stages such as from user preference understanding to content analysis. Typical recommendation algorithms were built on the following bases: (1) assuming users and items are IID, namely independent and identically distributed, and (2) focusing on specific aspects such as user preferences or contents. In reality, complex recommendation tasks involve and request (1) personalized outcomes to tailor heterogeneous subjective preferences; and (2) explicit and implicit objective coupling relationships between users, items, and ratings to be considered as intrinsic forces driving preferences. This inevitably involves the non-IID complexity and the need of combining subjective preference with objective couplings hidden in recommendation applications. In this paper, we propose a novel generic coupled matrix factorization (CMF) model by incorporating non-IID coupling relations between users and items. Such couplings integrate the intra-coupled interactions within an attribute and inter-coupled interactions among different attributes. Experimental results on two open data sets demonstrate that the user/item couplings can be effectively applied in RS and CMF outperforms the benchmark methods.

1 Introduction

Recommender systems (RS) become increasingly important as they deeply involve our daily living, online, social, mobile and business activities. Typically, a set of users and items are involved, where each user u rates various items according to his/her respective preferences (embodied by preference rates) [12]. A new rate or item is then recommended to a user based on the rating behaviors of similar users on existing items.

Often recommendation algorithms come up with the outcomes based on the aggregated understanding of individual commonality. A rate is then predicted for a new item to a given user or a new user for a given item. The performance of applying such algorithms for real-time recommendation for specific users and items is often not very impressive. There are two important aspect that have not been considered thoroughly in RS. (1) The heterogeneity between users and between items, namely users and items are personalized and thus rating needs to

© Springer International Publishing Switzerland 2015
T. Cao et al. (Eds.): PAKDD 2015, Part II, LNAI 9078, pp. 707–719, 2015.
DOI: 10.1007/978-3-319-18032-8_55

be tailored according to individual characteristics. (2) The coupling relationships between users, between items, and between users and items, namely users and items are coupled and hence rating needs to capture the underlying interactions. These two aspects together essentially bring the recommendation problem to a non-IID context [4] [3], namely users and items are not as independent and identically distributed (IID) as usually assumed in the existing RS.

The existing RS algorithms and systems such as collaborative filtering and matrix factorization have been mainly built on the IID context, consequently they may overlook or may not fully capture the intrinsic heterogeneity and couplings. For example, many researchers try to influence or precisely estimate the latent factors [7] [19] [1] [13] through considering the attributes or topics information of users and items for latent factor models. Nevertheless, most of the existing methods assume that the attributes are IID. This is a very fundamental and critical issue for the RS community, as the big recommendation data in online, social, mobile and business applications is essentially non-IID. For example, a user's preference may influence his/her friends, item attributes are often associated with each other via explicit or implicit relationships [17] [18] [5]. Therefore in this paper, we deeply analyse the coupling relationships between users and between items based on their attributes, and incorporate the coupling interactions into MF to improve recommendation qualities.

Table 1. A Toy Example

Age	ZipCode	Country	Sex		Director Actor Genre	Scorsese De Niro Crime	Coppola De Niro Crime	Hitchcock Stewart Thriller	Hitchcock Grant Thriller
						God Father	Good Fellas	Vertigo	N by NW
20	10081	China	M	u_1		1	3	5	4
40	2007	Australia	F	u_2		4	2	1	5
20	2008	Australia	M	u_3		-	2	-	4

To illustrate the coupling relationships in RS, we give a toy example in Table 1. There is a rating matrix consisting of three users and four movies with their attributes. Most existing CF methods utilize the rating matrix for recommendation but ignore the attributes of users and items. However, when the rating matrix is very sparse, the attributes within users and items may also contribute to solving the challenges. Specifically, we can infer the relationship of u_1 and u_2 from the "Age", "ZipCode", "Country" and "Sex" attribute space. Similarly, we can get the movies' relationship from the "Director", "Actor" and "Genre" attribute space. Intuitively, the existing similarity methods such as Pearson or Jaccard measures can be applied to compute the similarities within users or items, based on the IID assumption. In reality, however this assumption is not always held and there are more or less exist coupling relations between instances and attributes. One observation is that the similarity of two attributes values are dependent on other attributes, for example, two directors' relationship is dependent on "Actor" and "Genre" attributes over all the movies. This dependent relation is called the inter-coupled similarity between attributes. Alternatively, within an attribute, one attribute value will also be dependent on other values of the same attribute. Specifically, two attribute values are similar if they present the analogous frequency distribution on one attribute, which leads to another so-called intra-coupled similarity within an attribute. For example, two directors "Scorsese" and "Coppola" are considered similar because they appear with the same frequency. We believe that the coupled similarities between values and between attributes should simultaneously contribute to the relationships within users and within items, namely user coupling and item coupling. Incorporating the user coupling and item coupling into MF model may reach more satisfactory recommendations.

The incorporation of such couplings into RS is motivated by analysing the interactions between different attributes by disclosing IID assumption. A complete consideration of couplings may provide a practical mean for enhancing the effectiveness of RS. Especially, if we do not have ample rating data, the objective user coupling and item coupling can be utilized to make recommendations. To date, our study is the first work to simultaneously consider the user coupling and item coupling based on their objective attributes, and integrate them together into the matrix factorization model.

The contributions of the paper are concluded as follow:

- We propose a coupled measure to capture the relationships for users and items, namely user coupling and item coupling, which consider the coupled interaction between attributes from non-IID perspective.
- We propose a Coupled Matrix Factorization (CMF) model by accommodating the user coupling, item coupling and users' subjective rating preferences together.
- We conduct experiments to evaluate the superiority of couplings and the effectiveness of CMF model.

The rest of the paper is organized as follows. Section 2 presents the related work. In Section 3, we formally state the recommendation and couplings problems. Section 4 first analyses the couplings in RS, then details the coupled MF model integrating the couplings together. Experimental results and analysis are presented in Section 5. The paper is concluded in the last Section.

2   Related Work

Collaborative filtering (CF) is one of the most successful approaches taking advantage of user rating history to predict users' interests. As one of the most accurate single models for collaborative filtering, matrix factorization (MF) is a latent factor model [8] which generally effective at estimating overall structure that relates simultaneously to most items. MF approach tries to decompose the rating matrix to user latent matrix and item latent matrix. Then the estimated rating is predicted by the multiplication of the two decomposed matrices. With the advent of social network, many researchers have started to analyse social

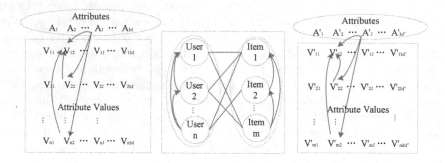

Fig. 1. Coupling Relations in Recommender Systems

recommender systems and various models integrating social networks have been proposed [20]. Social friendship is an outstanding explicit factor to improve the effectiveness of recommendation, however, not every web site have social or trust mechanisms. This explicit social gap greatly motivates us to explore the user and item couplings to improve recommendation qualities. Indeed, such couplings help to make reasonable recommendations when lacking valuable rating information.

Content-based techniques are another successful methods which recommend relevant items to users according to users' personal interests. Content-based methods often assume item's attributes are independent which is not always held in reality. Actually, several research outcomes such as [16] [17] [18] [9] [10] have been proposed to handle the challenging issues. However, user couplings are item couplings are still not completely considered for RS. This motivates us to completely consider the non-IID couplings and integrate the user couplings and item couplings into the MF model.

3   Problem Statement

A large number of user and item sets with attributes can be organized by a triple $S =< S_U, S_O, h >$, where $S_U =< U, A, V, f >$ describes the users' attribute space, $U = \{u_1, u_2, ..., u_n\}$ is a nonempty finite set of users, $A = \{A_1, ..., A_M\}$ is a finite set of attributes for users; $V = \cup_{j=1}^{J} V_j$ is a set of all attribute values for users, in which V_j is the set of attribute values of attribute $A_j (1 \leq j \leq J)$, V_{ij} is the attribute value of attribute A_j for user u_i, and $f = \wedge_{j=1}^{M} f_j (f_j : U \to V_j)$ is an information function which assigns a particular value of each feature to every user. Similar to S_U, $S_O =< O, A', V', f' >$ expresses the items' attribute space where $O = \{o_1, ..., o_m\}$, $A' = \{A'_1, ..., A'_{M'}\}$, $V' = \cup_{j=1}^{J'} V'_j$, $f' = \wedge_{j=1}^{M'} f'_j (f'_j : O \to V'_j)$ are all for items. In the triple $S =< S_U, S_O, h >$, $h(u_i, o_j) = r_{ij}$ expresses the subjective rating preference on item o_j for user u_i. User rating preferences on items are then converted into a user-item matrix R, with n rows and m columns. Each element r_{ij} of R represents the rating given by user u_i on item o_j. For instance, Table 1 consists of three users

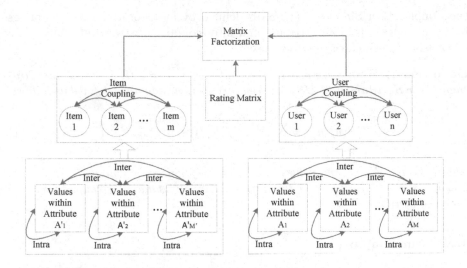

Fig. 2. Coupled Matrix Factorization Model

$U = \{u_1, u_2, u_3\}$ and four items $O = \{GodFather, GoodFellas, Vertigo, NbyNW\}$, $A = \{Age, ZipCode, Country, Sex\}$, $V_3 = \{China, Australia\}$, $f_3(u_2) = Australia$, and $A' = \{Director, Actor, Genre\}$, $V_3' = \{Crime, Thriller\}$, f_3' $(Vertigo) = Thriller$, and $h(u_2, Vertigo) = 1$. The existing similarity methods for computing the relationships assumed that the attributes are independent from each other. However, all the attributes should be coupled together and further influence each other. The couplings are illustrated in Fig.1, for users, within an attribute A_j, there is dependent relation between values V_{lj} and V_{mj} $(l \neq m)$. While a value V_{li} of an attribute A_i is further influenced by the values of other attributes A_j $(j \neq i)$. For example, attributes A_1, A_3, ... to A_J all more or less influence the values of V_{12} to V_{n2} of attribute A_2.

4   Coupled Matrix Factorization

In this section, we mainly introduce the coupled MF approach as shown in Fig. 2. CMF first computes the user coupling and item coupling which integrate the coupled interactions based on the objective attributes. Then, user coupling, item coupling and users' rating preferences are incorporated together into MF model.

4.1   User Coupling

Users are non-IID, as users share diverse properties but may also be inter-related for some reasons such as educational or cultural background. The user coupling can be calculated on top of the dependent relations for all attributes A_j by setting $S_{Ob} = S_U = <U, A, V, f>$. For two users described by the attribute space, the Coupled User Similarity (CUS) is defined by incorporating intra-couplings between values within an attribute and inter-couplings between attributes [16] to measure the similarity between users.

Definition 1. *Formally, given user attribute space $S_U = <U, A, V, f>$, the **Coupled User Similarity (CUS)** between two users u_i and u_j is defined as follows.*

$$CUS(u_i, u_j) = \sum_{k=1}^{J} \delta_k^{Ia}(V_{ik}, V_{jk})) * \delta_k^{Ie}(V_{ik}, V_{jk})) \tag{1}$$

where V_{ik} and V_{jk} are the values of attribute k for users u_i and u_j, respectively; and δ_k^{Ia} is the intra-coupling within attribute A_k, δ_k^{Ie} is the inter-coupling between different attributes.

4.2   Item Coupling

Similarly, items are non-IID. Each item owns different characteristics from others, and there may be coupling relationships between items such as the complementation between purposes. Similar to user coupling, the item coupling can be calculated by setting $S_{Ob} = S_O = <O, A', V', f'>$. For two items described by the attribute space, the Coupled Item Similarity (CIS) is defined to measure the similarity between items by integrating intra-couplings within an attribute and inter-couplings between attributes.

Definition 2. *Formally, given item attribute space $S_O = <O, A', V', f'>$, the **Coupled Item Similarity (CIS)** between two items o_i and o_j is defined as follows.*

$$CIS(o_i, o_j) = \sum_{k=1}^{J'} \delta_k^{Ia}(V'_{ik}, V'_{jk})) * \delta_k^{Ie}(V'_{ik}, V'_{jk})) \tag{2}$$

where V'_{ik} and V'_{jk} are the values of attribute j for items o_i and o_j, respectively; and δ_k^{Ia} is the intra-coupling within attribute A_k, δ_k^{Ie} is the inter-coupling between different attributes.

4.3   Coupled MF Model

MF approaches have been recognized as the main stream in RS through a latent topic projection learning model. In this work, we attempt to incorporate all discussed couplings into a MF scheme. Traditionally, the matrix of predicted ratings $\hat{R} \in \mathbb{R}^{n \times m}$, where n, m respectively denote the number of users and the number of items, can be modeled as: $\hat{R} = r_m + PQ^T$ with matrices $P \in \mathbb{R}^{n \times d}$ and $Q \in \mathbb{R}^{m \times d}$, where d is the rank (or dimension of the latent space) with $d \leq n, m$, and $r_m \in \mathbb{R}$ is a global offset value. Through this modelling, the prediction task of matrix $\hat{R}$ is transferred to compute the mapping of items and users to factor matrices P and Q. Once this mapping is completed, $\hat{R}$ can be easily reconstructed to predict the rating given by one user to an item.

In our proposed CMF, we take not only the rating matrix, but also the user coupling and item coupling, into account. All these aspects should be accommodated into a unified learning model. The learning procedure is constrained by three-fold: the learned rating values should be as close as possible to the observed rating values, the predicted user and item profiles should be similar to their neighbourhoods as well, which are derived from their coupling information. Specifically, in order to incorporate the user coupling and item coupling, we add two additional regularization factors in the optimization step. Then the computation of the mapping can be similarly optimized by minimizing the regularized squared error. The objective function is given as Eqn. 3.

$$
L = \frac{1}{2} \sum_{(u,o_i) \in K} \left(R_{u,o_i} - \hat{R}_{u,o_i} \right)^2 + \frac{\lambda}{2} \left(\|Q_i\|^2 + \|P_u\|^2 \right) + \frac{\alpha}{2} \sum_{all(u)}
$$

$$
\left\| P_u - \sum_{v \in \mathbb{N}(u)} CUS(u,v)P_v \right\|^2 + \frac{\beta}{2} \sum_{all(o_i)} \left\| Q_i - \sum_{o_j \in \mathbb{N}(o_i)} CIS(o_i,o_j)Q_j \right\|^2 \tag{3}
$$

As we can see in the objective function, the rating preference, user coupling and item coupling have been all incorporated together. Specifically, the first part reflects the subjective rating preferences and the latter two parts reflect the user coupling and item coupling, respectively. This means when we recommend relevant items to users, the users' rating preferences may take the dominant role. Besides this, another distinct advantage is that, when we do not have ample rating data, it is still possible to make satisfactory recommendations via leveraging the coupling information, e.g., one user will be recommended what his/her neighbours like or items similar to what he/she preferred before.

To optimize the above objective equation, we minimize the objective function L by the gradient descent approach:

$$
\frac{\partial L}{\partial P_u} = \sum_{o_i} I_{u,o_i}(r_m + P_u Q_i^T - R_{u,o_i})Q_i + \lambda P_u + \alpha(P_u -
$$

$$
\sum_{v \in \mathbb{N}(u)} CUS(u,v)P_v) - \alpha \sum_{v:u \in \mathbb{N}(v)} CUS(u,v)(P_v - \sum_{w \in \mathbb{N}(v)} CUS(v,w)P_w) \tag{4}
$$

$$
\frac{\partial L}{\partial Q_i} = \sum_{u} I_{u,o_i}(r_m + P_u Q_i^T - R_{u,o_i})P_u + \lambda Q_i + \beta(Q_i - \sum_{o_j \in \mathbb{N}(o_i)}
$$

$$
CIS(o_i,o_j)Q_j) - \beta \sum_{o_j:o_i \in \mathbb{N}(o_j)} CIS(o_j,o_i)(Q_j - \sum_{o_k \in \mathbb{N}(o_j)} CIS(o_j,o_k)Q_k) \tag{5}
$$

where I_{u,o_i} is the function indicating that whether user has rated item o_i, 1 means rated, 0 means not rated. $CUS(u,v)$ is the coupled similarity of users u and v, and $CIS(o_i,o_j)$ is the coupled similarity of items o_i and o_j. $\mathbb{N}(u)$ and $\mathbb{N}(o_i)$ respectively represent the user and item neighborhood filtered by coupled similarity.

Model Training. Through the above gradient descent approach, the best matrices P and Q can be computed in terms of the user coupling, item coupling and user-item coupling. The whole process of the coupled model starts at computing user coupling and item coupling based on their objective attribute space, then neighbors of users and items are selected from the couplings. Next, values of P and Q are randomly initiated followed by an iteration step to update P and Q until convergence according to Eqn. 4 and 5. After P and Q are learned from the training process, we can predict the ratings for user-item pairs (u, o_i).

5   Experiments and Results

In this section, we evaluate our proposed model and compare it with the existing approaches respectively using Movielens[1] and Bookcrossing[2] data sets.

5.1   Data Sets

Movielens data set has been widely explored in RS research in last decade. Movielens 1M data set consists of 1,000,209 anonymous ratings of approximately 3,900 movies made by 6,040 Movielens users who joined Movielens in 2000. Only users providing basic demographic information such as "gender", "age", "occupation" and "zipcode" are included in this data set. The movies also have a special "genre" attribute which is applied for computing the item couplings.

Similarly, collected by Cai-Nicolas Ziegler, Bookcrossing data set contains 278,858 users with demographic information providing 1,149,780 ratings on 271, 379 books. The ratings range from 1 to 10 and the users' "gender" and "age" attributes and the books' "book-author", "year of publication" and "publisher" have been used to form user and item couplings.

5.2   Experimental Settings

The 5-fold cross validation is performed in our experiments. In each fold, we have 80% of data as the training set and the remaining 20% as testing set. Here we use Root Mean Square Error (RMSE) and Mean Absolute Error (MAE) as evaluation metrics.

To evaluate the performance of our proposed CMF we consider five baseline approaches: (1) The basic probabilistic matrix factorization (PMF) approach [14]; (2) Singular value decomposition (RSVD) [2] is a factorization method to decompose the rating matrix; (3) Implicit social matrix factorization (ISMF) [11] is an unified model which incorporates implicit social relationships between users and between items computed by Pearson similarity based on the user-item rating matrix; (4) User-based CF (UBCF) [15] first computes users' similarity by Pearson Correlation on the rating matrix, then recommends relevant items to the

---

[1] www.movielens.org
[2] www.bookcrossing.com

given user according to the users who have strong relationships; (5) Item-based CF (IBCF) [6] first considers items' similarity by Pearson Correlation on the rating matrix, then recommends relevant items which have strong relationships with the given user's interested items.

The above five baselines just consider users' rating preferences on items but ignore the attributes of users and items. In order to demonstrate the effectiveness of our proposed method, we also compare it with other three hybrid models PSMF, CSMF and JSMF which respectively augment MF with Pearson Correlation Coefficient, Cosine and Jaccard similarity measures to compute the relationships within users and items based on their attributes. Simply put, we first respectively apply Pearson Correlation Coefficient, Cosine and Jaccard similarity to compute the similarities for users and items based on their attributes. Then we utilize the similarities within users and items to update and optimize the objective function to acquire the best P and Q as CMF. The objective function of PSMF, CSMF and JSMF is similarly constructed as Eqn. 3 by respectively replacing $CUS(u,v)$, $CIS(o_i,o_j)$ as the pearson similarity, cosine similarity, and jaccard similarity between users and between items.

5.3   Experimental Results and Discussions

We respectively evaluate the effectiveness of our CMF model in comparison with the above baselines and the other three hybrid methods PSMF, CSMF and JSMF by selecting the optimal parameters $\alpha=1.0$, $\beta=0.2$ for Movielens, and $\alpha=0.6$, $\beta=1.0$ for Bookcrossing.

Superiority over MF Methods. Because the users of Movielens data set have the basic demographic data for user coupling, and the movies also have natural genre attribute for item coupling, the experimental results compared to the MF methods in Table 2 can show the effect of user coupling and item coupling. In this experiment, we respectively compared the experimental results regarding different latent dimensions with MAE and RMSE metrics. When the latent dimension is set as 100, 50 and 10, in terms of MAE, our proposed CMF can reach an average improvements of 21.24%, 12.88% compared with PMF and RSVD approaches. Similarly, CMF can also improves averagely 58.77%, 39.93% regarding RMSE over PMF and RSVD approaches. Besides the basic comparisons, we also compare our CMF with the latest research outcome ISMF which utilizes the implicit relationships between users and items based on the rating matrix by Pearson similarity. From the experimental result, we can see that CMF can averagely improve 15.22% and 45.87% regarding MAE and RMSE respectively. In conclusion, the experiments on Movielens data set clearly indicate that CMF is more effective than the baseline MF approaches and the state-of-the-art ISMF method regarding MAE and RMSE when latent dimension is respectively set to 100, 50 and 10, due to the strength of user coupling and item coupling.

Similar to Movielens, Bookcrossing data set also has certain user demographic information, and rich book content information such as "Book-Title", "Book-Author", "Year-Of-Publication" and "Publisher". After removing all the invalid

Table 2. MF Comparisons on Movielens and Bookcrossing

Data Set	Dim	Metrics	PMF (Improve)	ISMF (Improve)	RSVD (Improve)	CMF
Movielens	100D	MAE	1.1787(28.09%)	1.1125 (21.47%)	1.1076 (20.98%)	**0.8978**
		RMSE	1.7111 (71.07%)	1.5918 (59.14%)	1.5834 (58.30%)	**1.0004**
	50D	MAE	1.1852 (18.43%)	1.1188 (11.79%)	1.1088 (10.79%)	**1.0009**
		RMSE	1.8051 (58.98%)	1.6103 (39.50%)	1.5835 (36.82%)	**1.2153**
	10D	MAE	1.2129 (17.19%)	1.1651 (12.41%)	1.1098 (6.88%)	**1.0410**
		RMSE	1.8022 (46.25%)	1.7294 (38.97%)	1.5863 (24.66%)	**1.3397**
Bookcrossing	100D	MAE	1.5127 (3.65%)	1.5102 (3.40%)	1.5131 (3.69%)	**1.4762**
		RMSE	3.7455 (0.76%)	3.7397 (0.18%)	3.7646 (2.67%)	**3.7379**
	50D	MAE	1.5128 (3.67%)	1.5100 (3.39%)	1.5131 (3.70%)	**1.4761**
		RMSE	3.7452 (0.74%)	3.7415 (0.37%)	3.7648 (2.70%)	**3.7378**
	10D	MAE	1.5135 (3.73%)	1.5107 (3.45%)	1.5134 (3.72%)	**1.4762**
		RMSE	3.7483 (1.20%)	3.7440 (0.77%)	3.7659 (2.96%)	**3.7363**

ISBNs, all the books in the data set are cleaned. Therefore the experimental results on the Bookcrossing data set can also demonstrate the impacts of user and item couplings. We depict the effectiveness comparisons with respect to different methods on Bookcrossing data set in Table 2. We can clearly see that, our proposed CMF method outperforms all the counterparts in terms of MAE and RMSE. Specifically, when the latent dimension is set as 100, 50 and 10, in terms of MAE, our proposed CMF can reach an average improvements of 3.68%, 3.70% compared with PMF and RSVD approaches. While CMF can averagely increase 0.98% and 2.78% regarding RMSE over PMF and RSVD approach. Furthermore, the prominent improvements compared with the baseline approaches regarding MAE and RMSE are resulted from considering complete couplings. Additionally, we also compare the CMF with the state-of-the-art method ISMF, the result shows that the improvements can reach to 3.41% and 0.44% regarding MAE and RMSE respectively. Therefore, we can conclude that our CMF method not only outperforms PMF and SVD which are basic MF methods, but also performs better than the state-of-the-art model ISMF in terms of MAE and RMSE metrics.

Superiority over CF Methods. In addition to the MF methods, we also compare our proposed CMF model with two different CF methods UBCF and IBCF. In this experiment, we fix the latent dimension to 100 for our proposed CMF model. On Movielens, the results in Table 3 indicate that CMF can respectively improve 0.49% and 2.42% regarding MAE, and 0.18% and 19.54% in terms of RMSE. Similarly compared with UBCF and IBCF, on Bookcrossing data set, the results show that the CMF can reach huge improvements respectively 33.02% and 31.03% regarding MAE, and 24.68% and 19.04% regarding RMSE. Therefore, this experiment clearly demonstrates that our proposed CMF performs better than UBCF and IBCF methods. The improvements are contributed by the full consideration of the couplings in RS.

Table 3. CF Comparisons on Movielens and Bookcrossing

Data Set	Metrics	UBCF (Improve)	IBCF (Improve)	CMF
Movielens	MAE	0.9027 (0.49%)	0.9220 (2.42%)	**0.8978**
	RMSE	1.0022 (0.18%)	1.1958 (19.54%)	**1.0004**
Bookcrossing	MAE	1.8064 (33.02%)	1.7865 (31.03%)	**1.4762**
	RMSE	3.9847 (24.68%)	3.9283 (19.04%)	**3.7379**

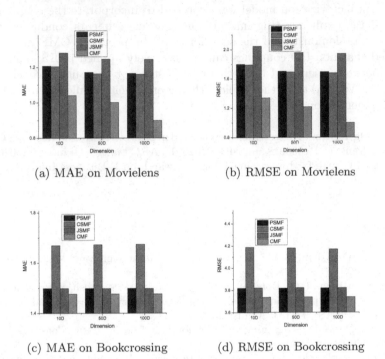

(a) MAE on Movielens   (b) RMSE on Movielens

(c) MAE on Bookcrossing   (d) RMSE on Bookcrossing

Fig. 3. Superiority over Hybrid Methods on Movielens and Bookcrossing

Superiority over Hybrid Methods. In order to demonstrate the effectiveness of our proposed method, we also compare it with other three hybrid methods PSMF, CSMF and JSMF. From the resultant Fig. 3 on Movielens data set, we can clearly see that the proposed CMF method greatly outperforms PSMF, CSMF and JSMF in terms of MAE and RMSE. Specifically, CMF can averagely improve PSMF, CSMF, JSMF by 20.14%, 19.63%, 27.78% regarding MAE, and by 54.58%, 53.45%, 79.50% regarding RMSE. Similarly on Bookcrossing data set, the results in Fig. 3 clearly indicate that the CMF method also perform better than PSMF, CSMF and JSMF regarding MAE and RMSE. The results show that CMF can respectively improve 2.24%, 19.57%, 2.2% in average regarding MAE, and 8.13%, 44.18%, 8.38% in terms of RSME compared with PSMF, CSMF, JSMF. From this experiment, we can conclude that our proposed CMF is more effective than these three hybrid methods.

6   Conclusion and Future Work

In this paper, we studied Recommender System from a non-IID perspective, specifically, we mainly focused on the significant non-IID coupling relations between users and between items to improve the quality of recommendations. The couplings disclosed the traditional IID assumption and deeply analysed the intrinsic relationships between users and between items. Furthermore, a coupled matrix factorization model was proposed to incorporate the coupling relations and the explicit rating information. The experiments conducted on the real data sets demonstrated the superiority of the proposed CMF method and suggested that non-IID couplings can be effectively applied in RS. The heterogeneity between users and between items is still not thoroughly considered in this paper. We need to further explore the heterogeneity challenge for enhancing our recommendation model in the future.

Acknowledgments. This research is sponsored in part by Australian Research Council Discovery Grants (DP1096218 and DP130102691), ARC Linkage Grant (LP100200774), and National Centre for International Joint Research on E-Business Information Processing (2013B01035).

References

1. Agarwal, D., Chen, B.-C.: Regression-based latent factor models. In: KDD, pp. 19–28 (2009)
2. Alter, O., Brown, P.O., Botstein, D.: Singular value decomposition for genome-wide expression data processing and modeling. Proceedings of the National Academy of Sciences (PNAS) **97**(18), August 2000
3. Cao, L.: Non-iidness learning in behavioral and social data. Comput. J. **57**(9), 1358–1370 (2014)
4. Cao, L.: Coupling learning of complex interactions. Information Processing & Management **51**(2), 167–186 (2015)
5. Cao, L., Yuming, O., Philip, S.Y.: Coupled behavior analysis with applications. IEEE Trans. Knowl. Data Eng. **24**(8), 1378–1392 (2012)
6. Deshpande, M., Karypis, G.: Item-based top-n recommendation algorithms. ACM Trans. Inf. Syst. **22**(1), 143–177 (2004)
7. Gantner, Z., Drumond, L., Freudenthaler, C., Rendle, S., Schmidt-Thieme, L.: Learning attribute-to-feature mappings for cold-start recommendations. In: ICDM, pp. 176–185 (2010)
8. Koren, Y.: Factorization meets the neighborhood: a multifaceted collaborative filtering model. In: Li, Y., Liu, B., Sarawagi, S. (eds.) KDD, pp. 426–434. ACM (2008)
9. Li, F., Xu, G., Cao, L.: Coupled item-based matrix factorization. In: Benatallah, B., Bestavros, A., Manolopoulos, Y., Vakali, A., Zhang, Y. (eds.) WISE 2014, Part I. LNCS, vol. 8786, pp. 1–14. Springer, Heidelberg (2014)
10. Li, F., Xu, G., Cao, L., Fan, X., Niu, Z.: CGMF: coupled group-based matrix factorization for recommender system. In: Lin, X., Manolopoulos, Y., Srivastava, D., Huang, G. (eds.) WISE 2013, Part I. LNCS, vol. 8180, pp. 189–198. Springer, Heidelberg (2013)

11. Ma, H.: An experimental study on implicit social recommendation. In: SIGIR, pp. 73–82 (2013)
12. Melville, P., Sindhwani, V.: Recommender systems. In: Encyclopedia of Machine Learning, pp. 829–838 (2010)
13. Menon, A.K., Elkan, C.: A log-linear model with latent features for dyadic prediction. In: ICDM, pp. 364–373 (2010)
14. Salakhutdinov, R., Mnih, A.: Probabilistic matrix factorization. In: Advances in Neural Information Processing Systems, vol. 20 (2008)
15. Xiaoyuan, S., Khoshgoftaar, T.M.: A survey of collaborative filtering techniques. Adv. Artificial Intellegence, 2009 (2009)
16. Wang, C., Cao, L., Wang, M., Li, J., Wei, W., Ou, Y.: Coupled nominal similarity in unsupervised learning. In: CIKM, pp. 973–978 (2011)
17. Wang, C., She, Z., Cao, L.: Coupled attribute analysis on numerical data. In: IJCAI (2013)
18. Wang, C., She, Z., Cao, L.: Coupled clustering ensemble: Incorporating coupling relationships both between base clusterings and objects. In: ICDE, pp. 374–385 (2013)
19. Wang, C., Blei, D.M.: Collaborative topic modeling for recommending scientific articles. In: KDD, pp. 448–456 (2011)
20. Yang, X., Steck, H., Liu, Y.: Circle-based recommendation in online social networks. In: KDD, pp. 1267–1275 (2012)

Complementary Usage of Tips and Reviews for Location Recommendation in Yelp

Saurabh Gupta[1]([✉]), Sayan Pathak[2], and Bivas Mitra[1]

[1] Department of Computer Science and Engineering,
Indian Institute of Technology Kharagpur, Kharagpur 721302, India
`saurabhgupta30@gmail.com, bivas@cse.iitkgp.ernet.in`
[2] Microsoft Corporation, Bangalore, India
`sayanpa@microsoft.com`

Abstract. Location-based social networks (LBSNs) allow users to share the locations that they have visited with others in a number of ways. LBSNs like Foursquare allow users to 'check in' to a location to share their locations with their friends. However, in Yelp, users can engage with the LBSN via modes other than check-ins. Specifically, Yelp allows users to write 'tips' and 'reviews' for the locations that they have visited. The geo-social correlations in LBSNs have been exploited to build systems that can recommend new locations to users. Traditionally, recommendation systems for LBSNs have leveraged check-ins to generate location recommendations. We demonstrate the impact of two new modalities - tips and reviews, on location recommendation. We propose a graph based recommendation framework which reconciles the 'tip' and 'review' space in Yelp in a complementary fashion. In the process, we define novel intra-user and intra-location links leveraging tip and review information, leading to a 15% increase in precision over the existing approaches.

Keywords: Location based social networks · Location recommendation · Mining graph and network data

1 Introduction

LBSNs have become a popular tool for users to share their locations with their friends. Services like Foursquare allow users to post 'check-ins' at their current locations to let their friends know about their whereabouts. The popularity of such LBSNs has encouraged the development of systems recommending new locations to users (henceforth referred to as location recommender systems). Typically, any such location recommender system leverages the rich check-in information about each user to suggest relevant locations. This is because the number of times a user checks-in to a particular location, indicates her preference for that location.

Apart from check-ins, there are other mechanisms through which users can engage with LBSNs and express their preference (positive or negative) for a

© Springer International Publishing Switzerland 2015
T. Cao et al. (Eds.): PAKDD 2015, Part II, LNAI 9078, pp. 720–731, 2015.
DOI: 10.1007/978-3-319-18032-8_56

location that they have visited. Two such alternate modalities are present in Yelp - namely, tips and reviews, capturing users' opinions about venues. Tips are short textual descriptions about a venue - typically describing the standout characteristic of a location or a set of things to do or to avoid at a particular location. Reviews, on the other hand, present a much more detailed account of a person's experience at a location. A major difference between tips and reviews as against check-ins is that the frequency of check-ins is much greater. A typical user might post only a handful of tips for a particular location, whereas she might check-in at a particular location 30-40 times. In this work, we look at the problem of location recommendation in a popular LBSN - Yelp. Specifically, we focus on the information provided by tips and reviews to generate location recommendations for users. To the best of our knowledge, this is the first work which looks at the problem of location recommendation from the lens of tips and reviews taken together, rather than check-ins. We propose a graph-based framework which seamlessly combines signals coming from tips and reviews. We consider users and locations as nodes in a graph with weighted edges constructed between user-user, user-location and location-location pairs, based on tip, review and friendship information in Yelp. Finally, a spreading activation technique is used to recommend locations to a target user. This work makes the following contributions in the context of location recommendation in LBSNs:

- We identify how tips are different from reviews in the way they are used to express preference for a visited location. We then propose a graph based recommendation framework which seamlessly combines the disparate spaces of tips and reviews, along with users' social network.
- In the graph-based framework, we propose novel intra-location and intra-user links based on factors like a) distance between locations (LLD), b) frequency with which users write tips and reviews (LLT/LLR), c) entropy of the common locations for which users either write tips/reviews (UUT/UUR) and d) the friendship relationship between users (UUF).
- We find that intra-location links like LLD and LLT/LLR affect the recommendation accuracy to a greater extent as compared to intra-user links like UUT/UUR and UUF. We also find that the way distance is incorporated in the graph has a huge impact on the recommendation accuracy.

2   Tip Space vs Review Space in Yelp

Yelp is an LBSN which helps users search for any kind of business in a city - from a restaurant to a hair-dresser. It is different from some of the other LBSNs like Foursquare in that it allows users to post both tips as well as reviews for the locations that they have visited. We run all our experiments on the Yelp dataset made freely available by Yelp as part of their data challenge contest.[1] The details of the dataset are mentioned in Table 1. This data is for Phoenix, Arizona covering the time period between Feb. 2005 to Feb. 2014. We can see from Table 1

---

[1] http://www.yelp.com/dataset_challenge

Table 1. Yelp Dataset Details

# unique locs	# unique users	# Tips	# Reviews	# users with at least 1 review	# users with at least 1 tip	# locs with at least 1 review	# locs with at least 1 tip
15,585	70,817	1,13,993	3,35,022	70,817	15,231	15,579	10,473

that all 70817 users have written at least one review whereas only 15231 users have written at least one tip. In addition to this, the number of locations for which at least one tip has been written is 10473, whereas the number of locations for which at least one review has been written is 15579. These numbers suggest that the number of users as well as the number of locations covered by reviews as an interaction medium, is much more than tips in Yelp. To study the difference between check-ins and tips/reviews as interaction mechanisms in LBSNs, we also considered Brightkite dataset [3] which was collected over the period of Apr. 2008 - Oct. 2010 and gives details about users' check-in behavior. We can see from Table 2 that not only is the average number of check-ins that a user makes is higher than the corresponding numbers for tips and reviews, but the maximum number of check-ins that a user makes for a venue is also significantly higher than writing tips or reviews. This is because the level of user involvement required by these diverse interaction mechanisms varies. While check-ins incur very little cognitive load on the part of the user (who just has to indicate whether she was present at a location or not); the textual nature of tips and reviews requires the users to expend more effort in writing them. Figure 1 shows that more than

Table 2. Comparison of check-ins(Brightkite) with Tips/Reviews(Yelp)

Avg. check-ins per user	Avg. Tips per user	Avg. Reviews per user	Max. check-ins by a user	Max. Tips by a user	Max. Reviews by a user
92.34	7.48	4.73	2100	104	10

80% of the users have no location for which they write both a tip as well as a review. Thus, if we consider that a user prefers a location if she writes a tip or review for that location, then the locations that a user likes through her tips are very different from the locations that she likes via her reviews, in Yelp. We also investigated three types of venues for each user : venues for which only tips were posted (Type T), venues for which only reviews were posted (Type R), venues for which both reviews and tips were posted (Type TR). After creating these 3 types for each user, we calculated the average distance between each pair of venues within a type. These values were then averaged over all the users and reported in Table 3. It can be seen that venues for which only tips are posted by a user are closer geographically to each other as opposed to venues for whom only reviews are written by a user. These observations provide a strong motivation to leverage the signals coming separately from tips and reviews in a complementary way so as to provide better recommendations to users.

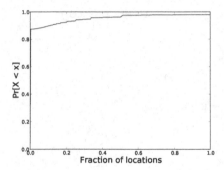

Table 3. Distance in kms.

Avg. Distance b/w venues of Type TR	0.623
Avg. Distance b/w venues of Type T	1.237
Avg. Distance b/w venues of Type R	6.015

Fig. 1. % of users writing both tips and reviews for same locations

3   Graph-based Recommendation Framework

We represent our framework as a multilayer graph $\mathcal{M} = (\mathcal{G}, \mathcal{C})$ where $\mathcal{G} = \{G_\alpha; \alpha \in \{U, L\}\}$ is a family of user and location layers. Each layer is individually represented with the help of a subgraph. Specifically, $G_U = (X_U, E_U)$ represents the user subgraph of $\mathcal{M}$ with $X_U = \{x_1^U, ..., x_{N_U}^U\}$ denoting the set of user nodes and $E_U = \{(x_i^U, x_j^U) | x_i^U, x_j^U \in X_U\}$ denoting the set of edges between users. Similarly, $G_L = (X_L, E_L)$ represents the location subgraph with $X_L = \{x_1^L, ..., x_{N_L}^L\}$ denoting the set of location nodes and $E_L = \{(x_i^L, x_j^L) | x_i^L, x_j^L \in X_L\}$ denoting the set of edges between locations. Finally,

$$\mathcal{C} = \{E_{\alpha\beta} \subseteq X_\alpha \times X_\beta; \alpha, \beta \in \{U, L\}, \alpha \neq \beta\} \tag{1}$$

is the set of edges between nodes of the user graph G_U and the location graph G_L representing inter-layer connections. E_U and E_L, on the other hand, represent intra-layer connections.

3.1   Edge Creation and Edge-weights

In this subsection we discuss how the weights for user-location, user-user and location-location links are chosen in our framework.

User-Location Links (UL) - Among the user-location links, we distinguish links either being formed based on tips or based on reviews.

Review Based Links (ULR): We create a link between a user a^U and location i^L if a^U has written at least one review for i^L. If a^U has written k number of reviews for i^L, then

$$w^R(a^U, i^L) = \frac{1}{1 + \frac{C}{k}} \tag{2}$$

where C is the maximum number of reviews that a user has written for any location and R denotes a review-based edge.

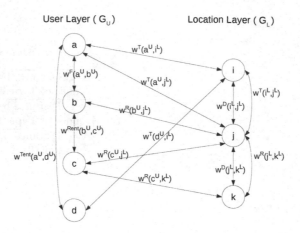

User Layer (G_U)   Location Layer (G_L)

Fig. 2. Graph-based framework represented as multilayer graph $\mathcal{M}$

Tip Based Links(ULT): Similarly, the weight of a tip-based edge between user a^U and location i^L is computed as:

$$w^T(a^U, i^L) = \frac{1}{1 + \frac{C}{k}} \tag{3}$$

User-User Links (UU) - Among the user-user links we make three distinctions with respect to the nature of such user-user links.

Friendship-based Links (UUF): Two friends a^U and b^U might have similar location preferences and hence a link is created if they are in each other's social network. Let $|\Gamma_{a^U}|$ denote the number of friends user a^U has, in her social network.

$$w^F(a^U, b^U) = \frac{1}{|\Gamma_{a^U}|} \tag{4}$$

where F denotes that this is a friendship based link. We can see from Eq. 4 that $w^F(a^U, b^U)$ would be different from $w^F(b^U, a^U)$.

Tip Entropy Based Links (UUT): Two users, a^U and b^U, can also have an implicit connection if they have written a tip for the same location i^L. Let $\Theta^T_{a^U b^U}$ denote the set of those locations for which both a^U and b^U have written a tip; and $\Phi^T_{i^L}$ denote the set of users who have written a tip for location i^L. Also, let $|t_{a^U i^L}|$ denote the number of tips a^U has written for i^L and $|T_{i^L}|$ denote the total number of tips that have been written for i^L. Tip-entropy of a location i^L in $\Theta^T_{a^U b^U}$ is then defined as

$$E_{i^L} = -\sum_{a^U \in \Phi^T_{i^L}} q_{a^U i^L} \times log(q_{a^U i^L}) \tag{5}$$

where $q_{a^U i^L} = \frac{|t_{a^U i^L}|}{|T_{i^L}|}$. Let us imagine a scenario where $i^L \in \Theta^T_{a^U b^U}$ and $j^L \in \Theta^T_{c^U d^U}$. If tip-entropy of location i^L is high and that of j^L is very low, we can infer stronger interest overlap between users c^U and d^U as opposed to users a^U and b^U. This is because a location with a high tip-entropy (for example - an airport) represents a location which has been visited by several casual visitors who have opted to write a tip for that location. Thus, the chances that two users who are bound by such a common location, have similar interests, are less. Hence, weight of the link connecting users a^U and b^U, based on tip-entropy is defined as

$$w^{T_{ent}}(a^U, b^U) = \sum_{i^L \in \Theta^T_{a^U b^U}} \frac{1}{E_{i^L}} \qquad (6)$$

where T_{ent} denotes that the link weight is based on Tip entropy.

Review Entropy Based Links (UUR): In a similar vein, the weight of the link between users a^U and b^U, based on review-entropy is computed as

$$w^{R_{ent}}(a^U, b^U) = \sum_{i^L \in \Theta^R_{a^U b^U}} \frac{1}{E_{i^L}} \qquad (7)$$

Note that previous works [9,10] have outlined the importance of place entropies based on check-ins in the task of friendship link prediction among users. Our use of tip and review entropies, on the other hand, uses them to connect similar minded users in a location recommendation setting.

Location-Location Links (LL) - We consider three types of links which can be created between two locations.

Review Based links (LLR): Locations i^L and j^L can have an implicit link via reviews written by a common user for both i^L and j^L. The motivation behind creating such an edge is that the similarity of reviewing behavior across i^L and j^L (exhibited by the common users who write a review for both i^L and j^L) can be used to infer similarity between i^L and j^L.

Let $\Pi^R_{i^L j^L}$ denote the set of those users who have written a review for both i^L and j^L. Also, let $|R_{a^U i^L}|$ denote the number of reviews that user a^U has written for location i^L; and let β_{a^U} denote the set of locations for which a^U has written a review for. Then, strength of connection between i^L and j^L is computed as

$$w^R(i^L, j^L) = \sum_{a^U \in \Pi^R_{i^L j^L}} \frac{1}{1 + \sum_{k^L \in \beta_{a^U}} |R_{a^U k^L}|} \qquad (8)$$

The rationale behind Eq. 8 can be explained through the following example. Let us suppose that users a^U and b^U have written a review for both locations i^L and j^L, i.e. i^L and j^L are connected because of users a^U and b^U. If the number of reviews that a^U has written for all locations (i^L and j^L included) is much greater than the total number of reviews that b^U has written for all locations, then the contribution of b^U in the link between i^L and j^L should be higher than the contribution of a^U. In other words, if a user is selective in her choice of venues for which she writes a review (evident by the lesser number of total reviews written by her), then her contribution should be weighed more heavily.

Tip Based links (LLT): Similarly, weight of the edge between locations i^L and j^L, formed on the basis of tips is given by

$$w^T(i^L, j^L) = \sum_{a^U \in \Pi_{i^L j^L}^T} \frac{1}{1 + \sum_{k^L \in \beta_{a^U}} |T_{a^U k^L}|} \tag{9}$$

Distance Based links (LLD): Prior work has established that people tend to visit locations which are close to the locations that they have already visited [13,14]. To account for this fact, we propose to create an edge between two locations with a weight which is inversely proportional to the distance between those locations. Instead of creating a distance based edge between each pair of locations, we create an edge between locations i^L and j^L, only if i^L and j^L have some common user who either has written a tip for both or review for both. The weight of such an edge is then computed as

$$w^D(i^L, j^L) = \frac{1}{1 + dist_{i^L j^L}} \tag{10}$$

where $dist_{i^L j^L}$ is the geographical distance between i^L and j^L.

All the weights are normalized in the same manner as described in [8]. For each node z, the sum of weights (S) for all the links going out from it is computed, and then each each weight of the link going out from z is divided by S.

3.2   Preference Propagation on the Graph

We use the graph described in the previous subsection for recommending locations to users. Let us denote the target user, for whom we want to generate recommendations, by a^U. We use a spreading activation algorithm (branch and bound serial, symbolic search algorithm (BNB)) in a manner similar to the one used in [2]. To initiate the recommendation process, we activate the target user with an activation value of 1, while all other nodes are initialized to 0. In the next iteration (next time step), a^U is 'expanded' in the sense that all its neighbors (locations as well as users) are activated by an amount proportional to the weight of the edge connecting the two nodes, and are added to a list Q. Therefore, the activation value that is assigned to an arbitrary neighbor z, of a^U equals

$$\mu_z(t+1) = \mu_{a^U}(t) \times w(a^U, z) \tag{11}$$

where $\mu_x(t)$ represents the activation value of node x at time t and $w(a^U, z)$ denotes the weight of the edge connecting a^U and z. Note that, depending upon whether z is a location or a user, there might be multiple edges connecting a^U and z. The activation values accumulated through all such edges becomes the new activation value of z. In the next iteration, a previously 'unexpanded' node

p with the highest activation value in Q, is 'expanded', with all its neighbors getting activated according to Eq. 11 (again accumulated over all kinds of edges between the two nodes). There might be some neighbors of p which are already in Q. Such nodes' values are increased by the value computed above. While other neighbors,which were not already in Q are added to Q with their corresponding activation values. This process of 'expanding' a previously 'unexpanded' node with the highest activation value in Q is repeated for a fixed number of iterations (50 in our experiments). Finally, the location nodes having the maximum activation values, that have not yet been visited by the target user, are recommended to her.

4   Experimental Methodology and Results

From the Yelp dataset described in Section 2, first we filter those users who have written tip/review for at least 30 locations. After this initial preprocessing step, we are left with 2076 fairly active users in terms of posting tips and reviews; and on whom we do all the further experiments. For each user we randomly mark off 20% and 40% of the locations that she has written a tip/review for, as the test set to test the effectiveness of our approach. The rest of the data for each user is treated as training data and used to build the graph with appropriate edge weights as described in Section 3. Then, for each target user, we run the preference propagation procedure outlined in Section 3.2 to generate top k recommendations. We repeat this experiment 5 times and report the average $precision@N$ and average $recall@N$.

$$precision@N = \frac{\sum_{u_i \in U} |TopN(u_i) \cap L(u_i)|}{\sum_{u_i \in U} |TopN(u_i)|} \tag{12}$$

$$recall@N = \frac{\sum_{u_i \in U} |TopN(u_i) \cap L(u_i)|}{\sum_{u_i \in U} |L(u_i)|} \tag{13}$$

where $TopN(u_i)$ is the top-n locations which have been recommended to the user u_i, $L(u_i)$ is the set of locations in the test set of u_i and U is the set of all the users. We do experiments for two values of N - 5 and 10.

We compare our approach with the following benchmark approaches:

- User-based collaborative (CF) filtering which takes similar users' opinion into account to recommend new locations to users. We refer to CF applied on the user-location space connected by tips as CF-Tip and that on the review space as CF-Review.
- Graph-based recommendation algorithm proposed in [5] referred to as G-Local henceforth; with the weights used for the links connecting users to locations being the same as used in our proposed framework.
- NMF [7], a state-of-the-art recommendation method which accounts for latent factors underlying user preferences. We mark the entry c_{ij} in the user-location matrix as 1 if the u_i has written either a tip or a review for

l_j. We also experimented with values of c_{ij} proportional to the number of times u_i writes a tip/review for l_j, but it produced inferior results and hence we report the results corresponding to c_{ij} values being 1. Both NMF and G-Local provide intuitive ways to combine the information coming from tips and reviews.

Table 4. Bold values indicate significant increase with p<0.01 w.r.t. NMF.

Testing %	Metrics	Our Framework	NMF	G-Local	CF-Review	CF-Tip
20%	$precision@5$	**0.07791**	0.0687	0.06211	0.01159	0.01191
	$precision@10$	**0.06729**	0.05837	0.05382	0.01103	0.01172
40%	$precision@5$	**0.12614**	0.11928	0.09637	0.02944	0.02903
	$precision@10$	**0.1144**	0.10099	0.08613	0.028	0.02913

Table 5. Bold values indicate significant increase with p<0.01 w.r.t NMF. Bold value with * indicates significant increase with p<0.05 w.r.t NMF.

Testing %	Metrics	Our Framework	NMF	G-Local	CF-Review	CF-Tip
20%	$recall@5$	**0.02736**	0.02413	0.02181	0.00316	0.00354
	$recall@10$	**0.04727**	0.04100	0.03781	0.00601	0.00697
40%	$recall@5$	**0.02187***	0.02068	0.01671	0.00368	0.00407
	$recall@10$	**0.03967**	0.03502	0.02987	0.00701	0.00817

Note that the low values of precision and recall reported in Tables 4 and 5 are not unreasonable because the accuracy of recommendations is usually not very high when dealing with sparse datasets (where density of user-item matrix is low) [11]. For instance, $precision@5$ reported in [11] for a dataset with density of 8.02×10^{-3} was 0.05. The sparsity of the preprocessed dataset used in our experiments is 5.61×10^{-3}. We can see from Tables 4 and 5 that the proposed framework outperforms other approaches significantly. Specifically, when 20% of the data is held-out as the test set, our approach outperforms NMF (the next best) by 13.4% and 15.28% w.r.t $precision@5$ and $precision@10$ respectively. Moreover, the robust performance of our approach vis a vis the other approaches is maintained even when the sparsity of the data is increased by marking 40% of the data as the test set.

We also constructed the user-location graph with links between users and locations created at random, respecting the density of user-location links formed either based on tips or reviews in the original dataset. Each of the weights are then computed as described in Section 3 and finally the spreading activation algorithm as used in our framework is used to recommend locations to users. The $precision@5$ and $recall@5$ (testing % = 20%) achieved by such an algorithm (averaged over 5 runs) is 0.034 and 0.0108 respectively; hugely under performing our framework as can be seen from Tables 4 and 5. This shows that the better recommendations generated by our framework are not by random chance.

5   Discussion

Our method differs from G-Local [5] in that apart from considering user-item links based on user preferences, our approach additionally considers intra-user and intra-location links, based on tip and review information. Hence, we wanted to see what is the effect of these extra links on recommendation accuracy individually. To that end, we selectively removed features from our framework and measured the recommendation accuracy. Table 6 shows the results of removing various features from our framework one at a time. As can be seen from Table 6, removing links of the type LLD decreases the precision and recall by the maximum amount, suggesting the importance of connecting locations based on distance. Links of the type LLR and LLT (where locations are connected based on the number of reviews/tips, the common users have written), have the next highest impact on the accuracy. Friendship links (UUF) seem to have the least impact on the recommendation accuracy.

Table 6. Effect of Individual Features (Testing % = 20%)

Metrics	All Features	LLD Removed	UUR & UUT Removed	LLR & LLT Removed	UUF Removed
$precision@5$	**0.07791**	0.06603	0.07664	0.07119	0.07734
$recall@5$	**0.02736**	0.02319	0.02692	0.02500	0.02716

Use of Distance: We know of one prior work [15] which incorporates distance in a graph-based setting for location recommendation. Our use of distance is different from theirs in that we only create a distance based link from l_i to l_j if there is a user who has written a tip/review for both l_i and l_j, rather than having links drawn from each location to its nearest k locations. We propose a

Table 7. Capturing Distance in an appropriate way (Testing % = 20%)

Metrics	G-Local	Our Framework	G-dist
$precision@5$	0.06211	0.07791	0.06107
$recall@5$	0.02181	0.02736	0.02145

framework, G-dist, that contains all the features that our proposed framework contains except for distance based links. The distance based links in G-dist are created in accordance with the approach outlined in [15]. In G-dist, distance based links are created from each location to its nearest 608 locations. Note that 608 is the average number of location-location links, that exist in the graph constructed according to our proposed framework. As can be seen from Table 7, having distance-based links to the nearest locations seems to take the recommendation process astray and produces poorer results than even G-Local which does not consider any intra-user or intra-location features. However, the different treatment of distance in our framework, streamlines the recommendation process to produce better results.

6 Related Work

While there has been a lot of interest in location recommendation in LBSNs, we think that interaction modalities other than check-ins have not featured prominently in majority of the studies. Collaborative Filtering (CF) [4] is a domain independent technique which uses past user behavior to compute user-user as well as item-item relationships to infer new user-item relationships [6]. Latent factor models like matrix factorization methods are a branch of CF methods that characterize users as well as items based on k number of latent factors [6].

Non-graph Based Location Recommendation: In [12], only user preference along with social (friend) and geographical factors were considered to recommend locations, whereas our approach considers other user and place properties derived from tips and reviews, apart from the social and geographical factors. In [1], the authors modeled users' preferences based on the categories of locations that they visited and identified experts in a city for a particular category; who were then used for location recommendation.

The graph based recommendation framework proposed here differs from the graph-based approaches discussed in [5,8,15] in the following significant ways:

- To alleviate the sparsity problem that besets collaborative filtering algorithms, a graph based recommendation framework was proposed in [5] where, only links connecting users to items based on the users' purchase history were considered. Our work on the other hand, also considers intra-user and intra-location edges based on tips and reviews.
- A model based on personalized random walk over user-location graph was proposed in [8] which made use of social network information along with check-in information. However, our work defines novel features about users and locations, which are in turn derived from tips and reviews. The user-user links of the type UUT/UUR, introduced in this work, are the first attempt to connect users in a graphical setting, with information other than social network information for recommendation purposes. Another point of departure of our work from [8] is the novel use of distance in our framework, not accounted for in [8].
- To the best of our knowledge, a recent work [15] is the first one to discuss the possibility of having edges between two locations for location recommendation. While the work in [15] connects locations l_i and l_j only based on distance, we create additional edges like LLR/LLT which capture semantic information other than distance. Also, while they create an edge from l_i to its nearest k locations, we create edges only between those 2 locations for which there is at least one user who has written tip/review for both of them.

7 Conclusions

In this paper, we describe a graph-based location recommendation system which makes use of signals coming from tips and reviews as opposed to check-ins.

We define novel features based on tips and reviews which can be used to connect users and locations. A simple preference propagation algorithm on the constructed graph is then used to generate recommendations. The recommendation accuracy of our proposed framework is better than the existing methods. Among the intra-user and intra-location features proposed, we conclude that links of the type LLD and LLT/R are better than UUR/T and UUF type of links. We also observe that having distance based links between k number of nearest locations may not be the best way to leverage distance based information and find that using distance based links between venues with at least one common user (based on tip/review) leads to much better location recommendations.

References

1. Bao, J., Zheng, Y., Mokbel, M.F.: Location-based and preference-aware recommendation using sparse geo-social networking data. In: SIGSPATIAL 2012, pp. 199–208. ACM, New York (2012)
2. Chen, H., Ng, T.: An algorithmic approach to concept exploration in a large knowledge network (automatic thesaurus consultation): Symbolic branch-and-bound search vs. connectionist hopfield net activation. J. Am. Soc. Inf. Sci. **46**(5), 348–369 (1995)
3. Cho, E., Myers, S.A., Leskovec, J.: Friendship and mobility: user movement in location-based social networks. In: KDD 2011, pp. 1082–1090. ACM, New York (2011)
4. Goldberg, D., Nichols, D., Oki, B.M., Terry, D.: Using collaborative filtering to weave an information tapestry. Commun. ACM **35**(12), 61–70 (1992)
5. Huang, Z., Chen, H., Zeng, D.: Applying associative retrieval techniques to alleviate the sparsity problem in collaborative filtering. ACM Trans. Inf. Syst. **22**(1), 116–142 (2004)
6. Koren, Y., Bell, R., Volinsky, C.: Matrix factorization techniques for recommender systems. Computer **42**(8), 30–37 (2009)
7. Lee, D.D., Sebastian Seung, H.: Learning the parts of objects by nonnegative matrix factorization. Nature **401**, 788–791 (1999)
8. Noulas, A., Scellato, S., Lathia, N., Mascolo, C.: A random walk around the city: new venue recommendation in location-based social networks. In: SOCIALCOM-PASSAT 2012, pp. 144–153. IEEE Computer Society, Washington, DC (2012)
9. Pelechrinis, K., Krishnamurthy, P.: Location affiliation networks: bonding social and spatial information. In: Flach, P.A., De Bie, T., Cristianini, N. (eds.) ECML PKDD 2012, Part II. LNCS, vol. 7524, pp. 531–547. Springer, Heidelberg (2012)
10. Scellato, S., Noulas, Mascolo, C.: Exploiting place features in link prediction on location-based social networks. In: KDD 2011, pp. 1046–1054. ACM, New York (2011)
11. Ye, M., Liu, X., Lee, W.-C.: Exploring social influence for recommendation: a generative model approach. In: SIGIR 2012, pp. 671–680. ACM, New York (2012)
12. Ye, M., Yin, P., Lee, W.-C.: Location recommendation for location-based social networks. In: GIS 2010, pp. 458–461. ACM, New York (2010)
13. Ye, M., Yin, P., Lee, W.-C., Lee, D.-L.: Exploiting geographical influence for collaborative point-of-interest recommendation. In: SIGIR 2011, pp. 325–334. ACM, New York (2011)
14. Yuan, Q., Cong, G., Ma, Z., Sun, A., Thalmann, N.M.: Time-aware point-of-interest recommendation. In: SIGIR 2013, pp. 363–372. ACM, New York (2013)
15. Yuan, Q., Cong, G., Sun, A.: Graph-based point-of-interest recommendation with geographical and temporal influences. In: CIKM 2014, pp. 659–668. ACM, New York (2014)

Coupling Multiple Views of Relations
for Recommendation

Bin Fu[1,2], Guandong Xu[1]([✉]), Longbing Cao[1], Zhihai Wang[2], and Zhiang Wu[3]

[1] Advanced Analytics Institute, University of Technology, Sydney, Australia
`Bin.Fu@student.uts.edu.au`, {`Guandong.Xu,Longbing.Cao`}`@uts.edu.au`
[2] School of Computer and Information Technology,
Beijing Jiaotong University, Beijing, China
`zhhwang@bjtu.edu.cn`
[3] Jiangsu Provincial Key Laboratory of E-Business,
Nanjing University of Finance and Economics, Nanjing, China
`zawuster@gmail.com`

Abstract. Learning user/item relation is a key issue in recommender system, and existing methods mostly measure the user/item relation from one particular aspect, e.g., historical ratings, etc. However, the relations between users/items could be influenced by multifaceted factors, so any single type of measure could get only a partial view of them. Thus it is more advisable to integrate measures from different aspects to estimate the underlying user/item relation. Furthermore, the estimation of underlying user/item relation should be optimal for current task. To this end, we propose a novel model to couple multiple relations measured on different aspects, and determine the optimal user/item relations via learning the optimal way of integrating these relation measures. Specifically, matrix factorization model is extended in this paper by considering the relations between latent factors of different users/items. Experiments are conducted and our method shows good performance and outperforms other baseline methods.

Keywords: Recommender system · Collaborative filtering · Matrix factorization

1   Introduction

Recommender system is a type of technology to overcome the information overload problem by estimating users' preferences and finding potentially desirable items for them [1]. For instance, online bookstore *Amazon* would infer users' preferences by analyzing the explicit ratings they have given, and then recommend to them some books that might catch their interests. Currently, recommender system has attracted extensive attention form both of academia and industry, and various methods have been proposed by researchers for recommending items of different types, e.g., movies [2], books [3], and music [4], etc.

© Springer International Publishing Switzerland 2015
T. Cao et al. (Eds.): PAKDD 2015, Part II, LNAI 9078, pp. 732–743, 2015.
DOI: 10.1007/978-3-319-18032-8_57

Among existing methods, collaborative filtering (CF) has been investigated and extended extensively [5]. The underlying assumption of CF is that similar users/items will give/receive similar ratings in future. Therefore, the success of CF essentially rely on how to learn the similarities, i.e., the extent of *relations*, between users/items. Roughly speaking, User/item relation can be measured using different types of auxiliary data, e.g., historical ratings, social networks [6–8], users and items' attributes [9], user-generated tags [10], and reviews [11], etc.

director			Robert	Tom	...	Jack
...			...	...	...	...
genre			*Musical*	*Crime*		*Thriller*
age	**gender**	**...**	**job**	v_1	v_2	...	v_m
25	F	...	lawyer	u_1	1	?	...	1
27	M	...	doctor	u_2	1	?	...	4
35	F	...	teacher	u_3	?	4	...	?
...	...	...	...	...	1	?	...	?
36	M	...	teacher	u_n	?	1	...	?

Fig. 1. An example of movie recommendation

Despite the improvments, existing methods mostly exploit only one type of data, e.g., historical ratings, to learn user/item relations. The potential issues include: (1) Each type of data might suffer the insufficiency problem. Figure 1 is a toy dataset for movie recommendation. It shows that: a) the rating matrix is extremely sparse, and b) an user/item is characterized by few attributes. Hence similarities and relations learned based on such kinds of data might be inappropriate. (2) Each type of data only offers a partial view of the underlying user/item relation, whereas multifaceted factors might be involved. For instance, if user u_1 follows other users' suggestions when rating items, there might be multiple possible factors: they have *similar attributes*, e.g., they are lawyers; they have shown *similar interest*, e.g., rating movies similarly, or they are *friends* in social networks, etc. Therefore, it is more advisable to couple multiple relations measured from different aspects to approximate the underlying user/item relation.

In this paper, we try to address above issues based on the matrix factorization (MF) framework [12]. The key tasks include: (1) Couple multiple views of relations measured from different types of data to approximate the underlying user/item relation. The objective of coupling them together is to capture a better understanding of user/item relation. Similar motivation has been employed in multiple kernel learning, in which different kernels correspond to different notions of similarity [13]. (2) Incorporate user/item relation in the matrix factorization model. To this end, we propose a new recommendation framework based on the MF model, and it can be characterized as follows:

(1) Firstly, multiple types of data are explored and the corresponding relations are coupled in a supervised learning framework. Specifically, for users u_i and u_j, k different data sources are exploited, thus we get k relation values. i.e., $A^{ij} = (a_1^{ij}, a_2^{ij}, ..., a_k^{ij})$, which are then treated as descriptive attributes of the underlying relation between u_i and u_j, i.e., s_{ij}. Next, we approximate s_{ij}

as $s_{ij} = f_\theta(A^{ij})$, thus more appropriate s_{ij} can be obtained via learning the optimal parameter θ of function f by minimizing a given loss function. Likewise, the same strategy is applied to items.

(2) Secondly, in order to utilize the user/item relation in MF framework, we assume that each user u_i latent factor vector $\hat{P}_i$ is determined by two components: his/her basic factor vector P_i, and other similar users' basic factor vectors $P_{neighbors}$. We set $\hat{P}_i = g(P_i, P_{neighbors}, S)$ to utilize the relations and dependencies between users. Here S is a matrix that indicates the relation values between all possible pairs of users. The same strategy is applied to items.

We finally compare our model with other state-of-the-art methods on datasets with different types of data to compute user/item relation respectively. To summarize, our main contributions are as follows.

- We propose an innovative model to exploit multiple types of auxiliary data and couple multiple measures of user/item similarity and relation in a supervised learning framework.
- We extend classical MF model to enable that relations between users/items can be exploited when learning each user's/item's latent factor vector.
- We conduct extensive experiments. The results validate the effectiveness of our model, and indicate its applicable scenarios.

The remainder of paper is organized as follows. Section 2 briefly reviews the related work. Problem statement is given in section 3. In section 4, we describe our model in detail. Experimental design and results are presented in section 5, followed by the conclusions in section 6.

2   Related Work

2.1   Compute User/Item Relation Using Different Types of Data

According to the types of data source have been utilized to compute user/item similarity, related methods mainly fall into the following categories.

(1) Utilizing historical ratings. The is the most common data have been investigated in traditional CF technique. Two basic methods are user-based approach and item-based approach [2,5]. User-based approach tries to compute the user relation based on their past ratings, and item-based approach tries to compute item relation base on the ratings they have received. *Cosine* similarity and *Pearson correlation* are the two main measures have been used to compute user/item similarity and relation. (2) Exploiting social networks. Social networks introduce explicit user-user relationship that can be represented as an asymmetric or symmetric matrix [8,14]. The basic assumption is that users connected in a social network should have similar preferences. Based on this assumption, various models introducing social networks have been proposed [7,15,16]. (3) Exploiting user contributed information. As summarized in [17], various types of data generated by users have been explored, such as tags, multimedia content, reviews and comments, etc. User relation is defined by the similarity between corresponding content contributed by the users.

2.2 Matrix Factorization (MF) Model

MF model has been extensively investigated and extended recently for recommendation task. It essentially treats the recommendation as a *matrix completion* problem, and solve it by transforming the original rating matrix R to the multiplication of two matrices $P^T Q$, thus r_{ij}, i.e., the rating assigned to item v_j by user u_i, can be approximated as $r_{ij} \approx P_i^T Q_j$. Here, P_i and Q_j can be treated as a latent factor vector representing user u_i and item v_j respectively. The optimal P and Q are learned by minimizing a particular loss function.

The relations between users/items' latent factor vectors are ignored in basic MF model. Some of above methods thus try to incorporate the use/item relation into MF model by assuming that similar user/item should have similar latent factor vectors. Specifically, social networks are utilized in [6,7,15], and social networks and historical ratings are explored in [16], etc.

Our proposed model is also based on MF model, and similar to these methods. The differences include: (1) Instead of determining user/item relation using a single type of data, our model utilize multiple data sources to measure different user/item relations and couple them together. (2) The weights of different measures and the ultimate user/item relation values are learned in a supervised learning framework, thus they could be optimal for current recommendation task.

3 Problem Statement

Based on MF model, we assume that each user/item's latent factor should be dependent on other similar users'/items' latent factors. Thus critical tasks include (1) Estimate the similarities between users/items, here similarity means the extent of user/item relation, we will use this term in following part. (2) Model dependencies between similar users/items in MF model. The targeted problem can be formulated as follows.

Let $S^U = [S_{ij}^U]_{n \times n}$ denote the similarities between users, where s_{ij}^U is the similarity between user u_i and u_j. As stated before, user/item similarity can be measured from different aspects, e.g., similarity based on ratings, similarity based on connectivity in social networks, etc. We then treat these values as descriptive features of the ultimate user similarity. Formally, S_{ij}^U should be determined as

$$S_{ij}^U = f_\theta(A^{uij}) \tag{1}$$

Here $A^{uij} = < a_1^{uij}, \cdots, a_z^{uij} >$ is a set of values consists of different similarity measures for u_i and u_j. f is a function that generates the final similarity and θ is a set of parameters that need to be estimated. Since different measures are computed on different data sources, thus multiple types of data are explored.

Likewise, let $S^V = [S_{ij}^V]_{m \times m}$ denote the similarities between items, where s_{ij}^V indicates the similarity of item v_i and v_j. Same strategy is applied to learn item similarity, we can get

$$S_{ij}^V = f_\vartheta(A^{vij}) \tag{2}$$

Here $A^{vij} =< a_1^{vij}, \cdots, a_z^{vij} >$ is a set of different similarity measures for item v_i and v_j, and ϑ are the parameters that need to be estimated.

To model the dependency relations between users/items, we let P and Q denote users' and items' basic latent factors as in traditional MF model. However, P and Q should be updated due to the influence between similar users and items. Thus, u_i's final factor vector $\hat{P}_i$ is $\hat{P}_i = g(P, S^U)$. Here g is the function determines the final factor vectors. It is clear that $\hat{P}_i$ is determined by other users' basic factor vector with respect to their similarities. Same strategy is applied to items, and we get $\hat{Q}_j = g(Q, S^V)$.

Finally, rating r_{ij} is estimated as $\hat{r}_{ij} = \hat{P}_i^T \hat{Q}_j$. Now, the specific task is how to design function f, g, and estimate the parameters $< \theta, \vartheta, P, Q >$, which is elaborated in next section.

4   Coupling Multiple Views of Relations

4.1   Multiple Views of Similarity

Given a particular type of data, there are various ways to compute similarities between users/items. For example, given the rating matrix $R_{m \times n}$, similarity between u_i and u_j can be defined as the similarity between corresponding row $R_{i.}$ and $R_{j.}$ in R. *Pearson correlation* can be used to compute the similarity between these rows. There are other ways of measuring similarity given more definitions or more types of auxiliary data. We couple these measures using the framework shown in (1). The results of different measures are then fed into function f to generate the ultimate similarity value. In this paper, we choose the sigmoid function for f, since it is a differentiable function which makes the following learning process feasible, and its output range is $[0, 1]$ that is natural for representing the similarity. The definition of f for determining user similarity is thus define as:

$$S_{ij}^U = f_\theta(A^{uij}) = \frac{1}{1 + e^{-\theta^T A^{uij}}} \tag{3}$$

Similarly, to learn the ultimate item similarity S^V, we get the following formula.

$$S_{ij}^V = f_\vartheta(A^{vij}) = \frac{1}{1 + e^{-\vartheta^T A^{vij}}} \tag{4}$$

The key problem is how to estimate the optimal values of parameters θ, ϑ.

4.2   Dependency Propagation Among Users/Items

Due to the similarity, an user's preference should be influenced by his neighbours, i.e., a set of similar users. Likewise, an item should also show common characteristics with other similar items. In other words, the latent factor vector of an users/item should be dependent on others' with respect to their similarities.

Let us focus on *user* firstly. In our model, each user u_i is assumed to have an initial factor vector P_i. Then, u_i will get an ultimate factor vector $\hat{P}_i$ due to the dependency propagation among similar users. We formulate this dependency propagation as follows:

$$\hat{P}_i = (1 - \alpha_u)\frac{\sum_{l\in\mathbf{Nu}(i)} S_{il}^U P_l}{|\mathbf{Nu}(i)|} + \alpha_u P_i \tag{5}$$

Where $\mathbf{Nu}(i)$ denotes u_i's neighbours, a set of users who are most similar with user u_i according to S^U. It is shown that $\hat{P}_i$ is determined by two components, i.e., his initial factor vector P_i, and his neighbours' factor vectors. Here α_u is an adjustable parameter to control the influence of dependency propagation. Since each user depends on his/her neighbours, a dependency graph is actually formed. The dependencies are propagated over the whole graph, thus each user essentially is influenced by all other users directly or indirectly. For items, we can get a similar formulation. Given an item v_j, its ultimate factor vector $\hat{Q}_j$ is:

$$\hat{Q}_j = (1 - \alpha_v)\frac{\sum_{l\in\mathbf{Nv}(j)} S_{jl}^V Q_l}{|\mathbf{Nv}(j)|} + \alpha_v Q_j \tag{6}$$

Here $\mathbf{Nv}(j)$ denotes item v_j's neighbours that are most similar with it according to S^V, and α_v is the parameter to trade off the two components.

Once the $\hat{P}_i$ and $\hat{Q}_j$ are determined, the rating r_{ij} can be approximated as $r_{ij} \approx \hat{P}_i^T \hat{Q}_j$. By now, the whole process of our model is outlined, and a graphical illustration is given in Fig. 2, in which $P_{i1...k}$ are factor vectors of user u_i's k neighbours, and $Q_{j1...k}$ are factor vectors of item v_i's k neighbours. $A^U = [A^{uij}]_{n\times n}$ is a matrix, and its element A^{uij} is the precomputed vector of similarity values for user u_i and u_j. Similarly, $A^V = [A^{vij}]_{m\times m}$ is also a matrix, and A^{vij} is the precomputed vector of similarity values for item v_i and v_j. Finally, $< \theta, \vartheta, P, Q >$ are the parameters need to be estimated.

4.3 Parameter Estimation

To begin with, we define the loss function which will be used in our model to learn the optimal parameters. Its definition is:

$$L(\theta, \vartheta, P, Q) = \frac{1}{2} \sum_{(i,j)\in\mathbf{D}} (r_{ij} - \hat{P}_i^T \hat{Q}_j)^2 + \lambda(||P||_F^2 + ||Q||_F^2) \tag{7}$$

here $\mathbf{D}$ is the training set that contains a set of (u_i, v_j, r_{ij}) tuples that record the already known r_{ij} for pair of (u_i, v_j). λ is the parameter to control the influence of the regularization term, and $|| \cdot ||_F$ denotes the Frobenius norm. $\hat{P}, \hat{Q}$ are defined in (5) and (6). Now, the task is to learn optimal parameters $\theta^*, \vartheta^*, P^*, Q^*$ that can minimize above loss function.

We use stochastic gradient descent method to solve this task. The key issue in the process is to compute the partial derivatives of a single training instance's loss with respect to the parameters. Without loss of generality, we show how to

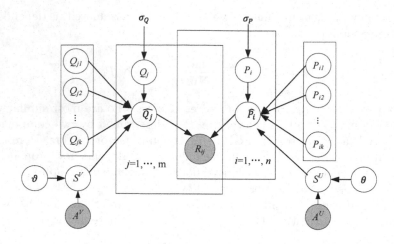

Fig. 2. Graphical illustration of our proposed model

compute the partial derivative for a tuple (u_i, v_j, r_{ij}). Specifically, for (u_i, v_j, r_{ij}), the loss function becomes:

$$L_{ij}(\theta, \vartheta, P, Q) = \frac{1}{2}(r_{ij} - \hat{P}_i^T \hat{Q}_j)^2 + \lambda(||P||_F^2 + ||Q||_F^2) \qquad (8)$$

For simplicity, we set $e_{ij} = (r_{ij} - \hat{P}_i^T \hat{Q}_j)$. The partial derivatives of all parameters are computed as follows.

According to (3), $\theta = < \theta_1, \cdots, \theta_z >$ is a vector of parameters used to determine user similarities S^U, so it only exists in $\hat{P}_i$. According to (5), for any element θ_t in θ, the partial derivative is

$$\frac{\partial L_{ij}}{\partial \theta_t} = -e_{ij}\frac{\partial \hat{P}_i^T}{\partial \theta_t}\hat{Q}_j = \frac{-e_{ij}(1 - \alpha_u)}{|\mathbf{Nu}(i)|} \cdot (\sum_{l \in \mathbf{Nu}(i)} \frac{\partial S_{il}^U}{\partial \theta_t} P_l)^T \hat{Q}_j \qquad (9)$$

Here $|\mathbf{Nu}(i)|$ is the number of u_i's neighbours, the remaining problem is $\partial S_{il}^U/\partial \theta_t$. According to (3), it can be computed as follows.

$$\frac{\partial S_{il}^U}{\partial \theta_t} = \frac{\partial \frac{1}{1+e^{-\theta^T A^{ujl}}}}{\partial \theta_t} = \frac{e^{-\theta^T A^{uil}}}{(1 + e^{-\theta^T A^{uil}})^2}A_t^{uil} \qquad (10)$$

As defined before, A^{uil} is a vector of similarity measures for user u_i and u_j, and A_t^{uil} is the tth element of A^{uil}.

Similarly, ϑ is a vector of parameters to determine item similarities S^V. According to (6), for any element ϑ_t in ϑ, the partial derivative is:

$$\frac{\partial L_{ij}}{\partial \vartheta_t} = -e_{ij}\hat{P}_i^T\frac{\partial \hat{Q}_j}{\partial \vartheta_t} = \frac{-e_{ij}(1 - \alpha_v)}{|\mathbf{Nv}(j)|} \cdot \hat{P}_i^T(\sum_{l \in \mathbf{Nv}(j)} \frac{\partial S_{jl}^V}{\partial \vartheta_t}Q_l) \qquad (11)$$

Here $|\mathbf{Nv}(j)|$ is the number of item v_j's neighbours. According to (4), $\partial S_{jl}^V / \partial \vartheta_t$ can be computed as follows.

$$\frac{\partial S_{jl}^V}{\partial \vartheta_t} = \frac{\partial \frac{1}{1+e^{-\vartheta^T A^{vjl}}}}{\partial \vartheta_t} = \frac{e^{-\vartheta^T A^{vjl}}}{(1+e^{-\vartheta^T A^{vjl}})^2} A_t^{vjl} \qquad (12)$$

Here A^{vjl} is a vector of similarity measures for item v_j and v_l, and A_t^{vjl} is the tth element of A^{vjl}.

For $P = [P_1, \cdots, P_n]$, only P_i and $P_l(l \in \mathbf{Nu}(i))$ are involved. Therefore, the partial derivative $\partial L_{ij} / \partial P_i$ is

$$\frac{\partial L_{ij}}{\partial P_i} = -e_{ij}\alpha_u \hat{Q}_j + \lambda P_i \qquad (13)$$

For any $P_l(l \in \mathbf{Nu}(i))$, $\partial L_{ij} / \partial P_l$ is

$$\frac{\partial L_{ij}}{\partial P_l} = \frac{-e_{ij}(1-\alpha_u)S_{il}^U}{|\mathbf{Nu}(i)|} \hat{Q}_j + \lambda P_l \qquad (14)$$

Similarly, for $Q = [Q_1, \cdots, Q_m]$, only Q_j and $Q_l(l \in \mathbf{Nv}(j))$ are involved. Therefore, $\partial L_{ij} / \partial Q_j$ is computed as

$$\frac{\partial L_{ij}}{\partial Q_j} = -e_{ij}\alpha_v \hat{P}_i + \lambda Q_j \qquad (15)$$

For any $Q_l(l \in \mathbf{Nv}(j))$, $\partial L_{ij} / \partial Q_l$ is computed as

$$\frac{\partial L_{ij}}{\partial Q_l} = \frac{-e_{ij}(1-\alpha_v)S_{jl}^V}{|\mathbf{Nv}(j)|} \hat{P}_i + \lambda Q_l \qquad (16)$$

We now have described the overall process of our model and how to compute all the necessary components. We call our model as CMR (Couple Multiple Relations). Note that each user/item's neighbours would change in each iteration, due to the dynamical updating of the similarities. Arguably, our model could learn more appropriate user and item similarities by explicitly optimizing the loss function. The validation is shown in next section.

5   Experimental Results

5.1   Evaluation Metrics and Datasets

The popular metrics, MAE and $RMSE$, are used to measure the performance of all the models. Let T denote a test dataset that contains a set of (u_i, v_j, r_{ij}) tuples, and $\hat{r}_{ij}$ denote the prediction of r_{ij}. These metrics are defined as belows.

– Mean Absolute Error (MAE).

$$MAE = \frac{\sum_{(i,j)\in T} |r_{ij} - \hat{r}_{ij}|}{|T|} \qquad (17)$$

– Root Mean Squared Error (RMSE).

$$RMSE = \sqrt{\frac{(r_{ij} - \hat{r}_{ij})^2}{|T|}} \tag{18}$$

For both of the metrics, a smaller value means a better performance.

We user 2 datasets, *Movielens1M* and *Yelp*, to train and evaluate the models. They include different types of data, enabling us to compute user/item similarity from diverse aspects.

– MovieLens1M is a dataset regarding recommending movies. Several types of data are included: (1) User attribute. Each user is represented by a vector of attributes, including age, gender, and occupation, etc. (2) Movie attribute. We just keep the category set of each movie as its attributes. (3) Rating set which consists of a set of $< user, item, rating >$ tuples. We remove users and items which have made/received less than 50 ratings, and finally there are 4297 users and 2514 movies left.

– Yelp is a dataset regarding recommending Point-of-Interest (POI), such as restaurant, hotel, etc. Several types of data are included: (1) Social networks. For each user, his/her friends are explicitly given. (2) POI attribute. We similarly keep the category set of each POI as its attributes. (3) A set of ratings. The original dataset contain POIs in several cities in United States. We just keep POIs in city "Las Vegas" and users who have rated any of these POIs and the corresponding ratings.

5.2   Baselines and Setting

We compare the proposed *CMR* model with other three classical models. These models include:

– *MF* model [12]. This model assumes that users/items are independent from each other, the similarities are ignored.

– *SoRec* model [6]. It takes one type of user relation into consideration. For dataset MovieLens1m, user relation is defined as the similarity between users' ratings. For dataset Yelp, relations between users are derived from the social networks, which means the similarity is 1 if two users are connected in the social networks, otherwise is 0.

– SR_{i+}^{u+} model [16]. It incorporates one type of user relation and one type of item relation into MF. Item relation is defined as the similarity of items' ratings in both of two datasets. To compute user similarity in the two datasets, we adopt the same approach adopted in *SoRec*.

– our *CMF* model. the aim of our model is to couple multiple measures of user/item relation. Specifically, two types of user relation are considered: a) similarity between users' ratings, and b) similarity between users' attributes (MovieLens1m) or connectivities in social networks (Yelp). Two types of item relation are also considered: a) similarity between items' ratings, and b) similarity between items' attributes.

Pearson correlation is used to compute the similarity between users' and item's ratings, and Euclidean distance is used to measure the similarity between users' and item's attributes. After trying different values, the length of latent factor vector is set to be 10 in all models. In *SoRec*, SR_{i+}^{u+}, and *CMF*, the number of neighbours for each user or item is also set to be 10. Since all models use stochastic gradient descent, we also experiment with different values of learning rate. Finally, the learning rate is set to 0.3 in the baseline models and 0.05 in out model. 0.005 is chosen for the regularization term weight λ in all models. For simplicity, α_u and α_v in our model are assigned with same value, and we use α to replace α_u, α_v in following part.

5.3   Results and Analysis

We begin with investigating the impact of α on our model's performance. Both of the two datasets are split into training set (80%), and test set (20 %). We vary α's value from 0 to 1, and the results on two datasets are shown in Fig. 3.

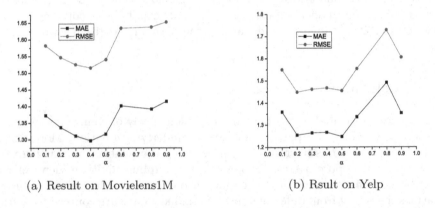

(a) Result on Movielens1M   (b) Rsult on Yelp

Fig. 3. The impact of α on our model

In our model, α is used to control the influence of similar users/items on current user/item. Seen from Fig. 3, experiments on two datasets show the similar results. That is, optimal performance is obtained when α is around $0.4 - 0.5$, which means the influence of user/relation should be exploited to an appropriate extent. Both of overemphasizing ($\alpha \to 0$) or ignoring ($\alpha \to 1$) the influence of user/item relation are inadvisable.

Next, we create another 4 training sets by sampling 20%, 40%, 60%, and 80% of original training set. Models are evaluated on these training sets, and the results are shown in Table 1-2. Each table is divided into two parts. The left part contains results in terms of MAE, and the right part contains results in terms of RMSE. Bold items indicate the corresponding models that performs best. α is set to be 0.4 in our model.

Table 1. Performance on MovieLens1M

Data set	MF	SoRec	SR_{i+}^{u+}	CMR	MF	SoRec	SR_{i+}^{u+}	CMR
100%	1.3947	1.369	1.3669	**1.2969**	1.6884	1.6485	1.6722	**1.5158**
80%	1.4199	1.4252	1.4252	**1.3229**	**1.4199**	1.6918	1.6918	1.5424
60%	1.4258	1.4245	1.4245	**1.2760**	1.6939	1.6919	1.6919	**1.4935**
40%	1.4024	**1.3580**	1.3915	1.3863	1.6755	1.6487	1.6624	**1.6061**
20%	1.3892	1.4167	1.4167	**1.3143**	1.6886	1.6849	1.6849	**1.5327**

Table 2. Performance on Yelp

Data set	MF	SoRec	SR_{i+}^{u+}	CMR	MF	SoRec	SR_{i+}^{u+}	CMR
100%	1.4599	1.4216	1.2207	**1.2661**	1.7542	1.7186	1.4823	**1.4671**
80%	1.4350	1.4042	1.3120	**1.1318**	1.7424	1.6970	1.5697	**1.3301**
60%	1.4873	1.4565	1.4146	**1.1852**	1.7586	1.7239	1.6752	**1.3853**
40%	1.4264	1.4167	**1.3661**	1.4219	1.7100	1.6964	**1.6239**	1.6254
20%	1.4614	1.4342	1.1677	**1.2130**	1.7814	1.7199	1.4263	**1.4159**

Tables 1-2 show similar results. They indicate that our model is significantly superior to the other models. Our model performs best on 4 out of the 5 datasets in terms of MAE and RMSE both on MovieLens1M and Yelp. Therefore, the results validate our model's effectiveness due to its capability of coupling multiple views of user/item similarity and learning the optimal user/item similarities for current task.

6   Conclusions

This work is motivated by the fact that each single way of computing user/item relation only captures a partial view of the underlying relation. Moreover, no explicit loss function is refereed to when learning the relations in existing work. Therefore, we propose a novel framework for coupling multiple views of relations between users/items in this paper. In our proposed framework, multiple relations measured from different types of auxiliary data are coupled to approximate the underling user/item relation. In conclusions, our model's advantages include: (1) Multiple similarity measures are coupled together, thus a more accurate and comprehensive approximate of the user/item relation could achieve. (2) Motivated by metric learning etc., we learn the user/item similarity by explicitly optimizing a particular loss function, thus the similarities acquired would be optimal for current recommendation task. (3) Both of the dependencies between users and between items are exploited based on MF model. The experimental results validate the effectiveness and superiority of our model. In following work, we aim to explore more types of auxiliary data to achieve more appropriate approximation of the underlying relations.

Acknowledgments. This research has been partially supported by Beijing Natural Science Foundation, China (4142042), and National Centre for International Joint Research on E-Business Information Processing, China (2013B01035).

References

1. Bobadilla, J., Ortega, F., Hernando, A., Gutirrez, A.: Recommender Systems Survey. Knowledge-Based Systems **46**, 109–132 (2013)
2. Sarwar, B., Karypis, G., Konstan, J., Riedl, J.: Item-based collaborative filtering recommendation algorithms. In: Proceedings of the 10th International Conference on World Wide Web, pp. 285–295 (2001)
3. Ziegler, C.-N., McNee, S.M., Konstan, J.A., Lausen, G.: Improving recommendation lists through topic diversification. In: Proceedings of the 14th International Conference on World Wide Web, pp. 22–32 (2005)
4. Kaminskas, M., Ricci, F., Schedl, M.: Location-aware music recommendation using auto-tagging and hybrid matching. In: Proceedings of the Seventh ACM Conference on Recommender Systems, pp. 17–24. ACM (2013)
5. Su, X., Khoshgoftaar, T.M.: A Survey of Collaborative Filtering Techniques. Advances in Artificial Intelligence **2009**, 4 (2009)
6. Ma, H., King, I., Lyu, M.R.: Learning to Recommend with Explicit and Implicit Social Relations. ACM Transactions on Intelligent Systems and Technology **2**, 20 (2011)
7. Jamali, M., Ester, M.: A transitivity aware matrix factorization model for recommendation in social networks. In: Proceedings of the Twenty-Second International Joint Conference on Artificial Intelligence, pp. 2644–2649 (2011)
8. Tang, J., Hu, X., Liu, H.: Social Recommendation: A Review. Social Network Analysis and Mining **3**, 1113–1133 (2013)
9. Singh, A.P., Gordon, G.J.: Relational learning via collective matrix factorization. In: Proceedings of the 14th ACM SIGKDD International Conference on Knowledge Discovery and Data Mining, pp. 650–658. ACM (2008)
10. Xu, G., Gu, Y., Dolog, P., Zhang, Y., Kitsuregawa, M.: SemRec: a semantic enhancement framework for tag based recommendation. In: Proceedings of the Twenty-Fifth AAAI Conference on Artificial Intelligence (2011)
11. Tang, J., Gao, H., Hu, X., Liu, H.: Context-aware review helpfulness rating prediction. In: Proceedings of the 7th ACM Conference on Recommender Systems, pp. 1–8 (2013)
12. Salakhutdinov, R., Mnih, A.: Probabilistic matrix factorization. In: Proceedings of the Twenty-First Annual Conference on Neural Information Processing Systems, pp. 1257–1264 (2008)
13. Gonen, M., Alpaydn, E.: Multiple Kernel Learning Algorithms. The Journal of Machine Learning Research **12**, 2211–2268 (2011)
14. Konstas, I., Stathopoulos, V., Jose, J.M.: On social networks and collaborative recommendation. In: Proceedings of the 32nd International ACM SIGIR Conference on Research and Development in Information Retrieval, pp. 195–202 (2009)
15. Ma, H., Yang, H., Lyu, M.R., King, I.: Sorec: social recommendation using probabilistic matrix factorization. In: Proceedings of the 17th ACM Conference on Information and knowledge management, pp. 931–940 (2008)
16. Ma, H.: An experimental study on implicit social recommendation. In: Proceedings of the 36th International ACM SIGIR Conference on Research and Development in Information Retrieval, pp. 73–82 (2013)
17. Shi, Y., Larson, M., Hanjalic, A.: Collaborative Filtering beyond the User-Item Matrix: a Survey of the State of the Art and Future Challenges. ACM Computing Surveys **47**, 3 (2014)

Pairwise One Class Recommendation Algorithm

Huimin Qiu[⊠], Chunhong Zhang, and Jiansong Miao

Beijing University of Posts and Telecommunications, Beijing, China
{Hermione.qiu.hm,zhangch.bupt.001}@gmail.com, miaojiansong@bupt.edu.cn

Abstract. We address the problem of one class recommendation for a special implicit feedback scenario, where training data only contain binary relevance data that indicate user' selection or non-selection. A typical example is the followship in social network. In this context, the extreme sparseness raised by sparse positive examples and the ambiguity caused by the lack of negative examples are two main challenges to be tackled with. We dedicate to propose a new model which is tailored to cope with this two challenges and achieve a better topN performance. Our approach is a pairwise rank-oriented model, which is derived on basis of a rank-biased measure Mean Average Precision raised in Information Retrieval. First, we consider rank differences between item pairs and construct a measure function. Second, we integrate the function with a condition formula which is deduced via taking user-biased and item-biased factors into consideration. The two factors are determined by the number of items a user selected and the number of users an item is selected by respectively. Finally, to be tractable for larger dataset, we propose a fast leaning method based on a sampling schema. At the end, we demonstrate the efficiency of our approach by experiments performed on two public available databases of social network, and the topN performance turns out to outperform baselines significantly.

Keywords: Implicit feedback · One class · Rank-oriented recommendation · Pairwise

1 Introduction

As the concept of personalization is becoming prevailing, recommender systems have been widely used in various Internet Service[4, 7]. Recommendation algorithms may differ as the scenario changes with service. Generally, there exist two categories: explicit and implicit scenarios. Under the former scenario, explicit ratings are given by users, such as a 1-5 scale in Netflix, and ratings is a measure of preference. Lots of researchs have been done on this issue, a representative work was done by [1] in Netflix prize. However, this kind of data is hard to achieve, for that it's a bad experience for user to be compelled to provide ratings. On the other hand, implicit scenario indicates the situation where only user's implicit feedback can be got. However, there are two kinds of implicit feedback. One kind of implicit feedback can be transferred into ratings, e.g., the counts of a

© Springer International Publishing Switzerland 2015
T. Cao et al. (Eds.): PAKDD 2015, Part II, LNAI 9078, pp. 744–755, 2015.
DOI: 10.1007/978-3-319-18032-8_58

user purchasing a brand, the counts of a user browsing a website, etc. Since that the counts of user behavior can be interpreted as a measure of preference, just as ratings do, this scenario is easier to tackle with. The other kind implicit feedback has no "count" information and is usually represented by binary relevance data, which indicates a user's attitude to an item, such as the followship between users in social network. Though Lots of research has been done for implicit feedback, few attention has been paid to the situation where only binary relevance data can be obtained, which is exactly we focus on in this paper.

For convenience, the implicit scenario where only binary relevance data is available is short for binary implicit scenario. To figure this scenario out, we take the followship in social network as an specific example. If a user follows someone else, he is assumed to like that person, which is also called a positive example. As for persons who are not in his follow list, it's not sure whether he like or dislike them, which are missing examples. Obviously, we can only know part of positive examples from user feedback, with the remaining hidden in the missing part, which is a mix of positive and negative examples. Hence, the dataset is sparse and has one class ambiguity, which are two main challenges in binary implicit scenario. Various solutions have been proposed to settle with this, such as considering it as a one class classification problem[5,12,14] or completing missing data according to a certain strategy[6]. Taking missing data as unknown, one class classification learn only from known positive examples and use EM-like algorithms to iteratively learn the classifier and then predict on the unknown. While the completing method assign fixed values to missing examples according to weighting or sampling strategies. However, it is too arbitrary to take all missing data as unknown or assign unchangeable values to them. In this paper, we try to figure out a compromise by dealing with training data in a new method.

Our proposed model is a constrained pairwise rank-oriented model, which is based on a rank-biased measure raised in Information Retrieval. We choose the rank-biased measure as the basis of our model because that we expect our model to achieve a better recommend list. As stated in [3,10], rank-biased measures penalize mistakes of top items more than other items, which leads to a better topN performance than traditional measures that take mistakes of all items as equally important. This is also proved by [9,13], etc. What's more, to be tailored for binary implicit scenario, we do two improvements to achieve the compromise we mentioned before, which are also contributions of this paper:

- We construct a new pairwise rank-biased measure, in which rank differences of item pairs are taken into consideration. We assume that a user prefer positive items that he has action on than items that he doesn't have any action on. The interpret of training data will be more accuracy, since that we consider all items in a proper method instead of ignoring missing data like one class classification[5,12,14] do.
- We add a constrained condition to refine the pairwise rank-biased measure. This constrained condition is deduced by considering user-biased and item-biased factors. In a social network, user-biased and item-biased factors are determined by the number of persons a user follows and the number of followers a user owns respectively.

The paper is organized as follows. In section 2 we review the related work. Model we propose is presented in section 3, which is followed by the optimization process and a fast leaning method which targets at reducing the computation complexity of optimization stated in section 4. In section 5, we empirically evaluate the efficiency of our approach by comparing with some baselines on two public datasets. Finally, we conclude the paper.

2   Related Work

The most popular optimization measure for recommender is error metric[1]. However, it is barely suit for explicit feedback or implicit feedback that can be transformed into ratings and not for the binary implicit scenario which we are targeted at. Recently, a new kind of measure has emerged, rank-biased measures [9,11,13]. This kind of measures optimize on rank information, and as the comparison done in [15] shows, they have a better topN performance. In addition, Mean Average Precision (MAP), a common rank-biased measure, is a more suitable measure for binary implicit scenario, which is shown in [2]. That why we choose MAP as the basis of our model.

Generally, recommendation work that targets at binary implicit scenario can be divided into several directions. One direction is one class classification. Machine leaning algorithms that optimize model only on limited positive examples are proposed, such as [5,12,14]. Another direction is one class collaborative filtering (OCCF)[6], where missing data are completed according to certain strategy. Then traditional CF model, such as Matrix Factorization and k-nearest neighbor (KNN), can be trained on the completed dataset. In contrast to these, our model is a pair-wise rank-biased model. Similar work have been done by [8,9]. [9] is based on smoothed mean reciprocal rank, a rank-biased measure, however, it only estimates the thought of ranking and didn't consider much about the sparse and ambiguity problems of dataset. [8] utilizes pairwise comparisons between item pairs, while its criterion is Area Under the Curve (AUC). However the AUC can't reflect the topN performance well[15] because it is not a top-biased measure. In our paper, we mainly go into the sparsity and one class ambiguity problems, and try to construct a tailored rank-biased measure.

3   Pairwise Rank-oriented Model

3.1   Problem Definition

The problem setting in this paper is as following: Suppose that we are doing recommendation for a social network, which is a representative binary implicit scenario, and we want to provide a list of persons who a user may interest most to him. Notice that for a user in social network, his items are all other persons except himself instead of common goods like books, radio or something else. Thus, if the number of user is M, we can got a M·N (N is equal to M) matrix Y by taking all user item pairs' relationship as entries. Obviously, this matrix is binary and diagonally meaningless. If user i follows item j, we indicate this user item pair $\langle i, j \rangle$ by $y_{ij} = 1$, otherwise we are not sure the relationship, $y_{ij} = 0$ is used to note it.

Rank-biased measure which we choose as the basis of our model is a concept drawing insight from Information Retrieval field, where various measures have been proposed, such as Normalized Discounted Cumulative Gain (NDCG), Mean Average Precision (MAP) etc.. However, some of them are not suitable for the binary implicit scenario, because graded relevance is needed. Take this into consideration, we choose MAP, which has been used in many paper, as our model's basis. A common form of MAP measure utilized in recommendation[9] is as following:

$$
AP_i = \frac{1}{\sum\limits_{j=1}^{N} y_{ij}} \sum_{j=1}^{N} \frac{y_{ij}}{r_{ij}} \sum_{k=1}^{N} y_{ik} \mathrm{I}(r_{ik} \leq r_{ij})
$$

$$
MAP = \frac{1}{M} \sum_{i=1}^{M} \frac{1}{\sum\limits_{j=1}^{N} y_{ij}} \sum_{j=1}^{N} \frac{y_{ij}}{r_{ij}} \sum_{k=1}^{N} y_{ik} \mathrm{I}(r_{ik} \leq r_{ij})
$$

(1)

Where r_{ij} indicates the of person j among all recommended persons provided for user i. $\mathrm{I}(r_{ik} \leq r_{ij})$ is a pre-defined function: if the condition in the bracket is true, it returns 1, otherwise 0. The maximum of AP_i promotes the relevant items of user i and depresses the irrelevant items, which optimizes a user's item ranks according to item relevance. Then MAP is the sum of all users' AP, which ensures the item ranks of all users is optimized. The detail of how to utilize MAP will stated in next two section.

3.2 Pairwise Ranking

In this section, a new measure, pairwise ranking (PWR), is proposed based on MAP, which alleviates the ambiguity problem of training matrix mentioned before. Traditional methods optimize directly on pointwise train data, and generally they assign fixed values to missing examples to complete the original dataset. However, an arbitrary assignment will cause the error of dataset, which will lead to an inaccurate model. To avoid this, we utilize item pairs as training inputs and construct a new pairwise rank-biased measure. This different approach liberates us from the assignment, and utilizes relationship between item pairs which is more reliable.

Here, we obey two basic hypothesis to decide the relationship between item pair: for a user, it is more likely that, 1). known positive examples are assigned higher values than missing ones. 2). known positive examples' values are closer. For the convenience of stating, we use N_i^+, N_i^- and N_i to denote known positive item set, missing item set and integral item set of user i. Eq. (1) can be divided into two parts: ranks of N_i^+ and comparisons between N_i^+ and N_i^-. Inspired by this equation, we derive our pairwise rank-biased measure, PWR, which is shown in Eq. (2).

$$PWR_i = \frac{1}{\sum\limits_{j=1}^{N} y_{ij}} \sum_{j=1}^{N} y_{ij} \sum_{k=1}^{N} I(r(j) - r(k)) \tag{2}$$

However, problem arises when it comes to optimization. Optimizations are done on continuous function, however, Eq.(2) depends on the rank and is piecewise constant and non-differentiable. There exists different approaches to deal with this, one of which is smoothing the function and preforming gradient descent optimization on the smoothed version. Borrowing thought from [2], we derive the approximation of $\frac{1}{r(i)}$, f_{ij}, which is used to denote a predicted value for user item pair $\langle i, j \rangle$, and rank can be gotten by sorting score ascending. With a sigmoid function $g(x) = \frac{1}{1+e^{-x}}$, a common monotonous function, the indicator function can also be transferred into continuous function $g(f_{ij} - f_{ik})$. And then, taking all users into consideration, we can get the integral function:

$$PWR_i = \frac{1}{\sum\limits_{j=1}^{N} y_{ij}} \sum_{j=1}^{N} y_{ij} \sum_{k=1}^{N} g(f_{ij} - f_{ik})$$

$$PWR = \frac{1}{M} \sum_{i=1}^{M} \frac{1}{\sum\limits_{j=1}^{N} y_{ij}} \sum_{j=1}^{N} y_{ij} \sum_{k=1}^{N} g(f_{ij} - f_{ik}) \tag{3}$$

3.3 Constrained Pairwise Ranking

With pairwise rank-biased model mentioned in last section, ambiguity and sparsity problems of training dataset is alleviated. However, we can still mine more useful information from the limited train matrix, and it 's promising to get a more accurate model with more information used properly. We all know that users and items are all different to some extent. We can also find that different users may have different selectivity and different items may own different popularity, which is a valuable information. This viewpoint has been proposed and proved by [6]. We approve the definition of selectivity and popularity in [6], and interpret and utilize it in a new method. To make concepts clear, we give the formula of selectivity and popularity first:

$$selectivity_i = \frac{\sum\limits_{j=1}^{N} y_{ij}}{N}$$

$$popularity_j = \frac{\sum\limits_{i=1}^{M} y_{ij}}{M} \tag{4}$$

Where M and N are the total number of users and items. It's a acknowledged common sense that a user who select more items has a higher selectivity, which also means that items he didn't select have higher confidence to be disliked by him, vice versa. It's also obvious that if an item is more popular, it will be more likely to be disliked by users who didn't select it. Therefore we can merge these two factors into a value C_{ij} which indicates the confidence of predicted value for user item pair $\langle i, j \rangle$. In the end, we can deduced a weighted measure function:

$$C_{ij} = \alpha * \frac{\sum_{j=1}^{N} y_{ij}}{N} + \beta * \frac{\sum_{i=1}^{N} y_{ij}}{M} \tag{5}$$

α and β, tunable parameters which depends on the influence of user and item factors, can be gotten from experiment. Since C_{ij} is the confidence coefficient, borrowing the thought of error metric, we can view C_{ij} as the weight of error and then get a weighted error metric $\gamma_{C_{ij}}$. To stay the same form with PWR, we continue to use $g(f_{ij})$ to indicate the predicted value of y_{ij}. Since y_{ij} can be 0 or 1, the error of predicted value is calculated respectively. Generally the more accurate the model is, the smaller the value of error metric is. To stay monotonic, $\gamma_{C_{ij}}$ is added to PWR as a negative item:

$$\gamma_{C_{ij}} = C_{ij} * ((1 - y_{ij})g(f_{ij})^2 + y_{ij}(1 - g(f_{ij}))^2)$$

$$cPWR = \frac{1}{M} \sum_{i=1}^{M} \frac{1}{\sum_{j=1}^{N} y_{ij}} \sum_{j=1}^{N} y_{ij} \sum_{k=1}^{N} g(f_{ij} - f_{ik}) - \sum_{i=1}^{M} \sum_{j=1}^{N} \gamma_{C_{ij}} \tag{6}$$

4 Optimization and Fast Learning

4.1 Optimization

The key task of optimizing is to approximate f_{ij} in cPWR. Note that matrix F is a predicted score matrix, which contains predicted scores for all user item pairs. Based on the latent factor model, we replace F matrix with two lower dimension matrixes: user matrix U and item matrix V. Thus, f_{ij} is the inner product of user vector U_i and item vector V_j. Usually, regularized penalty should be added to avoid over-fitting problem. After finishing these replacements, we can get the final objective function:

$$\ell(U, V) = \frac{1}{M} \sum_{i=1}^{M} \frac{1}{\sum_{j=1}^{N} y_{ij}} \sum_{j=1}^{N} y_{ij} \sum_{k=1}^{N} g(U_i V_j - U_i V_k) - \frac{1}{MN} \sum_{i=1}^{M} \sum_{j=1}^{N} C_{i,j} * ((1 - y_{ij})g(U_i V_j)^2$$

$$+ y_{ij}(1 - g(U_i V_j))^2) - \lambda(|U|^2 + |V|^2) \tag{7}$$

The main parameters to be approximated are matrix U and V. Here we choose stochastic gradient ascend to maximize this function. $\ell(U, V)$ is the average of all $\ell(U_i, V)$, which is determined only by history data of user i. Hence, With respect to U and V, we can derive gradient formulas of them by taking other parameter as fixed. For user vector and item vector respectively:

$$\frac{\partial L}{\partial U_i} = \frac{1}{|N_i^+|} \sum_{j=1}^{N} y_{ij} \sum_{k=1}^{N} g'(f_{ij} - f_{ik})(V_j - V_k) -$$

$$\frac{2}{MN} \sum_{j=1}^{N} C_{i,j}((1 - y_{ij})g'(f_{ij}) + y_{ij}(1 - g'(f_{ij})))V_j - 2\lambda U_i$$

$$\frac{\partial L}{\partial V_j} = \frac{1}{|N_i^+|} (\sum_{k=1}^{N} y_{ij}g'(f_{ij} - f_{ik})U_i - \sum_{k=1}^{N} y_{ik}g'(f_{ik} - f_{ij})U_i$$

$$- 2 \sum_{i=1}^{N} C_{i,j}(1 - g'(f_{ij}))V_j - 2\lambda_2 V_j)$$

(8)

Our model aimed at solving practical application problem, hence the scale of calculation must be limited. At first, we should analyze the complexity of current training procedure. Suppose that matrix Y has M·N entries, especially N is equal to M, and the dimension of latent factor is K. Note that $|N_i^+|$ is number of known positive examples for all user i. For each iteration, the learning process training across all users, and for each user, gradient is calculated for every item vector and this user's vector. Due to the comparisons between item pairs, the complexity of user-fact gradient formula is $O(|N_i^+| \cdot N)$ and the complexity of item-fact gradient formula is approximately O(N·N). Although matrix Y is sparse enough to make complexity of user-fact gradient calculation tractable. With every update of user vector, all item vectors are updated, which is exactly the computational bottleneck of our algorithm. In next section, we will propose a fast learning method to address this computational bottleneck.

4.2   Fast Learning

The key idea of speeding up the optimizing process is to update item vectors selectively and the main reason leads to the untractable calculation is that the times of item vectors update is quadratic of the number of items. To improve this, we propose a sampling strategy, which can sampling a small portion of items to update during each iteration. For the convenience of statement, we use two items pair sets: pos_pos and pos_neg pair set and what we want is that the difference between pos_pos pair to be smaller and pos_neg pair bigger. Hence, we can choose item pairs which have bigger differences from pos_pos pair set and item pairs which have smaller differences from pos_neg set. At last, representative items can be selected for next iteration loop. The detailed sampling strategy and the integral algorithm in presented in Algorithm 2 and 1.

Algorithm 1. Integral Learning Algorithm

Input: Train Matrix Y, constrained parameters α, β, regularization
parameter λ, learning rate γ and sampled item set n^-, n^+
maximum number of iteration max_iter, minimum error error_min

Output: two latent matrix U,V

initial U, V with random values and t

for every user, get his N_i^+ and N_i^-

loop

    % update user vector

    for i in M **do**

        $U_i^{(t)} = U_i^{(t-1)} + \lambda \frac{\partial \ell}{\partial U_i^{(t-1)}}$ according to Eq. (10)

    done

    sample n^- and n^+ from N_i^+ and N_i^- according to Algorithm 2

    % update item vector

    for j in n^+ **do**

        $V_j^{(t)} = V_j^{(t-1)} + \lambda \frac{\partial \ell}{\partial V_j^{(t-1)}}$ according to Eq. (11)

    done

    for k in n^- **do**

        $V_k^{(t)} = V_k^{(t-1)} + \lambda \frac{\partial \ell}{\partial V_k^{(t-1)}}$ according to Eq. (12)

    done

    $err = \ell_t - \ell_{t-1}$

    if err $\leq$ err_min then

        break;

until t $\geq$ max_iter

---

5 Experiment

5.1 Dataset Description

We evaluate our algorithm on two public available social network dataset, which all provide binary implicit feedback. The first Epinions dataset is the trust

---

Algorithm 2. Sampling Strategy

Input: training matrix Y_i, size of sampled positive items and missing items
$n_i^+\_max$ and $n_i^-\_max$ for all users

Output: sampled positive item set n_i^+ and missing set n_i^- for all users

for each user i

    get positive item set N_i^+ and missing item set N_i^-

    calculate all difference values set D_i^{++} for pos_to_pos pairs

    calculate all difference values set D_i^{+-} for pos_to_neg pairs

    sort D_i^{++} by descend order and sort D_i^{+-} by ascend order

    pick first serval item pairs from two sort list

    user items from picked pairs to fill n_i^+ and n_i^-

done

---

relationships interact for a who-trust-whom online social network Epinions[1], which contains 75879 users and 508837 directed friendships.The second dataset are collected on Slashdot[2], which is a technology-related new website allowing user to tag each other as friends or foes. The number of users is 77360 and the number of friendships is 905467, which makes it a less sparse dataset.

5.2 Evaluation

Both datasets are be processed in the same method. We adopt 10-fold cross validation method on our datasets. Detailed to say, we divide datasets for 10 times at the ratio of 9:1 from the perspective of every user randomly, and we take the bigger part as the train data and the remaining as the test data. Different datasets ensure that our model's performance is universal. In addition, the algorithm run across on different train set for 10 times, which make the average measure more compellent.

The measure we optimized for is a good substitute for top-N performance, but too complex to be intuitive. Hence the evaluation measure we adopt in our experiment is Precise@n (p@5, p@10), the precision rate of the topN recommend list. Precise@n is the proportion of positive examples in the topN recommend list.

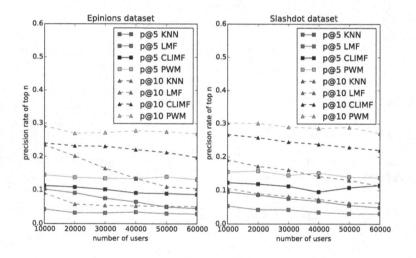

Fig. 1. Precision rate comparison of PWM and baselines on two dataset

5.3 Comparison

We evaluate our approach's performance by comparing with three representative baselines: Item-based KNN, LFM and CLiMF, which are described as following:

---

[1] http://snap.stanford.edu/data/soc-Epinions1.html
[2] http://snap.stanford.edu/data/soc-Slashdot0811.html

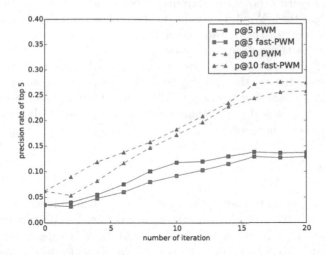

Fig. 2. The variation of precise@5 of PWM and fast-PMW along iterations

- Item-based KNN: First, the similarities between all items are calculated according to the intersection of two user sets, which constituted by users who have selected them. All these values are included by a similarity matrix. And then, taking a user preferred items in history as seed items, the similar items to these items will be taken as candidate for the user's recommendation. At last, Top-N list can also be picked out by ranking all candidates by corresponding weights.
- LFM: Latent Factor Model, is in fact an improved matrix factorization approach. It takes root mean square error as optimization measure, and performs gradient descent to optimize.
- CliMF: This is a similar algorithm with ours for it also adopt ranking-oriented measure[9], but it did not consider much about the sparsity and ambiguity existing in binary implicit feedback. The comparison with CliMF will verify that whether our schema of disposing missing data in implicit feedback is feasible.

The evaluation results are shown in Figure 1 and 2. First, we analysis the performance of all compared models showed in Figure 1. On one hand, we can see that when compared with traditional models, ranking-oriented models have an obvious superior precision rate. Especially when the size of training data get bigger, traditional models' performance are even worse. This proves that rank-biased measure is a more effective for topN performance. On the other hand, our model outperforms another ranking-oriented model CliMF. Different from CliMF, our model do more deep exploration on binary implicit scenario and construct a tailored pairwise ranking-oriented measure. The higher precision rate indicates that our exploration is valuable and feasible.

In addition, Figure 2 shows the computing speed and stability of our proposed fast learning algorithm. We did experiment with 50000 persons of Slatdot and keep trace of precise@5 (p@5) rate along the iteration process. It shows that two methods reach the convergence point almost at the same iteration, while the fast learning method takes less time in every iteration because the sampling strategy it adopts. Hence, the training procedure is speeded up. At the same time, we can see that the fast learning method only lower the precision rate to a tolerable level, which is still higher than other baselines. Generally speaking, the fast learning method makes our model more suitable for practical application.

6   Conclude

In this paper, we go into the binary implicit scenario. Confronted with one class ambiguity and sparsity problems, we find that a deeper comprehension of train data can achieve a better approach to fix it. On this basis, we derive our pairwise rank-biased model: First, we utilize the train data in the form of pairs. Second, we add a constrained condition to refine the optimizing process. For future work, we plan to address the problem of how to simply define parameters in our model.

Acknowledgments. This work is supported by the National Science and Technology Support Program that TV content cloud services technology integration and demonstration based on social network (No.2012BAH41F03).

References

1. Bell, R.M., Koren, Y.: Improved neighborhood-based collaborative filtering. In: KDD Cup and Workshop at the 13th ACM SIGKDD International Conference on Knowledge Discovery and Data Mining. sn (2007)
2. Chapelle, O., Mingrui, W.: Gradient descent optimization of smoothed information retrieval metrics. Information Retrieval **13**(3), 216–235 (2010)
3. Cremonesi, P., Koren, Y., Turrin, R.: Performance of recommender algorithms on top-n recommendation tasks. In: Proceedings of the Fourth ACM Conference on Recommender Systems, pp. 39–46. ACM (2010)
4. Davidson, J., Liebald, B., Liu, J., Nandy, P., Van Vleet, T., Gargi, U., Gupta, S., He, Y., Lambert, M, Livingston, B., et al.: The youtube video recommendation system. In: Proceedings of the Fourth ACM Conference on Recommender Systems, pp. 293–296. ACM (2010)
5. Liu, B., Dai, Y., Li, X., Lee, W.S., Yu, P.S.: Building text classifiers using positive and unlabeled examples. In: Third IEEE International Conference on Data Mining, ICDM 2003, pp. 179–186. IEEE (2003)
6. Pan, R., Zhou, Y., Cao, B., Liu, N.N., Lukose, R., Scholz, M., Yang, Q.: One-class collaborative filtering. In: Eighth IEEE International Conference on Data Mining, ICDM 2008, pp. 502–511. IEEE (2008)
7. Pera, M.S., Ng, Y.-K.: What to read next?: making personalized book recommendations for k-12 users. In: Proceedings of the 7th ACM Conference on Recommender Systems, pp. 113–120. ACM (2013)

8. Rendle, S., Freudenthaler, C., Gantner, Z., Schmidt-Thieme, L.: Bpr: bayesian personalized ranking from implicit feedback. In: Proceedings of the Twenty-Fifth Conference on Uncertainty in Artificial Intelligence, pp. 452–461. AUAI Press (2009)

9. Shi, Y., Karatzoglou, A., Baltrunas, L., Larson, M., Oliver, N., Hanjalic, A.: Climf: learning to maximize reciprocal rank with collaborative less-is-more filtering. In: Proceedings of the Sixth ACM Conference on Recommender Systems, pp. 139–146. ACM (2012)

10. Steck, H.: Evaluation of recommendations: rating-prediction and ranking. In: Proceedings of the 7th ACM Conference on Recommender Systems, pp. 213–220. ACM (2013)

11. Taylor, M., Guiver, J., Robertson, S., Minka, T.: Softrank: optimizing non-smooth rank metrics. In: Proceedings of the 2008 International Conference on Web Search and Data Mining, pp. 77–86. ACM (2008)

12. Ward, G., Hastie, T., Barry, S., Elith, J., Leathwick, J.R.: Presence-only data and the em algorithm. Biometrics **65**(2), 554–563 (2009)

13. Weimer, M., Karatzoglou, A., Viet Le, Q., Smola, A.: Maximum margin matrix factorization for collaborative ranking. Advances in Neural Information Processing Systems (2007)

14. Yu, H., Han, J., Chang, K.C.-C.: Pebl: positive example based learning for web page classification using svm. In: Proceedings of the Eighth ACM SIGKDD International Conference on Knowledge Discovery and Data Mining, pp. 239–248. ACM (2002)

15. Yue, Y., Finley, T., Radlinski, F., Joachims, T.: A support vector method for optimizing average precision. In: Proceedings of the 30th Annual International ACM SIGIR Conference on Research and Development in Information Retrieval, pp. 271–278. ACM (2007)

RIT: Enhancing Recommendation with Inferred Trust

Guo Yan$^{(\boxtimes)}$, Yuan Yao, Feng Xu, and Jian Lu

State Key Laboratory for Novel Software Technology,
Nanjing University, Nanjing, China
{gyan,yyao}@smail.nju.edu.cn, {xf,lj}@nju.edu.cn

Abstract. Trust-based recommendation, which aims to incorporate trust relationships between users to improve recommendation performance, has drawn much attention recently. The focus of existing trust-based recommendation methods is on how to use the observed trust relationships. However, the observed trust relationships are usually very sparse in many real applications. In this paper, we propose to infer some unobserved trust relationships to tackle the sparseness problem. In particular, we first infer the unobserved trust relationships by propagating trust along the observed trust relationships; we then propose a novel trust-based recommendation model to combine observed trust and inferred trust where their relative weights are also learnt. Experimental evaluations on two real datasets show the superior of the proposed method in terms of recommendation accuracy.

Keywords: Recommender system · Trust-based recommendation · Observed trust · Inferred trust

1 Introduction

Recommender system has been widely used in many real applications to find the most attractive items for the users. One of the most successful methods for recommendation is collaborative filtering which is based on the existing historical feedback from users to items. In addition to the user-item feedback, the trust relationships between users have been incorporated to improve the recommendation performance [4,7–9,20,21]. These methods, which are referred to as *trust-based recommendation*, mainly employ the user homophily effect which assumes that socially connected users may have similar preferences for the same item.

To date, the existing trust-based recommendation methods mainly focus on how to make use of the observed trust relationships. However, the observed trust relationships are usually sparse in many real applications. Consider the running example in Fig. 1. As we can see in the example, user A trusts user B, and we can take the rating from user B to item X into consideration if we want to estimate the rating from user A to item X. How to make use of the rating from

© Springer International Publishing Switzerland 2015
T. Cao et al. (Eds.): PAKDD 2015, Part II, LNAI 9078, pp. 756–767, 2015.
DOI: 10.1007/978-3-319-18032-8_59

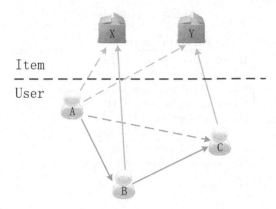

Fig. 1. The solid/dashed lines between users are the observed/inferred trust relationships. The solid lines from users to items are the observed ratings, and the dashed lines between users and items are the ratings we need to estimate (best viewed in color).

user B to item X and the connection from user A to user B is the main focus of the existing trust-based recommendation methods. However, these existing methods may become less accurate if we want to know the possible rating from user A to item Y in the example. Here, our key observation is that since user A trusts user B and user B trusts user C, user A may also trust user C to some extent; consequently, we can estimate the rating from A to item Y based on the inferred trust from user A to user C. Compared to the existing methods that use observed trust relationships only, the inferred trust may also be helpful for recommendation.

In this paper, we propose to infer some unobserved trust relationships for trust-based recommendation. By employing the transitivity property of trust, we first infer several types of unobserved trust based on the observed trust relationships. Next, we propose a novel trust-based recommendation model by incorporating both observed and inferred trust relationships, and we further propose a genetic algorithm to learn the relative weights of different trust relationships.

The main contributions of this work include:

- The proposed RIT model that infers unobserved trust and incorporates both observed and inferred trust for recommendation. The proposed method can tackle the sparseness problem of existing trust relationships, and thus achieve better recommendation performance especially for the cold-start users. To the best of our knowledge, we are the first to directly employ the inferred trust for recommendation.
- Extensive experiments on two real datasets showing the superior of the proposed method over several existing trust-based recommendation methods in terms of recommendation accuracy (e.g., up to 5.27% improvement over the best competitor in terms of MAE). Further, even greater improvement is achieved for the recommendation of cold-start users (e.g., up to 20.81% improvement over the best competitor in terms of MAE).

Table 1. Symbols

Symbols	Definition and Description
$\mathbf{R}$	the user-item rating matrix
$\mathbf{T}$	the observed trust matrix
$\mathbf{U}, \mathbf{V}$	the low-rank matrices for $\mathbf{R}$
$\mathbf{R}'$	the transpose of matrix $\mathbf{R}$
$\mathbf{R}(:, i)$	the ith column of $\mathbf{R}$
$\mathbf{R}(u, i)$	the element at the uth row and i^{th} column of $\mathbf{R}$
$\mathbf{I}^R$	the indicator matrix of $\mathbf{R}$
n, m	the number of users and items
r	the low rank for $\mathbf{U}$ and $\mathbf{V}$

The rest of the paper is organized as follows. Section 2 describes the proposed RIT model. Section 3 presents the experimental results. Section 4 covers related work. Section 5 concludes the paper.

2   The Proposed Model: RIT

In this section, we present the proposed RIT model for recommendation. The main notations we use throughout the paper are listed in Table 1. For example, we use capital bold letter $\mathbf{R}$ and $\mathbf{T}$ to represent the rating matrix from users to items and the observed trust matrix between users, respectively. Based on the notations, the trust-based recommendation methods take the rating matrix $\mathbf{R}$ and the observed trust matrix $\mathbf{T}$ as input, and aim to estimate the potential preference from a user to an item.

2.1   The Proposed Formulation

Inferring Trust. As mentioned in introduction, our key observation is that the inferred trust can help to improve the recommendation performance. Here, we first describe how we obtain the inferred trust. We consider three types of inferred trust [3,22] as shown in Fig. 2.

The first type of inferred trust is from *direct propagation*. As shown in Fig. 2(a), direct propagation means that if a certain user that is trusted by U_i (e.g., U_3 in Fig. 2(a)) trusts U_j, we can infer that U_i might also trust U_j to some extent. We can obtain this type of trust with the matrix operation $\mathbf{T}^2$.

The second type of inferred trust is called *co-citation*. The intuition behind co-citation is that if two users (e.g., U_i and U_j in Fig. 2(b)) are both trusted by some other users (e.g., U_1 in Fig. 2(b)), they may trust each other. Co-citation can be represented as $\mathbf{T}'\mathbf{T}$ in its matrix form.

The third type of inferred trust is *trust coupling* which is shown in Fig. 2(c). Similar to co-citation, trust coupling is based on the intuition that two users (e.g., U_i and U_j in Fig. 2(c)) may trust each other if both of them trust some other users (e.g., U_2 in Fig. 2(c)). We represent this kind of inferred trust as $\mathbf{T}\mathbf{T}'$.

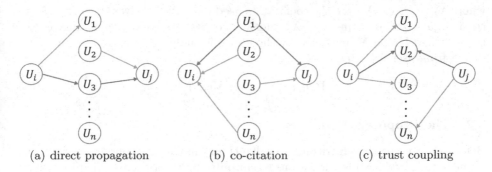

(a) direct propagation   (b) co-citation   (c) trust coupling

Fig. 2. Infer trust from U_i to U_j with trust propagation (best viewed in color)

Incorporating Observed and Inferred Trust. Next, we show how we incorporate both the observed trust and the three types of inferred trust. To incorporate them, the basic idea is to employ the homophily effect [10], which indicates that similar users tend to have similar preferences for items. In other words, if user u trusts another user v, they might have similar latent preferences for items. Taking observed trust as an example, we can add the following constraint

$$\min \sum_{(u,v)\in \mathbf{T}} \mathbf{T}(u,v)\|\mathbf{U}(:,u) - \mathbf{U}(:,v)\|_2^2 \tag{1}$$

where $\mathbf{U}(:,u)$ and $\mathbf{U}(:,v)$ indicate the preferences of user u and user v, respectively. In Eq. (1), we actually encourage the corresponding user latent preferences $\mathbf{U}(:,u)$ and $\mathbf{U}(:,v)$ to be closed to each other, and greater $\mathbf{T}(u,v)$ would encourage them to be closer. For the three types of inferred trust, we can substitute the $\mathbf{T}$ matrix with the inferred trust matrices such as $\mathbf{T}^2$, $\mathbf{T}'\mathbf{T}$, and $\mathbf{TT}'$.

To incorporate both observed trust and inferred trust, we adopt the commonly used matrix factorization framework [5] due to its flexibility. By considering all the four trust matrices (one for observed trust and three for inferred trust), we have our RIT model:

$$\min_{\mathbf{U},\mathbf{V}} \mathcal{L}(\mathbf{R},\mathbf{T},\mathbf{U},\mathbf{V}) = \min_{\mathbf{U},\mathbf{V}} \frac{1}{2}\|\mathbf{I}^R \odot (\mathbf{R} - \mathbf{U}'\mathbf{V})\|_F^2 + \frac{\lambda_u}{2}\|\mathbf{U}\|_F^2 + \frac{\lambda_v}{2}\|\mathbf{V}\|_F^2$$

$$+ \lambda_{g_1} \sum_{(u,v)\in \mathbf{T}} \mathbf{T}(u,v)\|\mathbf{U}(:,u) - \mathbf{U}(:,v)\|_2^2$$

$$+ \lambda_{g_2} \sum_{(u,v)\in \mathbf{T}^2} \mathbf{T}^2(u,v)\|\mathbf{U}(:,u) - \mathbf{U}(:,v)\|_2^2 \tag{2}$$

$$+ \lambda_{g_3} \sum_{(u,v)\in \mathbf{T}'\mathbf{T}} (\mathbf{T}'\mathbf{T})(u,v)\|\mathbf{U}(:,u) - \mathbf{U}(:,v)\|_2^2$$

$$+ \lambda_{g_4} \sum_{(u,v)\in \mathbf{TT}'} (\mathbf{TT}')(u,v)\|\mathbf{U}(:,u) - \mathbf{U}(:,v)\|_2^2$$

where $\lambda_{g_1}, \lambda_{g_2}, \lambda_{g_3}, \lambda_{g_4}$ are the coefficients used to control the weights of the four trust regularization terms, and λ_u and λ_v are used to avoid over fitting. Once the matrices $\mathbf{U}$ and $\mathbf{V}$ are learnt by the above formulation, we can estimate the rating from user u to item i as

$$\hat{\mathbf{R}}(u, i) = \mathbf{U}(:, u)' \mathbf{V}(:, i). \tag{3}$$

2.2   The Proposed Algorithm

Before we present the algorithms to solve the formulation proposed in Eq. (2), we need to introduce a few notations to simplify the descriptions. There are four types of trust regularization terms in Eq. (2), which can be re-written as:

$$\lambda_g \sum_{(u,v)} \left(\frac{\lambda_{g_1}}{\lambda_g} \mathbf{T} + \frac{\lambda_{g_2}}{\lambda_g} \mathbf{T}^2 + \frac{\lambda_{g_3}}{\lambda_g} \mathbf{T}'\mathbf{T} + \frac{\lambda_{g_4}}{\lambda_g} \mathbf{TT}' \right) \|\mathbf{U}(:, u) - \mathbf{U}(:, v)\|_2^2 \tag{4}$$

where $\lambda_g = \lambda_{g_1} + \lambda_{g_2} + \lambda_{g_3} + \lambda_{g_4}$.

For simplicity, we define $w_i = \frac{\lambda_{g_i}}{\lambda_g}$, and define matrix $\mathbf{Q}$ as

$$\mathbf{Q} = w_1\mathbf{T} + w_2\mathbf{T}^2 + w_3\mathbf{T}'\mathbf{T} + w_4\mathbf{TT}' \tag{5}$$

where $w_1 + w_2 + w_3 + w_4 = 1$. This $\mathbf{Q}$ matrix actually contains all the four types of trust (one observed trust and three inferred trust) and their relative weights.

With these new notations, the optimization problem in Eq. (2) can be re-written as

$$\min_{\mathbf{U},\mathbf{V}} \mathcal{L}(\mathbf{R}, \mathbf{T}, \mathbf{U}, \mathbf{V}) = \min_{\mathbf{U},\mathbf{V}} \frac{1}{2} \|\mathbf{I}^R \odot (\mathbf{R} - \mathbf{U}'\mathbf{V})\|_F^2 + \frac{\lambda_u}{2} \|\mathbf{U}\|_F^2 + \frac{\lambda_v}{2} \|\mathbf{V}\|_F^2$$

$$\lambda_g \sum_{(u,v)\in\mathbf{Q}} \mathbf{Q}(u,v)\|\mathbf{U}(:,u) - \mathbf{U}(:,v)\|_2^2 \tag{6}$$

Then, the task of the algorithm is to learn the weights ($W = [w_1, w_2, w_3, w_4]$) that are contained in the $\mathbf{Q}$ matrix and the two low-rank matrices ($\mathbf{U}$ and $\mathbf{V}$).

Learning Coefficients W. We adopt the so-called real-valued Genetic Algorithm [11] to learn the weights W. The basic idea is to treat the coefficients $W = [w_1, w_2, w_3, w_4]$ as a chromosome. After randomly generate a group of chromosomes, we can generate its offsprings by three basic operators, i.e., selection, crossover, and mutation.

For the selection operator, we can simply select several best chromosomes. Here, we need a fitness function to determine how good the chromosome is. In this work, we choose prediction accuracy and measure the fitness through cross validation. For the crossover operator, we select two parent chromosomes and generate the child chromosome as $W^{new} = \sigma W_1^{old} + (1 - \sigma)W_2^{old}$, where σ is a random number between 0 and 1 to control the proportion of each parent chromosome. For the mutation operator, we select two coefficients in

---

Algorithm 1. The *computeCoefficient* Algorithm

Input: The $n \times m$ rating matrix $\mathbf{R}$, the $r \times n$ user latent matrix $\mathbf{U}$, the $r \times m$ item latent matrix $\mathbf{V}$, the $n \times n$ trust matrices $\mathbf{T}$, $\mathbf{T}^2$, $\mathbf{T}'\mathbf{T}$, and $\mathbf{TT}'$, the number of coefficient groups κ, and the maximum iteration number l_c

Output: The coefficients $W = [w_1, w_2, w_3, w_4]$

1: randomly initialize κ groups of coefficients W;
2: **while** current iteration number $\leq l_c$ **do**
3:    select $\sqrt{\kappa}$ groups of best coefficients;
4:    **for** $i = 1$ to $\sqrt{\kappa}$ **do**
5:       **for** $j = 1$ to $\sqrt{\kappa}$ **do**
6:          crossover a new chromosome W^{new} from W_i^{old} and W_j^{old};
7:       **end for**
8:    **end for**
9:    **for** $i = 1$ to κ **do**
10:      mutate chromosome W_i^{new} from W_i^{old};
11:   **end for**
12: **end while**
13: **return** the best coefficients W;

W (e.g., w_i and w_j) and set $w_i^{new} = w_i^{old} - \tau$ and $w_j^{new} = w_j^{old} + \tau$. Notice that, both crossover and mutation will keep the sum of W as 1.

The algorithm is summarized in Alg. 1. We first randomly initialize a group of W's (step 1). For each iteration, we select several best W's (step 3), and generate their offsprings via crossover (step 4-8). Then, we adopt the mutation operator for each offspring (step 9-11). Finally, after several iterations, we return the best W.

Updating Matrices U and V. Next, we show how we update matrices $\mathbf{U}$ and $\mathbf{V}$. First of all, the fourth term of Eq. (6) can be written as

$$\sum_{(u,v)\in\mathbf{Q}} \mathbf{Q}(u,v)\|\mathbf{U}(:,u) - \mathbf{U}(:,v)\|_2^2$$

$$= \sum_{(u,v)\in\mathbf{Q}}\sum_{k=1}^{r} \mathbf{Q}(u,v)\|\mathbf{U}(k,u) - \mathbf{U}(k,v)\|_2^2$$

$$= \sum_{(u,v)\in\mathbf{Q}}\sum_{k=1}^{r} \mathbf{Q}(u,v)\mathbf{U}^2(k,u) + \sum_{(u,v)\in\mathbf{Q}}\sum_{k=1}^{r} \mathbf{Q}(u,v)\mathbf{U}^2(k,v)$$

$$- \sum_{(u,v)\in\mathbf{Q}}\sum_{k=1}^{r} 2\mathbf{Q}(u,v)\mathbf{U}(k,u)\mathbf{U}(k,v)$$

$$= \sum_{k=1}^{r} \mathbf{U}(k,:)(\mathbf{D}_1 + \mathbf{D}_2 - 2\mathbf{Q})\mathbf{U}'(:,k)$$

$$= Tr(\mathbf{U}(\mathbf{D}_1 + \mathbf{D}_2 - 2\mathbf{Q})U')$$

Algorithm 2. The *RIT* Algorithm

Input: The $n \times m$ rating matrix $\mathbf{R}$, the $n \times n$ observed trust matrix $\mathbf{T}$, and the latent factor size r

Output: The low-rank matrices $\mathbf{U}$ and $\mathbf{V}$

1: compute inferred trust: $[\mathbf{T}^2, \mathbf{T}'\mathbf{T}, \mathbf{TT}'] = computeInferredTrust(\mathbf{T})$;
2: initialize $\mathbf{U}$ and $\mathbf{V}$ randomly;
3: learn coefficients: $W = computeCoefficient(\mathbf{U}, \mathbf{V}, \mathbf{R}, \mathbf{T}, \mathbf{T}^2, \mathbf{T}'\mathbf{T}, \mathbf{TT}')$;
4: compute $\mathbf{Q}$ as defined in Eq. (5);
5: **while** not convergent **do**
6:     $\mathbf{U} \Leftarrow \mathbf{U} - \eta \frac{\partial \mathcal{L}}{\partial \mathbf{U}}$ where $\frac{\partial \mathcal{L}}{\partial \mathbf{U}}$ is defined in Eq. (8) and η is the learning step;
7:     $\mathbf{V} \Leftarrow \mathbf{V} - \eta \frac{\partial \mathcal{L}}{\partial \mathbf{V}}$ where $\frac{\partial \mathcal{L}}{\partial \mathbf{V}}$ is defined in Eq. (8) and η is the learning step;
8: **end while**
9: **return** $\mathbf{U}$ and $\mathbf{V}$;

---

where $\mathbf{D}_1$ and $\mathbf{D}_2$ are two diagonal matrices with $\mathbf{D}_1(u, u) = \sum_{v=1}^{n} \mathbf{Q}(u, v)$ and $\mathbf{D}_2(v, v) = \sum_{u=1}^{n} \mathbf{Q}(u, v)$. Since matrix $\mathbf{Q}$ may be asymmetrical, $\mathbf{D}_1$ and $\mathbf{D}_2$ may not equal to each other.

Eq. (6) can be re-written as

$$\mathcal{L} = \frac{1}{2} Tr[-2(\mathbf{I}^R \odot \mathbf{R})\mathbf{V}'\mathbf{U} + (\mathbf{I}^R \odot (\mathbf{U}'\mathbf{V}))\mathbf{V}'\mathbf{U}]$$
$$+ \frac{\lambda_u}{2} Tr(\mathbf{UU}') + \frac{\lambda_v}{2} Tr(\mathbf{VV}') + \lambda_g Tr(\mathbf{U}(\mathbf{D}_1 + \mathbf{D}_2 - 2\mathbf{Q})\mathbf{U}') \tag{7}$$

Then, we have the following derivatives

$$\frac{\partial \mathcal{L}}{\partial \mathbf{U}} = -\mathbf{V}(\mathbf{I}^R \odot \mathbf{R})' + \mathbf{V}(\mathbf{I}^R \odot (\mathbf{U}'\mathbf{V}))' + \lambda_u \mathbf{U}$$
$$+ \lambda_g(\mathbf{U}(\mathbf{D}_1 + \mathbf{D}_2 - 2\mathbf{Q})' + \mathbf{U}(\mathbf{D}_1 + \mathbf{D}_2 - 2\mathbf{Q}))$$
$$\frac{\partial \mathcal{L}}{\partial \mathbf{V}} = -\mathbf{U}(\mathbf{I}^R \odot \mathbf{R}) + \mathbf{U}(\mathbf{I}^R \odot (\mathbf{U}'\mathbf{V})) + \lambda_v \mathbf{V} \tag{8}$$

Finally, we summarize the overall algorithm to solve the problem defined in Eq. (2) as in Alg. 2. As shown in the algorithm, we first compute the inferred trust from observed trust (step 1) and then initialize the $\mathbf{U}$ and $\mathbf{V}$ matrices (step 2). Next, we learn the coefficients to control the weights of each trust regularization terms (step 3), and define the $\mathbf{Q}$ matrix (step 4). After that, the algorithm begins the iteration procedure. In each iteration, we alternatively update the user latent matrix $\mathbf{U}$ and the item latent matrix $\mathbf{V}$ with gradient descent method (step 6 and 7). The iteration procedure will continue until at least one of the following conditions is satisfied: either the Frobenius norm between successive estimates of both matrices $\mathbf{U}$ and $\mathbf{V}$ is below a certain threshold or the maximum iteration step is reached.

3   Experiments

In this section, we present the experimental results. The experiments are designed to answer the following questions:

- How accurate is the proposed method for recommendation?
- How does the proposed method perform for cold-start users?
- How robust is the proposed method?

3.1   Experimental Setup

In our experiment, we use two datasets: Epinions and Ciao [18]. These two datasets contain not only the user-item ratings but also the trust relationships between users. For the trust relationships, we assign $\mathbf{T}(u, v) = 1$ if user u trusts user v and $\mathbf{T}(u, v) = 0$ otherwise. Table 2 shows the statistics of the datasets.

Table 2. The statistics of Epinions and Ciao datasets

Data	Epinions	Ciao
users	33,725	6,102
items	43,542	12,082
ratings	684,371	149,530
trust relationships	405,047	98,856
sparsity of ratings	0.0466%	0.2028%
sparsity of trust relationships	0.0356%	0.2655%

For both datasets, we randomly select a set of ratings as the training set and use the rest as test set. We conduct two groups of experiments with different percentage of training set (50% and 90%). We fix the rank $r = 10$, set $\lambda_g = 0.1$, and $\lambda_u = \lambda_v = 0.1$ unless otherwise stated. We use cross-validation to learn the coefficients W on the training set.

For the evaluation metrics, we adopt the Root Mean Square Error ($RMSE$) and the Mean Absolute Error (MAE) to compare with the existing trust-based recommendation methods:

$$RMSE = \sqrt{\frac{1}{|\mathbf{R}_T|} \sum_{(u,i) \in \mathbf{R}_T} (\mathbf{R}_T(u, i) - \hat{\mathbf{R}}_T(u, i))^2}$$

$$MAE = \frac{1}{|\mathbf{R}_T|} \sum_{(u,i) \in \mathbf{R}_T} |\mathbf{R}_T(u, i) - \hat{\mathbf{R}}_T(u, i)|$$

where $\mathbf{R}_T$ is the test set, $\mathbf{R}_T(u, i)$ denotes the actual rating that user u gives to item i, $\hat{\mathbf{R}}_T(u, i)$ denotes the predicted rating from user u to item i, and $|\mathbf{R}_T|$ is the total number of ratings in the test set. Lower values of $RMSE$ and MAE mean better prediction performance.

The compared methods include RSTE [7], SoRec [8], SocialMF [4], FIP [20], and SR [9]. Notice that all these compared methods use the observed trust only.

Table 3. Effectiveness comparisons. Our RIT method outperforms all the compared methods.

Data	Training	Metrics	RSTE	SoRec	SocialMF	FIP	SR	Basic	RIT
Epinions	50%	$RMSE$	1.1975	1.2082	1.1905	1.1880	1.1897	1.1732	**1.1467**
		MAE	0.9548	0.9655	0.9484	0.9452	0.9469	0.9325	**0.9134**
	90%	$RMSE$	1.1154	1.1202	1.1140	1.1111	1.1137	1.1023	**1.0872**
		MAE	0.8768	0.8824	0.8748	0.8713	0.8739	0.8648	**0.8528**
Ciao	50%	$RMSE$	1.2022	1.2392	1.2051	1.2067	1.2054	1.1827	**1.1469**
		MAE	0.9180	0.9580	0.9215	0.9216	0.9213	0.9017	**0.8696**
	90%	$RMSE$	1.0899	1.1108	1.0889	1.0947	1.0908	1.0769	**1.0561**
		MAE	0.8157	0.8415	0.8159	0.8203	0.8166	0.8056	**0.7852**

Table 4. Effectiveness comparisons for cold-start users. Our RIT method can achieve even greater improvement for cold-start users.

Data	Metrics	RSTE	SoRec	SocialMF	FIP	SR	Basic	RIT
Epinions	$RMSE$	1.4315	1.4085	1.4073	1.4007	1.4012	1.3905	**1.2857**
	MAE	1.2124	1.1789	1.1876	1.1778	1.1791	1.1681	**1.0523**
Ciao	$RMSE$	1.4781	1.4771	1.4784	1.4904	1.4892	1.4420	**1.2398**
	MAE	1.2079	1.1983	1.2058	1.2141	1.2114	1.1635	**0.9489**

3.2   Experimental Results

Effectiveness Comparisons. We first compare the effectiveness of different methods in Table 3. In the table, we also report the results of the Basic case of our model where only the observed trust matrix is used. The Basic method is similar to the SR method except that it uses the trust as side information while SR uses the rating similarity. As we can see from the table, the proposed RIT method outperforms all the other trust-based recommendation methods wrt both $RMSE$ and MAE on both datasets. For example, on the Epinions data with 50% training data, our method improves the best competitor (FIP) by 3.48% wrt $RMSE$ and by 3.36% wrt MAE; for the Ciao data with 50% training data, our method improves the best competitor (RSTE) by 4.60% wrt $RMSE$ and by 5.27% wrt MAE. Overall, the results indicate that the inferred trust is indeed helpful to improve the performance of recommender systems.

Effectiveness for Cold-Start Users. The cold-start problem, which aims to recommend items to cold-start users, is still a remaining challenge in recommender systems. Intuitively, the proposed RIT model can better handle the cold-start problem as we infer many unobserved trust relationships. We conduct experiments to validate this intuition, and define users who have expressed less than 5 ratings as cold-start users. The results with 50% training data are shown in Table 4. As we can see, our RIT method achieves larger improvement in the cold-start scenario. For example, on the Epinions data, our method improves the best competitor (FIP) by 8.21% wrt $RMSE$ and by 10.66% wrt MAE; on

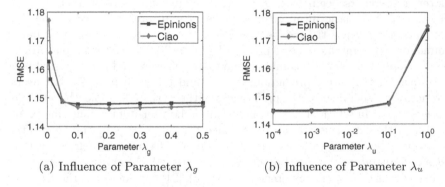

(a) Influence of Parameter λ_g (b) Influence of Parameter λ_u

Fig. 3. Influence of parameters. Our method is robust wrt the parameters.

the Ciao data, our method improves the best competitor (SoRec) by 16.07% wrt *RMSE* and by 20.81% wrt *MAE*. This result indicates that our RIT model can better handle the cold-start problem by inferring unobserved trust.

Influence of Parameters. Finally, we study the influence of the parameters in our RIT model: λ_g which controls the importance of trust relationships, and λ_u and λ_v which control regularization. The results are shown in Fig. 3 where we use 50% training data and only report *RMSE* on y-axis for simplicity. Similar results can be observed for 90% training data and the *MAE* metric. For convenience, we set $\lambda_u = \lambda_v$. As we can see from Fig. 3(a), on both datasets, *RMSE* decreases quickly and then stays stable as λ_g increases. For λ_u and λ_v, as we can see from Fig. 3(b), *RMSE* is stable when they are less than a certain threshold (e.g., 0.1) on both datasets. Overall, the proposed RIT model is robust wrt the parameters.

4   Related Work

In this section, we review some existing approaches for recommender systems, including traditional recommendation, trust-based recommendation, and the cold-start recommendation.

Collaborative filtering is widely used in many traditional recommender systems. These traditional recommender systems mainly focus on the mining of user preferences via the user-item rating matrix. Collaborative filtering can be divided into two categories: memory-based approaches and model-based approaches. Memory-based approaches aim to find similar users or items for recommendation, known as user-based methods [1] and item-based methods [15]. Some researchers also propose to combine user-based methods and item-based methods [19]. Model-based approaches focus on learning models from data with statistical and machine learning techniques, in which matrix factorization has been widely adopted. These matrix factorization methods [6,12,14] make predictions by factorizing the user-item rating matrix into low-rank latent matrices.

Later, researchers begin to take the trust relationships between users into consideration. The basic intuition is that users may prefer to accept the recommendations by their trusted people. With this intuition, many trust-based recommender systems have been proposed [4, 7–9, 20, 21]. For example. Ma et al. [7] develop a model which combines the tastes of users and their trusted people together for the rating prediction; Jamali and Ester [4] propose that users' preferences are dependent on the preferences of their trusted people ; Ma et al. [8] propose a model in which the trust matrix and the user-item rating matrix are connected with shared user latent preferences; Yang et al. [20] aim to use the inner product between two users' latent preferences to recover their social link; Ma et al. [9] introduce the social regularization term which serves as a constraint to regularize the user latent preferences; Yao et al. [21] consider both the relationships between users and the relationships from items. All the above trust-based methods employ the observed trust for better recommendation. Different from them, we also consider the inferred trust which might be useful for recommendation. In the recent work by Fazeli et al [2], they try to infer implicit trust into trust-based recommendation methods when explicit trust relationships are not available. In contrast, we infer the implicit trust from explicit trust and combine them together to improve the prediction accuracy of recommender systems.

One of the remaining challenges of recommender systems is the cold-start problem, i.e., how to recommend items for users that have expressed few ratings. Existing solutions can be roughly classified into two classes: interview based and side-information based. The interview based methods add an interview process to collect the preferences of cold-start users in their sign-up phase [17, 24]; the main problem of this type of methods is the additional burdens. The second type of methods incorporate the side information to enhance prediction accuracy for cold-start users, such as the content information of items [16], the attributes of users/items [13, 23], or the social relationships between users (e.g., trust-based recommendation). In this work, we enhance the trust-based recommendation by inferring unobserved trust from observed trust.

5   Conclusion

In this paper, we have proposed a trust-based recommendation model RIT based on matrix factorization and social regularization. In RIT, we first infer several types of unobserved trust based on the observed trust relationships by employing the transitivity property of trust. Next, we propose to incorporate both observed and inferred trust relationships for better recommendation. Experimental evaluations on two real datasets show that our method outperforms the existing trust-based recommendation models in terms of prediction accuracy, and greater improvement is observed for the recommendation of cold-start users.

Acknowledgments. This work is supported by the National 973 Program of China (No. 2015CB352202) and the National Natural Science Foundation of China (No. 91318301, 61321491).

References

1. Breese, J.S., Heckerman, D., Kadie, C.: Empirical analysis of predictive algorithms for collaborative filtering. In: UAI, pp. 43–52 (1998)
2. Fazeli, S., Drachsler, H., Loni, B., Sloep, P., Bellogin, A.: Implicit vs. explicit trust in social matrix factorization. In: RecSys, pp. 317–320 (2014)
3. Guha, R., Kumar, R., Raghavan, P., Tomkins, A.: Propagation of trust and distrust. In: WWW, pp. 403–412 (2004)
4. Jamali, M., Ester, M.: A matrix factorization technique with trust propagation for recommendation in social networks. In: RecSys, pp. 135–142 (2010)
5. Koren, Y., Bell, R., Volinsky, C.: Matrix factorization techniques for recommender systems. Computer **42**(8), 30–37 (2009)
6. Lee, D.D., Seung, H.S.: Learning the parts of objects by non-negative matrix factorization. Nature **401**(6755), 788–791 (1999)
7. Ma, H., King, I., Lyu, M.R.: Learning to recommend with social trust ensemble. In: SIGIR, pp. 203–210 (2009)
8. Ma, H., Yang, H., Lyu, M.R., King, I.: Sorec: social recommendation using probabilistic matrix factorization. In: CIKM, pp. 931–940 (2008)
9. Ma, H., Zhou, D., Liu, C., Lyu, M.R., King, I.: Recommender systems with social regularization. In: WSDM, pp. 287–296 (2011)
10. McPherson, M., Smith-Lovin, L., Cook, J.M.: Birds of a feather: Homophily in social networks. Annual Review of Sociology, 415–444 (2001)
11. Mitchell, M.: An introduction to genetic algorithms. MIT Press (1998)
12. Mnih, A., Salakhutdinov, R.: Probabilistic matrix factorization. In: NIPS, pp. 1257–1264 (2007)
13. Park, S.T., Chu, W.: Pairwise preference regression for cold-start recommendation. In: RecSys, pp. 21–28 (2009)
14. Salakhutdinov, R., Mnih, A.: Bayesian probabilistic matrix factorization using markov chain monte carlo. In: ICML, pp. 880–887 (2008)
15. Sarwar, B., Karypis, G., Konstan, J., Riedl, J.: Item-based collaborative filtering recommendation algorithms. In: WWW, pp. 285–295 (2001)
16. Schein, A.I., Popescul, A., Ungar, L.H., Pennock, D.M.: Methods and metrics for cold-start recommendations. In: SIGIR, pp. 253–260 (2002)
17. Sun, M., Li, F., Lee, J., Zhou, K., Lebanon, G., Zha, H.: Learning multiple-question decision trees for cold-start recommendation. In: WSDM, pp. 445–454 (2013)
18. Tang, J., Gao, H., Liu, H.: mtrust: discerning multi-faceted trust in a connected world. In: WSDM, pp. 93–102 (2012)
19. Wang, J., De Vries, A.P., Reinders, M.J.: Unifying user-based and item-based collaborative filtering approaches by similarity fusion. In: SIGIR, pp. 501–508 (2006)
20. Yang, S.H., Long, B., Smola, A., Sadagopan, N., Zheng, Z., Zha, H.: Like like alike: joint friendship and interest propagation in social networks. In: WWW, pp. 537–546 (2011)
21. Yao, Y., Tong, H., Yan, G., Xu, F., Zhang, X., Szymanski, B.K., Lu, J.: Dual-regularized one-class collaborative filtering. In: CIKM, pp. 759–768 (2014)
22. Yao, Y., Tong, H., Yan, X., Xu, F., Lu, J.: Matri: a multi-aspect and transitive trust inference model. In: WWW, pp. 1467–1476 (2013)
23. Zhang, M., Tang, J., Zhang, X., Xue, X.: Addressing cold start in recommender systems: A semi-supervised co-training algorithm. In: SIGIR (2014)
24. Zhou, K., Yang, S.H., Zha, H.: Functional matrix factorizations for cold-start recommendation. In: SIGIR, pp. 315–324 (2011)

Author Index

·

Printed in the United States
By Bookmasters